Introduction to Phytoremediation of Contaminated Groundwater

James E. Landmeyer

Introduction to Phytoremediation of Contaminated Groundwater

Historical Foundation, Hydrologic Control, and Contaminant Remediation

Dr. James E. Landmeyer
U.S. Geological Survey
South Carolina Water Science Center
720 Gracern Road
Columbia, SC 29210
USA
jlandmey@usgs.gov

This volume was produced under a cooperative research and development agreement between the U.S. Geological Survey and Springer Science+Business Media, B.V.

ISBN 978-94-007-1956-9 e-ISBN 978-94-007-1957-6
DOI 10.1007/978-94-007-1957-6
Springer Dordrecht Heidelberg London New York

Library of Congress Control Number: 2011937578

Printed on acid-free paper

Springer is part of Springer Science+Business Media (www.springer.com)

The deeper the roots, the higher the reach

To Lori, Kaylee, and Ava–
May our roots always find deep water.

Preface

"Nature is driven by water"
Leonardo da Vinci
(from Richter 1888)

Groundwater is one of our most important natural resources. Groundwater provides drinking water to more than 50% of the population of the United States. This fact may not surprise a domestic-well owner, but may surprise municipal water customers who drink tap water supplied by wells. A greater percentage of the population in other countries, particularly in the developing world, relies on groundwater. Groundwater usually requires minimal treatment prior to drinking as groundwater is filtered during flow through porous sediments. Moreover, groundwater can be decades to thousands of years old and, therefore, contain precipitation that fell long before the production and release of modern-day contaminants.

These facts do not imply groundwater is protected from contamination. On the contrary, groundwater resources can be more vulnerable to contamination than surface water. For example, contaminant sources may be located a short vertical distance above shallow groundwater that typically is discharged to surface water or pumped by wells after only a few years in the subsurface with correspondingly little time for contaminant cleansing to occur. Alternatively, the slow groundwater-flow rates characteristic of deep groundwater means that contamination will take more time to be discharged from aquifers. In both cases, contamination of groundwater should be avoided if at all possible. If contamination occurs, remediation of groundwater to pre-contamination conditions should be accomplished as efficiently, as quickly, and as cost effectively as possible to ensure that current and future demands on the groundwater resource can be met.

Phytoremediation is one of many potential alternatives that can be used to restore contaminated groundwater. In general, phytoremediation is the use of living organisms—plants—to restore contaminated environments to less harmful levels. It is commonly accepted that plants have the potential to cleanse the air, such as taking in carbon dioxide (CO_2) and releasing oxygen (O_2). But plants also have the potential to cleanse water. This is because water controls all aspects of plant survival; a glance at the waterless and plantless moon quickly confirms this statement. From seed to maturity, plants use water for food production, structural support, cell metabolism, and growth. It is this fundamental requirement for water that places plants in the unique position to be able to remediate contaminants dissolved in groundwater and to alter the local groundwater-flow system.

The application of plants to remediate contaminated groundwater may seem a recent and novel occurrence. Phytoremediation, however, has a much older and natural precedent. Terrestrial plants used for phytoremediation represent the culmination of evolutionary adaptations that provide a particular plant with a selective advantage for survival and reproduction. Adaptations include defenses against chemical threats and invasion by pests, as well as the production of chemicals solely to exclude the growth of other plants that compete for limited resources, essentially through constant involvement in a natural chemical warfare to ensure survival. These processes, along with photosynthesis, involve chemical reactions in plants that occur only if water is present. Moreover, the successful colonization of land by previously aquatic plants at least 400 million years ago possibly succeeded, in part, because early terrestrial plants remained close to areas of surface water, such as swamps and bogs, thus allowing the roots to be in constant contact with discharging groundwater.

The interaction between plants and groundwater has only recently become widely accepted for application to remediate contaminated groundwater. This is because the general observation of plants implies that precipitation is the source of water to plants, rather than groundwater, which is less readily observable. Moreover, plant physiology and hydrogeology have been treated as being separate disciplines and have produced hydrogeologists with little official training in plant physiology or plant–water interactions and plant scientists often receive little opportunity to study the fundamentals of hydrogeology.

This book is a synthesis of information from the fields of plant physiology, hydrogeology, and contaminant fate relevant to the phytoremediation of contaminated groundwater. This book can be used as a reference for researchers actively involved in phytoremediation. This book also can be used by environmental professionals who will either propose and implement a phytoremediation project or review its effectiveness in meeting environmental mandates. The inclusion of original-source references provides plant physiologists and hydrogeologists with fundamental information about each respective discipline that can be used as a primer or an introduction to each other's field. This book also provides numerous case studies that can provide a starting point for use in undergraduate or graduate courses in hydrology, hydrogeology, hydroecology, and environmental science. Moreover, this book is intended to be the introductory text I wish had been available in the early 1990s when I started working in the field of the phytoremediation of contaminated groundwater.

This book is divided into three parts. Each part covers a fundamental aspect of plant and groundwater interactions necessary to implement or evaluate the phytoremediation of contaminated groundwater. Part I is an introduction to the history of plant and water interactions that includes the first published record of this interaction (Chap. 1); the effect of plants on the hydrologic cycle and how this cycle can be used as a simple model toward understanding the effect of plants on groundwater at phytoremediation sites (Chap. 2); an introduction to the fundamentals of plant physiology that includes the entry and movement of water through plants and what processes facilitate and regulate water movement in plants (Chap. 3); an introduction to the fundamentals of hydrogeology that can be used to design phytoremediation plantings and to assess the effect of plants on groundwater levels and flow at phytoremediation sites (Chap. 4); and presents examples of plant and groundwater interactions that occur naturally and, therefore, support the application of plants to restore contaminated groundwater (Chap. 5). Because the chapters in Part I do not deal directly with the phytoremediation of contaminated groundwater some readers may choose to proceed directly to Part II. Much of the information presented in Parts II and III, however, is based on the fundamentals introduced in Part I.

The focus of Part II is the use of phytoremediation to affect groundwater recharge, discharge, levels, and flow. Part II demonstrates how sites characterized by groundwater contamination can be assessed for the feasibility of phytoremediation (Chap. 6); how to plant a site where trees can be used to achieve hydrologic control (Chap. 7); presents various conceptual models and frameworks that can be used to assess the interaction between plants and groundwater to meet regulatory goals (Chap. 8); how to monitor the effectiveness of phytoremediation for hydrologic control (Chap. 9); and an introduction to general economic and regulatory factors that often are encountered during the implementation of phytoremediation (Chap. 10). Relevant and original published case studies are provided to reinforce key concepts.

Part III presents the use of phytoremediation to affect contaminant concentrations in groundwater. Part III demonstrates how the evolution of plants as a part of natural biogeochemical environments provides the background for plant-contaminant interactions (Chap. 11); how the physical properties of contaminants control the interaction of contaminants with plants during uptake from groundwater (Chap. 12); how plants can be used for the phytoremediation of groundwater contaminated by polyaromatic hydrocarbons (PAHs), monoaromatic hydrocarbons (MAHs), gasoline, fuels that contain oxygenates such as methyl *tert*-butyl ether (MTBE), chlorinated solvents such as perchloroethylene (PCE), trichloroethylene (TCE), vinyl chloride (VC), and other less common groundwater contaminants, such as explosives, perchlorate, and tritium (Chap. 13); how various frameworks, including numerical models, can be used to assess the effectiveness of plant interactions with contaminated groundwater (Chap. 14); how to monitor the effectiveness of phytoremediation using a variety of field methods (Chap. 15); and an introduction to general economic and regulatory factors that can affect phytoremediation implementation (Chap. 16). An epilogue is provided to highlight novel research topics and potential areas of promising research (Chap. 17). Relevant published case studies are provided and readers are encouraged to consult these original sources.

The future of the phytoremediation of contaminated groundwater is the responsibility of individuals or groups of professionals who are tasked with the implementation of phytoremediation under field or laboratory settings. For phytoremediation to be a scientifically defensible remediation alternative for contaminated groundwater, the processes of groundwater uptake and contaminant detoxification by plants must be shown to adhere to fundamental principles and physical laws. Moreover, these processes of groundwater uptake and contaminant detoxification and how they are monitored should provide reproducible results over space and time.

While it may be possible someday to 'phytoremediate' a site characterized by contaminated groundwater simply by installing plants, we are not there yet. To get there, we need to ask ourselves, how did phytoremediation work at this site? Was the causative agent of remediation based on plant or microbial processes or both? Did the plants affect the water table, capillary fringe, or unsaturated zone? What did not work as planned, and why? Perhaps this scenario can be summarized as

"An expert is a person
who has made all the mistakes
that can be made in a very narrow field."
Niels Bohr (1885–1962; from Mackay 1991)

In the end, the use of the natural ability of some plants to take up groundwater, refresh it through natural processes, and provide us with clean groundwater is surely an indication of good stewardship of one of our most important natural resources.

Columbia, SC, USA James E. Landmeyer

Acknowledgments

Funding to prepare this book was provided by the taxpayers of the United States of America through the Toxic Substances Hydrology Program of the United States Geological Survey (USGS). Dr. Herbert T. Buxton, Director of the Toxics Program, supported the creation of a comprehensive overview of the interaction between plants and contaminated groundwater that would compile the numerous independent studies being conducted by a variety of governmental and private researchers. Thanks also are given to David W. Morganwalp and Michael J. Focazio, also with the USGS Toxics Program—this book could not have been completed without their support. Moreover, thanks go to Marjorie S. Davenport and Eric W. Strom, former and current Directors, respectively, of the USGS Water Science Center in Columbia, South Carolina, for their encouragement while various drafts of this book were being prepared.

I thank three reviewers who looked at different parts of early versions of this book; Dr. John J. Quinn of Argonne National Laboratory; Jeffery A. Kuhn of the Montana State Department of Environmental Quality, and; Richard S. Dinicola, of the USGS Water Science Center in Tacoma, Washington. As peers and colleagues in the field of phytoremediation of contaminated groundwater, I appreciate their insights and careful reviews. I thank Rebecca J. Deckard and John M. Watson of the USGS Enterprise Publishing Network (EPN), and Sandra C. Cooper of the USGS Southeast Area for their thorough editorial and USGS policy reviews of early drafts. Any errors that remain cannot be attributed to them. I thank James Ray Douglas, USGS EPN for his artful scientific illustrations.

The opportunity to prepare this book was provided in 2001 by Kenneth Howell, former Editor at Kluwer Academic Plenum Publishers, now part of Springer. His early support was essential and timely, and I thank him for his persistence, patience, and vision. I thank Mrs. Betty van Herk, Environmental Science Editor at Springer, for her patience, guidance, and useful comments as various drafts of this book were being prepared since 2004.

I thank the South Carolina Department of Natural Resources (SCDNR) for their support in 1998 of the author's early phytoremediation research conducted near Charleston, South Carolina. I thank Thomas Effinger, Director of the Environmental Services Division of South Carolina Electric & Gas Company, for the opportunity to research various methods of monitoring the phytoremediation of a petroleum-hydrocarbon-contaminated shallow aquifer. I thank Brad Atkinson, North Carolina Department of Environment and Natural Resources (NCDENR), Division of Waste Management, for his unwavering enthusiasm regarding the evaluation of phytoremediation as a potential reimbursable technology to restore groundwater contaminated by leaking underground storage tanks, and his persistence in the establishment of a phytoremediation demonstration site in North Carolina. At the demonstration site, discussions with Brad, Dr. Elizabeth Guthrie-Nichols of the North Carolina State University, and J.P. Messier of the Department of Homeland Security United States Coast Guard (USCG), helped guide the development of some parts of this book. I also am indebted to individuals from USGS Water Science Centers around the country whose questions, asked during the author's participation in the USGS National Research Program lecture series in 2007, unknowingly strengthened the discussion of some issues covered in the book.

I also extend thanks to Professor Samuel S. Harrison, Allegheny College (retired) whose enthusiasm for groundwater and his talent for teaching left an impression on this (then) undergraduate student in 1988–89 who had enrolled in one of the few undergraduate courses in hydrogeology available at that time. Posthumous thanks are extended to Drs. O.E. Meinzer, John S. Brown, Walter N. White, and Charles H. Lee, all with the USGS, for their innovative and pioneering research in the United States during the early 1900s into the effects of plants on groundwater—they unknowingly helped to provide a firm foundation for some of the current issues facing the phytoremediation of contaminated groundwater.

Finally, the task of putting together this book permitted the opportunity to be immersed in the excellent research that has been and continues to be published in the growing field of phytoremediation. Because of the almost daily release of new information on phytoremediation, apologies are offered in advance to those whose relevant work may not be cited in this edition.

Contents

Part I Overview of Plants, Groundwater, and Their Interaction

1 Historical Foundation of Plant and Groundwater Interactions 3
1.1 Plant Physiologic Contributions 4
1.1.1 The Natural Philosophers 4
1.1.2 Andrea Cesalpino and Water Absorption 5
1.1.3 Johann Baptista van Helmont, Water, and Plant Growth 5
1.1.4 Robert Hooke and Cells 6
1.1.5 Marcello Malpighi and Fluid Flow in Plants 7
1.1.6 Stephen Hales and Fluid Pressure in Plants 8
1.1.7 Carolus Linnaeus, Georges-Louis Leclerc, and the Organization of Plants 8
1.1.8 Plant Solute Uptake 10
1.2 Hydrogeologic Contributions 11
1.2.1 Water Quantity in the Western United States 12
1.2.2 The U.S. Geological Survey, O.E. Meinzer, and Phreatophytes 13
1.2.3 First Observation of Plant and Groundwater Interaction 15
1.2.4 Charles H. Lee and His Experiments 16
1.2.5 John S. Brown and the Salton Sea 17
1.2.6 Phreatophyte Facies 17
1.2.7 G.E.P. Smith, Plants, and Groundwater Fluctuations 18
1.2.8 W.N. White, Plants, and Groundwater Fluctuations 19
1.3 Summary 23

2 Integration of Plant and Groundwater Interactions 27
2.1 Historical Observations of Water Movement 27
2.1.1 Musings During the Early to Middle Ages 28
2.1.2 Renaissance and Observation 29
2.1.3 Experimentation and Testing: The Beginning of the Scientific Revolution 29
2.2 The Water Budget and Hydrologic Cycle 30
2.2.1 The Water Budget 30
2.2.2 The Hydrologic Cycle 32
2.3 The Hydrologic Cycle, Plants, and Groundwater 33
2.3.1 Surface Tension 33
2.3.2 Vapor Pressure 34
2.3.3 Evaporation 34
2.3.4 Measurement of Evaporation 35
2.3.5 Transpiration 36
2.3.6 Measurement of Transpiration 37
2.3.7 Evapotranspiration 37
2.3.8 Measurement of Evapotranspiration 38

2.3.9 Tank Experiments 40
2.4 Summary 42

3 Fundamentals of Plant Anatomy and Physiology Related to Water Use 43
3.1 Plant Cell Structure, Photosynthesis, Respiration, Growth, and Dormancy 43
3.1.1 History of Unraveling Photosynthesis 43
3.1.2 Prokaryotic and Eukaryotic Cells 44
3.1.3 Photosynthesis, Chlorophyll, and ATP 47
3.1.4 Carbon Fixation: C_3, C_4, CAM, and Aquatic Plants 50
3.1.5 Plant Respiration and Glycolysis 51
3.1.6 Growth and Hormones 52
3.1.7 Dormancy 56
3.2 Roots 56
3.2.1 Root Physiology, Growth, and Depth 57
3.2.2 Root Evolution 60
3.2.3 Adventitious Roots 62
3.2.4 Root Hydraulic Conductivity 62
3.2.5 Effect of Redox Condition on Roots 63
3.3 Roots, Rhizosphere, Bacteria, and Mycorrhizae 64
3.3.1 Rhizosphere Bacteria and Nitrogen Fixation 65
3.3.2 Mycorrhizae 66
3.4 Roots and Water Absorption 67
3.4.1 Diffusion and Osmosis 68
3.4.2 Solute Entry 69
3.4.3 Water and Solute Uptake by Root Cells 70
3.4.4 Effect of Rhizosphere Surfactant Release on Water Uptake 71
3.4.5 Hydraulic Lift and Water Redistribution 71
3.5 Vascular Tissues, Leaves, and Transpiration 72
3.5.1 Xylem and Phloem 73
3.5.2 Cohesion-Tension Theory and Water Movement 76
3.5.3 Stomatal Resistance 78
3.5.4 Factors Affecting Transpiration 80
3.5.5 Soil–Plant–Atmosphere Continuum 82
3.5.6 Radiation Balance 83
3.5.7 Compensating Pressure Theory 84
3.5.8 Stems, Bark, and Lenticels 84
3.5.9 Leaves 85
3.5.10 Vacuoles 88
3.5.11 Plant Life Forms 89
3.5.12 Dew 89
3.6 Plant Water Status 89
3.6.1 Water Potential 89
3.6.2 Measurement of Water Potential 92
3.7 Summary 93

4 Fundamentals of Groundwater Hydrogeology 95
4.1 Henri Darcy and Darcy's Law 95
4.1.1 Hydraulic Conductivity 96
4.1.2 Plants and Groundwater 98
4.1.3 Anisotropy 99
4.1.4 Porosity 99
4.1.5 Hydraulic Gradient 101

4.2 Static Water—Hydraulic Head 101
4.3 Flowing Water—The Bernoulli Equation 102
4.4 Aquifers 102
4.4.1 Unconfined Aquifers 103
4.4.2 Unsaturated Zone, Capillary Fringe, and Capillary Zone 104
4.4.3 Specific Yield and Specific Retention 105
4.4.4 Confined Aquifers 105
4.4.5 Confining Units 106
4.5 Aquifer Properties 106
4.5.1 Transmissivity 106
4.5.2 Storage Coefficient 106
4.5.3 Heterogeneity 108
4.6 Groundwater Flow 108
4.6.1 Equipotential Lines 108
4.6.2 Flow Net Analysis 108
4.7 Groundwater Recharge and Discharge Areas 108
4.7.1 Infiltration 109
4.7.2 Recharge and Discharge Areas 109
4.7.3 Steady-State Flow and Transient Flow 110
4.8 Groundwater and Surface-Water Interactions 110
4.8.1 Wetlands and Swamps 111
4.9 Wells 111
4.10 Groundwater Fluctuations 112
4.11 Groundwater Models 113
4.12 Summary 114

5 Plant and Groundwater Interactions Under Pristine Conditions 115
5.1 Plants and Groundwater Recharge 115
5.1.1 Precipitation and Potential Evapotranspiration 116
5.1.2 Reduction in Recharge 117
5.2 Plants and Groundwater Discharge to Surface Water 117
5.2.1 Reduction in Surface-Water Flow 118
5.2.2 Increase in Surface-Water Flow 121
5.2.3 Changes in Surface-Water Chemistry 122
5.3 Plants and Groundwater Levels 123
5.3.1 Plants and Groundwater Discharge 125
5.4 Plants and Groundwater Chemistry 125
5.5 Summary 126

Part II Plant and Groundwater Interactions for Hydrologic Control

6 Site Assessment and Characterization 131
6.1 Site History of Contaminant Release, Assessment, and Characterization 132
6.1.1 Contaminant-Release History 133
6.1.2 Contamination Assessment and Characterization 133
6.2 Site-Specific Hydrologic Goals 134
6.2.1 Reduction in Groundwater Flow Across Property Boundaries 135
6.2.2 Reduction in Groundwater Flow to Potential Receptors 135
6.2.3 Reduction in Groundwater Recharge at Source Areas 136
6.2.4 Supplemental Use with Other Hydrologic Strategies 136
6.2.5 Contingency Plans 136
6.2.6 Regulatory Approval 136

6.3 Site Visit ... 137
6.4 Plant Hydroecology Assessment and Characterization ... 137
6.4.1 Potential Water Sources and Plant Distribution ... 138
6.4.2 Plant-Nutrient Availability ... 138
6.4.3 Plant Laboratory Studies ... 139
6.4.4 Weather and Climate ... 140
6.4.5 Plant-Available Water ... 142
6.4.6 Soil Bulk Density and Water Content ... 144
6.5 Hydrogeologic Assessment and Characterization ... 144
6.5.1 Groundwater System Concept ... 145
6.5.2 Depth to Groundwater, the Unsaturated Zone, and Infiltration ... 145
6.5.3 Hand-Auger Method ... 146
6.5.4 Hollow-Stem Auger Method ... 146
6.5.5 Sonic Technology ... 146
6.5.6 Direct-Push Technology ... 146
6.5.7 Monitoring-Well Installation and Depth to Groundwater ... 147
6.5.8 Groundwater-Flow Direction ... 148
6.5.9 Hydraulic Conductivity and Aquifer Tests ... 148
6.5.10 Ground-Penetrating Radar ... 149
6.6 Available Phytoremediation Site-Assessment and Characterization Documents ... 150
6.6.1 U.S. Environmental Protection Agency (USEPA) ... 150
6.6.2 Interstate Technology & Regulatory Council (ITRC) ... 150
6.6.3 Remediation Technologies Development Forum (RTDF) ... 153
6.7 Summary ... 153

7 Plant Selection, Installation, and Management to Affect Groundwater ... 155
7.1 Plant Types and Selection Criteria ... 156
7.1.1 Herbaceous Plants ... 156
7.1.2 Woody Plants ... 157
7.1.3 Natural Plant and Groundwater Interactions as an Analogy to Plant Selection ... 163
7.1.4 Plant Succession ... 163
7.2 Site Preparation, Design, and Plant Installation ... 164
7.2.1 Site Preparation ... 164
7.2.2 Site Plants and Planting ... 166
7.2.3 Recharge-Reduction Design ... 171
7.2.4 Hydrologic-Barrier Design ... 171
7.3 Environmental Factors That Affect Plant Growth and Groundwater Use ... 173
7.3.1 Groundwater–Soil–Plant–Atmosphere Continuum ... 174
7.3.2 Soil Physical Composition ... 174
7.3.3 Soil Chemical Composition ... 176
7.3.4 Soil Moisture Composition ... 179
7.3.5 Soil Topography ... 180
7.3.6 Depth to Water Table ... 180
7.3.7 Semiconfined to Confined Groundwater Conditions ... 182
7.3.8 Groundwater Geochemistry ... 183
7.3.9 Plant Ecological Conditions ... 183
7.3.10 Climate and Vapor Pressure Deficit ... 184
7.3.11 Site Operation and Maintenance, Pruning, and Fertilization ... 185
7.3.12 Growing Season Length and Effect on Acceptance of Phytoremediation ... 187
7.4 Summary ... 188

8 Conceptual Frameworks for Phytoremediation to Achieve Hydrologic Goals ... 189
8.1 Initial Approaches to Assessment ... 189
8.1.1 Free-Surface Water Evaporation ... 189
8.1.2 Micrometeorological Data ... 190
8.1.3 Transpiration Well ... 190
8.1.4 Tank Experiments ... 191
8.1.5 Foliage Volume ... 191
8.2 Assessment of Potential Evapotranspiration, Recharge, and Groundwater Discharge for Phytoremediation Effectiveness ... 191
8.2.1 Precipitation ... 191
8.2.2 Recharge ... 191
8.2.3 Potential Evapotranspiration ... 192
8.2.4 Groundwater Discharge and Potential Evapotranspiration ... 193
8.3 Case Study: Reduction in Groundwater Flow Across a Property Boundary, Charleston, South Carolina ... 195
8.3.1 Potential Evapotranspiration Relative to Groundwater Discharge ... 197
8.3.2 Groundwater-Level Fluctuation Monitoring ... 198
8.4 Alternative Conceptual Frameworks for Groundwater Control ... 200
8.4.1 Water-Use Estimate Framework ... 200
8.4.2 Groundwater Flux Framework ... 203
8.4.3 Numerical Model Framework ... 204
8.4.4 Water-Budget Framework ... 206
8.4.5 Plant-and-Monitor Framework ... 207
8.4.6 Other Conceptual Frameworks ... 208
8.5 Summary ... 208

9 Monitoring Plant and Groundwater Interactions ... 209
9.1 Plant Physiologic Monitoring Methods ... 209
9.1.1 Water Potential ... 209
9.1.2 Root Hydraulic Conductance ... 210
9.1.3 Sap Flow ... 211
9.1.4 Stomatal Conductance ... 215
9.1.5 Leaf Area Index ... 216
9.2 Hydrogeologic Monitoring Methods ... 217
9.2.1 Groundwater Levels ... 218
9.2.2 Groundwater Flow ... 221
9.2.3 Vertical Groundwater Flow ... 221
9.2.4 Groundwater Volume Removed by Plants ... 221
9.2.5 Groundwater Discharge ... 222
9.2.6 Soil Moisture ... 222
9.3 Integrative Monitoring Methods ... 223
9.3.1 Geochemistry ... 223
9.3.2 Stable Isotopes ... 223
9.3.3 Meteorology and Plant Characteristics for Groundwater Uptake ... 227
9.3.4 Water-Balance Equation ... 230
9.3.5 Remote Sensing ... 230
9.3.6 Tracers ... 231
9.3.7 ET and Groundwater Models ... 231
9.4 Summary ... 232

10 Economic and Regulatory Factors That Affect the Phytoremediation of Groundwater 233
10.1 Economic Factors that Affect the Implementation of Phytoremediation 234
10.1.1 The Native Vegetation Versus Planted Vegetation Dilemma 234
10.1.2 Hydrologic Control: Phytoremediation Compared to Pump-and-Treat and Trenching 235
10.1.3 Use of Phytoremediation as a Supplemental Remedy 238
10.1.4 Operation, Maintenance, and Disposal Issues When Using Plants 238
10.1.5 Laboratory and Greenhouse Studies 239
10.2 Regulatory Factors That Affect Implementation of Phytoremediation 239
10.2.1 Time Required to Reach Hydrologic Control 239
10.2.2 Transgenics and Other Obstacles to Public Acceptance 240
10.3 Summary 241

Part III Contaminant Interaction, Partitioning, Uptake, Transformation, Metabolism, and Loss

11 Plant Interactions with Biogeochemical Environments 245
11.1 Plants, Ecology, and Biogeochemistry 246
11.1.1 The Flow of Energy and Electrons 246
11.1.2 The Flow of Oxygen 249
11.1.3 The Flow of Carbon 251
11.1.4 The Flow of Nitrogen 253
11.1.5 The Flow of Phosphorus 256
11.1.6 The Flow of Iron 257
11.1.7 The Flow of Sulfur 259
11.1.8 The Flow of Potassium 259
11.2 Plants and Natural Chemical Compounds 259
11.2.1 Allelopathy and Plant-Chemical Warfare 260
11.2.2 Plants as Environmental Indicators 264
11.2.3 Plants and Toxicity Assessment 265
11.2.4 Geobotanical Prospecting and Phytomining 266
11.3 Plants and Extreme Natural Environments as an Analogy for Groundwater Contamination 266
11.3.1 *Spartina* and Mangrove Monocultures 266
11.3.2 Extreme Temperatures 268
11.4 Plant Interactions with Contaminated Soil and Water 269
11.4.1 Natural and Constructed Wetlands 269
11.4.2 Plants and Biosolids 272
11.4.3 Natural Attenuation 272
11.4.4 Plants and Riparian Buffers 274
11.5 Summary 274

12 Chemical and Physical Properties That Affect the Interaction Between Plants and Contaminated Groundwater 275
12.1 Contaminant Partitioning in the Subsurface and Plant Uptake 275
12.1.1 Water–Soil–Contaminant and K_{ow} 276
12.1.2 Water–Soil–Air–Contaminant, and Henry's Law 277
12.1.3 Water–Soil–Air–Plant–Contaminant 278
12.1.4 Leaf and Tree Tissue Processes That Control Contaminant Loss 283
12.1.5 Root Uptake of Contaminants in the Volatile Phase 284
12.1.6 Uptake, Partitioning, and Transport Conceptual Models 285
12.1.7 Plants and Contaminant Interactions 286

12.2 Plant Rhizosphere Processes and Contaminant Fate 288
12.2.1 Microbial, Fungal, and Root-Zone Processes 289
12.2.2 Rhizotron Methodology 291
12.2.3 Root-Zone Changes in Subsurface Sediment Chemistry 291
12.2.4 Release of Root Exudates and Increased Bioavailability 291
12.2.5 Cometabolism 292
12.2.6 Endophytes 292
12.3 Reduction-Oxidation Processes Controlled by Plants and Contaminant Fate 292
12.3.1 Influence of Oxygen, Carbon Dioxide, and Methane on Plants and Contaminated Groundwater 293
12.3.2 Naturally Anoxic Aquatic Environments and Contaminant Fate 296
12.3.3 Plant-Induced Redox Changes at Contaminated Groundwater Sites 296
12.3.4 Water-Table Fluctuations and Contaminant Fate 297
12.4 Plant Biochemical Processes for Groundwater Contaminant Degradation and Detoxification 297
12.4.1 Phase I Reactions 299
12.4.2 Phase II Reactions 302
12.4.3 Phase III Reactions 304
12.4.4 Other Processes of Contaminant Fate 304
12.5 Summary 305

13 Plant Control on the Fate of Common Groundwater Contaminants 307
13.1 Early Evidence of Plant and Contaminant Interaction: Herbicides and Pesticides 308
13.1.1 Contaminant Half-Life Concept 309
13.1.2 Contaminant Bioavailability 309
13.2 Plant Interactions with Aromatic Hydrocarbons: BTEX 310
13.2.1 Plant Interaction and Uptake Pathways 310
13.2.2 Plant Transformation Reactions 312
13.3 Plant Interactions with Polycyclic Aromatic Hydrocarbons 312
13.3.1 Plant Interaction and Uptake Pathways 313
13.3.2 Plant Transformation Reactions 318
13.4 Plant Interactions with Fuel Oxygenates and Additives 320
13.4.1 Plant Interaction and Uptake Pathways 321
13.4.2 Plant Transformation Reactions 325
13.5 Plant Interactions with Chlorinated Hydrocarbons and Solvents 325
13.5.1 Plant Interaction and Uptake Pathways 326
13.5.2 Plant Transformation Reactions 332
13.6 Plant Interactions with Nitroaromatics, NDMA, Dioxane, Perchlorate, and Tritium 333
13.6.1 Plant Interaction and Uptake Pathways 334
13.6.2 Plant Transformation Reactions 335
13.7 Concerns About Plant and Contaminant Interactions 337
13.8 Plant Selection for Specific Groundwater Contaminants 338
13.9 Historical Trends in the Initiation and Continuation of Phytoremediation at Groundwater Contamination Sites 339
13.10 Summary 340

14 Conceptual Frameworks for the Phytoremediation of Groundwater Contamination 341
14.1 Contaminant Mass Reduction Framework 341
14.1.1 The Conceptual Framework 341
14.1.2 Case Study, Fort Worth, Texas 342

14.2 Framework That Accounts for Solute Transport and Plant Processes 343
14.2.1 Advection and Dispersion 343
14.2.2 Diffusion 343
14.2.3 Sorption 344
14.2.4 Volatilization 344
14.2.5 Aerobic and Anaerobic Biodegradation Processes 344
14.2.6 Plant Processes 344
14.3 Existing Conceptual Models 344
14.3.1 PLANTX 344
14.3.2 CTSPAC 345
14.4 Site-Characteristic Data Needed to Support the Framework That Accounts for Solute Transport and Plant Processes 345
14.5 Use of Existing Groundwater Flow and Solute-Transport Models That Incorporate Plant-Related Processes 346
14.5.1 Groundwater-Flow Models 346
14.5.2 Case Study in Florida 346
14.5.3 Groundwater Flow and Solute-Transport Models 348
14.5.4 Unsaturated-Zone Models 348
14.5.5 Guidelines for Model Evaluation 349
14.6 Summary 349

15 Monitoring for Phytoremediation of Groundwater Contamination 351
15.1 Plant Physiologic Methods 351
15.1.1 Rhizospheric Community Analysis 351
15.1.2 Plant Tissue Samples 352
15.1.3 Diffusion Traps 358
15.1.4 Gas Bags 359
15.1.5 Infrared Analysis 359
15.1.6 Plant Fluorescence 359
15.1.7 Compound-Specific Isotope Analysis 360
15.1.8 Stomatal Conductance 360
15.1.9 Leaf Area Index 360
15.2 Hydrogeologic Methods 360
15.2.1 Conventional Well-Sampling Methods 361
15.2.2 Low-Flow Well-Sampling Methods 361
15.2.3 Diffusion and Dialysis Methods 361
15.3 Integrative Methods 362
15.3.1 Root Zone Models 362
15.3.2 Push-Pull Tests 362
15.3.3 Stable and Radioactive Carbon Isotopes 363
15.3.4 Tree-Ring Chemistry and Aquifer Properties 363
15.3.5 Lysimeters 363
15.3.6 Passive Soil-Gas Methods 363
15.4 Toxicity Testing 364
15.4.1 Axenic and Nodule Analogs 364
15.4.2 Laboratory Approaches 364
15.5 Summary 364

16 Economic and Regulatory Factors That Affect the Phytoremediation of Contaminated Groundwater 365
16.1 Plant-Enhanced Contaminant Phase Transfer 365
16.1.1 Natural Plant Toxic Compounds 365
16.1.2 Plant Transfer of Subsurface Contaminants to the Air 366

16.1.3 Fate of Contaminants in Leaf Litter ... 366
16.1.4 Plant Detoxification Reactions ... 368
16.1.5 Contaminant Fate in Food Crops ... 369
16.2 Potential for Transgenic and Mutagen Activation at Phytoremediation Sites ... 370
16.3 Bioaccumulation Potential at Phytoremediation Sites ... 372
16.4 Technical Impracticability and the Role of Phytoremediation ... 372
16.5 Sustainability of Phytoremediation and When to Stop ... 373
16.6 Climate Change, Carbon Sequestration, and the Role of Phytoremediation ... 374
16.7 Summary ... 374

17 Epilogue ... 375
17.1 Phytoremediation Makes Evolutionary Sense ... 375
17.2 The Ideal Phytoremediation Plant ... 376
17.3 The Future of Phytoremediation of Contaminated Groundwater ... 376

References ... 379

Index ... 401

Part I

Overview of Plants, Groundwater, and Their Interaction

He who sees things grow from the beginning will have the best view of them.

Aristotle (384–322 BC)
(quoted in da Farina (1990))

1 Historical Foundation of Plant and Groundwater Interactions

Phytoremediation can be broadly defined as the use of living plants to restore contaminated media to regulatory-mandated levels. Phytoremediation can be used to restore contaminated air, soil, surface water, and groundwater. In general, phytoremediation processes can remove contaminants through direct uptake by roots or leaves and can decrease contaminant concentrations by biotransformation in the root zone or plant tissues or through volatilization or sequestration. Because of the wide range of processes involved in phytoremediation, many terms exist that are used to describe specific interactions between plants and contaminants. Some include phytovolatilization, rhizoremediation, rhizofiltration, phytoextraction, and phytostabilization.

In this book, phytoremediation is defined as the application of existing or planted vascular vegetation, such as trees, to remediate contaminated groundwater. Remediation occurs by natural, plant-mediated processes, internal or external to the plant that render contaminants to less harmful forms or concentrations, in a manner that is scientifically reproducible and defensible. The contaminants discussed in this book include those commonly detected in groundwater at levels that require state or federally mandated corrective action. Contaminant classes include petroleum hydrocarbons, fuel oxygenates, chlorinated solvents, and explosives. Phytoremediation of other contaminants, such as metals, trace elements, and nutrients is beyond the scope of this book.

That some plants can be used to restore contaminated groundwater is a result of plant evolutionary history. Terrestrial plants, for example, evolved in part by controlling the naturally occurring water-potential gradient between the subsurface and air driven by evaporation. This movement of water from the subsurface to atmosphere through plants also represents a potential vector for contaminant transport. For the phytoremediation of contaminated groundwater, contaminant transport in water, or air, through the plant enables a variety of processes to occur that can facilitate the transition of a toxic contaminant into less toxic intermediates, or into bound, immobile residues within the plant. Such specific processes of contaminant biotransformation in plants are part of a broader capability acquired by many terrestrial plants as they evolved in the presence of natural hazards over the past 400 million years.

Fundamental evidence to support the interaction of plants and contaminated groundwater can be traced to early investigations into natural plant–water relations by plant physiologists and hydrogeologists. Early observations made by researchers in each discipline have contributed to the understanding of water uptake and transport in plants, and led to early investigation of the interaction between plants and groundwater. Some of the fundamental questions currently (2011) being asked regarding the extent of plant and groundwater interactions at sites characterized by water *quality* issues, also were asked in the late 1800s and early 1900s in the context of groundwater *occurrence* and groundwater *quantity* issues. Relevant questions asked then and now, and the chapters where they will be discussed in this book, include:

- Can roots remove groundwater from the water table, the capillary fringe, or both? (Chap. 1)
- Can roots survive in saturated sediments often depleted of dissolved oxygen? (Chap. 11)
- How does the relatively stable root distribution of most plants respond to seasonal fluctuations in the water table due to differences in recharge? (Chap. 11)
- Does the uptake of groundwater by plants cause a measurable fluctuation in the depth to water table? Can this fluctuation be measured? Does the lack of a water-table fluctuation indicate the lack of groundwater uptake by plants? (Chaps. 1 and 7)
- What is the greatest depth from which plants can access groundwater? At what depth is plant and groundwater interaction no longer feasible to achieve remedial goals? (Chaps. 7 and 9)

J.E. Landmeyer, *Introduction to Phytoremediation of Contaminated Groundwater*,
DOI 10.1007/978-94-007-1957-6_1, © Springer Science+Business Media B.V. 2012

- What plants use groundwater? Do they use other sources of water? Can these plants be used at contaminated sites? (Chaps. 1 and 7)
- How do some plants survive in areas characterized by high concentrations of organic solutes? (Chaps. 12 and 13)
- What solutes can be taken up by plants? Can contaminants in the dissolved or gaseous phases be taken up by plants? (Chaps. 11 and 12)
- What happens to contaminants after they enter plants? Is contaminant volatilization from plant leaves a decrease or increase in exposure risk? (Chap. 13)
- How long will it take for plants to decrease groundwater contaminants to acceptable levels? (Chaps. 13 and 15)

The historical separation that has existed between the disciplines of plant physiology and hydrogeology may explain why many of these questions have not been answered to the full satisfaction of practitioners of either field, particularly with respect to the restoration of contaminated environments.

Ultimately, the success of the phytoremediation of contaminated groundwater must be based on scientifically defensible and reproducible data that indicate that plants fundamentally interact directly or indirectly with groundwater. The goal of Part I, which consists of five chapters, is to present data that document that plants do interact with groundwater. Without such observations, particularly of the effect of plants on groundwater hydrology, phytoremediation might be incorrectly considered an extension of bioremediation processes based on the use of non-photosynthetic heterotrophic microorganisms to degrade contaminants.

1.1 Plant Physiologic Contributions

The study of plant–water relations encompasses a long period of history that can be traced back to many early cultures. This is because plants have been and remain important throughout everyday life as sources of food, fodder, and medicine. The study of plants, or botany (from the Greek *botana*, meaning pasture, grass, or fodder), can be traced back to the times before Christ (BC). Prehistoric man through trial-and-error probably recognized which plants could be eaten with no ill effects, which plants might cause harm, and which plants could be used to alleviate suffering. For example, the use of plants for medicinal purposes was recorded by the Sumerians as early as 3,000 BC The selection of seeds of desirable plants to secure a more controlled supply of food, medicine, or source of fermented beverages safer to drink than ambient water supplies probably was not a conscious decision, but arose from observations of the germination and growth of undigested seeds in waste-disposal areas. This establishment of a stable supply of plants, or crops, has been implicated by historians as one of the prerequisites for the subsequent establishment of civilization and the development of the arts and sciences.

1.1.1 The Natural Philosophers

The importance of plants to human survival has led to their being a subject of speculation and investigation. Aristotle (384–322 BC) proposed that plants were totally dependent upon the soil based on observation rather than experimental testing. Aristotle deduced that plants grew in soil, or humus, and, therefore, plants must be dependent upon humus for their survival. Aristotle stated that plant roots took from the soil miniature versions of their organic matter and came to be known as the Humus Theory. The legacy of this theory can be seen today in the marketing of plant 'food' by companies that make various soil amendments. The Humus Theory fit Aristotle's view that terrestrial things consisted of a combination of the four elements—fire, water, wind (air), and earth. This idea of the four elements had also been proposed by Empedocles of Agrigentum (492–432 BC). Even though plants were later shown to synthesize their own food, Aristotle was not entirely wrong with his idea of humus uptake, as will be shown later.

A student of Aristotle, Theophrastus of Eresus (372–287 BC) also known as Ferguson, recorded more than 500 plant names in his *Enquiry into Plants* (Loeb Classical Library 1916) and *On the Causes of Plants* (Loeb Classical Library 1990), considered to be the first written record about plants. The name Theophrastus, which means god's speech, was given to him by Aristotle. Theophrastus is often referred to as the Father of Botany. In the first century AD, Pliny the Elder (23–79 AD) also known as Cais Plinius Secundus, became interested in plants while a soldier in the Roman army. During his tours of duty, he noted that the distribution of plants seemed to be related to differences in climate; was he perhaps the first plant ecologist? Again, these differences were based on observation rather than experimentation. Pliny the Elder went on to list up to 1,000 plants in his 37-volume *Natural History* (Loeb Classical Library 1938) in the year 50 AD. In 1596 the Swiss botanist Gaspard Bauhin published *Illustrated Exposition of Plants* which contained about 6,000 plants (Hobhouse 2004).

Plants also were recognized in ancient times for containing strong chemical substances. For example, toxic plant extracts have played a role in hunting and human conflict for thousands of years. In ancient Greece, for example, an extract of hemlock was the poison of choice, as those familiar with the death of Socrates remember. Theophrastus wrote in his *Enquiry into Plants* and *On the Causes of Plants* about plant-derived poisons. Later, under the Roman ruler Nero, the Greek physician Dioscorides classified poisons that included those derived from plants. On the other hand, the physician Hippocrates wrote that certain plant substances that today we classify as alkaloids, had beneficial qualities as long as the dosage was constrained. Therefore, plants have the potential not only to synthesize powerful chemicals but to remain unaffected by them. This interaction regarding plants and natural chemical substances and the analogy it provides for the phytoremediation of contaminated groundwater is discussed in Chap. 11.

1.1.2 Andrea Cesalpino and Water Absorption

The focus of Aristotle and his students on the interaction between plants and *soil* limited the investigation into the interaction between plants and *water*. Also, scientists tended to study what Aristotle had *said* about a particular subject, rather than to go out and observe these subjects directly. Some of the earliest work conducted to understand the absorption of water by plants did not occur until almost fifteenth centuries later by the Italian physician and herbalist Andrea Cesalpino (1519–1603). He concluded that plants absorb water similar to how a sponge absorbs water, perhaps because his observations were made before the development of the microscope. His work was published in 1583 as part of his treatise *Des Plantis libri XVI* (Kramer and Boyer 1995).

The importance of Cesalpino's hypothesis on water uptake by plants was that it was based on the idea of a *physical* process rather than simply conjecture. This is an important distinction because during the time of his studies, plants were regarded as possessing forces of magnetism and suction. In fact, mysticism and alchemy were often used to explain other processes. For example, during the fifteenth and sixteenth centuries, there was great interest in testing Aristotle's four-element theory of fire, water, air, and earth. Many independent discoveries during this time challenged this concept, such as the observations of comets in the early 1600s. Such observations shook ancient ideas about the physical world and, by default, the roles of alchemy and mysticism were put in doubt.

Cesalpino's interest in plants led him and others to collect, dry, and press leaves of various plants so that the leaf structure could be studied. This practice remains to this day and such herbaria are found in most colleges or universities around the world. Today, at least 180 million specimens are listed in the *Index Herbariorum*. Cesalpino had other interests in plants beside water interactions. His observations of fungi, for example, led him to believe that they did not produce seeds and were derived directly from decaying substances. Cesalpino also directed the botanical garden in Pisa, Italy, and became physician to Pope Clement VIII in 1592 (Boorstin 1983).

1.1.3 Johann Baptista van Helmont, Water, and Plant Growth

The Flemish alchemist and scientist Johann Baptista van Helmont (1577–1644; Fig. 1.1), perhaps most famous for his first use of the term 'gas' (from the Greek *chaos*) to describe matter in an airlike state, continued this search into the relation between plants and water and composed perhaps the first laboratory experiment that challenged Aristotle's Humus Theory. In retrospect, his experiment seems almost too simple to have been remembered some 300 years later. He placed a 5 lb (pound) (2.2 kg [kilogram]) willow tree in a pot filled with 200 lbs (90.9 kg) of dry soil to which he added only water. After 5 years (y) the willow weighed almost 170 lbs (77.2 kg), whereas the soil weight remained essentially unchanged. He reasoned that plants,

Fig. 1.1 Johann Baptista van Helmont added only water to a 5 lb (2.2 kg) potted willow tree for 5 years and concluded that the plant increased in weight some 165 lb (75 kg) to a total weight of 170 lb due to the plant's uptake of water, rather than soil, because the weight of the soil did not decrease. This simple experiment challenged the prevailing Humus Theory. The importance of atmospheric gases on plant growth was not known at this time, so van Helmont's oversight into the increase in biomass from CO_2 fixation can be forgiven, perhaps more so than his failure to keep track of the volume of water he added to the pot over time.

therefore, used water to support their sustenance, rather than soil.

Van Helmont reached this conclusion, however, without the benefit of having recorded the total amount of water added over the 5-year period. In fact, neither Aristotle nor van Helmont was entirely correct; plants use both water *and* substances from the soil. What they could not have realized at the time, however, was the role that air, which contained gases such as carbon dioxide (CO_2), played in plant growth and in the increase in the willow's weight. It was not until recent times that it was revealed that plants in fact do take in substances, both mineral and organic, from dead or decaying organic matter in the soil. Aside from the rather dubious connection to alchemy and lack of meticulous quantification, the process of turning water into plant matter provided the first evidence that plants could take up significant quantities of water from soils. Moreover, van Helmont's experiments confirmed similar speculation made earlier by Sir Francis Bacon (1561–1626) that plants used water.

A similar conclusion regarding the importance of water for plant growth was made in 1700 by John Woodward (1665–1728). Woodward grew plants in pots much like did van Helmont, but only in water obtained from various sources, such as rainwater, or river water. Woodward observed the best growth was achieved when soils were added to the water. Additional confirmation that water and soil were important to plant growth was provided by the experiments of Jethro Tull (1674–1741). Tull stated that plants were sustained on fine particles of soil that entered the plant through the roots. This idea gave rise to the then novel but now common practice of tilling, the goal of which is to produce very fine soil particles for nutrient release.

1.1.4 Robert Hooke and Cells

As has been the case throughout history, rapid advances in the sciences usually follow the introduction of a new tool, piece of analytical equipment, or a difference in experimental approach. For example, in the early seventeenth century, many technicians were grinding lenses to make compound microscopes in an attempt to observe physical processes at scales of ever increasing resolution. This had not been done before. All that had been studied was the appearance of everyday things.

One of the first observations made with a microscope was related to plants, and was made by Robert Hooke (1635–1703). Hooke was an innovative scientist who made major discoveries in many fields. Using a two-lensed microscope he made himself, Hooke observed that the seemingly solid structure of dead cork tissues from the pith of an alder tree was, in fact, not a homogenous solid mass but rather composed of multiple copies of smaller, empty structures that he called *cella* (from the Latin word for *cell*, or small room) (Fig. 1.2). He wrote

> *...I could exceedingly plainly perceive it to be all perforated and porous, much like a honey-comb, but that the pores of it were not regular...these pores, or cells...were indeed the first microscopical pores I ever saw...*
>
> Observation XVIII, Micrographia (Hooke 1665)

This microscopic observation of cork cells helped to explain the characteristics of cork perceived at the macroscopic scale: a light-weight, nonabsorbent, compressible material. Hooke stated that this behavior could be explained by the fact that cork was actually a very small quantity of a solid but was spread out over a large dimension. He also observed a similar structure in elder tree pith, the tissues at the center of stems composed primarily of what turned out to be parenchymal cells. It was fortunate that Hooke's microscope was powerful enough to resolve individual dead cork cells, because cork cells are relatively thin and light in weight.

These observations of cellular spaces in non-living tissues led Hooke to speculate not just on its effect on plant structure but also the novel role of air in plant growth. In 1665, Hooke performed a simple experiment that proved that plants need to be exposed to air to survive. He placed seeds in soil both covered and uncovered with a glass jar; the covered seeds did not germinate. Hooke published these and other observations, drawn by hand, in 1665 in *Micrographia*, which became a best seller and was widely read even by non-scientists. Hooke's compound microscopes could

Fig. 1.2 A representation of an etching of cork cells from an alder tree similar to that drawn by Robert Hooke in his *Micrographia* (Hooke 1665). As most of this tissue is composed of air space, his observations explained why cork was lighter in weight than most woody plant samples of similar volume.

magnify between 20 and 50 times, which pales in comparison to those that could magnify up to 300 times made in 1670 by Anton van Leeuwenhoek. Interestingly, when Leeuwenhoek reported in 1678 that he had discovered little animals with these microscopes, it was Hooke who was tasked to confirm these findings.

Hooke's observations of cells in plant tissues were confirmed by other scientists and were seen in animal tissues as well. Almost 200 year later, in 1838, the German scientists Matthias Jakob Schleiden and Theodor Schwann linked these observations of the presence of cells in both plants and animals, respectively, and stated the Cell Theory of Life—that all life consists of cells, and that all cells arise from other cells.

1.1.5 Marcello Malpighi and Fluid Flow in Plants

Recognizing that water played an important role in the growth of plants, scientists began to make attempts to answer the question of *how* water was transported from soils through plants and finally to the air. In the late 1600s, the Italian scientist and physician Marcello Malpighi (1628–1694) took this question to task. Although it may seem odd that a physician would be interested in the flow of water in plants, at that time there was great interest in the study of the hidden flow of blood in the circulatory system of mammals. This inquiry had been initiated by the hypotheses of William Harvey, who realized in 1628, that the human heart was a pump, not a source of heat as had been envisioned up until this time. Harvey stated that this pump pushed blood through veins, which had always been known to contain blood, as well as through arteries, which had long been thought to either be empty or a redundant source of blood. It was Harvey that hypothesized that the two vessels were connected in a closed loop that passed through the heart.

In any case, Harvey's hypothesis was missing a major part; he had not been able to see what connected, or closed, the loop between the arteries and the veins. In 1661, Malpighi used microscopes to investigate blood flow in frogs and observed the presence of the smallest blood vessels, or capillaries, in the lungs. These observations provided the experimental evidence needed to demonstrate that capillaries connected the flow of blood in the arteries with that of the veins. Although Leeuwenhoek had seen capillaries in the tails of fish placed under his microscope earlier than Malpighi, for whatever reason Leeuwenhoek's observations are often forgotten. Malpighi extended such a microscopic approach to the comparative anatomy of various organs in the body of mammals, including humans. He also discovered that insects breathed not with lungs, but through a row of holes on their bodies. For all his efforts, Malpighi is regarded as the Father of Microanatomy. Moreover, unlike many contemporaries, such as Hooke and Cesalpino, Malpighi did not think that fungi arose from spontaneous generation from decaying organic matter, but rather from seed or fragments of themselves.

Within this context of study and method of investigation, it was Malpighi's causal observation of the jagged part of bark that led him to study fluid flow in plants. In one of his experiments, he removed the bark from completely around a tree, a practice referred to as girdling. The bark contained the tissues composed of cells that transport the food made in the leaves, called phloem, but did not contain the tissues composed of cells that transported water, called the xylem. His experiment was designed to determine the role that these structures have on fluid flow in plants, and whether or not it was analogous to the closed-loop system in mammals. Over time, he noted that the bark just above the area that had been removed started to swell, and exuded a fluid, or sap, that was sweet upon tasting. At first, the whole tree above the girdle did not seem to suffer adversely from the experiment. Over a period of a few weeks or months, however, Malpighi noted that the leaves wilted and died, and death of the complete tree soon followed. Malpighi concluded from his experiment that water was transported up to the leaves from the soil through the xylem, which after all had remained intact to the tree after the bark had been removed. However, the tree died not from a lack of water but because the roots had died due to the lack of transport of the necessary sugars that were formed in the leaves down to the roots, which cannot make their own food.

Malpighi also designed an experiment in which he observed a squash seedling planted in moist soil. He noticed that the seedling stopped growing if its first leaf was removed. This result suggested to him that the leaves were important in some nutritive way to plants, much as Hooke had hypothesized. Malpighi published the results of his comparative anatomy of plants as *Anatome Plantarum* (Plant Anatomy) in 1675 (Kramer and Boyer 1995). Ironically, similar observations of the water-transporting structures of wood were made again by Leeuwenhoek in 1676. Malpighi will surface again in Chap. 3 as the discoverer of another important part of plants, which turned out to be directly related to the uptake of water by plants and, therefore, useful to phytoremediation of contaminated groundwater.

Another scientist, the medical doctor Nehemiah Grew (1641–1712), also was interested in the fluid flow of plants. He speculated that the circulation of fluids in plants may be similar to that of the circulation of blood in humans. Some of his anatomical observations were later published in *The Anatomy of Plants* (1682) (Kramer and Boyer 1995). Grew also advanced the idea that roots act as 'mouths' that ingest water from the soil along with air.

1.1.6 Stephen Hales and Fluid Pressure in Plants

A unique feature about the water-transporting xylem cells previously mentioned is that after their differentiation and growth, these cells die, so that the entire plumbing system necessary to keep a tree alive is composed of non-living tissues. Plants can, therefore, transport water using cells that do not require an energy supply. Also, water transport is not constrained by diffusion because adjacent cells walls are not present in the non-living xylem. Although the transport of water in the xylem requires no expenditure of energy by plant cells, the accumulation of ions in the roots as water is taken up from soils lowers water potentials (i.e., the water concentration) in the root cells and does require the input of energy by living root cells. These topics are described in detail in Chap. 3.

The question of how far water can move up the xylem against gravity was an intimidating one for scientists in the seventeenth century. A clergyman and physician named Stephen Hales (1677–1761) made the first direct measurements of the pressure of water in the roots of plants. He, like other contemporaries such as Harvey and Malpighi, had been investigating the flow of blood in mammals and wondered if fluid flow in plants also followed the model of closed circulation. Hales used a piece of bladder, from a goat perhaps, tied around a pruned grape vine to stop its loss of fluid at the cut and realized the bladder could be used as a device to measure fluid pressure in the plant. As he was making his measurements of the root pressure of the grape plants, he noted that the highest flow was measured in the spring before the leaves came out, and that flow (measured as a pressure) dropped off considerably later in the season when the leaves were fully out. Hales then proceeded to demonstrate that this seasonal reduction in flow, or pressure, resulted from the evaporation of water from leaves that pulled fluids from the roots up through the xylem. Hales demonstrated this by removing leafy branches from a tree, completely removing leaves from one branch and allowing increasing amounts of leaves to be left on other branches. Each branch then was set in a container with a known volume of water. Over time, Hales observed that the branches with the most leaves lowered the water level the most in the containers.

Questions remained, however, about whether this movement of fluids in plants was a closed-loop circulation, like that in animals, or simply separate flows in separate tubes in opposite directions. Hales devised an experiment to test the circulation hypothesis in which he took a branch (from an apple tree), attached a tube to the cut end of the branch, filled the tube with water, and cut away the bark and the last growth ring along 3 in. (inches) (7.6 cm [centimeters]) of the branch just above the tube (Fig. 1.3). He also made a cut in the bark about 1-ft (foot) (0.30 m [meters]) from the tube. As he observed the water in the tube (which was 22 ft (6.7 m) long) taken up by the branch, he noted that the first cut became moister, whereas the upper cut remained dry. Hence, he was able to prove that the movement of water in a plant could not be described as a closed-loop circulation. In fact, he showed that a branch could absorb water through either end; sap, however, would only move from the shoots to the roots. In 1727, Hales summed up his observations in the publication *Vegetable Staticks* (or physics) (Hales 1969, reprint).

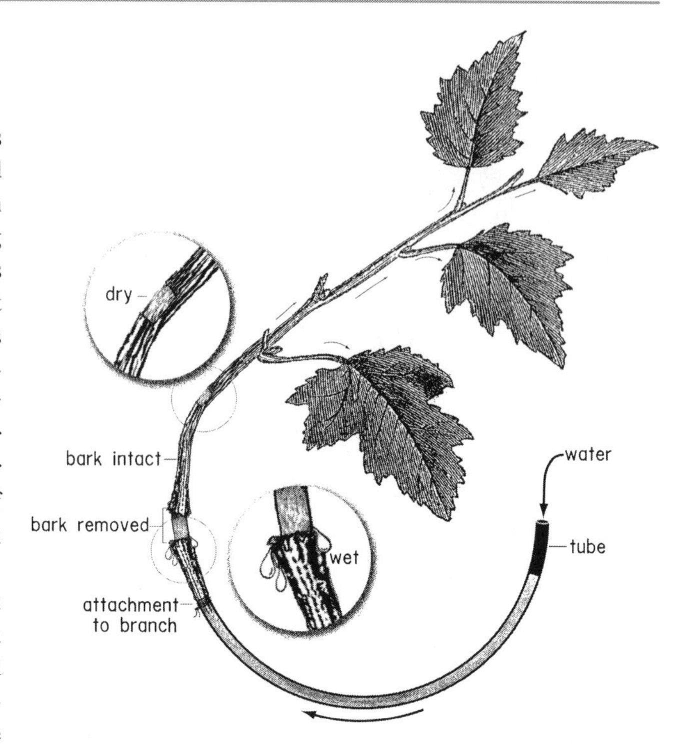

Fig. 1.3 A representation of Hales' experiment which determined that plant fluid circulation is open, not closed like the circulation of blood in mammals. Hales determined this when he attached a branch to a water-filled tube and noted that water flow (*large arrow*) did not occur beyond the first cut, and the upper cut remained dry. Hales also demonstrated that if the leaves were immersed in a bucket of water, water moved in the opposite direction, from leaves to the cut end.

1.1.7 Carolus Linnaeus, Georges-Louis Leclerc, and the Organization of Plants

Increased international trade and immigration characterized the sixteenth and seventeenth centuries and led to an increase in the types of plants being observed by Europeans and Middle Easterners. This increase in the number of plants made it imperative for a systematic classification system to be developed to handle the increasingly large body of known plants. For instance, take the common potato; each area had its own local name for this plant. The potato came to Spain from Peru by Jesuit missionaries, and from Spain, it spread throughout the world, and illustrated the need at that time for a more refined method of classification of plants, because the same plant was often referred to by a common name that also was used to describe many other very different plants. To further complicate things, plants were named by people in their native language. Moreover, the names of plants were cumbersome; for instance, catnip, known today as *Nepeta cataria*, was at one time called *Nepeta floribu interrupte spicatis pedunculatis*. Plant naming was clarified considerably when Latin was agreed upon as the official language for plant classification. Moreover, an additional task was to take this vast body of knowledge, based primarily on the spoken word, and compile it into written form for teaching and instructional purposes.

Even with this common language, problems still plagued early botanists and physiologists, because the Latin was applied to describe the gross physical characteristics of plants, such as flowers, bark, and leaf structure. Early attempts were made by many to improve upon these classification schemes, and the works produced by John Ray (1627?–1705), such as *Historia Plantarum* (1686, not to be confused with the book by Theophrastus of the same name), led to a general plant-classification system that is often used today. Ray divided the flowering plants, or angiosperms, that protect their seeds in fruits or nuts into two broad groups based on the number of proto-leaves a seed has in a small depression called the kotyle, or cotyledon. Seeds with one leaf were called monocotyledons and seeds with two leaves were dicotyledons. As we will see in later chapters, this single characteristic is useful on many levels; monocotyledons grow in height only, whereas dicotyledons also expand in girth. The Frenchman Antoine Laurent de Jussieu (1748–1836) also investigated plant forms based on the number of leaves that emerged from seeds and published his work *Genera Plantarum* (de Jussieu 1789), which was expanded upon in 1830 in *Botanicom Gallicum* by Augustin de Candolle.

The need for a more rigorous method of classification for plants (as well as animals) was ultimately met by the Swede Carl Ingemarsson (1707–1778). His father, Nils Ingemarsson, was a clergyman of the Lutheran Church. Nils loved plants and, to young Carl's delight, placed many unique ones around the church. Carl studied medicine rather than the clergy as had been the hope of his father, but he also studied botany—two fields that were linked by the common use of plants for medicinal purposes. During this time, Carl was asked to bring order to the messy personal collection of biological specimens of the nearby estate of Doctor Kilian Strobaeus (Hubbard 1916). Deemed beyond organization by others, Carl found his calling for turning such chaos into order, and began to show his talent for classification. Over time, Carl Ingemarsson developed a classification system for plants and animals termed binomial nomenclature which is still used today. His ideas were first released in 1735 in the 12-page *System Naturae* (Linnaeus 1735). Carl Ingemarsson is perhaps more widely recognized today as Carolus Linnaeus, as it was the custom for scholars at that time to take Latinized forms of their first names. In addition, the Ingemarsson family changed their surname to Linnaeus, a derivation of the root word for the linden tree, which is a plant, ironically, that can tap groundwater.

The classification system developed by Linnaeus is binomial, meaning two names, because it gave a generic (Genus) and specific (species) name to each plant or animal being classified. A genus refers to a group of related species. The species is the smallest unit of classification and can maintain its features through many generations. The concept of a species can be attributed to John Ray, based on his observations of many plant and animal specimens. Within the binomial system, the concepts of genus and species are related. For example, man is classified as *Homo sapiens* (L.), where the parenthetic L. stands for Linnaeus, to indicate that he was the first to describe and name the species for classification purposes. If a species has different appearances, say, spotted leaves versus plain leaves, this difference is denoted as a variety, such as the tree *Populus deltoides* var. DN-34, where a variety is a different type of species, much like a poodle is a special variety of dog. A cultivated variety, abbreviated as cultivar, is a variety grown for a particular unique feature. Most cultivars are reproduced by cuttings, rather than from seeds, and can be considered a clone of the parent. Such classification can continue on in the opposite direction to even larger groups of classification, such as family, order, class, phylum, and kingdom. Only the genus and species, however, is used when referring to specimens. This binomial system is analogous to how individuals are listed in phone books, where, for example, many people who share a common last name are differentiated by unique first or second names.

To ensure the adoption and use of his new method, Linnaeus coupled the binomial system of differentiation with another older classification method based on the readily observable differences in the reproductive organs of different plants. Previously in 1680, at a time when separate sexes were thought to be a trait characteristic of mammals only, Nehemiah Grew discovered that flowering plants (angiosperms) had both male (stamen) and female (pistil) reproductive structures. Pollen contained in the stamen was necessary for production of a seed in the pistil, a process called pollination. This idea also was studied in 1694 by the German botanist Rudolf Jakob Camerarius (1665–1721).

Linnaeus furthered these notions and used a simple microscope to assist his merging of the sexual classification and binomial schemes. This was a step ahead of differentiating plants based on larger more obvious structures, such as leaves or location of growth. With his microscope, Linnaeus observed that the number of stamen is constant for members of the same kind of plant. This makes sense today, because plants are parts of families that have similar genetic traits that were unknown during Linnaeus' time. Also, these features are less susceptible to changes induced by changes in environmental conditions. He placed plants into 24 classes based on their differences in the sexual parts of each plant.

That Linnaeus was corroborating his binomial classification system with the notion that plants had separate sexes was resisted and, in fact, considered by his peers to be immoral. As a result, Linnaeus, against his wishes, had to leave his teaching position at a university. In the end, it is the binominal system that has stood the test of time rather than the classification scheme based on sexual organs of plants. This is because some plants that have the same number of stamen are *not* related, and a constant number does not indicate genetic relation.

At the time of his publication of *Species Plantarum* in 1753 (Linnaeus 1753), Linnaeus had named and classified almost 6,000 species. This classification provided the basis for later taxonomists to perform more accurate classifications. For comparative purposes, today upwards of 400,000 different species of plants, from the smallest photosynthetic microorganism to the tallest redwood tree, have been classified according to Linnaeus' almost 260-year-old system. Moreover, recent efforts to catalog all living things on earth have reached over one million species. This number and overall diversity would be higher if fossilized plants and animals were included. The convention of using Latin names still is being used today. Moreover, because the binomial classification system is based on structural differences, Linnaeus' system turned out to be congruent with the Theory of Evolution offered later by Charles Darwin.

The classification of organisms was a topic of intense scientific investigation by many scientists during the time of Linnaeus. Another scientist in France also was investigating the relation of various plants and animals to one another. His name was Georges-Louis Leclerc (1707–1788), perhaps more widely recognized by his title Comte (Count) of Buffon, which was a small town in France near his home. He wrote his observations of the natural world into a large encyclopedia first published in 1749 called *Histoire Naturelle*. As a contemporary of Linnaeus and interested in answering similar questions about plant species and organization, he did not accept Linnaeus' system of plant classification based on the numbers of stamen and pistils. Rather, Leclerc noticed that crossing two species often resulted in sterile offspring, and that a species could be defined as a group that produces fertile offspring. Even though he published his book 4 years ahead of Linnaeus, it was the Linnaean system that survived, in part because the Count's book took more than 25 years for the translated version to reach widespread readership in England.

Upon wide acceptance of the binomial system, it became much easier for botanists to focus on other ways to study similarities observed across various plants. Some of the earliest studies by plant botanists involved solving the obvious problem of how plants take up and deliver water to leaves at great heights above the land. These botanists primarily focused on the structural components of plants that are associated with water transport, for example, the xylem and phloem tissues. Botanists also had to study perhaps the most important part of the plant, especially in terms of its relation to subsurface sources of water—the root system. In many ways, it was this study of root interactions with water that provides an early bridge between plant physiologists and hydrologists. These investigations provide a firm foundation of plant and groundwater interactions necessary to support the later application of plants for the remediation of contaminated groundwater.

The classification of plants has not been completed; it is an ongoing occupation, carried out by plant experts known as taxonomists. The task is the same as before: observe, compare, classify, generalize, and specialize. As good as the binomial nomenclature is as a classification system, it remains an artificial system based on some actual relations between genera of plants. It is not, however, a comprehensive taxonomic endpoint—the natural system, which would include all relations between all plant genera.

1.1.8 Plant Solute Uptake

After it was generally recognized that plants take up and transport water from the soil and release it to the air as vapor, scientists began to consider this process in the context of an adaptation for plant survival and reproduction. As early as 1676, the scientist E. Mariotte, whom we will meet again in Chap. 2, conducted experiments to demonstrate that the uptake of water by plants provided a process for them to obtain chemical elements from the earth. These experiments were a logical step from the Humus Theory of Aristotle and the experimental results of van Helmont. Advances in the understanding of water uptake in plants were less dramatic following Hales' work. In 1789, however, it is interesting to note that Samuel Williams calculated that 3,874 gal (gallons) (14,643 L [Liters]) of water passed through an acre of maple trees per day (d) in Vermont, for an unreasonably high flow rate of 6 gal/day/tree (Williams 1789).

Plant research in the early 1800s focused not only on the uptake of water but also of solutes. Remembering the experiment of Johann Baptista van Helmont and his willow tree, it was hypothesized that at least some minerals from the soil dissolved in the water were taken up by plants. In 1804, the

Swiss botanist N.T. de Saussure (1767–1845) found that the absorption of certain minerals by roots was not proportional to the absorption of water. Moreover, he determined that these solutes were not absorbed in proportion to their occurrence in soils, which suggested that roots have a selective permeability that allows entry of different solutes differentially.

The entry of water and solutes into plants was further investigated by H.H. Dutrochet in the 1830s, who explained the entry of solutes into plants as being driven along gradients in solute concentration. Known as osmosis (from the Greek *osmos*, meaning thrust or push) this provided a firmer physical-based process to explain water uptake by plants. This idea persisted for most of the nineteenth century. In fact, Charles Darwin investigated the relation between roots and the uptake of various solutes, such as ammonia carbonate, which resulted in the deposition of brown granules in the endodermal cells of plants, as reported for *Euphorbia peplus* (Darwin 1882). In the late 1800s, J. von Sachs offered the concept that the loss of water from leaves above ground, or transpiration, was related to water uptake in the root zone below ground. He also determined that clay soils contain water more available to plants than sandy soils, an observation that indicates that perhaps von Sachs was familiar with some of the fundamentals of hydrogeology, such as permeability, porosity, and hydraulic conductivity.

In the early 1900s, the idea of osmotic pressures as the metric for water status in plants was beginning to be challenged by the concept of water potential. This challenge arose from measurements of osmotic pressures that could not explain the movement of water throughout an entire plant. Additionally, the devices used to measure water potential became more accurate than the devices used to measure osmotic pressure. Today (2011), the generally accepted theory of water uptake by plants is that water is passively taken in as a consequence of the evaporation-driven process of transpiration from leaves along water-potential gradients; these concepts will be discussed in Chap. 3.

From this short history of the study of plants and their interactions with soil and water, it is evident that the *source* of water taken up by plants did not concern early plant physiologists. However, later plant physiologists did come close to making such a distinction when they developed a generalized system of plant classification based on the apparent source of water used by some plants. Early botanists recognized that certain types of plants could only be found when certain sources of water were present. This observation often is reflected in the common names of plants, such as bog asphodel, pondweed, and water lily. Plants that consistently are found in areas of constant water are called hydrophytes (from the Greek *hydro* and *phyte* meaning water plant)—these are the aquatic plants. Plants that compose the opposite end of the water spectrum, that is, plants that grow in the absence of constant supplies of water, are called xerophytes (from the Greek *xero* meaning dry). Other plants that can inhabit an area with available water but also high levels of salts are called halophytes (from the Greek *halo* meaning sea). As would follow, the xerohalophytes inhabit drier areas with high salt content, such as near evaporite deposits. The plants called mesophytes (from the Greek *meso* meaning middle), grow well in moist soils that are sufficiently aerated and use precipitation, when abundant, or groundwater during less frequent precipitation. The generalized root distribution of these plants and their relation to groundwater is shown in Fig. 1.4. Finally, observations made in the early twentieth century by a then young hydrogeologist, who we will meet soon, gave rise to the term phreatophyte (from the Greek *phreato* meaning well or well plant) to plants that remove water from the capillary fringe or water table. Plants that are totally dependent on groundwater are called obligate phreatophytes, and those that can use other water sources are called facultative phreatophytes. Perhaps this classification scheme can be viewed as the beginning of the integration between plant physiologists and hydrogeologists.

Finally, W.A. Cannon, a botanist with the Desert Botanical Laboratory of the Carnegie Institution of Washington, located in Tucson, AZ, made this statement:

> *The problems which deal with the presence of trees are primarily physiological and have mainly to do with the absorption and conservation of water. Each of these capacities varies with the species. Of the root relations, that of the root-and-water table is of prime importance, owing to the fact that the soil horizon tapped by the roots of trees derives by capillarity, from the level of the groundwater, its perennial supply of moisture.*
>
> W.A. Cannon (1923)

We will see how insightful about hydrogeology this botanist turned out to be.

1.2 Hydrogeologic Contributions

During the late nineteenth century when water potential was replacing osmosis as the metric for water studies in the field of plant physiology, the western part of the recently re-United States was being settled following the Civil War. It quickly became apparent that the largest constraint on potential settlement of this arid area was the lack of large quantities of surface water that were more prevalent in the humid eastern states. The need to examine alternative sources of water provided the necessary economic impetus to study the occurrence of alternative water sources, such as groundwater.

The study of groundwater occurrence and availability during this time was relatively new. So new, in fact, there

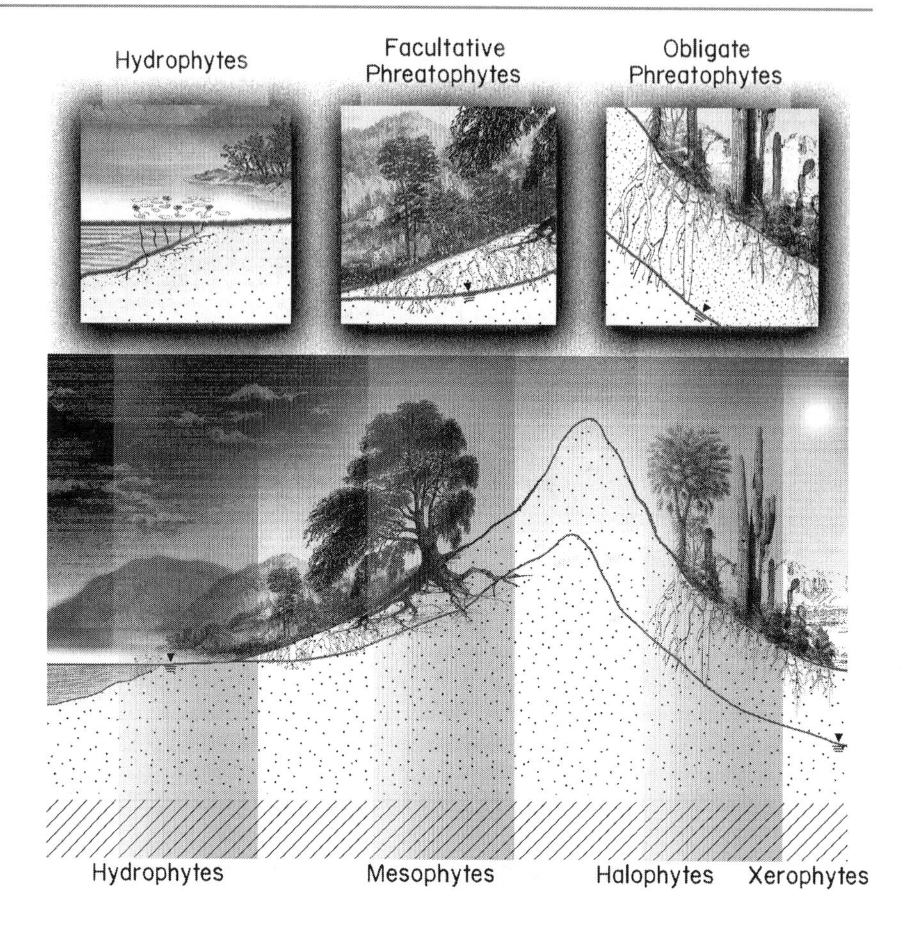

Fig. 1.4 Generalized root distribution with depth and respect to the water-table surface (indicated by the *inverted triangle*) for hydrophytes, phreatophytes, mesophytes, halophytes, and xerophytes. Facultative phreatophytes include plants belonging to the willow family, such as *Populus spp.* and *Salix spp.*, and obligate phreatophytes include plants such as mesquite (*Prosopis glandulosa*).

was no consensus on what name to give this topic of study. For example, the terms hydrogeology and hydric geology were used as early as 1802 and 1885 by Jean Baptiste Lamarck and John Wesley Powell, respectively. Lamarck is perhaps best known for his idea that all species were descended from other species, including man, and is widely regarded as the Father of Invertebrate Paleontology. John Wesley Powell fought in the Civil War, explored the Colorado River, and was the first Director of the U.S. Geological Survey (USGS), a federal agency created by an Act of Congress in 1879 to, in part, systematically study the water resources of the western United States. However, these men were describing geologic processes of erosion and sedimentation controlled by water, not groundwater. The first use of the term hydrogeology to mean the study of groundwater was in a report in 1880 by J. Lucas, and was later used by the USGS (Fuller 1906).

1.2.1 Water Quantity in the Western United States

During the late 1800s, it was clear that issues of water *quantity* were as important to managers then as water *quality* is today. New settlements needed to have an abundant and consistent supply of water to succeed. Because precipitation was neither abundant nor consistent in this area of the United States, and most surface-water sources were intermittent, isolated, and unreliable at best, water trapped deep in underground layers of rock and soil, called aquifers, became the focus of study. The early USGS hydrogeologists who studied these resources crossed the area on horseback. Consequently, some hydrogeologists noticed that certain plants persisted in arid areas in spite of inadequate precipitation. Moreover, plants that grew along often dry river flood plains were similar to those in more humid eastern areas.

Although these hydrogeologists were not trained in classical plant physiology, a few of them hypothesized that the survival of plants in arid areas must be linked to their ability to use groundwater. Understanding whether or not this was the case had not only scientific but economic implications. Such plants could be used as indicators of groundwater that may provide adequate supplies, when tapped, to support municipal usage. Others realized, however, that these same plants also would compete with man for water. Moreover, this groundwater use by plants was considered to be consumptive because the water left the basin after transpiration. Today, we realize that even though consumptive use occurs, these plants support a diverse ecological niche and are necessary to maintain part of the ecological system of riparian habitats; the study of these interactions is referred to as hydroecology.

It may seem ironic that initial observation of the interaction between plants and groundwater occurred in the desert regions of North America. Deserts, however, cover roughly one-sixth of the land area of the United States. Deserts are found in greatest abundance west of the Mississippi River between 15° and 30° latitude. This location is typical for most deserts around the globe where the weather is dominated by dry, falling air under high pressure. In contrast, humid areas are characterized by moist, rising air under low pressure.

The plant species that characterize deserts have developed at least two different strategies for survival in regard to water. Those that use water derived from relatively infrequent precipitation are the xerophytes introduced earlier. These types of plants usually become dormant between precipitation, and they have developed physical mechanisms to retain the water they take up, such as highly modified leaves that became spines, photosynthetic stems, and a unique photosynthetic process that is discussed in Chap. 3. Therefore, xerophytes adapt to limited water availability in deserts, or other areas with limited precipitation or to high infiltration rates through porous soils, by a strategy of water conservation. The second type of strategy is discussed in the next section.

1.2.2 The U.S. Geological Survey, O.E. Meinzer, and Phreatophytes

In the 1910s and 1920s, a hydrogeologist by the name of Oscar Edward Meinzer with the USGS traveled from the very dry lands of southern California to the valleys in adjacent Nevada. He observed that the apparent location of the water table in reference to land surface had a direct effect on plant occurrence and distribution. He summarized his ideas and observations in USGS Water-Supply Paper 577 (Meinzer 1927; the cover of which is shown modified in Fig. 1.5); some of this information was presented earlier (Meinzer 1926). From today's perspective, Water-Supply Paper 577 is regarded as more insightful than even the late author would have realized because his observations of plant and groundwater interaction help establish some of the first principles of the phytoremediation of contaminated groundwater.

Fig. 1.5 In the 1920s, Oscar Edward Meinzer of the U.S. Geological Survey related the presence of certain plants to the depth of groundwater in desert areas of the southwestern United States and summarized in his classic 1927 publication, Plants as Indicators of Groundwater, U.S. Geological Survey Water-Supply Paper 577.

O.E. Meinzer noted that certain plant species appeared to be associated with more consistent sources of water deep below ground rather than the infrequent, light precipitation that characterized the area. He apparently had been thinking about this for some time, as he had coined the term phreatophytes to categorize such plants in an earlier publication (Meinzer 1923). Meinzer's definition of a phreatophyte was a "plant that habitually obtains its water supply from the zone of saturation, either directly or through the capillary fringe" (Meinzer 1923). He was careful in stating that this new term was not designed to create a new category separate from other classifications of plants based on water source, but to overlap the existing terms already used by plant physiologists. In contrast to the previously mentioned strategy of water conservation, these plants that have deeper root systems that tap groundwater employed a strategy of drought avoidance. Today, the term phreatophyte is used routinely by both hydrogeologists and plant physiologists to refer to all plants that tap deep, perennial sources of groundwater.

Phreatophytes have root systems that tap groundwater from the capillary fringe or deeper as an ecological advantage in arid areas or in humid areas where precipitation is not constant from year to year or due to geological constraints such as the presence of lower permeability sediments near land surface. The distinction between obligate and facultative phreatophytes defined previously often is dependent on the initial root distribution with depth, the depth to water table, and the degree of water-table fluctuation; even non-phreatophytic plants can use groundwater if the water table rises into a shallow root zone.

In retrospect, the desert ecosystem turned out to be the best area for O.E. Meinzer to suggest that an interaction between plants and groundwater occurs. This is because the plants he believed to rely on groundwater often were restricted to discrete groups and were physically separated from the plants that relied on precipitation. To draw such conclusions about the interaction between plants and groundwater would have been more difficult to make, for instance, in humid regions, because humid areas offer more sources of water to plants.

As we will see in Chap. 9, phreatophytes can use alternative sources of water, such as precipitation or artificial irrigation, either simultaneous to using the water table or not. For example, although wild alfalfa (*Medicago sativa*) is classified as a phreatophyte, it can be irrigated for commercial purposes and apparently has not suffered from being removed from its native groundwater source (Robinson 1958).

Making the generalized observation between plant occurrence and distribution and the relation to groundwater apparently was not enough to satisfy O.E. Meinzer's curiosity about the effect of phreatophytes on groundwater resources. Meinzer realized the potential economic implications of his observations of phreatophytes as being (1) potential indicators of the presence of sustainable groundwater resources and (2) as sources of competition with man for limited groundwater supplies. Meinzer wanted to use phreatophytes to help determine the potential water yield of aquifers in arid areas and to produce maps showing the location of watering holes for the growing population in desert areas, one of the many tasks assigned to the USGS at that time.

As would be expected at the turn of the twentieth century with such economic issues at stake, O.E. Meinzer and a few other hydrogeologists sought to more fully develop the classification of phreatophytes in arid regions by relating the presence of phreatophytes to the known hydrogeologic properties of the desert. Although the distribution of desert plants with respect to water availability had received attention from plant physiologists before O.E. Meinzer's investigations, the discriminating factors they used to ascribe differences in plant distribution were changes in soil chemistry, such as alkalinity, not the depth to the water table.

A modern example of the linkage between plants, groundwater distribution, and groundwater supplies in the desert areas of the United States is provided by the aptly named resort town of Palm Springs. Located in the desert region of southeastern California, Palm Springs receives an average of 3–4 in. (7.6–10.1 cm) of precipitation annually in the valleys, with a little more in the adjacent mountains. Palm Springs lies within the Coachella Valley, formed between the Little San Bernardino and San Jacinto Mountains. The valley was formed when these mountains were uplifted during geologic activity that occurred along normal faults. These faults, or breaks in the rocks, created zones of impermeable, broken fragments of rock and clay. Groundwater recharged in the mountains tended to accumulate in these lower permeability zones at the faults and was discharged to land surface as springs. The water from these springs typically is hot, being geothermally heated deep underground. These springs were used by Native Americans for many centuries before the arrival of European explorers and then American pioneers. In 1920, J. Smeaton Chase, a resident of the then small town of Palm Springs, perhaps unknowingly wrote of this relation between geology, plants, and groundwater when he stated that Palm Springs was

> *the child of the mountain, for it lives in the mountain's protection and is nourished out of its veins*
>
> Chase (1920)

Geologically controlled springs in arid areas that offered a constant supply of groundwater also offered refuge from the heat underneath the large fan palms (*Washington filifera*, America's only native palm) that typically grow there. These palms have shallow root systems that are less than 20 ft deep. At many oases, palm distribution is aligned in the direction of the fault. This relation among geology, plants, and groundwater led to the rapid development of Palm Springs, and also brought unintended hydrologic changes. Prior to the 1960s, for example, people flocked to Palm Springs because the relative humidity of this area was a comfortable 3–5%. However, the increased numbers of people, landscape plants, gardens, irrigated golf courses, and swimming pools have increased the relative humidity to 20–30%!

Knowledge of the location of these fault-induced springs and subsequent oases was essential in order to survive in this arid area. The USGS recognized this need and was tasked to prepare maps of the springs, or water holes, in these areas. As an added benefit, the locations of springs also indicated the locations of major faults in the area—information needed for hazard assessment. The interactions among deep underground faults, upwelling groundwater, and phreatophytes are referred to as vegetated scarps (Figs. 1.6 and 1.7).

The relation among geology, plants, and groundwater is not unique to the deserts of the United States. The Negev Desert in the southern part of Israel, which is situated in the desert zone that extends from northern Africa, or the Sahara, to the Rub' Al Khali in Saudi Arabia, is characterized by rift valleys that occur between the Sinai Peninsula in Egypt and the Negev Desert, caused by the Syrian–African rift. Here lie the Gulfs of Suez and Elat and the Dead Sea. Along the many fault lines in these areas, upwelling groundwater

Fig. 1.6 An aerial view of the San Andreas Fault in Coachella Valley, California, seen here as a faint line running vertically in the middle of the picture. Groundwater flows from left to right in the alluvial sediments deep underground to discharge to the surface along the fault, thus supporting dense forests of phreatophytes, the dark masses along the left side of the fault in the picture, in an otherwise arid area (Modified from Proctor 1968). These clusters of growth related to geologic structure are called vegetative scarps.

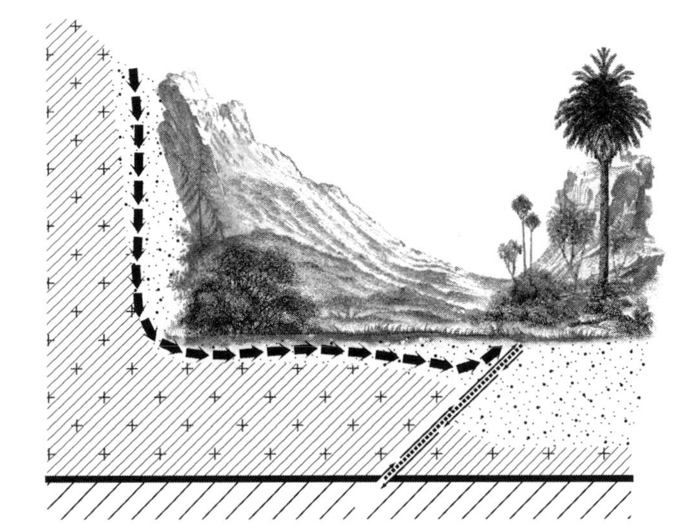

Fig. 1.7 An oasis, similar to that depicted above, is generally characterized by tall palm trees, a welcome sight unique to desert areas often devoid of other vegetation. Oases are present because geologic features such as faults in underlying rocks and overlying sediments create a zone of less permeable material (the *dashed line*), which causes horizontally flowing groundwater (*arrows*) to flow upward toward land surface. Plants such as shallow-rooted palms or deep-rooted cottonwoods, tap this groundwater and thrive even though little current precipitation exists.

supplies water to oases that characterize the area. These oases were instrumental in sustaining nomadic tribes (Issar 1990). Because the oases provided water and food from native date palms, they also were areas of the initial concentrations of many people, including today. In fact, the control of subsurface water, such as the Mountain Aquifer that receives recharge by precipitation that falls on the West Bank, is a continuing source of tension (Nativ 2004).

1.2.3 First Observation of Plant and Groundwater Interaction

O.E. Meinzer and another USGS scientist, Walter N. White, in the mid-1920s decided to observe more fully the interaction between plants and groundwater. To accomplish this goal, they investigated a crop of alfalfa (*Medicago sativa*) on a farm in the arid Escalante Valley of Utah that was planted in 1922 and irrigated for only 1 year. Meinzer and White installed observation wells during 1926 and 1927 to determine the depth to the water table, which varied between 6 and 15 ft (1.8 and 4.5 m) below land surface. On July 13, 1927, they excavated the complete root system of an alfalfa plant near an observation well. Upon excavation, they made a meticulous drawing of the relation of the root system to the measured depth to groundwater. The soil to a depth of 8.5 ft (2.6 m) was clay and peat loam, below which was mostly sand and gravel. With the highest water table at 6.5 ft (1.9 m), the roots grew in the direction of higher moisture content as the depth to water table increased to below 10 ft (3 m) (Fig. 1.8). During the period of high water table, the alfalfa growth in the immediate area was observed to be vigorous, and as the water table declined later in the season, the plants appeared withered.

Figure 1.8 depicts the first published evidence of the interaction between plants and groundwater, and this observation proceeds by more than 70 year the application of such plants for the phytoremediation of contaminated groundwater. More importantly, because Meinzer and White were hydrogeologists, they were perhaps the first scientists to relate readily observable differences in plant morphology, such as vigor, to less readily observable, hidden characteristics that compose some of the fundamentals of hydrogeology, such as depth to water table, thickness of the capillary fringe, and the role that soil properties have in controlling the bioavailability of water to plants. As such, their studies provide a fundamental basis for plant and groundwater interactions and, therefore, for the phytoremediation of contaminated groundwater.

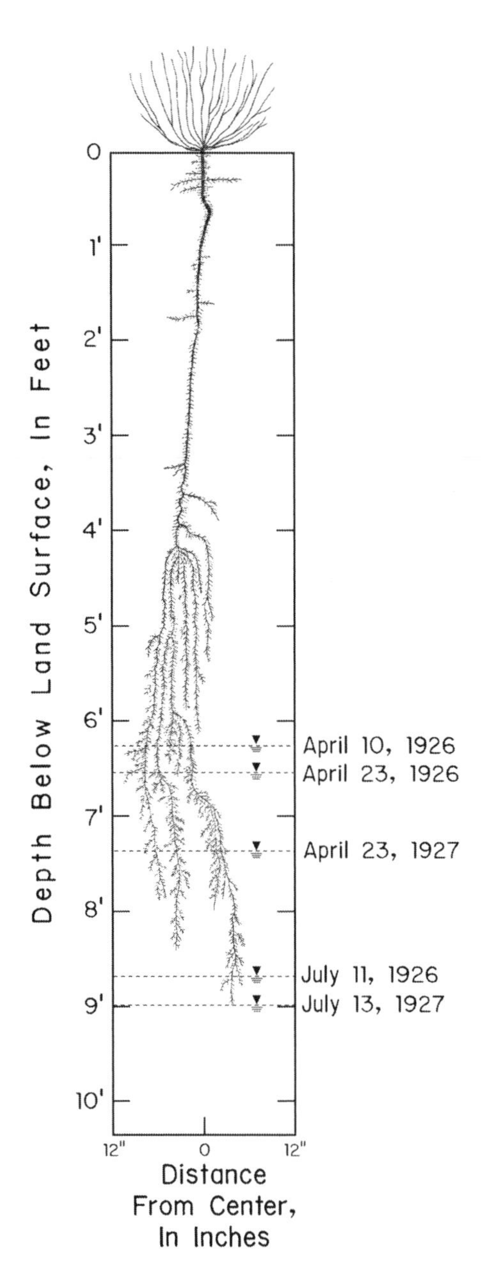

Fig. 1.8 The first published observation of the direct interaction between plants and groundwater, in particular the distribution of the root system of alfalfa (*Medicago sativa*), a wild phreatophyte, and depth to groundwater (the dated and dashed lines). The depth of the water table controlled the depth of the roots and the health of the plant (Modified from Meinzer 1927).

Such scientific observations also had economic implications. These investigations of Meinzer and White confirmed the belief that a profitable cash crop could be grown in arid conditions without costly, long-term irrigation. This notion, in part, led the way to economic growth in the western United States in the early twentieth century. Alfalfa is the third most widely grown crop in the United States (circa 2007) and used for hay, seed production, alfalfa meal (forage), and honey. This direct evidence of plant and groundwater interaction suggests a possible reason why alfalfa was grown and cultivated in Persia, the location of modern-day Iran, prior to the use of irrigation. Finally, alfalfa still can be found to grow wild in parts of arid Africa.

1.2.4 Charles H. Lee and His Experiments

The observations of Meinzer and White were built upon even earlier observations by another USGS colleague, Charles H. Lee. Because of the role of groundwater in controlling the types and distributions of plants in arid regions, Lee was interested in determining what component of the water budget was derived from the transpiration of groundwater by native plants compared to evaporation from surface soils. In his study of the water resources of Owens Valley, California, he recognized that

> *. . .the roots of vegetation, such as wild grass, penetrate the soil to groundwater and become the channels by which a large amount of moisture is conveyed into the atmosphere. Evaporation from bare soil combined with transpiration is in fact the most important element entering into computations relating groundwater for this region.*
>
> C.H. Lee (1912)

It is unclear whether or not he meant that the water was conveyed to the surface through the plants by transpiration, or outside of the roots through the channels.

In attempts to quantify the contribution of groundwater transpiration by plants to total water losses that included soil evaporation, Lee took a rather novel experimental approach. He placed large metal tanks in the ground, filled them with native soil, planted salt grasses, and created a constant artificial water-table surface in each tank through the addition of carefully measured quantities of water from adjacent reservoirs to account for losses by evaporation and transpiration over time. Although this experimental design was an excellent idea, the experiment turned out to be inconclusive, because the roots of the salt grass died before reaching the artificial water table. However, Lee's novel experimental approach did demonstrate an early recognition of plant interaction with groundwater and the potential effect of this interaction on the water budget. More importantly, Lee's research stood in stark contrast to other contemporary research being conducted in the western United States by others, including scientists in the USGS, who did not incorporate the potential use of groundwater by plants into their overall water budgets. As such, Lee's study heralded the beginning of similar investigations in the 1920s, such as those previously discussed by Meinzer and White. (Perhaps O.E. Meinzer and C.H. Lee should be considered the fathers of plant and groundwater interactions.)

Other desert plants were shown to interact with groundwater. Mesquite trees (*Prosopis spp.*) had long been known to have an extensive and deeply penetrating root system, even during the time of Meinzer's studies. Pictures of roots exposed along river banks in the early 1900s clearly depict the great depths reached by mesquite roots often exceeding 50 ft (15.2 m) (Fig. 1.9).

Fig. 1.9 The more than 50-ft (15.2 m) deep root system of a mesquite (*Prosopis spp.*) tree exposed along the eroded bank of a river (Modified from Gatewood et al. 1950).

1.2.5 John S. Brown and the Salton Sea

Other early investigations of the relation between groundwater and woody plants can be traced to work by John S. Brown of the USGS, as he studied the water resources of the Salton Sea region of California (Brown 1923). Brown's research included recording the depths to water table in areas where the occurrence and height of mesquite trees were measured, and the soil types recorded. As can be seen from a subset of Brown's data presented in Table 1.1, as the depth to groundwater approaches land surface, both the presence and growth of mesquite trees increases; when the opposite occurs, the abundance and growth of mesquite trees decreases. Brown's observations also reveal that there may be a limit to the depth that even mesquite roots can reach and support growth, because no mesquite trees were observed in areas where depths to groundwater, as measured in wells, exceeded 75 ft (22.8 m), particularly in well-drained porous sands. This maximum depth of mesquite growth supports observations described earlier that noted mesquite roots did not exceed 50–60 ft (15.2–18.2 m), as had been recorded in areas exposed along stream banks.

1.2.6 Phreatophyte Facies

As the study of the interaction between plants and groundwater became more numerous, O.E. Meinzer summarized the apparent relation between type of plant, both herbaceous and woody, that dominated in a particular area and the depth to water table, as measured in nearby wells. Part of his summary is depicted in Table 1.2.

The relation between the presence and growth of mesquite trees and depth to groundwater also was the subject of mapping activity that depicted the depths to water-table elevations in multiple wells, called contour maps, with respect to the dominant types of plant growth. As Meinzer and R.F. Hare of the USGS studied the Tularosa Basin in New Mexico, they produced maps of the types of vegetation that essentially grew in various clusters, or facies, and the relation of these phreatophyte facies to the depth to groundwater measured in wells (Meinzer and Hare 1915). Meinzer and Hare (1915) claimed the dominant controlling factor on plant distribution in arid areas was the occurrence of and depth to groundwater.

Meinzer and Hare (1915) also defined various phreatophyte facies as having different plants, but dominated by a particular species. These zones included the barren zone, the alkali zone, the mesquite zone, and the creosote bush zone. Other zones were described, but the zones listed above are characteristic of the most dominant ones. The barren zone, as would be expected, was devoid of vegetation, even though depth to groundwater was less than 25 ft (7.6 m) below land

Table 1.1 Relation of the depth to groundwater and occurrence and height of mesquite trees in the Salton Sea area of California, United States (Modified from Brown 1923).

Well location	Depth to ground water from land surface, in ft (m)	Character of mesquite growth	Nature of soil
Palen Mountains, Adams well	20(6.1)	One lone mesquite bush beside well, others not far away	Stream gravel
Eagle Mountains, Anshutz well	8(2.4)	Small clumps of mesquite in vicinity; trench cut in side of canyon shows roots of mesquite penetrating crevices of rock to water	Granite, somewhat jointed and sheared
Blair well	34(10.3)	Abundant mesquite, 10 to 12 ft high	Very porous sand, forms dunes
Chuckwalla well	7.5(2.2)	Mesquite abundant locally in bed of dry arroyo	Stream gravel and clay
Cook well	75(22.8)	None	Porous sand
Imperial	80(24.3)	None	Porous sand and silt
Indian wells, post office	34(10.3)	Abundant forests of mesquite 10 to 15 ft high	Porous sand, forms dunes
Stemberg well	45(13.7)	Scattering growth 2 to 3 ft high	Sandy silt
Palo Verde Valley	12(3.6)	Heavy timber over large areas	Porous sandy silt

Table 1.2 Relation of depth to groundwater and the occurrence of different herbaceous and woody phreatophytes; X indicates the plant was present (Modified from Meinzer 1927).

Depth to water table (feet)	Seepweed (*Dondia*)	Mexican salt grass (*Eragrostis obtusiflora*)	Alkali sacaton (*Sporobolus airoides*)	Chamiso (*Airiplex spp.)*	Mesquite (*Prosopis glandulosa)*
4	X				
4		X	X		
10		X	X		
20			X	X	
30					X

surface (Fig. 1.10). The alkali zone was dominated by salt grasses, such as those described in the experimental studies of Lee (1912), which are adapted to high concentrations of salts left in the upper layers of soils after the evaporation of shallow water. The mesquite zone was located on the sides of slopes composed of sediment deposited by streams, called tallus, that drained the adjacent San Andreas Mountains. Here, depths to groundwater approach 50–80 ft (15.2–24.3 m) because of the porous nature of the coarse materials. Finally, the greatest depth to groundwater, in excess of 100 ft (30.4 m), was characterized only by the creosote bush.

Cross-sectional diagrams made by Brown (1923) also were used to relate the influence of depth to groundwater on the types of plants observed in a particular basin, and he included a description of the vigor of the plants. For example, Brown drew cross-sectional diagrams in areas where mesquite growth appeared to be stunted as the depth to groundwater increased, which appeared to be more of a controlling factor than a result of differences in soil chemistry. He confirmed this observation by measuring the depths to groundwater in wells. His results are summarized in Fig. 1.11.

1.2.7 G.E.P. Smith, Plants, and Groundwater Fluctuations

Approaches to investigate the interaction between plants and groundwater systems that were undertaken by researchers in the early 1900s seem simple by today's standards. At that time, however, the hypothesis that plants interact with groundwater was novel. Use of simple methods during the 1920s provided direct evidence of the interaction between plants and groundwater in the arid southwestern United States. However, it remained unclear if this interaction could be measured more accurately than relating certain plants to the depth to groundwater.

The work of G.E.P. Smith in the mid-1910s started to provide a foundation for addressing this question that still is relevant today in the context of using phytoremediation to achieve hydrologic control of contaminant plumes in groundwater. Smith was an irrigation engineer with the University of Arizona, and he recognized the inherent difficulty in using soil-filled tanks to measure transpiration, as was made evident from Lee's investigations (Lee 1912). Smith used a more direct method that involved placing automatic water-level recorders in wells installed in a forest

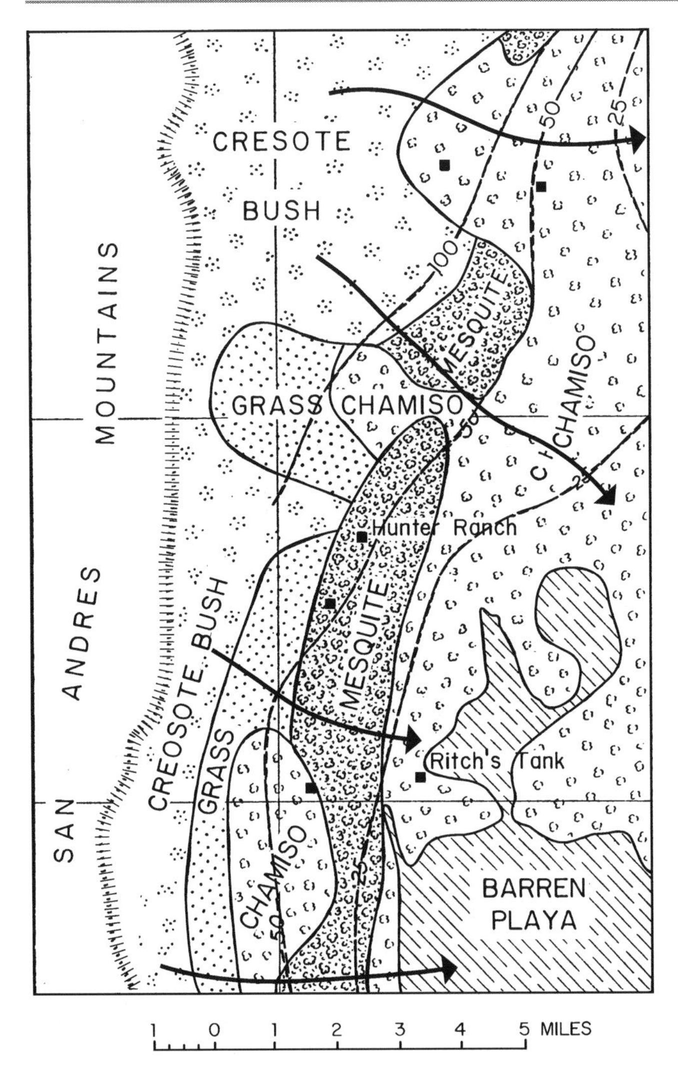

Fig. 1.10 Map showing an early depiction of a phreatophyte facies in relation to the depth to groundwater as measured in wells, shown as lines of equal groundwater level of 25-, 50-, and 100-ft (7.6-, 15.2-, and 30.4-m) below land surface. As the depth to groundwater increased, the plants changed from chamiso to mesquite to creosote bush as the root-depth increased (Modified from Meinzer and Hare 1915). The arrows depict the direction of groundwater flow.

of trees, such as mesquite or cottonwoods. The application of automatic water-level recorders in wells was a novel approach at the time. Previously, such recorders had been used by the USGS to measure surface-water elevation. Smith was able to demonstrate that a decline in groundwater level occurred on a daily basis as a result of the uptake of groundwater by trees. This process was confirmed over time as the water table declined only during periods of tree growth and no declines were observed at night or after the leaves fell. G.E.P. Smith described his observations in an unpublished paper given before the Geological Society of Washington in November 1922. It is not known whether Meinzer was aware of this early research.

1.2.8 W.N. White, Plants, and Groundwater Fluctuations

W.N. White observed a similar predictable, daily decline and rise in the water table that suggested removal of groundwater by plants (White 1932). He observed that during the growing season, automatic water-level recorders indicated a daily fluctuation in the water table in areas characterized by what he called groundwater plants, with only a slowly declining water level was observed in areas without such plants. Overall, he assigned the cause of the fluctuations to the fact that the water in the capillary fringe was being depleted by plants during the day at a faster rate than the capillary fringe could be resupplied with groundwater from hydrostatic or artesian pressures. Conversely, the water table increased at night because the evaporative demand on the water in the capillary fringe was eliminated by a decrease in transpiration, and local groundwater replenished what had been transpired.

Data from the automatic water-level recorders indicated that a groundwater fluctuation occurred daily and was initiated between 9 and 11 a.m., with the deepest groundwater levels measured between 6 and 7 p.m. later the same day after the time of highest transpiration. After 7 p.m., the groundwater levels began to return to levels similar to those measured in the morning. The lag time of a few hours between maximum daily groundwater demand and lowered water table is a result of the night-time replenishment of groundwater back into the cells of plants to meet structural needs, because much of the water originally there was removed during the day. A similar phenomenon can be readily observed in the typically wilted condition of most plants at night after a long hot day that is reversed in the morning before sunrise when the plants return to a non-wilted state.

Daily groundwater fluctuations caused by plants also were observed to have a seasonal pattern. Groundwater fluctuations were observed to begin in the spring after foliage emerged but were not observed to occur after a hard frost, or in other areas characterized by plowed fields or where plants grew but the water table was deep and beyond the reach of roots. White's observations of groundwater fluctuation also were correlated directly with changes in air temperature, wind movement, and sunlight intensity, and indirectly with humidity. These relations are similar to the factors that control evaporation. White was not surprised to note that the greatest drawdown, or declines in the groundwater table of up to 2.5 in. (6.3 cm), was observed on days that were hot and windy, and that even the amplitude of the daily groundwater fluctuation varied with the various stages of plant health and growth. An example of the effect of plant uptake on groundwater levels observed by White, in

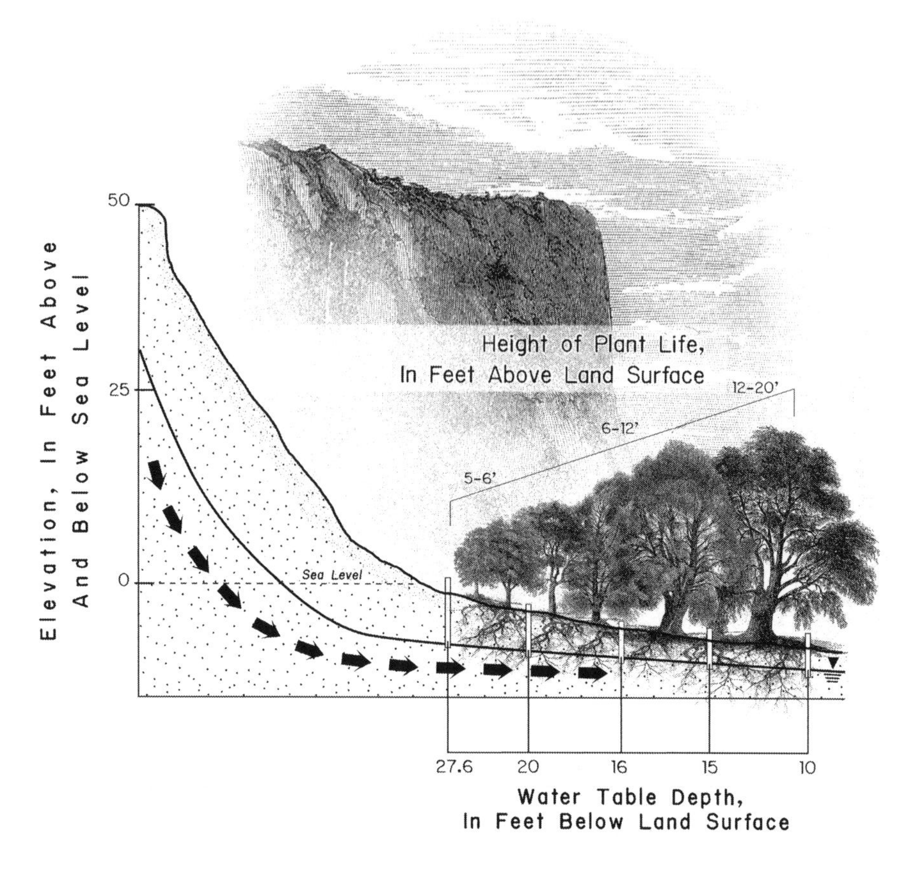

Fig. 1.11 Diagram of a cross section showing that as the depth to groundwater in wells decreased the height and size of mesquite (*Prosopis glandulosa*) trees increased (Modified from Brown 1923). The arrows represent groundwater flow.

this case at a nonirrigated field of alfalfa, is depicted in Fig. 1.12.

White also investigated plant-induced, groundwater-level fluctuations in separate fields composed of greasewood, shad scale, salt grass, and sedges and marsh grasses. The maximum groundwater drawdown observed ranged from 1.5 to 4.25 in. (up to 10.7 cm). As would be expected, the greatest drawdown occurred when the plant growth was the densest and the plants were growing rapidly. White also measured groundwater-level fluctuations in a thicket of willow trees, a plant more representative of the type often used in phytoremediation applications to address groundwater contamination. In such willow thickets, White noted maximum fluctuations in the water table of 3.75 in. (9.5 cm) during hot, clear weather in the summer, with no groundwater fluctuations in October after frost and leaf drop had occurred.

The lack of groundwater-level fluctuation in wells in cleared fields further supported White's conclusions that plant uptake of groundwater caused the observed daily rise and fall of the water table in planted areas. As shown in Fig. 1.13, groundwater-level fluctuations were observed in a field of greasewood but not in cleared land during the same monitoring period.

A similar method was employed by White to test the hypothesis that plants affect the water-table level, although he measured the groundwater fluctuation before and after some of the plants had been removed. To do this, water-level recorders were placed in wells in a field of alfalfa during the summer of 1926 and measured for a few days before and after the alfalfa was cut. The result on both the elevation and daily fluctuation in the water table is shown in Fig. 1.14.

These experiments were performed during periods of no precipitation, so the changes in the groundwater levels observed could be directly related to the presence or absence of phreatophytes.

White used similar field data to develop an empirical equation to describe groundwater use by transpiring plants during a 24-h period. This is an important contribution because it determines the water used by a plant that is attributed specifically from groundwater, not soil moisture. White's equation also can be used to compute the specific yield, or volume of groundwater that will flow due to gravity, of an aquifer based on plant-induced groundwater-level drawdown measurements (White 1932). White's equation is

$$Q = y(24r + s) \tag{1.1}$$

Where Q is the depth of groundwater transpired (inches or centimeters), y is the specific yield of the soil zone in which the observed groundwater-level fluctuation occurs (% by volume), r is the hourly rise in water table, or rate of groundwater inflow, in length per time from 12:00 to 4:00 a.m., the time of assumed zero transpiration when

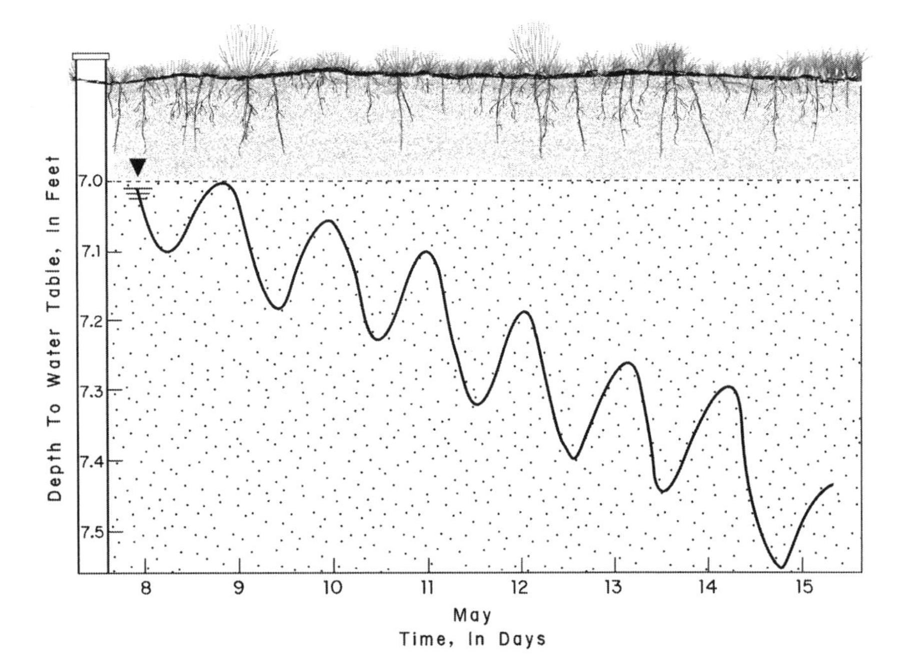

Fig. 1.12 One of the earliest observations of the daily fluctuation in groundwater over time from its highest level (*dashed horizontal line*) in a well installed in a nonirrigated alfalfa field in 1925. The daily decline occurred as groundwater was used by the plants. The groundwater-level rebound was sequentially lower each day, as no precipitation occurred (Modified from White 1932).

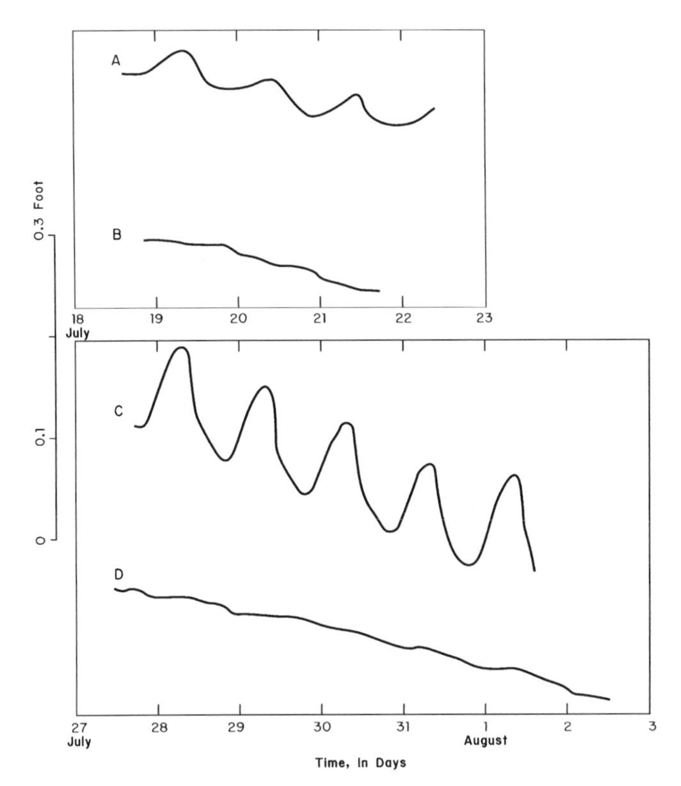

Fig. 1.13 Comparison of the daily fluctuation of the water table in a field of greasewood (*A*, *C*) and in a control plot of cleared land (*B*, *D*) showing the effect of plant and water use on groundwater levels relative to soil-water evaporation (Modified from White 1932).

groundwater levels recover by induced local convergent flow, and s is the net fall in groundwater level during the same 24-h period, in length. The terms r and s are derived from the water-table fluctuation data generated in a well at a site over at least a few days.

An important aspect of White's work was using laboratory approaches at the field scale to determine exactly how plant removal of groundwater was causing the observed groundwater-level fluctuations. White designed tanks that were filled with soil, water, and plants to reproduce the conditions of the water table in the field, similar to tanks designed by Lee. Figure 1.15 depicts a diagram similar to that White used to describe the effect of plant uptake of groundwater on a simulated water-table surface. At the bottom of each of six experimental tanks, a layer of gravel was placed and covered by a thicker layer of soil. Water was then added to the gravel layer through a recharge well in the center of the tank that was open at the bottom. Water displaced the air and saturated the highly permeable gravel and subsequently the less permeable soil above. When these two layers were completely saturated, with no air spaces, the addition of new water was stopped. Above the water level in the saturated soil, water was drawn up farther into the soil by capillary forces against gravity. This movement created a zone of water under tension.

White, among others, noted that most plants that relied on groundwater had roots within the capillary fringe, and some even had root growth below the water table in fully saturated sediments. White's experimental tanks contained plants ranging from grasses to woody plants common to the western United States, and the results of his experiments indicated that as the roots took up groundwater from the capillary fringe to meet transpiration demands, a hydraulic gradient was established that caused groundwater to move upward from the capillary fringe. The volume of groundwater removed from the capillary fringe by plants during the day induced groundwater to move upward from the saturated

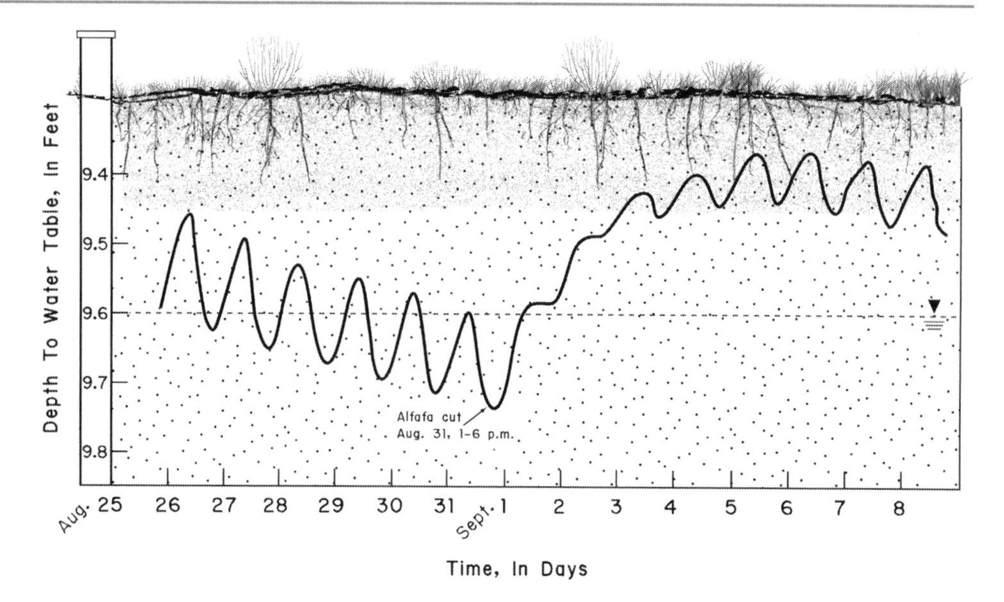

Fig. 1.14 Daily fluctuation of the water table beneath an alfalfa field from August 25, 1926 (*dashed horizontal line*) during plant-uptake of groundwater, until August 31, 1926, when the alfalfa was cut. When the alfalfa began to regrow, the groundwater-level fluctuation resumed (Modified from White 1932).

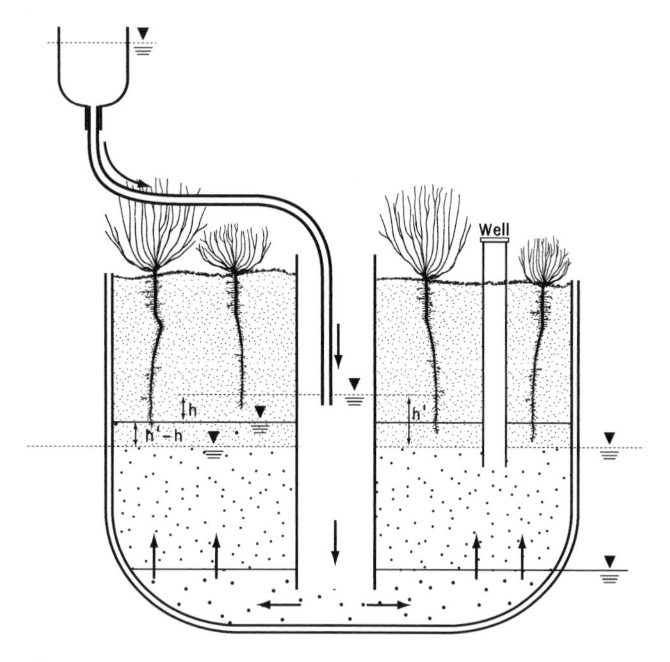

Fig. 1.15 Diagram of an experimental tank filled with an upward-fining profile of sediment, artificial groundwater (▼ depicts water-table surface) supplied by an external reservoir through a central pipe, and plants. Unfortunately, the plants used so much water that when it was depleted, the plants died (Modified from White 1932).

zone along this gradient to replenish water lost to the plants from the capillary fringe. This is purely a physical process. If the uptake of water by plants from the capillary fringe and replenishment by the upward movement of groundwater is faster than the rate of recharge of the aquifer by hydrostatic pressure, artesian flow, or lateral flow from upgradient areas, the water-table level declines. When the plants do not remove water from the capillary fringe, for instance at night when transpiration is lower or ceases altogether, the water table rises to replenish water that was lost.

A study conducted by the USGS during 1943–1944 by Gatewood et al. (1950) to determine the potential groundwater quantity that could be made available to the War Department (the original name of the Department of Defense) from the Safford Valley, Arizona, also provided additional, fundamental information on the interactions between plants and groundwater. As part of that investigation, the amount of groundwater being used consumptively by saltcedar trees (*Tamarix spp.*) was investigated before and after removing the trees in order to make available that volume of groundwater for other uses. These researchers used a variety of methods to determine the volume of groundwater used by plants, including tank, transpiration well, seepage-run, inflow-outflow, chloride increase, and slope-seepage methods. A major conclusion of their study was that of the 28,000 acre-ft (3.45×10^7 m^3) of water calculated to be used by vegetation along 9,303 acres (3.76×10^7 m^2) of a 46-mi (mile) (74 km [kilometers]) reach of the Gila River during the 1-year study, 23,000 acre-ft (2.83×10^7 m^3) was derived solely from groundwater. Hence, 82% of the water demand came from groundwater, and most of that was from saltcedar. Suffice it to say, this study was the first of its kind on a scale large enough to indicate that plant and groundwater interactions had significant hydrogeologic, water budget, and economic effects.

During this time period, the effect of plants on groundwater was included in textbooks such as *Hydrology* (Wisler and Brater 1956), although inclusion of these interactions fell out of favor in later textbooks. It took until the 1970s for the effect of phreatophytes on groundwater to once again be examined, although from the perspective of the effect of plants on flood control in open channels. Such work was done using analytical methods (Bouwer 1978). Still, questions as to the magnitude of the effect of these plants on groundwater, not just as indicators of the resource, remained unanswered. Such unanswered questions had serious economic implications for most of the western United States, because it had been hypothesized that a great proportion of natural groundwater discharge from a particular basin was due to the removal, through transpiration, of groundwater by plants. Hence, this amount of discharge had to be quantified before estimates could be made of the amount of groundwater that could be pumped at well fields without overstressing the aquifer system or affecting the ecological aspects of the area. Moreover, similar questions would be asked as the interaction of plants and contaminated groundwater was investigated, especially to achieve hydrologic containment or control.

1.3 Summary

Plants require water and, for some plants, this need can be met by groundwater. As early as the 1920s, it was clear that certain plants in arid regions of the United States possessed what was called the 'groundwater habit' prior to the introduction by O.E. Meinzer of the now widely-used term phreatophyte. It also was shown empirically that certain plants grow roots to the water table and capillary fringe and use groundwater to meet transpiration demands. This use of groundwater results in a measurable fluctuation in the water-level in wells.

Why is this information important to the phytoremediation of contaminated groundwater? These early observations provide an unbiased confirmation that plants and groundwater interact. This historical perspective is essential to those environmental professionals who either have to implement or regulate phytoremediation projects that involve contaminated groundwater. A chronology of important events in plant physiology and hydrogeology is provided in Table 1.3 to support the almost 100-year-old observation of plant and groundwater interactions.

Table 1.3 Timeline of important events in plant physiology and hydrology which provides a basis to support the fundamental interaction of plants and groundwater (References are found in the text of Chaps. 1 and 2).

Date	Significant events in plant physiology	Significant events in hydrology
300 BC	Aristotle stated that plants use only soil for growth (Humus Theory).	
	Theophrastus recorded more than 500 plant names in his *Enquiry into Plants* and *On the Causes of Plants*.	
250		Archimedes stated principles of floatation.
50 AD	Pliny the Elder lists up to 1,000 plants in his 37-volume *Natural History*.	
1398		English parliament passed laws to prevent pollution of English rivers.
1569		One of the earliest publications on water dowsing was written by J. Besson.
1580		Palissy's *Discourse Admirables* was published and discounted the popularly held notion that seawater returned to the highlands by underground passages.
1583	Andrea Cesalpino stated in his *Des Plantis libri XVI* that plants absorb water similar to a sponge absorbing water, rather than by forces of magnetism or suction.	
1620	J.B. van Helmont provided evidence that plants take up significant quantities of water rather than soil, which discounted the Humus Theory, and confirmed speculations made earlier by Sir Francis Bacon.	
1653		Pascal stated that a fluid exerts an equal pressure in all directions.
1665	Robert Hooke used a crude microscope (30X) and called empty holes of dead cork tissue "cells", in *Micrographia*.	
1667	John Ray classified 18,600 species of plants; divided angiosperms into monocotyledons and dicotyledons.	
1670	Anton van Leeuwenhoek, built the first useable microscope, which could magnify up to 300X.	
1674		Perrault determined the water budget, P-ET = R, in *Treatise On The Origin of Springs*.
1675	Marcello Malpighi experimentally determined that water was transported to the leaves through the xylem, as published in *Anatome Plantarum*.	

(continued)

Table 1.3 (continued)

Date	Significant events in plant physiology	Significant events in hydrology
1682	Nehemiah Grew, in *The Anatomy of Plants*, investigated plant sexuality.	
1687	John Ray penned *Historia Plantarum* in three volumes and helped define the concept of species.	E. Halley contributed to the concept of water budgets and published An estimate of the quantity of vapor raised out of the sea by warmth of the sun
1694	Rudolf Jakob Camerarius investigated plant sexual reproduction.	
1699	John Woodward described water loss through plants as passing through pores.	
1727	Stephen Hales, in *Vegetable Staticks*, measured root pressure but concluded that evaporation from leaves is more important in water transport in plants.	
1735	Carolus Linnaeus published *System Naturae*, only 12 pages long.	
1738		D. Bernoulli stated in *Hydrodynamics* that for a fluid to move faster, it must lose an equal amount of pressure; the converse also is true.
1749	Georges-Louis Leclerc published the first volume of 36 volumes of *Histoire Naturelle*.	
1753	Carolus Linnaeus introduced binomial nomenclature in *Species Plantarum*.	
1754	Charles Bonnet saw gas bubbles emitted from the leaves of underwater plants.	
1759	Caspar Wolff observed apical meristem cells.	
1761	German botanist Jakob Gottlieb Koelreuter produced the first documented hybrid plant.	
1772	Joseph Priestly demonstrated that plants take up carbon dioxide and release a gas, although he did not know at the time that it was oxygen.	
1779	Jan Ingenhousz stated in *Experiments on Vegetables* that the gas released by plants is oxygen and only occurs in the green parts of the plant during the day.	
1785	William Withering described the treatment of heart disease with digitalis from Foxglove leaves.	
1789	Samuel Williams calculated that almost 4,000 gal of water was transpired by 1 acre of maple trees, in *The Natural and Civil History of Vermont*.	
1790	The poet Göethe published *An attempt by J.W. von Göethe, Privy Councilor of the Duchy of Saxe-Weimar, to Explain the Metamorphosis of Plants*.	
1796		English farmer Joseph Elkington applied knowledge of groundwater and geology to drain wet lowlands.
1804	N.T. de Saussure suggested that the permeability of roots varies to explain solute uptake by plants.	
1817	Chlorophyll was isolated by P.J. Pelletier and J.B. Caventou.	
1827		English geologist William Smith, who published the first geologic map, published *On Retaining Water in Rocks for Summer Use*.
1830	H.H. Dutrochet explained the entry of solutes into plants by osmosis.	
1832	Robert Brown observed and named the nucleus of Orchideae.	
1838	Cell Theory of Life was promoted by M.J. Schleiden and T. Schwann, for plants and animals, respectively.	
1844	Hugo von Mohl discovered chloroplasts in cells of green plants, and used the term protoplasm to describe material within a living cell.	
1854		Mechanical windmill was first used to pump groundwater in the midwestern United States.
1856		Darcy described flow of water through porous media in *The Fountains of Dijon*.
1863		Dupuit ignored effect of vertical flow in regional groundwater flow.
1866	Gregor Mendel discovered the basis of the inheritance of physical characteristics.	
1872		The term acid rain was coined by Robert A. Smith.
1879		U.S. Geological Survey was established by Act of Congress.

(continued)

Table 1.3 (continued)

Date	Significant events in plant physiology	Significant events in hydrology
1880	J. von Sachs stated that transpiration of water from leaves was related to water uptake in the root zone.	J. Lucas first used the term hydrogeology to indicate water under the ground in *The hydrogeology of the lower Greensands of Surrey and Hampshire*.
1882	Charles Darwin reported root uptake of ammonia carbonate.	
1886		Forchheimer expanded on Dupuit's concept of regional groundwater flow using differential equations.
1897		First known use in the United States of the term aquifer appeared in W.H. Norton's *Artesian Wells of Iowa*.
1903	The Danish botanist Christen Raunkiaer introduced the classification of plants based on the location of buds.	First use of P-ET = R by the USGS, in *The Water Resources of Molokai, Hawaiian Islands*, by W. Lindgren.
1904	The rhizosphere was described by Lorenz Hiltner.	
1912	The first hybrid poplar, *P. deltoides* × *P. trichocarpa*, was produced.	
1922		G.E.P. Smith observed daily groundwater-level changes were related to peak periods of plant transpiration.
1923		O.E. Meinzer coined the term "phreatophyte" after observing survival of plants on groundwater in arid areas of the United States.
1927		O.E. Meinzer and W.N. White related root penetration to depth of water table. Meinzer authored USGS Water-Supply Paper 577.
1930	Van Niel's experiments indicated that oxygen released by plants comes from H_2O, not CO_2.	
1932		W.N. White measured groundwater-level changes related to grasses and trees and created the equation to calculate the volume of groundwater taken up by plants.
1940		Hubbert used flow nets to simulate groundwater flow between streams.
1941	Water was confirmed as the source of oxygen released by plants.	
1942		*Hydrology* textbook, edited by O.E. Meinzer, discussed transpiration with respect to groundwater.
1950	Georges Morel produced plant clones from undifferentiated cells.	J.S. Gatewood and others observed groundwater use by salt cedar in Arizona.
1954	Calvin cycle that depicts the path of carbon from photosynthesis was revealed.	
1966		Textbook *Hydrogeology* by Davis and DeWiest discussed using groundwater-level changes to determine water removed by plants.
1967		J. Hem related high chlorides in tree tissue to chloride-rich groundwater.
1975	R. Chaney reported the uptake of metals by plants grown in sludge.	
1978		*Groundwater Hydrology,* by Herman Bouwer, was published and specifically described the effect of phreatophytes on groundwater.
1983	First genetically modified plant was created which resulted in a tobacco plant resistant to an antibiotic.	
1986	The U.S. Environmental Protection Agency (USEPA) approved the release of the first genetically modified crop of herbicide-resistant tobacco.	
1989	Term phytoremediation was first used at the *Conference on Hazardous Waste Research*.	
1990	First phytoremediation project was funded by the Rocky Mountain Hazardous Substances Research Center.	
1991	Professor Ilya Raskin (Rutgers) used phytoremediation in a Superfund proposal.	
1992	H. Sandermann introduced the concept of the green liver as part of a pesticide-herbicide model that stated plant reactions are more like mammals than microbes.	
1993	S.D. Cunningham and W.R. Berti used phytoremediation in reference to plant and metal interactions.	
1998	S. Rock, USEPA, coined the term phytotechnology.	
2003	The book *Phytoremediation*, edited by S. McCutcheon and J. Schnoor, is published	
2006	International Phytotechnology Society founded.	

2 Integration of Plant and Groundwater Interactions

When he first came here with his divining rod,
he saw a thin vapor rising from the sward,
the hazel pointed steadily downward,
and he concluded to dig a well

Walden (H.D.Thoreau 1854)

The early investigations by plant physiologists who examined the interaction between plants and water were not concerned about the source of the water. Hydrogeologists, on the other hand, examined the effects that various demands had on groundwater resources in the water-poor region of the United States and reported the relation between depth to groundwater and plant distribution. For the most part, these investigations were carried out independent of each other over time and space (refer to Table 1.3). The lack of integration between the fields of plant physiology and hydrogeology continued throughout most of the twentieth century.

Evidence, however, of an early beginning of integration can be traced back to, of all things, the ancient act of dowsing for water using a divining rod. Commonly referred to as water witching, the diving rod conjures up images of a person under intense concentration pacing the ground holding a forked tree branch and waiting for the cut end to point downward under its own energy, thus indicating the presence of groundwater (Roberts 1951). Although water witching dates back to the sixteenth century, it was discredited in USGS Water-Supply Paper 416 because it was not considered to be scientifically defensible (Ellis 1917).

Whether or not one believes that water dowsers can find groundwater, it is interesting in terms of the history of plant and groundwater interactions to note the choices of wood used by dowsers. For example, the most common woods used are willow and witch hazel. Witch is from the Anglo-Saxon word *wicen* meaning to bend. This aptly describes the pliable nature of the wood of the witch hazel. Both witch hazel and willow often can be found where the depth of groundwater is shallow; these plants, therefore, can be classified as being facultative phreatophytes. Early dowsers likely chose the willow and witch hazel for use as a divining rod principally because the dowsers found these trees growing where water was located very near the land surface. Hence, even folkloric evidence of an interaction between plants and groundwater has existed since the advent of water witching. Fortunately, a more scientifically defensible process that unifies the fields of plant physiology and hydrogeology is found in one of the most fundamental concepts of hydrology—the hydrologic cycle.

2.1 Historical Observations of Water Movement

Of the total volume of water on earth of about 330 million cubic miles or about 360 quintillion gallons (1,360 quintillion L), 97% is in oceans and too salty for direct use by humans. The remaining 3% is freshwater, but most (2%) is ice. About 0.6% is groundwater, and less than 0.001% is surface water. Perhaps an easier way to envision the availability of freshwater is by analogy: if all of the earth's water were equal to a gallon, the amount of freshwater available for use would equate to a full tablespoon.

Under the influence of less than half of the sun's energy that reaches the earth, water is continually vaporized from the oceans and other surface-water bodies into the atmosphere where the water vapor condenses as precipitation and returns to the earth as freshwater. Because little new water has been created over time, the water in the oceans, rivers, lakes, atmosphere, and underground are continually exchanged back and forth between compartments. Some water is slowly returned into the cycle, as occurs when inland surface water evaporates and returns as precipitation in another basin. Groundwater discharge takes the longest amount of time to complete the cycle. Total exchange of global water occurs about every 3,000 years.

The exchange of a fixed, finite source of water between compartments is defined as the hydrologic cycle. The hydrologic cycle can be viewed as an account, at any given time, of the status of water in the various compartments of the

J.E. Landmeyer, *Introduction to Phytoremediation of Contaminated Groundwater*,
DOI 10.1007/978-94-007-1957-6_2, © Springer Science+Business Media B.V. 2012

earth. An understanding of the hydrologic cycle took time. The history of unraveling the hydrologic cycle depicts the evolution of thinking in many aspects of science that began with philosophical musings, led to direct observation of physical processes, and finally to quantification of these processes though direct experimentation and measurement.

2.1.1 Musings During the Early to Middle Ages

The first recorded musings about the movement and storage of water were most likely derived from observations of water that was readily available, such as precipitation and surface water (Fetter 1988). In some places, an apparent dilemma occurred between the observations of infrequent precipitation but the seemingly constant presence of lakes, streams, and springs. As early as 800 BC, the Greek philosopher Homer stated in Book 21 of the *Iliad* that

> *The deep-flowing Oceanus, from whose deeps every river and sea, every spring and well flows.*

Anaxagoras of Clazomenae (500–428 BC) wondered about the source of water seen in rivers, and stated that

> *Rivers depend for their existence on the rains and on the water within the earth, as the earth is hollow and has water in its cavities.*

These writings (cited in Fetter 1988) hint at an initial understanding of the hydrologic connection between oceans and rivers. At the time, however, this connection was believed to be made by the presence of large subsurface reservoirs and not precipitation, because the amount of precipitation did not seem to be able to support observed quantities of surface water. However, Plato (427–346 BC) wrote about a potential source of subsurface water as the connection, a cavern which he called Tartarus.

It has been shown, especially during droughts, that surface water does consist of groundwater, called base flow, so Plato and his contemporaries were not entirely incorrect in relating surface water to an underground source of water. In this case, however, they were referring to water being held in one large cave, which supplied all the water seen in the rivers. Water flowed from the oceans underground up to the tops of mountains and then flowed back down in rivers to the ocean. The fact that the landscape of Greece, which is dry most of the year and consists of limestone, fostered these observations is important in understanding the context in which the observations were made. The Greeks used the many sinkholes in the area as sources of drinking water and as locations to convey excess surface water to the ocean during flooding. Also, the Greeks noticed that there seemed to be more water in rivers than was supplied by precipitation and runoff, which supported the notion that the difference was made up by water from underground caverns.

In contrast to these ideas, in the first century BC, the Roman architect Vitruvius (90–20 BC) hypothesized, among other things in *De Architectura Libri Decem* (Fetter 1988), that precipitation and snow falling in the mountains reappeared as springs and streams in low-lying areas near the ocean. Seen from today's perspective, this notion was very insightful, in the sense that the source of water in the tops of mountains did not invoke the presence of subterranean reservoirs where water would have to flow uphill against gravity. Aristotle thought that springs and surface water were not connected and that springs were not caused by precipitation of rainwater from the sky but by vapors from underground cavities. In hindsight, Aristotle can be forgiven for his incorrect notion, because his conclusions were based on observations of the continual dripping of water from the roofs of karst caves in Greece, thermal springs, and the fact that some offshore springs flowed during high tide.

Most translations of the *Bible* contain a statement that reveals a beginning of an understanding of the hydrologic cycle:

> *For He draws up the drops of water, He distills His mist in rain, which the skies pour down, and drop upon man abundantly.*
> Job *36:27–28* (Revised Standard Version (RSV) 1971)

Another, similar statement in the *Bible* is attributed to King Solomon:

> *All streams run to the sea, but the sea is not full; to the place where the streams flow, there they flow again.*
> Ecclesiastes *1:7* (RSV)

The role of the sun in the movement of water from ocean to sky and back to land was noted in the writings of Aristotle. He stated,

> *Now when the sun in its circular course approaches [sic], it draws up by its heat the moist evaporation: when it recedes the cold makes the vapour that had been raised condense back into water which falls and is distributed through the earth. This explains why there is more rain in winter and more by night than by day: though the fact is not recognized because rain by night is more apt to escape observation than by day.*
> Aristotle, *Meteorologica*, 359.27-360.25 (Ross 1927)

The Chinese and Arabian cultures represented the height of civilization during the Middle Ages. Many advances were made by these societies in the fields of chemistry and astronomy. Therefore, it is ironic that they left little record of advanced understanding of the hydrologic cycle.

In Europe, Norse mythology reveals that the Viking universe consisted of nine worlds, and that the whole universe was held up by a large tree (Helfman 1972), a common theme in many ancient mythologies and often called the World Tree. These nine worlds (nine being the most significant number in Norse mythology) were located along the tree, called Yggdrasil from *Ygg* meaning the terrible one and *drasil* meaning horse, perhaps referring to the horse that the wise elder Odin rode. The tree just happened to be an ash tree. The branches held the stars and sky upward from the earth. The upper level was the realm of the gods, called the Asgard. The middle world was occupied by the Midgard, and the lower Hel. The World Tree was supported by three roots, and each root obtained water from three separate wells supplied by springs. Because these wells would have contained groundwater, perhaps it is coincidence that the ash tree is a phreatophyte. A similar legend circulates among the Buddhists in India, in which a spring runs at the base of a tree they worship; the tree, being a willow, also is a phreatophyte. Such images of a World Tree whose roots extend to groundwater also were used by the Aztecs of Mexico, who called their phreatophyte Tota. Even the cottonwood tree commonly used in sun dances by Native Americans is a phreatophyte.

2.1.2 Renaissance and Observation

Previous speculation during the Middle Ages about natural processes that followed theological proclamation began to be replaced by rational thinking and observational approaches that characterize the Renaissance Period in the fifteenth and sixteenth centuries (Durant 1953). Perhaps the best example of understanding the movement of water through direct observations rather than speculation is that provided by Leonardo da Vinci (1452–1519). He spent countless hours near streams and waterfalls, meticulously drawing the paths taken by seeds he threw into the water. Da Vinci, however, continued the notion of the Greeks that water rose from the oceans to the mountains through underground reservoirs, similar to observations he made of blood rising in the body to supply flow to a cut. Da Vinci had a fundamental understanding of the hydrologic cycle, however, and wrote in his notebooks (Curdy 1923)

> *Or do you not believe that the Nile has discharged more water into the sea than is at present contained in all of the watery element? Surely this is the case. If then this water had fallen away from the body of the earth, the whole mechanism would long since have been without water. So therefore one must conclude that the water passes from the river to the sea, and from the sea to the rivers, ever making the same-self round, and that all of the sea and the rivers have passed through the mouth of the Nile an infinite number of times.*

Because this process still invoked the transfer of water from oceans to mountains, Gregory Reisch (1467–1525) attributed such upward water movement to suction. Indeed, suction can move water against gravity, but not the great distances required to support these hypotheses.

Insightful explanations about the movement of water that did not include the need for underground transfer of ocean water to the mountains were made by the potter and geologist Bernard Palissy (1510–1590) in *Discourse Admirables* in 1580, such as

> *Rain water that falls in the winter goes up in summer, to come again in winter. And when the winds push these vapors the waters fall on all parts of the land, and when it pleases God that these clouds (which are nothing more than a mass of water) should dissolve, these vapors are turned to rain that falls on the ground.*

His conceptual model, although clear to us today, was not widely accepted by his contemporaries. This may have been a consequence of his being accused of heresy for insisting that fossils were the remains of once living creatures. He also has been considered the Father of Agricultural Chemistry on account of his work with manure application in cultivated fields to support plant growth over time. A short biography of this and other hydrological contributions can be found in Deming (2005).

2.1.3 Experimentation and Testing: The Beginning of the Scientific Revolution

The beginning of a more modern approach to understanding the movement of water in the hydrologic cycle was the quantification of observations that occurred in the seventeenth century. During this period, science emerged as a systematic method of inquiry about observations of the natural world with less influence from theological ideology (Durant 1953).

In the late 1600s, Pierre Perrault (1608–1680) wanted to determine if enough precipitation occurred to supply the flow observed in rivers that drained a basin. Until this time it was thought that precipitation amounts were insufficient to supply the flow in rivers, and that these flows were supplemented with water from underground caverns that, in turn, were supplied by ocean water that entered through holes in the ocean floor. This made sense, as previously discussed, because rivers typically flow even when no precipitation has recently fallen. Between 1668 and 1670, Perrault measured the precipitation that fell in the basin that drained to the Seine River. Then he made what may have been the first measurement of the discharge of water out of a basin. When Perrault multiplied the amount of precipitation by the drainage basin area, he found that the amount of precipitation, P, was six times greater than the discharge, Q, of water from the valley (Fig. 2.1), as stated in 1674 in *De l'origine des fountains* (*Treatise on the Origin of Springs*).

A more accurate measurement of the flow of water in the Seine River was perfected later in 1686 by Edme Mariotte (1620–1684). To measure flow, he measured the velocity of the river by using floats to calculate the distance that the float, being a surrogate for a particle of water, traveled per unit time and multiplied this velocity by the measured cross-sectional area, or the water depth multiplied by the river

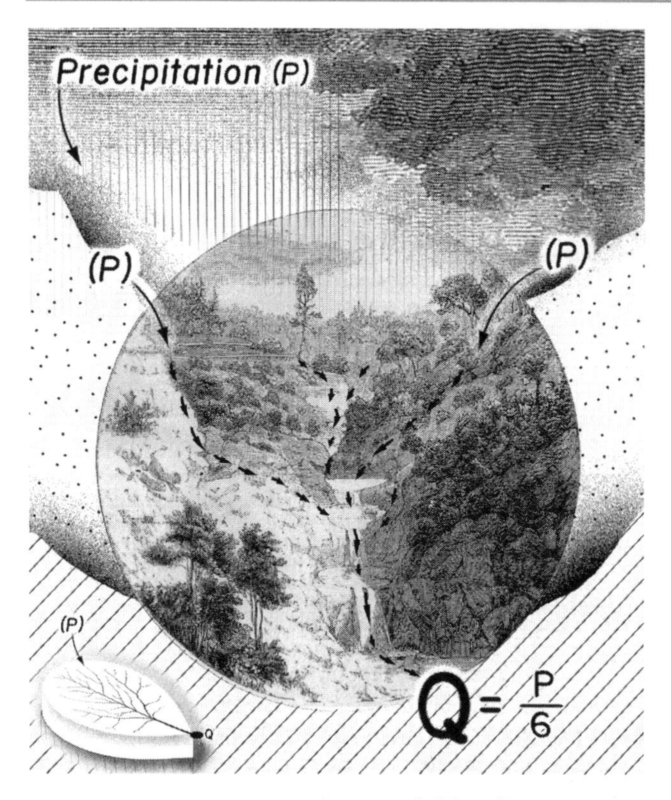

Fig. 2.1 A conceptual depiction of the Seine River basin where it was determined using a novel experimental method that the amount of water delivered to the basin by precipitation, P, was much greater (Up to six times) than the amount of water that left the basin by stream discharge, Q. This observation provided empirical evidence that a source of extra water existed to supply groundwater, evaporation, and ultimately transpiration.

width, to determine the discharge of water. He concluded as Perrault had stated that precipitation was sufficient to supply the flow of water in the river and to springs. More than 100 years later, similar observations of the relation between discharge and precipitation were made by J.F.D. Smyth in the Colony of Virginia (Rosenshein et al. 1986).

2.2 The Water Budget and Hydrologic Cycle

Although the measurements made by Perrault and Mariotte indicated that precipitation was greater than six times the river discharge, which refuted the need to invoke a source of water from massive subterranean caverns, the question remained: what happened to the balance of rainwater that remained in the basin? Moreover, what was the source of water to support the amount of precipitation being measured? Perrault (1674) stated that

> *... to cause this river (the Seine) to flow for one year, from its source to the place designated, and which must serve also to supply all the losses, such as the feeding of trees, plants, grasses, evaporation ...*

In 1687, the astronomer Edmond Halley (1656–1742) partly answered these questions by making accurate measurements of the amount of water being evaporated from the surface of the Mediterranean Sea. He concluded that the volume of evaporated water was sufficient to supply the water discharging to the oceans from local rivers. Halley's observation provided the foundation that water returned to the land from the oceans not though subterranean holes but as water vapor through the sky. Today, we know that about once every 10 day the moisture in the air falls and is exchanged with new water vapor.

Evaporation, therefore, became an important component to consider when investigating water flow in a basin, but this still only accounted for the removal of water from the rivers. Similar evaporation investigations also were performed in England and Wales. In 1802, for example, John Dalton (1760–1844) calculated a water balance for many counties in England by using the novel approach of a network of raingages and was able to state

> *Upon the whole then I think that we can finally conclude that the rain and dew of this country are equivalent to the quantity of water carried off by evaporation and by the rivers.*

Here is one of the initial records of the balance between the input of water to a basin and the removal of water from the basin, from which grew the concept of a water budget. These early scientists, however, had not accounted for all of the sources and sinks for water in a basin, especially from the standpoint of plants and groundwater.

2.2.1 The Water Budget

These early investigations into quantifying the balance between the inflow and outflow amounts of water in a basin provided the foundation for early conceptual models of the hydrologic cycle and the framework to quantify this cycle in terms of a water budget. Because the total quantity of water in the earth is finite, it can be handled mathematically using continuity equations. Thus, for any particular valley or basin, such as the Seine River valley studied by Perrault (1674), we can state the following equality shown in Eq. 2.1:

$$W_{\mathrm{Inflow}} = W_{\mathrm{Outflow}}, \tag{2.1}$$

where W_{Inflow} refers to water that enters a representative basin, such as precipitation per unit time, and W_{Outflow} refers to water that leaves a representative basin, such as by river discharge per unit time. Equation 2.1 represents steady-state or equilibrium conditions, such that the amount of water entering a basin during a certain period of time is completely removed from the basin during the same time, with no storage, S, of water in either lakes or groundwater. Mathematically, this can be expressed as the change in storage, ΔS, over time, Δt, equals zero, or $\Delta S/\Delta t = 0$.

This water-budget approximation can be considered valid when looking at the inflows and outflows of water for a particular basin over a period of many years, in which any changes in storage that do occur can be considered small relative to the larger quantities of inflow and outflow. When shorter timeframes are studied, however, $\Delta S/\Delta t$ will not be 0, and a transient form of Eq. 2.1 must be used, such as

$$W_{\text{Inflow}} - W_{\text{Outflow}} = \pm \Delta S/\Delta t. \qquad (2.2)$$

From Eq. 2.2 it can be seen that if water inflow is greater than outflow, storage of water will occur (the sign of $\Delta S/\Delta t$ will be positive), and if water inflow is less than outflow, then storage will be depleted (the sign of $\Delta S/\Delta t$ will be negative). Perrault's work showed that precipitation, or water inflow, provided six times more water than could be accounted for in surface-water flow out of the valley. Hence, there was either storage in the basin or water was leaving the system through an unaccounted process. Was this unaccounted water stored as groundwater or in lakes? Or was this water leaving the basin in other ways than by river discharge and evaporation? This imbalance of water must have troubled Perrault, for as part of his study he attempted to make what was probably the first measurement of water outflow by evaporation from surface water, even preceding the evaporation measurements made later by Halley.

In addition to losses of water from a basin by stream discharge and evaporation, there is the outflow of water through porous soils and sediments (infiltration), which recharges aquifers. Because of the interference of soil particles to infiltration, the time it takes to return this subsurface water to surface-water bodies is considerably longer than the time it takes water in rivers and streams to discharge to the ocean. Considering these additional processes of water outflow from a basin, Eq. (2.2) becomes

$$P - (Q + G + E) = \pm \Delta S/\Delta t, \qquad (2.3)$$

where P represents precipitation, Q represents river discharge, G represents groundwater flow or discharge, and E represents evaporation.

Although the removal of surface water by evaporation, E, was recognized by Perrault and Halley in the late 1600s, it wasn't until the experiments by Stephen Hales in the 1720s, however, that the evaporation of water through plant leaves was recognized and considered in water-balance calculations. Therefore, this process of the evaporation of water from a basin through plants, as transpiration, T, had to be included in the water-budget equation. Since both evaporation and transpiration are the conversion of water from a liquid to a vapor, the two processes often are combined into one term, called evapotranspiration, ET. As such, Eq. 2.3 becomes

$$P - (Q + G + ET) = \pm \Delta S/\Delta t, \qquad (2.4)$$

If we assume that the change in storage of water in a basin can be considered negligible over a long period of study, such as $\Delta S/\Delta t = 0$ in Eq. 2.4, and that we can further combine surface-water and groundwater return flow to the oceans, R, then Eq. 2.4 can be simplified to

$$P - ET = R. \qquad (2.5)$$

To summarize, Eq. 2.5 states that within a particular basin, the precipitation, P, not returned to the atmosphere as water vapor from evapotranspiration will flow as surface water in rivers or much more slowly as groundwater to the ocean, unless springs supplement surface flows. As was observed by Perrault (1674), because precipitation to the Seine River valley was six times the volume of surface water discharged from the basin, the balance could be attributed to groundwater flow, storage, and evapotranspiration. The removal of water by evapotranspiration occurs at a much faster rate than by subsurface flow and at a fairly constant annual rate, even though precipitation tends to vary widely. Therefore, knowledge of the role of evapotranspiration, in particular plants, becomes important. For example, in the continental United States the loss of water by evapotranspiration accounts for almost 70% of precipitation, or 2,871 bgal/d (billion gallons per day) (10,910 Mm^3/d [million cubic meters per day]). Of this total, a low estimate for transpiration by plants not used for economic purposes, called consumptive use, is about 106 bgal/d (403 Mm^3/d), or about 3% of total evapotranspiration, although consumptive use can be much higher (Moran et al. 2007). An excellent review of the impact of vegetation on the hydrologic cycle at a global catchment scale is presented by Peel et al. (2010).

In Alley et al. (2002), an example is provided that further emphasizes that evapotranspiration is a large component of the hydrologic cycle. In central Kansas, groundwater recharge by precipitation accounted for about 10% of total precipitation over a 6 year period. In addition, one explanation for recharge being such a small fraction of precipitation is that much of the precipitation that infiltrates into the upper layers of soil is rapidly used by plants. For example, Healy et al. (2007) stated that of the 76% of precipitation that infiltrated, up to 85% of this volume either evaporated or transpired.

2.2.2 The Hydrologic Cycle

The deceptively simple equality of Eq. 2.5 often is presented in the form of a variety of conceptual models that depict the flow of water between the various compartments of the hydrologic cycle. When it comes to the depiction of the direct removal of water by plants, or the indirect removal by evapotranspiration, there is considerable disagreement. Most graphs, flowcharts, or descriptions of the hydrologic cycle depict the outflow of water by plants in the form of precipitation, soil moisture, or surface water if the plants are aquatic macrophytes. Indeed, any subsurface water shown to be removed by plants in a particular schematic is most often assumed to be derived only from soil moisture in the unsaturated zone, not groundwater.

Although the method of presentation can differ widely, two representative models are presented in Figs. 2.2 and 2.3. In the first example, evapotranspiration is explicitly presented, but groundwater as a potential source for evapotranspiration is not (Fig. 2.2); in Fig. 2.3, plant use of groundwater is directly linked to water from aquifers. Ironically, Viessman et al. (1977) stated that transpiration, T, can be composed of both soil moisture and groundwater but did not display this in the figure. Also excluded are the interactions between surface water and groundwater.

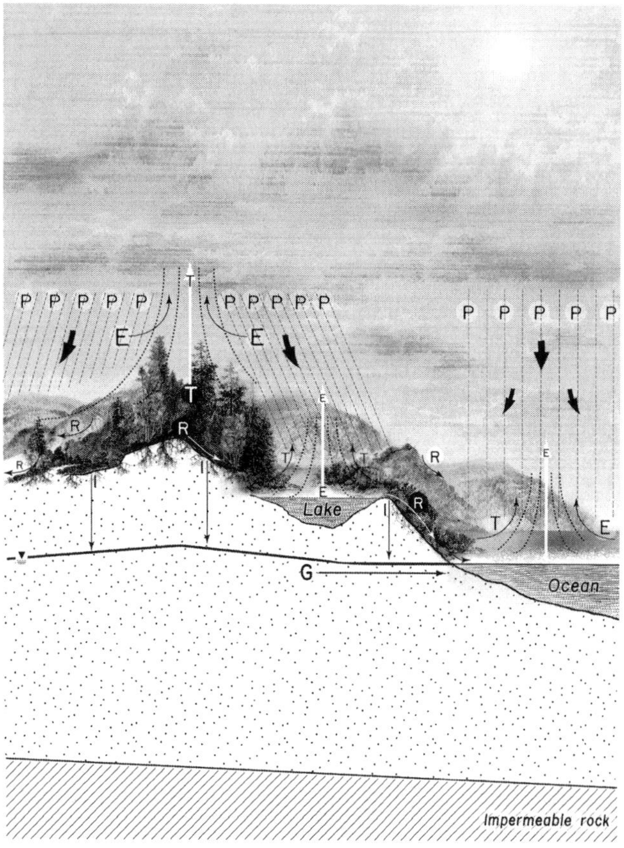

Water removal from a particular basin also is not depicted in the hydrologic cycle shown in Fig. 2.3. Although only two representations of the hydrologic cycle are reproduced here, even casual observation of the many textbooks available on water resources indicate that the removal of groundwater by plants is not uniformly depicted or often even considered.

The common lack of uniform depiction of plant uptake and transpiration of groundwater in the hydrologic cycle may be explained as follows. As shown in Eq. 2.5, the components of precipitation, P, and river and groundwater discharge, R, are more amenable to direct observation and, therefore, were the earliest components to be measured, as initiated by the field measurements made by Perrault and Mariotte. Even the process of evaporation from surface-water bodies could be measured after the work of Perrault and Halley, and others after them. So, the hidden nature of groundwater and the common misunderstanding that it can supply the water to plants led to its exclusion in many early, and even current, water-budget studies. Moreover, the effects of surface-water evaporation can exceed the effects of transpiration in certain areas; because evaporation of groundwater was considered to

P = Precipitation E = Evaporation G = Groundwater flow
T = Transpiration R = Surface Runoff I = Infiltration

Fig. 2.2 A common depiction of the hydrologic cycle that does not include groundwater as a potential source of transpiration (Modified from Viessman et al. 1977).

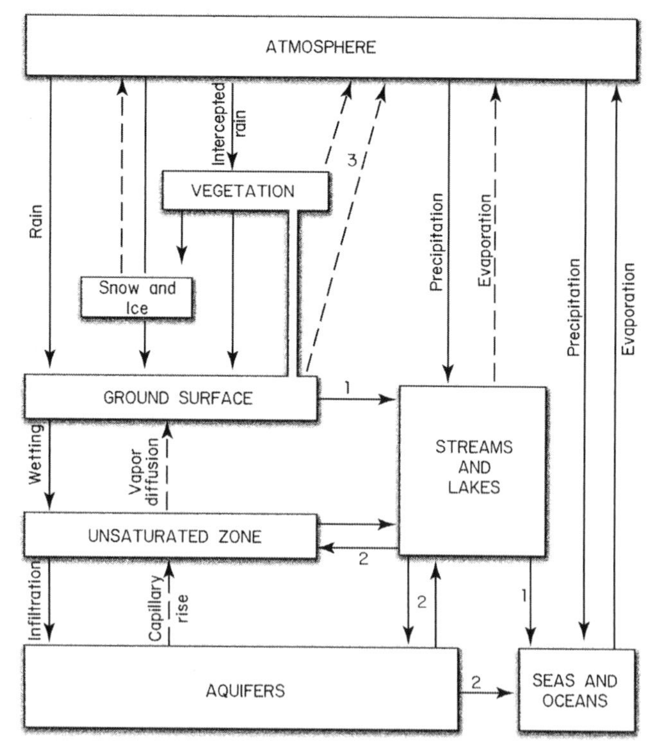

Fig. 2.3 A less common depiction of the hydrologic cycle that shows that vegetation can use groundwater for transpiration (Modified from Detay 1997).

be negligible, transpiration of groundwater was similarly assumed to be inconsequential.

In other studies, the magnitude of evaporation and transpiration often are estimated by using Eq. 2.5 following direct measurement of precipitation and river discharge. The accuracy of the values for evapotranspiration, *ET*, estimated this way, however, cannot exceed the uncertainties of the measurement of precipitation or discharge. Still, Eq. 2.5 can provide the hydrologist with much useful information. For example, an interesting conclusion made from measurements of precipitation, streamflow, and estimated *ET* is that *ET* has less variability from year to year compared to fluctuations that occur with precipitation and streamflow. Hence, drought conditions occur because of a lack of precipitation and because the removal of water by *ET* remains fairly constant over time and can exacerbate drought conditions. In this manner, *ET* acts like a constant tax on the water budget, even though the inflows and outflows of the hydrologic cycle increase or decrease.

Finally, the conditions necessary for the study of plant and groundwater interactions were not apparent in most of the humid areas of the United States, where groundwater often is not the sole source of water to plants, unless during times of prolonged drought; under these conditions, even deep aquifers can be affected. As such, the following quote from G.E.P. Smith, who was introduced in Chap. 1, seems to be a visionary, considering the time it was written:

> *The hypothesis has been held by the writer for a long time that the water drawn up through trees and transpired constitutes the principal loss from the groundwater reservoir, and that, in some cases, this loss is the total loss, while in all cases evaporation is an agency of less import.*
>
> G.E.P. Smith (1915)

2.3 The Hydrologic Cycle, Plants, and Groundwater

More than 70% of the annual precipitation in the United States is returned to the atmosphere as vapor through evaporation from lakes and rivers, soil, shallow groundwater, and by transpiration. These processes and their demand for water must all be met before the remainder can flow in streams or be stored as groundwater (Fig. 2.4). Both evaporation and transpiration are characterized by the physical change of the state of water from a liquid to a vapor. As such, the tendency for water to undergo this change can be explained in terms of the physical properties of water and how meteorological factors affect these properties. Finally, the evaporation of water through plant transpiration is controlled by these physical processes as well as by plant physiology that affect water movement, such as the regulation of vapor exchange at the surface of leaves. As a result of these constraints, *ET* rates can be affected by, and also can affect, the depth to groundwater.

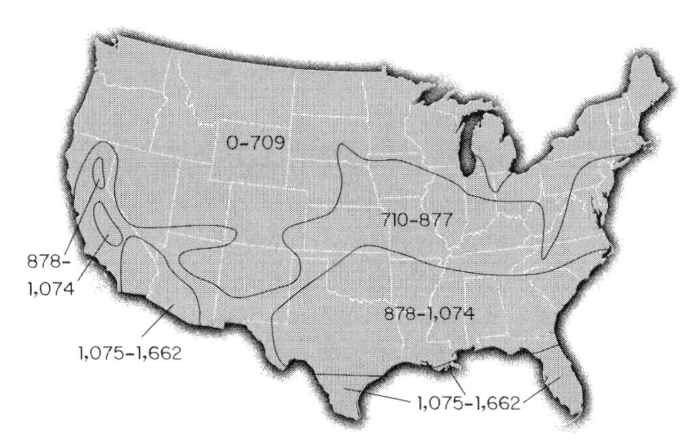

Fig. 2.4 Mean annual evaporation from shallow surface-water bodies, in inches, in the conterminous United States. The effect of evaporation on the water budget is more prominent in arid areas, such as the southwest relative to the southeast. One inch is equivalent to 2.54 cm (Modified from Viessman et al. 1977).

2.3.1 Surface Tension

From the time of Aristotle to the 1700s, water was thought to be an element, and composed of one thing only. The 1700s brought the realization that this was not the case, based on investigations by Karl Wilhem Scheele, in 1772, Joseph Priestly in 1774, Lavoisier in 1779, and Cavendish, in 1784. These scientists were all investigating the process of combustion; at that time, it was widely held that combustion resulted in the liberation of a separate entity into the air called phlogiston, derived from burned material.

The change in state of liquid water to a vapor requires energy. This energy is needed to overcome the strong intermolecular forces that exist between adjacent liquid water molecules. The strength of the intermolecular forces comes from the facts that the water molecule is characterized by a bent shape and that oxygen has a strong attraction for electrons. Both facts impart a polarity to the overall water molecule, or the separation of a slightly negative zone from a more positive zone. This polarity results in bonds between the hydrogen of one water molecule with the oxygen of another water molecule. In general, the energy to overcome these forces comes from the sun.

Because of these strong intermolecular forces, there is a higher tension between water molecules that are exposed to the atmosphere relative to the tension between adjacent water molecules. This tension, or attraction, at the surface occurs because the water molecules are more attracted to each other than to the air (Fig. 2.5). This phenomenon of tension explains the actual spherical shape taken by

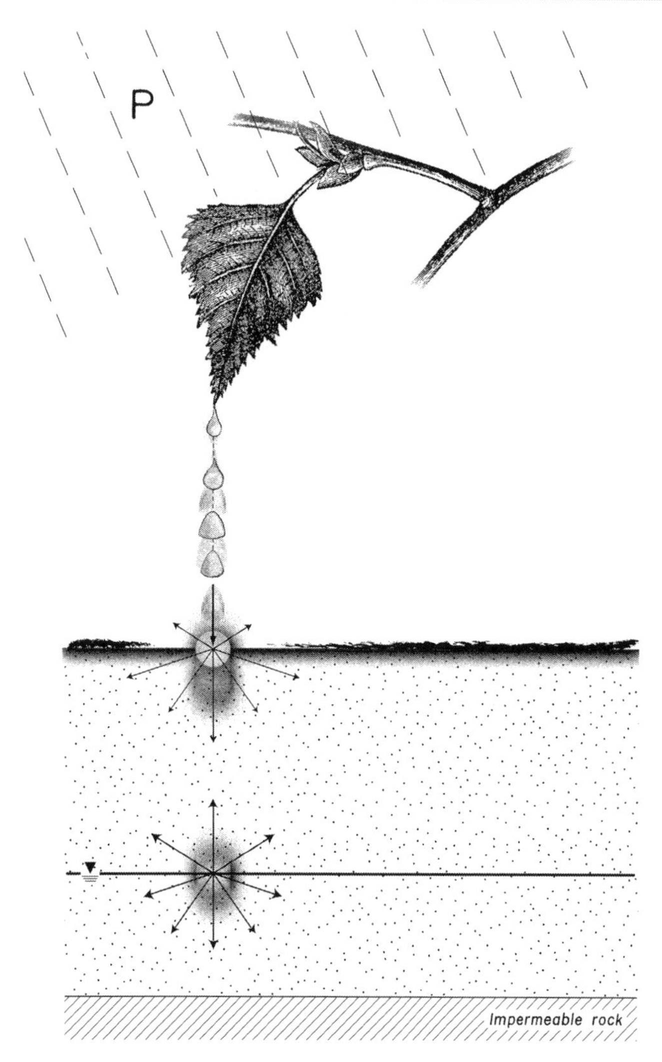

Fig. 2.5 Strong intermolecular forces of the water molecule lead to surface tension and control the shape of water droplets, such as precipitation on a leaf falling into porous media. The relative strength of surface tension is depicted by the length of the *arrows*.

raindrops rather than the teardrop shape commonly depicted. The spherical shape represents the smallest surface area a volume of liquid water can achieve while keeping tension to a minimum. This attraction also explains why isolated drops of water take a spherical shape; the surface tension pulls the water inwards in order to occupy the smallest volume. In plant and groundwater interactions, the surface tension of water also explains the capillary action of water in porous media as well as the cohesion theory of the ascent of sap, and why phreatophytes can use both capillary and groundwater, all which are described in Chaps. 3 and 4.

For this minimal volume of a sphere to be changed or increased, energy must be expended, so the shape of water droplets also has a thermodynamic explanation. As we will see in Chap. 3, this also explains why plants encounter a wilting point, where water present in soil cannot be removed because it is held under too great a tension. As such, the magnitude of the surface tension of water can be defined as the energy needed to increase this reduced surface area by a unit amount. The energy required to overcome these intermolecular forces is equivalent to about 540 cal/g (calories per gram) of water evaporated at 100°C, and up to 590 cal/g at 15°C. This is double the heat required to vaporize alcohols, such as methanol and ethanol, and four times the heat required to vaporize the organic compound benzene. Even a compound similar in structure to water, hydrogen sulfide (H_2S), is a gas at room temperature and has a boiling point of −60.7°C.

Under ambient environmental conditions, energy necessary to vaporize water is provided by the radiated or advected energy of the sun. The energy is conserved, because the temperature of the evaporated water does not change. Water also can be evaporated along gradients in vapor pressure, which will be discussed next.

2.3.2 Vapor Pressure

As liquid water molecules become energized and enter the atmosphere as vapor, the vapor molecules exert pressure on the remaining liquid water molecules. If this process is constrained inside a fixed volume, such as inside a glass test tube inverted into a dish of mercury that rises to a level in the tube balanced by the weight of the atmosphere, the pressure exerted by the vapor molecules can be measured. At equilibrium, vapor pressure on the liquid water molecules is 17.5 millimeters of mercury (mmHg) (0.694 in.) at room temperature (Fig. 2.6). The vapor pressure of water is not constant, however, for it decreases as solutes are added—solutes essentially dilute the water, and lower the vapor pressure. This explains why the solute ethylene glycol is added to the water used in pressurized cooling systems for internal combustion engines, because the water-coolant solution can exceed temperatures greater than the boiling point of pure water.

The rate of water evaporation from a free surface can be defined as the net exchange of water molecules across the water surface per unit time. Evaporation continues until the air becomes saturated with water vapor, which is the absolute humidity that can be held by a given quantity of air at a given temperature. Warmer air can hold more moisture than cooler air as long as adequate supplies of water are available. This relation to water availability explains the dry heat that characterizes arid areas.

2.3.3 Evaporation

Evaporation is fundamental to understanding the interaction between plants, water, and groundwater. Under equilibrium

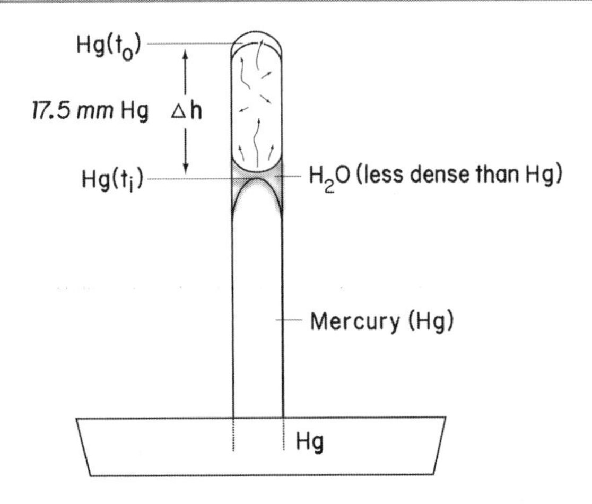

Fig. 2.6 The pressure exerted by water vapor, shown as the *arrows* above liquid water (*gray area*), is measured by inverting a tube filled with the vapor being measured and mercury in a dish. The difference in mercury level, Hg (t_o) and Hg (t_i) is equivalent to the vapor pressure. A higher vapor pressure indicates a greater tendency to become a gas. This figure also demonstrates the results of surface tension for two liquids of different properties. The top of the water has a meniscus shaped concave up, as the water molecules are more attracted to the walls of the glass tubing than to themselves. Conversely, the mercury meniscus is shaped concave down, since the mercury is more attracted to itself.

conditions, energy added to a closed system will be dissipated to the same magnitude somewhere else. If energy is added to a body of water, for example, the energy also is removed when liquid water changes state to a vapor.

The relation between evaporation and water vapor pressure was recognized as early as 1802 by John Dalton, known perhaps most commonly for his discovery of the partial pressures of gases, or Dalton's Law of Partial Pressures, as shown in Eq. 2.6

$$E = b(e_o - e_a), \tag{2.6}$$

where E is the evaporation rate, b is a coefficient that refers to the resistance of water vapor transfer to the atmosphere, e_o is the vapor pressure of the water surface, or supply, under saturated conditions, and e_a is the vapor pressure of water in the air at any given time, or demand. Hence, the rate of evaporation, E, is proportional to the difference between e_o and e_a, called the vapor pressure gradient. If e_a reaches conditions of e_o, evaporation stops. An alternative way to look at evaporation is in Eq. 2.7

$$E = C_{water} - C_{air}/R_{air}, \tag{2.7}$$

where E is evaporation rate, in grams per square meter per second (g/m^2/s), C_{water} is the vapor concentration of water at the surface, C_{air} is the vapor concentration in the air above the surface, in grams per cubic meter (g/m^3), and R_{air} is the resistance to the diffusion of water from liquid to vapor, in seconds per meter (s/m) (Kramer and Boyer 1995). As with the equation by Dalton, the net evaporation, E, is proportional to the difference between C_{water} and C_{air}. A similar relation between evaporation and water vapor is given by the equation of Thornthwaite and Holzman (1939).

The removal of water molecules entering and leaving the atmosphere above the water surface is controlled by many physical factors. Transfer of water from a liquid to a vapor will continue if a vertical gradient is established from the water to the atmosphere in which water vapor is removed and prevented from accumulating, as described in Eqs. 2.6 and 2.7. This vertical gradient is affected by the amount of moisture in the air above the water body. If the relative humidity, or ratio of absolute humidity to the saturation humidity of the air at a given temperature, is high, evaporation will be low, and the converse also is true. Evaporation also is affected by the amount of solar radiation, which warms both the water and air directly above it; cloud cover; air temperature; wind speed and, to a lesser extent, direction; and atmospheric pressure changes in relation to elevation that affect water vapor pressure.

Under conditions in which the water table is close to land surface, evaporation of groundwater can occur if the soil and sediment are coarsely textured, such as sands or gravels (Bouwer 1978), and the hydraulic conductivity of the soil and sediment, or the relative ease in which water moves through sediment pores, is conducive to water flow in the zone above the water table. Evidence of the evaporation of shallow groundwater is provided by the accumulation of salts at land surface where this process occurs. Areas associated with low agricultural output tend to have saline surface soils caused by the evaporation of shallow soil moisture and groundwater. Evaporation of water from soil when the water table is deep, however, also can result in soil that gradually dries out (Hillel 1998). More discussion on the potential for groundwater evaporation is discussed in Chap. 4. Moreover, because water continually is cycled between a liquid to vapor and back again, the amount of water in the atmosphere constitutes only a small portion of the total global supply of water at any given time, or about 0.04% of non-ocean water. The process of ice water being converted directly to water vapor is a special form of evaporation called sublimation.

2.3.4 Measurement of Evaporation

Measurement of evaporation can be made using indirect and direct methods. Evaporation can be estimated indirectly by using Eq. 2.3 if the other components of water inflow and outflow in a particular basin are known. Because the measurement of these factors can contain errors of between 5% and 10%, such estimates generally are not accurate.

Evaporation can be measured directly by using water-filled shallow metal Class A evaporation pans, and keeping records of the amount of water added to replenish the pan on a daily basis to maintain a predetermined maximum level. Because the water temperature in the pans tries to equilibrate with the temperature of the air, however, a conversion factor is required to relate evaporation rates in the pan to land or water evaporation rates. Typically, pan evaporation rates, in inches, are multiplied by 0.7 to get estimates of evaporation from water bodies much larger than the shallow pan. A precipitation gage should be placed near the pan to record rainwater added to the pan during the evaporation-measurement period.

2.3.5 Transpiration

In 1699, John Woodward (Kramer and Boyer 1995) observed plants he was growing in a liquid culture and stated that

> *... the greatest part of the fluid mass (water) that ascends up into plants does not settle there but passes through their pores and exhales up into the atmosphere.*

The process he was describing more than 300 years ago is transpiration. Transpiration is the evaporative loss of water from living plants. Transpiration produces about 75% of the water vapor over land surface and about 13% of the water vapor around the globe (Von Caemmerer and Baker 2007). Even dormant plants can lose water by transpiration. Transpiration of water from a leaf can be viewed as being similar to the evaporation of water from the free surface of exposed water, as described above, except that the water vapor must first travel through water-conducting structures within the plant to the leaf to reach the air (Fig. 2.7). At the most basic level, the same physical factors that influence evaporation from the surface of an exposed water surface also affect the transpiration of water from a leaf, such as the relative humidity of the air, and the amount of incoming solar radiation that impinges on the leaf. Unlike open surfaces of water, however, plant anatomy and water-conducting structures have the ability to resist evaporation, which is discussed in Chap. 3.

The rate of transpiration can be described by using Eq. 2.8:

$$T = C_{leaf} - C_{air}/R_{leaf} + R_{air}, \tag{2.8}$$

where T is the rate of transpiration, C_{leaf} is the vapor concentration inside the leaf tissue, and R_{leaf} is the resistance to vapor diffusion in the leaf (Kramer and Boyer 1995). Hence, transpiration, T, is proportional to the difference in vapor concentration between the leaf and air, normalized by any

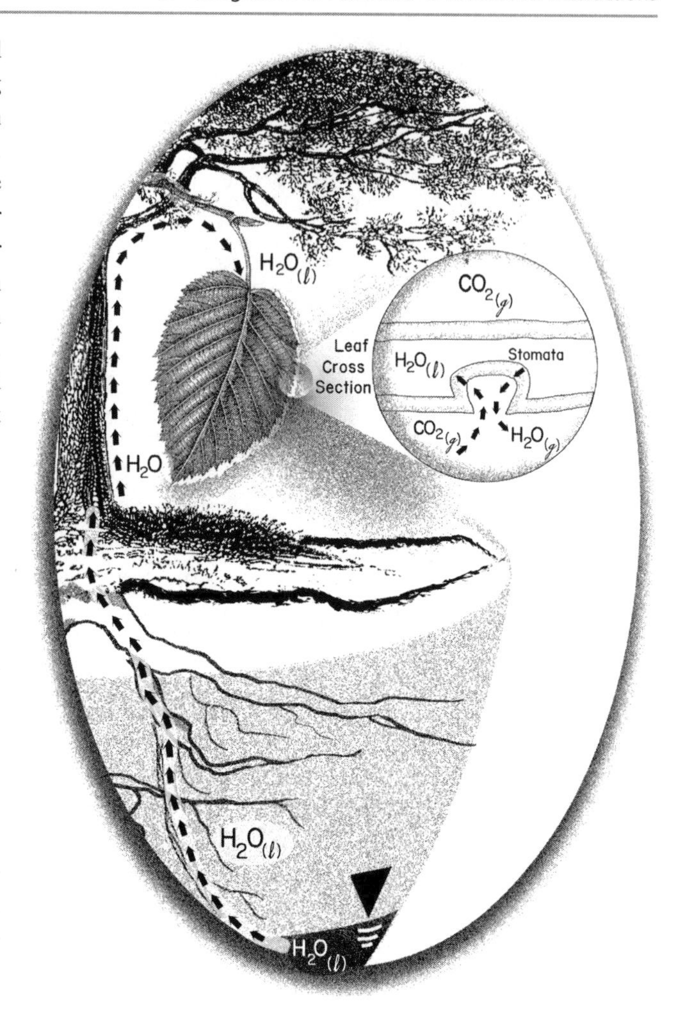

Fig. 2.7 The loss of water vapor, $H_2O_{(g)}$ by diffusion from leaves during transpiration following uptake by root hairs in the subsurface and translocation of $H_2O_{(l)}$ to the leaves. This is a consequence of the diffusion of $CO_{2(g)}$ into leaves which stimulates photosynthesis and the production of a stored source of energy.

resistance to this vapor transfer by the structure of the leaf. Ultimately, it is the transpiration of water along this vapor-pressure gradient that moves water from the roots to the atmosphere. Changes in physical factors, such as light intensity, will affect variables in Eq. 2.8, such as R_{leaf}. This can lead to a change in the physical status of the plant itself. For example, the loss of water will act to lower the leaf temperature and, therefore, under conditions of rapid transpiration, the leaf temperature of a plant may be lower than the air temperature.

In addition to the physical factors related to meteorological processes that occur above ground, physical properties of the soil, such as soil moisture content, porosity, and hydraulic conductivity, also play a role in controlling the maximum rate of transpiration. This is because water first must be moved from reservoirs of water in the ground to the root system. An example of the effect of soil properties on the transpiration of plants is the previously mentioned wilting point, a situation in which soil moisture is present in the pore spaces of subsurface sediment but is too strongly bound to these sediments to be bioavailable for plant uptake even though roots hairs may be in contact with the water. More on this subject will be discussed in Chap. 3.

The meteorological and soil-water factors influence transpiration, but so do plant physiological factors. A plant is not simply a passive straw through which water moves from the soil to the air. Plant-based factors include the total leaf area, size, location, and shape. Finally, as would be expected for plant and water interactions, biotic factors also play a role in controlling transpiration, such as the growth stage of a plant, its color and health, etc.

The upper limit of transpiration is determined by the total amount of solar energy available to evaporate water. In most cases, transpiration cannot exceed pan evaporation rates. For example, even though the surface area available for evaporation from the leaves of a tree can exceed that of unplanted bare ground by a factor of 6, the evaporation rate cannot exceed that of the total amount of energy available, regardless of whether the area is planted or not. However, if water is not limiting and plant transpiration is not regulated, or if winds move dry air continually past the leaf, plant transpiration may exceed pan evaporation.

2.3.6 Measurement of Transpiration

The first known attempt to quantify transpiration was performed by Stephen Hales in 1727 (Kramer and Boyer 1995). He related the amount of water transpired by individual potted plants to the weight lost from the pot; the greater the weight loss the higher the transpiration rate. His approach of measuring individual plant transpiration is still used today at many plant laboratories around the world, including those that are interested in the effect of various concentrations of groundwater contaminants on transpiration of plants used in phytoremediation. Hales also made field measurements and calculated that almost 4 tons of water per acre per day, or nearly 1,000 gal (3,780 L), was transpired from a field of cabbages.

The measurement of the use of water by larger stands of plants also is of interest, especially considering the need to group plants that have high transpiration rates close together for many phytoremediation applications. Similar to measuring the weight loss of a plant over time in the laboratory, a weighing lysimeter is used in the field to measure the water lost on a much larger scale. A weighing lysimeter is essentially a large pot placed in the earth, filled with soil, plant(s), and watered, and the changes in water content related to transpiration are quantified by weighing (Fritschen et al. 1973). The application of weighing-lysimeter tests and the transferability of results to large trees in natural conditions were investigated by Patric (1961). In Patric's study, 26 weighing lysimeters were constructed by placing soil in concrete tanks that measured about 10 ft by 21 ft by 6 ft deep (3.0 m by 6.4 m by 1.8 m). Initially, grass was grown in the tanks; later, the grass in some of the tanks was replaced with Coulter pine. Relative to the tanks that still contained grass, all soil moisture was removed from the tanks that contained the pines, and these tanks had higher *ET* rates as well.

Other methods that can be used to measure transpiration include those based on measuring the loss of water vapor or water weight (Lee 1942). To measure water vapor loss, an apparatus called the Freeman method can be used in which a cylinder is placed around part of a plant and a chemical is used to absorb the water vapor lost by transpiration. Another method is to place a chemical indicator strip, typically made of cobalt chloride, directly on a plant leaf to determine the amount of water lost based on the degree of color change in the chemical. To measure water-weight loss, plant-tissue samples either are removed and the loss of weight of the sample over time indicates the amount of water lost by transpiration or placed in a chamber called a potometer. The advantage of these methods is gained only if the period of time of the experiment is shorter than a few days.

Probably the most widely accepted methods to measure transpiration in plants is based on artificially applied radioisotopes or pulses of heat, both of which are used as tracers of water flow in individual trees. For example, tritiated water ($^1H^3HO$) was added to the soil around a Douglas fir tree (*Pseudotsugamenziesii spp.*) in the western part of the United States (Kline et al. 1976). This technique, as well as the heat-pulse method and others, are discussed in Chaps. 3 and 9.

2.3.7 Evapotranspiration

Because of the difficulties in measuring evaporation and the lack of knowledge about measuring transpiration, lumping them together as one term does not indicate that the combined rate is more accurate. In fact, it reveals the opposite to be true. This is because, in many instances, *ET* is not derived from careful measurement of evaporation or transpiration but represents the balance of flow that remains after other components in Eq. 2.5 are measured or estimated.

Evapotranspiration had been defined by C.W. Thornthwaite in order to estimate evaporative demands in large regions. The focus on larger areas tended to reduce the effects of large variations in climate, plant distribution, or changes in local topography. If water quantity is unlimited in a particular area, potential evapotranspiration, ET_p, can be determined. Potential evapotranspiration is essentially the maximum power of evaporation, or demand, of the atmosphere for water in a given area at a given time under given climatic conditions. Hence, ET_p cannot be considered a constant or a maximum for a given area if the area is characterized by significant fluctuations in climate. It is simply a concept to describe *ET* if water quantity is not constrained.

For the calculation of ET_p to be useful, plant transpiration is treated as one large leaf with no water stress that completely shades the soil and is growing maximally. In this manner, ET_p can be expressed in the following equation:

$$ET_p = C_{int} - C_{air}/R_{int} + R_{air}, \tag{2.9}$$

where ET_p is the rate of potential evapotranspiration (g/m^2/s), C_{int} and R_{int} are the vapor concentration of water at the surface of a moist leaf or soil, respectively, C_{air} is the vapor concentration in the air above the leaf surface (g/m^3), and R_{air} is the resistance to the diffusion of water vapor (s/m). Because both processes of evaporation and transpiration require energy from the sun to turn water into a vapor, varying the temperature of air by 10°C increments under equal atmospheric conditions will increase evaporation and transpiration by a factor of 2.

Because plants have access to subsurface sources of water along with exposed surface water, it is possible that transpiration can exceed free-surface evaporation, as was mentioned previously. This has been termed the oasis effect. The oasis effect was demonstrated by using water hyacinth [*Eichhornia crassipes* (Mart.) Solms] in reservoirs in Texas (Benton et al. 1978). The authors looked at lakes that had about 20% surface coverage by water hyacinth. This coverage was calculated to result in a transpiration loss of over 2,000,000 acre-ft (2.46 × 10^6 m^3) of water per year, about 20% of the annual yield of the lake. Moreover, when plants grow in isolated areas, either arid or humid, and receive warm, dry air from prevailing winds, the additional heat causes additional evaporation. Under these conditions, *ET* can exceed pan evaporation rates.

2.3.8 Measurement of Evapotranspiration

A variety of methods have been developed to estimate *ET* for individual or groups of plants. In general, these methods include quantification of (1) the mass flux of water, (2) mass balance methods in basins, and (3) energy balance methods. Weighing lysimeters, mentioned earlier, also can be used. Estimation of evapotranspiration also can follow the water-budget approach described for field studies. Evapotranspiration can be estimated from the rise and fall of the water-table surface if sufficient groundwater is drawn up to induce water release from storage; this method is described in detail in Chap. 4. If certain meteorological conditions are known, methods to use include the Thornthwaite method (Thornthwaite and Holzman 1939), the Penman method (Penman 1948), the Van Bavel method (Van Bavel 1966), and the Penman-Monteith method (Monteith 1965). These methods all are based primarily on an energy-budget concept, in which inflows and outflows of energy are balanced. The direct determination of evapotranspiration from large groups of trees is a difficult task because of the variability inherent to individual trees and the physical variables that affect *ET*. On the other hand, although estimates of *ET* on individual trees can be made fairly accurately, scaling these values up to the stand level is often problematic.

A fundamental approach to estimating *ET* is to solve the energy-budget equation. Under steady-state conditions, energy added to a system must be balanced by energy leaving the system. Mathematically, this can be expressed as

$$R_n = \lambda E + H + G, \tag{2.10}$$

where R_n is the net radiation, λE is the latent heat flux or the energy absorbed when the water evaporates or released when it condenses, H is the sensible heat flux or convective energy initiated by temperature gradients in the air, and G is the ground heat flux or heat that conducts into the soil, all in watts per meter squared (w/m^2). R_n can further be defined as

$$R_n = (\text{radiant energy}_{in}) - (\text{radiant energy}_{out}), \tag{2.11}$$

where radiant energy consists of long-wave, thermal and short wave, solar radiation, which can go into and out of the canopy.

The Penman method of estimating *ET* ironically requires no direct measurements of plant characteristics but instead relies on physical conditions, such as meteorological data, that are more routinely measured and, therefore, available to those interested in assessing the application of phytoremediation at a particular location. The Penman method is shown in Eq. 2.12 as

$$E = sR_{nw} + yE_a/s + y, \tag{2.12}$$

where E is the evaporation rate, s is the slope of the saturation-vapor-pressure curve at the wet-bulb temperature, R_{nw} is the net radiation over water, y is the psychrometric constant, and E_a is a function of the wind-speed and vapor-pressure deficits.

The Penman-Montieth method is a modification of the Penman method. It also is a physics-based model that links the energy budget with aerodynamic conditions, but, unlike the Penman method, also includes conditions designed to account for plant feedback on the rate of *ET*, such as terms that relate to the resistance of water movement through plant leaves through the stomata, r_c, and canopy, r_a. This model, consequently, is a measurement of actual ET, ET_a, rather than ET_p and is shown in Eq. 2.13:

$$L_v E = s(R_n + G) + paC_p[(e_s - e_a)/r_a]/(s + y) \times [(r_a + r_c)/r_a]. \quad (2.13)$$

Because this method is computationally cumbersome (for example stomatal-resistance data are difficult to determine), van Bavel (1966) presented a modified Penman-Monteith method.

The Bowen-ratio method also has been widely used to estimate ET. This method uses R_n, G, and the Bowen ratio, β, to calculate λE, the latent heat flux according to

$$\beta = H/\lambda E = C_p(T_2^* - T_1^*)/\lambda\varepsilon(X_2 - X_1), \quad (2.14)$$

where C_p is the heat capacity of air, λ is the latent heat of vaporization, T^* is temperature, ε is the ratio of the molecular weight of water relative to dry air, and X_1 is the mole fraction of the concentration of water as a vapor, or its partial pressure, in the presence of water, X_2. This method to estimate ET was compared with ET derived from constant-level weighing lysimeters in a study by Gay and Fritschen (1979) of a stand of saltcedar on the flood plain of the Rio Grande River in New Mexico during 5 days in June 1977. The results of both methods were in good agreement; ET was estimated to be 0.32 in./d (8.2 mm/d) using the Bowen-ratio method and 0.31 in./d (7.9 mm/d) using the lysimeter method.

Crop coefficients have been developed to relate the loss of water as ET by plants. The Blaney-Criddle formula was developed to depict the consumptive use of water by irrigated crops. In simple terms the relation between plant use and water availability is defined as

$$U = KF, \quad (2.15)$$

where U is the monthly water use in inches; K is the monthly or seasonal consumptive-use coefficient, and F is the sum of the monthly use factors. The F term is the product of average monthly temperature and monthly percentage of daylight hours. Corn for example has a coefficient, K, of 0.75 to 0.85, whereas alfalfa has a coefficient of almost 0.90 (Van der Leeden et al. 1990a; 1990b). Low values of K are representative of humid areas, and high values are representative of arid areas. As long as basic climate data are available at a particular site, estimates of consumptive use can be made using this method. As would be expected, an increase in temperature will increase K values (Blaney et al. 1942). The selection of K values was shown to be dependent on the plant species, growth density, and depth to water table by Rantz (1968).

Actual evapotranspiration, ET_a, for a particular area is the removal of water normalized by the very real limits placed on the availability and bioavailability of soil moisture. Actual evapotranspiration may approach ET_p in areas where precipitation is relatively constant and the vegetative cover is almost ubiquitous, such as in the jungles of South America. The relation between precipitation and ET_p and ET_a is depicted in Fig. 2.8. The interaction between plants and groundwater is greater in a groundwater discharge zone, where groundwater is flowing upward toward land surface, such that ET_a might approach ET_p.

An estimate of ET_p can be measured directly in the field by using atmometers (Worth et al. 1994). Atmometers can be used to measure the rate of evaporation from a wet surface to the atmosphere. Atmometers consist of a tube and a porous ceramic cup filled with water exposed to the

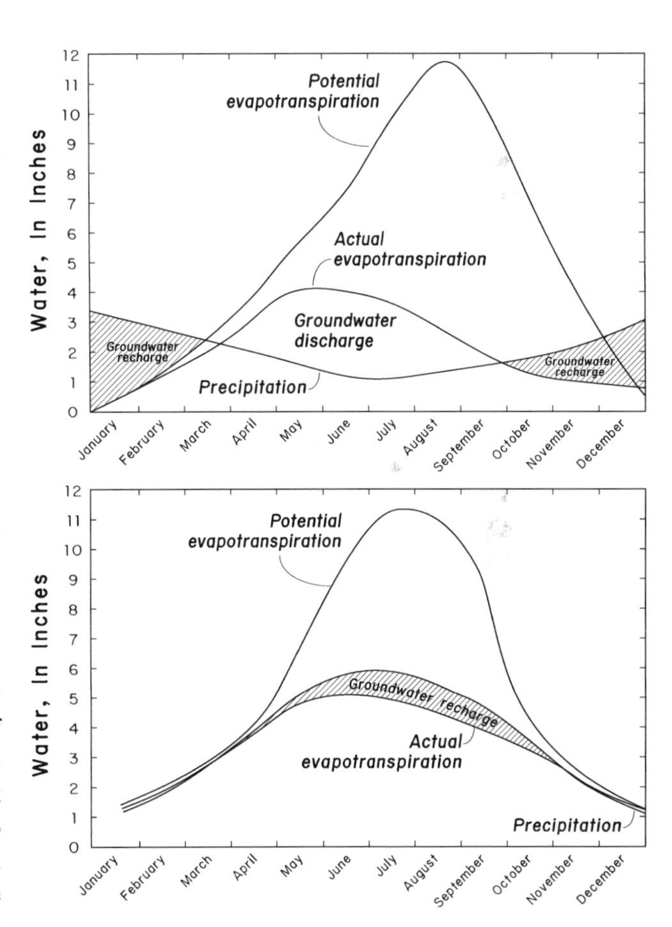

Fig. 2.8 Potential evapotranspiration, ET_p, compared to actual evapotranspiration, ET_a, in a semitropical area under conditions of (**a**) limited precipitation in the summer but abundant precipitation in the winter, so little moisture is stored in the soil, and (**b**) almost constant precipitation throughout the year and, consequently, little depletion of soil moisture (Modified from Fetter 1988). When ET_a is greater than precipitation, groundwater depletion will occur, and can be considered discharge if phreatophytes are removing groundwater. When ET_a is less than precipitation, groundwater recharge can occur. One inch is equivalent to 2.54 cm.

air, typically at a height of about 3 ft (1 m) above ground. Because this set up typically would measure only physical evaporation, the ceramic cup is covered by membranes to simulate the restriction of evaporation through leaves of plants.

As was discussed previously, between 540 and 590 cal are required to evaporate 1 g of water at 15°C. Because less than 400 cal/cm^2 are available on a clear sunny day, the maximum amount of evaporation from a free surface of water is equivalent to about 6 mm/d (Kozlowski and Pallardy 1997). Transpiration from planted areas cannot exceed this upper amount of evaporation and cannot be more than the evaporation from a free surface of water unless drier air is advected over the planted area.

In recent investigations of water budgets, the concept of growing degree days has been used in place of the more conventional methods used to estimate *ET*. As defined in Lorenz and Delin (2007), growing degree days, *GDD*, is the annual sum of the average temperature each day minus a base temperature for a particular area. The *GDD* is used rather than *ET* to estimate the amount of net precipitation remaining per year that may potentially recharge groundwater.

Most plants take up to 100 times more water than their own dry weight, indicating that water uptake is a wasteful process as a result of the need for leaves to contain pores in order to capture CO_2. In some cases, such as arid areas with low precipitation, low humidity, and high air temperature, plants lose to the atmosphere more than 90% of the water they take in. In other words, 10% or less water is retained for biomass. In addition, even though water predominantly exits a plant through leaf surfaces, the water content in leaves is small relative to the water content in stems or the trunk. For example, the leaf-water content is about 4% in 10–60 years-old trees, which is evident by the fact that the leaf-water content can supply transpiration demands for only a short period of time (Running 1979). Also, most deciduous leaves removed from a plant will be dry to the touch within 24 h. This also demonstrates that water exits most leaves rather than enters them. As we will see in Chap. 3, the water retained by plants is primarily stored in non-conductive trunk and stem materials and cellular components.

As *ET* continues, and the upper layers of soil dry out, plants with deep or more extensive root systems can continue to transpire. For these plants, if water is not available, then no evaporation or transpiration will occur, even if climatic conditions are favorable. An example presented by J.D. Hewlett (1982) using evaporation as a surrogate for transpiration helps explain this concept in terms of the availability of water to plants. Using a sealed bottle of water, if it is placed on a sunny windowsill, it will be warmed by heat from the sun, but no evaporation will occur. This scenario is analogous to the plant–water relations of a cactus or other desert plant. Removing the cap from the bottle will initiate evaporation, but even then, only a small portion of the added heat will result in evaporation because of the restriction of the neck of the bottle and the small surface area available for evaporation to take place. In this scenario the opened cap is analogous to the leaves of an herbaceous or woody plant. However, if water in the bottle is poured over a flat surface, heat will affect a now larger surface area causing evaporation to proceed more quickly. This is similar to plants that have large leaf surfaces.

The unsaturated zone and water table are affected by *ET*. Evapotranspiration by phreatophytes in Washington State was reported by Harr and Price (1972). Because they quantified changes in groundwater levels in the study area along with changes in soil moisture, they were able to calculate that up to 31% of the annual *ET* rate of 8.2–9.8 in./year (21–25 cm/year) was derived from groundwater. They also reported that the higher rates of groundwater transpiration were related directly to shallow depths to the water table, and that increased depth to the water table resulted in lower groundwater use. These results should not be surprising, because many of the factors that control *ET*, such as plant height and canopy coverage, are smaller if the water table is deep. At a study site in northern Kansas, between 5 and 20% of *ET* was determined to be derived directly from aquifers, using a groundwater-flow model in conjunction with an atmospheric model (York et al. 2002). A similar investigation in Nevada and Utah reported that groundwater *ET* ranged from 6 to 38% of total *ET* when shrubs were present and up to 70% of total *ET* when grasses were present (Moreo et al. 2007).

As stated earlier, the extent of evapotranspiration is controlled by water availability and energy input. Some areas of the United States can have similar values for ET_p but widely different water budgets because of differences in energy input. For example, parts of southern Arizona and South Carolina have similar ET_p (between 34 and 42 in./year (878 and 1,074 mm/year; Healy et al. 2007) but drastically different precipitation amounts, such that Arizona is water limited and South Carolina is energy limited. The use of ET_p as a master variable to determine the potential for plant and groundwater interactions to occur in a particular study area or site needs to be evaluated in terms of the difference between this ET_p and precipitation.

2.3.9 Tank Experiments

The innovative experimental approaches used by Lee (1912) of the USGS as he sought to understand the water budget in arid areas, such as Owens Valley, California, were discussed in Chap. 1. Essentially, Lee estimated that the *ET* discharge of water from his 54.59 mi^2 (141 km^2) study area with a water-table depth of less than 8 ft (2.4 m) was about 109 ft^3/s (cubic feet per second), or 3 $(ft^3/s)/mi^2$ (Lee 1912). Similar tank experiments also were conducted by Parshall (1937), but the results were inconclusive, and the *ET* rates reported were larger than what would be expected for field conditions.

Lee's results were used by Robinson (1958) to determine the relation between *ET* and temperature and depth to groundwater (Fig. 2.9). It can be seen that as the depth to groundwater increases, *ET* decreases, and as the temperature increases, *ET* increases. Moreover, the rates of *ET* were less than pan evaporation, in most cases from 68% to 75% of pan evaporation (Robinson 1958). A conclusion from this result is that the pan evaporation rate can provide an upper bound on potential *ET* at sites being evaluated for phytoremediation.

Tank experiments to investigate *ET* from riparian trees also were conducted and the results were reported by Robinson (1970). Twelve 30 ft^2 (2.78 m^2) tanks between 7 and 10 ft (2.1 and 3 m) deep were planted with willows, greasewood, and rabbitbrush, and one tank was left unplanted. Robinson (1970) reported *ET* as a volume per foliage or quantity of water per foliage, which was affected by depth to groundwater, length of growing season, and nutrient toxicity.

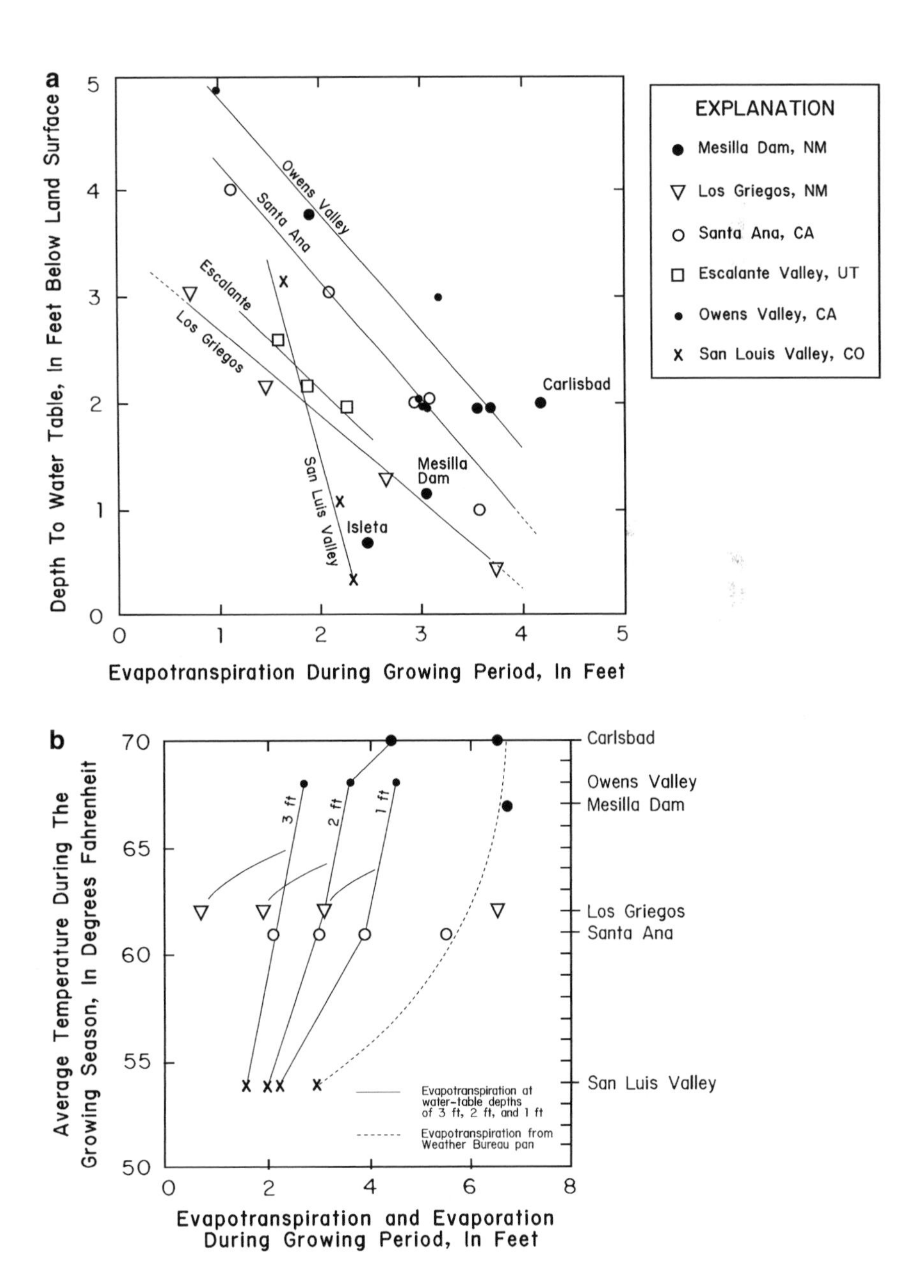

Fig. 2.9 Relation of saltgrass transpiration to (**a**) depth to water table at six sites in the southwestern United States where the shallow water table resulted in high transpiration, and (**b**) average air temperature, with high temperatures resulting in high *ET* rates, especially if the water table was closer to land surface (Modified from Robinson 1958). One foot is equivalent to 0.304 m.

2.4 Summary

Up to 70% of continental precipitation is returned to the atmosphere through evapotranspiration. In the extreme case of arid areas, transpiration from plants can exceed the amount of evaporation from the free surface of water; in many instances, consumptive use—the natural mining of groundwater resources by plants—can occur, and result in net water discharge from a basin. As such, plant water use has a major role in the global water budget, or hydrologic cycle, and this includes transpired groundwater.

Why is this information important to the phytoremediation of contaminated groundwater? The incident solar energy that reaches a site of groundwater contamination ultimately provides the energy that is used to affect the local water budget to either prevent recharge or to remove groundwater from shallow contaminated aquifers. This removal of energy in the form of evapotranspiration occurs and can be quantified, especially with respect to changes in depth to and fluctuations in the water table.

3 Fundamentals of Plant Anatomy and Physiology Related to Water Use

Water comprises between 75% and 90% of the mass of terrestrial and aquatic plants. Such high water content reflects an aquatic stage during evolution. Moreover, it indicates that the successful transition of plants from aquatic to terrestrial environments required the ability to retain and maintain a high water content. It should not be surprising, therefore, that water is one of two reactants in the deceptively simple equation of photosynthesis, perhaps the most important chemical reaction on earth:

$$CO_2 + H_2O \rightarrow \text{plant (chlorophyll), light} \rightarrow CH_2O + O_2. \quad (3.1)$$

All green plants regardless of size, location, or species use this reaction to synthesize food. Every day, plants react about 400 million tons of carbon as CO_2, with 70 million tons of hydrogen from H_2O and liberate 1.1 billion tons of oxygen (O_2). This reaction is so important that all CO_2 in the atmosphere eventually passes through the leaves of plants about every 300 years.

The photosynthetic reaction indicates the importance of water as a reactant, and provides an excellent starting point to describe the important role that water, including groundwater, plays in the life of plants, and ultimately, how this role affects the phytoremediation of contaminated groundwater. The objective of this chapter is to provide information on plant and water relations that can be used during the phytoremediation of contaminated groundwater.

3.1 Plant Cell Structure, Photosynthesis, Respiration, Growth, and Dormancy

Equation 3.1 reveals that plants are autotrophs and synthesize their own food by turning non-living, inorganic materials, such as CO_2 and water, into living, organic matter. The entire conversion rests on a speck of green, organometallic pigment called chlorophyll. This compound singularly harnesses the radiant kinetic energy of the sun into potential chemical energy to be used by plants and, in turn, all forms of heterotrophic life. For chlorophyll to work, however, water is required; therefore, it follows that various sources of water, including groundwater, can affect photosynthesis.

3.1.1 History of Unraveling Photosynthesis

The story of photosynthesis is revealed in the paths taken by carbon, hydrogen, oxygen, and light. Much as it took hundreds of years and the efforts of many individuals to understand the hydrologic cycle, understanding photosynthesis also took time and effort. In 1754 Charles Bonnet (1720–1793) observed gas bubbles on leaves of submerged aquatic plants. In the 1770s, Joseph Priestley (1733–1804) performed a series of simple experiments that were the first steps toward unraveling the source of the gas bubbles observed by Bonnet. When Priestley placed a burning candle or a live mouse in a closed airtight jar the candle flame extinguished and the mouse died long before the candle wax ran out. If, however, he placed a plant, such as a sprig of mint used in his experiment, in the jar with a candle or mouse, the candle remained lit and the mouse survived. (Priestley's experiment led to the beginning of the still popular tradition of giving flowers to hospital patients in the hopes that the plants would help cure the patient by purifying the air.)

In any case, the role of the plant in keeping the flame lit and the mouse alive remained a mystery and was the source of much dispute. Priestley stated in 1774 that

> *The putrefaction of such masses of both vegetable and animal matter, is in part at least repaired by the vegetable creation. And not withstanding the prodigious mass of air that is corrupted daily by the aforementioned causes, yet, if we consider the immense profusion of vegetables upon the face of the earth, it can hardly be thought but that it may be a*

J.E. Landmeyer, *Introduction to Phytoremediation of Contaminated Groundwater*,
DOI 10.1007/978-94-007-1957-6_3, © Springer Science+Business Media B.V. 2012

sufficient counterbalance to it, and that the remedy is adequate to the evil.

Joseph Priestley, *Experiments and Observations* (1774) (from Kramer and Boyer 1995)

A twist in this story is that prior to Priestley's experiment, the Swedish chemist Carl W. Scheele also discovered that oxygen was the gas given off by plants, but he is less recognized today because his paper was published after Priestley's paper.

In 1779, a colleague of Priestley, the Dutchman Jan Ingenhousz, demonstrated in *Experiments on Vegetables* that the mystery element must be a gas, because when he placed a plant under water the leaves, but not the stems, released bubbles; this gas production was observed only when the leaves were exposed to sunlight, and gas bubbles were not released in the dark (Leicester and Klickstein 1952). He also demonstrated that plants release CO_2 in addition to oxygen and can use it during photosynthesis, which will be discussed later.

In 1780, the pastor Jean Senebier observed that only the green parts of plants that contained chlorophyll released a gas which we now know to be oxygen, and took in gaseous CO_2. This helped to confirm Ingenhousz's observations of a lack of bubble production by stems. The Swiss botanist N. T. de Saussure, introduced in Chap. 1, showed that the release of oxygen by plants occurred only if the uptake of CO_2 occurred. In 1844, Hugo von Mohl discovered what he called leaf green or chloroplasts, in the green parts of plants. Finally, in 1845, the German biologist Robert Mayer wrote that plants absorb one form of energy as light and give off another form of energy as chemical bonds.

It may appear that all the parameters in the process of photosynthesis are accounted for. However, what remained unanswered was the source of the oxygen released. Was the oxygen from gaseous CO_2 or liquid water, both of which contain oxygen? Because it was recognized that oxygen also is a gas, it was assumed that the oxygen produced was derived from gaseous CO_2 rather than liquid water. Moreover, because two atoms of oxygen compose the oxygen molecule, the implication is that two molecules of water would be required.

In the 1930s, this question was examined by C.B. Van Niel who proposed that water, not CO_2, was the source of oxygen released by plants. Later, the different stable isotopes of oxygen were used to confirm Van Niel's hypothesis. As will be shown in Chap. 9 oxygen stable isotopes are useful in many studies involving water, because oxygen has a light isotope (^{16}O) and a heavy isotope (^{18}O) that consists of two additional neutrons. This difference in the number of neutrons results in a difference in the atomic mass of each isotope, and each isotope behaves differently in chemical reactions. For example, the lighter oxygen isotope (^{16}O) reacts faster than the heavier and kinetically slower isotope (^{18}O).

This difference in oxygen isotope mass and kinetics was exploited by Samuel Ruben and Martin Kamen, who applied water that contained the heavy oxygen isotope ($H_2{}^{18}O$) and CO_2 that did not ($C^{16}O_2$) to plants. They then measured the isotopic composition of the oxygen released by the plants and found that it contained the heavy isotope (^{18}O), thus confirming that the oxygen released by plants is derived from water. Hence, the important role of water in the photosynthetic equation in the creation of organic compounds from inorganic compounds was demonstrated. Moreover, groundwater can be an important source of water used in photosynthesis. This fact provides a rational basis for the use of plants that interact with groundwater at sites characterized by groundwater contamination. First, however, water entry must be discussed, and this means looking at plant–water interactions at the cellular level.

3.1.2 Prokaryotic and Eukaryotic Cells

Robert Hooke's observations of cork using a crude microscope revealed that the cork was not a homogeneous solid mass but composed of small chambers separated by walls that Hooke called cells, as described in Chap. 1. The idea that any organism or object was composed of cells was revolutionary. Hooke's observation led to the formation by Matthias Schleiden in 1838 of the Cell Theory, stating that all life consists of cells, and a single cell is the minimal unit that has the properties of life.

The Cell Theory also provides a framework to explain the structures within cells and their functions. Anatomy is the study of the structure of living things. From an anatomical basis, plants can be conceptualized as a fluid (liquid) that lives in another fluid (gas) above ground, and below ground lives in both fluids. Each interaction between these fluids and the plant is separated by membranes. Physiology, on the other hand, is the study of the function of these structures. Physiology tries to answer such questions as why is a leaf shaped the way it is, or why does one plant have a shallow root system and another plant has a deep root system?

Plants are composed of individual cells with groups of specialized cells that make up various organs, such as roots, stems, and leaves, each which have different functions. This is similar to most multicellular organisms. All plant cells are alive, or in the case of xylem cells were once alive, and work together to satisfy the needs of the plant as a total organism.

Cells are not equal in structure or function, however. Cells are classified into two types, prokaryotic and eukaryotic, based on structural differences. Prokaryotic cells, the simpler of the two types, are representative of bacteria, such as *Esherichia coli* (*E. coli*). They consist of a somewhat rigid cell membrane, or plasmalemma, that surrounds the cell cytoplasm both of which are contained within the cell wall. The cell wall helps provide rigidity to the cell, especially when the cellular components contain water and can exert internal pressure on the inside of the cell wall. The cell wall is composed of peptidoglycan, polysaccharides, and amino acids. The genetic material that dictates the metabolism and reproduction of the cell exists as a single strand of free-floating deoxyribonucleic acid (DNA) and is not bounded within the cytoplasm by a separate membrane.

In contrast, the eukaryotic cell is more complex and is characteristic of higher life forms, such as algae, fungi, plants, and animals. The generalized structure of a plant cell is shown in Fig. 3.1. Overall, a plant cell consists of the living cytoplasm, called protoplasm, that is contained within a cell membrane (plasmalemma or cytoplasmic membrane) similar to the prokaryotes. The cytoplasm is a colloidal, gel-like material where most of the metabolism of the cell occurs. In plants, the plasmalemma of each cell is the interface between the exterior of the cell, called the apoplast, which is composed of the continuous extracellular aqueous phase between cells, and the interior of the cell, or symplast. The plasmalemma controls the passage of various elements, organic compounds, and water into and out of each cell.

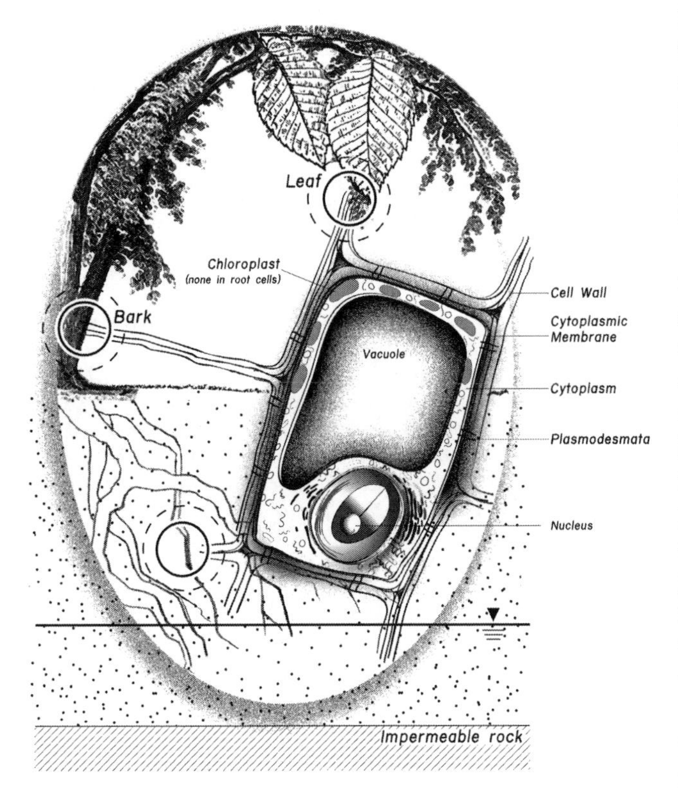

Fig. 3.1 Simplified structure of a eukaryotic cell in various parts of a plant, such as the leaf, bark, and root. The cytoplasmic membrane is synonymous with the plasmalemma. Chloroplasts are not found in root cells of most plants, except a few epiphytes like orchids. The importance of water is evident from the relative size of the vacuole, a water-storage organelle.

Individual cells are not isolated from each other after division and differentiation into tissues and various organs. The cell walls permit the exchange of cytoplasmic materials between adjacent cells through interconnections called plasmadesmata. It is the presence of such plasmadesmata that causes multicellular plants to be different from single-celled bacteria, which tend to be like little isolated factories that have to manufacture all the processes of life and metabolism. Individual plant cells also can do this to some extent, but generally act together as tissues that are differentiated to perform separate tasks to benefit the whole plant.

In plants, the cell protoplasm and plasmalemma are surrounded by a tough cell wall for protection just like for the prokaryotes. In the eukaryotic cell wall, peptidoglycan is absent; instead, the cell wall is made of polysaccharides, such as cellulose, which is a polymer of glucose molecules (120 in all) and is arranged in thin sheets called microfibrils that slide past each other to accommodate cell growth. The cell wall has to be rigid to provide physical support but also has to be porous and, therefore, semipermeable. As such, water and other dissolved compounds can pass through, but passage of smaller compounds or macromolecules essential to life is excluded.

The ability of plants to synthesize cellulose was of primary importance in their transition from the oceans to land, as its strength provided structural support that allowed the cell cytoplasm inside to remain bathed in water. As cells divide and elongate, the pressure exerted between adjacent cells induces the production of pectin that causes the cells to hold fast but remain pliable. Pectin, to those familiar with the process of canning, is what binds jelly together. When plants reach maturity, a secondary cell wall is made out of cellulose and an additional compound called lignin, composed of various aromatic compounds. When placed together, cellulose and lignin are known commonly as wood. Cellulose also is the principal component of paper, which is most likely made of wood pulp.

The structure of the cytoplasmic membrane allows control of compound entry and exit. The inner part of this membrane consists of hydrophobic compounds that resist interaction with water, and the outer part consists of hydrophilic compounds that readily interact with water. This membrane is, therefore, very thin, only two molecules thick (Fig. 3.2). This bilayer structure imparts a semipermeability to the cell to allow certain compounds, such as

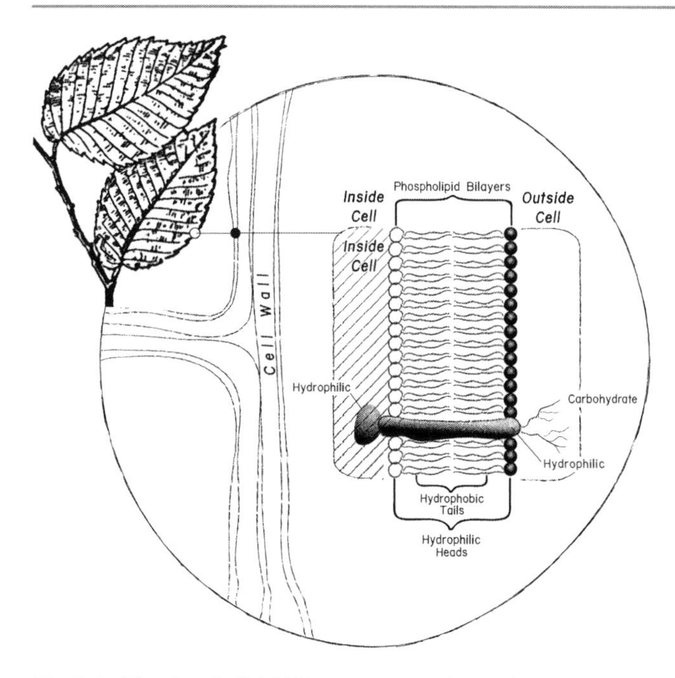

Fig. 3.2 The phospholipid bilayer structure of a section of a eukaryotic cell membrane. The hydrophobic ends of the molecules point inward and are overlain by hydrophilic heads. This structure helps explain the round shape of most cells (Modified from Curtis 1983).

water, to cross the membrane easily with no expenditure of energy by the cell, whereas other compounds are completely excluded from entry, and some gain entry only after energy is expended by the cell. Figure 3.2 depicts a linear segment of the cell membrane, with each end open to the environment. In reality, this cannot be the case; the ends link together to form the typical circular shape of most cells (Smith and Szathamáry 1999). In this manner, a complete membrane is created and water is retained inside the cell.

The cytoplasm also is filled with other structures called organelles that help transfer energy and perform other specialized functions. Mitochondria in the cytoplasm use the process of respiration, the reverse reaction of photosynthesis, to convert the energy stored during photosynthesis into work; more specifically, converting reduced organic compounds in the form CH_2O into adenosine triphosphate (ATP) the major form of energy used by the cells. In plants, mitochondria are concentrated in the actively growing cells of the root tips, which often have to penetrate impermeable material of the subsurface. The cytoplasm also contains other organelles, such as ribosomes, which function to create proteins from messenger ribonucleic acid (mRNA). Others organelles include the endoplasmic reticulum, a variable structure of membranes that serve as sites for chemical reactions or ribosomal attachment; golgi bodies or complexes for storage; and lysosomes that contain hydrolytic enzymes.

Areas of a plant where cells constantly divide are called meristems. These include cells at the root tip, or apex, as well as the tip of new shoots, called apical meristems. It is these areas that account for the growth in root and shoot length in most woody plants. In contrast, grasses have meristematic tissues located near the area of leaf attachment to the stolon or rhizome and are called intercalary, which explains why the tips of grasses can be removed during mowing or herbivory without damaging future plant growth. In either case, this growth pattern is called primary growth, a common characteristic of both herbaceous and woody plants, in which the plant increases in length.

Other areas of rapidly growing tissues that require energy in the form of ATP are the lateral meristems. Rather than increasing in length, these cells account for an increase in width, or girth, of a plant; such growth is called secondary growth. This growth is initiated in the 1–10 cell thick living tissue called the cambium. In woody plants, these cell layers are covered by another cell layer called the cork cambium. As this layer of cells grows, the transport of food and water is cut off to the epidermal cells causing them to die and be replaced by a peridermal layer of cells, which becomes the bark of most plants. In order to reduce water loss through dead cells that no longer have to the ability to regulate material flow, before they die these cells secrete suberin, a waxy compound that acts as a waterproofing agent, into the cell walls.

Compared to prokaryotes, the eukaryotic cell contains DNA in a separate organelle. The presence of this organelle in eukaryotic cells was observed through a microscope in 1781 by Felice Fontana. In 1832, this observation was confirmed and named the nucleus by Robert Brown (1773–1858), the Scottish scientist recognized more for his observation of the motion of tiny particles suspended in fluids, called Brownian motion. Using his microscope, in 1832 Brown said

> *In each cell of the epidermis of a great part of this family, especially of those with membranous leaves, a single circular areola, generally somewhat more opaque than the membrane of the cell, is observable, one to each cell. This areola, or nucleus of the cell as perhaps it might be termed, is not confined to the epidermis, being found also…in the parenchyma or internal cells of the tissue. The nucleus of the cell is not confined to the Orchideae but is equally manifest in many other Monocotyledenous families; and I have even found it, hitherto however in very few cases, in the epidermis of Dicotyledenous plants.*
>
> Robert Brown (1832; in Ford [1985])

It was thought that incorporation of the nucleus into the eukaryotic cell was the result of phagocytosis, the engulfment of a food particle or prokaryote by the cell, or the result of symbiosis between different prokaryotes that provided a selective advantage to both cells. It was later shown experimentally in the 1930s, using giant marine algae (*Acetabularia acetabulum*), that the nucleus contained the information that permitted cell growth and reproduction.

An interesting adaptation unique to plant cells is that all plant meristematic cells contain the DNA to make itself as well as all other parts of the plant. For example, the initial cells in the shoots of a plant above ground can turn into roots if placed below ground. Theophrastus observed the consequence of this fact, without knowing what caused it, as stated in *Enquiry into Plants* (Loeb Classical Library 1916)

> *A plant has the power of germination in all its parts, for it has life in them all. . .the methods of generation of plants are these: spontaneous, from a seed, a root, a piece torn off, a branch, a twig, piece of wood cut up small, or from the trunk itself.*

As we will see in Part II, the ability of plants to grow from various pieces has implications for the kinds of plants and methods of planting at sites where phytoremediation will be used to remediate contaminated groundwater. Moreover, it has a direct effect on the cost of remediation in that phytoremediation is often less costly than other remedial designs.

An important organelle in plant cells relative to plant and groundwater interactions at contaminated sites is the vacuole. A vacuole is a space within the plant cytoplasm (Fig. 3.1). The origin of vacuoles in cells may be a consequence of phagocytosis. In this manner, unlike prokaryotes that must release digestive enzymes extracellularly to digest food and then reabsorb the smaller particles back through cell membranes, the presence of a vacuole enables eukaryotes to process digestion within the cell itself, which increases efficiency. The vacuole can account for most of the volume of a cell and up to 90% of the total volume of some mature cells. The presence of large vacuoles and cell walls renders plants a much larger surface area than the same volume of cytoplasm without these structures. Vacuoles are bound to a membrane called the tonoplast, just like the cytoplasmic membrane. The space of the vacuole can be occupied by water, food in the form of starch, or pigments. Vacuoles are often used by plants to isolate defensive toxic compounds away from other parts of the cell. Whether or not vacuoles can be used to store toxic compounds from contaminated groundwater after uptake by a plant is unknown and is an area of promising research.

Groups of individual cells that perform similar functions are called tissues. In plants, tissues include the meristematic cells already discussed as well as epidermal tissue. The parenchyma cells contain large vacuoles used to store various compounds such as water, oil, crystals, and tannins. These cells make up the pith of many plants. Collenchyma tissues contain pectin to provide structural strength to plant stems and petioles and contain chloroplasts and, therefore, can conduct photosynthesis. Schlerenchyma tissues contain lignin, unlike the collenchyma tissues. Other tissues that have air pockets between adjacent cells are called aerenchyma tissues. As in mammals, each group of tissues functions separately but usually to the benefit of the larger organism.

3.1.3 Photosynthesis, Chlorophyll, and ATP

Of all the organelles contained in the cytoplasm within the plasmalemma inside the cell wall, the one unique to plant cells are the plastids. Plastids may contain food, in the form of starch, or pigments. Plastids also contain chloroplasts, which form the center of the photosynthetic process because they contain thylakoids, the membranes of which contain the seminal substance of chlorophyll. Multiple layers of thylakoids look much like a stack of coins and are called grana, and these are surrounded by a liquid called stroma.

Chlorophyll was first isolated in 1817 by the French chemists P.J. Pelletier and J.B. Caventou and named for its green color from the Greek *chloros*, and *phyllon*, for leaf. Chloroplasts are surrounded by an inner and outer membrane. Chloroplasts contain DNA, RNA, and ribosomes and can divide independently of the cell's nuclear division. The ability of chloroplasts to undergo division separate from the cell suggests that chloroplasts once were independent cells, such as prokaryotic cyanobacteria, that became engulfed by larger cells that perhaps could not perform photosynthesis (Gray 1993; Clegg et al. 1994). Also, the plastids move around in each cell independently in order to maximize their exposure to changes in the position and condition of light. The white color of most roots indicates the absence of chloroplasts in the cytoplasm of these cells (Fig. 3.1), but the cells do contain leucoplasts for storage of starch. Some flowering plants such as orchids, however, have chloroplasts in their root cells.

Chlorophyll is a cytochrome and is composed of proteins, amino-acid building blocks that are conjugated with non-amino-acid compounds. The presence of chlorophyll is the fundamental criterion that separates those that have it, the plant kingdom, from those that do not, the animal kingdom. Bacteria, mold, yeast, and some fungi, however, are considered part of the plant kingdom even though they do not possess chlorophyll.

Chloroplasts contain two forms of chlorophyll—the blue–green chlorophyll *a* and the yellow–green chlorophyll *b* pigments—that give the familiar color to most plants. Typically, most plants have three times more chlorophyll *a* than chlorophyll *b*. In general, a molecule of chlorophyll consists of a magnesium atom (Mg) at the center of a porphyrin ring (Fig. 3.3), similar to the structure of mammalian hemoglobin, which contains an iron atom centered in a porphyrin ring, suggesting that chlorophyll is simply a heme that mutated to contain magnesium. The magnesium and iron function as catalysts, or metalloenzymes, to increase the probability a reaction occurs to completion. The porphyrin ring of chlorophyll is the center of light capture; the tail of chlorophyll consists of a chain of phytol for linkage to the lipid layers of the thylakoid membranes (Fig. 3.3). Many

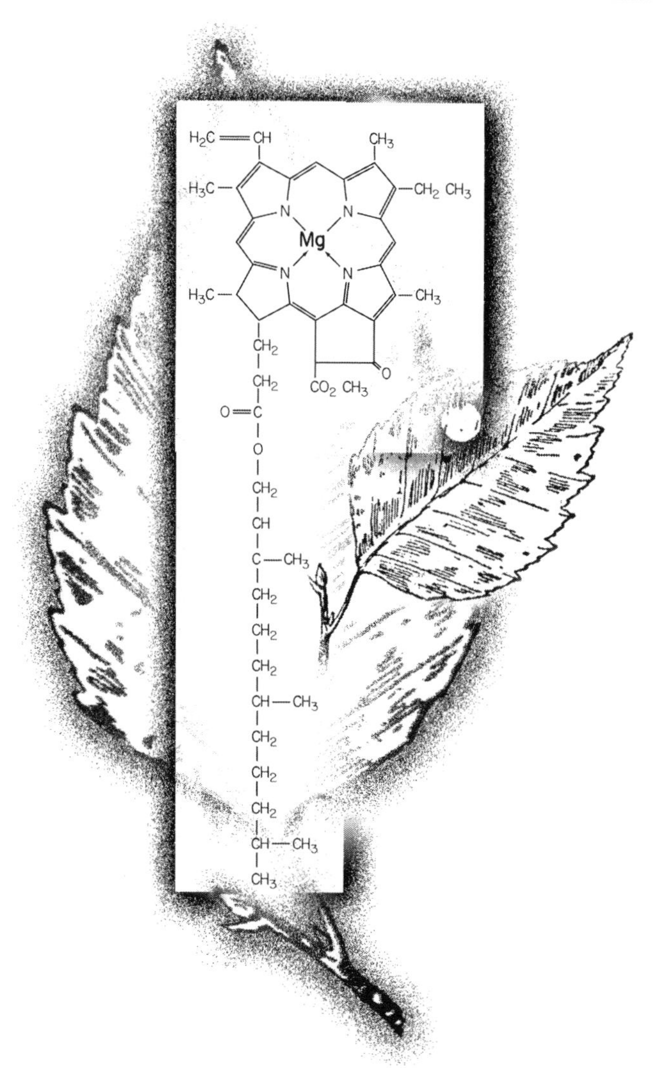

Fig. 3.3 The structure of chlorophyll *a*, showing the magnesium ion in the porphyrin ring, the center of light capture.

molecules of chlorophyll *a* are attached to each thylakoid. The thylakoid is the location where photons of light are processed. Chlorophyll is protected from being oxidized by too much solar radiation by another pigment, the carotenoid zeaxanthin, which acts as a heat sink (Fleming and Niyogi 2005).

The green color of most plants begs the question of why green, with all the visible colors of the rainbow? The incoming light that strikes the leaf surface appears invisible to the naked eye, but it contains all of the different wavelengths, or white light, that viewed individually are seen as different colors. Plants are green because they reflect light in the green spectrum. In general, during photosynthesis, the green-pigmented chloroplasts absorb incoming light energy. This source of light can be the sun or artificial lighting. Pigments have distinctive light absorption patterns, and chlorophyll absorbs the blue (400–500 nm [nanometers]) or red (600–700 nm) wavelengths of the solar spectrum. Plants also can use light in the red and violet parts of the spectrum for photosynthesis, however, which are less available than green light. Thus, plants usually have more light-sequestering structures than just chlorophyll. This is indicated by colors other than green in the fall foliage of many deciduous trees that result from the lack of chlorophyll *a* and *b* production in the chloroplasts and the appearance of accessory pigments in the chromoplasts, such as the carotenes and xanthophylls. Moreover, some leaves remain red year round because anthocyanins are present in the sap rather than in the chloroplasts, and they absorb the more plentiful green light.

The rate and magnitude of photosynthesis is dependent on many variables. As may be expected, the rate is related to the amount of light and air temperature of a particular area. Photosynthesis is directly proportional to light intensities greater than 25% of maximum sunlight, in terms of incoming radiation in kilocalories per square meter per minute ([kcal/m^2]/min). Light intensities less than 25% of maximum result in a decrease in the rate of photosynthesis for most plants. The rate of photosynthesis also is directly related to the amount of chlorophyll present. The conversion of light energy into sugar is relatively inefficient, usually between 1% and 2% for most plants. This is similar to the very small amount of energy actually converted to light, rather than lost as heat, in a typical incandescent light bulb. In reality, because the incident light on the surface of the earth is not a constant but varies with location and time and the summation of nights equate to about 6 months per year, the photosynthetic efficiency globally is about 0.1%.

Rates of photosynthesis increase as air temperature increases to a maximum and then decreases with further increasing temperature. The rate of increase is about 2–5 times per 10°C increase in temperature under conditions of normal light intensity such that it is not limiting. All plants respond differently to changes in temperature, and optimum rates of photosynthesis can occur when temperature ranges from 16°C to 40°C. As such, photosynthesis occurs in most plants in the morning before noon when temperatures are lower and evapotranspiration demands also are lower, and then later in the afternoon when light is still available but air temperatures decrease from peak levels. As will be discussed later in this chapter, this temperature effect is initiated by stomatal closing and opening—stomata tend to close as temperatures increase.

Because temperature and light affect rates of photosynthesis the rate of photosynthesis also is directly related to the rate of transpiration. If the air is dry, transpiration may be initially high until stomata close, which decreases both photosynthesis and transpiration rates. On the other hand, if the air is humid, photosynthesis can outpace transpiration. The transpiration efficiency, *TE*, of plants is expressed as the ratio of net production (from photosynthesis) to the transpiration of 1,000 g of water. As stated earlier, most of the water taken up by plants is released as water vapor and is not used in the photosynthetic equation; the *TE* of most plants, therefore, is extremely low, about 2 g of water used in photosynthesis per 1,000 g of water transpired.

Although not part of the structure of chlorophyll, one of the elements essential for its synthesis is iron. In areas where the earth's surface is exposed to oxygen, however, iron is in an oxidized form and unavailable for direct passive uptake by plant roots. Under anoxic conditions, however, iron is reduced and can be released into solution. This process of iron reduction under anoxic conditions primarily is mediated by iron-reducing microorganisms. Plants that have the ability to survive periods of flooding or conditions of a high water table and the attendant anoxic conditions have a higher probability of accessing the reduced, dissolved iron with little expenditure of energy. Too much dissolved iron, however, can be toxic to plants. The role of iron in plant growth is discussed in Chap. 11.

During photosynthesis, an electron-transport system similar to that used in mitochondrial cells also is present. The energy necessary to start the system is absorbed in the form of light radiation. Just as human skin is warmed by the sun's light radiation, this energy also is absorbed by the molecules in a plant. Energy absorption causes atoms to be raised to an excited level, whereby electrons orbiting the nucleus are pushed into a new, more distant orbit, causing these atoms to have more potential energy. Thus, incoming light energy is transformed into electrons and potential energy. Specifically, chlorophyll is raised to an excited state but also rapidly disposes of energy and returns to its pre-excited state. The energy is released as heat or light, called fluorescence, but the most important transfer of energy is in the formation of high-energy molecules. This light energy is absolutely necessary for the reaction of CO_2 and water to occur, because these reactants have an unfavorable Gibbs change in free energy in the absence of light.

The solar energy captured by chlorophyll in the chloroplasts in the leaves is used to split a water molecule into hydrogen and oxygen, and oxygen is released as a byproduct into the atmosphere. Four photons of light energy are required to split hydrogen from oxygen—the hydrogen then reacts with CO_2. In general, CO_2 enters the leaf through open stomata by passive diffusion along a concentration gradient. It enters the mesophyll cells, also by diffusion, along the watery boundary layer and then into the cytoplasm. To produce 1 molecule of sugar, as $CH_{2n}O$, requires 6 molecules of CO_2 to be reduced, which is the input of energy in the form of 24 electrons; this states the balanced form of Eq. 3.1. This equates to an energy requirement of 28.8 kcal/molecule, (kilocalories per molecule). The absorbed light, at the blue and red wavelengths of 400 and 700 nm, respectively, is about 40.5 kcal per 6×10^{23} photons; this is called photosynthetically active radiation (PAR). The conversion of light energy into reducing power is about 71% efficient.

The path from light radiation energy to chemical bond energy that occurs in the chloroplasts is complicated. There are two different light-driven reactions. The overall reaction is that water is split into hydrogen ions (H^+) and oxygen, which exits the plant through the stomata, and the evolved H^+ is used to reduce the coenzyme nicotinamide adenine dinucleotide phosphate (NADP) to NADPH, which is then used by the cell to reduce gaseous CO_2 into carbohydrates. The water molecule is split to remove 4 electrons by oxidization. The water-splitting reaction occurs inside the thylakoid membrane, and the reduction of NADP to NADPH occurs on the outside of the membrane. Hence, there exists a gradient in H^+ concentration from inside the membrane where it is produced to the outside where reduction occurs.

The H^+ used to reduce NADP to NADPH also converts adenosine diphosphate (ADP) and phosphate into ATP. The synthesis of living cell matter takes energy, and the form of energy used by all living organisms is ATP. The H^+ gradient established by the splitting of water and its use in NADPH also leads to the storage of ATP. The NADPH can reduce CO_2 to sugars, and the ADP used to run energy requiring reactions. Both are cycled back to the oxidized forms of NADP and ADP, ready for use again.

As stated previously, the sugar synthesized is used by plants not just for food but to construct other important organic compounds, from the waxy cuticle that covers many leaves to wood itself. As such, Eq. 3.1 also can be written as

$$6CO_2 + 6H_2O \rightarrow \text{plant, light, enzymes, minerals} \rightarrow C_6H_{12}O_6 + 6O_2 + H_2O_{(g)} \quad (3.2)$$

The formation of ATP from solar energy and then electron energy can follow two paths; cyclic or noncyclic phosphorylation. In cyclic, ADP is linked to phosphate using the electron energy released from light sorption of chlorophyll as the excited electrons drop back to ground state. These electrons are cycled continually. In noncyclic phosphorylation, however, electrons are not cycled but are passed along a series of electron carriers or transport systems. In doing so, ATP is formed. In the noncyclic case, the electron energy to drive the formation of ATP is derived from the splitting of water. This electron transport system of noncyclic phosphorylation in photosynthetic organisms is very similar to oxidative phosphorylation in the mitochondria of non-photosynthetic organisms. During photosynthesis, the input energy is light;

whereas in non-photosynthetic animals, the input energy is reduced organic matter such as plant matter.

An analogy can be made between ATP and a chemical storage battery in that both store concentrated forms of potential energy for later use. The use of ATP to drive life processes releases maximum amounts of energy—up to −8,000 cal—after hydrolysis. The high energy released is a function of the bond energy in ATP, in which the last 2 phosphates are anhydride linkages that contain repulsion energy between the closely spaced phosphate groups. This is in contrast to the simple sugars first fixed during glycolysis, such as glucose-6-phosphate, which yields less energy than ATP upon hydrolysis—only −3,300 cal.

In other words, the storage of energy from the sun in carbohydrates formed by plants is analogous to the energy stored in water at elevation (Fig. 3.4). Both forms of energy are considered potential energy. If water stored at elevation flows downhill and turns a wheel the potential energy is converted to kinetic energy that can be used to perform work. This is the same with the sugar molecule. The stored energy is not physical, like water elevation, but chemical, which is converted to kinetic energy when it is used to drive the flow of energy in a living organism.

From Eq. 3.2 it appears that the total reaction must proceed in the presence of light. This is in fact true; light is needed to initiate photosynthesis. Additional reactions take place in the absence of light, however. These dark reactions include the use of NADPH to reduce CO_2 to carbohydrates and sugars. This is why dark reactions are sometimes referred to as carbon-fixing reactions. Dark reactions take place in the stroma of the chloroplast. The rate of dark reactions is dependent upon temperature and will increase with increasing temperature. Moreover, other forms of hydrogen can act as a hydrogen donor during dark reactions, such as hydrogen sulfide (H_2S) by some photosynthetic sulfur bacteria, or even molecular hydrogen (H_2). It is likely that abundant quantities of H_2S in ancient, anoxic environments were used to reduce CO_2 and initially release sulfur rather than oxygen, until supplies were depleted. As concentrations of oxygen in the atmosphere increased, ozone (O_3) formed prohibiting ultraviolet radiation from entering the earth's surface and aiding in the invasion of land by plants. Once an oxic atmosphere was established, aerobic or oxidative metabolism was advantageous, whereas oxygen previously had been (and still is) considered a toxin. As a consequence, conditions existed such that captured chemical-bond energy in photosynthates could be harvested efficiently by organisms, including man, and oxidized back to CO_2. The abundant supply of water, however, for use not only in cellular processes but also as a ubiquitous hydrogen donor, makes it the preferred source of hydrogen for the reduction of CO_2 in dark reactions.

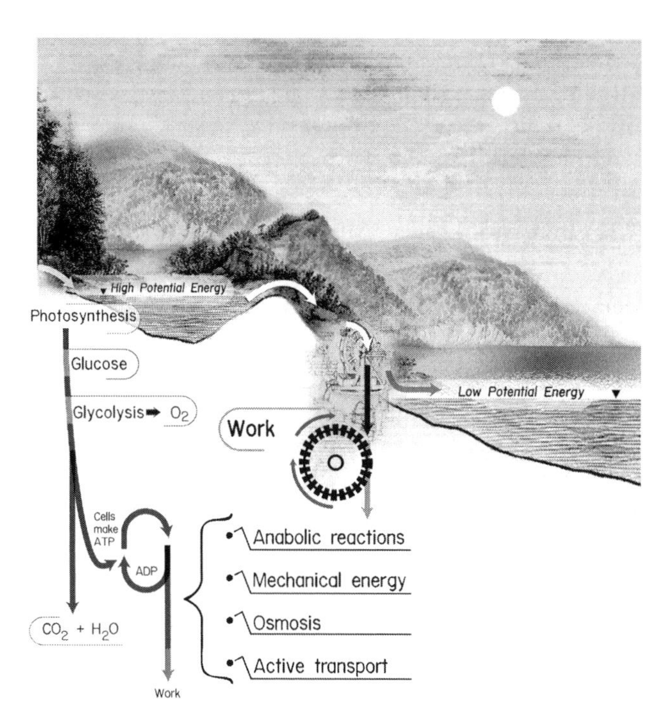

Fig. 3.4 The potential energy stored in adenosine triphosphate (ATP) is analogous to that of the potential energy associated with the elevation of water, where both can be used to perform work for making proteins or turning a water wheel, respectively.

3.1.4 Carbon Fixation: C_3, C_4, CAM, and Aquatic Plants

Light provides the energy to stimulate chlorophyll and split water and release H^+, and atmospheric CO_2 enters the cell passively, but these processes together do not make CO_2 available for subsequent reduction. Most plants reduce, or fix, CO_2 by using the enzyme ribulose biphosphate carboxylase/oxygenase, called either Rubisco, or RuBP. This enzyme is used by plants to take the carbon from CO_2 and add it to pre-existing sugars to create two molecules of a 3-carbon (C_3) sugar called phosphoglyceric acid (PGA). This process is referred to as the Calvin cycle, or the Calvin-Benson cycle, after the work of Calvin's chemistry group in California. Most C_3 plants tend to predominate in moderate climates where water is readily available. A 4-carbon (C_4) sugar also can result from carbon fixation by certain plants. Most C_4 plants tend to predominate in hot, dry climates or tropical ones with high light intensity. Photosynthesis is at a maximum around 30°C for C_3 plants, but temperatures must be near 45°C for C_4 plants. C_4 plants still use the Calvin-Benson cycle, but it occurs elsewhere in the plant. C_4 plants have to be more efficient in the conversion of light energy into reduced organic matter than C_3 plants that grow in more humid areas.

Carbon also can be fixed during the crassulacean acid metabolism (CAM) or Hack-Slack cycle. This type of reduction can be envisioned as an aerobic reduction of CO_2, unlike the anaerobic reduction of CO_2 that produces CO_2 and CH_4 (methanogenesis). Shallow-rooted plants in the desert under conditions of essentially permanent water limitation have to be more efficient in taking up CO_2 than plants with less water stress. Desert plants accomplish this by fixing CO_2 into a C_4 carbohydrate, such as oxaloacetic acid, malic, and aspartic acids, during the day before converting it to the glucose typical of the C_3 and C_4 plants at night when water stress is lower.

Unlike C_3, C_4, or CAM plants, plants that reside in aquatic environments fix CO_2 similar to single cell blue-green algae with respect to photosynthesis and gas exchange. Terrestrial plants fix CO_2 after it passively dissolves in the water of the outer layer of mesophyll cells; the products are then translocated throughout the plant. Conversely, aquatic plants passively absorb CO_2 dissolved in the water column, either as a dissolved gas ($CO_{2(g)}$) or as an ion such as carbonic acid (H_2CO_3), bicarbonate (HCO_3^-), or carbonate (CO_3^{2-}); this occurs directly as no or minimal stomata are present in aquatic plants. If air is in equilibrium with water, the amount of dissolved CO_2 is near 0.5 mg/L, compared to about 8 mg/L for oxygen. It is this minimum concentration of CO_2 that must then diffuse passively through the water before entering cells; in fact, concentrations need to be at least 10 times higher to overcome aqueous boundary layers of still water near leaf surfaces, as exist in air near terrestrial leaf boundaries.

Aquatic plants differ from terrestrial plants in other ways. Aquatic plants do not have a strict xylem in the manner that terrestrial plants have. As such, the presence of roots in most aquatic plants is for anchorage and food storage rather than for water or solute uptake, as evidenced by the lack of root hairs. Aquatic plants have very thin leaves with thin epidermal cuticles, and many are dissected. Such plants also have a more extensive network of pore spaces interconnected throughout the plant, similar to terrestrial plants that grow in waterlogged soils, such as phreatophytes that can be used for phytoremediation. These pore spaces also conduct CO_2 formed by methanogenesis in the sediments near the roots to the leaves for use in photosynthesis.

Aquatic plants also have adapted to their watery environment in terms of oxygen fate. Oxygen is a byproduct of photosynthesis and needed for respiration. Whereas terrestrial plants use stomata to emit oxygen produced by photosynthesis and to take up oxygen for passive transport to roots, aquatic plants use aerenchyma tissues for gas transport. Oxygen produced by photosynthesis is stored in these tissues after collection by diffusion. Once significant internal pressures are produced during gas collection, oxygen is transported to roots that often grow in anaerobic sediments where oxygen is limiting to root growth and survival.

3.1.5 Plant Respiration and Glycolysis

Photosynthesis results in the formation of food for plants, but this trapped light energy cannot be used directly by the plant. The trapped energy first needs to be unlocked and the key is the equally important process of respiration. Respiration in plants is similar to that in animals and is essentially the reverse of photosynthesis shown in Eq. 3.1. Energy is required to drive respiration much like energy is required to drive photosynthesis. The energy for respiration is derived from (1) the excitation of electrons in the chlorophyll molecule, and (2) the hydrolysis of water. In general, during respiration the trapped energy is transformed into the organic compound glucose, which can then be used by the plant. In doing so, glucose is converted back to CO_2 and is released, ready to be reduced again by plants. These sugars also can be used to synthesize the necessary chemicals of the plant structure. Individual glucose molecules, for example, can be polymerized to form cellulose that is used in the cell wall, as previously mentioned. Most of the carbohydrate products in the natural world are built from glucose molecules arranged in various manners to give rise to various compounds.

Most of the enzymes that participate in plant aerobic respiration come from the mitochondria located in the cytoplasm. Similar to respiration in mammals, plant respiration permits the oxidation of photosynthetically produced reduced organic matter, or photosynthate to produce energy, water, and CO_2, according to

$$C_6H_{12}O_6 + 6O_2 \rightarrow 6CO_2 + 6H_2O + \text{energy}. \quad (3.3)$$

This is an essential part of the global carbon cycle and is discussed in Chap. 11.

As is shown in Eq. 3.3, respiration produces CO_2 and the rate of production is variable. Plants cannot afford to resubmit the very CO_2 they fixed right back into the atmosphere during respiration, so respiration proceeds slowly, often resulting in the production of other compounds that will store this energy for later use. In part, slower reaction rates occur because diffusion is the process by which CO_2 is lost and O_2 is gained. This is perhaps best demonstrated by the extensive network of air spaces between cells of aquatic macrophytes. In these cases, the majority of the bulk plant consists of nothing but air space, surrounded by a 1-cell-thick wall that permits maximum oxygen and CO_2 diffusion.

In plants, as in animals, the conversion of stored chemical energy in glucose to useful energy is through glycolysis where photosynthate is broken down in a series of steps. Glycolysis leads to the conversion of sugars that contain six carbon atoms into two molecules of pyruvic acid, or pyruvate, that each contains three carbon atoms. This step occurs within the cell cytoplasm and does not require oxygen.

The sugar is then reacted with phosphate (PO_4) using energy from ATP to create glucose phosphate. Two molecules of ATP and two molecules of pyruvate are released per molecule of sugar undergoing respiration.

The pyruvic acid is then exposed to oxygen to yield energy, CO_2, and water as part of the Kreb's cycle, also called the citric acid cycle or tricarboxylic acid (TCA) cycle. The Kreb's cycle occurs in the mitochondria of each cell and consists of reactions of organic acids. The cycle begins when one molecule of pyruvate is converted to three molecules of CO_2, remembering that pyruvate had three carbons atoms. Not all three molecules of CO_2 come from the 3-carbon compound of pyruvate; rather, two molecules become associated with coenzyme A, or acetyl coenzyme A. The Kreb's cycle has to complete the cycle two times to convert each 6-carbon sugar into CO_2. The cycle has to have a constant supply of oxidized nicotine adenine diphosphate (NADP) and flavin adenine diphosphate (FAD). Each of these reactions is controlled by specific enzymes, and these enzymes are proteins that are encoded by specific genes. The bottom line is that the glucose is mineralized to CO_2 and its energy is converted into ATP.

Where does the oxygen in the Kreb's cycle come from? For terrestrial plants, it is gaseous oxygen from the atmosphere. Oxygen enters in the stomata and by pores in the stem and branches called lenticels, which are discussed later in this chapter. Because the gas-phase concentration of oxygen is much higher in the atmosphere relative to the lower partial pressures of oxygen in the soil, unsaturated zone, capillary fringe, and water table, oxygen can diffuse from leaves to roots along a concentration gradient. The transport through the plant and to the roots is through an interconnected series of air spaces between loosely packed cells that are collectively called the aerenchymal cortex. Much like how the xylem and phloem act as pipes to conduct fluid flow, these tissues act as pipes to conduct gas flow. The oxygen is consumed by respiration and the Kreb's cycle and the balance diffuses into the rhizosphere.

3.1.6 Growth and Hormones

The fundamental growth of plants has fascinated researchers, as well as laypersons, over the long history of the interaction between plants and man. The writer and poet Johann Wolfgang von Göethe, known today simply as Göethe, also did experiments in and wrote books on geology and optics, as well as plant life. He theorized that all varieties of plants could be reduced to a simple pattern of growth of basic parts or segments, and that mature plants were variations on this structural scheme. Moreover, he concluded from his observations that all parts of a plant above ground were modifications of a proto-leaf, except for the main stem. Göethe published this idea in *An attempt by J.W. von Göethe, Privy Councilor of the Duchy of Saxe-Weimar, to Explain the Metamorphosis of Plants* (1790; Durant 1953) (Table 1.3). He later composed a poem with the same central idea. Today this idea can be observed by the repetition a common angle between stem, branch, leaf main axis, and leaves of many plants.

What is plant growth, and why is a basic understanding of plant growth important to the phytoremediation of contaminated groundwater? The sugars produced by photosynthesis provide the energy needed to support metabolism and to drive plant growth. As may be expected, the selection of fast-growing plants often used for phytoremediation is directly dependent on the growth characteristics and water-use potential of the plants. Moreover, the different growth characteristics of various plants have important implications for the capital costs of phytoremediation and the time for remediation to occur at contaminated sites.

Growth is a prerequisite of a living entity. In general, for most woody plants, growth can be defined as active cell division at rates greater than cell death, and this typically can be found in both above-ground and below-ground meristematic tissues. The most obvious indication that plants grow is the transformation over time of a small seed, such as an acorn, to a mature tree such as a majestic 200-years-old oak. Cell growth is associated in plants with the production of food, respiration, and metabolism. Obviously, this type of cellular or tissue growth cannot occur without first having started with a fertilized female cell, or zygote, that started the subsequent chain of cell division; additional discussion of this topic is beyond the scope of this chapter.

There are two kinds of growth tissue in plants, the primary and secondary tissues, both of which arise from cells grouped into meristematic tissues, as introduced earlier in this chapter. Primary growth occurs in cells that can grow indefinitely and are found at the tips of shoots and roots, where growth results in an increase in length rather than thickness. The growth beneath the phloem and xylem that arises from the cambial tissues is called secondary growth, because this growth is lateral rather than vertical with a corresponding increase in girth. This is why initials carved in a tree trunk at chest height or a nail driven into a tree trunk is not displaced higher when revisited many years later. Both types of tissues are also found below ground. Shoots and roots elongate and expand in girth in dicotyledon plants, whereas monocotyledons increase in height only. This difference is regulated by many factors, both internal and external to the plant. External factors include the degree of sunlight, amount of precipitation, porosity of soil, degree of herbivory, etc. Internal factors include those similar to animals, such as hormone production and regulation. These internal factors can be affected, however, by changes in external variables.

The meristematic tissues and the production of new cells following division from existing ones is a process that needs to be elucidated, for it has implications regarding the types of plants that should be used for phytoremediation of contaminated groundwater. Meristem cells were first recognized by Caspar Wolff in 1759 (Table 1.3). The meristem is at the center of the stem, but the cells that divide occur at the tip, or terminal buds above ground or root tips below ground. These shoot and root meristems are characteristic of most plants but are more important in annuals that experience primary growth patterns. The meristems, whether in the tips of shoots and roots or in the cambial layer, are formed of undifferentiated cells that contain mostly cytoplasm and a few organelles along with a small vacuole, that later become differentiated into the various cells of the plant, such as epidermis, xylem, phloem, etc. These undifferentiated cells are the source of continued self-renewal in plants (Weigel and Jürgens 2002). As such, these meristem cells are analogous to human stem cells, which are undifferentiated cells that have the potential to be used for many purposes.

As the meristem cells divide, they move from the center of the shoot or root tip to the tip and then back and down around the original center. The meristematic cells divide rapidly at first and then stop. Additional increase in size is by cell-wall relaxation and elongation. In fact most increase in plant size is actually an increase in the size of existing cells rather than the addition of new cells. This is exactly opposite of most animals, in which growth is a result of cell division. After the elongation is over, the cell wall re-hardens.

The consequence of the division of meristem cells is hard to observe on a small scale, but large-scale consequences can more readily be observed. One example is the leaves of palms that are produced from the meristem at the tip of the plants and grow outward and then downward as younger leaves are generated. The unique bark of such plants is composed of older leaf petioles, with the older ones near the ground and progressively newer ones nearer the top. As these undifferentiated meristem cells divide, some become leaves where others become branches.

Growth also can be classified in plants as one of two main growth types—determinate or indeterminate. Determinate growth refers to a predictable growth cycle with the cessation of growth once a particular biomass is achieved, followed by death. Plants that exhibit determinate growth are typified by smaller plants, such as annuals. On the other hand, plants that tend to keep on growing over time for many years, even hundreds of years, exhibit indeterminate growth. This is exemplified by woody plants, such as bristlecone pines in the southwestern United States. In both cases, the overall biomass achieved by either type of plant is controlled by the genetic information contained in the cells.

Shoots grow in length from the terminal bud, where active cambium is located. Smaller buds are located farther down the stem and are called lateral buds. When the main tip is removed, by natural damage or pruning, the food and water that had gone to the removed tip is diverted to the lower remaining buds, and they will grow at increased rates. Anyone who has pruned a plant has experienced this phenomenon. On the other hand, removal of too many leaves after the buds have broken decreases potential food production. Many lateral buds remain dormant for the life of the plant. These dormant lateral buds combined with the fact that all buds, both leaf and flower, are formed during the previous growing season, help to ensure plant survival across a range of environmental stresses over time.

Growth also is enhanced by the osmotic uptake of water and resultant cellular swelling, or turgor. This stresses the cells by inducing the cell wall to stretch, and growth occurs by making this stretched cellular dimension a permanent part of the plant. As this is an increase in water pressure or potential, and water transport to leaves is along a gradient of decreasing water potentials (as we will see later in this chapter), a dilemma exists between water transport and cellular turgor maintenance. For example, in the spring when buds break, leaves will not enlarge and enable additional water transport unless water is initially available.

An increase in girth is accomplished by the cambium stem cells that create new xylem toward the center and new phloem toward the bark, with older xylem dying and providing support, and the phloem dying and becoming the cork and bark. Such cambial stem cells are characteristic of woody plants that have primary and secondary growth patterns. Less is known about these cambial stem cells than stem cells at the tips of shoots and roots.

An interesting feature of many woody plants is the relatively small proportion of the plant that grows compared to its overall biomass. Typically, less than 1% of a woody plant is, in fact, alive. This is because the nonliving parts of the vascular system can support the plant without the costs of metabolism. If this internal tissue rots, the tree will still survive, until it falls from lack of support; this explains why recently fallen trees were alive even though they had been hollowed out by decay. An interesting mental image to conjure is of a tree reduced to only its living cells; it would appear as a thin column of green cells linked to leaves above and white roots below, similar to the structure of a bubble, with little surface area but of high volume. In fact, even though plants are one of the longest living organisms on earth, such as the aforementioned bristlecone pine that can live to over 1,000 years of age, the actual growing cambium is only a few years old; the rest of the tree is composed of long-dead tissue. This is a consequence of the meristems, in which the growing cambial cells are replaced continually as other cells die.

Another characteristic of plants which has implications for phytoremediation is the rate of growth of certain plants. Kudzu (*Pueraria lobata*) can grow up to 1 ft (0.3 m) per day, as can some other vines. In the southeastern United States, kudzu is considered today to be an invasive plant, but was initially introduced by the federal government to reduce erosion in soils that were heavily planted during the cotton industry boom in the early 1900s, and to be used as a landscape plant, called porch vine. Kudzu can thrive in nutrient-poor soils because as a member of the legume family, kudzu creates nitrate in the root zone by the fixation of atmospheric nitrogen, as is discussed in Chap. 11. Some trees, such as poplars often used for phytoremediation, can grow 10 ft (3 m) or more in one growing season, if nothing needed to support this growth rate is limited. By comparison, some plants grow only a few inches per year.

What causes this difference in growth rate in plants? Is it difference in water use? Although water can increase cell turgor and result in cellular elongation, it does not necessarily affect the *rate* of cell division. The primary reason that some plants grow faster than others is because they possess a longer section of active meristem cells, up to 2-ft (0.6 m) long in some plants such as poplars, relative to those that have smaller sections of meristems and exhibit slower growth. As such, the individual cell-growth rates are the same for each plant, but plants that grow faster have more cells dividing across a greater length than in slower growing plants. Because massive amounts of energy are needed to support the enlarged area of growth, fast-growing plants typically are characterized by having larger leaves, and more of them, to make food or extensive and deep root systems for food storage and water and nutrient uptake. Both of these characteristics are why phreatophytes, such as willows and poplars, are successful for phytoremediation if groundwater is within reach of roots. In essence, the use of poplars to remediate contaminated groundwater is a result of poplars having large sections of stem cells to support, which drives the need to interact with groundwater.

Fast growth rates are more common for obligate phreatophytes along riparian habitats where water is not limited. Riparian plants also have a high rate of seed dispersal and rapid germination rates to take advantage of infrequent floods. In fact, the growth and seed-dispersal traits that ensure the survival of many generations of riparian trees are similar to those of plants considered to be weeds. Three examples of fast-growing riparian plants will be discussed here; tamarisk, eucalyptus, and melaleuca. The first two can be used in phytoremediation projects.

Tamarisk (*Tamarix* spp.*)*, or saltcedar, was introduced in the United States around 1860 for shade, wood, and flood control. It probably came from Europe, or the area around the Mediterranean Sea, because tamarisk also is the name of a river in the Pyrenees (Van Hylckama 1974). It also is mentioned in the *Bible*, and wood from these trees is found at many sites of antiquity in the Middle East. Alternatively, it is possible that Spanish conquistadors may have brought the plant with them when they invaded Mexico in the sixteenth century. Although originally intended to be a cultivated plant, tamarisk's ability to tolerate high-salinity conditions and its fast growth have allowed its range to be uninhibited since about the 1930s following a tree-planting campaign, called the shelterbelt project, to slow soil erosion after the Dust Bowl (Robinson 1965). Saltcedar trees produce large quantities of seeds; one tree can produce 500,000 seeds each year, with growth rates up to 10 ft (3 m) per year after germination.

The most aggressive of the *Tamarix* species are *T. pendantra* and *T. gallica*. In particular, these species have out-competed other trees to dominate the riparian corridors in at least 15 of the 17 western states, including Arizona, New Mexico, Texas, Oklahoma, Kansas, Colorado, Utah, California, Nevada, Oregon, Nebraska, Idaho, Montana, Wyoming, and South Dakota (Robinson 1965). In some cases, its establishment occurred after floods denuded large areas of native riparian species. Also, the damming of many rivers for water supply and flood control in this part of the United States has favored the tamarisk over native species, such as willow and cottonwood, which rely on the natural ebb and flow of flooding for seed dispersal.

Because saltcedar thrives along river banks in low topographic areas, the roots generally need to be no more than 25 ft (7.6 m) deep, because the depth to the water table is shallow in groundwater discharge areas. Because the saltcedar is fast growing, large volumes of groundwater are needed to support its growth. And because no economic benefit is derived from the wood or fruit, groundwater used by the trees and not returned to the basin is considered to be consumptive. Robinson (1965) reported that groundwater used by saltcedar can approach 9 acre-feet (11,097 m^3) per acre (4,047 m^2) in the southwestern United States. These are the trees used by Gatewood et al. (1950) in the tank studies of the effect of phreatophytes on groundwater resources discussed in Chap. 1.

The eucalyptus trees (*Eucalyptus globules*) found in the western United States came from Australia and may have been introduced by railway owners who planted these trees along the rights-of-way to provide a source of lumber for the ties as well as shade from the sun. It was soon found, however, that the wood was prone to splitting after the rails were attached. These trees grow fast, at about 10 ft (3 m) per year. Eucalyptus thrives not because it can grow along rivers and use shallow groundwater like the tamarisk but that it can utilize deep groundwater that is unavailable to most plants. The roots of eucalyptus are called lignotubers and can store starch and water for increased drought survival—the native people of Australia use the roots as a source of water.

Not all invasive plants that use copious amounts of groundwater are related to the western United States, however. Another non-native woody plant found in the United States that is characterized by fast growth rates and is, therefore, considered to be invasive is the melaleuca (*Melaleuca quinquenervia*). It also hails from Australia and grows about 6 ft (1.8 m) per year. Introduced into south Florida in 1906, it spread slowly but was greatly enhanced after planes were used to disperse seed throughout the Everglades. In 1993, it was estimated that melaleuca had spread to cover nearly 400,000 to 1.5 million acres (1.61×10^9 to 6.0×10^9 m^2) of the Everglades, spreading at a rate of 50 acres (202,000 m^2) per day. As would be expected, water losses also are higher in areas where melaleuca has invaded. They are hard to kill, and any threat to their survival leads the tree to release millions of seeds. It is perhaps ironic that the one person responsible for encouraging the use of melaleuca, as well as kudzu, in the United States, was the famous plant scientist David Fairchild, who spent his career introducing plants to the United States (see Fairchild (1938) for more information on kudzu and melaleuca introduction, as well as for compelling reading).

There are other reasons why some plants grow faster than others, and it has to do with survival, competition, and reproduction. To ensure survival over generations, however, seed production and dispersal are the main driving forces for accelerated growth and height. But the main one for some plants that use wind for seed dispersal, such as cottonwoods, oaks, and poplars, is height—they need to be tall to take advantage of the wind for a wide range of seed distribution and survival. Willow (*Salix spp.*) seedlings were found to have greater survival rates in fine-textured sediments high on point bars relative to coarse-grained sediments closer to surface-water fluctuations (Gage and Cooper 2004). The roots of seedlings need to reach the water table, which requires adequate water and nutrient availability, and may take several years.

If growth occurs in the tips of roots and shoots and the cambial layer where the meristem cells differentiate, what controls do the plants have, if any, on growth? In general, plant growth is controlled by hormones, which are organic molecules synthesized by the plant but not used directly for energy. Hormones can accomplish their work at very low concentrations. They can either increase or decrease plant growth, often in separate tissues of the plant at the same time. Hormones can be produced in one tissue and exert an effect in another tissue, sometimes a great distance away.

An excellent example of the influence of hormones on plant growth is the control of the color and lifespan of leaves of deciduous trees in temperate forests. As the growing season ends, the plant responds to lower levels of solar-radiation intensity and shorter days by stopping the production of chlorophyll and then disposing of the leaf that previously had provided the plant with food. This process is called photoperiodism and is controlled by a light-sensitive protein called phytochrome. Ethylene, essentially a plant hormone in gas form, promotes leaf drop by increasing the production of an enzyme that breaks down the cellulose in plant-cell walls.

Other environmental changes, or stresses, can induce the production of hormones in order to aid the survival of the plant. Abscisic acid (ABA) is a hormone that promotes plant dormancy and leaf abscission, and it is produced by leaf cells under conditions of water stress. The production of ABA, in turn, causes the stomata guard cells to close, which reduces transpiration and photosynthesis.

Another common example of the effect of hormones on plant growth is contained in a growing shoot on any landscape plant. In the spring, an increase in length occurs in the terminal bud, formed the previous year. Other buds, however, also are present along the stem behind the main terminal bud. The question, then, is what keeps these buds from growing as rapidly as the terminal bud or not at all in some cases? The terminal bud releases a hormone that suppresses the growth in other buds, a process called apical dominance. If the terminal bud is removed, the source of the hormone that has inhibited the growth of the other lateral buds is removed, and the lateral buds start to grow. This hormone is formed in the cell cytoplasm, but transported throughout the plant.

Another hormone that relates to cell elongation is gibberellin. It encourages cell elongation especially in stems. One of the possible reasons that it is advised to put aspirin in a vase of cut flowers is because the aspirin, or acetyl salicylic acid, acts as a surrogate hormone to induce the cut-end stems of flowers to root.

The hormone responsible for terminal bud elongation and root tip elongation is called auxin, or indoleacetic acid (IAA). Auxin is synthesized by the plant by conversion of an amino acid called tryptophan. Apparently, the purpose of the IAA hormone in a particular cell is to make the cell wall more pliable and to promote cell elongation. When exposed to light from one direction only, auxin production is induced in the shaded side of the terminal bud, and these cells elongate at a faster rate than the cells exposed to light with the net effect being that the bud bends toward the light source; this process is called phototropism. Other plant auxins include the synthetically derived naphthalenacetic acid.

Auxin also can be transported from the shoots through the phloem to the roots to stimulate root meristem growth. This process can be forced by gardeners or horticulturalists by addition of synthetic auxin, orindole-3 butyric acid at a concentration of 0.1%, to plant cuttings prior to installation in soil media. Such an approach also can be applied to cuttings used in phytoremediation projects, but it has some potential liabilities, which will be enumerated in Chap. 7—briefly, the use of too much auxin can actually inhibit root growth, much as the presence of auxin in the terminal meristem enhances its growth but inhibits that of adjacent lateral buds.

Another synthetic auxin was an impurity in the 2,4-D, or Agent Orange, used to defoliate tropical plants in Vietnam. In commercial applications, 2,4-D is a synthetic auxin that is widely used as an herbicide to kill dicots but not monocots. These hormones stimulate plants to essentially grow themselves to death through overstimulation of respiration, which rapidly depletes carbohydrate stores at rates faster than the production of new carbohydrate by photosynthesis. This occurs because synthetic auxins can influence RNA transcription and the production of proteins used for many growth purposes in the cells.

In addition to the effect of these hormones on cell elongation and respiration, there are other hormones that affect cell division. One group of such hormones is the cytokinins. In most plants, cytokinins are synthesized in the roots and reach the shoots after transport in the xylem. Their effect on cell division occurs between DNA replication and mitosis and is relatively unknown.

3.1.7 Dormancy

Dormancy in plants is a period of slow or no growth and is primarily a function of ambient air temperatures. If temperatures are too cold for a particular species, the enzymes that regulate the life of individual cells and operate within a range of temperature will be affected and growth will slow or cease.

As might be expected for multicellular organisms such as plants that grow exposed above ground as well as below ground, dormancy in different parts of a plant is achieved under different temperatures. In general, dormancy for above-ground buds occurs at temperatures around 45°F (7.2°C) for a time period between 4 and 8 weeks. Roots will still grow under these conditions, however, because changes in soil temperature generally lag behind changes in air temperature due to the insulating properties of soil and the high specific heat capacity of soil moisture. Roots continue to grow after leaf loss because of respiration of stored carbohydrate; growth will occur as long as oxygen in the subsurface is available. As we will see later in Chap. 7, in most parts of the United States, early fall is when many woody plants are typically installed, because the root systems are not dormant and continue to grow even while the shoots remain dormant.

Dormancy also is controlled by water availability. This fact is exemplified by dried seeds that can survive long periods of time and still remain viable. Water is important with respect to seed dormancy because water is the solvent in which cellular respiration takes place. Where there is growth, there is respiration; even non-growing cells respire slowly during dormancy. This is why trees retain water even during dormancy; the water was taken up during periods of previous transpiration, and loss of water to the atmosphere during the dormant period is limited by the thick, waterproof bark. The lenticels in the bark permit gas exchange to occur, but this exchange is limited to the cortex and cambium just beneath the bark and does not extend to the deeper xylem except possibly by diffusion when water flow rates are at a minimum.

Another example of dormancy and water availability is the common mistletoe (*Viscum album*) plant. The word mistel is the Anglo-Saxon word for dung, and the word toe or tan means twig, so the plant typically used during the Christmas holiday to promote amorous activity literally means dung on a twig. This made sense to early plant observers who could only fathom that the mistletoe in the high branches of trees were there on account of bird droppings that contained mistletoe seeds. In ancient times, the fact that a green plant grew during the bleak winter gave rise to its reverence by the Druids, who used it to ward off evil spirits at the start of a new year.

From a plant–water relation viewpoint, the presence of mistletoe in trees has important implications. Mistletoe makes it own food but with water taken by the roots from the host plant's xylem. It also uses the host plant for support, sun exposure, and protection from predators. Mistletoe can be found in most parts of a dormant tree, but its presence in treetops is unequivocal evidence that dormant trees not only contain water, but that water flow can support the mistletoe. Another feature of these plants is that they contain toxic, organic compounds called alkaloids. A few mistletoes do not harm an otherwise healthy tree, but massive infestations can lead to the tree's death.

Plants go dormant in colder weather but even in freezing weather most trees do not die. This is because cold temperatures induce genes to produce antifreeze compounds in the cells to decrease the freezing point of water. This process takes advantage of the colligative property of water and can protect cells from death down to −50°F (−45°C). Unfortunately, the production of antifreeze compounds by human cells does not occur, and such cold temperatures cause ice to form in cells, or frostbite.

3.2 Roots

The height of a plant and its crown of leaves are usually the most impressive features that catch our attention. The depth and distribution of roots below ground, however, are equally impressive but often less well recognized. Leaves can be removed and the plant will survive, but a loss of roots will lead to plant death even if the leaves were present. Examples of this fact are plants that survive after having been cut to the ground, as during coppicing, and plants that have shoots frozen but are replaced with new growth.

The root system has an equal if not greater role than leaves in the survival of vascular plants, such as phreatophytes used for phytoremediation. For example, roots take in water and store excess sugar produced during photosynthesis. Root cells lack chloroplasts, however, and need to be supplied with reduced organic compounds produced by the leaves. In turn, the shoots and leaves of terrestrial plants require water but are not in direct contact with it like roots. If most plants are removed from the soil and their roots are left exposed, the whole plant will rapidly die. The root systems of plants can be visualized as an underground forest—an inverted reflection of the above-ground forest we encounter everyday. The results of some studies indicate that the root systems of plants, such as grasses, can be more than 100 times the surface area of the leaves and stems.

The extent of root penetration and distribution in the subsurface can vary from plant to plant. For example, broom snakeweed (*Gutierrezia sarothrae*) is a woody perennial that can be found in the grass-dominated rangelands in areas west of Texas. One of the reasons that broom snakeweed can compete with grasses is the difference in root structure between the two species. Grasses have a higher root density nearer the soil surface than snakeweed, but snakeweed has a deeper taproot than grasses. This structural difference provides a selective advantage to the snakeweed when the upper layers of soil moisture dry out and cause the stomata in the grasses to close, whereas the snakeweed can remain open and continue photosynthesis on account of the taproot (Wan et al. 1993).

3.2.1 Root Physiology, Growth, and Depth

In general, roots are underground extensions of a plant's trunk and, therefore, retain many characteristics common to shoots. Roots contain a cambium layer of stem cells continuous with the trunk above ground. Similar to how shoots grow above ground, the elongation of roots is called primary growth, and the increase in root diameter is called secondary growth, caused by cell division from a procambium layer of root stem cells. This division ultimately will provide the root cells that form the epidermal layer, the cortex, and the vascular system. As these cells grow larger the percentage of cell volume occupied by the vacuoles increases to occupy most of the cellular space. Root growth, like most life processes, is controlled by temperature, enzymes, and the genes that control their synthesis. The hormones gibberellin and auxin produced in the shoot meristems also influence the growth of root meristems. The plant hormone auxin is used to regulate the formation of lateral roots from the main root stem meristematic cells, or pericycle.

All roots have a common denominator, which is that all plants, under natural conditions of sexual reproduction, start as seeds. From this seed, an embryonic root, or radicle, depending on the plant type, grows either into many small adventitious roots and develops into a fibrous root system, which consists of many roots from the base of the stem, or develops into a large taproot. Plants also can have both types of roots. Herbaceous plants, such as grasses, have fibrous root systems, and although some only penetrate shallow depths, others have very deep roots. These dense root systems are poised to capture water from infrequent precipitation, and they also act to stabilize the grasses and soils in areas such as the Central Plains of the United States where little topographic relief results in high wind speeds and subsequent high soil-erosion potential. When perennial grasses were removed from the Central Plains in the 1930s and replaced with annual agricultural plants, the less-extensive root systems resulted in the production of massive amounts of airborne dust, a time period commonly referred to as the Dust Bowl. This event is a classic example of a lack of appropriate land management and cycles of climate change. By comparison, many deciduous trees have taproot systems, particularly plants that are obligate phreatophytes.

Much as with stems, root elongation by cell expansion, rather than cell division, occurs in the root tip. The root apex, or tip, is hard in order to penetrate through the pore spaces in soil or rock. The pressure to push through the soil pores is provided by the growing cells just behind the root apex. The root apex moves through the pores in a spiral direction. The pressure of the roots moving through the soil is more like a wedge than a nail, in that a wedge, because of its shape, expands as it moves downward and exerts lateral forces.

Because cell elongation leads to a weakening of the root cell wall, the roots need protection from the solid matrix as they push through the soil pores. This protection comes from the root cap, which also secretes a mucigel that lubricates the soil to aid in soil penetration. This mucigel has to be continually produced, as it leaves a trail as the root advances. Although the meristem is located at the root tip, it is not exactly at the extreme end but rather behind it a bit and imbedded into the tip. The various structures of the root in relation to water uptake are shown in Fig. 3.5.

To support this growth and cell elongation, growing roots seek moisture in the soil. In doing so, shallow roots can grow a considerable distance through the air spaces of the unsaturated zone to reach moisture. This explains why roots can be found in caves, because the root tip can grow through the air, or essentially 100% porosity—if moisture is encountered, root hairs develop to extract the water for the plant. Certain

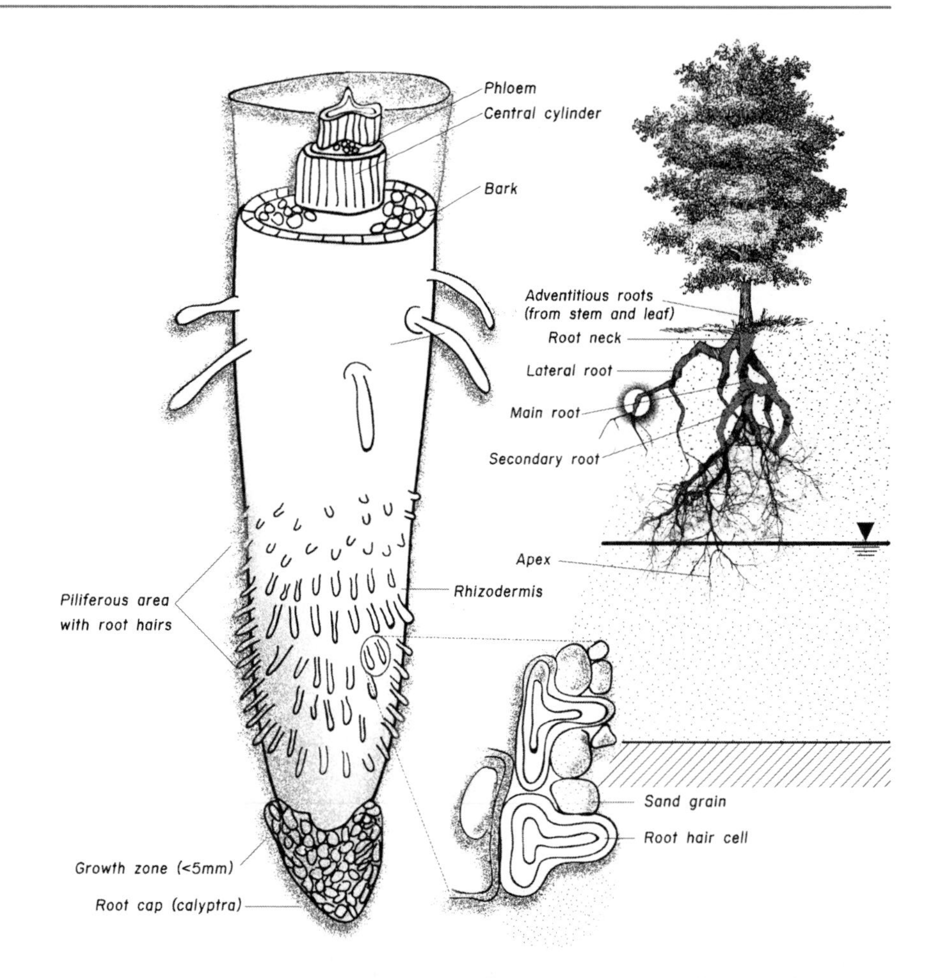

Fig. 3.5 The magnified structure of a growing root tip (*left*) in relation to the root distribution of a woody plant (Modified from Kozlowski and Pallardy 1997). Most water enters at individual root hairs by diffusion along a gradient in water potential from wetted sediments (*lower right*), the capillary fringe, or water table (*upper right*). Root hairs exist in the piliferous area, located just behind the growth zone of the root tip. Other important water-acquisition roots include those that are below as well as above ground, such as adventitious roots.

woody plants in Hawaii, however, have roots that, although similar in structure near the ground surface, become linear as they grow through the open areas of old lava tubes. These roots can be tens of feet long and grow through the moist atmosphere of the lava tube. This structure also enables them to act as conduits to deliver rainwater by gravity to the growing root apex far below ground.

The downward growth of roots is aided by the presence in the cells at the root tip of organelles in the cytoplasm called amyloplasts. These amyloplasts contain grains of starch that make the amyloplasts heavier at the bottom of each cell, and provide orientation with respect to gravity. The position of the amyloplasts is not fixed in the cytoplasm, so the amyloplasts are free to move to the lowest position in the cell as the whole root grows.

Root hairs form behind the root tip and are the main site of water entry (Fig. 3.6). They are an extension of the epidermal cells and arise from the secondary growth that occurs after elongation. The root hairs are not more than 1/250 to 1/3 of an in. (0.01–0.8 mm) long. The density of root hairs and their capacity for withdrawing water from the ground is enormous. Some grasses have been calculated to have almost 4,000 ft (1,219 m) of root hairs in 1 cubic in. of soil.

Fig. 3.6 Adventitious roots on hybrid poplar cuttings formed during shipment in plastic bags and elongated after immersion of the cut end into water (Photograph by author).

Between the root hairs and epidermal cells are the less closely spaced cells of the cortex, or parenchymal cells, which have air spaces between them. The root cortex is covered by the epidermis. Pore spaces between cortex cells permit oxygen to enter from underground sources or to pass to the cells from above ground. These pore spaces constitute the largest cellular volume of the root. Roots of plants that are exposed to standing water or periodic flooding conditions have adapted to the attendant decrease in oxygen content by using the cortex tissues to transport atmospheric oxygen by diffusion to the submerged roots—such plants include the cypress and mangroves. These adaptations are especially useful if the surface-water body is characterized by slow to stagnant flow.

On a larger scale than root-tip elongation, overall root growth is needed for the plant to reach a volume of soil that contains sufficient amounts of water to support its needs. A popular myth is that the distribution of roots below ground is a mirror image of the distribution of shoots above ground. In fact, the reality is more interesting, because for some plants, the below-ground mass of roots is greater than the shoot mass. For example, the production of root material exceeded that of above-ground wood production by almost three times for both pine and hardwood forests in the southeastern United States (Harris et al. 1977). In addition, the distribution of roots with depth for many trees is closer to one third the height of the tree, because lateral roots can extend beyond the plant crown. This root distribution probably results from access to rainwater and higher concentrations of oxygen in the shallow air spaces in the soil. However, another important control of depth and lateral distribution is the type of soil; root systems are able to extend farther laterally in sandy soils than in clayey soils. This may be due to the lower permeability of clays and the higher porosity and water content or the lower dissolved oxygen levels.

Most woody plants have the majority of roots close to the surface of the soil; this can be observed after many species of trees has fallen over. This root distribution permits plants to access water from infiltrating precipitation, be exposed to levels of oxygen near that of the atmosphere in the soil pores, and access nutrients from decaying leaf litter. It has been reported that 60–90% of the roots of trees are located in the upper 1–2 ft. of soil (Le Maitre et al. 1999). However, many trees have roots that penetrate great depths, as long as significant amounts of oxygen are available or can be transported from the air to the roots, and there is sufficient soil permeability to allow the roots to elongate. In soils that do not receive frequent precipitation, or if evaporation from the soil surface is excessive, the only available water for plant growth may be near the water table. Therefore, the maximum depth that roots can penetrate is important in determining a plant's ability to thrive with a lack of water from frequent precipitation. This phenomenon is discussed further in Chap. 5.

Much as stems branch off of larger growth above ground, the same branching also occurs in root systems. These root branches tend to grow at right angles to the parent root. The rate of root growth varies between species of woody plants, from a few hundredths of an inch per day to greater than almost 1 in./day (a few millimeters per day to greater than 25 mm/day). Essentially the existing root mass remains attached to the soil matrix and only the tip extends outward into the pore spaces of the soil matrix. Most root growth, as elongation, occurs in the spring and continues into the fall even after shoot growth has stopped. A controlling factor is the temperature of the soil and the availability of soil moisture.

Not all parts of root systems interact with water. The outer epidermal layer of cells, especially on older roots, can be replaced by a tough, bark-like layer of cells filled with waxes that are less permeable to water, a process called suberization. Suberin is the same substance used in the Casparian strip to block water transport into the vascular system, as described later in this chapter. Suberized roots comprise a higher percentage, in some cases more than 99%, of the total root system in older plants (Kramer and Bullock 1966). Limited water absorption can occur through suberized roots even though the permeability is reduced relative to unsuberized roots.

Speaking of root age, root longevity varies among species, so no exact lifespan for roots can be used to encompass all woody plants. As would be expected, the longer and larger the root, the older it typically is. For root hairs, however, considerable turnover occurs, such that 30–90% of the root hair mass is replaced annually (Fogel 1983).

The food made by the leaves moves throughout the plant by the phloem and is stored throughout the above-ground and below-ground structures. But how does this food get allocated? Do the stems and shoots receive more carbon because they are closer to the source than the roots, which are located farther away? In arid areas where water tables are deep and precipitation infrequent, more carbon is allocated to the deeper roots relative to the shallower roots. Snyder and Williams (2003) reported that the defoliation of such deep-rooted plants that access the water table also rely on shallower water and allocate carbon to the root system. Under such a scenario, the plants become carbon limited, and it appears that available carbon is shunted to shallower roots, perhaps because the water present near the deeper roots cannot be extracted by a decrease in the water potential gradient after defoliation.

Whereas the maximum height of a tree is constrained by the physical forces of gravity and friction between water and xylem, these are not constraints on root penetration below ground. In fact, McElrone et al. (2004) and Jackson et al. (1999) measured tree-root penetration up to 20 m (66 ft) below ground surface and was able to make measurements of some of these roots exposed in caves. These researchers reported that deeper roots have wider diameters in xylem tissue than shallower roots from the same plant, perhaps in order to overcome increased hydraulic resistance at depth (McElrone et al. 2004)—this is similar to the need for thicker electrical wiring as its length increases in order to minimize frictional losses of electron flow. In other words, because water absorbed by deeper roots has a longer path to travel than water absorbed by shallower roots, the diameter of the deeper roots must be larger to overcome limitations imposed by resistance to flow.

3.2.2 Root Evolution

A discussion of roots, their function, relation to water use, and relevance to phytoremediation cannot be considered adequate without some background on the evolutionary history of terrestrial plants. The common theme between plant evolution and phytoremediation is how plants obtain the necessary elements from the environment for survival and reproduction. Much of this information is based on fossil and molecular clock evidence; molecular, or gene, clock determination uses the constant rate of change in DNA mutation caused by random drift to understand evolutionary development.

Terrestrial plants, such as those used for phytoremediation purposes, began as unattached, one-celled photosynthetic organisms, such as algae, that floated near ocean or freshwater surfaces. Algae include both unicellular organisms, such as the blue-green algae, and multicellular organisms, such as green algae. Most contain chlorophyll and, therefore, are photosynthetic and considered to be aquatic plants. These are not true plants, however, because they do not contain cells organized into tissues or organs, such as stems and leaves, nor can they reproduce sexually like higher plants. The unattached unicellular blue-green algae bask in freshwaters near the surface to obtain sunlight and have multiple sources of CO_2 to sustain photosynthesis, including atmospheric CO_2, dissolved CO_2 as bicarbonate ions, and CO_2 as minerals. Under such conditions a vascular system is simply not necessary.

The transition of plant life from water to land could have occurred only through the adaptive formation of structures that provided selective advantages to these early plants for survival. Plants went from structures that encouraged gas and water exchange by diffusion, as seen in green algae, to structures that resisted diffusional water loss, except at highly regulated locations, such as the stomata. Plants went from the lack of a rigid support system, as none was necessary in the water which supported the plants, to the development of internal woody fibers that provided rigidity even after the cells died. Whereas water absorption occurs by diffusion over the length of the body of an aquatic plant, it is restricted to the subsurface root zone that often remains moist, especially for phreatophytes with roots in the capillary fringe or water table.

The multicellular forms of green algae may have evolved from the tendency of unicellular algae to attach to each other. These algae are found in both freshwater and saltwater. Green algae, which contain chlorophyll *a* and *b* as well as the accessory pigment betacarotene, for example, is found in many forms, but one of the simplest is the attachment of cells end-on-end to form strings of cells of variable length. These algae evolved from a common uralgae, the first oxygen-producing plants that started turning the originally reducing atmosphere of the earth into an oxidizing one. The interrelation of these cells is more than simply a grouping of independent cells, which technically would be considered a colony, but rather the reproduction of cells that have independent cell walls that remain attached following division. For example, the freshwater green algae *Spirogyra*, familiar to most primary-school biology students, has some division of labor among cells. This is clearly indicated by the formation of a holdfast on the last cell used to anchor the algae and is considered to be a protoroot. Another example is seaweed, or sea lettuce, called *Ulva*, which is commonly found on beaches along the northeastern United States.

The initial transition from aquatic to terrestrial plant life probably started when some algal cells remained exposed after being washed up onto a shoreline but resisted drying out completely because their epidermal layer contained wax and the other parts of the epidermis contained holes that permitted the introduction to the cell of atmospheric, rather than dissolved or aqueous, CO_2. Alternatively, algal cells could have been washed up and then covered with a fine layer of sediment that protected them from desiccation. In either case, this transition occurred about 500 MYa (million years ago). The capability of cells to prevent desiccation and to essentially bring water with them as they colonized the landscape was an important development that led to the colonization of land by plants. Moreover, this ability to resist drying out provided a selective advantage, because it provided access to minerals in the soil matrix where competition was less intense than in water due to the vast numbers of algae and phytoplankton.

Mosses represent the next evolutionary step from aquatic algae toward the establishment of terrestrial plants. Although moss clumps, or beds, look like a single plant, it is actually composed of many thousands of closely spaced, single-stranded plants. Each strand, or filament, has structures that grow down into the soil for anchorage, called rhizoids—again, a type of protoroot. The green part above ground has scales that function as leaves, although they technically are not, that also can absorb moisture from the air. Mosses probably were some of the first plants to contain organized chloroplasts. This provided an advantage over algae because CO_2 is more readily available in air than in water, even at its low atmospheric concentration of 350 ppm (0.035%). Botanists classify mosses as bryophytes because they do not have vascular tissues. Fossil evidence suggests the presence of terrestrial bryophytes as early as 350 MYa during the Devonian Period. They represent part of the sequential steps toward vascular plants because, like the water-conductive parts of terrestrial plants, or xylem, that no longer are living cells and, therefore, lack cytoplasm so that water can be transported at rates that exceed diffusion, most mosses have cells that store water that also are no longer living cells. That the adaptation of mosses to terrestrial environments was successful is evident in their common presence today.

In order to become what are recognized today as land plants, several additional structural changes had to occur. First, single cells had to group together into tissues specialized to hold the plant upright at some distance above land surface. This came about with the Psilophyta, which contained the first true stems, characterized by the club mosses and horse tails. The younger Carboniferous Period, some 300 MYa, was a time of globally warmer temperatures and lush plant growth that included large, tree-sized club mosses and ferns. These plants formed thick deposits of peat after death and, following burial and time, comprise the thick coal beds that we mine and use today for energy. It is ironic, perhaps, that today the burning of such fossil fuels is believed by many to be a contributing cause of global warming, when, in fact, even warmer temperatures than today were necessary to support such extensive plant growth during the Carboniferous Period to produce the fuels being burned. Moreover, it also is interesting that the decayed plant remains that constitute the source of most fossil fuels today consist of plants found in damp areas near surface water and were, most likely, phreatophytes that used groundwater.

Within this vertical support structure, other cells had to align themselves to transport water from the soil to other parts of the plant and to harvest sunlight while exposed to the air; this later process required leaves. The first true leaves were found in the *Filicophyta*, or ferns. These structures provided more surface area for gas exchange to occur than that offered by stems alone. It is these structural changes that resulted in the classification of vascular plants. The exposed epidermal cells had to synthesize compounds that would render them waterproof against loss of moisture, but, at the same time, provide some permeability to permit the entrance of CO_2. Fossil evidence indicates the rise of vascular plants at least from 400 MYa, during the Silurian Period. The fossil evidence includes one of the earliest vascular plants, *Rhymia major*, which had a separate set of cells shaped in a cylinder inside the body that permitted water to move upward and downward inside the plant.

This evolutionary progression from aquatic, single-celled algae to multicellular, vascular land plants is repeated today during the colonization and subsequent succession of barren land by plants. For example, rocks exposed at land surface originally contain no topsoil, so it is essentially devoid of moisture and available nutrients and, therefore, any life forms. Under such conditions lichen thrives, which are both algae, typically blue-green, and fungi. The lichen survives by extracting the bound nutrients from the rock or barren soils of abandoned fields by using acidic excretions. Mosses become established on the detritus left behind by the lichens and then, when sufficient accumulations of organic matter are deposited, seeds from annual vascular land plants with high rates of seed production, dispersal, and germination are established. Once a fertile soil is developed, and soil moisture and groundwater accumulate, slow-growing vascular plants, such as conifers, dominate, and then fast-growing deciduous trees. The result of such succession is the development of a hardwood forest ecosystem, often called the climax community, and may take many years. In many ways, the ecological succession of plants, one on top of each other, is similar to the development of human culture and civilization in the same area over time; across the Middle East, for example, there are many nongeologic topographic highs, called tells, which represent the succession of invading cultures that then built upon the wreckage of the invaded.

The reference to seed production, dispersal, and germination above gives rise to another chapter in the history of plant evolution; the earliest plants did not reproduce sexually, but asexually, without flowers. The link of current plants to an aquatic past is reflected in the need for lichen and mosses to have enough moisture available, usually supplied by dew, to permit the sperm to swim to the ovary of the lichen or moss. These non-flowering plants included the mosses and ferns, which reproduce using spores. Later, the inefficiency of spore production compared to germination led to seed production by these and other plants, such as cycads and conifers. The seed is a fertilized egg surrounded by a source of food and protected from the environment by a tough seed coat. These seeds, because they arose without a flower, are called gymnosperms. The pines today are relatively unchanged from those of 300 MYa. The development of pollen as a means of transporting genetic information was first seen in gymnosperms. Although pollen is an irritant for allergy suffers, it marked a profound step forward to the terrestrial habitat for plants, because it freed plants from relying on a watery medium for reproduction. Plant success on land is not only a consequence of the advantage of seed-based reproduction and the absence of water as a vector of fertilization, as is necessary for mosses, but also, as we will see later in this chapter, the advantage of structural adaptations to survive in environments where water is a limiting factor.

The flowering plants are called angiosperms, or enclosed seeds, and they can be found in the fossil record as far back as the end of the Cretaceous Period some 65 MYa. These plants are divided into monocots and dicots, as was discussed in Chap. 1, based on the number of leaves, either one or two, respectively, attached to a seedling. In general, the dicots have highly differentiated vascular systems that formed during secondary growth from the meristem, whereas monocots have more random and diffuse systems and no secondary growth; any increase in girth is caused by cell elongation or enlargement, and not division.

Some plants have returned to an aquatic habitat after a period of land colonization, and some highly developed plants never fully left the watery environment in the first place. One such type of aquatic plant is the water lily, in which the root, or in this case the rhizome or underground stem, is submerged in the bed sediment of a lake or pond, the stem is in the surface-water column, and the leaves and flowers float on the water surface exposed to sunlight and air. Air diffuses along a concentration gradient in the cortex cells or aerenchymal tissues to the root zone to support root respiration even in sediments that often do not contain oxygen. Water lilies produce flowers and, therefore, had once in the past made the transition to land but arrived back in the aquatic environment.

Some plants live entirely under water, such as *elodea* or *hydrilla*. These plants contain extra amounts of chloroplasts to deal with the dilution of light in the water column and have large air spaces in their tissues. They obtain water by direct diffusion into the leaf cells and the CO_2 from the atmosphere or bicarbonate. The roots of such plants do not have vascular tissues for water transport because they lack xylem and, therefore, the need for root hairs, but they do have phloem for food transport and storage and a root for anchorage. Other plants float about on the surface of the water without attached roots; these are truly hydroponic plants, such as duckweed (*Lemna spp.*), which essentially is a free-floating leaf.

3.2.3 Adventitious Roots

Roots also can develop above ground and are called adventitious roots. Adventitious roots arise from the root initials located in or near lenticels along the stem (Ginzburg 1967; Hook et al. 1970). They are found especially in riparian facultative phreatophytes that grow in areas where surface-water levels and sediment elevations fluctuate as is characteristic of a flooding regime. Adventitious roots have vascular connections that terminate within the annual growth ring of the tree's current year. Most adventitious roots arise under conditions when water is not limiting, whether from periodic flooding or high humidity levels. In some cases, adventitious roots can become the main tap root, due to water limitations, infestation or animal attack, or impermeable geologic strata. Adventitious roots often can be short lived, however, and thrive only if abundant moisture is available. A cutting of a poplar tree placed in a plastic bag, for example, that contains high humidity will form adventitious roots (Fig. 3.7), but these roots will shrink and die after the cutting is exposed to drier air. The formation of adventitious roots on stock used to install phytoremediation systems and its implications for successful planting is discussed in more detail in Chap. 7.

3.2.4 Root Hydraulic Conductivity

Hydraulic conductivity is a term that describes the characteristic of fluid movement through a porous media and is, somewhat fortuitously, common to both plant physiology and groundwater hydrology. With respect to plants, hydraulic conductivity describes the diffusive movement of water from one cell to another cell through the cell membranes and plasmadesmata, called the symplastic pathway. The central question here regarding the flow of water from cell to cell is what the initial rate of water movement is and what controls this rate. The resistance to water movement by the cell membrane of roots can be quantified and is called root hydraulic conductivity, L_p. As we will see in Chap. 4, when used in groundwater hydrology, the term hydraulic

Fig. 3.7 Adventitious roots on hybrid poplar cuttings formed during shipment in plastic bags and elongated after immersion of the cut end into water (Photograph by author).

conductivity, K, also describes the flow of a fluid, groundwater, through porous sediments.

As water passively enters the root hairs by diffusion, the water can only reach the xylem after first passing through multiple cell walls, the symplastic pathway, or simply through cell walls and spaces, the apoplastic pathway. Water encounters these cell membranes, which provide a large resistance to flow, along its path. Once in the xylem, however, this resistance is no longer present, as the xylem contains no cytoplasm or cell membranes. In roots, this resistance is quantified as root hydraulic conductivity, L_p, and can be estimated using

$$L_p = W_v/\Delta\psi \quad (3.4)$$

where W_v is the velocity of water transport from one cell to another and $\Delta\psi$ is the difference in water potential between cells; the concept of water potential will be more fully described later in this chapter.

Quantification of root hydraulic conductivity provides a meaningful way to compare the magnitude of the degree of resistance to diffusional water flow between cells, in volume of water per unit area of membrane per unit time per unit driving force ($m^3/m^2/s/MPa$). The velocity of water transport, W_v, from one cell to another can be described by

$$W_v = L_p(\Delta\psi). \quad (3.5)$$

Over time, intracellular water movement will decrease as the difference in water potential, $\Delta\psi$, between the cells decreases and additional water movement ceases.

Tap roots often are characterized by much higher root hydraulic conductivities relative to shallower roots. The tap root of most plants may grow to greater depths but are fewer in number relative to the more abundant roots in the shallow parts of the soil horizon, especially in temperate, well-drained nutrient-poor soils. This is the case for longleaf pines (*Pinus palustris*) in the Coastal Plain geophysical provinces of the United States, where tap roots average about 60 ft (180 m) in length (Heyward 1933; Pessin 1939). The higher root hydraulic conductivity may result from a difference in physiology of these deeper roots compared to shallow roots, where the tap roots have long and continuous xylem (Le Maitre et al. 1999). The higher root hydraulic conductivity also may reflect the increasing potential to encounter the water table, and subsequent increase in water potential. Therefore, the fact that fewer roots are at depths nearer the water table is compensated for the roots having higher root hydraulic conductivities (Fig. 3.8).

There also is a relation between root hydraulic conductivity, root age, and hydraulic conductivity of the soil. For a given plant, older tap roots tend to have lower root hydraulic conductivities than younger tap roots, as the former contain more suberin. Moreover, vigorous tap-root growth actually can increase the hydraulic conductivity of the soil, both aerially and vertically. For example, in Australia, researchers measured the hydraulic conductivity of silty-clay soil in the unsaturated zone beneath and just outside of a tree plantation and found that the hydraulic conductivity of the soil was lower outside than inside the plantation (Rural Industries Research and Development Corporation 2000). The researchers also measured the hydraulic conductivity at various depths beneath the plantation. Although there was a gradual decrease in hydraulic conductivity with depth, the highest hydraulic conductivity values were associated with the root zone (Rural Industries Research and Development Corporation 2000).

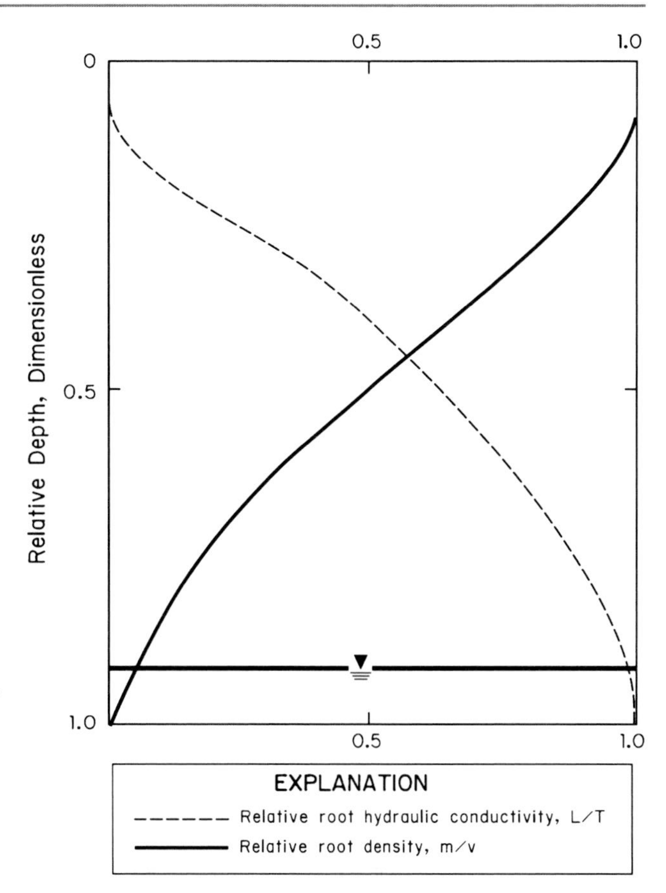

Fig. 3.8 The generalized relation among depth, root density, and root hydraulic conductivity common to many phreatophytes. The lower root density near the water table is compensated for by increased root hydraulic conductivity.

3.2.5 Effect of Redox Condition on Roots

Plants are autotrophic, aerobic organisms; plants require oxygen during respiration to release the energy stored in the food made during photosynthesis. In most terrestrial environments with little sedimentary organic matter, oxygen is not limiting. In aquatic or subsurface systems that contain either natural or contaminant inorganic and organic matter that can interact with oxygen, oxygen can become limiting. Such oxygen-depleted, anoxic conditions also can be an advantage to some plants. This is because under anoxic conditions certain essential and micronutrients, such as iron and phosphorus, can be passively taken up by plant roots with little energy expended.

As we saw in the discussion of photosynthesis, the splitting of water produced hydrogen and is used to reduce CO_2; essentially a transfer of hydrogen atoms or electrons. The reduced carbohydrate produced, such as glucose, can be oxidized, such that electrons are removed, either by the plant, an herbivore, or omnivore, through respiration to derive energy in the form of ATP. The most common oxidant used during respiration is molecular oxygen. The coupling of sequential reductions and oxidations involves changes in energy, too, because the electrons flow downhill as the energy to drive life is released.

Although for survival most terrestrial plants need to maintain constant contact with water, either in the form of precipitation, surface water, soil moisture, or groundwater, too much water adversely affects root growth and overall plant health because it restricts the availability of oxygen. Ready access to the water table by phreatophytes suggests robust plant growth, as water would not be limiting for these plants. However, even phreatophytes will suffer from oxygen limitations in the presence of adequate water supply. Water physically displaces air from soil pores, and oxygen has a low solubility in water (8 mg/L at 25°C). Plants have adapted to low oxygen levels by forming interconnected gas passageways, the cortex or aerenchyma discussed previously, that permits the diffusive transport of air to the roots. The diffusion of atmospheric oxygen from the leaves to the roots was first investigated in willows, a tree often planted at phytoremediation sites.

Even though plants produce oxygen during photosynthesis, as described at the beginning of this chapter, this process occurs primarily in the above-ground portion of the plant. Conversely, because roots do not photosynthesize, roots consume oxygen during growth through respiration—if oxygen is used at a faster rate than it is replaced, root cells will die from asphyxiation. This interaction between plant growth, water availability, and oxygen levels is particularly evident in recently flooded areas before the water recedes or in areas where the water-table level has increased into the root zone. On the other hand, the frequent vertical fluctuation of the water-table level in response to precipitation events and(or) changes in plant use of groundwater may only affect part of the total vertical extent of root mass. A falling water table may actually increase the amount of oxygen in the subsurface—this fact will turn out to facilitate the oxidation of organic xenobiotics released to groundwater. A notable exception to these extremes is the use of aquaculture methods to grow some terrestrial plants for commercial or hobby purposes. In this case, the seemingly stagnant water is kept fully saturated with dissolved oxygen by using containers that have a large surface area, shallow depth, and mechanical aeration.

One of the most extreme examples of how trees respond to large changes in redox conditions caused by changes in water levels is the rain forests of the Amazon. There, trees can be submerged under 45 ft. (13 m) of water for as long as 6 months each year (Kubitzki 1989); the forest essentially becomes a lake. Fernandez et al. (1999) investigated the influence of prolonged submergence on water potential, photosynthetic rate, and leaf conductance in the submerged trees. Some trees retained their leaves after submergence, and others lost their leaves, which regrew after the water level subsided. Water potentials increased in the trees as the water level rose, as water was not limiting. Submergence decreased photosynthesis by 50% from levels measured when the leaves were not submerged, indicating that the leaves could still photosynthesize during submergence.

The presence of trees in anoxic aquatic environments has given rise to many hypotheses about how terrestrial plants can oxygenate these anoxic systems. Cypress trees (*Taxodium distichum* L.), for example, have outgrowths on their lateral surface roots, called knees, that grow upward and above the mean high-water line. The knees are absent on cypress that grow in water that remains at a constant level, and appear only where the surface-water level fluctuates. The location of the knees suggests that they do not arise from the terminal meristem of the root tips. Rather, cypress knees are a localized growth of the cambium on the upper surface facing the water that produces multiple layers of xylem in a small, focused location. Although cypress knees are widely believed to be used by the trees to increase root aeration, this is contradicted by experimental evidence (Kramer et al. 1952). The oxygen that enters the roots does so by diffusion through the fluted part of the trunk, comprised of an increased volume of aerenchymal tissue between the bark and the phloem (Fig. 3.9).

3.3 Roots, Rhizosphere, Bacteria, and Mycorrhizae

In fact, man lived in a sea of bacteria. They were everywhere—on his skin, in his ears and mouth, down his lungs, in his stomach. Everything he owned, anything he touched, every breath he breathed, was drenched in bacteria.

The Andromeda Strain, Michael Crichton (1969)

Just as mammals, such as ourselves, carry other life forms on and in our bodies, plants also harbor other life forms. For instance, the roots of terrestrial plants increase the organic

Fig. 3.9 The wider base than trunk of this baldcypress (*Taxodium distichum* L.) in Congaree National Park, near Columbia, South Carolina, illustrates the adaptation of this particular species to the low oxygen levels that characterize the saturated sediments in which these trees grow. The increased surface area of the wider base permits oxygen diffusion at rates to equal root respiration (Photograph by author).

content of the soil and, therefore, create a microniche for heterotrophic bacteria. The zone immediately surrounding the roots that supports these bacteria was called the rhizosphere in 1904 by Lorenz Hiltner (in Anderson et al. 1993). It was later discovered that this rhizosphere exists first to benefit the plant but also benefits the bacteria.

The rhizosphere can extend up to 0.397 in. (10 mm) from individual roots. Another definition of the rhizosphere is that it comprises the thickness of soil that remains attached to plant roots if they are exposed and the plant is shaken. The rhizosphere is occupied by a variety of bacteria that make available atmospheric nitrogen to the roots for uptake by reducing, or fixing, atmospheric nitrogen, a process that most plants cannot do by themselves. The rhizosphere also is occupied by a variety of fungi, or mycorrhizae, that aid the plant in the uptake of minerals and water; this is in contrast to other fungi and bacteria that often can infect plant roots. The heterotrophic bacteria and fungi in the rhizosphere, in turn, receive excess organic matter from the plant as a source of carbon and energy. In many ways, the symbiotic relation between plants and microorganisms was an evolutionary negotiation between the heterotrophic bacteria intent on damaging the plant and the plant intent on decreasing this threat to survival. One consequence of the beneficial relation between plants and root microbes is reflected in the habit of many older gardeners that pour sugary soft drinks on the ground near roots of some plants to facilitate this interaction. (Probably the most famous organism that inhabits the rhizosphere is the truffle.)

The number of microbes in planted soils can be at least an order of magnitude greater than the number in unplanted soils. In the upper soil layers, such as the O and A zones, bacteria can approach 10^5 to 10^8 cells/g of dry soil because of the presence of roots. As a result, concentrations of CO_2 in the rhizosphere are higher relative to areas without roots. The root uptake of water physically concentrates water near roots, and roots produce organic compounds that act as surfactants to lower water surface tension; this reduction in water tension provides an advantage to rhizospheric microbes that seek to avoid strongly negative water potentials, which would stop the diffusional uptake of nutrients necessary to microbial life. Such a relation between plants and increased bacterial numbers in response to water availability is particularly noted during drought conditions, such that rooted areas act as oases of water for microbes.

3.3.1 Rhizosphere Bacteria and Nitrogen Fixation

Probably the most investigated relation between plants and the rhizosphere involve the soil bacterium *Rhizobium*. These bacteria interact with plants after entrance through root hairs during the seedling stage and extend into the cortex. The *Rhizobia* are beneficial to plants because the bacteria reduce gaseous nitrogen (N_2) in the soil air (although present at concentrations near 80%, nitrogen is not bioavailable for plant uptake) into nitrogen that is available to plants. *Rhizobia* are associated with the roots of a class of plants called the legumes, such as clover, beans, and alfalfa. Prior to the manufacture of inorganic sources of nitrogen in fertilizers, the addition of manure and interplanting of alfalfa and clover among other crops were the principal methods of delivering nitrogen to non-leguminous crops. Moreover, today it is common practice to rotate fields of leguminous crops with non-leguminous crops.

This interaction between plants, bacteria, and nitrogen availability raises the question of how did this interaction develop? After all, if plants are able to fix atmospheric CO_2 why didn't plants also develop the ability to fix atmospheric N_2? The answer, in part, lies in the fundamentals of chemical bonds. Atmospheric N_2 is held together by a triple bond between the two nitrogen atoms—much energy is needed to break a triple bond. Plant-associated bacteria provide the source of energy needed to break the triple bond and make the nitrogen available to plants. Moreover, the bacteria that can break this triple bond can do so only in the absence of oxygen; in the presence of oxygen the enzyme that the bacteria use to beak the triple bond and fix N_2 is oxidized and rendered unfit. This creates a dilemma: plants require nitrogen to survive, but plants also require oxygen to support respiration. Thus, a compromise evolved. The nitrogen-fixing bacteria form large nodules on the roots of most plants, which results in the formation of a microanaerobic zone inside the nodule where N_2 fixation occurs after the oxygen is consumed (Hardy and Havelka 1975). In return the plant supplies the bacteria with organic compounds which the bacteria need for energy and cell growth.

Let us look more closely at the bacteria that live inside these plant-root nodules. *Rhizobia* are free-living bacteria in the soil and must gain entry into plant roots before nitrogen fixation occurs. The evolution of plants permitted infection avoidance by disease-causing bacteria and viruses, so entrance of any particular bacteria, even potentially beneficial ones, is highly regulated. The root hairs of plants, such as legumes, release lectin, a protein that binds with a sugar released by the *Rhizobia*. Upon such surficial linkage, the root hair curls around the bacterium and the bacterium then purchases entry into the root hair cell by the insertion of a tube, called the infection thread, which causes the cell to divide and forms a nodule around the site of bacterial entry. This inversion and convolution produces the ideal structure for subsequent nitrogen fixation: an anoxic central core where the bacteria reduce nitrogen into ammonia, and an oxic outer rind where nitrification can occur and also supply the root cells with oxygen.

Another problem quickly arises, however. Both bacterial and plant cells require nitrogen so how is nitrogen rationed among these competing needs? It turns out that the bacteria fix more nitrogen than they require for themselves, so some becomes available to the plants. In turn, in order to ensure a supply that is greater than the nitrogen needs of the plant, the plant provides excess glucose to the bacteria. Part of the rationale behind this production of excess nitrogen by the bacteria may be found in the fact that plants that have root nodules carry the gene for part of a compound called leghemoglobin, which is needed by the *Rhizobia* to utilize oxygen for respiration, and the *Rhizobia* contain the other gene (Hardy and Havelka 1975). This is an important process for plant nutrition and survival, because elements that a plant needs other than nitrogen are derived from the dissolution of the soil matrix. Moreover, the bacteria recycle some of the nitrogen back into the atmosphere during denitrification. The nitrogen cycle is discussed in greater detail in Chap. 11.

What about plants that do not have an association with nitrogen-fixing bacteria? How do these plants meet their nitrogen needs? One example of a non-leguminous nitrogen-fixing plant is the alder tree (*Alnus spp.*). Alder roots have filamentous bacteria, such as *Frankia*, that perform nitrogen fixation. But not all non-leguminous plants have *Frankia*. In Chap. 11, the relation between such plants and the use of groundwater as a source of nitrogen is discussed; it will be shown that most of these plants possess the phreatophytic habit.

3.3.2 Mycorrhizae

The presence of plant roots increases not only the numbers of bacteria in the soil but also the number of fungi. In general, fungi are primarily responsible for the decomposition of dead plant matter. The fungi of many plant roots in the rhizosphere are called mycorrhizae, from the Greek, *mykes*, meaning fungus, and *rhiza*, meaning root. More than 80% of all vascular plants have mycorrhizae. This is another example of a symbiotic relation that benefits both the fungi and the plant. Plants can grow in the absence of mycorrhizae, but enhanced growth always is observed where fungi are present. The fungi gain entry into plant tissue through insertion of hyphae into open stomata or wounds on the bark or suberized roots. The linkage between tree wounds and potential fungal entrance has implications for sample-collection methods pertaining to tree-tissue monitoring during phytoremediation projects, which is discussed in Chaps. 9 and 14.

The association between plants and fungi is not a recent phenomenon. The fossil record contains evidence of root and mycorrhizal connections. The role that this relation played in the transition of plants to land is unclear, but the association with fungi undoubtedly would have been an advantage. Not all fungal and root interactions are beneficial to both organisms, however. In fact, one of the largest impediments to a stable food supply is the attack of various fungi on food crops. Moreover, some of the commonly used trees and planting methods, such as monoculture, used for phytoremediation purposes are vulnerable to widespread fungal attacks.

If most plant diseases and infections are caused by fungal entry and growth, how did the beneficial symbiotic relation between plants and mycorrhizae develop? At first, bacteria probably were associated with plant infection, until the growth and reproduction of each was enhanced by their mutual interaction. Keep in mind that fungi are heterotrophs and require reduced organic carbon as a food and energy source. Plant roots while alive provide a carbon source much like after death. Perhaps plants shed organic matter into the rhizosphere while alive to satisfy the organic carbon needs of the fungi in order to remain negatively unaffected by fungal root colonization. Roots secrete the organic substance mucigel, as well as organic chemicals that could ward off potential threats, as is discussed in Chap. 11. Plant roots also can contribute to feeding these rhizospheric bacteria and fungi as a consequence of root turnover, the annual shedding of dead root matter.

Many different kinds of mycorrhizae are associated with plant roots depending on the site of colonization. Mycorrhizae that exist on the inside of the roots are called endotrophic mycorrhizae or vesicular arbuscular mycorrhizae (VAM) (Safir 1987) and are found within the cells of the root cortex. Mycorrhizae that exist on the outside of roots are called ectotrophic mycorrhizae (Bowen 1984). Endotrophic mycorrhizae are more common than ectotrophic mycorrhizae, and they are found in more than 90% of the world's herbaceous plants (Dhillion and Zak 1993). Endotrophic bacteria rarely are associated, however, with woody plants, which tend to be dominated by ectotrophic mycorrhizae. In fact, some tree species, such as oaks, maples, and hickories, have few to no root hairs and, therefore, rely solely on ectotrophic mycorrhizae for water uptake.

Mycorrhizae are believed to be useful to plants in that the mycorrhizae's mycelium, or hyphae, increase the total surface area of the roots for increased uptake of water and minerals (MacFall 1994). Ectomycorrhizae form a mantle around the root or between cells in the cortex so that they actually inhabit the space between cells of the plant root. In this manner, plants use ectomycorrhizae to access previously unavailable minerals in decaying leaf litter at the ground surface. The fungi commonly seen on the forest floor and called toadstools or mushrooms provide visual evidence of the presence of mycorrhizae in the rhizosphere of a nearby tree, as these mushrooms are the fruiting bodies of fungi (Fig. 3.10). These fungi are carpophores, the reproductive part of the underground mycellium.

The mutually beneficial interaction between plants and fungi can probably find its earliest beginning in a humble plant that many pass by today unaware—the lichen discussed previously and shown in Fig. 3.10. The part of lichen visible to the eye is the fungal part; it is believed that this acts to keep the sequestered algal cells inside from drying out. The fungi provide the algae access to water and minerals, which can be absorbed from sources other than the ground, such as precipitation, rocks, and bark, in a manner analogous to the mycorrhizae of tree roots.

Mycorrhizae benefit plants because of enhanced water uptake as well as other processes necessary for plant survival. As we saw earlier in this chapter, dissolved iron is essential for the production of chlorophyll. The iron encountered by most plant roots is in the oxidized form and, therefore, not bioavailable. To increase iron bioavailability and to facilitate iron uptake, plants and mycorrhizae produce organic acids that can chelate iron and increase iron solubility; this process is discussed in Chap. 11. Also, it is possible that fungi degrade organic matter in the soil and release stored nutrients for uptake by trees. In return, the fungi essentially hitch a ride toward new sources of water and soil organic matter as the roots grow.

Some rhizosphere bacteria can release substances to decrease the germination of certain seeds relative to other seeds for which they are better suited. Such allelopathic relations and the implication for plant exposure to groundwater contamination is discussed in Chap. 11. Fungi may

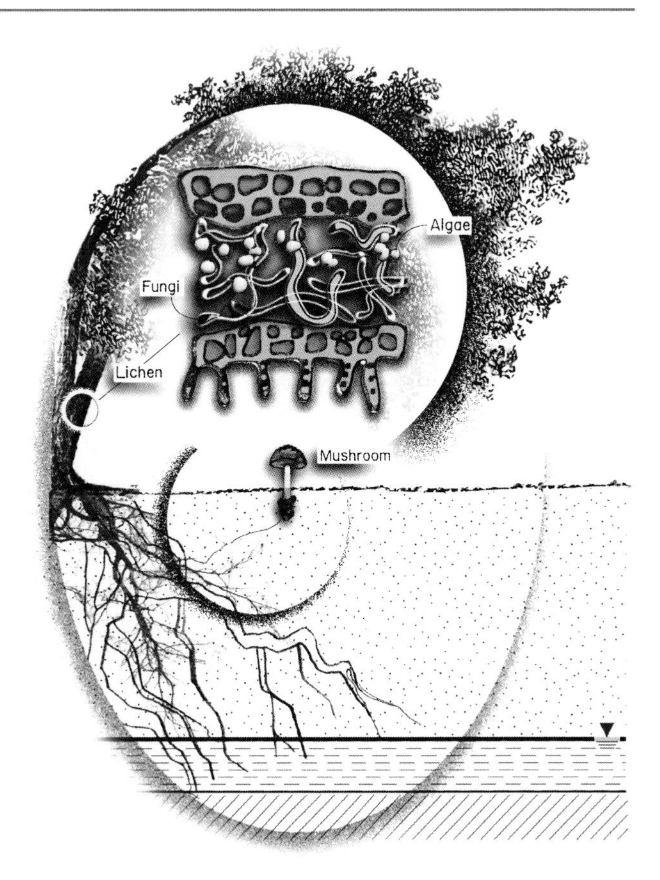

Fig. 3.10 The mushrooms often seen on the ground at the base of trees are the fruiting bodies of mycorrhizae. A similar symbiotic association is provided above ground by lichen that contains algae and fungi.

also protect the tree from bacterial root pathogens, as they are rendered inactive by the antibiotic effect of root fungi.

Another geochemical consequence of plant and mycorrhizae interaction is seen in flooded soils characterized by little dissolved oxygen. Although oxygen from the atmosphere can diffuse through the plant cortex to support root respiration, the production of toxic hydrogen sulfide (H_2S) from sulfate reduction in anoxic soils can be a detriment to plant growth. Sulfur-oxidizing bacteria such as *Beggiatoa*, associated with plant roots in these environments can oxidize the H_2S to harmless levels and, therefore, remove the threat to plant growth. This relation has important implication for phytoremediation plantings where the trees will interact with anoxic, contaminated groundwater.

3.4 Roots and Water Absorption

Plants, on average, are greater than 80% water by volume. A germinating seed will send a single meristem upward to capture light and CO_2, but the seed would die if it didn't also send downward a single root meristem to capture water. The same seed could have remained dormant, sometimes for centuries for some seeds, but once exposed to water and oxygen, the process of cellular respiration and growth begin. On the other hand, once a leaf is removed from a plant, it is isolated from its water source and will dry out according to its transpiration characteristics.

Plants need water to survive but cannot simply abstract it from the humidity in air in most cases. It may appear as a dilemma that plants require water but limit uptake to only one main area, the roots. Whereas all of the other parts of the plant are designed to minimize the loss of water, the roots are designed to maximize the gain of water. Plant structures have specifically evolved to interact with water above and below ground; in the latter, roots can interact with the water stored in the pore spaces of sediment, where the water is less likely to be removed by evaporation. For roots to interact with water it must be present in the soil in the shallow root zone or deeper, such as groundwater.

Unlike the capability of plants to react to light sources from great distances, roots cannot react to the presence of moisture more than a few millimeters away from each root. In this sense, roots do not grow toward water but expand and lengthen until water is reached. After a root encounters water, root hairs grow quickly until the water is depleted. If a root encounters no water, or if water has been removed and is not recharged, the root hairs become desiccated and die. This is why phreatophytes that have root masses in the capillary fringe or water table are able to tolerate drought conditions even though shallower sediments dry out.

3.4.1 Diffusion and Osmosis

In general, all life can be defined as a dynamic steady-state condition of the acquisition, processing, and excretion of various chemicals and substances with the surrounding environment. These processes occur on a cellular level continually until death, and, in most cases, are driven by diffusion.

Diffusion is the process in which molecules of a substance tend toward an even distribution throughout a given volume. Movement is in the direction of concentration gradients. If a drop of food coloring is placed in still water, the molecules of pigment will bump into each other at a greater probability in the center of the drop. This random interaction propels the other food-coloring particles away from the center. As the particles leave the center, there is less probability that adjacent food-coloring particles will interact. Once an equilibrium distribution has been reached, each particle will have the same probability of interaction with other particles. It follows, then, that the rate of diffusion of a particular particle is proportional to the original concentration of the particle.

Diffusion is a scale dependent process. Take for example the human need for oxygen. Atmospheric oxygen cannot directly enter our bodies because the distance across which diffusion would have to occur would be too great. It has been calculated that for oxygen to diffuse from our head to our toes would require 100 years. With the advent of lungs and a circulatory system connected to every cell, however, oxygen diffuses rapidly into the capillaries in the lungs, where the cell thickness is about 1.5 microns. This relation between diffusion and distance is one of the reasons that plants have thin leaves and thin root hairs, in order to facilitate steep concentration gradients of essential reactants over a short diffusional distance. Life simply could not exist without diffusion.

On an individual cell-by-cell basis, the diffusive movement of water molecules in response to a water-concentration gradient across a selectively permeable membrane is called osmosis (from the Greek, *osmos*, for push). Water in contact with a root cell that contains solutes essentially enters the cell to lower the water concentration inside and will continue to diffuse until equilibrium conditions are established. The greater the solute concentration across such a membrane relative to pure water, the greater the tendency is for water to cross that membrane, and the higher the osmotic potential. Conversely, the higher the osmotic potential of a solution, the higher the tendency for water transport into the solution. As the water moves to the place of higher solute concentration, the pressure will increase if in a fixed volume.

An alternative definition of water transport into cells relates to the minimum osmotic pressure applied to a solution of water and solute to keep it from gaining more water. Here, osmotic potential is operationally defined as the negative of the osmotic pressure. Hence, if a solution is placed in contact with pure water separated by a membrane, the greater the tendency for water to be transported across the membrane to equalize water concentrations. The higher the osmotic pressure is, the lower, or more negative, is the osmotic potential. These measurements are done with an osmometer, derived from a much earlier simple one made from a pig's bladder by the French physician Joachim Henri Dutrochet in the early 1800s. These experiments further illustrate that osmosis is essentially simple diffusion but through a semipermeable membrane.

Polar water can enter individual cells by diffusion because of the water molecule's small size and weight. Unlike the solute molecules that enter the cell under the direct regulation imposed by the physical structure of the cell membrane, water enters and exits the cell with no apparent regulation or discrimination by the cell membrane and its semipermeable structure. This is similar to how a GORE-TEX® jacket prevents liquid water from entering from the outside but allows water vapor to exit to the outside. Diffusion is a slow process, even for a small molecule like water, but if a plant can grow a large enough volume of roots, the diffusion of water into the root hairs at least will equal the removal of water by evaporation from the leaves.

These physical processes of diffusion and osmosis are passive processes, because no energy is spent by the plant to facilitate the entry of water; the solute concentration in the cell increases, the water concentration decreases, and water enters the cell by osmosis. Increased pressure in the cell results from the water entry—during osmosis, plant cells can obtain pressures up to 15 atms (atmospheres), which is great enough to explain why sidewalks and driveways near trees often are lifted.

When an individual plant cell encounters waters and it passes through the cell wall and cell membrane, water and its solutes, which will be discussed later, can then enter the storage compartments of the cell vacuoles. The vacuoles themselves are contained within a selectively permeable membrane called the tonoplast. Vacuoles contain a dilute concentration of sugars and salts, also known as sap, which is about 98% water. Because of its solute concentration and lower water concentration, or high osmotic pressure and low osmotic potential, dilute water can enter the vacuole. In general, other plant cells contain a more concentrated solution inside separated by a semipermeable cell membrane from the external, more dilute solution of the cytoplasm.

Water entering through the plasmalemma creates an internal pressure, called turgor, in the vacuole, that is exerted on all parts of the cell structure. This makes the plant rigid; plants that are in a water deficit appear limp and wilted because they lack turgor. Turgor is analogous to tire inner tubes that hold air; upon inflation of the inner tube with air, or the plasmalemma with water, pressure is exerted outward on the inside surface of the inner tube, or cell wall, which make both rigid.

Wilting also can occur, however, when cells are exposed to a higher concentration of solution external to the plant relative to the internal solution. The water potential is higher in the plant and lower outside, and water exits the plant cells, a process called plasmolysis. Moreover, the removal of water by transpiration at faster rates than water uptake in the roots causes the same wilting phenomenon, because it reduces the hydraulic pressure. Water may be bioavailable, but it is being removed at a faster rate than it can be taken up. Conversely, the water may be held onto soil grains at too high tension levels for removal by plants.

Once sufficient osmotic pressure to produce turgor has taken place and the cell membrane presses against the cell wall, excess water is pushed into the cortex at the same rate that it is taken in by osmosis. This is the same cortex penetrated by the hyphae of ectotrophic mycorrhizae. Because the cortex has a greater concentration of solutes, it also has a lower, more negative water potential than the epidermis or area outside the root. Thus, water flows to the endodermis, and from there to the xylem where the flow creates a slight pressure, called root pressure. Although this pressure can move water to low heights, this process does not supply water to the tops of most plants. This pressure is responsible, however, for the occurrence of water droplets on the tips of leaves, a process called guttation, when sufficient water supplies are available to meet and exceed the transpiration demands of a plant (Fig. 3.11). Typically, water exits plants as vapor from the leaves, but here water exits as a liquid.

Fig. 3.11 The loss of liquid water, rather than water vapor, by plants is called guttation and is a result of root pressure caused by osmosis when soil water is not limiting and relative humidity is high (Photograph by author).

3.4.2 Solute Entry

The mechanism that drives the passive uptake of water by osmosis in the root zone is primed by the initial intake of solutes by root cells and the resultant increase in intra-cell solute concentration. This is important especially in terms of water limitation, because a cell that has a higher solute concentration will be able to take up water at greater negative water potentials than a cell with a lower solute concentration.

As stated earlier, the semipermeable structure of the plant cell membrane is the key component to the survival of individual plant cells and, hence, the total plant. Plants have the disadvantage of not being able to move, at least great distances at appreciable rates, to reach new water sources, and most water sources are characterized by an extremely dilute solution of solutes. Hence, the concentration of these substances, such as salts, is much lower outside of the plant cell relative to inside the plant cell. Even in areas where the salts are more concentrated in solution, such as near the ocean, the cells of plants in such areas, for example the cordgrass *Spartina*, have even higher salt concentrations in order to continue osmosis.

The physical structure of the bilayer of the cell membrane allows it to act as a selective membrane, which allows some molecules to pass through while keeping out others, as shown in Figs. 3.1 and 3.2. The first line of discrimination is focused on molecule polarity with respect to the cell membrane. Water has been called the universal solvent for good reason. The dipolar nature of water allows it to orient around individual ions in solution. Take for example what happens when a crystal of sodium chloride (NaCl) is added to water. The negative end of the water molecule, the oxygen, engulfs the sodium ion, and the positive end of the water molecule, the hydrogen, engulf the chloride ions. This causes separation of the original sodium chloride into Na^+ and Cl^- ions.

For non-polar, non-ionic materials, the initial entry into the cell membrane occurs by passive diffusion from areas of high solute concentration to areas of lower solute concentration. The rate of this initial diffusion is determined in part by the solubility of the molecule in lipids, for the inner part of the cell membrane is non-polar (Fig. 3.2) and part by the cell-membrane selectivity. Non-polar molecules that dissolve easily in non-polar lipids enter easily, which supports the old adage, like dissolves like. Although smaller non-polar compounds enter the fastest, even large molecules with high molecular weights can enter. On the other hand, polar molecules, either organic or inorganic, of larger weight and size, enter more slowly. One reason that larger polar organic molecules, including many compounds that can become groundwater pollutants, are excluded entry into the cell is because the cell membrane must hold onto polar molecules already in the cell and keep them from leaving.

Other molecules can be transported through the cell membrane, from the external side of the cell membrane to the internal side, by carrier molecules called permeases. Although this process is driven by the diffusive gradient of the molecule's external concentration relative to its concentration inside the cell, permeases increase the rate of diffusion, even though no energy expenditure is required by the cell; this is called facilitated diffusion or transport.

In some situations, however the rate of diffusive or facilitated transport entry of a compound is not fast enough to meet cellular demands. When this occurs, energy in the form of ATP must be used by the cell to increase the rate of transport, sometimes against diffusion-based, concentration gradients, where concentrations inside the plant are higher than outside the plant; this process is called active transport. The rate of flux into the cell is limited to the occupation of sites by the carrier molecules, such that a maximum rate of uptake cannot be exceeded. The compounds that enter the cell through the membrane by either simple diffusion, facilitated diffusion, or active transport obtain a concentration in the cell cytoplasm that is regulated by the cell and its water content. In this case, the water already in the cell acts as a solvent.

The establishment of an electrochemical gradient, based on ion pumps, rather than the expenditure of ATP, also can be used to drive active transport. Cells of plants need high internal concentrations of potassium (K^+); this is one of the big three ingredients present in commercial, inorganic fertilizer. As is the case within animal cells, potassium is used to make proteins. Concentrations of potassium in the cell are much higher than that outside the cell; how can more potassium be taken in as it is used when the potassium inside is not being lost? The answer is that hydrogen ions (H^+) are pumped out of the cell, thus creating a net negative potential inside the cell. This electrochemical gradient causes positively charged ions, such as K^+, to enter the cell. Although this process is considered passive, energy is spent in the form of ATP to expel the hydrogen (H^+) from the cell. Another monovalent ion, sodium, typically is more concentrated outside the cell than inside. This differential gradient is maintained by the sodium-potassium pump, where potassium enters and sodium exits the cell simultaneously.

3.4.3 Water and Solute Uptake by Root Cells

Leaves have evolved to retain water vapor while roots have evolved to enhance water uptake. How exactly, then, does the water move from the root hairs to the xylem? We saw in Chap. 2 that one of the characteristic properties of water is the ability to generate significant tension between water molecules exposed to a non-water surface. Therefore, water constrained by small pore spaces is under tension and can produce movement based on capillarity forces alone—no biological process is necessary to move water under such circumstances, just surface tension. Capillary action can cause water to move through plant-cell walls and follow the intercellular spaces of the cortex. In this process, however, water would not have to enter the protoplasm of the cortex cells. The endodermis blocks further water movement because the lateral cell walls are impermeable to water. Water must first go through the protoplasm of cells before being allowed to gain entry into the xylem.

Water first diffuses into root hairs located behind the root meristem (Fig. 3.5). Each root hair is an individual epidermal cell. They are called hairs because they are small, almost hairlike in appearance. As such, they increase the surface area of the outer layer of root cells, or the epidermis (Fig. 3.5) by several hundred-fold (Cailloux 1972). The roots of some grasses can have as high as 14 billion root hairs. The large number or hairs means that the distance traveled by water and dissolved substances is decreased (Itoh and Barber 1983). Root hairs also are useful to plants in that they can penetrate smaller pore spaces between soil grains that may contain large volumes of water, such as when substantial amounts of clay are present as either aggregates or layers in the soil.

The density of root hairs varies with the type of plant and prevailing environmental conditions, such as water availability and oxygen content. For example, trees can have from 20 to 500 root hairs/cm^2, and grasses as much as 2,500 hairs/cm^2 (Kramer 1983). When exposed above ground, root hairs are covered by cutin, and the root cap is covered by mucigel; this polysaccharide-rich material acts as a lubricant for root-hair penetration into tight soils. As the root increases in length, the zone of root hairs also moves, with new hairs forming near the growing tip and old hairs dying at the end of the root-hair zone.

Because individual root hair cells contain water and solutes, the water potential in the root hairs is lower than outside the root hairs; consequently, water enters the root hair by osmosis as previously discussed (Fig. 3.5). After this initial entry of water, the root hair cell is now at higher water potential than adjacent cells, and water moves from the root hair to these cells and ultimately into the cellular vacuole and xylem. The root hair becomes depleted in water after this transfer and is, therefore, receptive to more water entry from outside the root hair as long as water is bioavailable. If this occurs along a series of interconnected cells, such as the xylem, a mass flow of water will occur.

In general, water that enters the root hairs can enter the internal xylem tissue of the plant through two mechanisms: (1) through the cell walls and membranes of the individual cells that compose the outer layer of epidermal cells in the roots, or (2) between the cell walls of individual epidermal cells (Fig. 3.12). As previously discussed, the first mechanism requires water to go through the living cytoplasm of cells and is called the symplastic pathway. The second method has water going between intracellular spaces and is called the apoplastic pathway. The apoplastic pathway of water is similar to finding one's way through a maze. For the apoplastic pathway, soil water enters the root between adjacent cell walls of epidermal cells (Fig. 3.12). Water passes from cell wall to cell wall of the epidermis and cortex cells by diffusion. Upon reaching the endodermis, however, water must pass through the Casparian strip similar to the symplastic pathway. The Casparian strip cells are not permeable to water. As such, the membranes of these cells regulate the entry of water, solutes, and gases into the plant as a whole. Within the Casparian strip and outside the stele is the pericycle cells, which give rise to branch roots that grow through the cortex into the sediment.

Various analytical and numerical models have been developed in attempts to simulate the relation between plant roots and water and solute uptake. Somma et al. (1998) developed a three-dimensional model of root water and solute uptake. Water uptake is simulated as being affected by water potential and osmosis and solute uptake is simulated by passive and active uptake. In essence, plant transpiration and solute assimilation are coupled through water-use efficiency data.

3.4.4 Effect of Rhizosphere Surfactant Release on Water Uptake

The production of organic exudates along root sheaths and tips increases the solute concentration in the soil water and, therefore, can affect water surface tensions. Water is under considerable tension in the unsaturated zone because of the presence of air in the pore spaces. As the concentration of organics increases in the soil water from plant roots, such as sugars and phosphatidylcholines, the surface tension of water becomes lowered and may facilitate the entry of water into plants (Passioura 1988; Read and Gregory 1997). Compared to pure water, a surfactant and water solution will decrease surface tension between 10% and 50% (Read et al. 2003).

For plants, the surfactant property of root exudates is especially beneficial as the soil profile dries, for the exudates will have a greater ability to wet the remaining water under higher tensions and thereby lower the tension to facilitate plant uptake. Surfactant addition to the unsaturated and saturated zones of the subsurface has been shown, however, to decrease the diffusivity and hydraulic conductivity of the sediments.

3.4.5 Hydraulic Lift and Water Redistribution

The vertical movement of water from roots deep in the unsaturated zone, capillary fringe, or saturated zone and redistribution of this water to shallower roots is called hydraulic lift (Richards and Caldwell 1987). The physical basis behind hydraulic lift is best explained in terms of the various components of water potential in the root zone. In areas where transpiration and evaporation remove soil moisture from shallow surface soils, the total water potentials decrease or become more negative. Deeper soil with higher, less negative water potential then moves upward toward the surface to replenish the lost water. Much of this occurs at night when transpiration and evaporation cease. This upward movement of water from deeper wetter zones to a drier unsaturated zone is a principal mechanism of unsaturated flow in some arid regions (Andraski et al. 2005). Of course, this process only continues as long as there are deeper sources of water in the capillary fringe with less negative water potential or groundwater at atmospheric pressures. If this source of water is beyond root growth or dries up, the plants wilt and will not recover the following day.

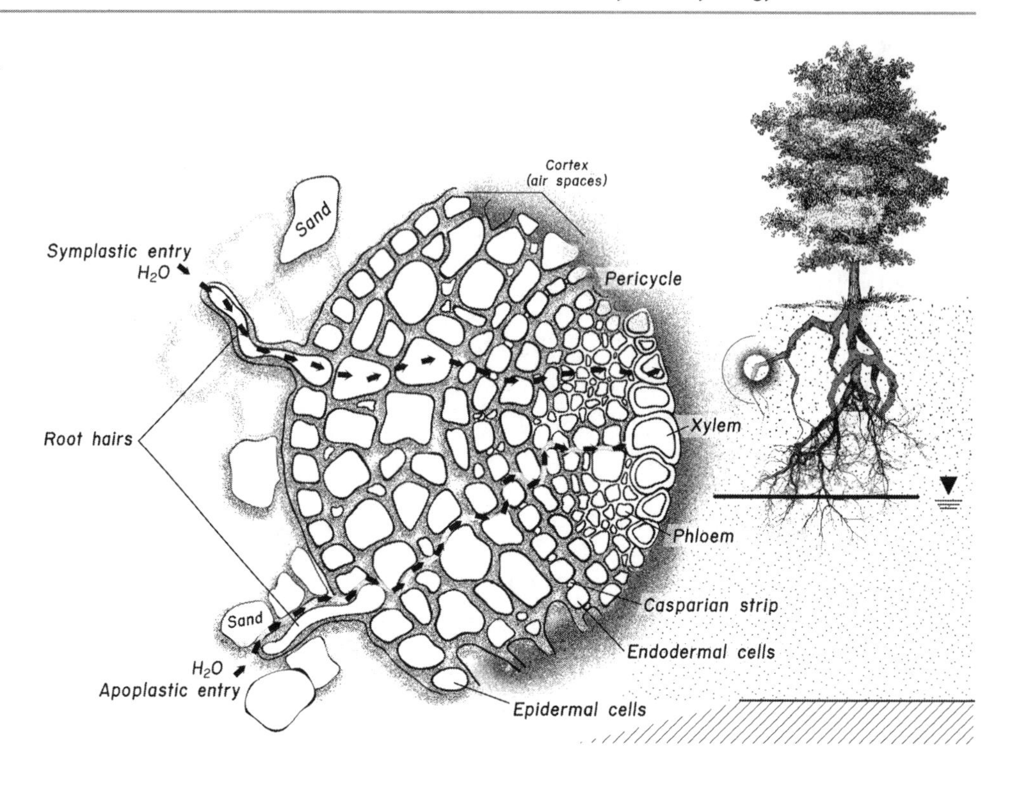

Fig. 3.12 The entry of water and solutes into the vascular system of a plant can occur by two pathways. Water and solutes follow a path from the sediment or pore space to a root hair by diffusion, through the intracellular space to the cell walls of the cortex to the endodermis through the Casparian strip and finally the xylem (Modified from Curtis 1983).

Evidence exists in Hultline (2003) and Leenhouts et al. (2006) that water in roots in the shallow soils also can move in the opposite direction of hydraulic lift; that is, downward from wetter soils to roots that exist in deeper, drier sediments. This process, called hydraulic push, may explain how the roots of phreatophytes follow the fluctuation of the water table and how such plants can become established after germination, grow deeply into relatively dry sediments, and reach declining water tables over time.

Brooks et al. (2009) present evidence that the subsurface water used by some plants is not necessarily from recent precipitation and infiltration but rather from residual water that has been entrained and, therefore, represents water from multiple past precipitation events. In essence, their data of water stable isotopes indicates that all of recent infiltration does not necessarily move rapidly from the land surface to the water table, also called translator flow. Plants that grow in the dry Mediterranean climate of the study area have adapted to these dry conditions by essentially mining this trapped water and, therefore, would not be classified as phreatophytes that rely on groundwater.

3.5 Vascular Tissues, Leaves, and Transpiration

At its simplest, a plant takes advantage of the radiant energy of the sun to make food, and can control to some extent the rate of water evaporation. This movement of water through plants from soil to air is called transpiration, from the Latin *trans* for across and *spiro* for to breathe. Moreover, from the entry of water into root hairs to the release of water vapor from the leaves, no to very little energy is expended by the plant—plants can be considered, therefore, as efficient parasites of the sun, air, and water.

But how can water be transported from the roots to the leaves at the top of the tallest trees? Some ancient coast redwoods in California's Redwood National Park are over 370 ft (112 m) tall, weigh about 1.6 million pounds, and are about 600 years old; in fact, the oldest was 2,200 years old but was logged in 1933. The coast redwood is *Sequoia sempervirens*, whereas other sequoias can be found in the Sierra Nevada (*Sequoiadendron giganteum*) and China (*Metasequoia glyptostroboides*). These are not only the tallest but the fastest growing conifers in North America, which is significant because conifers primarily are slow growing as they have adapted to low-moisture climates. Fossils of similar trees have been found in sediments that date back to the Jurassic Era some 160 MYa in areas much more widespread around the world than today.

The current, more isolated, distribution of redwood trees can be explained primarily as the result of abundant soil moisture (Rundel 1972; Preston 2008). Moreover, these trees were classified by Robinson (1958) as phreatophytes. Rundel (1972) reported that during seasonal dry periods when precipitation was low, relatively high soil moistures were measured beneath redwood tree roots, supplied by groundwater. This groundwater had to have been recharged elsewhere during times of higher precipitation. In fact, groundwater may play a role in the maintenance of tall redwoods over 2 millennia.

The transport of water to the top of plants cannot be solely explained by root pressures that result from the osmotic uptake of water by roots, as was discussed previously. In fact, the pressures of water within the xylem have to be less than atmospheric pressure in order to follow a decreasing pressure gradient from root to leaf. Because atmospheric pressure tries to enter a vacuum, it is possible that water is moved upward as a vacuum is created inside the xylem. However, the maximum lift generated by such a process would be no greater than 33 ft (10 m) at sea level, the weight of the atmosphere, and would decrease at higher elevations. The aquatic plant royal water lily (*Victoria amazonica*) has very large leaves, as much as 8 ft (2.4 m) across, and grows from a rhizome located on the bottom of a pond. A petiole connects the rhizome to the leaf at the water surface. This petiole can approach lengths up to 22 ft (6.7 m) but has not been found to be longer, presumably because of the inability for evaporation to pull a vacuum greater than the atmospheric pressure that holds it down.

In the nineteenth century, it was thought that the plant cells acted as small pumps to push the water up to the leaves. This hypothesis was dispelled, however, by the German botanist Eduard Strasburger after he killed the cells of a 40-ft (12 m) length of wisteria vine by boiling them but left the upper leaves alive. The leaves did not wilt and die, which indicated that water transport did not require the cells in the stem to be alive. Strasburger observed a similar response after he immersed the stems of plants in toxic solutions of heavy metals. The conclusion was that living forces were not necessary to raise the water to great heights.

The amount of water transpired relative to the amount of water used to produce biomass can be described as the transpiration ratio of a particular plant. This transpiration ratio, *TR*, is the weighted value in mass of water used to produce a mass of a particular crop. A *TR* of 1 indicates that all water taken up by the plant was used to produce biomass, an unlikely scenario. A *TR* greater than 1 indicates that extra water was taken up but not used to support biomass synthesis. Corn, for example, grown in Colorado has an average transpiration ratio of 1,405 lbs (638 kg) of water per pound of crop (Van der Leeden et al. 1990a, 1990b). Watermelons have an average *TR* of 1,102 lbs (500 kg) of water per pound of crop. On average, more than 2,000 tons of water must pass through the roots of a crop plant to result in only 20 tons of biomass (1%), and even after the crop is dried, only 5 tons will remain, of which 3 tons will be from water. Hence, only 0.15% of the total water used is actually incorporated into the plant biomass; the remainder is returned to the atmosphere by *ET*. Of this *ET*, the proportion that is derived from groundwater may approach 100% in dry climates (Moreo et al. 2007).

Transpiration ratios exist for other plants as well. As might be expected, many native weed species in North America can tolerate drier conditions and have *TR*s under 500. Transpiration ratios also are available for woody plants that may be used at phytoremediation sites. Because trees are not considered to be crops in the same manner as corn, the *TR* is computed as the pounds of water needed per pound of dry-leaf matter. Maple trees that thrive in moist soils have a *TR* of 1,281 whereas conifers have *TR*s less than 250 (Van der Leeden et al. 1990a, 1990b).

3.5.1 Xylem and Phloem

We have seen how the cellular structures of plant tissues relate to water uptake at the level of the individual cell. Liquid water enters individual root hair cells by capillary action and osmosis, diffuses to the xylem, and rises to the leaves against gravity to exit as water vapor. But how exactly are the two phases of water connected in a plant? Part of the answer lies with the transition from the oceans and lakes of one-celled photosynthetic algae to the land. Successful transition resulted only after a set of interconnected cells was evolved to transport water and it's dissolved constituents under negative pressure. As stated previously, lichens and mosses do not have such conducting tissues, whereas ferns do. Ferns contain substantial amounts of lignin that enable them to stand erect off the ground in search of light and provide protection from desiccation and herbivory. Ferns grew to great heights in primeval forests when the atmosphere was warmer than today. Currently, ferns are much shorter and tend to be present as understory plants in temperate forests. Ferns still need water to reproduce, however, as a medium to transmit sperm to the ovule.

It is the linkage between cellular structure and the laws that govern the physical properties of matter that interact together in the movement of fluids in vascular plants through the xylem and phloem. As we will soon see, these tissues are present from the growing root tip to the stems to leaf petioles to the leaf veins.

3.5.1.1 Xylem

The xylem is composed of thick cells connected end on end, thereby forming continuous tubes that allow the passage of water from the site of entry in root hairs to the leaves. Xylem is produced by the vascular cambium stem cells on the side facing the center of the plant and only function to carry water after they have died and lost their cytoplasm. The cambium is the 2 to 10-cell thick layer of living tissue that gives rise to the xylem and phloem in dicots. It is absent in monocots that have only primary growth.

The xylem is composed of tracheid cells and vessel cells. The tracheids are essentially single, long, pointed cells that have holes or pits that permit exchange of fluid between adjacent tracheid cells. Their construction is similar to how perforated pipe laid in drains can move water along hydraulic gradients both vertically and horizontally. This construction provides strength but at the cost of increasing the resistance to water flow. Tracheids are more characteristic of gymnosperms, such as conifers, although they also are present in angiosperms. Angiosperms possess vessels, which are not as narrow as the tracheids and are connected end-to-end to form a continuous pipe; therefore, water flow is not as impeded as with tracheids. Both tracheids and vessels consist of cells that die after they form tubes; this is a consequence of their intended use to transport fluids for the whole plant rather than just within their own cytoplasm. Vessels have rows of shorter cells that have thick cell walls on the inside in the form of rings or spirals, similar to the rings within the walls of mammalian trachea, or windpipe. These cells cannot tap into the transpiration stream and place a burden on the plant. Conifers have tracheids and resin canals in the xylem. The xylem in pines are not hollow tubes supported by spirals but instead are pitted. For example, the xylem cells lined up end to end are not continuous, as in the hardwoods, but pinch off to points at both ends, where adjacent cells overlap. At this juncture, water can move through the thin cell walls.

The xylem cells die each year as they are replaced by new cells. The plant hormone auxin that is present in the cambial cells regulates the growth (Tuominen et al. 1997) and death (Moreau et al. 2005) of xylem cells. Higher auxin concentrations are found closer to the cambium, and concentrations decrease in the xylem cells with age. The use of rings to document the life of a tree was first documented by Leonardo da Vinci. He also observed that the width of each ring was an indication of the relative amount of moisture available during each year since growth began.

Suction in the xylem can cause gases to enter the fluids, called cavitation. Cavitation can occur due to excessive tension, embolisms from air exchange with the cortex, freeze and thaw cycles, or disease. Perhaps the most likely source of cavitation is water limitation because of drought conditions. Cavitation of the water column in the xylem can decrease water transport and result in decreased hydraulic conductivity, dehydration, and even plant death. Because of the deleterious effect of cavitation on water transport in the plant, and because water movement is induced by evaporation in the stomata, plants must maintain stomatal conductance below the maximum that leads to cavitation (Sparks and Black 1999). This is especially true in more drought-resistant species, which have decreased hydraulic conductivities and more control over stomatal conductance.

The presence of gases in xylem fluids is not just from tension breaks but also from the entrance of atmospheric oxygen and respiration production of CO_2 by surrounding live tissues (Kramer and Kozlowski 1960). Both oxygen and CO_2 can be present from 1% to 20% and typically are inversely related. In most cases, the concentration of oxygen is lower in the xylem than in the atmosphere, and the concentration of CO_2 is higher in the xylem because of the presence of the respiring cambium. In the cortex and bark outside of the cambium, however, gas content is similar to that of ambient air by plant-atmosphere exchange through lenticels.

Some xylem structures are considered adaptations to lower water availability, much in the manner that stomata are regulated by limited water or strongly negative water potentials. For example, in areas of California that experience droughts, plants have high conductivity xylem when water is available but narrower vessels and tracheids when water is limiting. Such physiological adaptations are related to avoidance of embolism formation, which would result in plant death (Kolb and Davis 1994). Another survival approach for deciduous trees is to replace embolized xylem tissue each year with new xylem.

3.5.1.2 Phloem

The tissue that forms on the outside of the cambial stem cells is similar to xylem in that it transport fluids, but that is where the similarity stops. The cells that comprise the phloem are relatively thin and connected end-to-end to form tubes that permit the passage of water and solute, such as sugars, from the shoots to roots and back again. They are present closer to the surface of the plant than the xylem. The walls of these cells are thinner than the xylem and, unlike the xylem, are composed entirely of living cells and, therefore, retain their cytoplasm. These phloem cells do not, however, contain nuclei at maturity. The phloem consists of different cells, depending upon whether the tree is a hardwood or conifer. In hardwoods, the phloem cells consist of the sieve-tube cells that are not entirely open end-to-end, and the companion cells.

As discussed previously, sap is a dilute solution of water and organic molecules such as sugars. Sap also contains proteins, such as amino acids, and hormones, such as auxin, that are produced in the shoot meristems and travel through the phloem to influence root growth in the root meristems. The phloem and cambium of the Scotch pine (*Pinus sylvestris*) was collected, dried, crushed, by Laplanders and turned into bread during times of scarce game, as was noted in 1732 by Linnaeus who traveled throughout Lapland. The name Adirondack, commonly associated with the mountain range in New York, actually means tree eater, which was the custom of the Native American Adirondack tribe that lived in the Adirondack mountains. These examples underscore the primary role of phloem, to transport the products of photosynthesis throughout the plant.

The phloem predominately moves fluids downward from plant-produced dissolved organic substances in the leaves to the roots. The driving force for this movement of sap is the concentration gradient that exists between the zone of production in the leaves and the zone of consumption. Because the concentration of sugars is high in the leaves, the concentration of water is low. Water enters from the xylem to dilute this solution and creates high osmotic pressures in the leaf phloem. Conversely, in the roots these sugars are oxidized for growth, and the phloem concentration is rendered diluted, which causes water to exit these cells and create a decrease in osmotic pressure (Fig. 3.13). These relations lead to the downward movement of sap from the leaves to the roots. The direction can be reversed during the spring in deciduous trees, such that the roots contain more sugar than the new leaves and the sap flows up the phloem, called the pressure-flow hypothesis. However, some sap flow is not controlled by either pressure or gravity, as some conifers yield sap during daylight to heal wounds.

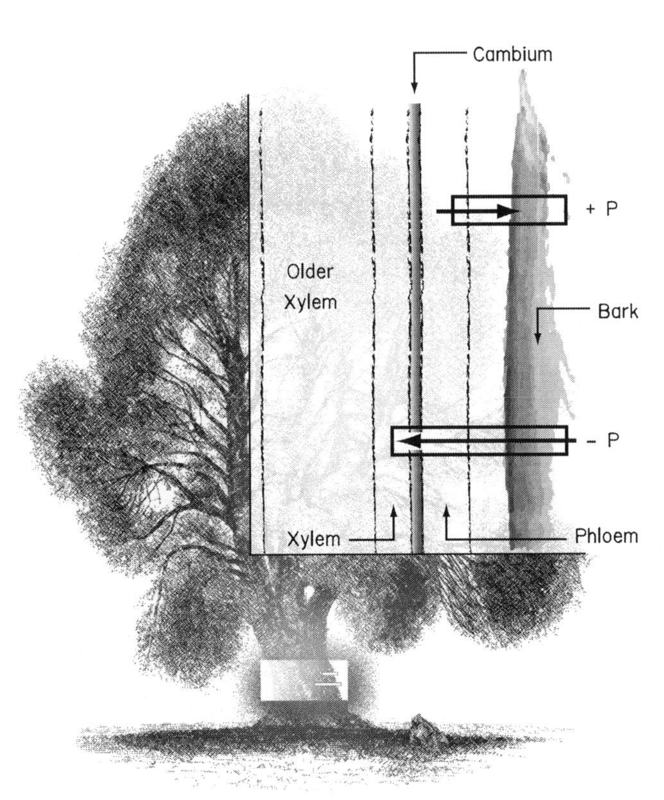

Fig. 3.13 Advancement of an increment corer (shown as the *rectangles*) into a tree reveals a horizontal pressure gradient, such that the phloem is pressurized (+P) and the xylem is often under tension (−P). If the older xylem is decayed, it is possible that it also is pressurized, however, and can emit gases, such as methane produced by methanogenic bacteria in the rotting, anoxic wood.

3.5.1.3 Law of Laplace

The diameter of the conducting vessels in vascular plants is very small, from the root hairs to the xylem. These cell walls must be thick enough, however, to withstand the strain when these vessels are under tension or pressure, but remain thin enough to ensure intracellular exchange by diffusion. The existence of such an apparent contradiction is best explained in terms of the relation between the radius of a vessel and its wall strength, called the Law of Laplace:

$$T = PR, \tag{3.6}$$

where T is the tension on the vascular wall (dynes/cm^2), P is the pressure (mm Hg), and R is the radius of the vessel (cm). If the radius is large, the wall thickness of the vascular tube also must be large; if the radius is small, the wall thickness is small. This explains how even in the largest trees, the vascular system can be composed of many narrow, thin-walled tubes.

3.5.1.4 Rays

As we have seen from the previous discussion of the cellular components of the plant tissues, most are vertical in orientation because they arise from the vertically growing cambium. Some horizontal tissues also can arise, however. Lateral movement of fluid from the outer rings of phloem to the inner non-living xylem tissues can occur through pith rays. A ray consists of parenchymal cells produced by the cambium. In softwoods, some ray cells have ducts that carry resinous compounds. Rays can be present in young plants as a result of primary growth of the terminal meristem of the bud and function to store or translocate food.

3.5.1.5 Ring and Diffuse Porosity

The location, size, and number of xylem vessels are characteristically different for trees based on the presence of water-conducting structures of the vessels, tracheids, fibers, and parenchymal cells. In coniferous plants, the xylem does not contain true vessels, and such trees are called non-porous trees. In deciduous plants, the wood is characterized by vessels having different structural appearances. These trees, called diffuse-porous trees, have both tracheids and vessels and a more uniform distribution within each annual ring, where the vessels are smaller in radius and farther apart from each other. Conversely, ring-porous trees have vessels distributed in the early wood and are larger and more closely spaced. In diffuse-porous trees, water can move in the vessels in multiple rings; in poplar trees, for example, water movement has been observed to occur as deeply as 4–8 cm into a tree. In ring-porous trees, the water is conducted only in the outer annual ring.

3.5.2 Cohesion-Tension Theory and Water Movement

Even though the flow of water in plants has been pondered since the days of Aristotle, the cause or causes behind the movement of water against gravity from the roots to leaves remains a topic of considerable debate (Zimmermann et al. 2004; Brooks et al. 2004). As was discussed in Chap. 2, a molecule of water consists of two hydrogen atoms bonded to an oxygen atom. There is a bond between each hydrogen atom and the oxygen atom of an adjacent water molecule. Hence, a series of adjacent water molecules is linked by this cohesive force. This intermolecular force is what drives the tension found in water. Therefore, a molecule of liquid water that exits a leaf after being turned into water vapor is linked to a molecule of liquid water that is still in the leaf, and these water molecules all are linked to water in the plant down to the root hair. This can be envisioned as being similar to the bucket brigades used to fight fires in the seventeenth and eighteenth centuries. As such, water molecules that exit the plant pull up additional water molecules through the plant, which causes additional water molecules in the root hairs to rise after entry from the soil by osmosis.

Near the turn of the twentieth century, this cohesive force between adjacent water molecules was offered as the reason for fluid flow in trees against gravity; this came to be known as the Cohesion-Tension Theory (Dixon and Joly 1895). Indication that this theory is not fully accepted is given by textbooks in the early twentieth century. For example, Gager (1934) describes the movement of water upward through a plant as "not perfectly understood, but the pulling force resulting from transpiration is, perhaps, the main factor."

Because much of the soil moisture present in the subsurface above the water table is under tension, including water in the capillary fringe, the cohesion of individual water molecules to each other has to be stronger than the tension holding water molecules to soil particles. Energy has to be input, or spent, for water to move against these forces. The force that initiates this process is the evaporation of water and lowered water potential and the energy is from the sun. Water movement is upward in the direction of decreasing water potentials, or increasing negative values of water potential, ψ. If no evaporation occurs, the water column does not move. Unlike the plumbing system of most homes in which water is supplied through the pipes under pressure and flow to faucets is created when the pressure is relieved to a lower level, movement of water in plants from roots to leaves in the direction of lower tensions, and flow through the plant is initiated by decreasing the leaf water potential.

Experimental evidence of the high negative tensions necessary to support transpiration streams 10s of ft (1s of m) above ground was provided by P.F. Scholander, using the pressure bomb that he devised to measure water potentials (Scholander et al. 1965). He stated that water potentials of cut shoots could provide an estimate of the tension of the water in the transpiration stream prior to the cut. The tension that a thin column of water can withstand is immense, between about 3,000 and 150,000 lbs/in.^{2}, and is much stronger than steel.

Because the transpiration stream is under considerable tension, or low water potentials, in the Cohesion-Tension Theory, there is a point when the tension is overcome and the water column breaks, which can lead to the cessation of water flow. The water in the sap can turn into vapor to block water flow through cavitation. As stated previously, cavitation can be the result of the lower pressures of water under tension causing gases to come out of solution as the pressure is lowered. It reduces the hydraulic conductivity of the xylem. Cavitation in tall redwoods was reported to occur once leaf water potentials exceeded −1.9 MPa (Koch et al. 2004). Some hybrid poplars have been shown to have cavitation in the xylem during prolonged droughts, assuming that no water is removed from the water table, and that this cut-off of water supply increases the hydraulic resistance in the stems resulting in premature shoot death (Tyree et al. 1994). Conversely, the water in the transpiration stream can freeze and break. Both processes lead to gas-filled spaces, or embolisms, in the xylem that transports the water. These embolisms have been shown, however, to be repairable, with water entering the gas-filled cells and continuing to flow (Canny 1997). These results place into question the need to explain sap flow solely in terms of measurements of cohesion and tension, as will be discussed below. Moreover, some plant physiologists claim that water will break when tension reaches −0.6 MPa, much lower than that required by the Cohesion-Tension Theory to lift water in plants more than 300 ft (91 m) tall.

We have seen that the movement of water in the xylem follows a decrease in water potential and generally is upward against gravity. Some water exits through the bark, however, and this loss is discussed later in this chapter. But what path does an individual molecule of water follow from the root hair to the stomata? Is the path straight up the side of the tree, or does it follow a different flow path? Some indication of the path is suggested by the arrangement of leaves along stems or the pattern of fissure appearance and lenticels on the outside of some trees. The arrangement of leaves along a common axis, in this case, a stem, is called phyllotaxis, and the relation to a spiral pattern was introduced in the mid-1700s by C. Bonnet in which leaves were related to a helix winding around a cylinder, which was challenged later by a logarithmic pattern suggested by A.H. Church (Kramer and Boyer 1995). The spirals can be either left- or right-handed.

Much less obvious is the sometimes spiral construction of xylem beneath the bark. Waisel et al. (1972) reported a study of the path that water that had been stained with dye takes in various shrubs and trees. The source water was stained with acid fuschin to a concentration of about 1 g/L. This tracer was either injected directly into the plant or exposed roots, or the roots were immersed in a solution of water that contained the dye. The path taken by the water as seen by the tracer then was recorded at its position in cross sections of the plant taken from above the tracer injection and immersion site. Arrival in above-ground branches by the appearance of dye in the cross sections provided evidence for Waisel et al. (1972) to categorize the flow in trees as belonging to two main types: spiral and straight ascent (Figs. 3.14a, b, respectively). It also was observed that, in some cases, the point injection of the water dispersed into a wider zone around the circumference of the tree as the water ascended, termed ring ascent. Ring ascent was revealed in some trees to progress in either a clockwise or counterclockwise pattern. Waisel et al. (1972) looked specifically at trees that today are candidates for use at many phytoremediation sites—species of the *Populus* genera, in this case, *Populus euphratica*. They examined water and dye flow in a tree from a wet site that reportedly had diffuse porosity compared to a tree from a drier site having ring porosity. In both cases, straight ascent that turned into ring ascent was noted.

The rationale behind the spiral pathway of water ascent from the roots to the leaves is not directly intuitive. The answer, however, is contained in the efficiency of the plant world as influenced by selective pressures. Tree trunks and stems can be envisioned as cylinders. The shortest path between two points on a cylinder is described by a helix, or spiral pathway. The pattern of flow also is related to the number of leaves. As was first recorded by Leonardo da Vinci, a correlation can be made between the cross-sectional area of the branches and their attendant leaves and the cross-sectional area of the trunk. This indicates that, in essence, the

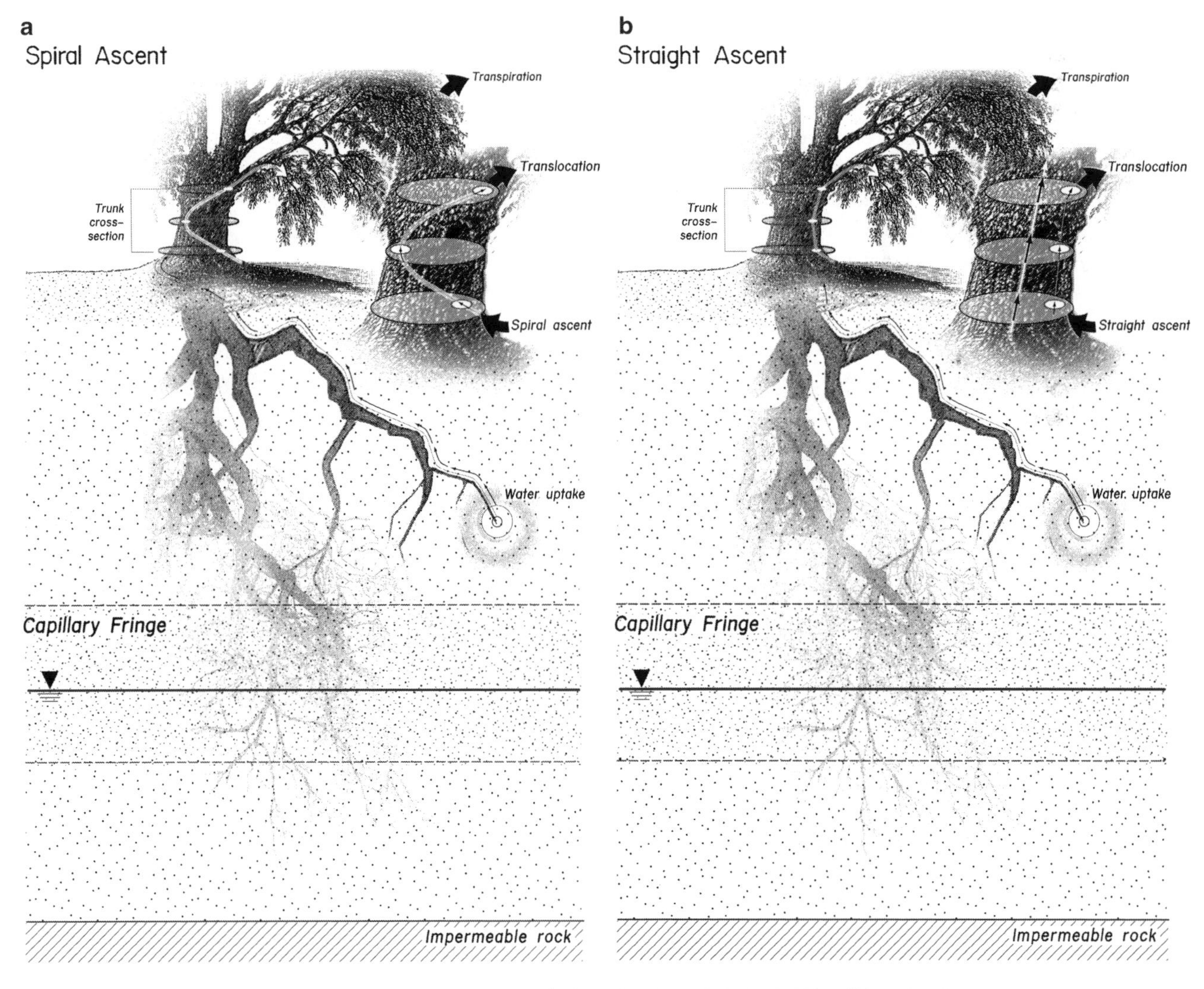

Fig. 3.14 (**a**) Water ascending a tree in a spiral vertical pathway in the xylem beneath the bark. This will have implications for core collection and analysis at contaminated sites, as discussed in Chap. 15. (**b**) Water ascending a tree in a straight vertical pathway in the xylem beneath the bark.

trunk is a large pipe of a given cross-sectional area that provides only enough water to support the summed cross-sectional area of the branches above it.

The same correlation between flow and total cross-sectional area would have to apply below ground for the root system as well. Shinozaki et al. (1964) brought this relation into focus with the unit-pipe analogy; that a group of leaves on a branch was fed by a section of conducting tissue fed from underground by a set of roots. The movement of water from the roots to the leaves, therefore, is not random but highly specific. This often can be seen in trees struck by lightning, in which a few roots are killed and the branches that these roots supply dry up and die, but the main tree survives.

The flow of water in a well-watered plant has a budget such that the water uptake equals the water lost. This flow of water in the plant, Q, is analogous to the flow of an electrical current in a circuit, such as

$$Q = \text{Water potential difference/sum of the resistance.} \quad (3.7)$$

As water moves though the xylem, resistance to flow is encountered. This happens to the flow of any substance through a conduit, such as electrons through a wire, where the resistance is felt as heat. To overcome resistance to flow, the tension of water permits resistance to occur without the water becoming vaporized into the gas phase, or for embolisms to form. The resistance to flow is about 1.5 atm/ 32 ft (10 m) in height, so for a 459-ft (140 m) tree, the resistance is about 35 atm; no trees on earth are higher.

Perhaps the best example of how useful the Cohesion-Tension Theory is in supplying the water needs of tall plants is provided by the existence of the coast redwoods (*Sequoia sempervirens*) in California mentioned previously. These trees can reach heights of more than 350 ft (106 m). Is this the limit that water can be transported along a negative water potential gradient? Would even more negative potentials be possible, since plant cells need some water pressure to maintain turgor? If such negative water potentials induce stomatal closure in most plants, how do the upper leaves of the *Sequoia* photosynthesize? To help answer these and other questions, Koch et al. (2004) took water-potential measurements, both predawn and midday, of the leaves of *Sequoia* at different heights and found them to correlate directly, from about −0.7 to −1.3 MPa (predawn) and from −1.2 to −1.84 MPa (midday) at 131 ft (40 m) and 354 ft (108 m), respectively. Turgor was measured also with respect to height and found to decrease with height from 0.93 MPa at 164 ft (50 m) to 0.48 MPa at 360 ft (110 m). Pressures are low but still remain positive, even though the leaf water potentials are −1.84 MPa at 354 ft (108 m).

An alternative explanation of the ascent of water in plants that does not rely on the large negative pressures required by the Cohesion-Tension Theory is based on data collected by Balling and Zimmermann (1990). They used a pressure probe to assess the water pressure of intact xylem cells. They found that moderate negative pressures existed, rather than extreme negative pressures, between −1 to −10 MPa, required by the Cohesion-Tension Theory, which the authors described as placing water in an extremely unstable, metastable state akin to superheated water. Also, their explanation of flow does not require the xylem vessels to contain a continuous column of water. As such, Zimmermann et al. (2004) suggest that water movement in plants occurs in a series of complicated lifts of various volumes of water.

Finally, no discussion of theories of water movement in plants can be considered complete without mention of the possible effect of the moon's gravitational pull on water bodies on earth, including water in plants. The phenomenon of tides in oceans and large lakes caused by the moon's gravitational pull during its phases is well known. Less well known, but just as verifiable by observation, are groundwater earth tides in deep, confined aquifers that is discussed in Chap. 4. The reason that the moon influences oceans and deep groundwater is because both are large volumes of water. In plants, water is present in small tubes that seemingly would not be affected by gravitational differences. There are many gardeners, however, who believe in planting based on the moon's phases.

3.5.3 Stomatal Resistance

Carbon dioxide must enter the leaves for chlorophyll to help make plant food. Entry is gained through stomata, pores in the leaves (from the Greek, *stoma*, meaning mouth) that regulate this gas exchange—they control CO_2 uptake as well as water vapor loss. The stomata typically account for less than 1% of the total upper and lower surface area of leaves and are small (about 15 μm), so that even on a small leaf there are between 13,000 and 100,000 stomata/cm^2; an average-sized leaf can contain millions to 10s of millions of stomata. Stomata present on one surface of the leaf are called hypostomatous plants, and stomata present on both leaf surfaces are called amphistomatous. Many plants, such as *Populus* used for phytoremediation, are amphistomatous, and these plants tend to be characterized by higher rates of growth and transpiration. Different types of stomata can be found on plants and classified based on the structure of the guard cells and adjacent cells. The different types of stomata are the anomocytic, anisocytic, paracytic, and diacytic.

The movement of water through a leaf is regulated by stomata, for they act like a valve that balances the input of water from the root hairs to the demands of water by the atmosphere by evapotranspiration. Stomata can partially or fully close, based on the lower water potentials of adjacent guard cells, which are specially shaped epidermal cells. When water is limiting, guard cells shrink and remain closed. When water is available, the guard cells swell and open, permitting gas exchange to occur. The uneven length of guard cells on adjoining sides of the stomata means that when the water potential is high, the cells expand, and pull apart, creating a pore that permits gas exchange to occur. When water is not available, the guard cells shrink and the stomata close, causing gas exchange to cease. This feedback loop means that even if water is available to a plant, if transpirational forces are great, then the plant will close the stomata and there will be no photosynthesis even on days when it would appear that conditions were appropriate.

Plants also tend to have the majority of their stomata on the undersides of leaves, such that the probability of the stomata being clogged by particulate matter is reduced. As we saw before, plants also control indiscriminate water losses by the presence of a waxy cuticle and short, hair-like growths from the epidermal layer. These hairs act to increase the relative humidity near the leaf surface by reducing the rate of air flow over the leaf surface; a higher relative humidity decreases transpiration.

The closure of stomatal guard cells to limit transpiration when water potentials become increasingly more negative also decreases the potential for food production. Although more water can be taken up by the larger surface area of root hairs than lost by leaves if water is unlimited, these conditions are not typical and other mechanisms are needed to regulate water losses from the leaves by transpiration during photosynthesis. If the driving force of water uptake in the roots is roughly equal to water lost in the leaves, assuming that water is not limiting, the question becomes what are the various resistances to the transpiration of water and what are their magnitudes? Regulation of transpiration occurs through changes in stomatal conductance. For many plants, if the leaf water potential drops to -1 MPa, the stomata close. The water potential difference between the soil and air is what drives water use by plants. A simple estimate of the magnitude of this potential for flow is called stand conductance, in which the daily water use is divided by the vapor pressure gradient (Rural Industries Research and Development Corporation 2000).

Transpiration from leaves is controlled by at least two main factors—the vapor pressure gradient or vapor pressure deficit (VPD, Pallardy and Kozlowski (1979)) that exists between that in the leaf air spaces relative to that of the surrounding air and any resistance to this diffusional transfer of vapor that might be present. Such resistances include changes in the stomatal aperture, or stomatal resistance, where resistance is the inverse of conductance, such that resistance = 1/conductance. A high conductance is equal to a low resistance and vice versa. Resistances also include the boundary of air near the surface of the leaf that receives emitted water vapor. Transpiration, T, is related to the resistances by

$$T = P_{wv(\text{leaf})} - P_{wv(\text{air})}/r_s + r_b, \qquad (3.8)$$

where the vapor pressure gradient is $P_{wv(\text{leaf})} - P_{wv(\text{air})}$, in kilopascals (kPa), r_s is the stomatal resistance, and r_b is the boundary layer resistance present at the surface of the leaf. Because diffusion is a slow process, transpiration is limited by the magnitude of the diffusional release of water from the leaf air to the external air.

In most cases the highest resistance to flow from the leaves is the loss of water vapor. Leaves are attached to the stems of plants, either directly (from the Latin *sessile*, meaning sitting on) or indirectly by a leaf stalk, or petiole (from the Latin *petiolus*, meaning stalk). The petiole permits the leaf to move in response to winds without being torn and also to change the position of the flat surface of the leaf in reaction to the changing position of the sun with respect to the relatively fixed location of the plant. Leaves can be arranged as single leaves from each petiole, called simple leaves, or as multiple leaves on a single petiole, called compound leaves. Plants with compound leaves permit sunlight to reach lower leaves.

Most plants balance the need to have a maximum surface area exposed to sunlight for photosynthesis and the uptake of gaseous CO_2, while reducing the loss of water vapor. In between the outer layer of the leaf, or epidermal cells, are the mesophyll cells (from the Greek, *meso*, meaning middle). The mesophyll cells contain chloroplasts where photosynthesis occurs following photon absorption. The chloroplasts are mobile within the cell cytoplasm and orient toward the sun (Fig. 3.15). These photosynthetic cells, or parenchyma, are present in two layers—the palisade and spongy parenchyma. As the names imply, the palisade cells are long and tightly spaced and closest to the upper epidermis, and the spongy layer below the palisade cells is composed of less tightly spaced cells surrounded by air spaces that, as we will see, include water vapor. This is why for most leaves, the upper surface is a darker green then the underside, for there are more chloroplasts in the upper mesophyll cells. The palisade layer is more photosynthetically active than the spongy layer because of its location.

As might be expected, not all leaves of a plant are exposed to sunlight under the same intensity. Hence, it would appear that shaded leaves do not transpire to the extent that sunlit leaves do. Does this indicate that more water will become available to the leaves exposed to sunlight at the expense of decreased water availability to shaded leaves? Experiments done in the field by Brooks et al. (2003) indicate that no such response occurred, and that stomatal conductance in exposed leaves did not increase.

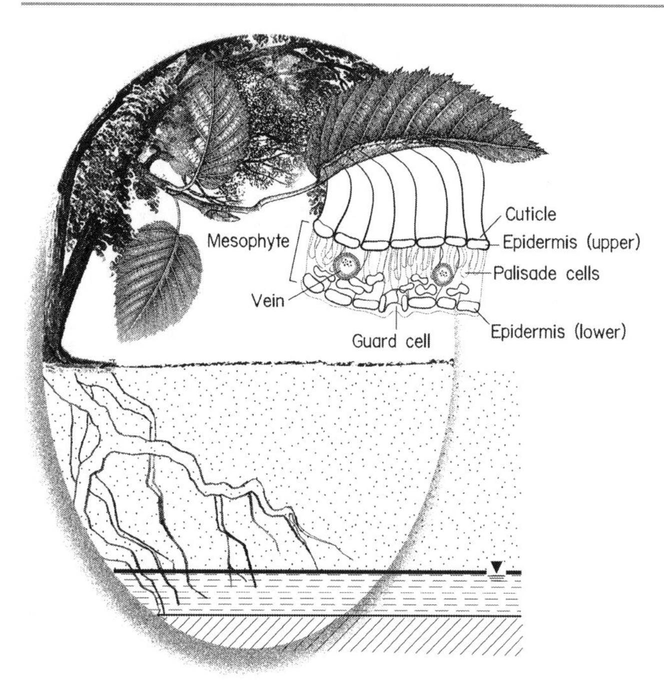

Fig. 3.15 The basic structure of a leaf that depicts the location of the guard cells that surround the stomata on the underside of leaves. Some phreatophytes, such as poplars, have stomata on both sides of the leaf.

Water transported to the leaf through the xylem enters the leaf through vascular bundles, more generically known as veins. The veins are primarily found in the spongy parenchyma, because that is where the spaces between cells facilitate gas exchange. Conversely, photosynthetic products are removed from the leaf and transported to other parts of the plant by the phloem.

Water can be translocated up the xylem to leaves and within the various parts of a tree. This internal translocation of water often is seen in trees with branches that are exposed more strongly to the east. Water stored in the plant during the night moves from the western part of the plant to meet the earlier *ET* demand in the eastern part of the plant (Daum 1967).

But let us return to the movement of water from the soil to the atmosphere through a plant. Because the palisade layer of cells in the mesophyll is the location of photosynthesis, these cells have a high concentration of sugars and, hence, low water concentration or potential. The water potential in the veins, attached to the mesophyl and xylem, is higher than the mesophyll. Hence, water enters the mesophyll cells and the water that is not used to support turgor or used in photosynthesis is removed as water vapor. These vaporized water molecules are replaced continually according to the Cohesion-Tension Theory.

Changes in the water potential of a leaf not only changes stomatal conductance but also the shape of the leaf itself. Many plants respond to water deficits or high atmospheric temperatures or low humidity by leaf rolling. This acts to reduce the surface area of the leaf by up to 50%. Although this is a common adaptation to thin-bladed plants, such as grasses, broad-leafed plants respond in a similar manner but typically wilt the entire leaf surface at a steeper angle to incoming radiation. Moreover, these processes are reversible once water potentials are restored.

Even if water is unlimited, transpiration has been shown to not be constant because of changes in the atmospheric conditions that may cause the stomata to close. Van Hylckama (1968) presented data from a study site near Buckeye, Arizona, where if windspeed and temperatures increased each day, even if water was available, *ET* decreased. He concluded from these observations that widely held assumptions of the constant rate of water use by riparian vegetation cannot be supported. These results have implications for the assumption of constant water-use rates by trees when simulated as part of some phytoremediation models.

For tall trees, very negative leaf water potentials increase stomatal resistance or decrease conductance, and this affects the stable isotopic signature of the carbon being fixed as sugars. Koch et al. (2004) analyzed the stable carbon isotopes of leaf tissues collected from the upper part of redwood trees and found that relative to the stable carbon value of the lower leaves, the upper leaves were enriched in the heavier carbon isotope (^{13}C). For most plants, there is enrichment in the light carbon isotope (^{12}C), because the lighter carbon isotope reacts faster in the carbon fixation reaction during photosynthesis. At higher elevations, however, where stomatal resistances also are higher, less CO_2 is available, and all is used regardless of whether it is isotopically light or heavy. The tissue in the upper leaves of the redwood samples was −22.2 permil; by comparison, most C_3 plants have tissue stable carbon isotope values near −27 permil. These differences in carbon isotopes and their use in monitoring the phytoremediation of contaminated groundwater systems is further explained in Chap. 15.

Stomatal regulation also is under some control of plant-produced hormones. For example, the hormone ABA accumulates in plant shoots deprived of water. This accumulation, in turn, causes the stomata to close (Nilson and Assmann 2007). These researchers go on to review some of the insights gained from the recently sequenced genome of *Populus*.

3.5.4 Factors Affecting Transpiration

There is a saying in Iowa that on a hot summer's night, you can hear the corn grow. Although at first glance this seems to be a mere exaggeration, there is some truth to this statement. For example, growth is a biologically mediated reaction that has higher rates at higher temperatures, and the energy for such growth in plants, the respiration of food produced by plants, is dependent on temperature-controlled processes, such as gas exchange, water evaporation, and transpiration. The rapid growth under high summer temperature under non-limiting nutrient conditions can result in cells rapidly stretching until they break.

The two factors that are the largest regulators of transpiration are water availability and stomatal closure. When sunlight is available, leaf stomatal resistance is low. When sunlight is not available, leaf stomatal resistance increases. For example, the optimum rates of photosynthesis occurred in *Tamarix* when the air temperature was 23–28°C, which occurred in the morning when *ET* demand remained low (Anderson 1982). Photosynthesis decreased by 35% as the air temperature increased.

The factors that affect transpiration include those related to the pathway that water takes from the soil to the atmosphere through the plant. This includes the soil in which the roots grow, the availability of water in the capillary fringe or saturated zone, the structure of the root and its hydraulic conductivity, the type of xylem in the trunk and branches, stomatal resistance, and ultimately the moisture content of the air.

Many factors affect the removal of water vapor from leaves. These include (1) climatic factors, such as the amount of sunlight energy, the driving force in initiating evaporation, the relative humidity of the air near the leaves, the vapor pressure deficit, and wind speed; (2) water factors, such as its availability in soil, and the hydraulic conductivity of the soil to move the water to the roots, and (3) the resistance of water flow through plants. Evaporation doubles for every temperature rise of about 10°C, as we saw in Chap. 2. If the air is drier than the gas spaces in the spongy parenchyma, water loss from the leaf is greater than when the air is more humid. Wind speed affects the humidity near plants; higher wind speeds tend to remove humidity near the leaf more quickly than calm winds and results in increased transpiration.

If all these factors are held constant, the predominant factor controlling transpiration is the bioavailability of water in the subsurface. If water is not limiting, then the maximum amount of water transpired will be constrained by the solar energy input and VPD. Too little water for plant uptake, however, can result in xylem cavitation in certain trees, as has occurred in poplars exposed to drought conditions. Essentially, this blocks water flow through the xylem and results in shoot death (Tyree et al. 1994).

In a gross sense, the more biomass a particular plant has, such as number and size of leaves, the higher its transpiration is if water is not limiting. This was demonstrated in a study by de Wit (1958) in which the amount of water transpired by a crop during the growing season was linearly related to the production of harvested biomass. Moreover, the slope of the line was specific for different plants.

Tsao (2003) presented a range of transpiration rates for grasses and woody plants that often are chosen for the phytoremediation of contaminated groundwater. Grasses, such as buffalo grass, winter rye, and alfalfa, ranged from 0.02 to 0.55 in./day (0.5 to 14.1 mm/day), and for woody plants transpiration ranged from 0.3 to 100 gal/day/tree (1.1 to 378 L/day/tree). Nagler et al. (2003) and Wilcox et al. (2006) report that saltcedar, cottonwood, and willow used about 13.2 gal/day/tree (50 L/day/tree). These values do not discriminate whether the transpired water was precipitation, soil water, or groundwater, however.

Contrary to popular belief, transpiration by plants can occur during the night and when plants are dormant. There may be no leaves and, therefore, no stomata, but the stems contain lenticels that are open to the atmosphere from at least as deep into the tree as the phloem. Transpiration at night is possible if certain climatic and water conditions occur. Decker et al. (1962) reported for a stand of woody phreatophytes, that evapotranspiration measured during the night (8:30 p.m to 5:30 a.m.), was not zero but approached 11% of the daily evapotranspiration rate.

As stated in Chap. 2, transpiration can be estimated by a number of different methods. Ferro et al. (2003) suggested

$$T = ET^{*}\theta_{f}^{*}LA, \tag{3.9}$$

where T is the water use (volume per tree), ET is a reference evapotranspiration, θ_f is the water-use multiplier, and LA is the leaf area of the tree. This equation is for an individual tree but can be assessed at a forest scale by multiplying the value for an individual tree by the area planted. The use of this equation assumes no limitations on transpiration, such as stomatal closure and disease.

As we saw above, transpiration is affected by factors that determine the resistance of the diffusion of water from the leaf to the air. The water potential of the air, Ψ_{air}, is

$$\Psi_{air} = RT/V_w ln(RH), \tag{3.10}$$

where R is the gas constant, T is temperature (°K), V_w is the partial molar volume of liquid water, and RH is the relative humidity of the air, the fraction of current saturation relative to total saturation at a given air temperature:

$$RH = C_{wv}/C_{wv} \text{ at saturation}, \tag{3.11}$$

such that relative humidity ranges between 0 and 1 and, if multiplied by 100, provides the percentage of relative humidity.

For leaves to perform gas exchange, the relative humidity must be near 100%. As the air is often less than this value, a gradient exists for water vapor to exit the leaf. This continues as the wind removes saturated air, and drier air moves over the leaf. This is why houseplants have a hard time surviving the dormant period in heated homes unless water is added on a frequent basis.

From this it follows that the temperature of the leaf also becomes a factor that affects transpiration. As the leaf temperature increases, so does the temperature in the leaf air spaces. Transpiration occurs until the relative humidity of the leaf air increases to nearly 100%. This explains why, in arid areas, desert plants have fewer stomata and open them only during the night, and why, in humid areas, maximum transpiration occurs when the relative humidity of the air is lowest and little transpiration occurs when relative humidity of the air is higher, even though water may be available.

Photosynthesis is at a maximum on sunny days with relatively cool air temperatures, because the leaf temperature remains low. The converse also is true. Still, the driving force for water movement from the leaf to the air is the water-vapor deficit, or VPD, between the leaf and the air. As can be seen from the preceding discussion, the need for plants to open stomata to expose cells to atmospheric CO_2 also results in the loss of water vapor. This balance of water inputs and outputs can be described in terms of water-use efficiency, *WUE*, which is the ratio of net photosynthesis, in grams of CO_2 up taken per gram of transpired water. Desert plants have higher *WUE* than plants in humid settings.

The age of a tree also affects transpiration. Older trees tend to have a higher transpiration rate. Tsao (2003) reported that the transpiration rate for poplars increased with age; for trees 2, 5, and 30 years, the transpiration rates were a maximum of 10, 53, and 200 gal/day/tree, (37.8, 200, and 756 L/day/tree), respectively. Two-years-old cottonwoods had transpiration rates of 3.8 gal/day/tree (14.3 L/day/tree), and a 19-years-old cottonwood had a rate of 95 gal/day/tree (359 L/day/tree); for these studies, the source of the water transpired was not delineated. This trend tended to be observed for other woody plants as well. This observation may be explained by the increased root density of older plants, which generally are in contact with additional water sources.

The depth to groundwater also can affect transpiration. This is more important for plants that have shallow roots than deep roots when the water table is decreasing, and the reverse also is true if the water table rises. Gazal et al. (2006) investigated the transpiration of cottonwood trees growing along the San Pedro River in Arizona by using sap-flow methods. Trees growing in areas characterized by different depths to the water table were compared. Where streamflow was perennial, groundwater was shallower compared to deeper groundwater where the streamflow was intermittent. During drought conditions, the cottonwoods at the intermittent site underwent water stress with little increase in transpiration even though the VPD increased throughout the day. When recharge from precipitation occurred, however, the water table rose and these trees increased transpiration.

The hydraulic characteristics of the soil and aquifer sediments also affect transpiration, as is described in Chap. 8. The higher the soil porosity the more readily the soil is accessible by root growth, water infiltration, and air penetration. Hultline et al. (2006) hypothesized that plants growing in coarse soils where the water table declined rapidly would be more likely to experience xylem cavitation than plants growing in a more clay-rich soil where the water table might be less likely to fluctuate. The plants that grow in coarser soils will be affected more by drought conditions than those growing in more compacted soils.

3.5.5 Soil–Plant–Atmosphere Continuum

Plants represent an interface between water in the subsurface and water in the atmosphere. Surface-water bodies, such as streams and ponds, also link subsurface water with the atmosphere, but only at specific locations where groundwater is above land surface. Water used for transpiration and photosynthesis represents a more widespread transfer of water from the soil and subsurface reservoir to the atmospheric reservoir. Even during conditions of no flow, plant biomass represents a standing volume of stored water.

Soil contains at least three major components—solids, gases, and solutions. The solid is the material that composes the sediment, such as inorganic sands or organic material such as peat or lignin. The gases are from the mixing of the atmosphere in the soil pores spaces with gases generated abiologically and biologically in the soil itself, such as methane, CO_2, or H_2S. The solution is typically the water that contains dissolved nutrients and other elements.

The soil is an important part of the plant–water relation because soil is a source of nutrients and micronutrients and provides storage for water. We have seen how water in soil in contact with root hairs enters the plant and travels through the xylem to be evaporated at the leaf–air interface and driven by higher to lower water potentials, and how this removal of soil water into the atmosphere greatly influences the hydrologic cycle. In fact, this interaction of soil water with the atmosphere through the water-conducting vessels of vascular plants has been described generally as the Soil–Plant–Atmosphere Continuum (SPAC). As was discussed in the section on plant transpiration, an observable lag occurs between the removal of water from the leaves and the uptake of water by the roots, which leads to midday wilting in plants even when the soil contains water. This occurs because the plant resists water flow, much like a copper wire resists the flow of electrons or soil resists the flow of water.

Up to this point, we have discussed the plant and atmospheric components of the water balance of plants with little mention of the role that soils play, other than as a system to provide anchorage for the plant and a source of water and minerals for dissolution and uptake. After all, many flowering plants can be grown hydroponically, where the water is essentially a liquid culture. While certain terrestrial plants can be grown successfully solely in water containing the right concentrations of minerals, this situation occurs only commercially. The only exception is the common duckweed (*Lemna spp.*), which actually spends its entire life as a free-floating, hydroponic plant. We also have seen that individual disciplines have focused on water from the perspective of plant physiology, hydrogeology, and meteorology, which has led to various terms to describe the status of water. This was partly handled by the introduction of the integrating concept of water potential.

For plants to survive in soils, a balance is needed between the ability of the soil to retain water, transmit water, and to permit aeration. These somewhat mutually exclusive goals are described by the relative amounts of capillary and noncapillary pore spaces in a soil sample. Assuming that half the volume of soil is solid material, the balance is made up of pore spaces. If the space is less than 60 μm, water can be held against gravity by the capillary forces of adjacent water molecules as previously discussed in Chap. 2. This water usually is referred to as field capacity, or the water left after drainage caused by gravity. Pore spaces larger than 60 μm do not permit water retention, but although this may seem problematic it actually is essential to provide the introduction of air to respiring roots. Equal proportions of capillary and noncapillary pore spaces are best for water retention and aeration. As would be expected from just casual observations, the proportion of capillary to noncapillary pore spaces differs dramatically for different soil types. This is because of the difference in soil composition and texture.

Root hairs can take up water through diffusion and osmosis as long as the water potentials are lower inside the root hairs relative to the water potential of soil water. However, the continual intake of water to supply transpiration demands also is dependent, ultimately, on the supply of water in the soil as moisture, capillary fringe water, or groundwater. As would be expected, if the resupply of water to representative soil is eliminated, the water potentials of the soil decrease. For a given soil type, the water potentials can decrease during midday, and plant cells lose water pressure and wilt. This situation reverses at night when transpiration decreases as the atmospheric demand for water decreases, and the plant water potentials return to equilibrium with soil water potentials. If water potentials decrease to a maximum negative level, usually −1.5 MPa, however, the wilting can be permanent, even at night. The exact maximum water potential varies from species to species but also is dependent on the soil type.

The type and texture of soil where a plant grows is important and should be investigated as part of phytoremediation site-assessment activities, which are discussed in Chap. 6. For example, clay-rich soil may have higher water content than sandy soil, but less water availability, whereas the sandy soil may have less water content but more water availability. The availability of water to move though soils to roots is based on the hydraulic conductivity of the soil or sediment. The magnitude of this hydraulic conductivity is a function both of the soil type and given water potential. As soil dries out, the water potential becomes more negative as air replaces water in the soil pore spaces, which results in decreased hydraulic conductivity.

The concept of soil-moisture storage is a central concept to most plant physiologists. Even plants with shallow roots that rely on precipitation obtain water not from direct infiltration, because it is too much to use all at once, but from what remains in the soil after precipitation ceases. If the amount of water input is immediately lost, there is no soil-moisture storage at all. If the amount lost is less than the amount of water input, then storage can occur, similar to the relation presented in Chap. 2:

$$\text{Soil moisture storage } (SMS) = \text{inputs} - \text{outputs}$$

or

$$SMS = \text{Precipitation} - (\text{Evaporation} + \text{transpiration} + \text{runoff} + \text{drainage to groundwater}). \tag{3.12}$$

As seen in Eq. 3.12, drainage to groundwater is not considered a source of water to soil moisture, but represents a loss. In contrast, however, plants can use water from the capillary fringe and water table, not only soil moisture. This is crucial in the phytoremediation of contaminated groundwater, as we will see.

3.5.6 Radiation Balance

The energy used to type this sentence as well as to read it ultimately came from the sun. Fusion reactions transmit energy from the sun, in the form of light and heat, to the earth. This light energy is captured as chemical-bond energy in the structural parts of plants. This energy is released perhaps most obviously when the parts of a plant, like wood, are burned. The stored energy is released as light and heat as the chemical-bond energy is broken down in the presence of oxygen. At much slower rates, organisms that consume plant material use chemical-bond energy to drive their own metabolism and release nutrients back to the plants to support their growth.

3.5.7 Compensating Pressure Theory

The observation that embolism damage can be repaired and transpiration resumed (Canny 1997) places in doubt the Cohesion-Tension Theory of transpiration. Canny (1997) showed that breaks in the transpiration stream and subsequent repairs may happen to a plant several times a day. In fact, the very tension that Scholander et al. (1965) reported is regarded by Canny (1997) to be the compensating pressure applied to the embolized vessels.

3.5.8 Stems, Bark, and Lenticels

When one looks at plants closely, it becomes apparent that for most plants, all parts of their structure are derived from modified leaves, as was written about by the poet and naturalist Göethe. There are many similarities between the function of leaves and stems. Stems are the above-ground extension of the root neck that provides the support structure for the leaves. Stems contain the vascular tissues for sugar and water transport. However, like leaves, some stems contain chlorophyll and can undergo gas exchange and photosynthesis. For example, cacti and other succulents carry out photosynthesis not in their leaves, which have been reduced to spiny projections, but in their stems, as a consequence of low water availability and high vapor pressure deficits. In some cases, more food is produced than consumed in some herbaceous plants with green stems, such as sugar cane.

Additional gas exchange can occur from the atmosphere to the cambium through the lenticels (Fig. 3.16). Here, oxygen can diffuse in and drive cellular respiration, especially in the growing phloem and cambium. Lenticels also have been shown to be the source of adventitious root development in some phreatophytes (Ginzburg 1967). If the stems of such plants are exposed to an increase in water content, such as submersion during flooding or increased air humidity, the parenchymal cells fill with water and expand and then grow and divide, sometimes producing multiple roots from one lenticel. It is the genes of leaf cells that give rise to other types of cells to form other tissues.

Branches are secondary stems that depart from the main stem. These secondary stems produce additional branches and contain leaves and buds. This growth is called ramification. If the main stem continues its upward growth throughout the plant's life and lateral stems grow from it, this is termed monopodial ramification. This is characteristic of most conifers. On the other hand, deciduous trees tend to stop growth in the main stem, and the lateral stem growth predominates. This is called sympodial ramification.

Bark is the outermost layer of cells of woody plants. It is produced by the cork cambium beneath it, and it grows outward and then dies. Bark consists of non-living (no

Fig. 3.16 Lenticels on a 3-years old hybrid poplar tree at a phytoremediation site near Charleston, SC (Photograph by author).

metabolism) cells that are filled with lipids. Crevices created by the expansion of the underlying and growing cambium are used to supply oxygen to the growing and respiring cells through the lenticels. One of the purposes of the bark is to protect the growing cambium from environmental assault from the outside and to prevent the loss of sap and water from the inside. It tends to be more waterproof than water permeable. One example of the excellent waterproofing traits of bark is provided by the birch tree, which was used for thousands of years by the native people of North America in the construction of canoes. The bark typically was gathered during the summer and, after removal from the tree, was placed inside out on the canoe's frame.

Following winter leaf drop, the effect of stems on transpiration can be substantial. This is caused by the lenticels. Much as humans loose their outer layer of skin cells periodically, so do trees loose their outer layer of cork. The lenticels allow oxygen to reach the cells in the growing and, therefore, respiring layer of cambium cells beneath the bark and cork, mainly as diffusive transport through the intercellular space in the cortex. Because of the presence of the continuous cambial sheath, however, it is unlikely that gas exchange occurs between the outer phloem and inner xylem by way of the lenticels. The cork cambium cells that do not contain suberin can extend through the periderm to the atmosphere. They are similar to the stomata in that they are sites of gas exchange. After all, plants are aerobic organisms and require oxygen in order to respire the food they make, and this process occurs in all cells. Many plant species that grow in swamps or other areas inundated by standing water and anoxic conditions maintain an oxic halo around their roots in the rhizosphere by transporting oxygen from the lenticels to the roots (Hook et al. 1970).

Lenticels are necessary in order to supply the living cells with oxygen to supply respiration of previously fixed CO_2. This is important during times of no light or for parts of the plant that do not contain chlorophyll. Bark is not impermeable, but it is too thick to permit diffusion of oxygen into the phloem and cambium beneath it; at a 21% oxygen concentration in the atmosphere, oxygen could diffuse less than a millimeter into the bark. This is why pores are necessary to facilitate oxygen diffusion, at least to balance the oxygen consumed by cellular respiration. This also is why oxygen availability below ground in the root zone is important, and why waterlogged soils that do not dry out can result in the death of plants that are not adapted to these conditions of oxygen limitation.

There are at least two different types of lenticels—transverse and longitudinal—based on the orientation of the lenticel opening along the stem. Longitudinal lenticels are more involved in gas exchange than the transverse lenticels. Under flooded conditions, the intercellular space of the layer of cells beneath the lenticels, or phellogen, increases to facilitate the diffusion of gas. The phellogen look like raised white bumps on the outer bark. Lenticels also affect transpiration by being a portal of water-vapor escape (Kozlowski and Pallardy 1997). During the summer when leaves are out, most water vapor is eliminated through the stomata, as leaves offer a much lower resistance to water loss. After leaf drop, however, the resistance to water-vapor elimination through the lenticels offers a lower resistance to water movement.

Stems are a variable storage compartment of water but a continual storage compartment of carbon (Chiou et al. 2001). For example, chemicals that have a greater octanol: water partition coefficient, K_{ow}, in a plant will be sorbed onto tissue rather than become transpired; this is described further in Part III.

3.5.9 Leaves

The general shape of many plant leaves reflect a compromise between light and CO_2 capture and water loss. All leaves can restrict short-term daily water limitation by stomatal closure. Longer term seasonal water restrictions, however, are not always handled in this manner. Deciduous trees, for example, shed their leaves in response to drier conditions. Evergreens retain their leaves longer before shedding, because they generally have fewer stomata, less leaf surface area, and lower rates of water usage. As a tradeoff, however, they tend to grow at slower rates than deciduous trees even when water is not limiting.

The structure of leaves has to accommodate a variety of processes; light capture, capture of CO_2, formation of food, transportation of food from the leaf, and be a site of water exchange. These jobs are done by three primary tissues—the epidermis, the mesophyll, and the vascular system. As in the case of most epidermal layers of the plant, the epidermal layer of leaves is one-cell thick and the outermost layer of the leaf surface, save for the presence of wax on the cuticle of the uppermost leaf surface. The lower epidermal layer is broken up by the presence of the stomata. The guard cells contain the chloroplasts. The mesophyll cells are sandwiched in between the upper and lower epidermal layers.

The vascular system is contained in the veins, which probably is one of the more recognizable parts of a leaf. Veins can run throughout a leaf. The narrow leaves of grasses have veins that run parallel to the direction of the leaf. Most other plants have a strong central vein fed by numerous smaller secondary veins that spread out to the leaf edge, and these are interconnected by smaller veins. The central vein is attached to the petiole, which provides connection to the stem or branch. These veins are locations where the xylem and phloem are interconnected.

As the leaves form from the growing meristematic cells, there is an overall shape to grow forward as well as cells that will become the upper part of the leaf and cells that will become the lower part of the leaf. Moreover, certain cells will become the vascular tissues, some will become the palisade tissue near the top surface, and some will become the mesophyl near the lower surface. After most of this cell division occurs, the leaf cells then can expand to the size of their potential, without much additional division. Winter buds, for example, contain encapsulated small leaves and even flowers. Cellular division and growth have occurred already even before the bud opens. All that then remains is the expansion of these tissues by the uptake of water.

The function of leaves as sites of gas exchange is analogous to mammalian lungs or fish gills, where O_2 is taken up and CO_2 is released into fluids, such as air or water, respectively. With lungs, every cell requires a fresh supply of oxygen and must rid itself of CO_2, but neither can diffuse into or out of the cell to sources in the atmosphere. That is the job of the circulatory system, which is the site of cell-to-cell gas exchange with fluids and lungs where the gas exchange with the atmosphere occurs. The mesophyll cells of plants are in contact with air over 90% of their surface area. This increases the diffusion of CO_2 into the leaves.

Water from the liquid phase is transferred to the air in a vapor phase by diffusion. In general, an amount of water equivalent to a leaf's fresh weight is vaporized every 20 min. on a sunny day (Canny 1990). Leaves resist water loss in at least three ways—the resistance to water-vapor loss in the intracellular air spaces, *IAS*, connected to the stomata, r_{IAS}, resistance to water loss by the aperture of the stomata, r_{ST}, (previously discussed), and even after water vapor escapes, it enters the layer of saturated air, called the boundary layer, next to the leaf surface, r_{BL}. The total resistance to water vapor loss from the leaf, l_T, is

$$l_T = r_{IAS} + r_{ST} + r_{BL}. \tag{3.13}$$

In contrast to leaf resistance, the flow of water vapor can be viewed in terms of the leaf conductance of water, which is simply the inverse of the resistance, where leaf conductance is 1/leaf resistance. The units of measurement are L/T (cm/s). Leaf conductance increases with the amount of light available and with higher, less negative water potentials in the leaves, but decreases when CO_2 concentrations are elevated or the vapor pressure deficit is high. In this way, leaves can be viewed as nozzles adjusted to such a fine spray that only water vapor emanates from the openings.

Not only does the flat outer structure of leaves suggest its role as the site of gas exchange, but so does the inner structure. The cross section of a leaf reveals that while its outer surface is protected from excessive water loss or other liquid entry by the waxy cuticle, the side or sides with stomata contains more void space than cells; this is the place for gas exchange, with CO_2 dissolving in water along a concentration gradient and water evaporating along a concentration gradient in response to the VPD.

The air space within leaves of various plants has been estimated. In pine needles of conifers, air spaces represent about 5% of the total volume of leaf tissue, and this tends to increase with leaf size, such that corn is 10% and tobacco is 40%. Much like our lungs, the total surface area of these pore surfaces may exceed the external leaf surface area by up to 30% to ensure that gas exchange occurs.

Leaves often are covered with small projections of tissue that resemble animal hair. These tissues, or trichomes (from the Greek *trichome*, meaning growth of hair) are often found on the underside of leaves. Each trichome is an individual cell that arose from the meristem, and its rise above the leaf surface is indicative of its rigid cell wall. Some trap air and act to insulate the leaf from extreme drops in temperature. On the other hand, trichomes can reduce the influence of solar radiation and water loss by transpiration. Plants that have trichomes have cells that contain the genes for such structures, and this gene is absent in plants that do not contain trichomes. Specialized leaf hairs also can directly control the water status of plants. Such hairs that can release water are called hydrathodes, and this occurs so that excess water can be removed from the plant.

Close inspection of many woody plants reveals that a characteristic angle of departure of leaves, as well as branches, is repeated throughout the plant. The angle is the same for the lateral branches from the main trunk and from the branches coming off the lateral branch and finally the petioles from the branches. This is done by the plant to maximize the amount of light and air available to all leaves of the entire plant. This form is accomplished by the reaction wood cells that arise from the live xylem. This can be seen in examining the cross section of a piece of fallen limb; there will be more wood, or xylem, on the underside of the branch than on the top side. This is to provide structural support and, in a way, acts as a cantilever.

The position of leaves with respect to each other also is not random. Branches from the main trunk can be depicted as lines coming from a circle, which would be the cross section of the tree trunk. As you go up the trunk, successive branches come off the circle at the same angle, be it 90° or 120° or another angle (Fig. 3.17). This succession of trunk to branch is continued from branch to twig and from twig to leaf petiole. Leaves can be arranged in an opposite pattern, as is found in the maples, such as *Acer rubrum*. The arrangement of leaves on a single stem or branch also maximizes exposure to light; most leaves radiate outward from their nearest neighbor.

Even the arrangement of stems from the main stem is at a fixed angle and is designed to decrease the interference from other stems for light capture. In fact, the angle between branches and the main stem, between leaf petioles and branches, and between small branches and large branches tends to be the same for each species, a mathematical

Fig. 3.17 The three branches of this small fern radiate from the main stem at 120° (Photograph by author).

solution to the issue of light capture that needed to be solved as plants went from water to land.

When leaves are arranged in a spiral in order to not shade adjacent leaves, the pattern follows a geometric series called the Fibonacci series, or numbers, after Leonardo of Pisa, also known as Leonardo Fibonacci (1175–?), a mathematician from the thirteenth century. In 1202, Fibonacci was interested in the potential reproductive prowess of rabbits. Rabbits can mate at the age of 1 month. If a male and female rabbit are placed together, and the female gives birth to only two babies, one male and one female, at the end of the first month there will be one pair, the original rabbits. At the end of the second month there will now be two pairs, the original pair plus the new pair (the two offspring), and at the end of the third month, there will be three pairs, and so on. Although these reproductive assumptions are rarely met, Fibonacci realized that this progression had an interesting numerical order of 0, 1, 2, 3, 5, 8, 13, 21, 34, 55, 89, 144, 235, and so on, for each number is the sum of the two proceeding numbers. Such sequences are seen across many natural and physical sciences.

The manner in which some plants produce lateral shoots from the main stem also can follow a Fibonacci series. For instance, examination of the number of lateral branches from the main stem of a tree above ground can follow the pattern of one, and then another branch above this one, so that a horizontal line drawn through this node will intersect it and the previous branch, so now there are two, and then a line drawn though this reveals three branches, and so on through the next branch, five, and then the next, eight. This is in the vertical plane, and there also is a sequence of growth that follows the Fibonacci numbers for the lateral plane around the main stem, expressed as the arrangement of branches, or leaves for herbaceous plants around the main stem. To arrive at this conclusion, the leaves are counted from the bottom, and the number of leaves per complete 360° revolution around the plant is counted. For example, if branches, twigs, and leaves can be arranged at 180°, 120°, and 144° intervals, this also is ½ (1 revolution = 2 leaves), 1/3 (one revolution = 3 leaves), and 2/5 (two revolutions = 5 leaves), and so on. Therefore, these spirals all have a uniform ratio regardless of the plant type. This is believed to be a result of the natural selection of branch and leaf arrangement to maximize the leaf exposure to sunlight. The study of this relation is called phyllotaxis, and even though most plant species grow in some manner following these patterns, it is not a universal phenomenon and can change for each plant with respect to environmental variables.

What do Fibonacci numbers have to do with the plants selected for phytoremediation purposes? The most common arrangement of branches or leaves from a stem is the angle 137.5° (Fig. 3.18). This angle provides maximum sunlight exposure to the leaves as well as downward penetration of light. If this angle is changed only slightly, the amount of leaves that can be packed most efficiently into the smallest area or space will be decreased. The arrangement of leaves for maximum sunlight exposure is related to the transpiration rate of plants and, therefore, water use. The plants that have higher transpiration rates have more leaves, as would be expected. Both poplar and willow trees are characterized by high *ET* rates and often are selected for use in the phytoremediation of contaminated groundwater. These trees have 8 leaves per 3 revolutions, and some willows have 13 leaves per 5 revolutions. Again, this arrangement helps explain why the trees are found where the water table is shallow or where water is not limiting; the leaf surface area, therefore, does not become a factor in growth and reproduction.

Fig. 3.18 The repetition of a common departure angle (137.5°) that most plants exhibit is in response to the most efficient exposure to sunlight and, therefore, food and energy production.

Leaves of plants differ in many respects, such as shape, but the largest classification criterion is the timing of leaf drop. Deciduous (from the Latin *decidu-*, for falling off) trees make and drop their leaves in one growing season, whereas evergreen trees hold onto their leaves for multiple growing seasons. Contrary to popular belief, not all evergreens are coniferous plants. Live oaks (*Quercus spp.*), wax myrtle, ligustrum, and many hollies are not coniferous but retain their leaves. Like the conifers, these plants tend to be found in drier soils, so once leaves are produced, dropping them and regrowing new ones each year would be too costly. Also, conifers are dominant in colder areas and maintain their leaves because they have a very small surface area, are covered by a thick cuticle, and are insulated by wax. The needles, or leaves, of spruce trees go so far as to dehydrate their leaves to reduce freezing. Deciduous trees are more likely to be found in moist soils, but water availability is reduced when it is bound up as ice or snow. Hence, leaf drop occurs as water availability declines.

Leaf drop also reduces evaporation from the upper layers of the soil and provides organic matter that can be reused by the plant after recycling by detritovores. One of the fastest rates of such recycling is exhibited in the tropical rainforests, where high temperatures, high precipitation amounts, and poor soils cause a rapid recycling rate. Conversely, the needle shape and thick bark of most conifers are attributes that enable them to live in dry or cold climates.

Earlier in this chapter the interrelations among plant roots and bacteria and fungi were discussed. There also is a relation between bacteria and the leaves of some plants. For example, bacteria are present in the fluted margins of the leaves of *Ardisia crispa*. These bacteria are heterotrophic rather than photosynthetic and live in the apoplast area between the cells. It is believed that they provide growth-promoting substances because if the plants are heat treated to kill the bacteria, the plants survive but grow at a reduced rate with less vigor (Margulis and Sagan 2002). There also are cyanobacteria (*Nostoc spp.*) that inhabit the leaves of *Gunnera manicata* at higher elevations in Ecuador. Inside the leaves, cyanobacteria reduce atmospheric nitrogen to nitrate for plant uptake.

Many phreatophytic plants, either obligate or facultative, have large leaf surface areas. This can be accomplished by having fewer but larger leaves, as is characteristic of *Populus*, or by having many smaller leaves, as is characteristic of *Betula* or *Salix*. In both cases, the high rate of growth of these plants requires the additional carbon to be supplied by more fixation of CO_2 in the leaves, more water as a reactant in photosynthesis, and increased gas exchange. It is this total volume of tissue that undergoes gas exchange that ultimately determines the size of a particular plant. As a plant grows larger, the surface area must increase as a square function. For example, doubling the radius of a sphere from 2 to 4 in. (5–10 cm) will result in a four-fold increase in surface area and an eight-fold increase in volume.

A close analogy to the role that leaves play is given by the solar panel. Both capture solar energy and convert it to other uses (Fig. 3.19). Leaves absorb the sun's electromagnetic radiation when it impinges on chlorophyll in chloroplasts and the exited electrons are used to produce ATP from ADP, essentially electrical energy. Solar panels take the same solar radiation energy and convert it into electrical energy; for example, a 4 ft^2 (0.37 m^2) solar panel can produce about 35 W of energy. Essentially, the sun is the source of electrons and both leaves and solar panels distill these electrons. Much like leaves, solar panels are thin and have a large surface area. Solar panels consist of thin sheets of semiconductor material, such as silicon, to which impurities,

Fig. 3.19 A leaf and a solar panel essentially perform the same task; harnessing the sun's energy for storage in a more useable form. Here, the solar panel provides power to a battery that runs lights for a parking lot at night. The tree, a palo verde (*Cercidium microphyllum*) near Tucson, Arizona, uses the sun to generate ATP to support life (Photograph by author).

such as phosphate, have been added. The semiconductor material is connected to metal contacts below the silicon. Solar radiation through photons of light strike this material and some of the energy is absorbed by the semiconductor, which knock loose electrons from the silicon-phosphate semiconductor, and this causes electron flow; this is why solar panels are called photovoltaic cells. To direct this flow into a current, electrical fields are used. The efficiency of a typical solar panel is about 15%. Often, the sunlight energy so obtained during the day is collected and stored in a battery for use at night.

The local amount of sunlight across the United States varies with latitude. Typically, there are about 4 h of peak sunlight from Washington to Maine, and this number increases farther south, which has 4.5–5 h of peak sunlight. Maps of average solar radiation, called insolation maps, can be used for a particular area and are discussed as part of site-assessment activities presented in Chap. 6.

3.5.10 Vacuoles

Vacuoles serve many purposes for plant and water relations as previously described, but also can be considered to be the cellular dumping ground of the plant. As described in

Chap. 11, vacuoles aid in the survival of a plant when exposed to changes in the geochemistry of the soil or water it encounters. For example, although calcium is necessary for plant growth, excess calcium can be eliminated as calcium oxalate in specialized vacuoles called crystal idioblasts. Calcium oxalate crystals in plants were observed as early as the seventeenth century, by Antonie van Leeuwenhoek. Calcium oxalate crystals are perhaps best known by people afflicted with kidney stones, which are hard crystals of calcium oxalate. The source is often a diet rich in plants, such as spinach, which contains these minerals in the vacuoles. The formation of these crystals is thought to impart a wide range of advantages to a plant, from defense against herbivores, by a form of crystals called raphides, which are discussed in Chap. 11, to calcium regulation and structural support.

3.5.11 Plant Life Forms

Even casual observation reveals the role that the location and environment have on predominant plant life, distribution, and diversity. The establishment of monocultures across a wide area is the exception, such as *Spartina* or *Mangrove* or a cultivated field, rather than the rule. But even in an area of varied plants, certain distributions occur. The distribution can be one affected by space (location) and time (season). For example, some trees in the same forest lose leaves in the fall and some do not. One of the earliest systems of classifying this distribution of plants was performed by the Danish botanist Christen Raunkiaer in 1903, and is based on the occurrence and position of buds relative to ground surface. It contains five different classifications:

1. *Phanaerophytes*, where the buds are 9.8 in. (25 cm) above ground surface and include most trees, shrubs, and vines
2. *Chamaephytes*, where the buds are closer to the ground, below 9.8 in. (25 cm), and include the herbaceous and some woody plants
3. *Hemicryptophytes*, where the top growth dies but a bud persists at or below the ground surface, such as grasses
4. *Cryptophytes*, where the bud is beneath the ground surface or water; and
5. *Therophytes*, where the bud is not present but the plant persists by seeds.

The main variable here is the extent of protection that different plants offer to the buds to ensure survival of the next generation. This is affected directly by the environment and climate of an area. In humid areas that do not have frost, for example, the plant communities are dominated by Phanaerophytes. Conversely, desert areas of limited moisture typically discourage such plants in favor of Therophytes. Although this classification scheme is not widely used, it is included here because it indicates the importance of the interaction between plants and sources of water.

3.5.12 Dew

Water found on plant leaves early on a cool morning after a previously hot, humid day is called dew. As the ground cools faster than evaporation occurs, water in the atmosphere condenses. One source of this water can be from plant transpiration during the previous day or night. Although the formation of dew recaptures some of the water that would have left by evaporation, it is a source of water that is typically important only in very arid climates. Lichens, however, use dew as a water source in most climates.

3.6 Plant Water Status

Water can compose up to 90% of most plants. To briefly summarize what was discussed previously, water is used as the solvent in the cell cytoplasm, the source of hydrogen in the carbohydrates made by photosynthesis, a vector of entry into cells and movement throughout the plant, and provides support and cell elongation. Water entry into cells by osmosis occurs when the concentration of water inside the plant cell is lower than the concentration of water outside the cell. Because the osmotic potential of cells varies, an alternative measurement of plant water status is desirable; this parameter is called water potential.

3.6.1 Water Potential

A Sisyphean task is one in which just as the task is about to be completed, everything falls back to the level at which the task was started. Its derivation is from the Greek myth where Sisyphus, as part of his punishment by the god Zeus, was tasked with pushing a heavy boulder up a hill, only to have it roll back down the hill as he approached the top.

Although it may be a useful metaphor for some aspects of life, it also is a useful way to describe the energy contained in matter based on its relative position. In the example above, the elevation of the hill provided the potential energy for the boulder, which was turned into kinetic energy as the boulder rolled downhill. Water can assume similar potential energy in an elevated storage tank which is released as kinetic energy when a garden sprinkler or water fountain is turned on.

The first law of thermodynamics states that energy may be transferred but the sum remains constant or is conserved: energy can be changed but not destroyed. An excellent example is the internal combustion engine, which takes chemical energy from liquid fuel, converts it into mechanical energy, which is transferred to the coolant and then radiator as heat energy, which transfers it to the atmosphere. Moreover, sunlight hitting a paved parking lot gets transferred into heat, whereas the same sunlight hitting the leaf of a plant is, in part, stored as energy in compounds made by the plant.

The second law of thermodynamics states that the disorder of a system, or entropy, always increases or entropy is not conserved. For example, using photosynthesis, converting simple molecules, like CO_2 and H_2O, into more complex organic molecules may decrease the entropy of the plant initially, going from disorder to order, but this is more than balanced by the increase in entropy that results from the input of solar energy.

The exchange of energy from one compartment to another can take two forms—heat and work. In the example of the internal combustion engine, the chemical energy in the fuel is exchanged into the mechanical energy of the pistons moving in the cylinders to turn a crankshaft. Hence, the fuel energy is converted into both heat and work. Work can be defined as force multiplied by the distance covered by the force or against a resisting force.

Water also contains a certain level of energy inherent to its atoms and bonds of its molecular structure. Water molecules in motion can exchange this energy with their surroundings, such as a cell wall. The speed of molecular motion is what determines the magnitude of its energy. At normal temperatures and pressures, the ability of molecular interaction to occur spontaneously is referred to as Gibbs free energy, after work done by J.W. Gibbs in the 1930s (Kramer and Boyer 1995). This free energy is proportional to the number of molecules on a free energy per mole basis and can be called chemical-free energy, or chemical potential, u. This potential is essentially a measure of the tendency of a chemical to undergo transformation, or the potential energy state of a chemical. Water, for example, can be evaporated, advected, or diffused.

Because u is not an absolute, an interesting aspect of chemical potential is the difference between the u of an initial and final state. If the difference is negative, the change occurs; if the difference is positive, or zero, no change occurs. Zero chemical potential represents the system at equilibrium. For example, the conversion of liquid water into gaseous hydrogen and oxygen does not occur spontaneously because the change in u is positive. But how then, can we explain the similar conversion of liquid water into gaseous oxygen during photosynthesis? Although the hydrogen is used to reduce CO_2, the primary driving force to overcome the increase in u is provided by solar energy.

The chemical potential, or free energy, of water that arises from the random movement of water molecules in contact with each other will change if temperature or pressure changes. As might be expected, boiling water that releases water vapor causes the chemical potential to increase because of the increased force at which water molecules collide with each other. The same conversion process that turns water to vapor also occurs in evaporation and transpiration, although the chemical potential is lower because of lower temperatures. Similarly, increased pressure also causes the chemical potential of water to increase as if the temperature has increased, because the water molecules are so tightly packed together—this is why squeezing an ice cube can cause it to rapidly melt.

For a sample of pure liquid water at 1 atm (1 bar) and 20°C, the reference standard for u is arbitrarily assigned as $u_o = 0$ Megapascals (0 MPa). This value is the highest water potential possible, by definition, so all other potentials will be a negative number less than zero, because differences in water potential are more useful than actual values. The more negative a value of u is for water in soil, for example, the lower is the energy state and more energy is required to remove water from soil, where the energy ultimately is derived from the sun.

The unknown u, u_w, of water can be compared to the standard,

$$u_w - u_o. \tag{3.14}$$

When u_w is pure water,

$$u_w - u_o = 0. \tag{3.15}$$

As previously discussed, the chemical potential of water can be increased by increasing temperature and pressure. On the other hand, the chemical potential of water can be decreased by the reverse—decreasing temperature and pressure—or by adding a solute to the water to form a solution. When u_w is not pure water, as is the case when a solute is added and the number of molecules of water decreases, or, the presence of the solute molecules, which often are larger than the water molecules, interferes with water molecule movement, the end result is that the chemical potential of water, u_w, decreases

$$u_w - u_o = \text{ a negative number.} \tag{3.16}$$

The negative number indicates that the movement of water would occur spontaneously from a less negative (water with less solute), to a more negative number (water with more solute).

The units of chemical potential are energy units/mole, or Joules/mole. To convert to pressure, in the case of water, the chemical potential can be converted to a water potential, ψ, by normalizing the difference in chemical potential by the partial molar volume of water (V, in m^3/mole), or

$\psi = (u_w - u_o)/V$, in Joules/m^3 = Newtons/m^2 = Pa (pressure). If pure water is placed in contact with a water-solute solution of less chemical potential, it is possible to measure the chemical potential of the water or the solute molecules. The potential of water, however, typically is investigated.

Water potential, ψ, is the force of water in a system and the ability of water to perform work, or in the case that we are interested in, the potential for water to enter or exit a plant cell. In other words, water potential is the potential energy of the water per unit mass of water. Water moves from higher ψ to lower ψ. Similar to the example of the boulder moving freely from an area of high potential to lower potential, the flow of water in a plant is from areas of high water potential, or negative ψ, to areas of lower water potential, or more negative ψ. The advantage offered by using water potential as a measurement of the water status of plants is that it is based on physical, rather than biological, reference standards.

Water potential is composed of many other factors or subcomponents that affect the movement of water in soils, such as presence or absence of solutes, called the osmotic potential, ψ_o; the weight of the water present, or pressure potential, ψ_p; the affect of gravity on the water, or gravitational potential, ψ_g; and the extent of water adhesion to the soil by surface tension forces and hydrogen bonding, called the matric potential, ψ_m:

$$\psi = \psi_{pie} + \psi_p + \psi_g \qquad (3.17)$$

Figure 3.20 represents water potential and how it relates to the potential for water to flow from soil to plant to the atmosphere.

As shown in Fig. 3.20, the flow of water is from the least negative water potentials, or the wettest soils (but not groundwater, because it is at atmospheric pressures rather than under tension), to the area of most negative water potential, the area of least water, such as the plant leaves. Plants, in general, have negative water potentials because the amount of free energy in the plants is less than free water under the same conditions. When transpiration is not occurring, the standing water potential as affected by gravity should equate to about −0.01 MPa/m increase in height (Woodruff et al. 2004). In other words, a 10-m tall tree would have leaf water potential 0.1 MPa more negative than the lower leaves. The water potential from the plant's perspective is the measure of energy in each component of water transport, from outside in the subsurface, to inside the plant, and into the atmosphere. A decrease in water availability results in even more negative water potentials.

Because water potential can be used to examine water flow, there is a relation between soil water potential, ψ, and soil hydraulic conductivity, K. As the soil water potential in

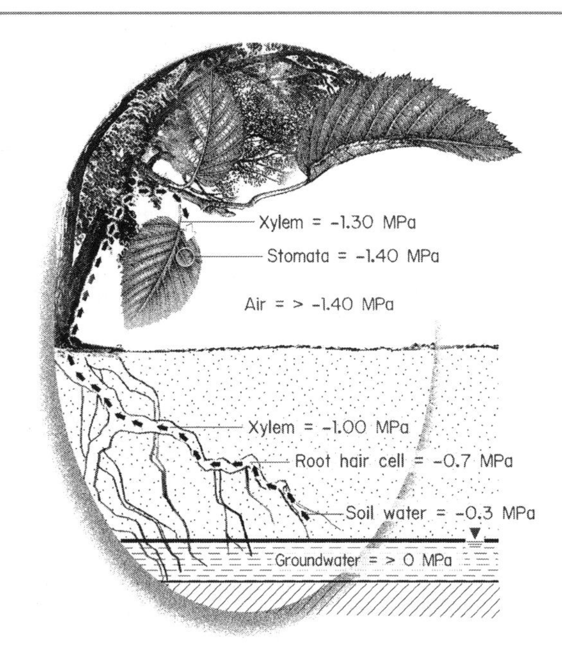

Fig. 3.20 Typical water potential (ψ, in MPa) along the flow of water from the subsurface to the atmosphere (Modified from Kozlowski and Pallardy 1997). Plants that tap groundwater do not have to overcome the stronger tensions in unsaturated soil.

the unsaturated zone becomes more negative and approaches field capacity, the hydraulic conductivity decreases as more air is added to the system, which increases the soil tortuosity encountered by the water. Conversely, the hydraulic conductivity of saturated sediments does not change, as long as the sediments remain saturated. This fact provides an additional advantage to facultative phreatophytes when soil moisture decreases, as flow is not impeded under saturated conditions.

This status of water potential affects the rate of photosynthesis and, therefore, transpiration. Too-negative water potentials cause stomata to close in response to increased resistance, which reduces transpiration and photosynthesis. Under conditions of unlimited water, however, turgid cells that have positive ψ and overcome the negative ψ_o can lead to near-zero values for total water potential. This is because the pressure potential, ψ_p, represents the hydrostatic pressure in the system, which becomes negative as turgor is lost during transpiration.

Relative humidity and water potential also are related and have an effect on plant water status. The water potential of the atmosphere is strongly negative throughout most ranges of relative humidity, and water is lost from less-negative water potentials in the leaves and to the air. When relative humidity approaches 100%, however, the water potential of the air decreases rapidly, becomes less negative, and approaches zero. Therefore, little water loss by transpiration occurs even though the stomata are open. This fact supports what drives transpiration—the large difference in water potential between the soil and the atmosphere.

3.6.2 Measurement of Water Potential

As important as water potential is, it would be meaningless as an indicator of plant water status if it could not be measured readily. Fortunately, water potentials can be measured using instruments that measure pressures or tensions, such as psychrometers, hygrometers, and tensiometers. In most cases, a device that measures pressure used to stop the movement of water, the hydrostatic pressure, is used to measure water potentials.

With the psychrometer method, a piece of plant material of unknown water potential is placed in a sealed chamber that also contains a droplet of a solution of known water potential. As the name of the method implies, if the plant material has a lower water potential than the droplet of solution and, hence, a lower vapor pressure by way of a higher solute concentration, preferential evaporation from the droplet cools its surface. Conversely, if the droplet of solution has a lower water potential than a less-concentrated plant-material sample, the sample will evaporate and warm the droplet. Hence, if the water potential of a particular solution is known and it results in no net movement of water to cool or warm the droplet, the sample of plant material must have the same water potential. Because a change in temperature can cause a change in water potential (for example a change of 0.01°C = 0.1 MPa, 0.1 MPa = 1 bar, 1 bar = 14.5 psi), the chamber must be kept at constant temperature and, therefore, is primarily a laboratory method. This method has been used extensively by Boyer and Knipling (1965).

Another method involves placing a piece of plant material in a chamber that can be pressurized in order to restore the distribution of water potential between living and nonliving xylem cells in the plant material. Because the act of taking a biopsy of plant material releases tension when the water column in the xylem is broken, water initially flows into the living cells by osmosis. Pressurization of the chamber, however, can reverse this flow of water back to the xylem. This method has been used since the mid-1960s, after widespread use by P.F. Scholander and others (1965), and can be used in the field. The main advantage to water potential measurements with the pressure chamber approach is that it incorporates the linkage of the atmospheric demand for water, the soil's or sediment's supply of water, and the plant reaction to both. Another method applies pressure to the whole plant rather than to tissue samples, for example: the branch or trunk is cut, sealed off, and pressure is applied. The amount of pressure applied is related to the water-flow characteristics of the plant.

Because the components that compose water potential can vary, the total plant water potential is not a constant. Reference times for collecting water potential, however, have been established and include a daily low at dawn (pre-dawn) when the plant roots and soil moisture are in equilibrium and a daily high at noon. A plant sample, such as a leaf, is taken by cleanly cutting it with a sharp knife. Because the water in the xylem is always under tension, or negative water potential, no water will flow from the cut; however, sap can flow because it is under positive pressure in the phloem. To exude the water and to record the pressure, the sample is placed in the chamber but with the cut end outside at ambient pressure. When the water is exuded under pressure, it equates to the plant's original tension. This can be done in the field with portable instruments.

The pre-dawn water potential of plants is made under conditions of closed stomata and, hence, the plant water potential will be equal to the soil water potential. In the afternoon, however, stomata are open and water flows from the soil to the atmosphere through the plant, and measurement of the water potential of the plant is an indicator of water demand in the air. Phreatophytes tend to have a more constant supply of water and, therefore, have a higher and more consistent pre-dawn water-potential value over time (Snyder and Williams 2000).

The turgor pressure of individual cells can be directly assessed by using a pressure probe. An air-filled glass tube sealed at only one end can be inserted into a cell. The pressure in the cell compresses the gas in the glass tube, and the pressure calculated using the ideal gas law. The hydrostatic pressure of individual cells also can be measured by using a similar approach in which a glass microtube is filled with incompressible oil. This oil can be readily distinguished from the sap that flows into the tube, and the sap flow can be offset by depressing a plunger, which can indicate the hydrostatic pressure in the cell.

Other field instruments include tensiometers that can be installed directly in the field to measure the water potential of the water in the soil near plant roots. A tensiometer consists of a tube with a porous ceramic cup attached to one end. It is filled with water and capped with rubber septa prior to installation in the field. If the soil is drier than the water-filled porous cup, water will flow out of the tensiometer, and the change in pressure in the headspace above the water level in the tube can be measured with a transducer.

Why go to all this trouble to define water potential? Why doesn't a simple measurement of just the water content of soil, or even the percentage of soil moisture, suffice? This is because soil water content and moisture percentage can indicate the relative difference in water amounts between two soil samples, but neither measurement can indicate in which direction water will flow. For example, the water content of soil can be the same as the water content of roots, but no indication of flow direction is suggested. Moreover, some water is tightly bound to soil particles so a high soil moisture content does not necessarily indicate that the water is bioavailable. The percentage of soil moisture at least can indicate whether or not water is present, even though it may not be bioavailable. This is the main reason why tensiometers are more useful and can rapidly indicate if water is at tensions near to, equal, or greater than the wilting point.

Water potentials can be used to estimate the velocity of water flow in the xylem. The velocity, W_v, of water moving upward through xylem of a constant radius, r, driven by a difference in water potentials, $\Delta\psi$, measured between two elevations that are separated by distance, Δx, can be estimated by this version of Poiseuille's equation:

$$W_v = ((r)2/8(\text{viscosity}))(\Delta\psi/\Delta x) \tag{3.18}$$

Water velocities between 1 and 45 m/h have been reported; the low rates are for plants with small-diameter xylem, and the high rates are for plants with large-diameter xylem.

3.7 Summary

From the landward transition of the earliest single-cell blue-green algae to the complex multicellular land plants used for the phytoremediation of contaminated groundwater, the requirement for water to sustain life remains a common denominator. Whereas the single-cell plant could float freely about in the upper layers of surface water and be exposed to sunlight, the multicellular plants had to develop a more complex anatomical adaptation to secure the plant's needs for water—the root system. Although underground, many plants have extensive lateral roots that take in precipitation. Even many of these plants, however, have tap roots that go deeper into the subsurface to collect the more perennial supply of groundwater in the event that precipitation is infrequent or soils are porous. Hence, even though plants have moved beyond being completely immersed in surface water, terrestrial plants remain connected to water, and phreatophytes primarily connected to groundwater.

Why is this information important to the phytoremediation of contaminated groundwater? The fact that water is present is not as important to plants as is the bioavailability of water—water has to be present at tensions above the wilting point to be accessed by roots. Plants that have roots that access groundwater, however, essentially eliminate this physical constraint on water availability.

4 Fundamentals of Groundwater Hydrogeology

Groundwater composes more than 98% by volume of the available freshwater on earth. Available does not mean readily accessible, however, because more than half of the available freshwater is held too tightly onto subsurface sediments by molecular attraction to flow to wells. In general, groundwater is the water that completely fills the void spaces in rocks, sediments, and soils and can move under the influence of gravity. The ultimate source of groundwater is precipitation as was described in Chap. 2.

This connection between precipitation and groundwater was not always clear. For example, Seneca (4 BC–65 AD) wrote that precipitation didn't soak deeply into the ground but only penetrated the upper layers of soil because

> *First of all, being a diligent digger among my vines, I can affirm from observation that no rain is ever so heavy as to wet the ground to a depth of more than 10 feet. All the moisture is absorbed in the upper layer of earth without getting down to the lower ones.*
>
> (Kramer and Boyer 1995)

Ironically, what Seneca perhaps was observing in his vineyard was the rapid uptake of precipitation by his vines before it could become recharge, an observation he may have made had he dug where there were no plants.

How, then, could the water collected from much deeper wells be explained if water did not penetrate beyond the upper soil? Seneca had suggested that the element earth was being converted to the element water or that air in the earth was condensed into water.

Bernard Palissy, introduced in Chap. 2, was one of the first to deduce the correct flow of water in the hydrologic cycle and provided perhaps the first definition of the relation between groundwater and springs.

> *... these waters (rain), falling on these mountains through the ground and cracks, always descend and do not stop until they find some region blocked by stones or rock very close set and condensed. And they rest on such a bottom and having found some channel or other opening, they flow out as fountains or brooks or rivers according to the size of the opening and receptacles...*
>
> (Palissy 1580)

Water moves through porous sediments because water molecules in contact with soil particles lose their cohesiveness and, therefore, can flow around the surface of soil particles. This relation between water and geology forms one of the foundations of the discipline of hydrogeology.

Although a few scientists and engineers in the 1800s had begun to expand on the relation between precipitation, geology, and groundwater, such as Joseph Elkington and William Smith (Stephens and Ankeny 2004 and references therein), the first to explain the movement of groundwater in aquifers in a useful manner was an engineer named Henri Darcy. His work is widely used in the hydrogeological and hydrological sciences, and can be applied to address hydrologic issues regarding the phytoremediation of contaminated groundwater.

4.1 Henri Darcy and Darcy's Law

The concept that has become the fundamental basis of groundwater flow and quantity investigations actually started as part of an investigation into surface-water quality. In 1856, Henri (Henry) Darcy (1803–1858), a hydraulic engineer, was tasked by his native city of Dijon, France, to investigate the city's drinking-water supply system. The city had been diverting surface water and having it flow through large beds of sand. The sand filtered out the larger particles and thereby greatly improved the quality of the surface water used for drinking water; this process is still being used by many municipalities around the world for water treatment. Being an engineer Darcy set up a laboratory-scale experiment of the sand beds in the basement of a local hospital to examine how different sizes of sand affected the flow of

J.E. Landmeyer, *Introduction to Phytoremediation of Contaminated Groundwater*,
DOI 10.1007/978-94-007-1957-6_4,

water (Freeze 1994). Darcy passed water through well-sorted sand grains in a column of known dimensions (Fig. 4.1).

The column or cylinder had a known cross-section area, A, and was stoppered at both ends, with the exception of a tube at the elevated end that allowed water to enter, Q_{in}, and a tube at the bottom that allowed the water to exit, Q_{out}. To measure the water properties inside the cylinder Darcy fitted two mercury-filled tubes, or manometers, through the cylinder so that they penetrated into the sediment-filled column. The two tubes were separated by a distance, L. He filled the column completely with water, turned it on end, and then added a flow of water, Q_{in}, at the top to equal the flow that left the bottom. In other words, flow conditions were at steady state, where $Q_{in} = Q_{out}$.

As the water flowed through the column, it also rose inside the manometers. The elevation to which water rose in each manometer above a common datum is called the head. Darcy observed that the head in each tube was different and decreased in the direction of flow. For example, the water rose higher in the tube closest to the inflow, h_2, relative to the water level in the tube closest to the outflow, h_1. The difference between these two water-level elevations was, therefore, h_2-h_1, or Δh. As we will see later in this chapter, Darcy's measurement of head in each tube represent the sum of the pressure head of the column of water above the manometer inlet, and the elevation of the manometer inlet from a common datum.

Darcy was not breaking new ground with this observation that water flow through a sediment-filled column resulted in a head difference that could be measured. Water flow had been studied, albeit in open channels, prior to Darcy. The flow of fluids through pipes also was an area of great scientific interest, particularly in the time of Newton, when laws describing solids were just becoming unraveled. The scientists Poiseuille, of poise fame, and Hagen in 1841 (Freeze and Cherry 1979) both studied the flow of fluid through pipes with the diameter of capillary tubes and noted that fluid flow, Q, was proportional to the cross-section area, A, of the pipe and rate of discharge, v

$$Q = vA. \tag{4.1}$$

Darcy must have known of these studies, because Darcy's observations are a corollary to Eq. 4.1.

Darcy expanded upon his results of water flow through columns and demonstrated that the specific discharge, or velocity, v, of water through the cylinder of cross-section area, A, was directly proportional to the difference between h_2 and h_1, an observation not considered by Poiseuille and Hagen since they were using pipes with no manometers. Darcy observed that a greater head difference between two measuring points gave rise to higher specific discharge if the distance between the two measurements, ΔL, remained constant. Also, v was indirectly proportional to L ($1/\Delta L$) if Δh was constant. Combined, Darcy's observations state that

$$v = \Delta h/\Delta L, \tag{4.2}$$

where h is the hydraulic head and $\Delta h/\Delta L$ is the hydraulic gradient. In much the same way that changes in height of a mercury-filled capillary tube, or thermometer, indicate changes in air temperature, changes in the height of water-filled tubes indicate changes in the water gradient and, therefore, discharge. Darcy's studies were important in that until his work the flow of water had been investigated only in streams and rivers.

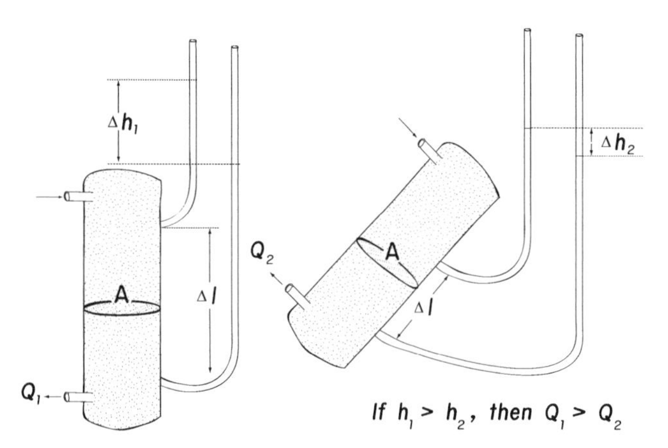

Fig. 4.1 Darcy's miniaturized sand filter, or column, used in his laboratory experiments to more easily observe the effect of differences in head, h cross-sectional area, A and hydraulic conductivity, K on water flow, Q.

4.1.1 Hydraulic Conductivity

Up to this point of Darcy's experiment, the statement could be made that the flow of water through the sand-filled columns between two points was controlled solely by the head gradient. But what if the column contained gravel instead of sand? Although Darcy packed his columns carefully with uniform sand grains characteristic of those that were in the sand filters of Dijon, he hypothesized that the flow of water, as specific discharge, v, through a column of gravel would be different from flow through sand. To account for the flow of water through different soil types that might affect the flow rate through the column, Darcy introduced a constant of proportionality, K, such that Eq. 4.2 becomes

$$v = -K\Delta h/\Delta l. \tag{4.3}$$

The term K in Eq. 4.3 is called hydraulic conductivity and has units L/T; some hydrologists refer to K in units of gal/day/ft^2. The minus sign is used to indicate that the flow of water is in the direction of decreasing head, or a negative head gradient. By convention, this fact usually is omitted when addressing large-scale field problems. This term represents how different sediments affect water flow: as Darcy stated, K is a coefficient of the permeability of a bed of sediment. If the porous media is poorly sorted or has a small grain size, then K decreases. On the other hand, if the porous sediment, or media, has a larger grain size, or is spherical in shape, the K increases.

The hydraulic conductivity of a particular sediment also is affected by the properties of the fluid. For example, if Darcy had passed oil rather than water through a sand-packed column he would have gotten different results for K. An additional term, therefore, is required to cover the effect of the properties of different fluids on flow. This is referred to as the intrinsic permeability, k, with dimensions of L^2. Because intrinsic permeability is more often used by the petroleum industry, and the effect of plants on pristine and contaminated groundwater involves water only, the reader is referred to other sources for more information on k (e.g., Tindall and Kunkel 1999).

Darcy observed in his laboratory that the specific discharge, v, also was directly proportional to the flow, Q, in L^3/T, and indirectly proportional to the cross-section area of the column, A, in L^2, or

$$v = Q/A, \tag{4.4}$$

When Eq. 4.4 is substituted in Eq. 4.3, the result is

$$Q = -K\Delta h/\Delta l A, \text{ or } Q = -KiA. \tag{4.5}$$

Even though Darcy's experiments were designed to test the use of various materials through which Dijon's surface water could be filtered, Eq. 4.5 also can be used to examine the flow of groundwater through aquifers and is a fundamental principle of hydrogeology.

The primary result of Darcy's experimental design or model was that the volume of water, Q, that could flow through the sand beds was related directly to the cross-section area of the bed, the hydraulic conductivity of the sands, the difference in head between the unfiltered surface water above the sand bed relative to the head leaving the bottom of the sand bed, and indirectly proportional to the length, or thickness, of the sand bed (Fig. 4.2). He now could use this model to estimate what effect that changes in each variable would have on the flow of surface water through the sand without actually having to test it out by using the massive sand beds themselves.

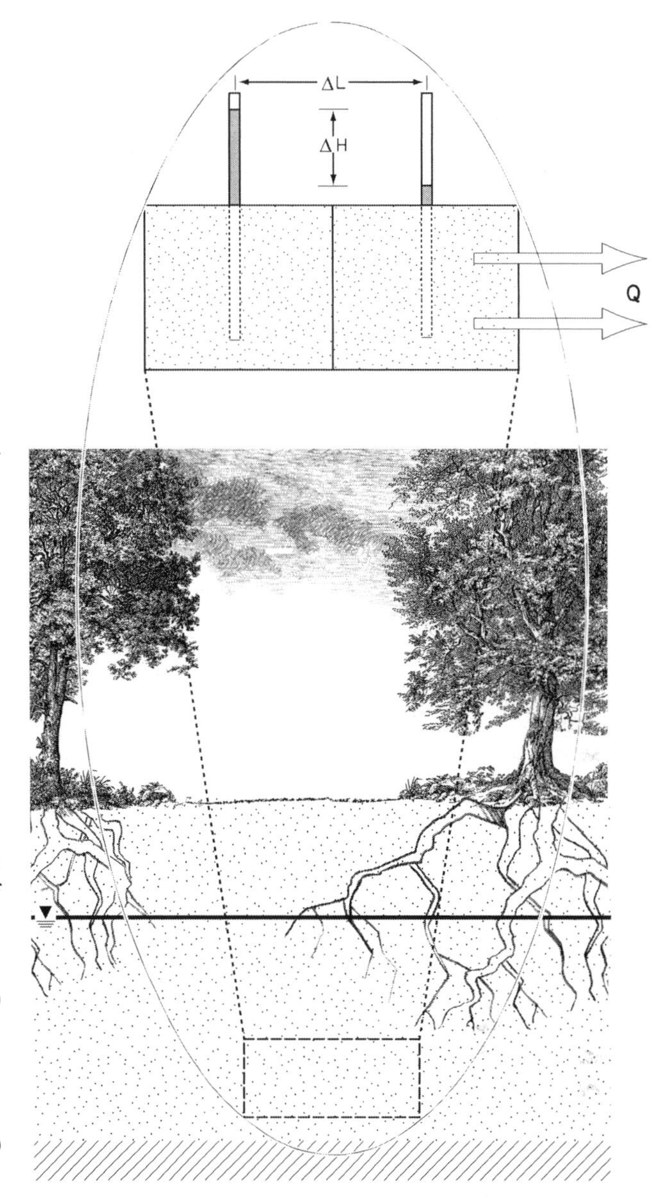

Fig. 4.2 The relation of groundwater flow, Q, to the head gradient; $\Delta H/\Delta L$.

The effect of various materials on water discharge under different hydraulic gradients is plotted in Fig. 4.3. There is a direct relation between the hydraulic gradient, $i = \Delta H/\Delta L$, and discharge, Q. The slope of each straight line represents the hydraulic conductivity, K, for each sediment-type tested and captures the relation between the two variables on the discharge of water through porous media. Again, although Darcy was concerned with the flow of surface water through sand beds, it also is applicable to groundwater flow through aquifers (Fig. 4.2). Measured values of K range from 10^{-4} to 10^{-9} cm/s for clays, 10^{-3} to 10^{-5} cm/s for loam, and 10^{-2} to 10^{-3} cm/s for sands (Tindall and Kunkel 1999).

One of the strengths of Darcy's Law is that it was derived by observation rather than from physical law. As such, Darcy's

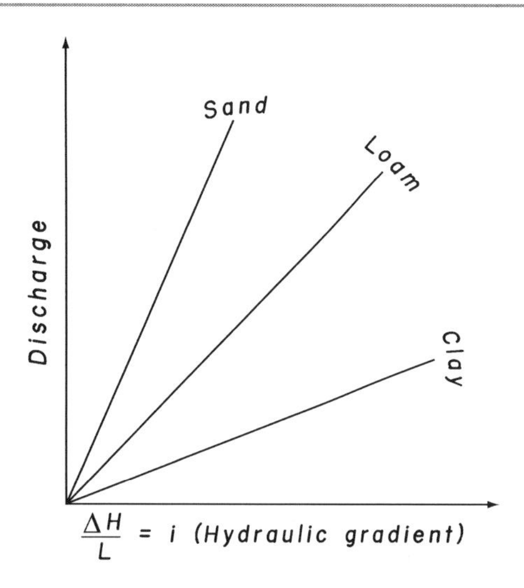

Fig. 4.3 Relation between discharge and hydraulic gradient for different geologic media. Under a given hydraulic gradient, greater flow occurs in sandy sediments or aquifers relative to aquifers with finer sediments. To support a constant rate of flow a steeper head gradient is required in aquifers with finer sediments.

Law has a few limitations, particularly from the perspectives of mathematics or fluid mechanics. First, the equation is more useful to describe the large-scale properties of flow relative to describing flow on a small scale. This is because Darcy's Law assumes that an aquifer acts as a homogeneous continuum of porous material, much like one big column. This in fact is in contrast to most aquifers, which tend to be heterogeneous on a smaller scale because of the sedimentary processes that lead to aquifer formation, such as layers of sand adjacent to layers of silts and clays. This approach averages all the variations of hydraulic conductivity inherent in a particular aquifer into one hydraulic conductivity value that can be used to calculate discharge.

As a result, even though the hydraulic conductivity value in Darcy's Law has units of velocity (L/T) and Q has units of flux (L^3/T), it does not indicate a velocity of groundwater but rather a volume of water moving through a cross-section area. This is because only part of the cross-sectional area, A, is available for fluid movement; the rest of the area being filled with solid matter. Darcy's Law also cannot be used to investigate water flow in the unsaturated zone or if flow is turbulent. To further investigate these shortcomings of Darcy's Law, attempts have been made to derive a similar equation for fluid flow through porous media from physical laws, such as the Navier–Stokes equation (see Gray and Miller 2004). Although many groundwater scientists are well versed in applied mathematics, most are more interested in expressions that consist of parameters than can be readily measured—a nod to Darcy's original experimental approach.

Even with these concerns the usefulness of Darcy's Law to solving complex groundwater-flow problems is its simplification of parameters that control flow, in particular

$$v = iK/n_e \tag{4.6}$$

such that the average linear velocity of groundwater, v, can be estimated with knowledge of the head gradient, i, as measured using at least two wells, the hydraulic conductivity, K, and effective porosity, n_e, from laboratory measurements or reference values.

Darcy's Law is similar to an earlier empirical expression that describes the flow of electrons in conductive materials—Ohm's Law. In 1827 Georg Ohm stated that the flow of current, I, in amps is directly proportional to the potential difference in voltage, V, across two points and is inversely proportional to the resistance, R, between these two points. Moreover, the hydraulic conductivity term Darcy used is analogous to that discussed in Chap. 3 with respect to soil hydraulic conductivity, root hydraulic conductance, and even stomatal conductance, although this latter instance deals with the movement of water vapor. In all cases, however, the central issue is the rate of water movement either through a cell membrane or through a porous media.

4.1.2 Plants and Groundwater

Not only did Darcy experimentally determine a fundamental relation between water and its movement through porous media, Darcy also may have been the first hydrologist to recognize the important relation between certain plants and groundwater. That Darcy was aware of at least a general relation between plants and groundwater is evident in quotes made by a Chief Engineer in Dijon, who stated that in some Morvan forests, springs could be found were alder trees were growing (Darcy 1856). Darcy himself also discussed the observations made by the French abbot Paramelle and his observations about springs and plants. Darcy stated that Paramelle could find groundwater discharging as springs by the following method:

> *As Father Paramelle slowly walks through a valley or continuous depression to find a spring there, it is obvious that he looks carefully at plants and at the ground, from which he seeks to infer the nature of and the strength of the plants, the consistency of the soil, the probable presence of water, and even the approximate depth of the water below the ground surface.*
>
> Darcy (1856), translated by Bobeck (2004)

Paramelle obviously had knowledge about plant and water relations and must have been aware of the previous work related to the movement of water through plants, as described in Chap. 1. Paramelle stated

> *Plants draw the material that feeds them from the soil and the atmosphere. Roots withdraw from the earth water, salts, and organic substances provided by manure. We know that this*

property of absorption occurs at the ends of the roots. When the water present in the soil is full of soluble materials and enters into the rootlets, the water becomes part of the plant juice and this substance is called sap, properly speaking. The rising sap reaches the leaves and undergoes several modifications with which we shall not concern ourselves here. We will say only that it gives up a large part of its humidity there, which is released into the atmosphere as aqueous vapor through all the green parts, and especially through the pores that cover the lower side of the leaves. Sometimes this transpiration is so abundant that it becomes noticeable as sweat in the form of droplets. The measurement of the product of this transpiration, or of the excess of total aqueous volume absorbed over the amount that the plant assimilates, gives us an idea of the importance of the first volume. The famous physiologist Hales found that the average transpiration of a sunflower was 20 ounces (1.25 pounds) during the 12 hours of a dry and hot day, and up to 3 ounces during a dry, hot night without dew. He also found that a dwarf apple tree can exhale 15 pounds of water during 10 hours of the day. In Sologne, I have seen very wet and, as a result, very unhealthy land completely desiccated and drained by planting green trees.

Darcy (1856), translated by Bobeck (2004)

Darcy also recognized that transpiration of groundwater by certain phreatophytes was the cause not only of making wet ground dry but behind the rapid tree growth he had observed near springs that subsequently went dry. In his own words, Darcy stated that

A spring called the Fountaine des Suisses (Swiss Spring) in Dijon had almost entirely disappeared; it produced only 1/5 liter per minute. Two lines of poplars planted along the small valley where the spring occurred showed the following phenomenon. The poplars had been planted at the same time; however, the first of each line showed more than double the growth of the following ones. I had a trench dug in the place where the spring was thought to emerge, and I noticed the roots of the two first trees had already advanced 8 to 10 meters toward this spring in the middle of the natural basin where they had grown, and were in the process of taking it over entirely. After some work to modify the course of the spring, its volume again equaled 12 to 13 liters per minute.

Darcy (1856), translated by Bobeck (2004)

Darcy stated that this relation between plants and groundwater could be used to locate springs (Sharp and Simmons 2005). Interestingly, Darcy foretold of future investigations into the consumptive use of water by plants in arid areas of the United States, which were examined more than 50 year later by the USGS. Darcy's interest lay in using plants to find springs that could then be diverted to supply man's needs. Darcy stated

...when brought to the surface this water has real usefulness rather than contributing to making naturally rich terrain even richer in vegetation as a result of its mineral composition and its location in a swamp.

Darcy (1856), translated by Bobeck (2004)

4.1.3 Anisotropy

The distribution of hydraulic conductivity is never uniform as required by Darcy's Law. Even in Darcy's column experiments, the sand grains are not all uniform in size and water flow that occurs through the middle of the column will be different from water flow that occurs through sand in contact with the sides of the container. As can be imagined, this difference in hydraulic conductivity and its effect on flow in the space of a sand-filled column can be much larger when expanded to the field scale.

Under ideal, homogeneous conditions hydraulic conductivity would remain uniform throughout the aquifer. Because of the orders-of-magnitude variation inherent in the grain sizes of complex geologic settings, it follows that the hydraulic conductivity of a particular area varies with respect to space. This variation is called aquifer heterogeneity, from the Greek *hetero* meaning different and *gene* meaning birth. If the variation in hydraulic conductivity does not impart a preferential groundwater-flow direction, conditions are called isotropic. If hydraulic conductivity varies along a direction, this variation is called anisotropy, from the Greek *an* meaning without, and *iso* meaning equal.

Because most unconsolidated aquifers are geologic units created by sedimentary processes that result in layered depositional sequences, most aquifers have some degree of anisotropy as a result of the orientation of sediments during original deposition. The vertical hydraulic conductivity, K', is often much lower than the horizontal hydraulic conductivity, both of which may vary directionally. Aquifer systems that have undergone some degree of weathering since deposition or formation, such as karst in limestone aquifers, tend to have anisotropic conditions. Anisotropic flow can be enhanced by continued groundwater flow to wells during pumping, especially if the influence of the well causes surface water of a different geochemistry to enter groundwater and cause dissolution or precipitation reactions to occur along flow paths.

4.1.4 Porosity

In Darcy's experiments, the volume of water added to the sand-packed column represented the gross porosity of the saturated material. In other words, the total porosity, n, of a particular volume of the column media can be described as the ratio of the volume of voids, V_v, to the total volume, V_t, where $V_t = V_v + V_s$, or $n = V_v/V_t$ (Fig. 4.4). This also holds true for most geologic media. The porosity of a porous medium is directly proportional to its degree of sorting,

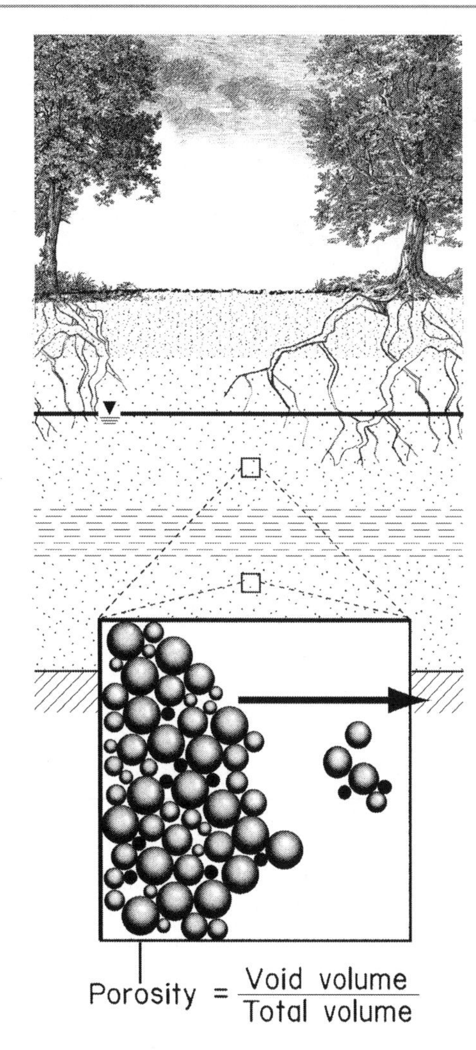

Fig. 4.4 Porosity for a porous media is the volume of the void spaces in a total volume of sediment. The effective porosity is the fraction of porosity that is interconnected and can support groundwater flow.

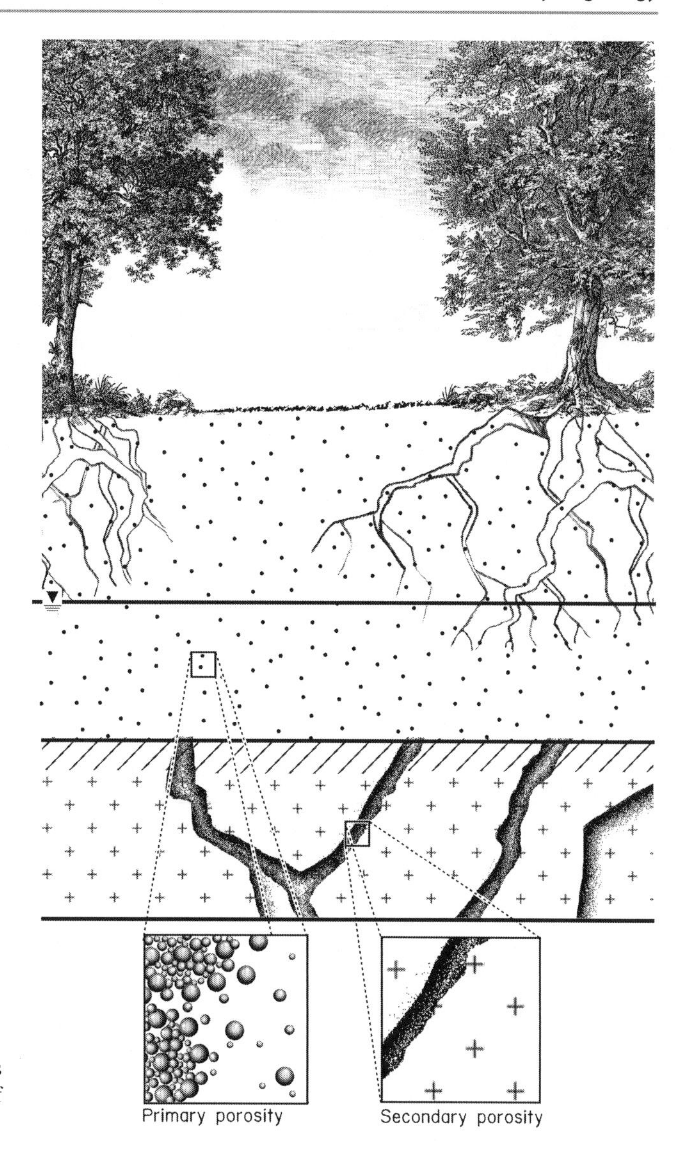

Fig. 4.5 Generalized views of primary and secondary porosity in porous media and fractured rock, respectively.

such as grains of uniform size distribution relative to grains of non-uniform size, inversely proportional to its degree of compaction, and not dependent on grain size. In other words, a unit volume of ball bearings of the same diameter has the same porosity as a unit volume of bowling balls.

This definition of porosity suggests that the volume of voids accounts not only for water content but also for the volume of flowing water. The latter does not hold, however, because not all pores are interconnected to the extent that water travels from one to the other through the material. For example, an individual particle of water introduced at the inlet, Q_{in}, in Darcy's column filled with sand will travel a longer path through the column length, l, filled with a porous material relative to the flow path taken if no media were present. This longer flow path, l_t, relative to the straighter path, l, is called the tortuosity, T, where $T = (l_t/l)^2$. The porosity that results in interconnected flow through porous media is called effective porosity, or n_e. Higher values of hydraulic conductivity usually mean higher values of n_e. Clays, however, that have higher porosity than sands actually have much lower values of hydraulic conductivity and n_e.

The porosity of a porous media may not be constant over time. Porosity can be a function of the pore spaces that have existed since the sediments were deposited or a function of events that have happened since deposition. The first instance, already discussed, is referred to as primary porosity (Fig. 4.5). The second instance is called secondary porosity and can occur to geologic materials such as limestone that are weathered over time by the flow of groundwater that contains carbonic acid from precipitation. Igneous and metamorphic rocks that essentially have no primary porosity can develop secondary porosity if the rocks become fractured; it is such fractures that are tapped for water supply.

4.1.5 Hydraulic Gradient

As previously noted in the derivation of Darcy's Law, the hydraulic gradient, $\Delta h/\Delta l$, is important in driving groundwater flow through porous media. The head loss between two wells in the direction of flow represents the energy needed by groundwater to overcome inertial forces and the friction of water particles in voids relative to those attached to sediments by hydroscopic force or tension (Fig. 4.6). Stearns (1927) performed laboratory investigations to determine whether small hydraulic gradients could be responsible for groundwater flow. He reported that, as expected from Darcy's Law, groundwater flow could occur even under a hydraulic gradient of a few inches per mile.

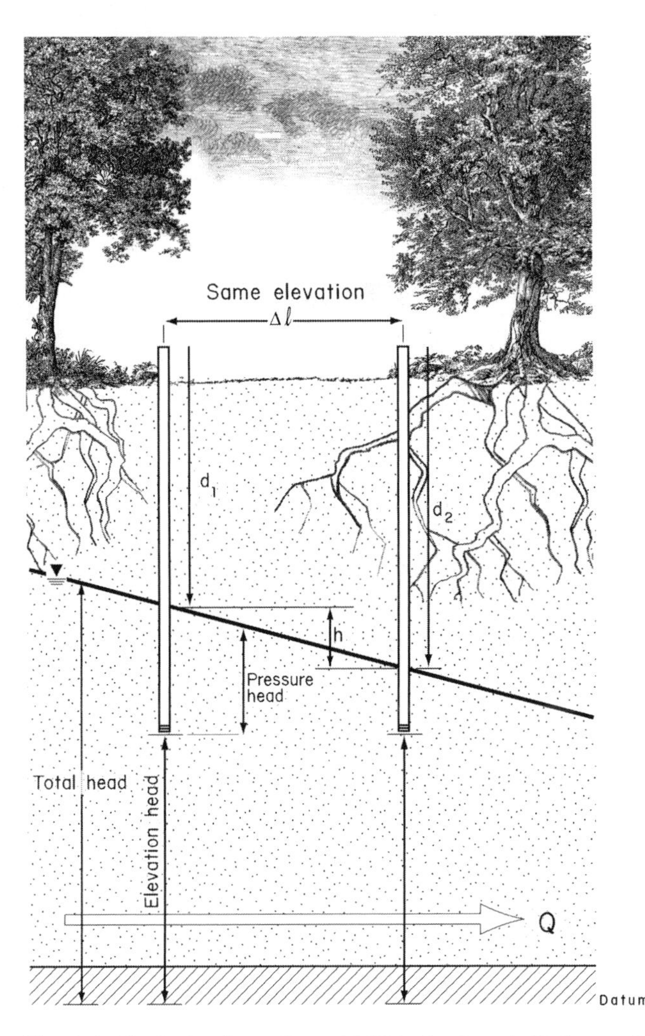

Fig. 4.6 The hydraulic gradient, $\Delta h/\Delta l$ is the force behind groundwater flow, Q. *MSL* is mean sea level.

4.2 Static Water—Hydraulic Head

A fluid is a substance that does not maintain a shear stress; in contrast, a solid has a definite volume and shape and does maintain a shear stress, whereas a fluid has a definite volume but not shape. In other words, fluids continually deform. For example, if you hit a baseball with a bat, the baseball goes in a predictable direction. If you hit a balloon filled with water with a bat, the direction of the water droplets upon the balloon bursting is far less predictable. Because fluids cannot maintain shear stresses, the surface of a liquid at equilibrium is flat, or horizontal. As we saw in Chap. 2, the effects of surface tension negate this equilibrium if water is placed in a thin-diameter tube, which usually results in a concave upward surface.

The pressure, P, at any point on the flat surface of a static fluid is composed of the force, F, exerted by the fluid on the unit area of the point, in ML/T^2. In other words, pressure is the force per unit area, or $P = F/A$. The pressure stress in a static fluid is the same in all directions. This was observed by Pascal as early as 1653. Isaac Newton defined force as F (a vector) $= ma$ (a vector), where $m =$ mass and $a =$ acceleration. Therefore, a body of mass, m, under static conditions accelerates solely because of gravity. Mass should not be confused with weight; mass is the same regardless of location, whereas weight is unique to location because it equals the force of gravity on a particular object.

But the pressure in a fluid can vary from the pressure at its surface. For example, in a column of water, the pressure changes with respect to the height of the column because of the difference in the weight of the water at various heights in response to gravity. In other words, the pressure at a particular location is directly proportional to the depth of the column of water above it. Generally, the pressure based on the weight of water is related to the density, ρ, of the fluid, where $\rho = M/L^3$ times the depth, h, of the water. The pressure, P, at any height, h, in that column is

$$P = \rho g h. \tag{4.7}$$

In other words, the pressure at a given depth below the water surface is greater than atmospheric pressure and can be quantified by $\rho g h$.

Pressure is often defined in pascals (Pa) or Newton per square meter where $1\ \text{Pa} = 1\ \text{N/m}^2$. In hydrogeology the effect of atmospheric pressure on water is usually considered to be negligible under the assumption that it can be considered constant over time. Under some short-term conditions, however, this assumption is not always valid, especially for semi-confined to confined unconsolidated and fractured-rock aquifers (Landmeyer 1996).

4.3 Flowing Water—The Bernoulli Equation

The flow of fluids is the result of the driving force of gravity to overcome inertia and the resisting force of friction as the fluid flows; flow is constant once these two opposing forces are equal. In groundwater, the hydraulic gradient is what imparts the gravitational component of flow. Flow is essentially an energy gradient such that water at a higher elevation has more energy than water at lower elevation. Energy here means the force multiplied by the distance, or the amount of work done. The initial energy for flow is attained by elevation and is called potential energy and by flow called kinetic energy. Because the rate of groundwater flow is relatively slow and laminar compared to turbulent surface-water flow, the kinetic energy is small. Heat as a form of energy in groundwater normally varies little over space and is considered a constant.

This understanding of fluid flow as a special condition of static fluids was not gained easily, however. As late as the early 1700s, during the time when Harvey was concerned with the circulation of blood in humans (Chap. 1) and Hales was concerned with the circulation of water or sap in plants (Chap. 1; Table 1.3), no device existed to measure the pressure exerted by the flow of these fluids against their vessels. Harvey noticed that when a blood vessel was ruptured, the level of blood issuing would rise and fall in rhythm with the contractions and relaxations of the heart. The French scientist Edmé Mariotte (Chap. 2) was experimenting with measuring the pressure of water flowing out of a pierced pipe. His experimental device was novel and simple: the water let out of the pipe was allowed to push against an instrument that contained lead, and when the amount of lead added was equal to the force of the water against it Mariotte was able to calculate the pressure of the water (Guillen 1995). Although useful for measuring fluids that could be permitted to leak, it would have been a disastrous way to measure the blood pressure of humans.

This quest for understanding fluid pressure and its measurement did not go unnoticed by a Swiss mathematician named Daniel Bernoulli (1700–1782). He published his work on his theories of fluids called *Hydrodynamics* in 1738. As Darcy did almost 100 years later, Bernoulli conducted physical experiments with water flowing through pipes of different sizes and recorded the changes in pressure. Although the behavior of fluids seems trivial today, during Bernoulli's time the behavior of fluids was unknown, compared to Isaac Newton's contemporaneous revelations about the behavior of solids. When Bernoulli added a glass tube to the side of a pipe that contained flowing water, the water rose up into the tube until it stopped at a certain elevation: could this perhaps be where Darcy got his idea to measure the head in his columns using thin tubes more than 100 years later?

However, if more than one glass tube is measured in order to determine the direction of flow in the pipe, the heights must be comparable to a common datum, called the elevation head. Hence, after Bernoulli's experiments, fluids could be characterized by knowing their pressure, density, and elevation. In the mid-1800s, Darcy measured levels of mercury in thin tubes and converted them to equivalent levels of water with respect to a common surface, the bottom of the column. These measurements reflected the elevation of the water level, or head, above the column bottom added to the elevation of the head above the sand. In other words,

$$H = \text{elevation head} + \text{pressure head, and} \tag{4.8}$$

$$H = z + (P/\rho g), \tag{4.9}$$

where ρ = mass density (M/L^3), g = gravity (L/T^2), and P = pressure (ML/T^2). Density, ρ, is the mass of a unit volume of a substance with dimensions (M/L^3).

In groundwater investigations, Bernoulli's equation can be used to deduce that the potential of a fluid at any point in a porous medium can be measured by the head above a common datum (Fig. 4.7). This is because gravity, g, can be considered constant among measurements. Instead of using manometers, hydrogeologists use piezometers, which essentially are pipes with two open ends, or observation wells consisting of 1- to 4-in. diameter pipes with no more than 10 ft (3.3 m) of slotted pipe, or screen, at the end penetrating the aquifer. For the measurements to be comparable the depth from land surface to the end of the pipe, or midpoint of the slotted section, must be constant. Hence, wells screened with larger slotted sections or a section that crosses two aquifers may not be comparable.

4.4 Aquifers

The sand- and water-filled columns used by Darcy enabled water to flow from the inlet to the outlet pipes. On a much larger scale in nature, aquifers are sediments that contain water available to flow to a well in useable volumes. The study of groundwater flow through sediments, geologic media, and aquifers is called hydrogeology. Although the term hydrogeology was first used in a report by J.B. Lamarck (1802), he used the term to refer to sediments that had been deposited by water flow. The first use of the term hydrogeology to describe groundwater flow in aquifers was in a paper by J. Lucas (1880). About the same time, the newly formed USGS adopted the term in published results of hydrogeologic investigation studies.

Even though the term aquifer is widely used today, C.V. Theis of the USGS used the term hydropher to describe the part of the geologic unit that was saturated with groundwater and the groundwater in the hydropher the groundwater body (see Clebsch 1994). In the time before Darcy, the water

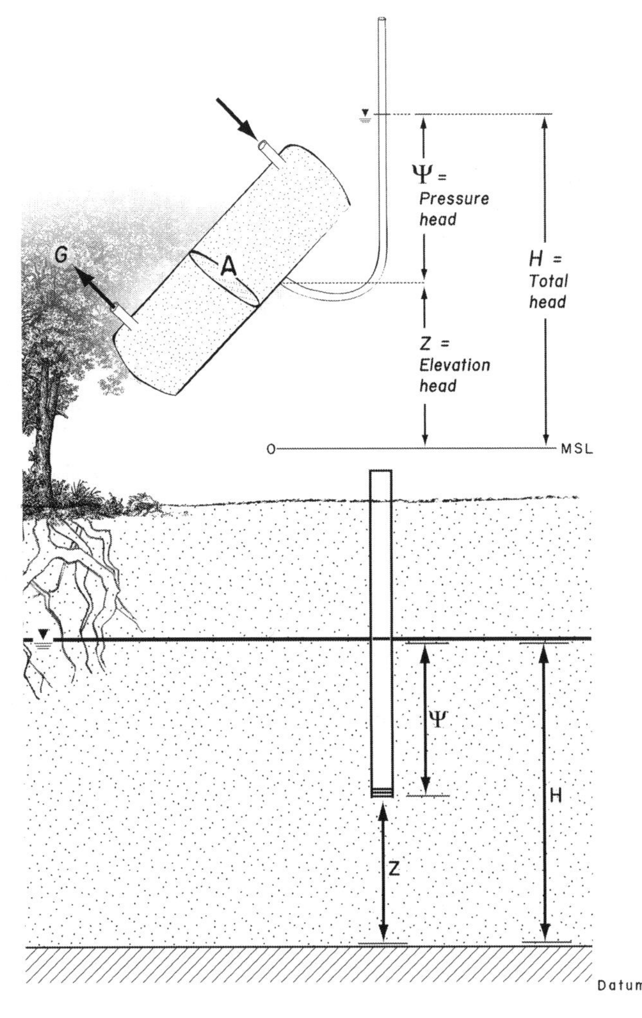

Fig. 4.7 The relation between total head, *H* or elevation head, *Z* and pressure head, *p*, or ψ, for laboratory settings such as those used by Darcy (*top*) and a monitoring well in the field (*bottom*).

produced from wells was described as originating from water bearing strata, as described in an early book on geology by Hitchcock (1840). Moreover, the term aquifer often is relative; in humid areas, for example, fine-grained silts or fractured rock are considered to be low yielding but may be considered an excellent source of water in more arid areas. Regardless, the groundwater in aquifers can be classified under two major conditions with respect to Bernoulli's Law and geologic conditions—unconfined and confined aquifers.

4.4.1 Unconfined Aquifers

In unconfined water-table conditions, the uppermost surface of the saturated zone not under capillary tension represents the special case of Eq. 4.8 where water is at atmospheric pressure. The pressure head in Eq. 4.8 is zero; thus, the total head is simply the elevation head of the water above a common datum (Fig. 4.8).

Unconfined aquifers are often near land surface, especially in low-lying areas. Because the water-table surface is exposed to the atmosphere and infiltrating water, the position of the water table can rise and fall. The water table is essentially a free surface that can react to a stress by moving upward during recharge and downward during discharge (Fig. 4.8).

Because the water-table surface is often near or at the land surface in the case of some surface-water features, it has long been believed that the water-table surface follows the topography of the area. This is partially true in that water flow in unconfined aquifers is in response to differences in elevation head and gravity, and this situation can be initiated by higher topographic elevations. However, topography can be an overestimate of the shape of the water-table profile. For example, in the steep topographic gradients common to

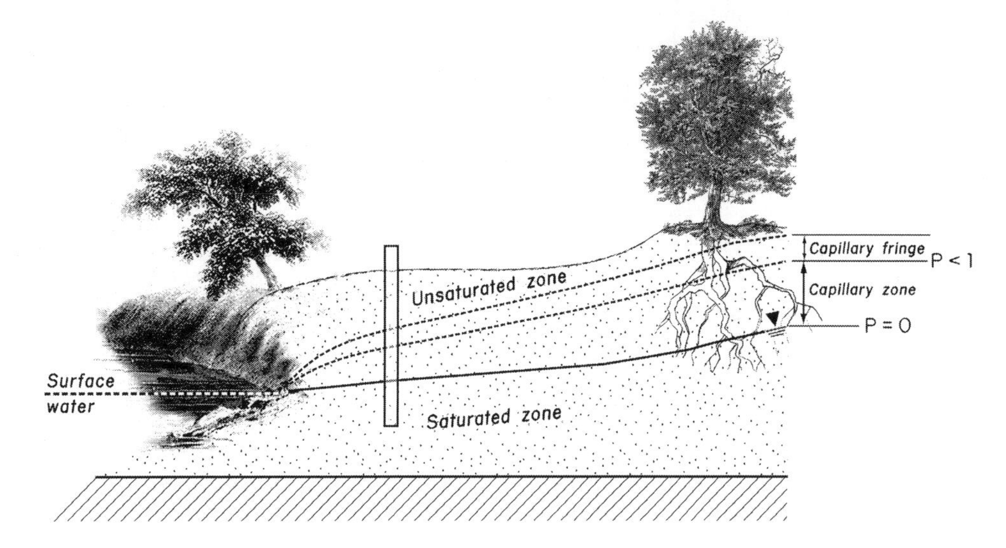

Fig. 4.8 The unsaturated zone, capillary zone and fringe, and water table. The distinction is made between a capillary zone and fringe such that the capillary zone represents where the voids may be completely filled with water but held under tension (*saturated tension*) and does not, therefore, flow freely to a well. The capillary fringe is where water is under tension but does not completely saturate the voids (*unsaturated tension*).

the sand hills in the Coastal Plain geophysical province of the eastern United States, the hydraulic gradient of the water is less pronounced because the higher permeabilities of the sandy materials lead to less frictional head losses such that groundwater flow can occur with lower hydraulic gradients (see also Fig. 4.3).

To simplify the definition of groundwater flow under such conditions, Dupuit (1863) and Forchheimer (1930) assume that the head at a vertical location is constant, all groundwater-flow lines are horizontal across the entire thickness of the water-table aquifer, vertical flow is eliminated, and the velocity of groundwater flow is directly related to the slope of the hydraulic gradient for each flow line. Although not relative for small-scale simulations of groundwater flow, the Dupuit-Forchheimer model is used widely for larger-scale simulations. An alternative conceptualization is that the water-table surface is controlled by the recharge potential of the sediments (Haitjema and Mitchell-Bruker 2005).

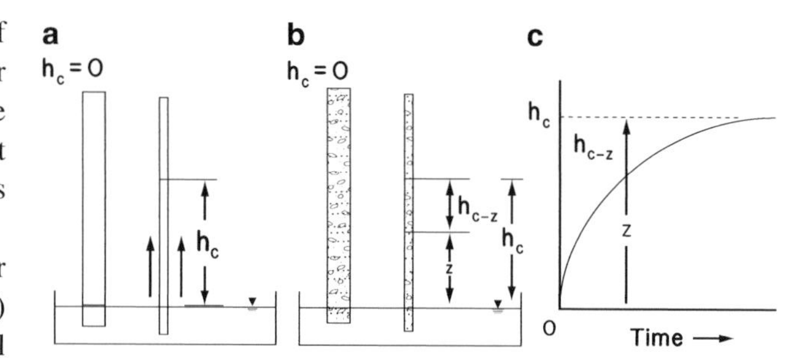

Fig. 4.9 Representations of how water behaves in the capillary zone above the water table. In both (**a**) and (**b**), representative large and small diameter pipes with an open end were placed in a pan of water. The larger pipe represents a well, and the smaller pipe represents interconnected pore spaces. In (**a**), the water in the larger diameter pipe is at the same level as the water in the pan. In the smaller diameter pipe, the water level rises in the direction of the arrows to a maximum level, h_c, above the water level in the pan. This is caused by water cohesion and adhesion to the inner wall of the pipe and reflects the capillary zone. In (**b**), the same two pipes are filled with porous media, but the water does not rise immediately in the smaller diameter pipe to h_c. Rather, it rises to some lower elevation, z, initially (**c**), and then over time reaches h_c. (Modified from Heath 1983).

4.4.2 Unsaturated Zone, Capillary Fringe, and Capillary Zone

The previous description of the water table often gives rise to a common misconception that the surface of the water table can be defined as a sharp interface between completely saturated and completely dry sediments. This is not the case at all, because the surface tension of water in contact with porous media will cause water to rise under tension above the fully saturated area where water is at atmospheric pressure (Fig. 4.8). The thickness of sediments nearest the water table where water completely saturates the pore spaces but is held under tension, or tension saturated, is the capillary zone. The thickness of the capillary zone is controlled by the size of the pore spaces (Fig. 4.9); increased porosity results in a thinner capillary zone and decreased porosity a thicker capillary zone. As the pore spaces begin to be occupied by more air than water above the capillary zone, water is still under tension and is called the capillary fringe (Fig. 4.8). Above this is the unsaturated zone, where water may or may not be present (Fig. 4.8). In some sub-disciplines of hydrogeology, the capillary zone and fringe are called either one or the other term and meant to be synonymous.

One of the reasons the concept of the capillary zone and fringe are often misunderstood is because they generally are not encountered during groundwater investigations that employ conventional site assessment techniques. For example, if a 2-in. diameter observation or monitoring well is installed through the capillary zone to the water table, the water that flows into the well is from the saturated zone where water is under pressure. If, however, a well of smaller diameter is installed next to the first, conventional well, groundwater will rise above the water table due to tension to a final height balanced by gravity and the smaller well's inner diameter.

Even though the capillary zone is hard to directly observe, it can be measured with certain devices. To assess the water potential under tension, a tensiometer can be used. The measurement of tension is crucial to understanding the source of water to plants, because the initial entry of water into root hairs is by capillary action, which then is continued by osmosis. The very small diameters of root hairs are designed to maximize diffusion and osmosis as well as take advantage of the surface tension of water.

The thickness of the capillary zone and fringe, or height above the saturated zone, can be estimated by using qualitative and quantitative approaches. Qualitatively, fine-grained sediments, such as silts and clays, have a thicker capillary zone and fringe than coarse-grained sediments, such as sands and gravels, assuming that soil moisture from previous precipitation is not confused with the uppermost location of the capillary fringe. This is because water is found in the small pores due to surface-tension attraction to the sediments; the large pores do not permit this to occur and are mostly filled with air. In most cases, however, the sediments are not homogeneous. If pore sizes are uniform, the capillary zone and fringe are larger than if the pore sizes were not uniform.

The use of a hand auger also can supply a reliable estimate of the thickness of the capillary fringe during field studies. The upper layer of the capillary fringe can be detected when the removal of the auger from the hole coincides with resistance, as well as accompanied by an audible sucking sound. The energy needed to overcome the tension and remove the auger from the borehole provides direct physical evidence of the forces required by plants to remove water from the capillary zone and fringe. It should also be noted that no water will enter the augered borehole, at least at first. Augering deeper will define the bottom of the capillary zone when water enters the borehole from the water table.

The height of the capillary zone can be estimated by using a more quantitative approach than a hand auger. The height, H, that a liquid column of water will obtain is described by

$$H = 2T\cos\theta/\rho g r \quad (4.10)$$

where T is the surface tension (N/m), θ is the angle of contact between the surface and the liquid, ρ is the density of the liquid (kg/m^3), g is the acceleration of a body due to gravity (m/s^2), and r is the radius of the capillary tube (m). For example, relative to sea level, T is 0.0728 J/m^2, θ is 20° (0.35 rad), ρ is 1,000 kg/m^3, and g is 9.8 m/s^2; therefore, a 1-m diameter well can allow water to obtain a height of about 1.4×10^{-5} m, or 0.014 mm, at sea level, which would be impossible to measure relative to other processes that affect this surface. If the well diameter were decreased to 1 cm, the water would rise 1.4 mm; if the well diameter were decreased to 0.1 mm, the water would rise 14 cm. The weight of the liquid column of water, therefore, is proportional to the square of the tube diameter. This equation can be used to calculate the height that water would rise above the water table using the largest effective porosity that is known for a particular site's soil or sediment.

An interesting and perhaps counterintuitive observation of contaminant transport processes that occur in the unsaturated zone is that the maximum transport rate, regardless of soil type, is near 13 m/d (Nimmo 2007). This indicates that rate-limiting processes that affect the speed of falling objects in the atmosphere may be similar to those in the pore spaces of the unsaturated zone.

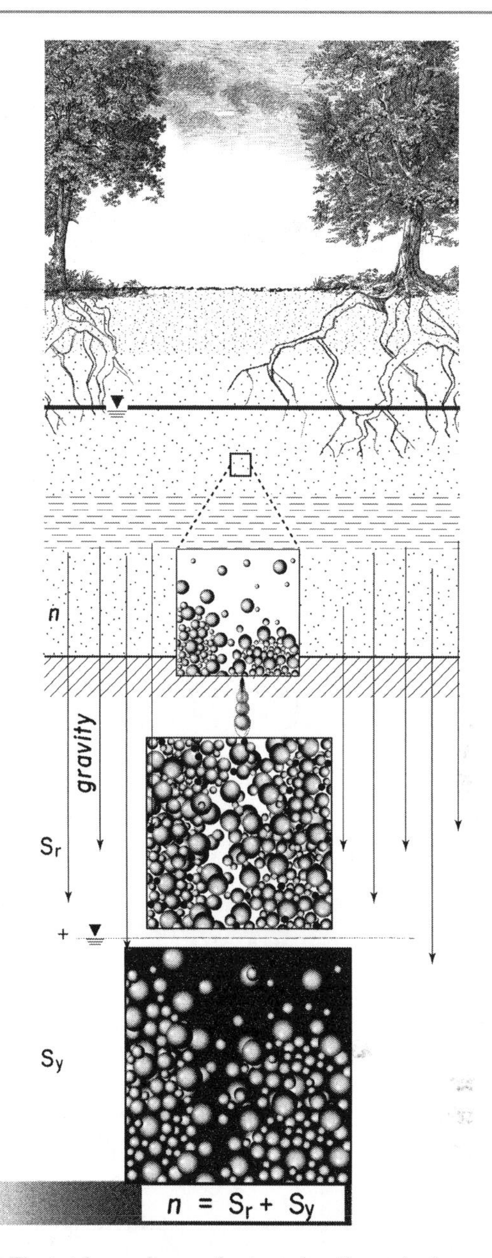

Fig. 4.10 The total porosity, n of saturated sediment is the sum of the specific yield; S_y or water removed by gravity, and specific retention; S_r or water that remains after gravity flow.

4.4.3 Specific Yield and Specific Retention

Because the water-table surface can fluctuate across the thickness of the aquifer, it follows that, based on porosity, the fluctuation could provide a direct indication of the change in water volume. However, this is not the case; because of gravity, only part of the fluctuating water is actually available for removal. This volume of water that drains from a water-table aquifer by gravity is called specific yield (S_y, Fig. 4.10); the balance left behind that is retained on media by tension against gravity is called specific retention, S_r. Hence, previously saturated sediments can essentially exhibit capillary fringe characteristics after water removal, and this provides a more useful definition of porosity. That is, total porosity, $n = S_y + S_r$. These terms are similar to wilting point and field capacity commonly used by plant physiologists and soil scientists, as discussed in Chap. 3.

4.4.4 Confined Aquifers

The Darcy column experiment discussed previously does not represent water-table conditions, as first might be imagined. Rather, it represents confined aquifer conditions, because the water levels in the tubes rise above the top of the elevated column. In other words, in the field, water levels in wells installed in an aquifer that underlies less permeable sediments will rise above the aquifer under confined conditions (Fig. 4.11). The surface of the groundwater level is not exposed to the atmosphere. If the groundwater level rises above ground surface, such as in valleys, groundwater will flow under its own pressure without being pumped. Such a condition was probably first observed in 1126 AD in Europe when wells drilled in the valley sediments in the Artois region gushed groundwater above land surface—as a result, freely flowing wells became associated with this region, and are called artesian wells (De Weist 1965). All groundwater in confined conditions is at pressures greater than atmospheric. Groundwater levels measured in wells in confined aquifers satisfy the Bernoulli equation and the total head measured represents the elevation head and pressure head.

Unlike the water-table surface in an unconfined aquifer where the head is simply the elevation head, the water surface defined by groundwater levels measured in wells in confined aquifers is less clear. It is a potentiometric surface, which represents the level to which groundwater will rise in a hypothetical well installed in a confined aquifer. If a well, or natural spring, does not penetrate a confined aquifer, the groundwater level will not rise above the aquifer even though it has the potential to do so.

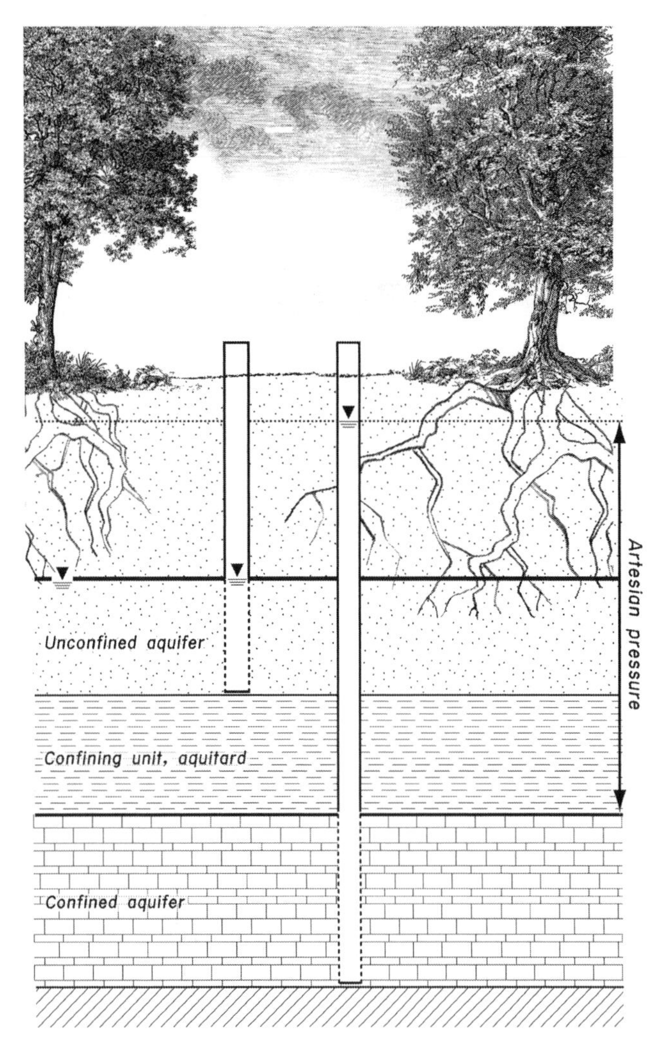

Fig. 4.11 A well installed through a confining unit into the underlying aquifer can be used to measure the pressure of the aquifer at that point. If the pressure is higher than land surface, a flowing well results.

4.4.5 Confining Units

Aquifers are considered confined when they are overlain by less permeable geologic strata. Geologic units that do not transmit useful quantities of groundwater to wells are called confining beds or confining units (Fig. 4.11). These units typically are composed of shale or unconsolidated silts or clay. Ironically, some confining units consist of clay minerals in which the porosity, n, is higher than in adjacent sand aquifers, but because the effective porosity, n_e, is lower, little water can be transmitted to wells. An important characteristic for confining units is the hydraulic conductivity of the material in the vertical direction, rather than in the horizontal position as is the case for aquifers.

4.5 Aquifer Properties

Unconfined and confined aquifers have many properties that can be measured. These include aquifer transmissivity, storage coefficient, and heterogeneity, which are discussed here.

4.5.1 Transmissivity

How can groundwater flow be estimated in a confined aquifer? First, the hydraulic conductivity, K, of the sediments that compose the confined aquifer is multiplied by the aquifer thickness, b. The resulting term is called transmissivity, T, such that $T = Kb$, in units L^2/T. Transmissivity can range from 1,000 ft^2/d (92 m^2/d) in poor aquifers to greater than 100,000 ft^2/d (9,200 m^2/d) in excellent aquifers. The usefulness of transmissivity is evident after substitution in Eq. 4.4, such that $Q = -KiA$, then $Q = -Kbwi$, where b is the thickness and w is the width of the aquifer, and finally $Q = Twi$ (Fig. 4.12).

4.5.2 Storage Coefficient

It was once thought that aquifers transmitted water through porous media only in the presence of a decreasing head gradient. This may have been a result of Darcy's column

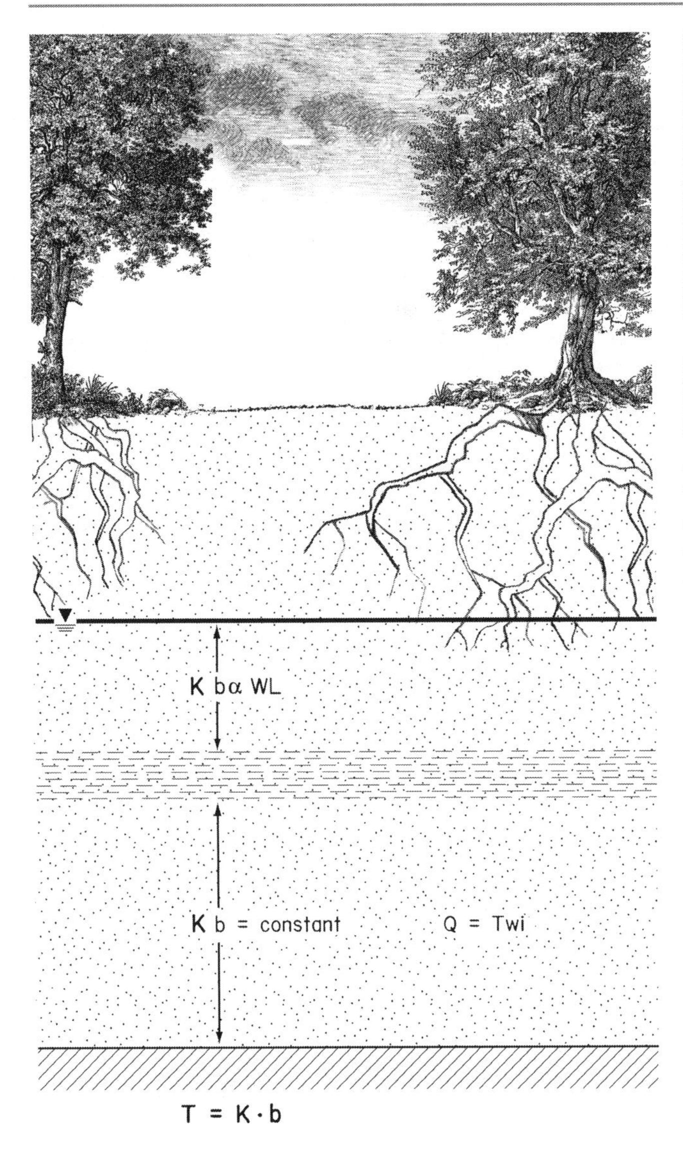

Fig. 4.12 Transmissivity, T of a generalized water-table aquifer is simply the hydraulic conductivity, K, times the thickness, b, thickness, which can vary over time. The transmissivity of a confined aquifer is the constant thickness, b, times the hydraulic conductivity, K, and, therefore, is unchanging.

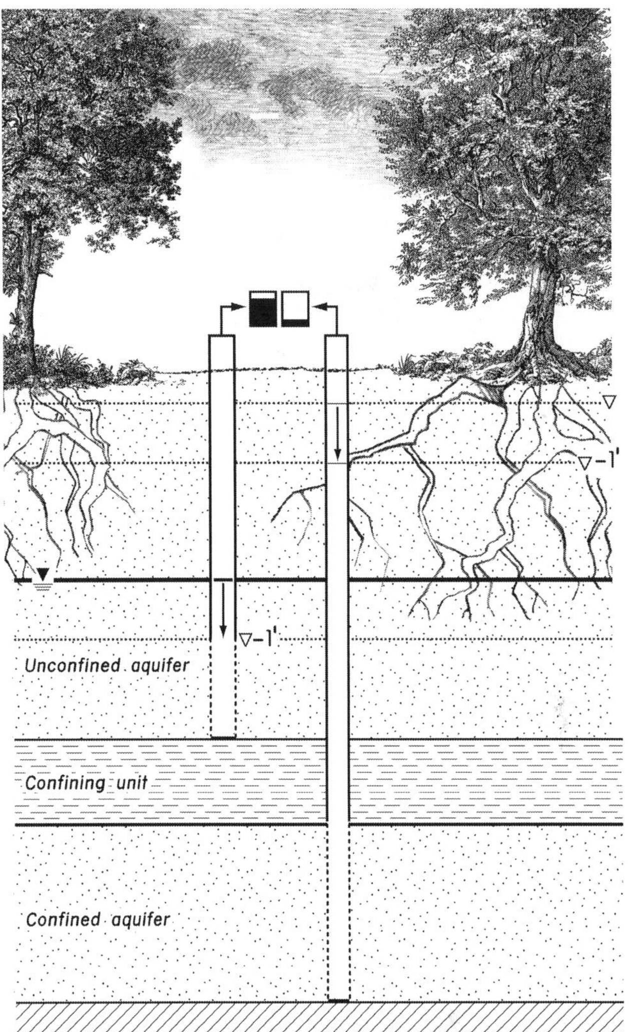

Fig. 4.13 The difference in how groundwater is stored in a water-table aquifer and in a confined aquifer under conditions of a unit decrease in head and the relative difference in the volume of groundwater released under this unit decrease in head is indicated by the black shaded area in the containers between the two wells.

experiment, in which steady-state conditions were used to induce flow. However, groundwater also can be yielded from aquifer sediments to wells by other processes. When a 1-ft groundwater-level decrease occurs in a water-table aquifer, the groundwater yielded from sediment pore spaces is by gravity, and the water removed from the sediments in the 1-ft zone is replaced by air. In a confined aquifer, however, groundwater is under pressure greater than the atmosphere. A similar 1-ft decrease in the potentiometric surface can be measured when groundwater is removed, but the sediments do not become dewatered. This is because confined aquifers not only transmit water but also store groundwater over time. Much like the relation between water flow and water storage in the water pipes in a house, when a faucet is opened, water flows, and when the faucet is closed, flow stops but water remains stored in the pipes. The amount of water stored in these pipes is dependent on the diameter of the pipes and the total length of all of the pipes. An analogous scenario occurs in confined aquifers. Groundwater can flow in response to a head gradient just as for a water-table aquifer, but groundwater also can be stored. Groundwater is released from storage during a release of pressure which causes the stored water to expand and the aquifer material to compress under the increased pressure from overburden following the removal of groundwater (Fig. 4.13). This is similar to the way dissolved gases are released from pressurized, carbonated beverages; the can is opened, causing the pressure to decrease and the gas to expand and be released from solution.

The volume of groundwater released from or taken into storage per unit surface area of aquifer per unit head decrease or increase is

$$S = \text{volume of water}/L^2(1\text{-ft head change}) \qquad (4.11)$$

To estimate the amount of groundwater potentially removed from storage, the storage coefficient, S_c, is multiplied by the thickness of the aquifer, or

$$S = S_c b \qquad (4.12)$$

The storage coefficient in confined aquifers ranges from 10^{-5} to 10^{-3} and for unconfined aquifers is the specific yield and ranges from 0.1 to 0.3.

4.5.3 Heterogeneity

Hydraulic conductivity can vary in an aquifer in the vertical and horizontal directions and gives rise to the conditions of aquifer heterogeneity. These differences in hydraulic conductivity reflect the absolute differences in consolidated or unconsolidated rock or sediments through which groundwater moves. Also, heterogeneity is a reflection of the depositional characteristic of the sediments, for example the depositional trend of a meandering river where coarse gravels and sands fine upward to silts and clays.

Aquifer heterogeneity is an important condition that can be evaluated by using several approaches. One approach is called hydraulic tomography (Yeh and Liu 2000). Whereas conventional aquifer tests produce a non-unique average of aquifer properties over space and time, hydraulic tomography tests use multiple wells that are discretized vertically by packers, such that screened intervals remain flowing and are considerably smaller than the entire well-screen interval. One of the multiple wells is pumped at the packed depth interval, and groundwater-level changes are measured over time. Subsequent pumping tests are done by lowering the pump to different levels in the well. The data are then evaluated by a mathematical model.

4.6 Groundwater Flow

So far we have looked at the flow of groundwater through porous media from the perspective of one or two measuring points only. In a laboratory experiment like Darcy's original column tests, two measuring points were needed to determine the head gradient that caused water to flow. If we place a number of such measuring points in Darcy's column, it becomes possible to generate a three-dimensional surface or contour of measuring points that have equal head, or potential. These lines depict equal head or groundwater pressure and are called equipotential lines, similar to the contour lines on topographic maps to depict elevation or the lines that depict atmospheric pressures on weather maps.

4.6.1 Equipotential Lines

The groundwater levels in a minimum of three wells should be measured in order to depict a three-dimensional groundwater-level surface. In general, the well locations are marked on a map, and the groundwater-level measurement of each well is listed with respect to a common datum for all wells. Lines can then be drawn, manually or by software, to connect the wells of equal groundwater potential and create an equipotential map.

A useful property of equipotential line maps is that they not only depict the contour of equal pressures in an aquifer but can indicate the direction of groundwater flow. This is because the groundwater-flow direction will cross equipotential lines at a right angle because groundwater follows the steepest hydraulic gradient along a path of least resistance. The direction of groundwater flow is not always constant over time, because groundwater potentials change in response to changes in recharge and discharge.

4.6.2 Flow Net Analysis

The use of equipotential lines to create maps of groundwater-flow direction is referred to as flow net analysis. Only a few equipotential lines are necessary to perform such analysis. By convention, a common contour interval between lines of equipotential is used, similar to the constant contour interval used for topographic maps. In practice, the number of wells usually is limited. Groundwater-flow lines of a similar interval are then drawn at right angles to the equipotential lines. The volume of groundwater can now be estimated from such a map, because two adjacent lines of groundwater flow form a flow tube and this prescribes a cross-sectional area. This is yet another extension of Darcy's Law: the product of aquifer thickness, b, and width, w, between adjacent equipotential lines can be substituted into the A of $Q = -KiA$ to estimate the discharge of groundwater.

4.7 Groundwater Recharge and Discharge Areas

The strength of flow net analysis is that a few measurements of head in an unconfined or confined aquifer can reveal much information about groundwater-flow direction and volume. For example, the equipotential lines of larger magnitude often indicate recharge areas. Equipotential lines of decreasing magnitude indicate the direction of groundwater flow toward springs, streams, ponds, wells, or the ocean.

4.7.1 Infiltration

Groundwater is derived from precipitation, either directly or through infiltration of surface waters derived from precipitation. Therefore, the factors that permit or exclude infiltration of water through soil to the water table are important. Infiltration is the process where water enters the soil zone but does not penetrate to the depth of the water table. In an ideal soil with no removal of water by plant roots the distribution of infiltrating water often is delineated by measurements of moisture content over space and time. An increase in moisture content from initially dry to increasingly wet soil results in the development of a wetting front: we can infer from Seneca's comment mentioned at the beginning of this chapter that he probably was observing such a wetting front.

The idealized case for a wetting front assumes that the upper soil layer is permeable enough to allow infiltration. Over time, surface soils can become impermeable because of the movement of fine particles into spaces between coarse particles from surface impaction. Air entrained in the pore spaces of dry soils also can decrease water infiltration. The air is displaced after a wetting front compresses it to the extent that the air pressure increases. This phenomenon can be seen when houseplants or lawns are watered and bubbles appear or can be heard as a popping sound at the soil surface.

Seneca had observed soil conditions after precipitation. But we also know from direct observation that the upper layers of loam soils can be dry and have low moisture content. This results from evaporation, transpiration, or lack of infiltration. The wetting and drying and re-wetting properties of a soil represents a characteristic of particular soils that is related to sediment texture and pore size. This relation between soil type and wetting and drying typically is expressed as a hysteresis curve (Fig. 4.14). More energy is required to lose water from a soil than gain water due to the surface tension of water.

Moisture in the upper layers of soil in the unsaturated zone that flows under the influence of gravity is the field capacity, as mentioned previously. Field capacity is analogous to the specific yield of an aquifer. As might be expected, water that is bioavailable to plants is present under moisture conditions that are above the wilting point but below field capacity (Table 4.1).

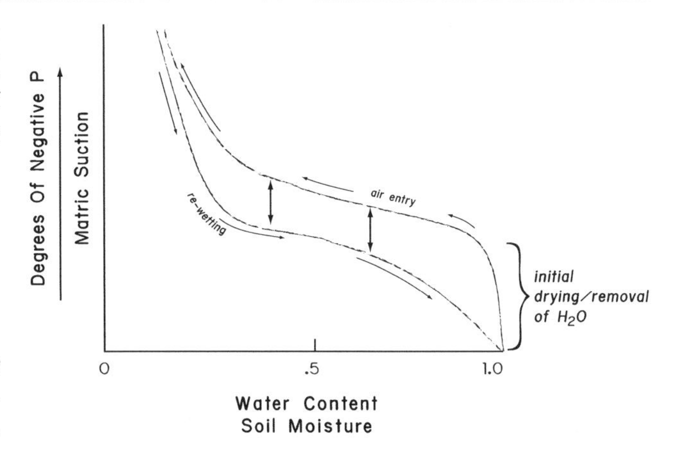

Fig. 4.14 A representative soil-moisture hysteresis curve that shows the cyclical wetting and drying of a representative soil (Modified from Hillel 1998).

Table 4.1 The range of plant bioavailable water for different soil types.

Soil type	Field capacity	– Wilting point (% Dry weight of soil)	=Bioavailable water
Sand	5	2	3
Loam	19	10	9
Clay	36	20	16
Peat	140	75	65

4.7.2 Recharge and Discharge Areas

Precipitation or surface water of higher elevation than the groundwater will accumulate in the soil as a wetting front, overcome the resistance of the soil, and, under gravity, infiltrate through the unsaturated zone, capillary fringe and zone, to reach the water table. The process of infiltrating water that becomes groundwater is called recharge. In recharge areas, a deep well exhibits a lower groundwater level compared to a shallow well (Fig. 4.15). Conversely, in discharge areas, the deep well will have a higher groundwater level compared to a shallow well (Fig. 4.15).

Recharge can occur to an unconfined aquifer along its entire surface and to the updip exposure (unconfined portion) of a deeper confined aquifer, called an outcrop area. Most recharge is comprised of precipitation infiltration in higher elevation areas with subsequent discharge in lower elevations. The location of discharge is controlled either by differences in geologic strata, or presence of surface-water bodies. When precipitation infiltrates into porous media that overlies a less permeable layer, and that layer continues aerially to intersect land surface, then groundwater is discharged to the surface as a seep or spring. Recharge also can occur to confined aquifers away from outcrop areas by leakage from adjacent confining beds. The occurrence of recharge is common in humid areas of the world, where

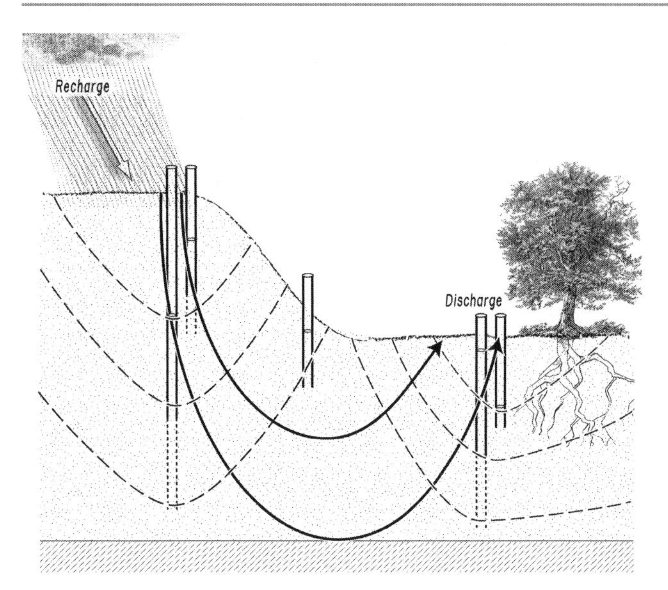

Fig. 4.15 Recharge and discharge areas defined using vertically discretized or nested well pairs. In the recharge area, the groundwater level is lower in the deep well relative to a shallow well. In the discharge area, the groundwater level is higher in the deep well relative to a shallow well.

the unsaturated zone is often less thick relative to the thickness of the water-table aquifer. In arid areas where precipitation is less than 10 in./year, the most recent recharge of extremely deep water tables may have happened more than 100,000 years ago (Alley et al. 2002).

Compared to the range of groundwater-recharge rates that tend to fluctuate over time due to differences in precipitation, groundwater discharge tends to remain relatively constant. This is why streams and rivers have relatively constant flow rates between precipitation events or during the early part of a drought. Also, recharge tends to be aerially distributed over a much larger area than more localized discharge areas, whereas discharge areas integrate groundwater flow lines from across the saturated aquifer thickness.

The rather consistent supply of groundwater discharge is one reason that phreatophytes are competitive with other plants, especially in areas that have frequent droughts. For example, trees that grow along streams have been observed to obtain greater heights and develop larger trunk diameters than upland plants of the same species (see Chap. 1). Conventional thought was that these riparian plants used surface water as their water source. However, it was shown using variations in the stable isotopes of water that exist between surface water exposed to evaporation and groundwater, that the large trees, such as maples (*Acer negundo*), contained water that had isotopic values similar to those of the groundwater, but not surface water (Dawson and Ehleringer 1991). Small trees contained a greater proportion of surface water, as shown by the isotopes, presumably because small trees have shallower, less developed roots. As the trees grow, they switch from surface water to groundwater as a source to meet water demands. The use of stable isotopes of water for monitoring the effect of phytoremediation on groundwater is discussed in Chap. 9.

4.7.3 Steady-State Flow and Transient Flow

In an aquifer system where the amount of recharge is equal to the amount of discharge, groundwater levels do not change over time and groundwater-flow conditions reach steady state. We saw this in Chap. 2, in Eq. 2.1 where $water_{inflow} = water_{outflow}$. In reality, however, steady-state flow conditions usually are not attained, because evapotranspiration demands for water limit infiltration and recharge.

4.8 Groundwater and Surface-Water Interactions

Groundwater not in storage tends to discharge to sinks such as surface-water bodies. During time of seasonally lower precipitation or drought, many streams can continue to flow at full stage for extended periods of time. Such baseflow is caused by groundwater discharge. This relation between groundwater and surface water was not widely recognized, however, until relatively recently (Winter 1999; Conant 2004, references therein) even though early hydrology textbooks hinted at the interaction (Wisler and Brater 1956). As we saw in Chap. 2, after plant transpiration and evaporation demands are met, the balance of groundwater enters surface-water sinks.

Many techniques have been developed to investigate the interaction between groundwater and surface water. These range from being as simple as using a shovel or boot heel to dig a depression, called a piezopit, in the stream bank near the level of surface water, allowing the piezopit to fill with water, and then placing small particles on the water surface in the hole and observing their movement (source: Dr. Samuel S. Harrison, Professor of Hydrogeology (ret.), Allegheny College, oral commun. 1988). Alternatively, simple iron or polyvinyl chloride (PVC) pipes can be pushed into the streambed sediment at various depths to determine the magnitude and direction of the vertical head gradient. More complicated devices called seepage meters also can be installed to determine the volume of groundwater discharge. The simplest of these devices, a cut-off 55-gal drum, is placed in the bed sediments of the surface-water body as described in Lee (1977); refer to Rosenberry and Menheer (2006) for a history of this meter. A valve is attached to the top of the drum lid, and a plastic bag filled with a known volume of water is attached to the valve. As groundwater flows through the open end of the drum inserted into the bed sediment, groundwater increases the volume of water in the bag. Conversely, if the drum is located in a groundwater discharge area, the initial bag of water will decrease in volume as it recharges the bed sediments. Sampling for differences in water temperature or chloride concentration also can be performed to determine the presence of groundwater discharge (Conant 2004).

The majority of the downgradient part of most surface-water systems in humid areas tends to be at low topographic elevations. When the surface-water level is lower than that of the groundwater in adjacent stream bank or streambed sediments, groundwater discharge occurs. As can be seen in Fig. 4.16, groundwater recharge that occurs over a broad area and far upgradient tends to discharge in smaller areas, such as lakes, ponds, wetlands, seeps, springs, or streams. Even when groundwater discharges to larger bodies of surface water, the location of groundwater discharge typically is restricted to near the shoreline.

4.8.1 Wetlands and Swamps

Although groundwater tends to discharge to surface-water bodies, it is possible for surface-water bodies to recharge groundwater systems under natural conditions in certain types of geologic environments. When this occurs, the surface-water bodies are called losing streams. Losing streams often are found in arid areas where infrequent precipitation rapidly enters dry river channels but then infiltrates into deeper sediments before substantial open-channel flow develops. Losing streams also occur in humid areas. For example, most wetlands or swamps are located in low-lying areas, which receive both runoff and groundwater discharge, and are, therefore, considered to be gaining. Some swamps, however, like the Okeefanokee Swamp in Georgia, are located on higher ground than surrounding areas and have a higher surface-water level than the surrounding wetland areas. Such swamps contain water supplied by precipitation rather than an inflowing stream, and the water will leak into shallow aquifers.

4.9 Wells

Because groundwater is not readily observable, any direct connection to it provides not only a source of water but also further insight into the subsurface. A well is an excellent means to achieve both goals; a well truly provides a looking glass into the subsurface. A well is a hole created into the earth that intersects saturated sediments or bedrock. A well is a point of artificially created discharge when pumped, if not drilled into an artesian system. As groundwater is removed from a well, the groundwater-level surface in the well becomes lower than the groundwater surface in the surrounding aquifer. As a result, groundwater flows to

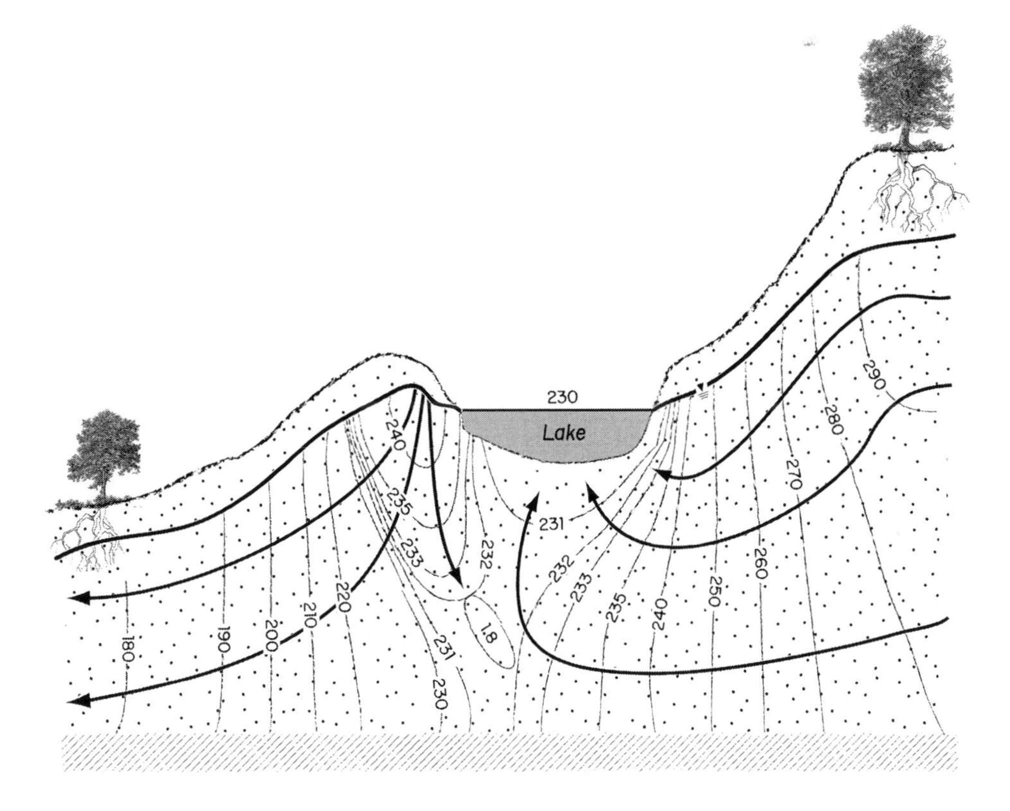

Fig. 4.16 Deep groundwater-flow lines (*arrowheads*) across the full thickness of an aquifer can converge near the surface at focused locations of discharge, such as shallow surface-water bodies (Modified from Fetter 1988). The contour interval for the equipotential lines is in feet, and is variable. One foot is equivalent to 0.304 m.

the well in response to the induced hydraulic gradient toward the well (Fig. 4.17). If the pumping rate, Q, in the well is greater than the rate of groundwater flow to the well, the groundwater level will continue to decline until the rate of groundwater flow equals the rate of pumpage, the groundwater level goes below the pump intake, the well goes dry, or groundwater is released from storage. In plan view, groundwater can converge to the well from all directions. From a flow net analysis, this convergence and head decline at the location of a pumped well is called the cone of depression (Fig. 4.17).

The groundwater resource has been accessed by wells by many civilizations throughout history in many areas of the world (Bennison and Bollenbach 1947). Perhaps the earliest wells were constructed in Israel during the Pre-Pottery Neolithic period around 6,000 year BC (Galili and Nir 1993). Some of the earliest attempts to drill deeply to access groundwater were made by the Chinese and Egyptians nearly 5,000 years ago (De Weist 1965). The Chinese used a churn drill made of wood to go through solid rock; deepening these wells took generations of laborers. Joseph's well in Cairo is almost 300 ft deep, through solid rock, and supplied groundwater to many civilizations including the Greeks and Romans until about 300 BC. A well was used by Eratosthenes (276–195 BC) to provide the earliest estimate of the circumference of the earth. In the United States, one of the oldest drilled wells is located in St. Augustine, Florida, drilled in the seventeenth century.

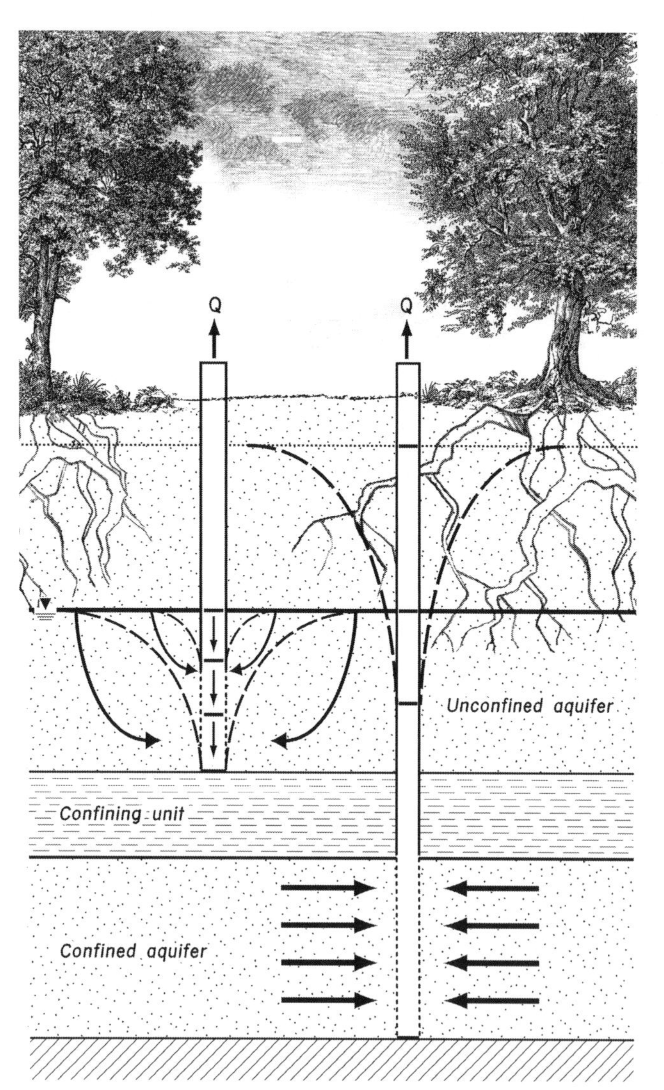

Fig. 4.17 A cone of depression (*dashed line*) for a pumped well in a water-table aquifer and confined aquifer (Modified from Heath 1983).

Unconfined and confined aquifers characterized by differences in groundwater yields also have differences in the cone of depression. In unconfined aquifers, pumped groundwater is derived from the specific yield of the aquifer sediments by gravity flow. As the groundwater level in the aquifer decreases the transmissivity, T, decreases, because $T = Kb$. Moreover, the rate of the development of the cone of depression is slow because drainage is by gravity only. In confined aquifers, pumpage reduces the pressure surface rather than dewatering the aquifer sediments. Pumping causes a decrease in groundwater pressure and can produce expansion of the remaining groundwater and compression of the aquifer sediments. As such, the cone of depression for confined aquifers expands more rapidly than for unconfined conditions.

4.10 Groundwater Fluctuations

Measurements of the groundwater level at a particular well over time reflects the basic water budget of Eq. 2.2; $\text{water}_{inflow}\text{-water}_{outflow} = \Delta S$. The groundwater-level surface in an unconfined aquifer is not fixed in space or time; it is exposed to atmospheric conditions. If water inflow exceeds water outflow, such as when it rains and recharge occurs, the groundwater level increases. If water outflow exceeds water inflow, the groundwater level declines. If water inflow equals water outflow, the groundwater level does not change; however, groundwater may be removed or added to the system but will not be measurable because the flows occur at equal rates. On the other hand, it is possible for the groundwater level to increase not because of an increase in inflow but because an outflow has stopped.

For unconfined water-table conditions, the relation between the amount of precipitation and the potential for the groundwater level to increase by recharge can be determined by using aquifer porosity. For example, if 1 in. (2.5 cm) of precipitation is added to an aquifer that has 50% total porosity, the groundwater level can rise 2 in. (5.1 cm). The rate of rise will be faster if the soil is already wet, and slower if it is dry, due to that particular soil's hysteresis curve. One inch of precipitation across 1 acre of land surface, or 43,560 ft^2 (4,007 m^2), is equal to a volume of 27,154 gal (102,642 L) of water. The weight of this water is more than 113 tons.

A well pumped at a rate equal to that of groundwater flow with no additional change in groundwater level over time represents steady-state flow conditions. If, however, the pumping rate is increased, the groundwater level will decline over time and represents transient conditions. This subtle difference is important when it comes to understanding the source of groundwater removed from a well, or in the case of a phytoremediation system, the source of groundwater transpired by a plant. Going back to the water budget equation introduced in Chap. 2, we can state that

$$Inflow - Outflow = \Delta S. \tag{4.13}$$

The groundwater taken up in a pumped well, or potentially by a tree with roots that reach the capillary fringe and water table, can come from an increase in groundwater inflow, a decrease in groundwater outflow, or from a drop in storage. For a water-table condition where storage is minimal, ΔS approaches zero such that

$$Inflow - Outflow = 0. \tag{4.14}$$

Thus, groundwater removed can be a result of either an increase in inflow or a decrease in outflow. This fact can be useful, as we will see in the next section, in attempting to apply phytoremediation concepts to reduce the flow of contaminated groundwater to adjacent surface-water bodies.

Other factors can affect groundwater levels, such as changes in pressure over the aquifer. One of the first records of groundwater fluctuations caused by an increase in pressure above the aquifer was in a report of a study in New York where an increase in the groundwater level in a well located near a railroad occurred after a train passed. Other causes include tidal fluctuations, barometric pressure fluctuations, and even earth tides. Van Hylckama (1968) reported a diurnal fluctuation in groundwater levels in plastic-lined tanks that contained soil, plants, and an artificially controlled groundwater level. When observed in bare tanks that contained no plants, the groundwater fluctuation was correlated to changes in atmospheric fluctuations in barometric pressure acting upon air trapped in the tanks during filling. Conversely, in tanks that contained transpiring plants, diurnal groundwater-level fluctuation was not related to changes in barometric pressure. Van Hylckama (1970) later reported that the depth to water table was an important factor in controlling how much groundwater could be used by plants such as saltcedar, and noted that even small differences in depth to groundwater, from 4.92 to 6.8 ft (1.5–2.1 m), for example, can reduce groundwater use even though the plants remain alive.

In Van Hylckama (1974), the relation between depth to water table and plant and groundwater use was investigated further. At a site near Buckeye, Arizona, plastic-lined evapotranspirometer studies revealed that for planted *Tamarix*, the groundwater use was 85 in./year (215 cm/year) when the depth to water table was set at 5 ft (1.5 m) below land surface. When the depth to water table increased to 7 ft (2.1 m), groundwater use decreased to 60 in./year (152 cm/year); when the depth to water table decreased to 9 ft (2.7 m), groundwater use decreased to less than 40 in./year (100 cm/year).

4.11 Groundwater Models

As described previously, the flow of groundwater in porous media results from a combination of independent and dependent variables. Before Darcy used his innovative column experiments to look at flow through porous media in order to solve a surface-water-quality problem, no predictive tool had been developed for use in assessing the effects of changes in these variables, such as head gradients, on the flow of groundwater. Darcy's Law provided hydrologists with such a tool, which enabled subsequent hydrogeologists to test various hypotheses without having to perform such tests in the field. As such, Darcy's Law may be considered one of the first groundwater-flow models and is still being widely used every day.

Since 1856, many physical, electrical, analytical, and numerical models that describe the flow of groundwater under various aquifer and flow conditions have been developed. These models were created to answer fundamental questions about the potential quantity of groundwater available in a pumped well field. Models that address the effect of various solutes dissolved in groundwater, such as salt or petroleum-based or halogenated contaminants have been developed to address questions regarding groundwater quality. In either case, it is important that model output be used to test hypotheses or to refine a conceptual model of the hydrologic system under investigation, such as establishing reasonable ranges for particular parameters, rather than expecting it to be the unique model that will address accurately all current and future questions.

The interactions described previously, such as groundwater and surface-water interactions, the water-table surface, recharge, discharge, and evapotranspiration, can be explained and modeled best by using the concept of a groundwater system. This system is defined by boundaries and conditions of groundwater flow or no flow and can be defined mathematically at each boundary. Examples include constant flux boundary, specified head, and the magnitude of evapotranspiration related directly to the depth to water table.

In many cases, vertical groundwater flow to tree roots from groundwater flowlines beneath the water table surface need to be simulated. This is especially important to determine if hydrologic control can be achieved, which is discussed in Chap. 8. Analytical models that are idealized and describe the removal of groundwater from wells that fully penetrate the thickness of a particular aquifer cannot be used in this situation. However, numerical models account for variations in the spatial hydraulic conductivity, K, in an aquifer, using ratios of vertical-to-horizontal hydraulic conductivity to account for anisotropy and the ability to create multiple layers with model cells or nodes discretized to the appropriate scale. It may be possible to ignore the effect of anisotropy on groundwater flow to plant roots, but the results will underestimate the effect of a potential plantation. Three-dimensional numerical models can be used to understand this phenomenon, because they account for groundwater flow in the lateral and vertical directions.

Numerical groundwater models can be used to account for all of the variables in a water budget for a particular site. As was stated previously, evapotranspiration can be the largest loss of water in a basin. In the groundwater-flow model MODFLOW (McDonald and Harbaugh 1988), loss of water from an aquifer is estimated by using a linear relation between evapotranspiration and water-table depth. Evapotranspiration occurs from the water table linearly until a predetermined depth is reached and evapotranspiration is assumed to be zero. An alternative approach was taken by Matthews et al. (2003) in which they simulated evapotranspiration in MODFLOW by using the recharge module, which specified recharge as a negative value. This approach was an improvement because the magnitude of evapotranspiration was not related to the water-table depth. This alternative approach does not represent flow of water in the unsaturated zone, however. Currently, there is no explicit way to simulate interactions of groundwater with vegetation that link root growth and groundwater use.

The MODFLOW groundwater model can be coupled to a conservative tracer package called MODPATH (Pollock 1994), which can be used to visualize individual groundwater flow paths following the release of artificial particles of water into the calibrated groundwater-flow model. This can be done to evaluate the effect of anisotropy on the deflection upward of groundwater flow lines in a water-table aquifer toward tree roots at or near the water table and capillary fringe. MODPATH also can be used to represent a plume of contaminant released from a source area and to determine the final size a phytoremediation planting must be to capture all the contaminant mass and provide maximum hydrologic control (Matthews et al. 2003); this is discussed further in Chap. 15.

Models can be used to further understand groundwater-flow systems. Models are more effective at revising conceptual models, rather than being actual representations of nature. This is especially true as the uncertainty of the parameters used and simulated time from calibrated conditions increases.

4.12 Summary

Groundwater is not readily observable but is readily quantifiable. Monitoring wells can be installed in unconfined or confined aquifers to measure groundwater levels and this information used to determine the volume and flow direction of groundwater by using Darcy's Law. Moreover, monitoring wells can be used to observe the effects of plants on groundwater, and is more fully presented in Part II.

Why is this information important to the phytoremediation of contaminated groundwater? The fundamental interaction that exists between plants and groundwater can be observed using these same fundamental concepts of groundwater hydrology, such as wells, measurements of groundwater levels, and Darcy's Law.

5 Plant and Groundwater Interactions Under Pristine Conditions

An initial understanding of plant and groundwater interactions did not follow a straightforward path. For example, it took many years and developments in forensic chemistry to elucidate that the oxygen released by plants during photosynthesis was derived from water absorbed by roots rather than from atmospheric CO_2 absorbed by the leaves. Also, geochemical techniques that involved stable isotopes revealed that trees that grow on the banks of rivers tap groundwater rather than the seemingly more available source provided by surface water. Moreover, the facts that groundwater is not readily observed and that plants release invisible water vapor makes it easy to forget that plants move enormous volumes of water on a daily basis, a process that is essentially hidden in plain sight.

The consequences of such hidden interactions between plants and groundwater can be understood, however, because the mass of water in a particular basin has to be conserved such that $P - ET = R$. The foundation of phytoremediation of contaminated groundwater is part of this fundamental interaction as revealed in the parameter of T in ET.

Some of the best evidence to support the application of plants to interact with contaminated groundwater is provided by plant and groundwater interactions that occur under natural, pristine environments. Some of these examples, including perhaps the first recorded observation of plant and groundwater interactions in 1926, were discussed in Chap. 2. More recently, the study of the interaction of groundwater, plants, and other ecological systems has been called by various terms such as ecohydrology, hydroecology, among others (Bond 2003; Lubczynski 2009; Lowry and Loheide 2010). The focus of this chapter is to provide additional examples of naturally occurring interactions between plants and groundwater and how these interactions can affect recharge, surface-water flow and geochemistry, and groundwater hydrology and water quality. This information on natural interactions provides the foundation for the application of these interactions at sites characterized by contaminated groundwater, and are discussed in Parts II and III.

5.1 Plants and Groundwater Recharge

Plant interactions with water are an important part of the hydrologic cycle as described in Chap. 2. Transpiration and evaporation return to the atmosphere up to 70% of the average annual precipitation in a particular basin. Of the 70% of water removed by evapotranspiration, the component of this total driven by transpiration alone can range from 5% to 80% (Larcher 1983; Moreo et al. 2007). Part of the transpired water can be composed of recent precipitation, precipitation that infiltrated into the upper soil layers, water from deeper within the unsaturated zone, or shallow or deep groundwater. In many areas, recharge is less than 10% of annual precipitation. From a mass-balance perspective, the processes of evaporation and transpiration limit the amount of water available for recharge. Moreover, the allocation of precipitation to evapotranspiration decreases the amount of groundwater available for discharge to either natural areas, such as springs, lakes, or rivers, or even to wells.

The effect of water availability on plant growth is at least anecdotally recognized by many laypersons through personal experience. For example, the relative width of tree rings, revealed after cutting down a tree, indicates the gross effect of water availability on annual tree growth. Within each annual growth ring, the springwood is lighter and thicker because the water-transporting xylem grew rapidly in response to the higher availability of water and other resources. The winterwood is denser and occupies less space, because water is less available when formed. These observations have been used recently by biologists to understand past and future climate conditions and effects on plant and water use.

J.E. Landmeyer, *Introduction to Phytoremediation of Contaminated Groundwater*,
DOI 10.1007/978-94-007-1957-6_5, © Springer Science+Business Media B.V. 2012

5.1.1 Precipitation and Potential Evapotranspiration

The effect of plant processes on recharge is no more evident than during comparison of the occurrence of precipitation relative to maximum evapotranspiration. Recharge can occur only where and when precipitation exceeds evapotranspiration. Some areas of the United States, for example, receive more precipitation during the summer and fall months, such as the southeastern United States which is in the path of hurricanes from across the Atlantic and the Gulf of Mexico. However, recharge during this time is low, because evapotranspiration is high. In the northeastern United States, more precipitation occurs during winter and spring (Fig. 5.1).

Plants affect the timing and volume of recharge through different processes. The presence of plants leads to a thick layer of organic leaf litter that increases infiltration rates. Extensive root systems increase the hydraulic conductivity of the soil and unsaturated zone around the roots, which can lead to increased infiltration amounts and rates. Johnston (1987) observed that infiltration rates increased from 0.20 to 3.97 in./year (7.20–100 mm/year) in a planted field and the time for infiltration to reach the water table decreased over time. Soil porosity and infiltration rates were 9 times faster for planted soils relative to bare soils (O'Conner 1985). However, increased infiltration rates support increased transpiration rates which also may decrease recharge. For some deep-rooted plants, infiltration rates may not increase because of higher vertical root hydraulic conductivities relative to lateral roots and fewer roots with depth (Pate et al. 1995).

Arid areas that have sparse precipitation and high evapotranspiration rates when combined lead to an interesting recharge pattern. In some parts of Arizona in the southwestern United States, for example, the depth to water table can be in excess of 300 ft (91.2 m). Recharge from recent precipitation is almost nonexistent. It has been shown, instead, that groundwater moves upward toward the land surface in response to high evapotranspiration (Andraski et al. 2003). Not only is no current recharge occurring, the evapotranspiration demand causes ancient recharge that occurred thousands of years ago to be brought to near the land surface. Deep water-table conditions also occur in humid areas, such as the Sand Hills regions of the Atlantic Coastal Plain, but the occurrence of a similar vertical movement of groundwater in response to evapotranspiration has not received much attention.

Jordan and Fisher (1977) examined the relation between precipitation and evapotranspiration and recharge on the island of St. Thomas in the Virgin Islands south and east of the tip of Florida. The island is about 14 mi (22.5 km) long and 2 mi (3.2 km) wide and has only two perennial streams. Average precipitation is about 40 in./year (101 cm/year) but high evapotranspiration rates permit only about 1–2 in. (2.5–5 cm) of recharge each year. Another effect of high evapotranspiration is the enrichment of salts in the remaining groundwater, up to 20 times more concentrated than precipitation. Conversely, Long Island, New York, an island of similar size but located in a more temperate climate, has recharge of more than 50% of the 40 in. (101 cm) of annual precipitation. This is due to the highly porous sediments of the underlying aquifers, high rates of groundwater flow, lower levels of solar radiation for shorter annual periods, and less potential evapotranspiration.

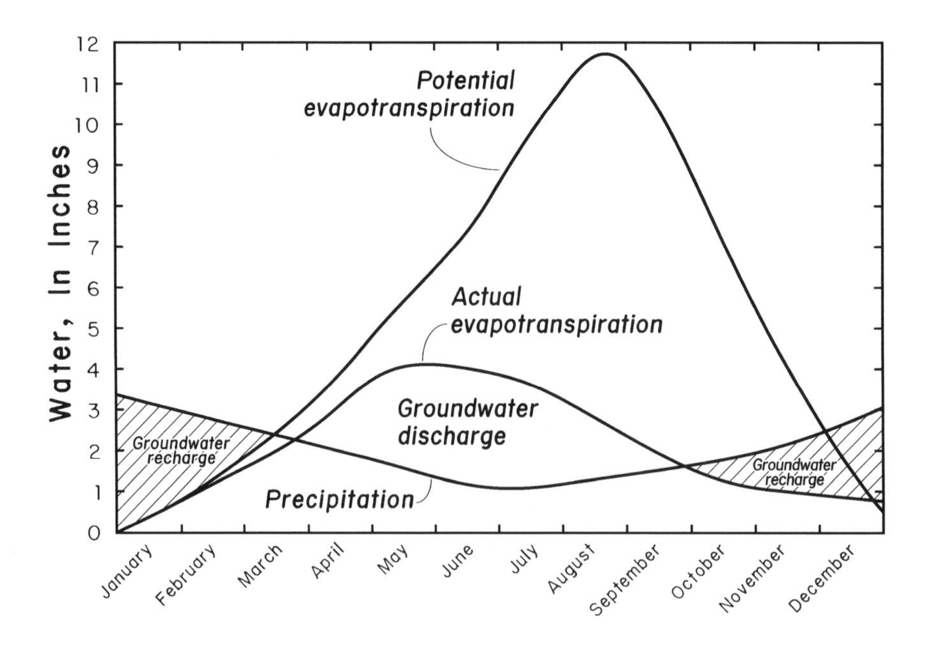

Fig. 5.1 The relation between precipitation and potential evapotranspiration, ET_p, actual evapotranspiration, ET_a, and recharge for a typical area in the humid northeastern United States, such as Pittsburgh, Pennsylvania (Modified from Fetter 1988). One inch is equivalent to 2.54 cm.

5.1.2 Reduction in Recharge

Phreatophytic plants under natural settings often provide little or no economic value. On the other hand, such plants can be of economic value when used as short-rotation coppice (SRC) for wastewater treatment, pulp production, or biomass production as an alternative energy supply. The effect of SRC on the water budget of two crops grown in different areas with different plants and, therefore, potentially different water-budget implications were examined by Allen et al. (1999), and their monitoring data are applicable to the phytoremediation of contaminated groundwater. The site was a 1.8-ha plantation in England. Six poplar clones were planted 3 ft (1 m) apart. This planting was useful for studying not only the effect of plants on water budgets but the effect of different plant physiologies, because one clone (Beaupré, *P. trichocarpa* Torr & A. Gray x *P. deltoides* Bartr. *ex* Marsh) had twice the leaf area and was taller than the other clone (Dorschkamp, *P. deltoides* x *P nigra* L.). As a result, the authors were able to show that the Beaupré clone had higher transpiration rates during June 1994—a mean of 5.0 mm per day compared to 2.4 mm per day for the other clone. Because of these high transpiration rates, Allen et al. (1999) concluded that such a rate of removal of soil water would adversely affect water resources in the United Kingdom.

In some cases, the plant-facilitated reduction in recharge was used to solve rather than create water-quantity problems. Recharge reduction was observed, for example, following the introduction of mesquite (*Prosopis julifora*) trees planted in the 1980s by local government officials around the city of Khartoum following a drought in the mid-1970s. The idea was the drought-tolerant, groundwater-using mesquite would slow the advance of the surrounding desert sands, which were encroaching at a rate near 10 mi/year (16 km/year). Because mesquite seeds are spread easily by birds and animals, however, the species invaded areas used for agriculture and has spread about 1,000 acres per year (4.0×10^6 m^2/year). As a result, recharge has declined and groundwater levels in wells have decreased from levels prior to those before the spread of the mesquite.

In some ephemeral river systems typical in arid climates, recharge areas often are constrained to dry river beds that contain the most permeable sediments. Phreatophytes located in these dry river beds, however, often intercept infrequent infiltration before it can become recharge (Fig. 5.2).

5.2 Plants and Groundwater Discharge to Surface Water

Trees located along streams use groundwater and not necessarily streamwater to meet evapotranspiration demands (Dawson and Ehleringer 1991) and at first may seem counterintuitive. After all, the density of trees increases near most surface-water bodies. This occurs in arid areas along stream channels and in humid areas in flood plains, swamps, and oxbow lakes. From an ecological standpoint, it is

Fig. 5.2 Phreatophytes, such as these saltcedar (*Tamarix spp.*) growing in a dry river bed in Tucson, Arizona, intercept infrequent infiltration before it can become recharge (Photograph by author).

advantageous for plants to use surface water when it is available but also to have deep roots to tap groundwater when surface water is less or even unavailable. These deep roots serve to allow the plant to survive drought conditions when surface water derived from runoff is scarce.

Some plants thrive only if groundwater is the water source. Harvey et al. (2007) investigated plant and groundwater interactions at peatland fens, or wetlands, in central Nebraska. This area of the United States is hot and arid, where potential evapotranspiration is 2–3 times annual precipitation, and is perhaps the last place that one would expect to find peatlands, which are more common in the humid southeast states such as Florida. The fens of Nebraska exist, however, because they are sites of groundwater discharge. Plants that can be found in these fens include rush aster (*Aster junciformis*), mud sedge (*Carex limosa*), and water sedge (*Carex aquatilis*).

At least two reasons why fens rely on groundwater were suggested by the work of Harvey et al. (2007). First, a microclimate of humid conditions in the wetlands, surrounded by the arid area, is created from the transpiration of groundwater. Second, plant growth is supported by nutrients in the groundwater. This same phenomenon of plant-transpiration-mediated nutrient transport and uptake from groundwater was seen in the Florida Everglades by Ross et al. (2006), who investigated the possible reasons why the groundwater beneath the many Tree Islands in the ridge and slough landscape typical of the Everglades is characterized by high phosphorus concentrations, even though the surrounding surface and groundwater have primarily low phosphorus concentrations under ambient conditions. Ross et al. (2006) measured groundwater levels in shallow wells on Tree Islands and observed that the lowest groundwater levels occurred during the summer months when evapotranspiration was the highest and the plants were using groundwater. Moreover, the deeper wells had higher groundwater levels than the shallower wells, which indicated a net upward flow of groundwater to roots.

The relation between *ET* and streamflow in the humid southeastern United States was investigated by Palmroth et al. (2010). They report for conditions in North Carolina that long-term records of streamflow, or discharge, can be used successfully to estimate regional values for *ET*. They show that the sum of annual stream discharge and *ET* was equal to annual, and independently, measured, precipitation; this result confirms the relation expressed in Eq. (2.5).

5.2.1 Reduction in Surface-Water Flow

Plants that grow along rivers that use groundwater affect surface-water quantity and flow by default. In short, plant transpiration of groundwater intercepts water that, given the opportunity, would become surface water. Riparian, from the Latin *ripari* meaning the bank of a stream, plants have been shown to affect the level and volume of surface waters that receive groundwater discharge. The effect of the riparian removal of groundwater is apparent for most rivers in the United States that are characterized by the lowest water levels and discharge during the growing months even though precipitation is often the highest during these times. An excellent review of the effect of riparian plants on hydrological processes is provided by Tabacchi et al. (2000).

The effect of phreatophyte uptake of groundwater on surface-water flows was first investigated in the early 1900s by White (1932) of the USGS, and in the 1940s by Robinson (1958) also of the USGS. In 1944, Robinson (1965) instrumented wells near the Gila River in Arizona in a large stand of saltcedar (*Tamarix*). During March 1944, before the growing season for saltcedar, no daily fluctuation of groundwater level was recorded in the well. During the growing season in June, however, Robinson observed up to 0.19 ft (0.05 m) in daily groundwater-level fluctuations. In October, after the saltcedar went dormant, little to no daily fluctuation was observed. During this same period, the level of the water in the Gila River also was monitored with an automatic recorder. During June, the daily fluctuations observed in the stage closely followed the groundwater-level fluctuations in the well. During the week-long period that daily stage fluctuations were observed, the river stage was lower at the end of the period, which indicted that not only was there less flow in the river each day because of transpiration, but the transpiration of groundwater also decreased the volume of surface water.

An additional effect of phreatophytes is a reduction in gradient from the originally higher groundwater levels to the river (Robinson 1958). Thomas (1952) reported that the Green River, which is characterized by valleys dominated by up to 40,000 acres (1.61×10^8 m^2) of phreatophytes, loses an average of 552 acre-ft (680,616 m^3) annually, roughly 278 ft^3/s, because of the uptake of groundwater by phreatophytes.

Other examples of the interaction between plants, groundwater, and surface water exist. Peak stream flows observed in the Rio Grande River decreased following an increase in non-native species in riparian areas (Roelle and Hagenbuck 1995). Another example is the removal of groundwater by trees prior to discharge to surface water in the Rio Grande and Rio Bravo Rivers at the Gulf of Mexico. A narrow border of trees and tall grasses flanks the shoreline on both sides of the bottom of steep cliffs that lead down to the river. A similar reduction in stream flows that resulted from the uptake of groundwater by phreatophytes was observed in the Northern Great Basin by Nichols (1993).

In his study, Nichols (1993) estimated that greasewood (*Sacrobatus vermiculatis*) could remove up to 9 in. (24 cm) of groundwater per year that otherwise would have discharged to the river.

Other studies of consumptive use of groundwater by plants and the effects of reduced surface-water availability can be found in Mower and Nace (1957), Mower et al. (1964), and Mower and Feltis (1968). Mower and Nace (1957) reported that phreatophytes use from 2 to 7.5 acre-ft/year (from 24,670 to 92,475 m^3/year) of groundwater within a 13,000 acre (5.2×10^7 m^2) part of the Malad Valley, Idaho. In Mower et al. (1964) different methods were used to determine consumptive use of water by phreatophytes in the Recos River basin in New Mexico. In this area of New Mexico, plant roots reach no deeper than 20 ft (6 m) below land surface. Mower and Feltis (1968) reported that between 135,000 and 175,000 acre-ft (1.6×10^8–2.1×10^8 m^3) of groundwater was consumed by phreatophytes in a 440,000-acre (1.78×10^9 m^2) portion of the Sevier Desert in Utah, more than four times the amount withdrawn by wells!

Bond et al. (2002) reported that a daily pattern in streamflow at an Oregon study site could be explained by the direct transpiration of groundwater, especially during the summer months when transpiration rates were higher. Although the decrease in stream base flow was small (between 1% and 6% of maximum measured base flow), it was an order of magnitude larger than water losses from evaporation. The importance of this study in terms of previously recognized seasonal relations between plants, groundwater use, and streamflow is that it demonstrates that this relation is based on the continuous, daily competition between plants and surface water for groundwater.

Another example of the well-studied interaction between phreatophytes and groundwater uptake and surface-water flow patterns is provided in the upper San Pedro basin in Arizona (Leenhouts et al. 2006). This river changes from an interrupted perennial stream to a continuous perennial stream along its course and has a dense riparian community along most of its length. Cottonwood and willows are the most abundant tree species in the flood plain. The presence of cottonwood was correlated to the median annual maximum depth to groundwater of 6.5 ft (2 m), and for willows 5.9 ft (1.8 m). The researchers showed that the continually decreasing groundwater levels from 30.8 to 31.8 ft (9.4–9.7 m) below land surface were caused by increased evapotranspiration.

In addition to the removal of groundwater by plants adjacent to rivers, groundwater withdrawals by man also affect riparian environments. In the San Pedro River valley in Arizona, the numbers of native riparian trees, such as the cottonwood (*Populus fremontii*) and Goodding willow (*Salix gooddingii*), have decreased and been replaced with tamarisk (*Tamarix ramosissima*) as groundwater withdrawals have increased to support agricultural, urban, and industrial needs (Lite and Stromberg 2005). As the depth to water table increased beyond 11.4 ft (3.5 m) below land surface, for example, Goodding willows had the highest rate of mortality, relative to areas where the depth to water table was more constant. This resulted in a reduced quality of wildlife habitat and increased potential for flood peaks and erosion. Results of a study on riparian cottonwoods along the Mojave River in California also showed higher tree mortality rates, between 60% and 95%, when groundwater-level declines were greater than 4.5 ft (1.4 m) (Scott et al. 1999).

Similar effects of groundwater withdrawals and riparian plants have been observed by using repeat photography. Some of the best examples come from the southwestern United States where mesquite and cottonwood trees were growing in the riparian area of the Mojave River in 1917 but after almost 80 years of groundwater development, all native species are gone, and tamarisk has replaced the native cottonwoods.

Lines and Bilhorn (1996) reported a decrease in surface-water flow in the Mojave River in southern California as a result of groundwater uptake by riparian vegetation. Based on the annual depletion of surface-water base flow, the riparian vegetation was estimated to have removed about 600 acre-ft (739,800 m^3) of water per year along a 2-mi (3.2 km) long stretch of the Mojave River.

As was outlined in Chap. 1, most of the original studies of phreatophytes occurred in the arid western United States where the distinction between phreatophyte and non-phreatophyte is clear and unambiguous. However, phreatophytes also are present in more humid, eastern areas of the United States. In more humid conditions, phreatophytes often are ignored, because their effect on water supplies is less drastic, especially if surface water is readily available. Phreatophytes in humid areas are predominately found where the depth to water table is shallow, but also can exist where the water table is more than 100 ft (30 m) below land surface. Along most flood plains in the eastern United States, phreatophytes can occupy large areas of point bars. The Congaree National Park near Columbia, South Carolina, for example, has the largest stand of bottomland hardwood trees in the United States, including willow (*Salix nigra*) and poplar (*Populus deltoides*) along the banks of the meandering Congaree River. In other low lying spots that receive groundwater discharge massive loblolly pines (*Pinus taeda*) exist that are 3 times older than most long-lived loblollies elsewhere, supporting the usage of the common name loblolly, which means moist depressions. In the Sand Hills regions of the eastern United States Coastal Plain, longleaf pine (*Pinus palustris*) have tap roots that can reach the water table deeper than 80 ft below land surface.

The effect on surface-water flow by plant uptake of groundwater in humid areas becomes more apparent during times of drought. At the Coweeta Experimental Forest in the Appalachian Mountains of western North Carolina, for example, the effect of trees on streamflow was investigated as early as 1947 (Dunford and Fletcher 1947). Similarly, the effect of transpiration on the daily fluctuations of groundwater adjacent to the North River near Annapolis, Maryland, during the summer of 1954 was observed as daily fluctuations in surface-water discharge. During a study period of 5 days in July 1954, the discharge of surface water varied from about 2.74 to 3.29 ft^3/s each day, for a change of about 0.55 ft^3/s/day. If this difference in surface-water discharge was attributed primarily to water lost daily to transpiration, it would be about 0.56 acre-ft/day (690 m^3/day). In a study of two small watersheds in the Coastal Plain of Georgia, Bosch et al. (2003) reported that increased evapotranspiration and plant uptake of groundwater affected the discharge of groundwater to streams by affecting the hydraulic gradient between the shallow aquifer and the surface water. During winter months and reduced evapotranspiration, the ET_p ranged from 48% to 74%; during the summer growing season, ET_p exceeded monthly average precipitation (Bosch et al. 2003). As a result, the water table was higher during winter months and lower during summer months. Gradients of the water table toward the surface water approached 3%, or the slope of the topography, during winter months. During drier periods, the lack of recharge or uptake of groundwater by plants dropped the hydraulic gradient to 1% or lower.

Fluctuations in the water table caused by phreatophytes were used to determine the effect of various approaches to control use of groundwater by phreatophytes in the southwestern United States (Butler et al. 2005). Phreatophyte consumption of groundwater prior to discharge is the cause of decreased surface-water flows in the Cimarron basin in Kansas. Many different control measures to increase surface-water flow by reducing the phreatophyte population have been done, and the metric of effectiveness for increasing streamflow is to monitor groundwater fluctuations. In simple terms, the presence of phreatophytes causes a diurnal fluctuation in monitoring wells; as these plants are removed, the diurnal fluctuations decrease.

In order to affect surface-water flows, the flow rate of water through the plant, as transpiration, has to be substantial, there has to be many plants, or both. Van der Leeden et al. (1990a, 1990b) published a table of the consumptive use of common phreatophytes in the western United States. The values are reported as use of groundwater in acre-ft per acre. Rates range from 3.3 (*Prosopis velutina*) to 7.8 (*Juncus balticus*) acre-ft per acre (4,068–9,617 m^3/acre). Measurements of sap flow, a surrogate for transpiration, in riparian trees were correlated to fluctuations in surface-water flows in a basin in Oregon (Bond et al. 2002). The daily variations in surface-water flows were recorded during summer drought periods, when flows are sustained by groundwater discharge (base flow) and are a function of the daily fluctuations in sap flow of streamside trees, such as red alder (*Alnus rubra* Bong.) and the evergreen, Douglas Fir (*Pseudotsuga menziesii* (Mirb.) Franco). The measured surface-water flows during summer were lower daily peak flows, and Bond et al. (2002) used this difference to estimate the amount of water that the riparian plants would have to transpire from the groundwater to account for this flow difference; essentially, the tree uptake of groundwater truncates that water which would have become surface water. Refer to Table 5.1 and 5.2 for additional details from the above studies.

Table 5.1 Groundwater use by phreatophytes growing along riparian systems.

River or State	Date	Acre-ft/acre[a]/year (m^3/m^2/year)	Study area (acres, mi or km)
Green river	1950s	552 (168)	40,000 acres (1.6×10^8)
Utah	1958	175 (53)	440,000 acres (1.7×10^9)
Idaho	1957	7.5 (2.2)	13,000 acres (5.2×10^7)
AZ	1950s	9 (2.7)	NA
MD	1954	204 (62)	NA
CA	1996	600 (182)	2 mi (3.2 km)
CA	1995	6,000 (1,828)	

NA means no data available
[a]The term acre-feet per acre (*acre-ft/acre*) is equivalent to the depth of water in feet (*ft*), such that 3 ft refers to 3 acre-ft/acre

Table 5.2 Percentage of surface-flow reduction by phreatophyte uptake of groundwater.

Study area	Groundwater removed, in acre-ft/year (m^3/year)	Total of surface-water flow, in percent
Cottonwood wash, AZ		
(with plants)	80 (along 4 river mi)	18
(after plants removed[a])	42	12

[a]Bowie and Kam (1968)

5.2.2 Increase in Surface-Water Flow

Plants can decrease surface-water flow, but can their removal also increase flow? The effect of clear-cutting hardwood forests on stream hydrology and ecology was studied at the Hubbard Brook Experimental Station in the 1970s. Overall, removing trees, whether by logging, infestation, or fire, increases the amount of water available for recharge, and the discharge to local streams increases. This increase in discharge is assumed to be a result of the increase in groundwater discharge, or base flow (Bosch et al. 2003; Verry 2003).

Riparian plants in the western United States, which is characterized by little annual precipitation, can affect the direction of flow between surface water and groundwater. For example, the Gila River in Arizona flows into the San Carlos Reservoir. The area is arid averaging less than 4 in. (10 cm) of precipitation per year. Melt water from adjacent mountains is the major source of water to streams in Arizona, including the Gila River. Because of the coarse characteristics of the surficial deposits across the lowlands of Arizona, segments of streams commonly act as sources of recharge to groundwater, because the total head is higher in the streams than in the shallow aquifers. Many phreatophytes along the banks of these losing-stream segments send roots to the shallow water table.

A 10-year study, conducted by the USGS and called the Gila River Phreatophyte Project, began in 1962. The study area was along a 15-mi (24 km) reach of the Gila River. Up to 6,000 acres (2.4×10^7 m^2) were covered by phreatophytes, mostly saltcedar and mesquite. The major component of flow to the shallow aquifer was from the Gila River which was monitored by using wells and streamflow measurements. To test the hypothesis that phreatophytes were causing the surface water to leak into the shallow aquifer because the plants were depleting local groundwater levels through transpiration, the plants were removed from the study area during 1966–1967. This resulted in the losing segments of the stream to becoming gaining segments (Culler et al. 1982; Winter et al. 1998).

As would be expected, the greatest effects of phreatophytes on surface-water resources are in small rivers or lakes where the amount of surface water is small relative to the amount of groundwater discharge. Bouwer (1975) presented a procedure for determining the effects of increased surface-water flow that resulted from the removal of phreatophytes that grow in flood-plain sediments. The removal of the phreatophytes permitted more groundwater to discharge to the surface-water body.

Bowie and Kam (1968) investigated the change in water use that would occur after riparian plants were removed along a section of the appropriately named Cottonwood Wash, Arizona. Along a 4-mi (6.4 km) reach, it was estimated that phreatophytes consumed 80 acre-ft/year of groundwater that, therefore, was not available to supply the adjacent surface-water body. This represents about 18% of the total surface-water flow (Table 5.2). After the phreatophytes were eradicated, a 50% reduction in groundwater removal by plants was observed, but stream flows increased by only 6%. In a different study, the water-budget method was used to determine evapotranspiration for saltcedar that grew along the Gila River flood plain in Arizona (Hanson and Dawdy 1976). The study area was about 5,500 acres (2.2×10^7 m^2). Measurements were made of the various components of water inflow and outflow both before and after the phreatophytes were removed. Following plant removal, the measured evapotranspiration decreased as much as 45% along one reach of the river; the effect on surface water was not given.

The relation between tree removal and increased surface-water flow is not always directly related. In a study area in Australia, a setting with eucalyptus trees, which are deep-rooted phreatophytes, produced little stormwater runoff to a nearby stream relative to a comparable area cleared of the same trees (Le Maitre et al. 1999). Other studies also indicate that eradication of phreatophytes does not always lead to increased surface-water flows (Collings and Myrick 1966). Researchers concluded following modification of the basin vegetation, that no statistically significant difference occurred in streamflow before or after plant removal. However, this may be because the coniferous plants studied were not directly linked in the first place to discharging groundwater and would have affected water flowing to the surface-water body only by changes in runoff.

A lesson could be learned about the importance of these naturally occurring, plant and groundwater interactions, and their affect on water resources from the experiences of ancient Greece. A then prosperous city called Ephesus was located near the mouth of the Cayster River that emptied into the Aegean Sea. This location provided easy access by way of Ephesus' port, to the trade and commerce that could occur with the then known world. The city enjoyed prosperous times even after becoming part of the Roman Empire and even after the sackings of the city by the Goths in 260 AD and by Arabic tribes in the early 700s. But the city still prospered. The city finally fell, however, to a seemingly inconsequential factor; the port had filled in with sediment carried to it by the Cayster River, and the city had been effectively cut off from the Aegean Sea and its link to trade (Freely 2004). What caused this siltation after so many years of prosperity? In one word–growth. As the city population increased, more land came under cultivation and large tracts of native forests were cut down. Not only did this loosen up the soil and increased erosion, the removal of the forests caused the water table to rise, as it was no longer being used for transpiration. In such a low-lying area, this caused additional water to discharge to the river and increased the volume of water, and sediment load, that spread out over the flood plain of the area and, ultimately, into the port.

5.2.3 Changes in Surface-Water Chemistry

The documented interaction between phreatophytes and surface-water flows also may affect the geochemistry of surface water. This would especially be the case if the geochemistry of the groundwater is different than surface water. Perhaps one of the first studies that observed the linkage between the uptake of water by phreatophytes and decreased surface-water flow and changes in surface-water chemistry was conducted in the Moshiri basin on the island of Hokkaido, Japan. The basin is characterized by oaks with a bamboo understory. During 1989, streamflow measurements were made at the basin outlet, and discharge decreased during the warm, dry summer months as a result of decreased precipitation and increased evapotranspiration (Fig. 5.3a). During July, the period of lowest streamflow, warmest weather, and least precipitation, the authors noted a relation between the fluctuation in streamflow and water chemistry on a daily, or diurnal, basis (Fig. 5.3b). Each day, streamflow was highest in the morning and decreased in the evening, the difference a result of removal of groundwater by riparian phreatophytes rather than by evaporation of surface water. This is because the dew point was higher than the stream temperature. Specific conductance measured in the streamflow also followed a daily pattern, but concentrations were low in the morning and higher in the evening. Groundwater in the area had lower total ions as total dissolved solids than the stream water, so a decrease in groundwater discharge would lead to increased influence on water-chemistry by the surface water.

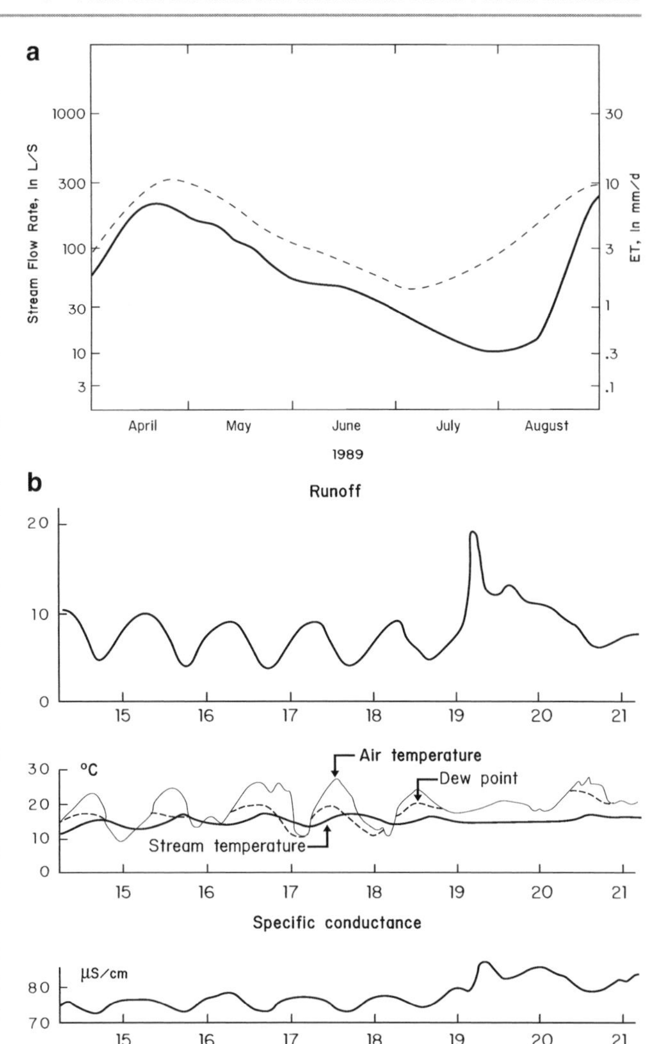

Fig. 5.3 The influence of plant uptake of groundwater on (**a**) surface-water flow and (**b**) geochemistry for a site in Japan. One millimeter is equivalent to 0.039 in., and one centimeter is equivalent to 0.39 in.

Clear cutting both riparian and upland forest trees can increase the rate of nutrient runoff, which can be measured by increased nutrient concentrations in stream water. For example, when nutrients, such as nitrate, no longer are taken up by plants, they become available for transport by runoff or groundwater. The net loss of nutrients from a clear-cut area can exceed by a factor of 8 the loss of nutrients from forests that are not clear cut. This is especially evident as increased nitrogen loading to streams near clear-cut areas (Bormann and Likens 1967).

A diurnal variation in trace-metal chemistry in streams was reported by Nimick et al. (2003). Rather than exhibiting a constant level of trace-metal concentrations over time, the streams exhibited a changing cycle of concentrations. The lowest concentrations of trace metals, such as manganese, cadmium, and zinc, occurred near the end of each day. Conversely, the highest concentrations occurred in the morning. Potential reasons for these fluctuations include sorption, diurnal uptake by riparian plants, and decreased input of groundwater that contains these trace elements during the day (i.e., a reduction in streamflow; Nimick et al. 2003).

Speiran (2010) reported that groundwater uptake by riparian phreatophytes effected nitrate concentrations in groundwater at two sites in Virginia. Nitrate concentrations in groundwater decreased from 10 to 2 mg/L after flow through a riparian zone. The riparian forest also focused local groundwater discharge to wetland systems that led to significant mass loss of nitrate through denitrification. Moreover, the study reported diurnal changes in groundwater levels near 0.25 m.

The beneficial effect of riparian plants on surface-water chemistry has been employed to decrease the discharge of groundwater contaminants such as nitrate to surface-water bodies (Tabacchi et al. 2000). In fact, many local municipalities enforce riparian buffers defined as a fixed width of vegetation near surface-water bodies that cannot be destroyed.

5.3 Plants and Groundwater Levels

The effect of fluctuations in the water table on the distribution of plants in a basin can be explained in areas where natural fluctuations of groundwater occur. For example, in marshes and swamps adjacent to tidally influenced surface-water bodies, the groundwater table can rise and fall in response to the daily tidal highs and lows, as well as to changes in barometric pressure. In such areas, the plant distribution is related closely to the mean depth to the water table, rather than the surface-water level. This relation of groundwater to plant distribution is primarily a result of less variability in the mean groundwater table in tidal swamps compared to nontidal swamps. Even a small difference in mean groundwater level results in a noticeable difference in dominant plant species distribution. For example, in a tidal stream in coastal Virginia, a groundwater table elevation difference of only 1.9 in. (5 cm) produced ash-blackgum dominated areas relative to maple-sweetgum areas associated with lower water-table conditions (Rheinhardt and Hershner 1992). Moreover, these researchers concluded that contrary to conventional thought, a more appropriate biological measure of wetness in tidal swamps should be the mean depth to water table, not the flooding duration, flooding height, or hydroperiod, as are more commonly cited.

The relation between plants and groundwater levels in individual wells also has been a focus of study. G.E.P. Smith (1915) indicated that the water table declined in wells installed in areas covered with trees during the growing season except at night or during dormancy. Other examples of a similar relation between groundwater table fluctuations and tree uptake of groundwater were shown by tank experiments by Lee (1912) and White (1932). The effects of phreatophytes on groundwater, as evidenced by daily fluctuations in groundwater levels in wells, were recorded in the Safford Valley in Arizona in 1944. Little groundwater-level fluctuation was noted before plant growth began in March. A few months later, however, after the trees, in this case saltcedar, had leafed out, the cyclical daily fluctuation observed earlier (as reported in Chap. 1) became evident, with a maximum observed decrease in groundwater level, or drawdown, of 0.19 ft (0.06 m). After the growing season ended in early winter, the fluctuations decreased (Robinson 1958).

A similar daily fluctuation in groundwater levels near a forested stream occurred in Michigan (Ferris 1949), which indicates that phreatophytes in the more humid eastern United States can affect water supplies, though less obviously. In the mid-1950s, results of a study at an experimental site in North Carolina, called the Bigwoods Experimental Forest, revealed that the groundwater level in shallow wells in a stand of loblolly pine declined during the summer and increased during the winter following the clear cutting of a 200-ft (61 m) long stand of pines (Trousdell and Hoover 1955). Before cutting, the observed groundwater level declined about 9.5 ft (2.8 m) during the summer. Clear cutting in late July reversed the decline to the point that the groundwater level rose 8.8 ft (2.6 m). In comparison, the groundwater level in a nearby stand of uncut trees remained low.

The relation between plants and fluctuations in groundwater levels also was noted in a humid area by Meyboom (1966), who was studying the poorly drained glacial moraine areas of the Canadian Plains. The Canadian Plains are dotted with numerous, small, water-filled depressions, called sloughs (pronounced sloo). Meyboom noted that these sloughs were surrounded by willow trees, such as basket willow (*Salix petiolaris*) near and in the water, and aspen poplar (*Populus tremuloides*) on higher ground (Meyboom 1966). To investigate the relations of surface water, groundwater, precipitation, and water use by the willows, Meyboom employed the nested well approach described in Chap. 4. The observation wells he installed were essentially 1- to 2-in. (2.5–5 cm) pipes with slots on the bottom and were installed at various depths. Wells with larger diameters were drilled and equipped with automatic water-level recorders, similar to those used by USGS hydrogeologists such as Meinzer, Brown, and White, as described in Chap. 1. When evapotranspiration was low in the winter, the level of water in the sloughs was higher than the water table, and water moved vertically down to the water table beneath the slough (Fig. 5.4). As evapotranspiration increased during the summer month of July, the flow direction in the water table reversed to one of higher hydraulic head beneath the slough, and water moved vertically upward (Fig. 5.4). This change in flow direction was caused by the seasonal removal of groundwater by the willows along the banks of the slough (Meyboom 1966). During the summer months, Meyboom concluded that one-fifth of the flow beneath the slough was diverted by willow uptake.

Subsequent explanations of the interactions of riparian plants, shallow groundwater, and resultant effects on surface water were given by Winter (1999), and examples are provided in Winter and Rosenberry (1995). For example, groundwater transpired by plants results in a decrease in the discharge to surface water, and surface water acts as a source of water to meet later evapotranspiration demands on groundwater.

Mower et al. (1964) used the method developed by White (1932) to determine the amount of groundwater taken up by transpiring plants. Again, this is the amount of water taken up by the plant from groundwater—a significant advance in understanding compared to alternative plant physiology

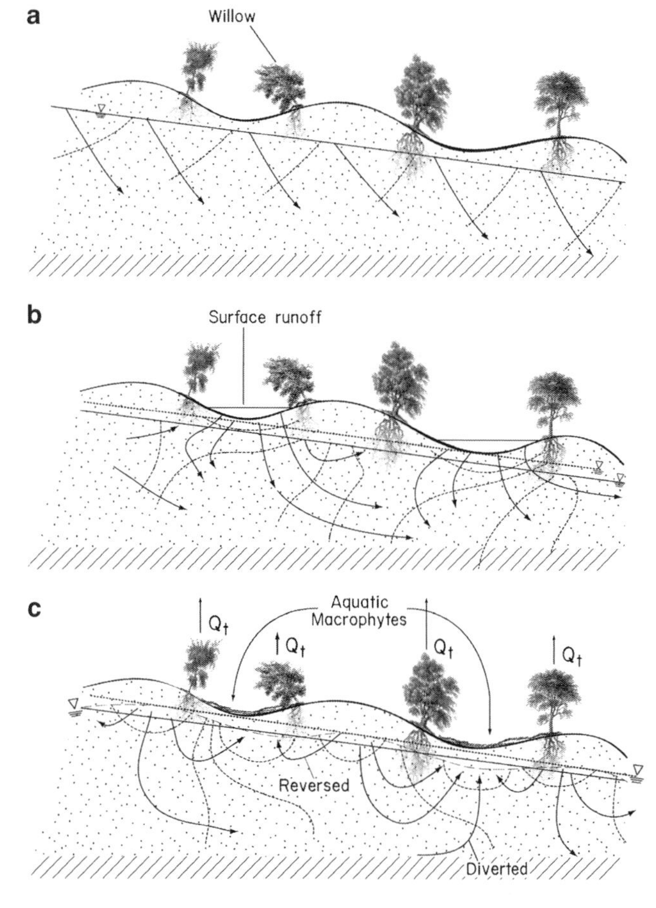

Fig. 5.4 Effect of groundwater uptake by (**a**) willows near surface water. The dashed lines in all figures represent equal head. (**b**) The dashed water table represents a higher water table following infiltration of surface water. (**c**) The lines with arrowheads indicate the direction of groundwater flow, and Q_t represents water removed by transpiration (Modified from Meyboom 1966).

methods that determine total water used, such as sap flow, regardless of the source of water. Mower et al. (1964) placed three wells in stands of phreatophytes such as saltcedar and grasses. Evapotranspiration from the water table at the three sites ranged from 2.5 to 5 ft/acre/year (0.76–1.52 m/acre/year), with an average depth to water table near 7 ft (2.1 m).

Kluitenberg et al. (2005) performed an investigation to determine the effect of phreatophyte evapotranspiration demand on groundwater levels in a shallow aquifer at two locations in Kansas—the Larned Research Site and the Ashland Research Site. Both sites are riparian groundwater and surface-water systems; the first is dominated by cottonwoods and willows and the second is dominated by *Tamarix*. At each site, groundwater fluctuations were measured in wells, and changes in soil moisture above the water table were measured by neutron probes. The researchers reported that the diurnal change in groundwater level was related to temporal changes in the following: the source of water to the plants, such as precipitation as opposed to groundwater; meteorological events, such as photosynthetically active radiation (PAR) and temperature; plant growth and density, such as vigorous as opposed to cut; proximity to surface water; and depth to water table.

A variety of studies to examine groundwater fluctuations either in the field or using controlled-experiment tank studies was conducted by the USGS in the 1960s. These include work by McDonald and Hughes (1968) in the flood plain of the Colorado River between California and Arizona and north of Yuma, Arizona, and downgradient of the Imperial Dam. Diurnal groundwater fluctuations ranged from 0.1 to 0.4 ft (0.03–0.12 m). The monitored well was located about 1,500 ft (456 m) from the gaining Colorado River in the flood plain that contained predominantly arrowweed (*Pluchea sericea*) from 1962 to 1965. These researchers made the observation that the initial decrease in groundwater level in the monitored well, actually called a transpiration well, from the rebounded higher levels during nighttime was seen in less than 15 min after the sun's rays hit the study area. The depth of groundwater fluctuation was directly related to air temperature and inversely related to relative humidity.

The groundwater-level fluctuations observed in these studies indicate the possible interactions that occur between components such as plant-root uptake rates, groundwater movement rates, and precipitation and recharge frequency, among others. The explanation behind the diurnal fluctuation observed is that the roots respond to a *VPD* by removing water from the capillary fringe or water table during the mid-morning to afternoon period when solar radiation is not at a maximum and the stomata remain open. The water table decreases, if and only if the rate of removal is greater than the rate of replenishment. Replenishment can be derived from storage, lateral movement of groundwater from upgradient areas, or recent recharge (Fig. 5.5). These factors are controlled by the climate and the hydraulic conductivity of the saturated- and unsaturated-zone sediments. Once the stomata close and evapotranspiration ceases, as occurs during the hottest part of the day, the rate of groundwater removal by plants is less than the rate of replenishment, and the water table rises. If this occurs over a period of time characterized by no precipitation, especially in unconfined aquifers of low hydraulic conductivity, the nighttime rebound groundwater level over time becomes progressively lower, and the maximum daily decline also becomes progressively deeper. The amount of change remains the same each day, however, as this is controlled by the VPD and plant-resistance characteristics, as long as the plant is healthy.

Szilágyi et al. (2008) report that groundwater-level fluctuations in wells located near gaining streams occurred 1–1.5 h after a fluctuation in the surface-water level in a small forested watershed. The explanation given for the lag is the decrease in the hydraulic gradient, *i*, between the wells in

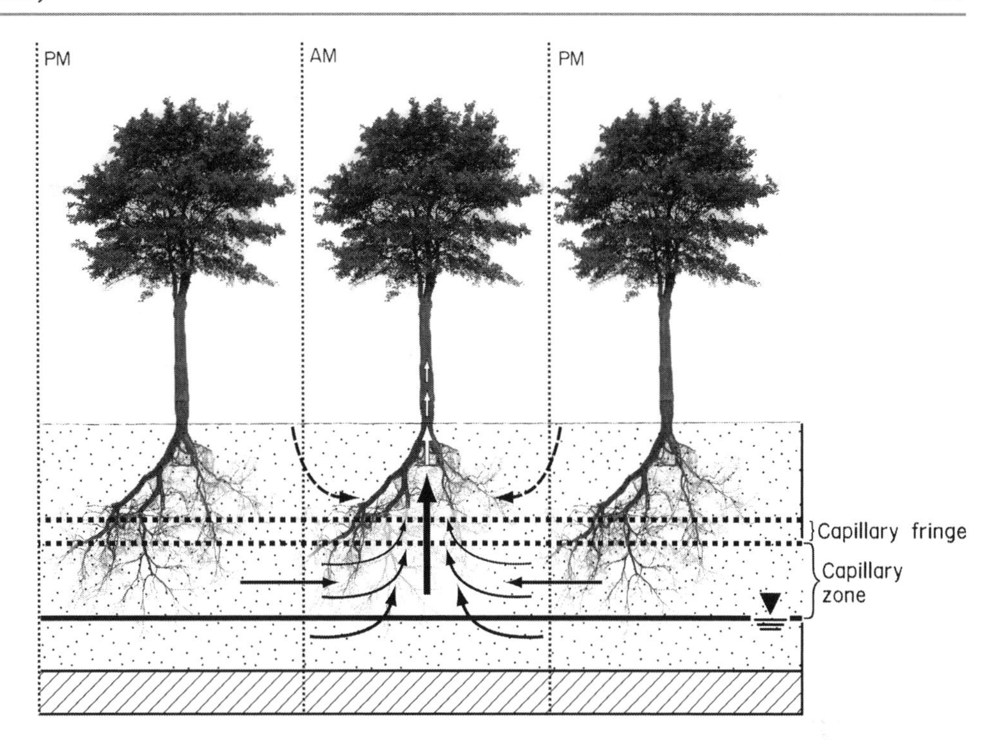

Fig. 5.5 Possible sources of water to plant roots in the capillary fringe or zone (shown) and water table and results compared to time of day.

the riparian zone and the surface-water body on account of water removed by *ET*. This decrease in gradient decreases the flow of groundwater to surface water as indicated by a more rapid decrease in surface-water levels. Gribovszki et al. (2010) provide an excellent review of the interaction between diurnal groundwater-level fluctuations and streamflow.

Other evidence that groundwater is used by trees can be inferred by looking at the width of tree rings over time. Although the relation to depth to groundwater is not apparent initially, if a tree has tap roots its rings tend to be a consistent width even during periods of little precipitation. On the other hand, the width of tree rings is not consistent in plants that use a less-consistent source of water (Douglass 1924).

5.3.1 Plants and Groundwater Discharge

Trees affect groundwater levels by decreasing the pressure head in the water table by direct root uptake or by abstracting water from the groundwater table along a water potential gradient. This can result in groundwater flow lines in the saturated sediments some depth from the water table to converge upward in response to the lower water levels and pressures. If the soil and sediment hydraulic conductivity were high enough and enough plants were present with high transpiration rates under meteorological conditions that would support evapotranspiration, such a plantation could become a zone of groundwater discharge.

Heuperman (1999) investigated the relation between phreatophytes and groundwater discharge at two sites in Australia. Using groundwater-level monitoring data, Heuperman was able to demonstrate that groundwater levels were lowered between 6.5 and 13 ft (2 and 4 m) relative to control sites that did not contain plants. The groundwater levels in a deeper aquifer, however, were not influenced. Whereas groundwater flow was vertical downward previous to planting, the transpiration of groundwater caused deeper water to flow upward to the shallow aquifer and reversed the vertical gradient. The hydraulic conductivity of the shallow aquifer was about 1.6 ft/day (0.5 m/day). At these sites, the deeper groundwater had higher salinity, and the upward flow of saline groundwater toward the root zone resulted in a zone of concentrated salts near the water table and capillary fringe.

Up to 80 sites in Australia were evaluated for plant and groundwater interactions by George et al. (1999) who reported that at the majority of sites, plants had no effect on the water table; however, the sites that did affect the water table were characterized by a considerably large planting. At some sites investigated, the decreased water table could not be solely attributed to the uptake of groundwater by plants, a decrease in recharge by plant uptake, or both. A study by Sánchez-Pérez et al. (2008) also found that hardwood trees growing in the riparian zone of the Rhine River in France did not directly take up water from the shallow water table at 3.3 ft (1 m) bls but, rather, took up water in the unsaturated zone.

5.4 Plants and Groundwater Chemistry

The interaction between surface water and groundwater has not always been clearly understood, as was demonstrated in Chap. 2. It was probably not until the 1960s that it was more widely recognized that most of the flow in rivers and streams during periods of low precipitation consists of groundwater discharge. In many cases, because the geochemistry of groundwater often is different than surface water, the types of plants that can grow in these areas provide a visible surrogate of the occurrence of groundwater discharge to rivers.

Rosenberry et al. (2000) were interested in the determination of groundwater discharge to a lake in Minnesota in order to estimate the lake's water budget. The researchers used both the presence and absence of certain plants to indicate specific locations of groundwater discharge. For example, the location of groundwater discharge to the lake was confirmed based on cooler water temperatures than the prevailing surface water temperature at the same location and absence of floating leaf and emergent vegetation but the presence of marsh marigold (*Caltha palustris* L.). The yellow flowers of the marsh marigold made it a visual indication of zones of groundwater discharge.

Most groundwater in discharge areas tends to be more mineralized than dilute infiltration water in recharge areas. In reporting results of a wetland study in Spain, Bernaldez and Benayas (1992) indicated that xerophytes were found in recharge areas in topographic highs. Conversely, in the lowland discharge areas characterized by groundwater of higher mineral content, both in salinity and alkalinity, the plants that predominated tended to be able to adapt to the higher salinity of the groundwater. Hence, a key control on the distribution of plant types can be related to the *supply* of groundwater and differences in the *geochemical* composition of groundwater, depending on its relation to the overall flow path from recharge to discharge areas.

Benayas et al. (1990) also investigated the interaction between riparian plant distribution and groundwater geochemistry. The researchers reported that mineralization of older groundwater along flowpaths in the aquifer system controlled the distribution of riparian vegetation in the study basin in central Spain. Groundwater samples collected farther downgradient from recharge areas were more mineralized; that is, the samples had higher specific conductance, pH, sodium and chloride concentrations, and the vegetation tended to be composed primarily of halophytic plants. In areas where the groundwater flow path was shorter the groundwater was characterized by lower concentrations of specific conductance, pH, and minerals, and the plants tended to be glycophytic and adversely affected by salts.

An interesting relation between a plant's need for dissolved chemicals and its ability to acquire them is illustrated in the Tree Islands in the Everglades of southern Florida. These Tree Islands, mentioned previously, form ridges and the surrounding lower areas result in sloughs. Tree Islands are geologically relatively recent, circa 5,000 year BP (Gleason and Stone 1994). The woody tree growth on these tree islands is extensive where land elevations are high and flooding is not continuous enough to kill roots by lack of oxygen. But where do these isolated trees get their necessary minerals? The slash (tree parts that fall to the ground) and old leaf litter are a potential source but are depleted of nutrients. Ross et al. (2006) hypothesize that transpiration decreases the underlying groundwater table, which focuses the discharge of adjacent surface water that, although contains naturally low concentrations of phosphorus, can have elevated concentrations of phosphorus from agricultural land-drainage canals—this process provides phosphorus to the root zone. This may provide a partial explanation for the observation that Tree Island soils are higher in phosphorus than the surrounding marsh soils.

Groundwater and surface-water data from a basin in South Australia also confirm the strong relation between transpiration and groundwater chemistry (Poulsen et al. 2006). Areas where massive amounts of groundwater discharge leave behind increased soil salinity are referred to as dryland salinity. In most cases, this increased salinity in shallow soils occurred following the removal of deep-rooted, perennial vegetation that was native to the area in order to plant shallow-rooted, annual, agricultural crops. Because the deep-rooted, native vegetation kept the water table low, the salinity was concentrated in the vadose zone by transpiration and evaporation enrichment was immobilized at depth (Barrett-Lennard 2002). Following the planting of agricultural crops, however, the increased water table from increased recharge and decreased transpiration mobilized the salts into solution: groundwater discharge to surface water increased the salinity of the surface water as well.

5.5 Summary

It is evident from a wide range of naturally occurring systems that phreatophytes can affect groundwater and, as a result, surface-water bodies. Phreatophytes can reduce the amount of recharge by uptake of either infiltrating water or groundwater. Conversely, the removal of phreatophytes can increase the amount of recharge. Phreatophytic use of water also can cause the water table to rise, such as when the unsaturated zone is large and depth to water table is great, or by direct uptake from the capillary fringe. These processes can occur in arid areas or in humid areas that are characterized by riparian plants, which can use surface water and groundwater to meet evapotranspiration demands. Phreatophytic use of groundwater prior to discharge to surface waters can affect surface-water flows, levels, and chemical composition. Conversely, some riparian systems use surface water that has entered the local groundwater-flow system, and upstream changes in surface-water flows can affect the ecology of these riparian systems. This precedent of natural plant and groundwater interactions can be extended toward the restoration of contaminated groundwater, the focus of Parts II and III.

Why is this information important to the phytoremediation of contaminated groundwater? The natural existence of plant and groundwater interactions in pristine areas that result in measurable changes in groundwater levels, flow, and surface-water discharge and chemistry provides unequivocal and fundamental evidence to support the use of phreatophytes in an engineered manner to interact with contaminated groundwater.

Part II

Plant and Groundwater Interactions for Hydrologic Control

. . .the soil horizon tapped by the roots of trees derives by capillarity, from the level of the groundwater, its perennial supply of moisture.

W.A. Cannon (1913)

Site Assessment and Characterization 6

The applied use of plant and groundwater interactions (as described in Part I) to achieve remediation goals at sites characterized by contaminated groundwater is a direct extension of these long-observed natural interactions. The specific application of plants to achieve remedial goals is, however, relatively new. The installation of plants at sites to affect the flow of contaminated groundwater in response to regulatory-driven site-restoration mandates was initiated in the late 1980s and early 1990s. Relative to the total number of sites in the United States that have documented groundwater contamination and require some type of corrective action, the number of published case studies of phytoremediation that specifically addresses groundwater flow issues is few. In most cases at sites characterized by contaminated groundwater, the chosen corrective action involves conventional pump-and-treat of contaminated groundwater or groundwater flow interception by trenching. This is despite the efforts made by various state and federal regulatory agencies, such as the U.S. Environmental Protection Agency (USEPA), as well as other federal agencies, to promote phytoremediation as an alternative corrective action.

Since the 1990s, the amount of information about phytoremediation of contaminated groundwater has increased with the release of numerous reports, journal articles, and the formation of the PhytoSociety, which publishes an international journal devoted to all aspects of phytoremediation. Many of these publications focus on the use of plants to remediate sites where soil contamination is the regulatory concern, however, rather than contaminated groundwater. As a result, the scientific and regulatory communities often are not in the position to unequivocally state how best to determine when phytoremediation will work at sites characterized by contaminated groundwater, how to implement phytoremediation correctly, and how to determine if the outcome is beneficial for site-remedial goals in a reasonable amount of time and in a cost-effective manner. A goal of Part II, therefore, is to present useful information that addresses concerns that surround the assessment, implementation, and verification of the effect of phytoremediation on the hydrology of contaminated groundwater.

In general, at most sites characterized by contaminated groundwater, phytoremediation can be used to achieve three main hydrologic goals:

- Prevention of contaminated-groundwater flow to cleaner offsite areas;
- Prevention of contaminated-groundwater flow from reaching regulated receptors, such as surface-water bodies, that may be located on or off site, and;
- Reduction in leachate formation and subsequent groundwater contamination near source areas.

Although each hydrologic goal is different, each is based on the successful interaction of plants and groundwater. Throughout Part II, the assumption is made that for a site where phytoremediation is being evaluated, the groundwater is either contaminated and the first two hydrologic goals are important or groundwater contamination is to be avoided and, therefore, the third goal is important. Also, the terms hydrologic and hydraulic will often be used interchangeably in Part II to emphasize the importance of the wide range of environmental factors that control groundwater flow.

The development of rigorous assessment and characterization approaches in order to successfully apply phytoremediation to achieve the three hydrologic goals is important for many reasons. First, site characterization activities may indicate that phytoremediation will *not* successfully affect the flow of contaminated groundwater, such as for conditions where the water table is too deep to promote groundwater and root interaction. Implementation of phytoremediation at such sites will lead only to the impression that phytoremediation, as a whole, is not a defensible technology. Second, inappropriate site-assessment activities may lead to the installation of a remedial technology that not only is more expensive but also may be less effective than phytoremediation. Third, if site-assessment activities

J.E. Landmeyer, *Introduction to Phytoremediation of Contaminated Groundwater*,
DOI 10.1007/978-94-007-1957-6_6, © Springer Science+Business Media B.V. 2012

indicate that plants can be used to control groundwater and site hydrology to contain a plume of dissolved-phase contaminants, then the complete remediation of recalcitrant contaminants by phytoremediation or other technologies may not be warranted. In this case, the influence of plants on the hydrology of contaminated groundwater can be considered to be a contaminant-independent process, as long as the contaminant concentrations are below levels toxic to plants. Finally, site assessment activities may reveal that existing or native vegetation may be appropriate to meet site-specific hydrologic goals and an engineered phytoremediation system need not be installed.

In Part II, methods and approaches are presented for phytoremediation evaluation that can be used during typical site-assessment and characterization activities at contaminated sites. These methods and approaches can help to determine the viability of a particular site for hydrologic containment or control by phytoremediation, either by uptake of groundwater or through a decrease in recharge. Such approaches are logical extensions of work conducted by early researchers as documented in Part I. The selection of plants, correct site characterization, design, installation, and monitoring of a phytoremediation system to achieve the three hydrologic goals also is discussed. In many cases, the burden of proof that phytoremediation will achieve the hydrologic goals is the responsibility of those proposing to use phytoremediation, and this proof can be achieved through proper site monitoring. Planting trees at a site typically does not require regulatory approval. However, the application of plants to achieve site-remedial goals within a regulatory context does require regulatory sanction and approval.

Chapter 6 provides techniques used to investigate sites with known or potential groundwater contamination. Some of this information will be a review for hydrogeologists who implement such techniques in their daily work. These same hydrogeologists, however, may be surprised to discover how commonly collected groundwater data can be used to assess plant and groundwater interactions. To others, this information will provide a starting point from which investigations can precede at individual groundwater contamination sites where the application of phytoremediation to achieve hydrologic goals is being considered as one of many remedial options, for instance, as part of feasibility studies. Chapter 6, however, does not compare costs of different phytoremediation approaches, or advocate a particular approach in order to realize cost savings; this is left to the individual project manager. Thus, no recommendations are given, for example, regarding the amount or type of fertilizer to use or which would be more cost effective.

Chapter 6 can be used by the following, depending on site-specific needs:

- Environmental consultants or others involved in site restoration to determine the range of remedial options for a particular site, including hydrologic control by phytoremediation;
- Environmental regulators who review proposals for projects in which phytoremediation would be used for hydrologic control;
- Environmental professionals and educators interested in the interaction between plants and groundwater.

The first two users often are at cross purposes regarding the fate of a particular site as a consequence of, in part, a lack of defensible published data on the effectiveness of phytoremediation of contaminated groundwater. Hopefully, Chap. 6 will provide common ground between these two users for discussion of phytoremediation for hydrologic control at specific sites to answer the general question, "Can phytoremediation be used successfully at a site to decrease the risks associated with contaminated groundwater?"

6.1 Site History of Contaminant Release, Assessment, and Characterization

To comply with federal laws, state regulatory agencies typically compile an annual list, or inventory, of sites where regulated compounds have been detected. States are given this responsibility pursuant to 40 Code of Federal Regulations (CFR), part of the Clean Water Act legislation passed in the 1970s. Based on the types of contaminants found or the type of contaminated source area, the sites are categorized under various laws. Recent spills or releases within the workplace are regulated under the Occupational Safety and Health Act (OSHA) under 29 CFR 1910. The USEPA regulates spills or releases to the environment under laws outlined in the Resource Conservation and Recovery Act (RCRA) and Comprehensive Environmental Response, Compensation and Liability Act (CERCLA, or Superfund). Most sites inventoried in the United States require an assessment of the contaminant distribution in groundwater, the groundwater-flow rate, and potential effects on human or ecological receptors, as well as a plan and eventual implementation of corrective action(s) to meet remedial goals.

The restoration of contaminated groundwater to compliance levels is documented in two general types of legal documents—a Corrective Action Plan (CAP) for sites regulated under RCRA, or a Record of Decision (ROD) for sites regulated under CERCLA. As part of the verbage in each of these documents, regulatory agencies hold responsible parties to the two following goals at a minimum, regardless of the site-specific contaminant:

1. Hydrologic containment and(or) control of the contaminant(s), and

2. Restoration of aqueous or dissolved-phase plumes, or zones of contaminant mass, to pre-contaminant or site-specific permissible levels.

Achievement of these goals is the restoration objective at most sites. At each site, specific contaminant-reduction goals must be met, and the restoration goal may be different for the same contaminant released in different areas. For example, in cases of small dissolved-phase plumes located in isolated areas with little potential for human or ecological impact, restoration goals often are implemented according to a risk-reduction standard, where permissible levels of contamination can be higher than if the contaminant was released in a more populated area. Overall, however, these two goals are the most common reasons for taking remedial actions at a site and, therefore, are important to an evaluation of phytoremediation.

6.1.1 Contaminant-Release History

All contaminated sites have unique histories of contaminant release and environmental effect. Contaminant releases can be characterized in terms of differences in space and time. Contaminant releases can be acute, such as a spill or accident that releases a substantial volume of contaminant over a larger area, or chronic, such as a slow low-volume release, as when a pipe joint or fitting leaks underground. Either type of release may have occurred in the past, over time, or more recently, but the contaminant remains in the ground. To complicate matters, the party responsible for the contaminant release may be the current owner of the property, a newer owner who assumes the liability of releases by previous owners, an owner of property adjacent to or downgradient from a contaminant source, or an unknown or disputed owner, as when contaminant sources are located in adjoining properties. Moreover, although zones of contaminated media can be delineated in most cases, the volume of the material(s) released is rarely, if ever, completely known.

Thanks, in part, to newer reporting regulations and more effective enforcement actions, contaminant releases that may have gone undetected 20 years ago now can be more rapidly detected. As a consequence, more recent releases generally have more specific information regarding the times and volumes of contaminant released as part of the site history. The importance of rapid-release detection systems required at most gasoline stations in the United States, for example, cannot be overstated.

Following most contaminant releases and detection, local or federal regulations stipulate the commencement of a formal site assessment and characterization of the extent of the environmental effects, if any. These site-assessment activities can occur voluntarily by the responsible party or in response to regulations that typically require any spill greater than the federal reporting limit of 25 gal (95 L) to be assessed for potential corrective action. As part of this formal response to a contamination event, the site assessment and history also can be supplemented with anecdotal evidence from former or current employees.

6.1.2 Contamination Assessment and Characterization

An important part of the initial determination of the history of contamination at a site is the assessment and characterization of the extent of contamination in the subsurface. This includes delineation of the vertical and horizontal dimensions of contamination to a point where uncontaminated, or background conditions, are known, and to identify potential human or ecological receptors. For example, a site with a known point-source release, such as a leak from an underground storage tank (UST) located within a populated area, will have to be surveyed for the presence of domestic or public drinking-water wells. If such a receptor already has been affected by contamination, such as water from a well that starts to taste or smell different, the process of site assessment is reversed; the source(s) of the contaminant release will need to be identified.

A contaminant release to the subsurface may affect the physical components that comprise the subsurface, such as the air, soil, water, and microbial ecology. The extent of a contaminant release on each component must be determined. For example, determination must be made about whether or not free-phase contamination exists, such as gasoline floating on the water table, because such free product is a long-term source of contaminant release to the water table. Moreover, the removal of free-phase contamination should be the initial goal of site-remediation plans that include phytoremediation because free-phase contamination typically will be toxic to plants. If the free-phase contamination is located some depth below the root zone, however, plants can be installed directly over the area to decrease groundwater recharge, increase subsurface oxygenation, and to decrease additional dissolved-phase contamination.

To assess the affect of contamination on the different components in the subsurface, the appropriate samples need to be collected and analyzed. One of the most efficient methods of assessing and characterizing the distribution of contaminants below the surface at a site is to drill boreholes. Drilling a borehole permits the collection of samples, both contaminated and uncontaminated, to determine soil physical and chemical properties, the collection of groundwater samples, the collection of soil-gas samples from the unsaturated zone, and the measurement of the thickness of any free-phase contaminant that enters the borehole. The occurrence and distribution of root material from existing vegetation also can be assessed during borehole drilling. These boreholes then can be used as locations to install temporary or permanent monitoring wells.

With respect to phytoremediation, one of the goals of site assessment and characterization is to map the extent of the contamination in context to the surrounding above- and below-ground features. Identification of these features can allow the general direction of groundwater flow to be determined, even before wells are installed, and guide the installation of monitoring wells. Because samples will be collected from below ground as part of site assessment and characterization, the locations of buried utilities, such as electric, gas, fuel, water, fiber optics, etc., must be identified as soon as possible. Even if these underground utilities have been previously located on a map, many local and state laws require that a private location service be contacted prior to subsurface investigation; otherwise, the party conducting the subsurface investigation may be held liable for damage. Moreover, sufficient space must be allowed on either side of located underground utilities to account for uncertainties inherent to the accuracy of these locations.

The source area or suspected contaminant-release must be delineated as accurately as possible within given site constraints. Source areas usually are the first part of a contaminated site that is delineated; a burst pipeline that spews raw product or the presence of objectionable odors are hard to attribute to natural causes. Conversely, if dissolved-phase contaminants have been detected in wells but the source area is unknown, the area of contaminant release often can be delineated by plotting the contaminated wells on a map and searching in upgradient areas—this approach is discussed further in Chap. 15.

The vertical extent of contamination at a source area from land surface through the contaminated zone(s) and to uncontaminated media also must be identified. This identification is facilitated where contamination tends to be closer to land surface and the contaminant is more likely to be concentrated over a relatively smaller volume of the subsurface. In areas downgradient from the source area, however, vertical delineation of the dissolved-phase contaminant is less likely to be accurate.

This decrease in assessment accuracy in downgradient areas is not a result of a lack of approach or technology; rather, it is an artifact of the conventional approach used to assess contaminated sites, in which monitoring wells are installed with screens located only across the water table. This conventional approach has led to at least two problems. First, the water-table surface is not a constant, fixed surface, but fluctuates sometimes substantially over time. Thus, the location of the water table during site-assessment activities may not be representative of the seasonal mean water-table surface. Wells installed with screens across the water table can go dry during droughts, periods of low recharge, or by unanticipated seasonal fluctuations. Second, groundwater in areas away from the source area may be recharged by uncontaminated water derived locally from above, which tends to push contaminated groundwater deeper below the surface of the water table. In this specific scenario, downgradient monitoring wells installed using the conventional approach of screening wells across the water table can produce uncontaminated samples while deeper, contaminated groundwater flows underneath the well and remains undetected. Examples of this scenario, effect on contaminant plume migration and detection, and solutions on how to avoid it are provided in Landmeyer et al. (1998b) and Wilson et al. (2005).

Within the context of contaminated sites that have a known history of release and some initial contaminant delineation, the following sections describe a potential approach that can be used to evaluate existing site- assessment information for the purposes of phytoremediation to achieve the three hydrologic goals. States may require, however, additional information be collected to comply with local regulations not covered here.

6.2 Site-Specific Hydrologic Goals

Prior to any discussion of the application of phytoremediation at a site characterized by contaminated groundwater, the question must be asked and answered, "Can the site be hydrologically controlled to decrease contaminant movement in groundwater?" Although a seemingly simple question, the answer is deceptively complex, because the answer often is site specific.

As described at the beginning of this chapter, phytoremediation can be implemented at sites characterized by groundwater contamination to achieve one or all of the following goals:

- Prevention of contaminated-groundwater flow to cleaner offsite areas—a reduction in groundwater flow across property boundaries;
- Prevention of contaminated-groundwater flow from reaching regulated receptors, such as surface-water bodies, that may be located on or off site—a reduction in groundwater flow to potential receptors, and;
- Reduction in leachate formation and subsequent groundwater contamination near source areas—a reduction in groundwater recharge at source areas.

The three hydrologic goals share a similar approach that can be used as part of comprehensive site-assessment and characterization. The following discussion will focus on the effect of plants on groundwater and how their effectiveness can be evaluated in terms of the three hydrologic goals.

The effect of plants on groundwater contaminants is covered in Part III.

In general, phytoremediation to achieve hydrologic control will be most effective in aquifers characterized by a depth to water table that is 30 ft or less. Water-table aquifers tend to be the most susceptible to contamination, and typically are the least cost-intensive to remediate using phytoremediation. Phytoremediation has been used for contaminated confined aquifers, with the remedial goal focused on the decrease in contaminant concentrations rather than to affect groundwater hydrology. The use of phytoremediation for this special case of confined aquifers is discussed in Chap. 14.

6.2.1 Reduction in Groundwater Flow Across Property Boundaries

Although groundwater flow occurs in the subsurface, its flow may affect management decisions made with respect to the land surface. Unlike surface water, such as rivers that provide a natural boundary between, for example, counties and states, groundwater flow does not recognize these or many other types of property boundaries. The relation between groundwater flow and property boundaries becomes relevant, however, when groundwater is contaminated. This scenario has different implications for different people, depending on the role that each person plays in groundwater contamination sites. Much like the fable about the seven blind men who each tried to determine what kind of animal an elephant was although only touching a separate part, each person who is involved as regulator, environmental consultant, responsible party, or a third party often see this relation between property boundaries and groundwater contamination from a unique perspective. For example, regulators are concerned with the spread of dissolved-phase contaminants in groundwater to off-site areas and may define a property boundary as a line in the sand that the migration of groundwater contamination will not be permitted to cross. This viewpoint is supported with respect to notable cases of third-party effects, such as occurred in Woburn, MA, where industrial solvents seeped into groundwater and migrated off site to reach downgradient municipal drinking-water wells. This incident was described in the book *Civil Action* (Harr 1995).

On the other hand, an environmental consultant hired by the responsible party may see a property boundary as a location that legally permits the maximum extent that permissible levels of contamination can exist in groundwater; such areas between a contaminant source and a property boundary are called mixing zones, where natural attenuation processes can be used to assimilate the contaminants. The site owner sometimes can use the defined property boundary to advantage by purchasing additional land downgradient of groundwater flow and perhaps at a lower cost than the option of more expensive site remediation. Moreover, a third party can view the property boundary as a line that, if crossed by groundwater contamination, can result in decreased property values or, conversely, as a potential source of revenue from a successful resource damage suit against the upgradient property owner.

In any case, the importance of property boundaries and groundwater contamination needs to be considered during any site assessment and characterization of a site for phytoremediation. At a minimum, the physical boundaries of a site provide an upper limit on the size of a phytoremediation planting. This size will ultimately affect the quantity of groundwater that can be hydrologically controlled by trees as is described in Chaps. 7 and 8.

6.2.2 Reduction in Groundwater Flow to Potential Receptors

Potential receptors of groundwater contamination include wells, lakes, streams, or springs. These features can be located on the same property as the contaminant release or maybe be located off site. A reduction in the flow of groundwater to these receptors can be a goal of the property owner who is concerned about future land use, of trustees responsible for the stewardship of public lands, or of regulators tasked with the protection of groundwater and surface water used for water supply and ecosystem resources, or recreational use.

Groundwater flow was conceptualized in Chap. 4 at three generalized physical scales—local, intermediate, and regional. Some knowledge of the scale of groundwater flow that might be expected at a contaminated site can be obtained prior to a site visit by determining the location of the site relative to the aerial or vertical extent of previously delineated major aquifer systems. This information is available from most USGS Water Science Centers across the United States. Because the root systems of the most commonly used phytoremediation plants rarely exceed 30 ft (6–9 m) below land surface, most sites that will benefit hydrologically from a phytoremediation system are those sites characterized by contamination in the water table and, therefore, represent predominately local flow systems. Local groundwater flow can be connected, however, to deeper, more intermediate and regional aquifer systems, depending on the hydrogeological framework of the site.

The presence of a receptor, such as a lake, pond, or wetland, at or near the contaminated site usually indicates a location of shallow groundwater discharge. As long as it can be documented that groundwater flow beneath a surface-water body does not occur, then most local groundwater-flow lines can be assumed to terminate at a surface-water feature.

6.2.3 Reduction in Groundwater Recharge at Source Areas

In general, localized source areas of potential contamination may seem as the last place to implement phytoremediation to achieve hydrologic control because these source areas often contain separate-phase contamination at levels toxic to plants. Contamination at source areas often extends completely from land surface to the water table. Moreover, source areas tend to be inaccessible, often being covered by buildings or paved surfaces.

Phytoremediation, however, can be used successfully to remediate such source areas. For example, installing plants in source areas can decrease recharge because of plant uptake of infiltration prior to flow through contaminated soil and sediment. This process of plant-facilitated reduction in recharge decreases both gross and net annualized recharge.

As discussed in Chap. 5, a decrease in the formation of leachate from source areas is a legitimate goal of phytoremediation. In fact, source areas such as landfills can be covered with plants rather than remain unplanted as is often the case (see Rock 2003 for a comprehensive review). Conventional landfill covers act to decrease soil permeability, and act as a water-percolation barrier, to achieve the goal of leachate-formation reduction, and are designed to comply with RCRA regulations promulgated with their creation in 1976. The goal of this approach is one of waste isolation. A reduction in soil permeability is achieved with mechanical soil compaction or with geotextiles. Such decreased soil permeability often is not sustainable, however, as a result of animal burrowing activities, erosion, earthquakes, etc., and often leads to excessive runoff and Horton overland flows. Conversely, vegetated covers, essentially a special case of phytoremediation where plants are installed on the landfill cover, rely on plants to reduce infiltration by first removing the water by evapotranspiration before the water can leach contaminants to the water table.

6.2.4 Supplemental Use with Other Hydrologic Strategies

The use of phytoremediation to achieve any of the three noted hydrologic goals also can be evaluated at sites where engineered hydrologic-control systems are in place, such as pump-and-treat or air stripping/air vacuum extraction (AS/AVE). In fact, a strong case can be made to integrate phytoremediation with conventional water and air control technologies, in order to accelerate site remediation during the time when roots have not yet interacted with groundwater. Conversely, at sites where long-running pump-and-treat systems have become ineffective or cost intensive due to biofouling or low specific yields, phytoremediation can be an alternative that is not affected by these obstacles. Moreover, real or perceived concerns regarding a potential lapse in water-removal processes during plant dormancy can be alleviated by using a strategy that integrates source-removal activities, such as the occasional operation of a pump-and-treat or AS/AVE system, with phytoremediation.

One particular scenario in which phytoremediation can be used successfully in conjunction with other remedial strategies is as an additional component of monitored natural attenuation (MNA) strategies. At sites where MNA has been implemented as a corrective action, a principal concern is the movement of the plume of contamination at rates faster than the rate of prevailing natural attenuation processes. A phytoremediation system applied as a downgradient hydrologic barrier could alleviate these concerns.

6.2.5 Contingency Plans

Contingency plans are required for all forms of site remediation in case the specified mechanisms do not work to the full satisfaction of the responsible party or regulator. Because of the uncertainty inherent in all decisions regarding site assessment, characterization, and phytoremediation, it is important to establish a contingency plan. Even if plants are shown to meet the goals of reducing recharge or off-site contaminant migration, the health of the phytoremediation system can be negatively affected by perils not related to groundwater contamination. This can include infestation by leaf-eating insects, molds, voles, shrews, beavers, and large quadrupeds such as deer; lighting strikes; fires; early frosts; snow; hail; drought conditions; inundation; and high winds, just to name a few. If the planting area is compromised by one or more of these threats, the contingency plan can be implemented.

6.2.6 Regulatory Approval

Planting trees at a contaminated site generally does not require regulatory approval. However, the purposeful application of plants to affect a reduction in contamination level, to decrease recharge, to alter the site hydrology, or even transpiration of volatile organic compounds generally does require regulatory contact and approval. Regulatory officials at the local, state, and federal levels involved with a specific site should be included during preliminary discussions of the potential implementation of a phytoremediation strategy, such as during remedial investigations and feasibility studies

(RI/FS) required by RCRA and CERCLA regulations. The types of plants to be used may require some oversight by local officials concerned with invasive species or plants that are perceived to create aesthetic concerns.

6.3 Site Visit

The amount of information that is available about the contaminant release and delineation history at a particular site will vary considerably. For some sites, a lot of information may be available, but it may not apply to the areas that require phytoremediation. Alternatively, the contamination may be delineated in the source areas at some sites, because the release event may be apparent, as with a ruptured pipeline, but contaminant delineation may not be as comprehensive in downgradient areas.

Such site information often can be found at the state or federal regulatory agency that oversees initial site-assessment activities that follow a known or suspected release. The amount of information available may vary from state to state and from contaminant to contaminant, such as fuels released by USTs, fuels released from aboveground storage tanks (ASTs), or solvents released by various industries; each type may be covered by separate assessment and corrective-action programs. Assessment data often are generated by a private consulting firm if a responsible party exists or by state or federal agencies if the responsible party cannot be determined. Data generated is available through the Freedom of Information Act (FOIA). Alternatively, discussions with state or federal project managers offer a more direct route to this type of site information, with the added benefit of initiating a discussion about a potential phytoremediation project.

After perusal of existing site data, a visit to the site is warranted. A site visit pays dividends by revealing much that cannot be observed simply by going over files of data. Prior to the investment of limited time and resources commensurate with any proposed phytoremediation project, a site visit may reveal that conditions of demographic issues, human or wildlife receptors, and extent of the source area are not conducive to phytoremediation. For example, consider the scenario in which a UST leak from a gasoline station is characterized by a large volume of gasoline floating on the water table that is also located upgradient from a reservoir used as a sole-source of drinking water. Because of the large volume of gasoline in the subsurface and the short distance to a surface-water receptor, more aggressive remedial actions other than phytoremediation likely will be needed. If the reservoir were not a source of drinking-water supply, such a scenario could be acceptable for hydrologic control by phytoremediation. In any event, a site visit helps to more rapidly form remedial hypotheses and make informed decisions on whether or not phytoremediation should be proposed to achieve hydrologic control. A site visit also permits the opportunity to interview others involved or neighboring residents; a 5-min conversation with a long-time resident near the site often reveals more information than contained in a site report.

A visit also can reveal areas within the property boundary of a contaminated site that cannot be used for phytoremediation. The presence of buildings or parking lots or other inaccessible areas may preclude the installation of a phytoremediation system. Also, areas that receive high foot or automotive traffic may have areas of reduced soil permeability because of compaction. Airports or other businesses may require specific lines-of-sight to remain unobstructed to ensure safe business operation, thus preventing the installation of a phytoremediation system in those areas. Moreover, the presence of buildings on site may negatively affect the amount of solar radiation available to grow and sustain the phytoremediation system.

A site visit also can permit a conceptual model of groundwater flow to be visualized. The local topography can be used to approximate the direction of groundwater flow and where recharge may occur. Steep topography often results in steep hydraulic gradients and fast rates of groundwater flow, which may limit the uptake of groundwater by plants at rates high enough to affect groundwater flow and contaminant transport. Conversely, low-lying areas that contain surface water throughout the year or during certain periods indicate groundwater discharge, which can potentially limit the flow of contaminated groundwater to off-site areas. Areas of standing water at high elevations, particularly for a short time after precipitation events, can indicate areas of groundwater recharge. For limitations to this approach, however, see Chap. 4 and Haitjema and Mitchell-Bruker (2005). Regardless of how much information can be gleaned about the depth to water table from reports and site visits, however, it still is useful to collect depth to groundwater data if only to confirm or refute the data previously collected by others or to establish the range of seasonal fluctuations that occur.

Finally, perhaps the most important reason to justify a site visit is to examine the presence and extent of existing vegetation. The existence of vegetation indicates that the minimum requirements for plant life are provided. The type of plant also may provide a proxy on the depth to groundwater or soil moisture content in the unsaturated zone as was described in Part I.

6.4 Plant Hydroecology Assessment and Characterization

Visits to contaminated sites around the world generally would lead to observations of some type of plant growth, even in arid desert areas. Plants, after all, represent more than 97% of the biomass on earth (Campbell et al. 2006). In fact, observations by researchers of plant growth at contaminated sites, such as made by Cunningham and Ow (1996), Fletcher (1991), and Chaney et al. (1997) indicated the possibility of using plants to restore contaminated sites—the genesis of phytoremediation. Two central questions arising from these observations were "What is the source(s) of water to the plants?" and "How do plants survive in a contaminated area?"

At a minimum, plant growth at sites where subsurface and groundwater contamination have been detected is a general indication that the necessary requirements to support plant growth are present. Growth indicates that plants are able to find sufficient amounts of nutrients, micronutrients, and support to withstand erosion or toppling by winds. It also indicates that water is available from precipitation, soil moisture, or even groundwater.

Much useful, preliminary information relative to the establishment of phytoremediation can be gained by an inspection of the plants at contaminated sites during a site visit. Previously existing site-assessment reports not focused on phytoremediation often indicate the presence of plants on a site map but in generic terms, such as "wooded area". A site visit would enable defining the distribution and types of such mapped vegetation, such as deciduous or conifer, at the site. This observation is important in the initial stage of a phytoremediation evaluation, to help in answering the questions "Is the site conducive to support plant growth?" and "Will extensive soil amendments be required?" and "Can existing vegetation be used as part of the overall phytoremediation strategy?" During a site visit, it is important to realize that some plants may not be indigenous to the area. For example, a site undergoing construction or revitalization may have had the native plants removed and replaced with landscape plants.

6.4.1 Potential Water Sources and Plant Distribution

A visit to a contaminated site provides a general view of the plant and water relations. Healthy plants indicate that water must not be a limiting factor. This indication is at least valid for the type of plants present under the prevailing weather conditions and conditions of the subsurface. The issue of natural water availability is important, because even if native vegetation is not part of a future phytoremediation design, its presence can indicate whether a supplemental water supply is needed.

Close inspection of the types of plants encountered during a site visit can provide an idea of the source of water being used by the plants. At one extreme, the presence of swamps or wetlands near the site indicates shallow depth to the water table. Plants in these locations use predominantly either surface-water runoff or groundwater. Conversely, in upland areas, the depth to water table typically is greater, and the plants typically rely solely on precipitation and soil moisture. Between such extremes, plants can rely on water from all sources, including groundwater.

6.4.2 Plant-Nutrient Availability

Plants at a contaminated site indicate not only water availability but the presence of other environmental factors necessary for plant establishment, growth, and survival. These factors include the availability of essential and trace nutrients, the status of plant health with respect to infestation of insects or other pathogenic organisms, and perhaps the presence or absence of subsurface contamination.

A few simple methods can be used to assess this plant-nutrient availability at a contaminated site as part of a site visit. Soil samples can be collected from the surface by using a variety of methods, and the samples can be analyzed for geochemical properties important to plant growth. This includes assessment of the concentrations of nitrogen, phosphorus, and potassium, and percent organic content. A geologic classification, such as particle-size analysis, can provide information about the relative grain-size distribution of the soils and the percentages of sands, silts, and clays which are directly linked to water movement and bioavailability as described in Chap. 4. In addition, samples can be analyzed for cation-exchange capacity, pH, and alkalinity. The important concept here is that some of these soil characteristics that affect plants can be changed by the addition of appropriate amendments. This is in contrast to the many site-specific variables that cannot be changed, such as precipitation amounts, solar radiation intensity, and local weather patterns.

The geochemical condition of groundwater also can be assessed during a site visit to determine if plants installed for phytoremediation can be sustained on the ambient water quality. The salinity of the groundwater, for example, as defined by measurement of total dissolved solids, needs to be assessed, because levels that are too high may not warrant the installation of even salt-tolerant species. The dissolved-oxygen (DO) content of groundwater also can be assessed, because roots require oxygen to support cellular respiration, and lack of oxygen in groundwater can often lead to death of the entire plant. The DO content generally is higher in moisture in the unsaturated zone relative to groundwater because the exchange of soil air with the atmosphere occurs at a more rapid rate than the interaction of the groundwater and the atmosphere through precipitation events.

A release of reduced organic matter to groundwater, however, will result in lowered concentrations of DO.

The organic matter, such as gasoline or other fuels, exerts a demand on DO and often leads to anoxic groundwater. This scenario does not mean that once contaminated, groundwater is isolated from additional sources of oxygen. As shown in Fig. 6.1, groundwater in a shallow gasoline-contaminated aquifer rendered anoxic by contamination received DO during recharge and was observed by using a DO-sensor placed in a monitoring well (Landmeyer and Bradley 2003). After recharge, DO concentrations returned to pre-precipitation conditions because of oxygen demands in the aquifer exerted by aerobic bacteria or the chemical oxidation of reduced inorganic species.

The presence of gases in the unsaturated zone, such as CO_2, also can be assessed during a site visit. If high concentrations of fuel are known or expected to be present in the groundwater, either as residual contamination or as a separate-phase product, the soil air may contain CO_2 at a concentration that can inhibit the respiration of plant roots. One simple method for determining the presence of high CO_2 concentrations is to dig a hole about half-way to the water-table depth and slowly lower a flame source, such as from a butane lighter or torch, into the hole. If the flame is extinguished, it may indicate the presence of CO_2. This should not be done for an extended period, because oxidation of the butane produces CO_2, and the presence of explosive gases, such as methane or hydrogen, also may be present in the hole. A safer approach to assess the presence of multiple gases in the soil zone during a site assessment would be to use a hand-held gas analyzer that measures the percentages of either oxidized or reduced gases.

Plant growth is related to the presence of root-associated fungal and bacterial communities. Hence, some basic microbiological work as part of site assessment can lead to a greater understanding of the microbial health of the contaminated area. Soil samples can be analyzed for various aerobic and anaerobic bacteria and the presence and distribution of mycorrhizal communities in areas that contain native vegetation. Alternatively, the establishment of baseline conditions in areas containing no native plants but designated for planting can help determine if plant addition will result in rhizosphere formation. Collection of this microbiological data also will help to determine if supplementation by root inoculation may be needed during plant installation, discussed in Chap. 7.

6.4.3 Plant Laboratory Studies

The presence of vegetation at a contaminated site does not always indicate that plants interact with contaminants or groundwater in a manner analogous to phytoremediation. One way of determining if native plants are interacting with groundwater, or if the plants proposed for use at the site can interact with the water table, is to conduct laboratory-scale experiments. Even if the native plants are not evaluated, site soil and groundwater samples can be used

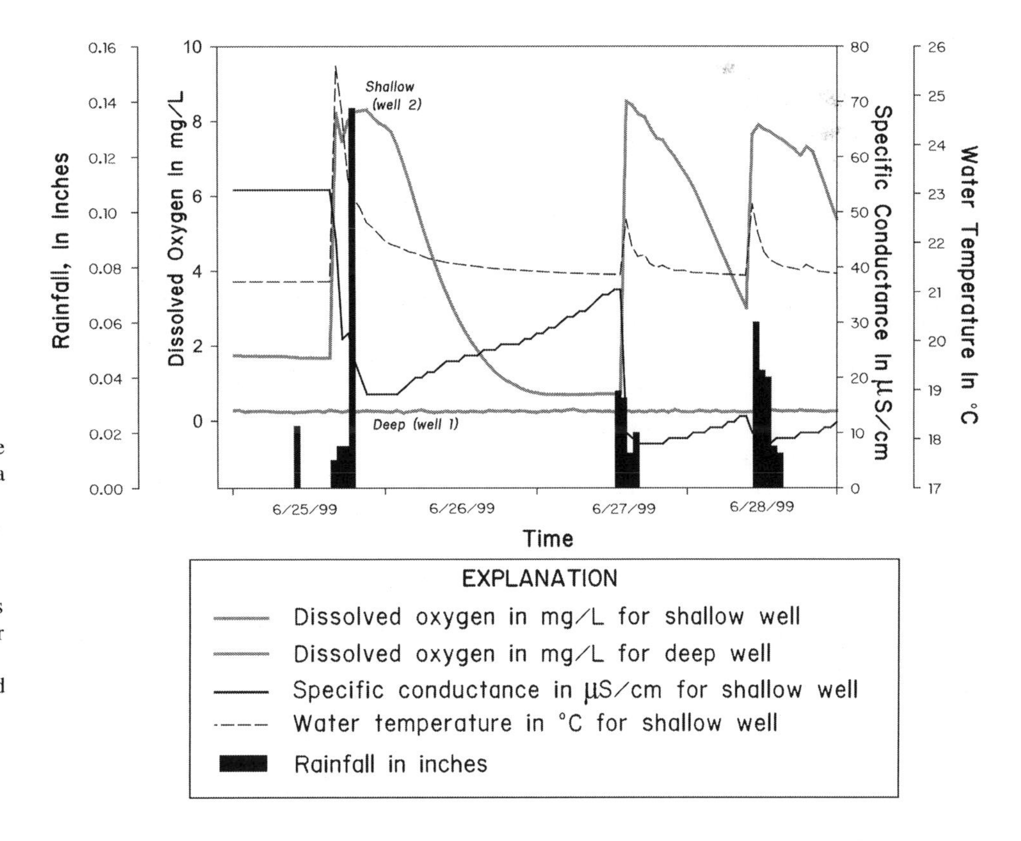

Fig. 6.1 The delivery of dissolved oxygen during recharge events to anoxic groundwater at a gasoline-contaminated aquifer, Laurel Bay, South Carolina. The depletion of oxygen was caused by microbial oxygen reduction rather than physical processes, as the DO-depletion slope is steeper than that of the slopes showing changes in water temperature and specific conductance (Modified from Landmeyer and Bradley 2003). One inch is equivalent to 2.54 cm.

as media for samples of the plants proposed for use at the site. In this manner, the level of contaminant that is toxic to a particular plant can be assessed before an entire phytoremediation system is installed correctly but fails to thrive as anticipated.

Most laboratory studies that are performed at a site are based on questions about the interactions between plants, water sources and availability, and groundwater contaminants. Part of this interaction leads to questions regarding the potential increase in risk exposure to human and wildlife populations caused by planting trees at a contaminated site. Laboratory-scale experiments are an approach to provide answers to these questions, and are discussed specifically in Chaps. 12 and 16.

6.4.4 Weather and Climate

The interactions between weather and plant distribution and growth are known generally to be inseparable, as was recorded as early as the first century AD by Pliny the Elder (see Chap. 1). The relation of periods when light, air temperature, soil moisture, humidity, and precipitation are optimal for plant growth defines the growing season for specific plants in specific regions of the world. For example, air temperature affects the rate of plant metabolism. In general, most plants can maintain metabolism between 35°F and 110°F (1.6–43°C). Short periods of exposure to sub-optimum temperatures, such as nighttime freezes, can result in the death of actively growing parts of a plant but the whole plant usually survives. Longer exposure to such conditions, however, generally results in plant death.

Under natural conditions, different plants have acclimated to a unique range of temperatures. The widespread natural distribution across North America, Asia, and Europe of the Genus *Populus*, for example, is one reason why poplar trees are commonly used for phytoremediation projects. The distribution of *Populus* and its affect on phytoremediation are discussed in Chap. 7.

The acclimation of different plants to different ranges in air temperature led the U.S. Department of Agriculture (USDA) to make maps that depict zones of plant hardiness, or tolerance to low-temperature extremes. These maps divide the United States into ten hardiness zones, located roughly horizontally across the country, as a function of temperature changes related to changes in latitude. Zone one is in the far northern United States and zone ten is in the far south. Each zone contains plants adapted to the average minimum air temperature in the zone. The determining factor of the success of a particular plant is the lowest temperature that it can survive without death of the roots. South Carolina, for example, is in hardiness zone eight, characterized by average minimum air temperatures between 10°F and 20°F (−12 to −6°C). Within each zone, plants more characteristic of a higher numbered zone usually can survive near bodies of surface water or large buildings, which broadens the lower range for a particular zone. Trees that grow best in zone eight, however, likely would not survive if planted in a lower zone, such as zone five. Table 6.1 presents the hardiness zones for some phreatophytes that can be used for phytoremediation.

Table 6.1 Preferred hardiness zones for trees typically used in phytoremediation.

Tree	Preferred hardiness zone, USDA
Birch, white	2–7
Birch, river	3–7
Willow oak	6–9
Willow, weeping	2–10
Sycamore	5–10
Populus spp.	2–10
Eucalyptus spp.	9–10
Baldcypress	5–10

6.4.4.1 Precipitation Maps

Precipitation amounts that tend to fluctuate annually for a particular area generally approach a fairly stable long-term average amount. Because of the reliance of plants on water, plant distribution closely follows precipitation abundance. Humid areas typically have more than 20 in. (50.8 cm) of precipitation per year, and areas with less than 10 in./year are considered arid. Between these two extremes are semi-arid areas. The average precipitation in each area constrains plant growth, like air temperature, especially in terms of establishing a phytoremediation planting (Fig. 6.2).

Not only is the amount of precipitation important, but also the timing, duration, and intensity of precipitation. The relation between precipitation and soil characteristics is important to plant distribution, because even high precipitation amounts that suggest *a priori* plant growth may not be

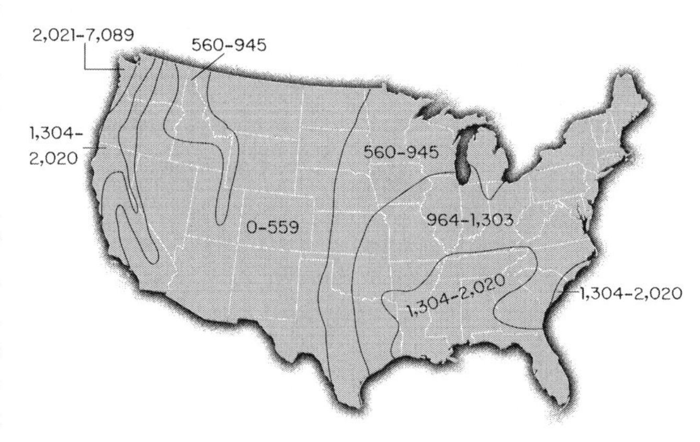

Fig. 6.2 Average annual precipitation, in millimeters, from 1980 to 1997, for the conterminous United States (Modified from Healy et al. 2007). One millimeter is equivalent to 0.039 in.

available to plants if the soil is extremely sandy and porous and readily drains away. This is the case in the upper Coastal Plain of many states, such as South Carolina, which receives 40–60 in./year (101–152 cm/year) of precipitation but supports only slash pine, stunted turkey oaks, and cacti—this area is essentially a desert in the rain. A similar condition also can be found in the Pine Barrens of New Jersey. In both areas, only longleaf pine (*Pinus palustris*) can thrive as they have a deep tap root.

Conversely, abundant precipitation does not always lead to lush growth if air temperatures are low, as is exhibited by conditions on the Aleutian Island chain of Alaska. There, the vegetation is dominated by grasses; no native trees can be found, although the Adak National Forest, which is part of the Alaska Maritime National Wildlife Refuge, consists of 33 spruce trees. Actually, these trees were planted in 1944 to boost the morale of the military personnel stationed in Adak, which had no trees at that time. This forest is so small, that the sign at the entrance states, "You are now entering and leaving the Adak National Forest."

6.4.4.2 Solar-Radiation Maps

All green plants require electromagnetic energy from the sun for photosynthesis as was outlined in Chap. 3. Less light over short periods results in a lower and slower rate of photosynthesis and, therefore, growth and reproduction potential. More light over long periods results in more food production, growth, and reproduction. It follows, then, that plant distribution is related to light conditions in a similar manner as to precipitation and air temperature. Hence, any phytoremediation effort is related to the light conditions of an area.

Insolation maps provide data about the length of time a particular location has solar radiation. In the United States, the average daily solar radiation per month, in kilowatt hours per square meter per day—essentially the average hours of sunlight energy input per day—ranges from 4 to 5 h/day (hours per day) for most of the northeastern and central plains states, 3–4 h/day on the coast of Oregon and Washington, and 5–6 h/day in the southeastern United States, including Florida (Fig. 6.3). Areas of California and some states along the front range of the Rocky Mountains down to Texas have between 6 and 7 h/day.

The solar data presented on insolation maps are spatial interpolations of solar radiation measurements taken between 1961 and 1990 and stored in the National Solar Radiation Data Base (NSRDB). The average values described above for solar radiation duration are averages of the 30 years of data from up to 239 sites that comprise the NSRDB (Fig. 6.3). Because the value given is the average value for each site, the number of hours of sunlight typically is lower during the winter months and higher during the summer months. For example, Columbia, SC, can have as much as 6–7 h/day of solar radiation in June compared to 4–5 h/day in December. In general, the minimum value of solar radiation can be determined for an area that is a candidate for phytoremediation by measurements taken on the shortest day of the year, December 21. The values for solar-energy input are important to know for a particular contaminated site, because all phytoremediation processes are based on the establishment of plant photosynthesis, growth, and transpiration, all of which are dependent on solar radiation for energy or water transport.

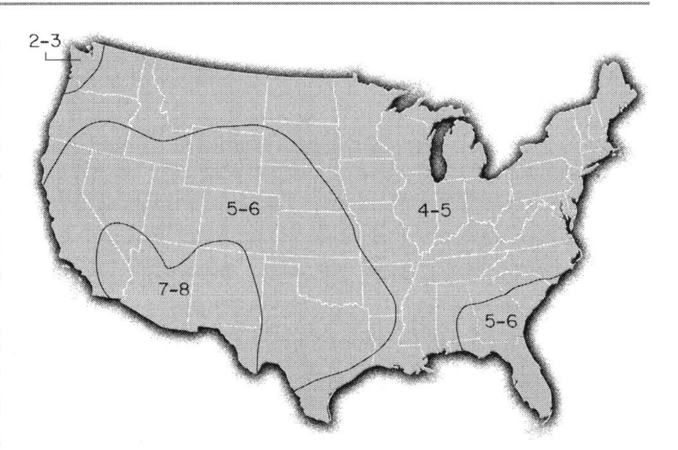

Fig. 6.3 Average annual daily solar radiation per month, in kilowatt hours per square meter for the 30-year period 1961–1990, for the conterminous United States (Modified from Marion and Wilcox 1994).

6.4.4.3 Growing Season Length and Potential Evapotranspiration

Solar radiation and the fluctuation of air temperature are the primary environmental factors that control the annual cycle of growth and dormancy in plants used for phytoremediation. The annual cycle of growth determined by favorable solar radiation and air temperatures is called the growing season. In the northern United States, in areas not affected directly by the Great Lakes, the growing season can approach about 140 days. Nearer the Great Lakes, the growing season increases to almost 200 days, because of the high heat capacity of water and its ability to moderate air-temperature fluctuations. Even when these areas receive snowfall, the snow acts as a good insulator to keep the soil temperatures in the root zone warm enough to encourage root growth. In the southern United States, the growing season can range from about 190 days to as much as the entire year, as the ground rarely is frozen.

The length of the growing season for a particular area can be highly variable, however, based on the topographic elevation and frequency of precipitation. For example, Atlanta, Georgia, and Lubbock, Texas, have similar average maximum and minimum air temperatures, about 90°F (32°C) and 28°F (−2°C), respectively, but different precipitation amounts. Atlanta is at a higher elevation and has 29 in./year of precipitation, whereas Texas is at a lower elevation and has only 15 in./year. This results in a longer growing season of 240 days in Atlanta relative to 200 days in Lubbock. Moreover, the western United States has many microclimates that range from the dry deserts of Nevada and southeastern California to the moist rainforests of Washington. Therefore, the average growing season length for a particular area must be used with care when assessing a site for phytoremediation.

Potential evapotranspiration maps have been constructed for the conterminous United States and also are useful for site-assessment and characterization purposes for potential phytoremediation projects (Fig. 6.4). This map indicates that higher potential evapotranspiration, ET_p, rates are related to the south and western states where the growing season length is longer. In general, higher ET_p rates indicate greater potential for hydrologic control by phytoremediation. Moreover, the greater the proportion of an area's water budget allocated to *ET* relative to precipitation suggests an even stronger potential for hydrologic control by phytoremediation to occur (Fig. 6.5). For example, values less than zero indicate water-limited conditions and a high potential for groundwater control and values greater than zero indicate energy-limited conditions (modified from Healy et al. 2007).

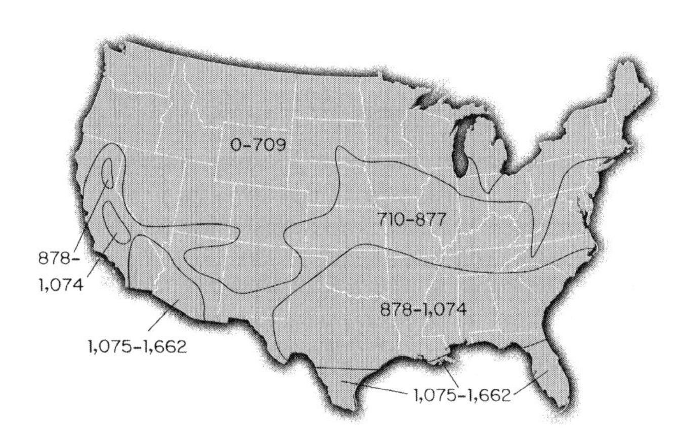

Fig. 6.4 Ranges in average annual potential evapotranspiration, in millimeters, for the 30-year period 1961–1990 for the conterminous United States (modified from Healy et al. 2007). One millimeter is equivalent to 0.039 in.

6.4.4.4 Vapor-Pressure Deficit

Vapor pressure is the gas solubility of a specific compound at a specific temperature and pressure, as described in Chap. 2. The higher the vapor pressure, the more likely the compound is to enter a gas phase. Because transpiration involves water vaporization controlled by diffusion, the concept of a vapor-pressure deficit, *VPD*, is used when describing plant and water interactions, both liquid and gas.

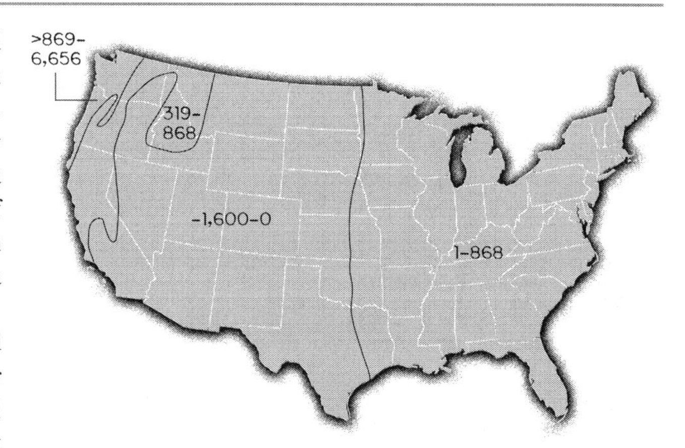

Fig. 6.5 Relation of precipitation to potential evapotranspiration in areas of the conterminous United States. Values less than zero indicate water-limited conditions. Values greater than zero indicate energy-limited conditions (Modified from Healy et al. 2007).

The magnitude of the *VPD* represents the difference between the vapor pressure at a given air temperature at saturated conditions and the actual vapor pressure at the same temperature (Kucera 1954). Vapor-pressure deficit can be determined by using common meteorological parameters, such as air temperature and relative humidity. Although the atmospheric pressure remains fairly constant at any given location, the lack of water and high specific heat characteristics lead to great fluctuations in air temperature on a daily basis in arid areas. Also, the relative humidity of a location changes on a daily basis and with respect to the measurement location. The *VPD* is important for plant transpiration and, therefore, phytoremediation because at a constant air temperature, with no direct input of radiant energy, water evaporates along vapor-pressure gradients, until equilibrium conditions are reached.

6.4.5 Plant-Available Water

Most plants thrive in well-drained soils for a number of reasons. First, loose soils permit root penetration. Second, well-drained soils permit the infiltration of precipitation. Third, well-drained soils permit atmospheric influx, including oxygen needed for respiration and nitrogen needed for fixation. Fourth, water from specific retention is held to the soil with less tension in soils with smaller pores and, therefore, is more readily available to the plant roots. Soils with too large pore spaces, however, drain quickly and retain little water, even though root penetration and gas distribution can be high.

The sediments in the subsurface consist of solids with pore spaces between them as was described in Chap. 4. The pore spaces can be filled with water, water and air, or air. As pores that are totally filled with water (fully saturated) drain by gravity, air enters the soil to replace that volume lost. These changes occur when the water table fluctuates, which directly affects groundwater use by plants.

As we saw with the description of Darcy's Law, groundwater is affected by gravity. Water movement in the unsaturated zone is called gravity drainage. If the hydraulic conductivity of the soil is higher than the rate of water entry into roots, the water will drain past the root zone before plant uptake. After gravity drainage occurs, the remaining water adheres by tension to the surfaces of soil particles. This water, called field capacity, has a tension of about −0.33 bar (roughly −0.33 atm). Although not affected by gravity, water held by tension is available for entry into plant roots, called plant available water, until tensions exceed −15 bar. As the soil dries out and tensions increase, plant wilting occurs, as plants cannot take up water even though water is present. At the wilting point, surface tension exceeds osmosis, the root water potential, ψ_r, is equal to the soil water potential, ψ_s, and no water flow occurs. The wilting point typically occurs when the volumetric water content in loams equals 30% (Fig. 6.6). The wilting point in clays is reached when the soil moisture decreases below about 15%, and in sandy sediments is nearer 5%.

The amount of water in soil or sediment between the wilting point and field capacity is referred to as specific retention, also called gravitational water or water-holding content by plant physiologists. The amount of water under specific retention that can enter plant root hairs is referred to

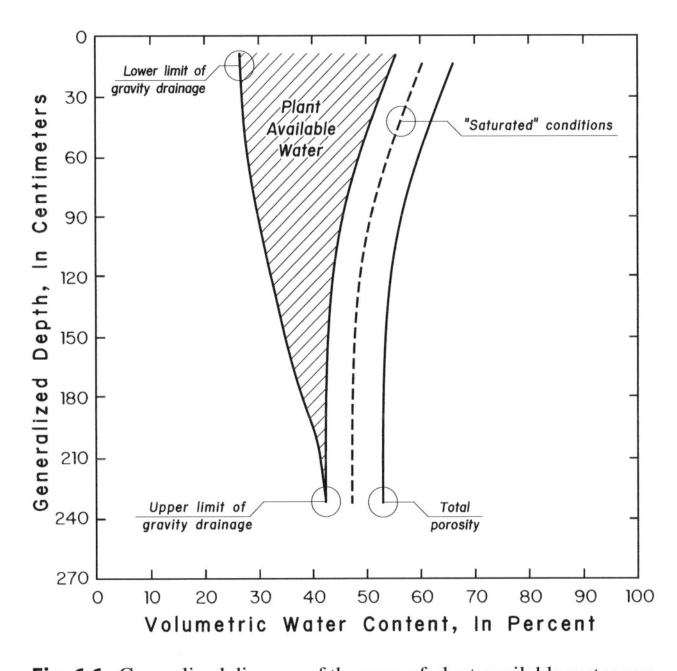

Fig. 6.6 Generalized diagram of the zone of plant-available water as a function of soil moisture, which is represented as the volumetric water content and depth below land surface. This relation explains why many plants, including some phreatophytes, have extensive shallow roots. One centimeter is equivalent to 0.39 in.

as plant-available water. Typically, plant-available water is half of the specific retention. The range of plant-available water is between −0.33 and −15 bar (−0.033 and −1.5 MPa [megapascal]). For conversion, 1 kg of mass on a unit surface area is equivalent to 9.8 Pa, where 100,000 Pa is equivalent to the pressure of the atmosphere, 100 Pa is equivalent to 1 mb (millibar), and 1 atm is equivalent to 1,013 mb.

The relation between water tension and plant-water availability can be envisioned by a simple analogy using a sponge. The water absorbed by a sponge after immersion represents water saturation. The amount of water that drains by gravity after immersion represents field capacity; when the sponge is squeezed, the water removed is the holding capacity, and the water that remains in the sponge represents water held by tension. Plant-available water is represented in this analogy, therefore, by the water available that is half of the water-holding capacity.

Fluctuations in water availability to plant roots produce different effects on plant growth and overall metabolism and health. A decrease in water availability, from drought conditions or insufficient irrigation, can result in stomatal closure and a reduction in CO_2 uptake. This decreases production of carbohydrate and, if continued for a long period, leads eventually to plant death by starvation. Too much water, on the other hand, especially if it is stagnant, can lead to the expulsion of air from the root zone and decreased root growth from lack of oxygen necessary to support root respiration. If the inundation lasts too long, the roots will die from a lack of oxygen. From this discussion, we can see that the water held by tension in the capillary fringe provides a major advantage as a source of water to plant roots, especially in terms of plants used at phytoremediation sites, because air is available in the pore spaces to support root respiration, the water will not drain by gravity, and the plants can overcome water tensions.

The thickness of the capillary fringe changes in space in different soil types and over time because of variations in infiltration amounts and rates. Sands have an average plant-available water thickness of 0.75 in./ft (1.9 cm/0.3 m) of soil column. This means that for a vertical soil column of 1 ft (0.3 m), an average of 0.75 in. (1.9 cm), or about 6% of the water in the sand pores, is available for plant use. Conversely, clays have an average of 2 in./ft (5 cm/0.3 m) of soil column, or about 16% of the total available. This is why plants grown in sand either need continuous water from frequent precipitation or a very deep root system that is exposed to soil moisture throughout the unsaturated zone down to the capillary fringe and water table; more of these factors are discussed in Chap. 7.

The widely used concept of the wilting point just discussed is not valid, however, for the roots of phreatophytes that are in contact with surface water or groundwater. For example, groundwater, by definition, is water present in the pores of sediments equal to or greater than atmospheric pressure as described in Chap. 4. The roots and root hairs of phreatophytes in contact with groundwater do not have to overcome water tension present in the capillary fringe. The implication for plants whose roots interact with the water table is that the concept of wilting point does not hold true for these roots and water will be available for uptake on a continual basis, as long as it can be accessed by the roots and as long as sufficient oxygen is available to support root respiration.

The selective advantage that phreatophytes have in terms of adapting to a reduction in soil moisture from decreased precipitation and drying surface soils and increasing negative water potentials is their ability to remove water under less negative tensions than the wilting point. The energy required to overcome the tension of water is equal to the osmotic water potential—the roots and the energy to lift the water through the plants to the atmosphere by the sun. By comparison, other forms of life in the subsurface, such as soil and aquifer microorganisms, can tolerate tensions no less than −0.01 MPa (Table 6.2).

To summarize for the purposes of site-assessment activities, the *presence* of water, or water content in unsaturated soils above the water table, or the water table itself, does not necessarily guarantee that plants will thrive. This is because the presence of water does not mean that water can move through the soil pores into the root hairs. This is determined by the *bioavailability* of water, or the water potential, which indicates whether or not water can be removed. Because plants derive water from the soil by osmosis and transport this water to leaves along a vapor-pressure gradient driven by evaporation, the remaining water is removed from the soil surface by enough energy to overcome surface tension (Fig. 6.7). At the point where surface tension is too high for plants to extract additional water molecules from the soil surface, plant-water uptake stops, even though water may still be present.

6.4.6 Soil Bulk Density and Water Content

The concept of bulk density familiar to soil scientists can provide useful information regarding the relation of plants to soil and water and, therefore, should be determined as part of site assessment. The bulk density of soil also can be measured easily. Bulk density refers to the ratio of dry soil to the total soil volume. In an uncompacted, humus-rich peat deposit, for example, the bulk density is lower than that of a similar volume of tightly compacted silty sand.

Conversely, the water that occupies the pore spaces of a volume of soil can be quantified using the concept of water content. Water content is determined by subjecting a weighed, moist, soil sample to oven drying and then reweighing the sample. The mass lost relative to the wet mass is the water content. Water content does not indicate either water availability to plants or the potential for water flow to occur, as was stated previously.

6.5 Hydrogeologic Assessment and Characterization

Perhaps the most important hydrogeologic factors that will determine the success of a phytoremediation project for hydrologic control at a site characterized by contaminated groundwater are the thickness of the capillary fringe and depth to groundwater. Simply put, if the minimum depth of the water table at its seasonal highest is beyond the reach of the deepest roots, then alternative remedial strategies may

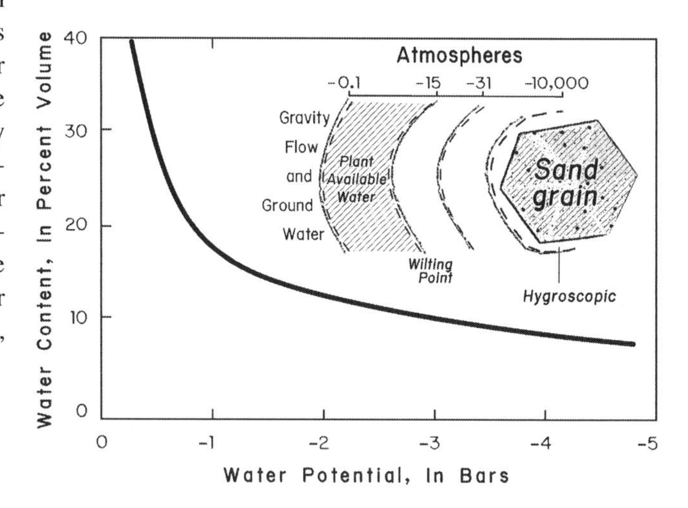

Fig. 6.7 As the water potential becomes more negative as sediments dry out, it is more difficult for water to enter root hairs. 1 atm is equivalent to 0.10 MPa.

Table 6.2 The water potential of various soil conditions encountered at phytoremediation sites.

Water potential (MPa)	Classification	Moisture characteristics
−30	Unsaturated	Hygroscopic water–bound to sediment.
−1.5	Unsaturated	Wilting point–tension limit at which plants can remove water.
−0.03	Unsaturated	Field capacity–water left after movement by gravity.
0.99 to −0.001	Capillary	Capillary zone and fringe–water that moves due to tension.
>0	Saturated	Water table–water flows to wells.

need to be investigated. Phytoremediation could still be used but would require more costly deep-planting techniques, such as trees installed in vertical holes using Tree-wells®, or the installation of an irrigation system.

Some of the methods that can be used to assess the hydrogeologic properties of a site for possible phytoremediation are similar to conventional groundwater remedial strategies. One such approach applied to achieve hydrologic containment and (or) control has been and continues to be the mechanical pump-and-treat approach. In brief, the pump-and-treat approach is based on the installation of relatively large-diameter wells that are screened or open to the part of the aquifer that is contaminated, and the groundwater is pumped and treated to acceptable levels prior to discharge, usually into the sanitary sewer system. At best, experience from many sites undergoing pump-and-treat remediation indicates that a time period on the order of decades is necessary to remove enough pore volumes of groundwater to decrease contaminant levels to acceptable levels (National Research Council 1994), and that the operation and maintenance costs associated with such long-term remediation are prohibitive. At worst, the wrong placement of the wells used for pumping, such as near property boundaries downgradient of the source area, can accelerate the spread of contaminants to previously uncontaminated areas. Finally, this approach of using groundwater as a vector to remove contaminants would have obvious limitations at sites characterized by aquifers of low transmissivity and low specific yield.

6.5.1 Groundwater System Concept

A common mistake made during site-assessment activities is to focus on the hydrologic conditions within the site-specific boundaries to the exclusion of that of the surrounding areas. Many sites assessed are only a few acres or less. Before site-specific characteristics are evaluated for the potential application of phytoremediation for hydrologic control, the site-specific conditions should be evaluated in context with the overall regional hydrogeologic framework, of which the site conditions are but a small part. For example, the locations of regional groundwater recharge and discharge areas must be evaluated in relation to the site location. It is possible that the site may be located in the recharge area of an aquifer used as a sole source of drinking water farther downgradient where the aquifer has become confined.

Information on regional groundwater systems can be found in either previous site-investigation reports, or in more generalized publications about the study area. Potential groundwater discharge areas that can receive contaminated groundwater, such as lakes, ponds, streams, and wetlands, must be evaluated in terms of their effects on local site hydrogeology. Although locations of such groundwater discharge areas generally are depicted on maps, their effect on local groundwater-flow paths, flow directions, and site-specific water budgets usually are not clear at this level of investigation. For example, in many cases the elevation of these surface-water features may not have been determined. This lack of information makes an *a priori* determination difficult as to whether a surface-water feature is, indeed, a site of localized groundwater discharge or recharge. In other cases, the water level of the surface-water body may be known and shown on maps in site reports, but how contours of equal groundwater levels interact with the surface-water body may not be depicted or known.

6.5.2 Depth to Groundwater, the Unsaturated Zone, and Infiltration

In general, shallow depth to the water table provides many advantages for the hydrologic control of groundwater by phytoremediation. When the water table is shallow, plant installation is relatively easy, and growth is more rapid. This is because groundwater quickly becomes the primary water source of the plants. However, a shallow water table also has disadvantages for phytoremediation. One disadvantage is that many phreatophytes are facultative and can use sources of water other than the groundwater that needs to be controlled.

This does not mean that sites where groundwater is at a considerable depth cannot be controlled successfully by using phytoremediation. At such sites, typically in arid areas or in humid areas characterized by porous sediment, the effects of plants and plant selection on the site hydrology may be more important in terms of water removal from the capillary fringe, a reduction in leachate formation by interception of infrequent precipitation, or induction of the upward movement of groundwater toward the land surface. Between these two water-table extremes are the cases of deeper semiconfined or confined aquifers.

The characteristics of the components of the unsaturated zone at a site can be evaluated without collecting sediment samples, but by observation of what happens to rainwater during or after precipitation. Areas of standing water after precipitation often indicate a low infiltration rate. Such 'recharge refusal' typically occurs in areas subject to compaction by foot or other heavier traffic. Too much compaction leads to runoff or ponded water. In the latter case, the water eventually infiltrates to the unsaturated zone and water table relative to the properties of the soils, such as porosity and permeability or hydraulic conductivity. In general, slow rates of infiltration range from 0.2 in./h to less than 0.05 in./h (0.5–0.12 cm/h), whereas rates above 5 in./h (12.7 cm/h) are rapid. If precipitation rates exceed infiltration rates, the water table can rise to intersect land surface.

Infiltration tests can be conducted readily in the field. For example, single- or double-ring infiltrometers can be installed; water is added to the infiltrometer, and the rate of water-level drop is measured to determine the rate of infiltration. Alternatively, a jug or carboy filled with water and turned on end in a collar of PVC three-way pipe can be used. A hole typically is dug 2–3 ft (0.6–0.9 m) deep above the water table, and 1–3 in. (2.5–7.6 cm) of gravel is added at the bottom to reduce turbidity when the water is added, and the rate of water decrease can be measured with a stopwatch.

A quick assessment of infiltration also can be accomplished by conducting a percolation test. A shallow hole is dug and the soil removed. A volume of water, such as 1 gal (3.7 L), is added to the hole and allowed to drain. The time to completely drain is recorded. Drainage is faster for sands and gravels relative to clays and silts. This also provides evidence of the ability for air to mix in different soil types, determined by the presence of bubbles in the water-filled hole.

6.5.3 Hand-Auger Method

A rapid method used to collect sediment samples, delineate the extent of soil contamination, and to describe the hydrogeologic properties of sediments at a site is the hand-auger method as described in Chapelle (1993). Frequently overlooked or undervalued because of its simplicity, a hand auger should be considered an essential component of any site investigator's toolbox. A hand auger consists of a metal cylinder that either is completely enclosed or contains openings that can be used when clayey soil is encountered. Attached at one end of the cylinder, or bucket, are two cutting blades that help the bucket to advance into the soil. After advancing the bucket to a desired depth, the auger is removed from the hole and emptied of its contents. Interestingly, perhaps this design was first employed by Bernard Palissy, whom we met in Chap. 2, as he examined the soils in France in the sixteenth century (Darcy 1856).

The hand-auger method also can be used to reveal the thickness of the capillary fringe, even if wells are present, because the sediment samples removed from the bucket can be examined for moisture content. Moreover, the approximate extent and thickness of the capillary fringe can be denoted by an increase in difficulty in removing the auger from the borehole as a result of water held under tension in the pore spaces.

The hand-auger method has some limitations for site assessment, however. One is that the sediment sample is disturbed during collection. Also, the method is not applicable in paved areas, in areas where subsurface obstructions exist, in coarse sediments, in areas with extensive roots, below the water table, or in areas known to have underground utilities. For these reasons, a hand auger should be used with caution.

6.5.4 Hollow-Stem Auger Method

A hollow-stem auger is essentially a large hand-auger bucket that is advanced into the soil by the power of an engine rather than manually and is used for deeper drilling than possible with a hand auger. The hollow-stem auger method also is called the rotary drilling method. Typical auger lengths are 4–8 ft (1.2–2.4 m); holes can be drilled deeper by attaching separate augers together. Similar to the hand-auger method, soil samples are compromised by substantial disturbance of the retrieved sample. This can be overcome, however, by attaching a hollow metal tube called a Shelby tube (after its inventor), or by attaching a metal tube that can be split into two sections, called a split-spoon sampler, to the end of the first auger. A benefit of using the hollow-stem auger method is that a monitoring well can be constructed and installed within the hollow stem of the auger after the sediment has been removed from the borehole and the auger sections then removed.

Boreholes also can be created in sediments by using air under high pressure. Called air-rotary drilling, high-pressure air from the end of a pipe is used to blow unconsolidated soil particles out of an advancing borehole and to bring them to land surface. This method does not permit the collection of intact soil samples, may cause volatile compounds in the unsaturated zone to be transported greater distances by the air pressure near the borehole, and presents an air-quality risk to drilling personnel.

6.5.5 Sonic Technology

Sonic drilling methods advance the drill rod into the subsurface by agitation, or mechanical vibration, rather than with mechanical rotary action. The hollow drill rods are vibrated at 50–150 Hz (hertz), or cycles per second. The hollow rods will retain the undisturbed sediment, which can be analyzed for aquifer properties or contaminant concentrations. Sonic, or vibracore, methods are suitable to unconsolidated aquifers although the total drilling depth may be less than with other methods.

6.5.6 Direct-Push Technology

Direct-push technology is based on using an engine-driven hydraulic pump to hammer a hollow rod with a solid tip into the ground. No cuttings are generated. The borehole created can be used for placement of a monitoring well, some which can be pre-assembled for rapid installation. Alternatively, groundwater samples can be taken during direct-pushing through a slotted or screened metal pipe attached to the end of the first drill rod. In this manner, groundwater can be collected rapidly and analyzed for various contaminants on site by using appropriate field-laboratory equipment. A variety of probes can be added to the end of the drill rod such as electrical resistivity probes and sensors that can detect VOCs. Moreover, aquifer properties such as hydraulic conductivity can be assessed using direct-push rods and traditional slug testing. This technology has made site assessment and characterization almost real-time activities.

6.5.7 Monitoring-Well Installation and Depth to Groundwater

Surface-water features, such as streams, ponds, or lakes, can provide an indication of the general depth to groundwater at most sites. This is because most surface-water systems receive discharge from groundwater, especially between precipitation events, as discussed in Chap. 4. For the purposes of site assessment and characterization activities, however, a more controlled method of documenting the depth to water table and its fluctuation is required. This can be accomplished by installing monitoring wells.

The construction of the ideal monitoring well to determine the groundwater level in porous media would be similar to the manometer Darcy used in his laboratory-column experiments (Darcy 1856). That is, a hollow tube open only at the end so that the water level inside the tube above the open end represents the pressure head at that open point. The pressure head elevation added to the elevation of the open end above a common datum would represent the total pressure head (see Fig. 4.6). In practice, however, monitoring wells tend to have an open, or more likely a screened interval that can be 5–10 ft (1.5–3 m) in length. The larger open or screened interval is used by convention in unconsolidated sediments because it integrates groundwater across a larger part of the aquifer, permits groundwater samples to be collected even as the groundwater level changes, and permits easier removal of groundwater samples.

Because the installation of a monitoring well requires a borehole to be created, the presence of the borehole and, subsequently screened interval, results in an artificial condition that permits vertical groundwater flow to occur within the saturated zone that previously did not occur. If not properly sealed, surficial contaminants can enter the aquifer by downward leakage. In fact, a poorly installed monitoring well can short circuit normal infiltration. This vertical transfer of contaminants also can occur below ground, where contamination at one depth in the aquifer can spread to shallower or deeper parts of the aquifer, which previously were uncontaminated. For this reason, most state or federal regulations require the part of the well above the water table to be sealed with an impermeable material, such as grout or cement, that impedes such short-circuit flow, and with appropriate filter pack sediments near the well screen. A poured-concrete base at land surface also helps prohibit the entry of surficial contaminants to depth.

Accurate measurement of the depth to and fluctuation of the water table across the total suspected contaminated area of a site is warranted, because a uniform depth to the water table may not indicate uniform contaminant flow beneath the site. For example, recharge through uncontaminated parts of the site will deflect horizontal groundwater-flow paths, and any dissolved contamination in these flow paths will be pushed deeper below the water-table surface. A consequence for a phytoremediation project planted in such an area where the contaminated groundwater-flow paths are deflected would be that roots would interact with a water table that is not, however, contaminated.

The depth to the water table can be determined rapidly in monitoring wells by measuring the depth using either a steel tape or electronic water-level meter. In areas where few or no monitoring wells exist, such as at uninvestigated sites, the hand-auger method can be used to determine an approximate depth to groundwater, where practicable. Where the water table is shallow, a steel pipe or steel pipe with slotted screen called a drive point, can be driven into the ground and used as a temporary well to measure the groundwater level. Finally, a series of monitoring wells installed with screens located at increasing depths below land surface, or nested wells, should be added at a phytoremediation site in order to document the presence of vertical gradients or the inducement of vertical gradients by tree uptake of water. This approach is discussed in Chap. 9.

The depth to groundwater ultimately is the major factor that controls the success of the installation of a phytoremediation system to hydrologically contain and (or) control contaminated groundwater. In most cases, the shallower the depths to the water table the better for phytoremediation establishment and hydrologic interaction. However, there are exceptions to this rule. For example, because the water table is a planar surface that moves up and down in response to the balance between input by precipitation and removal by *ET*, its depth below land surface is not constant. In fact, the depth to the water table can vary considerably over a year, especially in areas containing natural vegetation, or at sites near tidally influenced surface-water bodies. An implication of these groundwater fluctuations for phytoremediation systems is that trees initially established with a root mass above the water table can die if inundated by a rising water table for long periods, which varies from species to species.

Alternatively, trees in humid areas may rely on precipitation as much as on groundwater to meet *ET* demands, so shallow depths to groundwater may not have a positive effect on phytoremediation of contaminated groundwater. Even though tree growth at a site may be successful and the trees documented to transpire water, very little of the transpired water will consist of groundwater. On the other hand, the water table may be too deep, as in arid areas or in higher, well-drained elevations in humid areas, to support the rapid establishment of trees at the surface installed by using conventional methods, as is discussed in Chap. 7.

A metric that can be used in evaluating phytoremediation projects is the mean depth to high water table calculated from seasonal groundwater-level fluctuations. This mean depth is affected by additional factors, such as tides or changes in barometric pressure that can be accounted for by using equations discussed in Chap. 9.

The above problems notwithstanding, phytoremediation for hydrologic containment or control of contaminated groundwater generally tends to be applicable and cost effective in areas where the depth to water table below land surface is 30 ft (9 m) or less. This would indicate that phytoremediation tends to be restricted to shallow aquifer systems. It is exactly these shallow aquifer systems, however, that are most vulnerable to contamination because of the proximity to land surface. Deeper groundwater can be candidates for phytoremediation for hydrologic control, however, by using a variety of approaches that are beyond the scope of this book. In brief, these approaches include installation of a phytoremediation planting in recharge or discharge areas of deeper, more regional aquifers where groundwater flow is at or near the surface of the ground. Moreover, even shallow and deeply confined contaminated aquifers can be accessed by plants, either under natural or engineered conditions; some examples are provided in Chap. 8.

6.5.8 Groundwater-Flow Direction

Groundwater flow that occurs at a contaminated site is a vector of potential dissolved-phase contaminant transport. This is because many contaminants released to groundwater are soluble in water, and groundwater in porous media moves under the influence of the hydraulic head gradient, as demonstrated by Darcy's Law in Chap. 4. Because groundwater flow is relatively slow compared to the flow of surface water, individual particles of water tend to move by laminar rather than turbulent flow. If the direction of groundwater flow can be determined, such as by taking water-level measurements in monitoring wells, the compass direction of contaminant transport can be assessed with a high degree of certainty. Conversely, location of a dissolved-phase plume in groundwater may reveal a potential source area using the direction of groundwater flow.

6.5.9 Hydraulic Conductivity and Aquifer Tests

Many methods can be used to measure the hydraulic conductivity, *K*, of aquifer sediments. The magnitude of hydraulic conductivity is the rate-limiting step that controls the occurrence of groundwater flow and the potential uptake rate of groundwater by plants from the unsaturated or saturated zones. The easiest but least accurate method to estimate hydraulic conductivity is by using grain-size analysis. Problems of accuracy of grain-size analysis are related to the assessment being performed on a sample that is not in its original setting and was disturbed during collection. Another method that also is used on disturbed sediments and water are the tests done in the laboratory where the hydraulic conductivity is determined from the rate that water moves through a vertical column of saturated sediment.

Tests for hydraulic conductivity done in the field provide more accurate results, although they require more time to perform properly. A common method used is the single-well slug test, in which a volume of water is added to a monitoring well and the rate that the affected water level returns to static conditions provides an estimate of *K*. However, a slug test can sometimes provide a *K* value of the sand filter material used to pack the well screen, rather than the *K* of the aquifer sediments. These differences in *K* generally show up in the groundwater-level-change data plotted over the time of the test and result in two distinct slopes of change with respect to time. For example, the slope of the initial water-level change typically is the *K* of the well-screen filter pack material, and the latter slope of water-level change represents the aquifer hydraulic conductivity.

If water cannot be added to a well to conduct a slug test, as often is the case at a contaminated site, an artificial slug of equal volume made from a pipe or other material can be used to displace the water in the well and then rapidly removed. The limitation with both slug-test methods is that *K* is determined for the immediate area around the well only; slug tests in many wells need to be conducted to assess the distribution of *K* at depth across the site.

6.5.9.1 Aquifer Tests

In order to determine a hydraulic conductivity value for an entire site more efficiently than using slug tests on individual wells, an aquifer test can be done. Because a well creates an artificial zone of 100% porosity in an aquifer and is open to the atmosphere, a nonpumped well can induce groundwater flow toward it. Under pumped conditions, the removal of groundwater from a well is controlled by many factors that can be used in a diagnostic manner to understand the hydrogeologic properties of an aquifer. As depicted in Chap. 4, the groundwater pumped from a well is derived initially from stored groundwater that previously entered the well. The time for this stored water to accumulate can be a few days in low-permeability aquifers to a few minutes for wells in high-permeability aquifers. Once stored water has been removed, the groundwater level in the well declines to lower levels than the groundwater level in the adjacent saturated sediments of the aquifer outside the well. This scenario creates a hydrologic gradient from the aquifer to the well, and groundwater flows into the well.

The water level in the pumped well continues to decline until the rate of groundwater that enters the well screen from the aquifer is in equilibrium with the rate at which it is pumped. The difference between the groundwater level in the well before pumping and at any time after pumping is called drawdown. The relation between pumping rate, time, and groundwater-level drawdown reveals much information about the hydrogeologic properties of an aquifer. The groundwater level can be measured in the pumped well, but usually it is more desirable to measure the groundwater-level response in a nearby unpumped monitoring well.

As discussed in Chap. 4, in the 1930s the first person to link the relation among pumping, groundwater-level drawdown, and aquifer properties for confined aquifers was C.V. Theis of the USGS (Theis 1935). His equation shows the relation between pumping, Q, and drawdown, s, over time, t, as

$$T = QW(u)/4\pi s \tag{6.1}$$

where the term $W(u)$ is the well function and represents an infinite series, and the value of $W(u)$ is obtained from a table of values of u. Theis developed this equation to describe the flow of groundwater through confined porous media by adapting existing equations used to explain the flow of electrons through conductors.

The time required for an equilibrium to be established between the pumping rate and groundwater level will be different depending on the sediment composition of the aquifer being pumped. For example, a coarse-sand aquifer may reach equilibrium conditions much more rapidly than an aquifer composed of fine silts and clays, because water stored in a well completed in fine silts and clays takes longer to accumulate than it does in a sandy aquifer.

6.5.9.2 Flowmeter Tests

An aquifer test and a slug test are conducted differently, but they share the common requirement of artificially moving groundwater through a well after measurement of the static groundwater level. The static groundwater level, however, is static only in the sense that the groundwater level is not moving vertically. The flow of groundwater may continue to occur into the well screen from upgradient areas and out of the well to downgradient areas under ambient unpumped conditions. The rate of groundwater flow can be determined by using various approaches called flowmeter tests. Older flowmeter methods used an impeller placed in the well and the rotations around a fixed shaft were counted over time. Newer flowmeter methods measure flow by using the rate of the heat dispersal from a heat source placed into the well.

6.5.10 Ground-Penetrating Radar

As discussed in Chap. 3, the root systems of small or large trees that can be installed at phytoremediation sites, or of most plants in general, mostly are hidden from view. Roots often are seen at erosive features, because the presence of plant roots limits the further advance of erosion. However, for most tree roots to be observed, such as when measuring root penetration to the water table, takes considerable effort. A common method involves collecting sediment cores in a grid pattern near trees but away from the tree trunk, or bole, and observing the presence of roots in the recovered core material. Trenches can be dug alongside the tree to be examined, and examples are described in case studies in Chap. 8. Roots also can be unearthed by using water pressure and lifting the entire plant from the ground, although this is rarely done. Roots and other rhizosphere materials also are viewed by using specially manufactured digital cameras, called rhizotron cameras, that can be lowered down boreholes in the root zone. Drawbacks to these methods include time, expense, and the fact that the root tips and hairs may be lost.

Geophysical techniques have advanced enough to allow basic geophysics to be applied at many groundwater contamination sites and, therefore, have some application to site assessments for phytoremediation. They were developed primarily for finding oil in subsurface strata in the early 1960s. It came to be recognized that in many instances, these same technologies have applications for groundwater. The most commonly used geophysical method that can be applied to sites with contaminated groundwater where phytoremediation is being assessed is ground-penetrating radar.

The application of ground-penetrating radar (GPR) to identify and map below-ground root structure and distribution provides a non-invasive technique to study roots. The use of GPR for this purpose was investigated by Hruska et al. (1999). The site studied consisted of humus-rich surficial soils that graded to more loamy soils at depths less than 1 m from land surface where weathered bedrock was encountered. The forest trees at the site included 50-years-old oaks (*Quercus petraea* (Mattusch.) Liebl.). The GPR unit was run by two of the trees along straight transects, along which are placed a transmitter of electromagnetic waves and a receiver some distance away. Reflection of the transmitted energy as it impinges on various subsurface contacts is what is received and plotted. The data presented by Hruska et al. (1999) indicated that the GPR provided resolution of roots between 1.1 and 1.5 in. (3–4 cm) in diameter. Although these trees had massive lateral root systems that were entirely dependent on capture of precipitation because the bedrock precluded the formation of an unconsolidated water table, the application of GPR to sites where phytoremediation is applied warrants further investigation.

6.6 Available Phytoremediation Site-Assessment and Characterization Documents

Currently, a few public documents are available that outline various approaches, or protocols, regarding the assessment, characterization, implementation, and monitoring of phytoremediation. In general, these documents focus to a greater extent on the phytoremediation of soil and sediments relative to groundwater. The documents discussed in this section are grouped according to the major agency associated with the protocol and the summaries presented here are not meant to be comprehensive.

6.6.1 U.S. Environmental Protection Agency (USEPA)

As one of the original proponents of the assessment of phytoremediation in restoring contaminated environments, the USEPA has produced many documents on the subject of phytoremediation. The two discussed here are (1) Introduction to Phytoremediation (U.S. Environmental Protection Agency 2000a), and (2) Brownfields Technology Primer: Selecting and Using Phytoremediation for Site Cleanup (U.S. Environmental Protection Agency 2001).

The Introduction to Phytoremediation (U.S. Environmental Protection Agency 2000a) is indeed that, an introduction. Topics range from an overview of the technical, economic and regulatory aspects of phytoremediation, to the evaluation of the different plant-based technologies that compose phytoremediation processes, to design criteria, and to performance evaluation. Case studies also are provided. Perhaps because of the general nature of the document and because nine different phytotechnologies are covered, the information provided on the application of phytoremediation to hydrologic control by the reduction of off-site flow is limited to two pages. The document rightly states that for plants to control site hydrology, the depth to water table cannot exceed the maximum rooting depth of the particular plant(s) selected for phytoremediation. Moreover, to achieve containment of groundwater, the rate of groundwater flow must equal the rate of plant uptake of water to keep the contaminated groundwater from flowing past the plants. In some cases, the USEPA document states that success can be had at such sites if trees are initially deeply planted to be as close to the water table as possible, such as through use of methods similar to the patented Tree-well® method reported in Gatliff (1994).

The application of vegetated covers to affect site hydrology is discussed thoroughly. The USEPA document outlines a list of decision-making processes that, if followed, will lead to the recommendation or refusal of the application of phytoremediation at a specific site (Table 6.3).

The USEPA Brownfields document (U.S. Environmental Protection Agency 2001) provides introductory material on the potential application of phytoremediation at sites designated as Brownfields sites, which are abandoned, idle, or under used, and have long histories of commercial or industrial practices but also are located in areas undergoing revitalization. In many cases, the federal regulators involved at such sites encourage the use of innovative technologies, such as phytoremediation. The document describes a strategy to control contaminated groundwater flow by surrounding a contaminant plume with plants and thereby creating a hydrostatic barrier of tree roots so that groundwater is captured by the roots and does not flow past them.

6.6.2 Interstate Technology & Regulatory Council (ITRC)

Because of the typically lower installation cost of phytoremediation relative to conventional pump-and-treat methods to control groundwater, the potential arises that phytoremediation may be selected as a remedial option when, in fact, site data do not support implementation. To help determine if phytoremediation can or cannot be applied at a site, a decision tool was developed by members of the Interstate Technology & Regulatory Council (ITRC). The ITRC is a coalition of state, federal, public, and industrial representatives with shared interest and expertise in remedial technologies and activities. The goal of ITRC is to advance innovative remedial strategies that are cost effective.

The ITRC developed a protocol similar to that of the USEPA discussed previously that assists in the decision-making process (Interstate Technology and Regulatory Council 1999) and has been revised (Interstate Technology Regulatory Council ITRC 2009). The decision-making tool is built on a flowchart framework, where data collected sequentially are used to answer questions regarding the use of phytoremediation at a particular site (Fig. 6.8).

Table 6.3 Decision-making list modified from Introduction to Phytoremediation (U.S. Environmental Protection Agency 2000a).

Decision	Task
Define problem	Conduct site characterization
	Identify contaminant and media
	Identify regulatory needs
	Identify remedial objectives
	Establish success criteria
Evaluate site for phytoremediation	Perform site characterization to include data for phytoremediation
	Identify phytotechnology that addresses site need
	Review all available information
	Select plants
Conduct preliminary studies	Screening studies
	Optimization studies
	Conduct field plot trials
	Reevaluate plant selection, if needed
Evaluate full-scale system	Design system
	Install system
	Maintain and operate system
	Evaluate system
Achieve objectives	Take quantitative measurements
	Meet criteria for success

Three separate frameworks are specific to the source of contamination—groundwater, soils, or sediments. In the groundwater flowchart, it is realized that hydrologic control is a form of contaminant containment, as previously stated earlier in this chapter. However, the flowchart provides only two courses of action: (1) hydrologic control by prevention of recharge through the contaminated area, similar to the goal of a vegetative cover at a landfill; and (2) the contaminated groundwater potentially being mechanically pumped and re-applied back to the planted area as irrigation water. Although hydrologic control can be defined as the reduction of leachate formation by reducing recharge to the water table, hydrologic control also can be used to capture plumes in downgradient areas prior to transport off site. This application is not included in this version of the ITRC groundwater decision tree. However, the groundwater decision flowchart supports the reduced feasibility of phytoremediation of contaminated groundwater at sites where the depth to groundwater is greater than about 20 ft (6 m). This is not to say that plants cannot use groundwater at depths greater than 20 ft, but that deep-planting methods usually increase the installation cost and(or) remediation time.

The ITRC released a second document that more comprehensively addresses the issue of hydrologic control by phytoremediation and the processes for proceeding from site assessment to site closure (Interstate Technology and Regulatory Council 2001). This document restates the importance of plants in reducing recharge to groundwater. It extends the previous document by acknowledging that plants can be used to create hydrologic barriers to prevent groundwater flow and contaminant transport off site. This ITRC document addresses some of the potential limitations of a proposed phytoremediation system, such as the potential lack of performance during periods of presumed plant dormancy. The ITRC document acknowledges that a conventional pump-and-treat system could be used during dormancy. Of course, the concern regarding the effect of dormancy on groundwater can be answered only if the rate of groundwater flow is known to enable predictions of groundwater movement downgradient during dormancy to be estimated.

The 2001 ITRC document includes a section on the technical requirements for phytoremediation to assist in determining if a site is appropriate for hydrologic control by phytoremediation. The process begins with assembling members of a phytoremediation team and generating a checklist, similar to Table 6.3, that includes the following: baseline site characterization, review of existing site data, agronomic site assessment, site visit, definition of remedial objectives, and answers to the question of how phytoremediation can be used to meet the objectives at the site. Once these data have been collected, a proposal that outlines the goals is drafted and submitted to various stakeholders for review, comment, and approval.

The document acknowledges a common problem at sites being analyzed for phytoremediation—plant physiologists can be concerned solely with plants, and hydrogeologists can be concerned only with groundwater. These specialized interests can result in little interaction between the scientists. This indicates that other specialists should be involved, such as risk assessors, environmental engineers, etc., if resources permit. The document continues with other activities, such

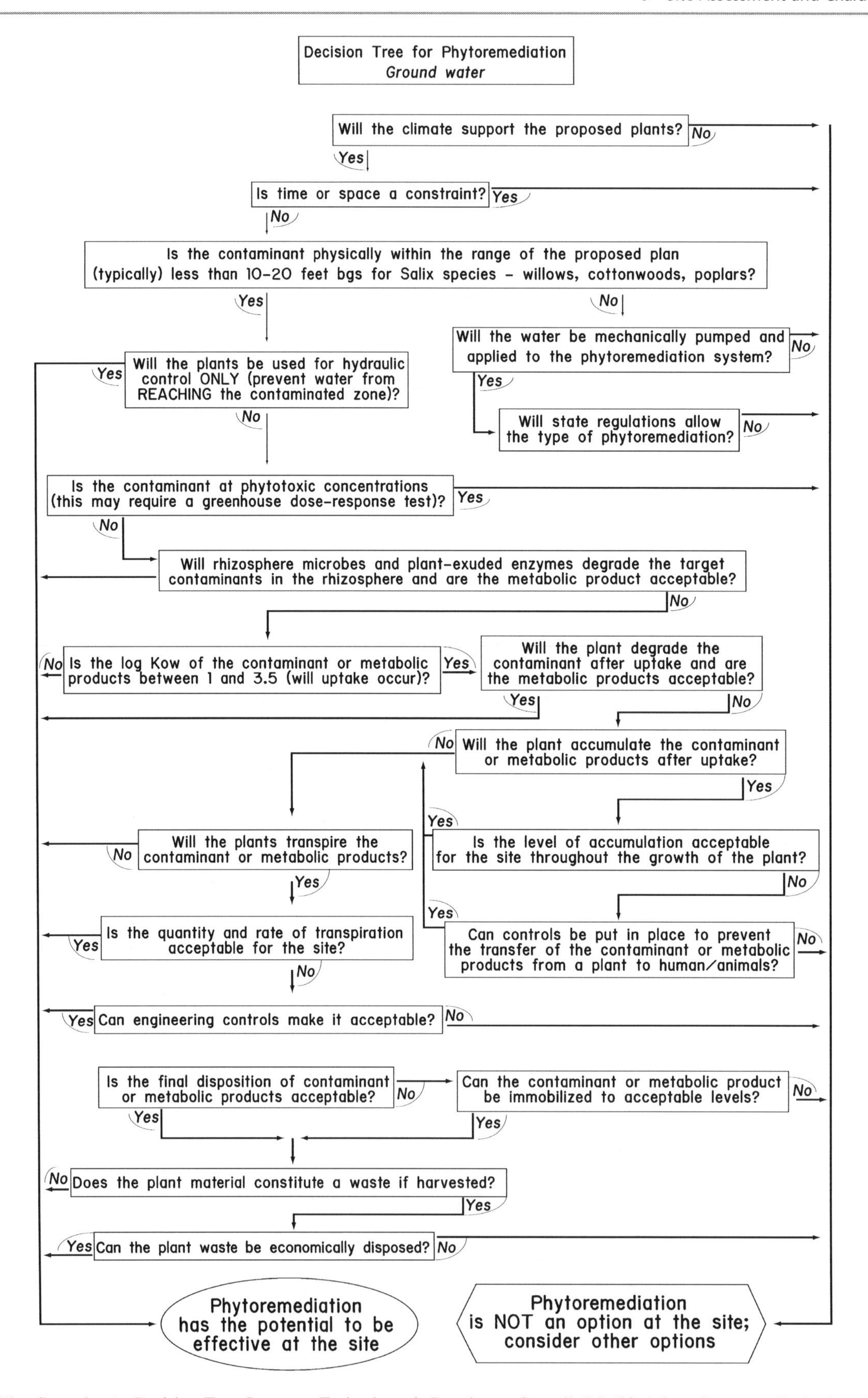

Fig. 6.8 The Groundwater Decision Tree, Interstate Technology & Regulatory Council (Modified from Interstate Technology Regulatory Council ITRC 1999).

as a feasibility study in the field or laboratory; an operation and maintenance plan; a monitoring strategy; and a contingency plan.

6.6.3 Remediation Technologies Development Forum (RTDF)

The Remediation Technologies Development Forum (RTDF) represents collaboration between the USEPA, Office of Research and Development and Technology Innovation Office, and public and private sectors. Established in 1992, its goal is to mutually seek innovative solutions to the growing problem of hazardous wastes in the United States. Moreover, to assist in widespread implementation, the solutions have to be cost effective. Funding for the RTDF is provided by the USEPA, with additional support by technology customers and other federal agencies, such as the Department of Defense (DOD) and Department of Energy (DOE). Additional funds are contributed by private concerns.

Because of the wide range of contaminants and clean-up strategies, the RTDF is divided into different action teams. The Phytoremediation of Organics action team produced a report that describes the development of a field protocol for determining the efficiency of certain plants in degrading various groundwater and soil contaminants (Remediation Technologies Development Forum 2005; Fig. 6.9). Little is said in the report regarding the hydrologic control of groundwater by plants. However, the team also has produced a large bibliography that contains many relevant articles on phytoremediation, and some address the issue of hydrologic control by plants.

Although the main focus of the report by RTDF (Remediation Technologies Development Forum 2005) is to present the application of phytoremediation for the remediation of chlorinated solvents at contaminated sites, the document acknowledges the importance of assessing the hydrogeology of a site and its influence on the extent of remediation of contaminants. The document states that certain trees can extract groundwater and depress the water table, which causes groundwater to flow to these trees. As a result of the change in groundwater flow, the risk of off-site migration of contaminants, in this case chlorinated solvents, is decreased.

As part of a protocol displayed in the document and reproduced with modifications in Fig. 6.9, one of the initial questions in the flowchart is whether the plant roots can meet the requirement for depth to water table and whether the results of modeling the site water balance with the addition of plants are promising. The protocol emphasizes the importance of using computer models to help define the water balance in order to determine if a potential phytoremediation system will affect the water budget. The primary simulation objective is to determine if the plant removal of groundwater will be at a rate fast enough to create a water-table depression to alter groundwater contaminant transport. Perhaps the best rationale for using models is to determine the time to reach such conditions where, at a minimum, the rate of groundwater flow equals the rate of uptake by plants.

One approach offered in the RTDF document (Remediation Technologies Development Forum 2005) is to perform groundwater capture-zone modeling. A capture zone is a delineation of the source of groundwater being pumped from a well. Capture-zone analysis usually is conducted by water managers to understand the source of water pumped by drinking-water supply wells and to determine if potential sources of contamination are located in the capture zones of these wells (see Landmeyer 1994 for an example). For phytoremediation purposes, pumping wells rather than transpiring trees are simulated.

6.7 Summary

A thorough review of the history and pre-existing data at sites is warranted to determine if phytoremediation is applicable and can be successful to remediate contaminated groundwater. Thankfully, a growing body of information exists regarding how to conduct, in a methodical and logical manner, site assessments and characterization to determine if conventional remedial practices or phytoremediation can be used. However, these guidelines and protocols often are limited in not being able to address site-specific concerns. Moreover, the protocols often omit the use of phytoremediation to achieve goals of hydrologic containment or control.

Why is this information important to the phytoremediation of contaminated groundwater? The time invested using conventional site-assessment and characterization approaches will either be returned in the form of a successful phytoremediation system and, therefore, a successful phytoremediation project, or in the knowledge that an alternative remedial strategy can be used if phytoremediation cannot be supported.

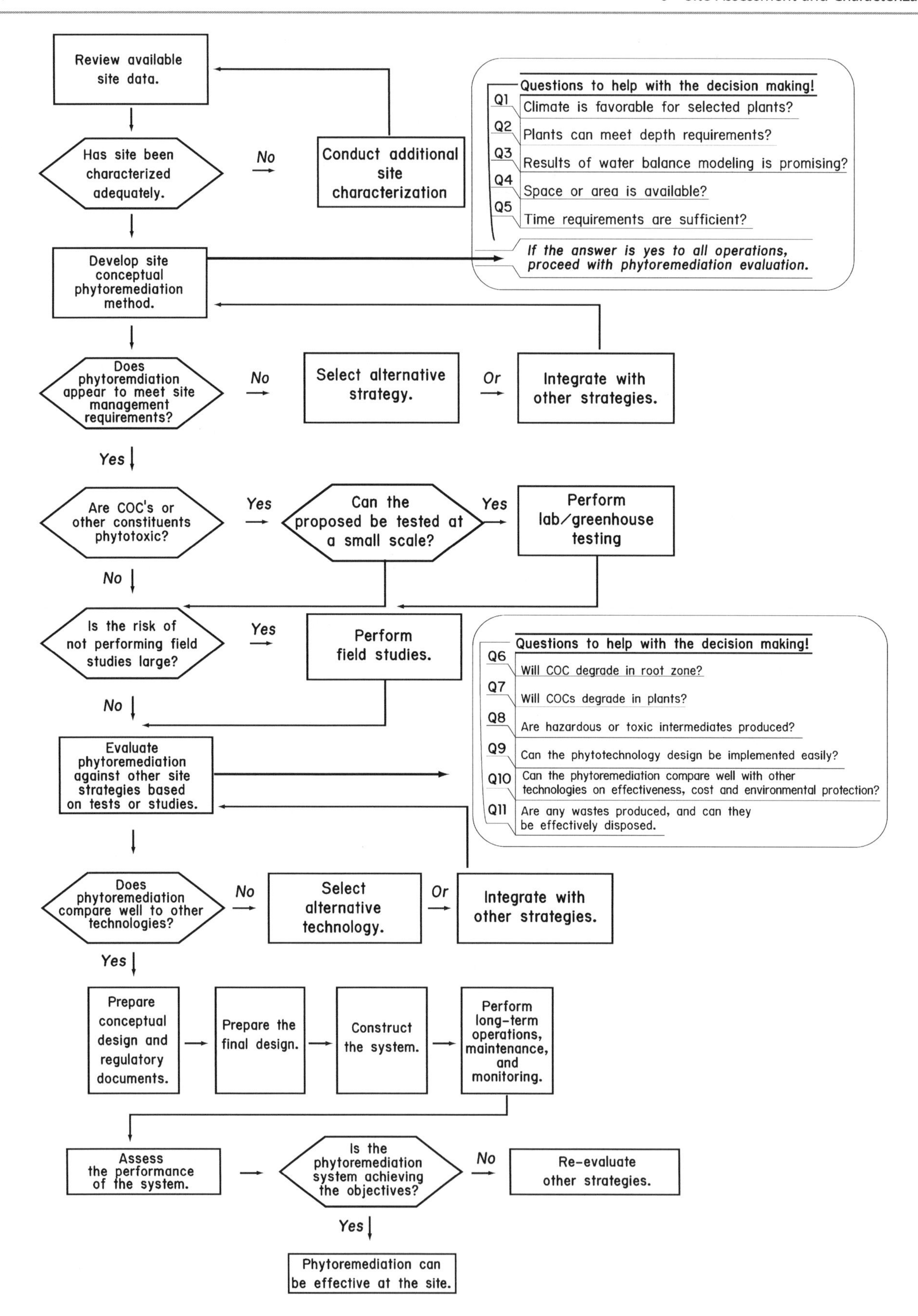

Fig. 6.9 The phytoremediation decision-making flowchart. COCs are contaminants of concern (Modified from Remediation Technologies Development Forum, 2005).

7 Plant Selection, Installation, and Management to Affect Groundwater

> *...like a tree planted by water, that sends out its roots by the stream,*
> *And does not fear when heat comes, for its leaves remain green,*
> *And is not anxious in the year of drought,*
> *for it does not cease to bear fruit.*
>
> Jeremiah 17:8 (RSV)

This quote from the *Bible* states what, up until recently, was thought to be a long-held observation—that trees growing near streams use surface water. We now know that trees near streams often use groundwater. Conversely, early observations indicated that a connection between plants and groundwater near surface water may exist, and that the occurrence of certain plants indicated the relative depth that groundwater could be found below land surface. For example, the Roman architect Vitruvius stated that (with emphasis by the author),

> *Besides these there are other indications of places where water can be found—namely, the presence of small rushes, willows which are not planted, alder trees, vitex, reeds, ivy, and all other such plants which occur and thrive only in places where there is water. One must solely rely on these plants, however, if they occur in marshes, which, being lower than the surrounding country, receive and collect and for some time retain waters that fall on the near-by fields in winter;* ***but if these plants occur naturally in places that are not marshes, one can seek for water in these places.***

Even in the Saharan and Sahel desert regions of northern Africa, it has long been known that some trees, such as acacia and tamarisk, prefer groundwater such that

> *Trees and plants sometimes afford invaluable assistance in locating successful wells, the position of master joints or belts of fissured or decomposed rock, along which underground water percolates, being not infrequently indicated at the surface by lines of trees or shrubs, known as 'aars.'*
>
> Wagner (1916)

The use of groundwater rather than surface water by plants was confirmed by the discovery that the stable isotopic values of water in trees that grow along streams are more similar to the values of groundwater than of surface water (Dawson and Ehleringer 1991). Riparian plants thrive near surface water because they tap groundwater discharge. Many of the plants used for phytoremediation of contaminated groundwater can be found naturally near streams, such as cottonwoods, poplars, birch, and sycamore. Incidentally, the Hebrew version of Jeremiah 17:8 suggests that the trees near the stream actually were using groundwater discharge. As revealed by Ross (2007) with emphasis by the author, the hydrologic importance of the passage is related to one word

> *They shall be like a tree planted* ***above*** *water,*
> *sending down its roots by a stream.*

This translation suggests that the trees indeed used groundwater rather than surface water, which would have been infrequent in the arid area of the story. It took almost 2,000 years and the use of stable isotopes of water to verify that groundwater can be a source of water to trees, as we will see in Chap. 9.

Since the term phreatophyte was introduced in 1927 by O.E. Meinzer, at least 70 species of plants have been characterized as phreatophytes. Such plants are not from one family but cross many different families, from herbaceous grasses to woody trees. Plants that rely solely on groundwater are called obligate phreatophytes, and those that rely on groundwater and other sources of water are facultative phreatophytes, as described in Chap. 1.

Much of what currently is understood about phreatophytes and water use, especially in regard to contaminated groundwater, is based, somewhat ironically, on efforts to control these plants or rid areas of them where they use large volumes of groundwater that otherwise could be used by man, as was introduced in Chap. 5. Therefore, the planting of phreatophytes at sites characterized by contaminated groundwater is actually a beneficial aspect of consumptive use that could, perhaps, be termed consumptive restoration.

J.E. Landmeyer, *Introduction to Phytoremediation of Contaminated Groundwater*,
DOI 10.1007/978-94-007-1957-6_7,

7.1 Plant Types and Selection Criteria

The establishment of biomass is a fundamental criterion for success at phytoremediation sites. It is reasonable, therefore, to investigate the factors related to plant establishment and to discuss the parameters that limit potential growth and how to take the appropriate measures to ensure growth over the life of the project. These issues must be addressed during the design stage of a phytoremediation project in conjunction with or following the site-assessment and characterization activities discussed in Chap. 6.

Based on the knowledge of plant and groundwater interactions presented in Part I, a few well-known herbaceous and woody plants that use groundwater are candidates for use at phytoremediation sites where hydrologic control is a goal. Examples of plants of each type are alfalfa and hybrid poplars, respectively. Many phreatophytes can have a large range of distribution but may not thrive in all climates where needed. Conversely, some plants are restricted to one type of hydroecology and can be used only when specific conditions are available or engineered.

A map was developed by the Department of Energy (DOE) biomass program that depicts the types of relatively fast-growing species that do well in certain areas of the country (Fig. 7.1). This map also is applicable to predict areas where the installation of a phytoremediation project may be warranted. Alternative plants also may be considered but will be discussed in less detail because less is known about their success at phytoremediation sites. Moreover, the lists of various herbaceous and woody plants presented here are to suggest which native or indigenous plants at a site can be used to either assess the presence of groundwater or to act as part of an overall hydrologic-control system.

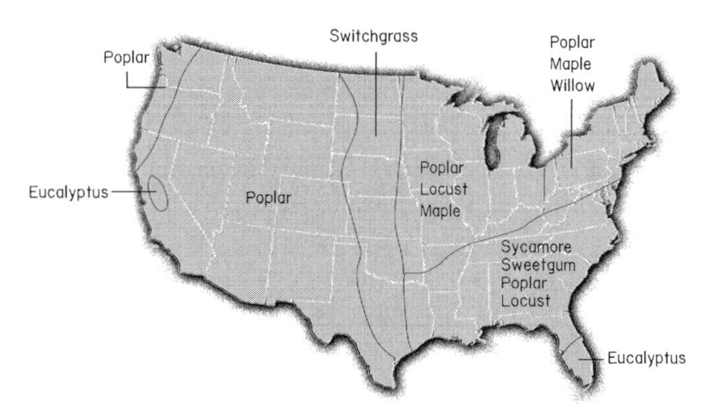

Fig. 7.1 Identified plant species that do well in certain parts of the United States for fiber production. These plants also have the potential to interact with groundwater because they are fast growing and have deep root systems.

7.1.1 Herbaceous Plants

Herbaceous plants do not have woody material in their structure. Herbaceous plants grow from meristematic tissue in a primary pattern, with no secondary growth. In fact, the derivation of herbaceous is *herb*, which is Latin for grass. Examples of herbaceous plants are given here in order of their relation to the depth to groundwater. This relation between plant type and depth to water table was given the term phreatophytic facies in Chap. 1. Most herbaceous plants typically use sources of water other than groundwater. In some cases, the depth to water table in areas of herbaceous growth may be greater than 4–8 ft (1.2–2.4 m); however, in areas that have a higher percentage of clays, soil moisture generally extends from the capillary fringe to land surface.

Perhaps the best known herbaceous plant that also is a phreatophyte and has the potential, therefore, to be used for phytoremediation to control groundwater is alfalfa (*Medicago sativa*) (Table 7.1). As we saw in Chap. 1, O.E. Meinzer used alfalfa to report what is probably the first published interaction between plants and groundwater. Alfalfa was probably imported to the United States through Georgia in the early 1700s (Meinzer 1927) and can be found today abundantly in every state in the United States. The roots of alfalfa can exceed lengths of 60 ft (18 m) if the depth to water table is that deep, especially in upland areas. Perhaps the most notable evidence that alfalfa is a phreatophyte is that it can grow, but perhaps not thrive, as a cultivated plant with no irrigation. Transpiration rates between 1.7 and 10.5 mm/day have been reported (Interstate Technology Regulatory Council 2009).

In low-lying areas that are consistently wet, intermittently wet, or submerged at least part of the year is where the herbaceous cattails, rushes, and sedges, of the genera *Typha, Juncus,* and *Scirpus,* respectively, grow. During

Table 7.1 Representative herbaceous phreatophytes native to the western United States (Modified from Robinson 1958) that may have phytoremediation potential to influence groundwater.

Common name	Scientific name	Relation to groundwater	
		Depth to water table below land, ft (m)	Remarks
Bermuda grass	*Cynodon dactylon*	na	Subtropical plant, from India.
Alfalfa	*Medicago sativa*	4 (1.2)	
Sacaton	*Sporobolus airoides*	5–25	Deep root system.
Vanadium bush	*Cowania stansburiana*	(1.5–7.6)	Indicates vanadium-uranium deposits and can absorb uranium.
Saltgrass	*Distichlis spp.*	2–12 (0.6–3.60)	Grows near ocean or in desert.

na not applicable

moist seasons, these plants indicate only obvious sources of water but are more useful as indicators of groundwater when precipitation is low and groundwater discharge meets their needs. If the moist area is simply a low area that collects surface-water runoff and where the supply of water is not constant, these plants are not likely to be found. Fossils of ancestral cattails have been found in rocks dating back to 400 MYa (Meinzer 1927), and their connection to groundwater perhaps is the key to their long-term success. Transpiration rates between 8.5 and 28.2 mm/day have been reported (Interstate Technology Regulatory Council 2009).

Reeds (*Phragmites*) also can be found in low-lying areas of groundwater discharge, either with or without the presence of surface water. Reeds particularly are useful as an indicator of fresh groundwater when found adjacent to saline surface-water bodies. Perhaps the oldest recorded instance of reeds growing adjacent to saline surface water is in the *Bible* in reference to the Red Sea. The term sea refers to standing surface water and the Red Sea was more likely called the Reed Sea, in reference to the reeds near its shoreline. Because these plants cannot use salty water, their presence indicates that fresh groundwater discharges to such a saline surface-water body (Issar 1990). Transpiration rates between 1.4 and 8.5 mm/day have been reported (Interstate Technology Regulatory Council ITRC 2009).

Because the transpiration rate of alfalfa and reeds is low, it is unlikely that stands of these plants would be used to control groundwater flow or decrease recharge at a contaminated site. A more likely scenario would be their presence to enhance the rhizosphere in areas of groundwater discharge to prevent contaminants from affecting surface-water quality—this is discussed in Part III.

Most grasses native to the United States are drought-tolerant plants that have deep roots to reach the water table or extensive lateral roots to access soil moisture. The presence of rye grass (*Elymus condensatus*), for example, indicates a shallow depth to groundwater in arid to semiarid areas. These plants tend to be found in valleys rather than in uplands. In the early 1900s, the location of the water supply for the city of Tonopah, NV, was based solely on observations that rye grass grew in the valleys along the predominately dry, desert terrain. A similar relation of shallow depth to water table and rye grass exists for more humid areas, as was reported by O.E. Meinzer (1927) for the Washington, D.C. area. Transpiration rates between 4.1 and 9.2 mm/day have been reported (Interstate Technology Regulatory Council 2009).

Another grass that uses shallow groundwater is saltgrass (*Distichlis spicata*). It grows naturally in areas where the depth to groundwater generally is less than 8 ft (2.4 m). As described in Chap. 1, in the early 1900s Charles Lee of the USGS used tanks filled with soil, water, and saltgrass to estimate that the removal of water by evapotranspiration in a 55-mi^2 (140-km^2) area in Owens Valley, CA, was equivalent to a continuous daily flow of 109 ft^3/s. Sacaton (*Sporobolus airoides*) is a grass found primarily in the western United States and is perhaps best known for indicating groundwater in dry places, such as Death Valley, CA.

The depth and fibrous nature of the lateral and vertical structure of grass roots supports the use of herbaceous plants in applications where the primary goal of phytoremediation is to reduce recharge. The ability to more closely space individual grass plants than is possible for larger plants increases the density of root distribution in the subsurface. Also, grasses can be planted between woody species that are planted at larger intervals to rapidly take up precipitation before newly planted trees are established, which will force the trees to seek deeper sources of water. Once the trees are established and closed canopy conditions are reached, the grasses will most likely die in response to a reduction in light penetration. By this time, the role of the grasses to reduce recharge is no longer needed.

Herbaceous plants have either an annual or perennial life cycle. Perennial herbaceous plants have many advantages over annuals for use at phytoremediation sites. Perennial plants need to be installed only once, whereas annuals need to be reinstalled on a yearly basis. The root systems of annuals, therefore, are not as hardy as the root systems of perennials, which must be large enough to seek water and supply food to the roots to support root respiration during the winter months. Typically, annuals require more intensive agricultural management because of limited root systems, so these plants require more fertilizer and water than perennials. More importantly, the extensive and long-lasting root systems of perennials contribute to the development of an organic-rich O-horizon that often is lacking when annuals are planted. This O-horizon contains the rhizospheric microbial community that not only supports itself and plant life but also plays a key role in global biochemical cycles, such as nitrogen and carbon, as is discussed in Chap. 11.

7.1.2 Woody Plants

Woody plants are characterized by a primary growth pattern that occurs from meristematic tissue as in herbaceous plants, but also increases in thickness from true cell division from the cambial tissue, called secondary growth. Herbaceous plants increase somewhat in girth, but this thickness is derived from cell elongation rather than division.

7.1.2.1 Poplars

The genus *Populus* is one of the most widely studied woody plants. Poplars compose about 10% of the willow family

Salicaceae. The *Salicaceae* and the *Flacourtiaceae* and 29 other families are grouped under the Malpighiales of the angiosperms and are native to and widely distributed throughout the northern hemisphere. Poplars can be found in hot and cold climates. There are about 35 species of *Populus* in the northern hemisphere, and about eight species of *Populus* are native to North America (Fig. 7.1). Its wide distribution is evident from the fact that the wood of poplar was widely used, rather than canvas, by Renaissance painters in western Europe, with perhaps the most famous example of art created on a poplar panel being the Mona Lisa (Durant 1953). Poplars are dioecius, which means poplars do not have male and female reproductive parts on the same tree; rather, separate male and female trees are required to reproduce through wind pollination. Because of this characteristic, they can easily form natural hybrids.

Poplars typically are found along waterways for it is in such settings that the many pollinated cottony seeds can fall to moist soil conducive to successful germination. This life cycle may be an advantage for the survival of poplars across the Great Plains of the United States, where frequent prairie fires from lightning strikes occur. The most widely distributed tree in North America is the quaking aspen, or *Populus tremuloides* (Robinson 1958). They also are the most widely studied. Other important species found worldwide include *P. acuminata*, *alba*, *angustifolia*, *balsamifera*, *deltoides*, *fremontii*, *heterophylla*, *nigra*, *texana*, *trichocarpa*, and *welizeni* (Robinson 1958). Poplars also were present and widely distributed as far back in the geologic record as the Cretaceous Period, more than 90 MYa.

The most well-known and broadly-used poplars, both for commercial and phytoremediation applications, are not the native specimens but hybrids. A hybrid is the result of the cross pollination of two different species of the same genus. Such cross pollination, called hybridization, was first recorded in 1761 by the German botanist Jakob Gottlieb Köelreuter (1733–1806) when he crossed *Nicotiana paniculata* pollen with *Nicotiana rustica* (Gager 1934).

A hybrid should not be confused with a clone, which is a vegetative propagation of an individual genotype that maintains the genetic character of the original plant. A clone is designated by the suffix *cl*. A clone also can be artificially produced from the undifferentiated cells of plants. This was discovered in the 1950s by Georges Morel, who agitated undifferentiated plant cells to which he added growth hormones. These cells then were separated and grew into individual plants that were genetically identical to the source of the cells. Cultivars are clones derived from closely related plants and given the suffix *cv*. The production of hybrid poplars follows the process traditionally used to produce desirable traits in other plants, such as roses. The ultimate goal of any attempt at crossing different parent plants of the same genus is to have the resultant hybrid seedling contain the best genetic traits of each of the parents, called heterosis.

The development of a hybrid poplar tree for phytoremediation purposes, for example, is similar to the long-practiced method of plant domestication. The best examples of plant domestication come from the development of cereal grain crops, such as wheat or rice. The best forms of these wild plants and others, such as sunflower, were intentionally selected by Native North Americans over many thousands of years to produce larger and larger seeds. In doing so, they also ended up with a plant of annual ecology rather than the perennial history it originated from. Detriments of this change are discussed in Chap. 16.

The first hybrid poplar was created in 1912 as a cross between the eastern cottonwood (*Populus deltoides*) from the Midwest to eastern United States and native western black cottonwood *Populus trichocarpa*. The result was a sterile hybrid poplar, or a cottonless cottonwood. It contained the rapid secondary and primary growth habits of each species, respectively. Since that time, many hybrid poplars have been created. To ease confusion, hybrids are referred to by the initials of the first letter of each species in the cross pollination. For example, the hybrid from the cross pollination between *Populus trichocarpa* and *Populus deltoides* was given the prefix TD; the female parent is listed first. The prefix of the cross between *Populus deltoides* and *Populus nigra* is, therefore, DN. Hybrids also are denoted by numbers assigned by their crossers, such as 49–177, where the first number refers to the cross number and the second number refers to the seedling number. The numbering convention refers to plants produced by the Washington State University Poplar Research Program.

Much information is available about the genus *Populus*, generated by those who have an economic interest in trees, such as the short-rotation wood culture (SRWC) industry, the paper and pulp industry, or state-based experimental stations interested in the reforestation of strip-mined land or landfills. This interest is economically driven, as 1 acre of poplar trees can produce 10 dry tons/year within a 6- to 8-year rotation, relative to 4 dry tons/year with a 50-year rotation for Douglas fir. Recently, these parties, the private sector, and federal government agencies such as the DOE, have become interested in using biomass derived from hybrid poplars as an alternative energy source, a practice common in Europe. Probably one of the first agencies to offer hybrid poplars to the general public in the United States was the Northeastern Forest Experiment Station in the 1920s.

The economic interest in poplars also drove the recent work in describing the entire genome of poplars. In this sense, poplars can be considered analogous to the importance of fruit flies to geneticists. As reported in Sterky et al.

(2004), the poplar is the internationally accepted model for molecular studies of tree biology. A Web-based database, called Populusdb, contains expressed sequence tags (EST) for 18 tissues that represent various plant organs. Such work led to the insight that angiosperms and gymnosperms have a high degree of similarity in genetic information. The genome of the black cottonwood, *Populus trichocarpa*, was found to contain more than 45,000 genes (Tuskan 2006).

The widespread distribution of poplars provides the best evidence that water-loving phreatophytes are not simply limited to groundwater in arid areas with little precipitation (Dickmann and Stuart 1983). For example, *Populus deltoides* grows to very large heights east of the Mississippi River. Also, *Populus heterophylla* can be found in most low-lying areas near springs or streams east of the Appalachian Mountains. Because all of these species tend to grow near surface water where the depth of the regional water table is shallow, they are dependent on groundwater for survival. These plants predominately use groundwater, but can also use rainwater and, therefore, are an example of facultative phreatophytes. These plants are facultative in that the depth for most cottonwood roots is no greater than 35 ft (10 m; Meinzer 1927; Robinson 1958). The examination of the use of groundwater by poplar trees began in the mid-1940s, as we saw in Chap. 1. For example, Gatewood et al. (1950) grew cottonwood trees (*P. fremontii*) in tanks and determined that the water use was about 7.6 ft (2.3 m) between October 1, 1943, and September 22, 1944, when the depth to water table was constant at 7 ft (2 m).

A possible explanation for why phreatophytes, such as poplar trees, have come to rely on groundwater and have established such a large range of growth may be related to events that occurred some 65 MYa. In general, dinosaurs were around as recently as the Cretaceous Period, some 90 MYa, as is evident from the fossil record. Such fossils are abruptly absent, however, in the younger Tertiary sediments directly overlying the Cretaceous sediments. This change is accompanied by an abrupt change in plant fossils as well. Geologists have identified this dramatic change in fossil assemblages, especially the lack of large dinosaur fossils in the Tertiary, for many years. It wasn't until the 1980s that Luis and Walter Alvarez detected high concentrations of iridium in the sediments at this interface between the Cretaceous and Tertiary rocks, called the K-T boundary, at many sites around the world. While iridium is present in the earth's crust, the stable isotope signature of the iridium measured at the interface was similar to the stable isotope signature of iridium present in asteroids. Alvarez suggested that the lack of dinosaur fossils in the Tertiary sediments could be explained by a mass extinction that followed an asteroid impact with the earth. The impact must have sent ash into the atmosphere which blocked light, changed the pattern of precipitation, and caused plants to die and the animals that fed on these plants and the animals that fed on these animals to die as well. Up to 50–80% of all plant and animal genera perished and is called the K-T extinction (Plummer and McGeary 1985).

What does this event have to do with the phytoremediation of contaminated groundwater? Interestingly, some plants and animals did make it through this mass extinction event, and they all share a common trait. Examples of the life forms that survived the K-T extinction are present today, essentially in an unchanged form, in swamps and wetlands. For example, it is possible to trace the lineage of turtles and alligators back to before the K-T extinction event. Plants that co-inhabited low-lying areas near swamps, streams, and lakes, also survived the extinction. These plants were phreatophytes or had characteristics of phreatophytes. So, the assumption is that these plants and animals used low-lying areas as a source of water and protection during the K-T extinction event, and the ultimate source of water in low-lying areas is groundwater.

Many studies across a diverse range of natural sciences indicate that poplar trees use groundwater, which provides a firm foundation for their application to contaminated groundwater as is stated throughout this book. Zhang et al. (1999) measured sap flow in poplar trees growing in a riparian setting in England and reported that 15–60% of the sap flow was composed of groundwater. At the study site, the aquifer material consisted of sandy loam, and the water table was about 4 ft (1.25 m) below ground. The trees investigated were *Populus trichocarpa* Torr. & A. Gray x *P. tacamahaca* L. (Clone TT32), which were about 6 years old and roughly 18 ft (5.5 m) tall. Sap flow was found to be directly related to solar radiation and vapor pressure deficit, with little sap flowing on cloudy and rainy days. Total water use by the trees approximated ET_p, as calculated by using the Penman equation. Transpiration rates between 13 and 200 mm/day have been reported for poplars (Interstate Technology Regulatory Council 2009).

With respect to the source of water in the sap in the trees, Zhang et al. (1999) recorded the drying out of the upper soil layers to a depth of 39 in. (100 cm). Because the trees continued to transpire, as indicated by positive sap flow, the authors suggested that water must be derived from deeper, more saturated soil layers near the water table. In fact, water removed directly from the capillary fringe or water table was estimated to increase from 15% in June to between 45% and 63% in July and August, respectively.

7.1.2.2 Willows

The genus *Salix* of the family *Salicaceae* also has examples of phreatophytes that can be found in humid climates. These trees, commonly known as willows, inhabit low-lying areas near streams where the depth to groundwater is shallowest, even in arid areas. Such habitats have been observed for some time, as exemplified by the following:

> *By the waters (streams) of Babylon,*
> *there we sat down and wept...*
> *On the willows (poplars) there*
> *we hung up our lyres.*
>
> Psalm 137:1–2 (RSV)

All 250-plus willow species are native to the United States. Like cottonwoods, willows are most likely to be found along streams or in flood plains where the depth to groundwater is 10 ft (3 m) or less. Unlike poplars, willows are pollinated by insects rather than wind.

Based on their location with respect to regional, intermediate, and local groundwater-flow systems, it should not be surprising that evidence of the interaction between willows and groundwater exists. For example, in the nineteenth century, this relation between willows and groundwater was noted in the following:

> *When the thalweg [middle] of a valley is uncultivated and one sees there growing naturally willows, poplars, alders, osiers, rushes, reeds, wild mint, silver weed, ground ivy, and other water-loving trees or plants, one should presume that the course of water is not deep in that place. However, as these kinds of plants thrive in all humid terranes they can only serve to indicate the presence of groundwater in so far as they are on a thalweg or at the bottom of a hollow.*
>
> (Paramelle 1856).

Observations such as these were later confirmed by experimental testing. For example, White (1932) recorded the water-table fluctuation in a well near willows to be 0.3 ft (0.09 m) during August 1926, where depth to groundwater was 5–6 ft (1.5–1.8 m) below land surface. Another study on the use of groundwater by willows was performed by Blaney et al. (1933) in California. They transplanted a willow (*S. laevigata*) from the field into a 6-ft (1.8 m) diameter, 3-ft (0.9 m) deep tank, where a depth to water table was maintained at about 2 ft (0.6 m). They recorded the water use to be equivalent to 4.4 ft (1.3 m) between May 1930 and April 1931. A similar experiment was conducted in New Mexico, but multiple plants were transplanted into a similar sized tank. These researchers recorded water use near 2.5 ft (0.7 m) between June 1936 and May 1937 (Young and Blaney 1942).

Willows have an advantage over poplars with respect to the water quality of groundwater. Willows tend to be more tolerant of salt stress than cottonwoods and are the predominant riparian plant along the Colorado River (Busch and Smith 1995). This information may be useful in designing a phytoremediation planting if site-assessment and characterization activities indicate high salt concentrations in the soil or groundwater. Transpiration rates between 10 and 45 mm/day have been reported for willows (Interstate Technology Regulatory Council 2009).

In the western United States, one of the most widespread phreatophytes is the greasewood (*Sarcobatus spp.*). It grows where the depth to water table is shallow or as deep as 60 ft (18 m). As stated in Chap. 1, USGS hydrologist W.N. White set up experimental tanks in Utah to determine the proportion of groundwater used by greasewood (White 1932). He determined that the seasonal use of groundwater ranged between 0.08 and 0.38 ft (0.02–0.1 m) when measured in natural stands of greasewood, whereas in his tank experiments with an artificially controlled water table the use was greater, nearly 2 ft (0.6 m) of groundwater.

7.1.2.3 Saltcedar

Whereas greasewood is native to the western United States, another phreatophyte that has become widespread but is not indigenous is saltcedar, or tamarisk (*Tamarix spp.*). The saltcedar is native to western Europe and Asia, and charcoal from *Tamarix* plants has been found in caves near Mount Carmel, Israel, that date between 12,300 and 10,500 BC (Ley-Yadun and Weinstein-Evron 1994). Saltcedar probably was brought to the United States in the 1800s. For example, Robinson (1958) cites an observation in Bowser (1957) that saltcedar was found to be thriving in the San Jacinto River in Harris County, TX, in 1884. The flood plains of rivers throughout the arid southwestern United States characterized by shallow groundwater apparently have provided the imported saltcedar a niche in which to outcompete other riparian plants. Saltcedar can tolerate both wet years and dry years, because of its deep, branched root system, produced during dry years, and adventitious roots from the bark produced during wet years, and a prodigious amount of seeds produced regardless (Robinson 1958). The photosynthetic organs of saltcedar are not true leaves but cladophylls, which are cylindrical stems that look like whirled leaves, perhaps more readily recognized to exist in asparagus. Saltcedar has replaced up to 90% of the native cottonwoods and willows in the lower Colorado River valley (Sala and Smith 1996).

There are differences between species of this genus. Whereas *T. aphylla* does not have to reproduce by seed and can retain its leaves throughout the year *T. gallica* produces seed and drops its leaves annually. Because of the massive seed production of *T. gallica*, it is more widespread throughout the western United States. This fecundity coupled with its copious use of groundwater make saltcedar a management problem for water managers in that area. For example, measurements taken in Carlsbad, New Mexico, indicate that the average use of groundwater by saltcedar grown in tanks was 5.48 ft/year (1.6 m/year; Blaney et al. 1942). Tank experiments performed by Gatewood et al. (1950) indicated that as the depth to water table increased, the amount of groundwater use decreased but was still high.

Similarly, high groundwater use was observed in natural stands of saltcedar by the same researchers.

7.1.2.4 Mesquite and Rabbitbrush

A phreatophyte native to the western United States is mesquite (*Prosopis juliflora*). Mesquite is a deep-rooted plant with roots that penetrate at least 50 ft (15.2 m) in search of the water table. It also uses precipitation or surface water if available. Interestingly, even though mesquite is considered to be an arid plant, its capacity to take up and retain groundwater is evident from accounts that upon cutting down a mesquite tree, its green wood is heavy, similar to the wood of eastern hardwood trees that are not water limited (Spalding 1909). Moreover, Spalding (1909) observed a relation between the distribution and depth of mesquite roots and groundwater:

> *The root system of these plants consists of a taproot which grows rapidly downward and when developed is always within reach of a permanent, deep water supply, and a system of widely spreading lateral roots which are in relation to more superficial layers of the soil...The contrast between this and the shallow root system of many of the great trees of eastern mesophytic forests, familiar to everyone who has seen them uprooted by heavy winds, is instructive.*
>
> (Spalding 1909)

The depth to water table also controls, to some extent, the vigor of mesquite trees. In areas where the water table is at shallow depths, mesquite can reach a height of 50 ft (15.2 m); in areas where the depth to water table is greater, the height of the tree is considerably shorter (Cannon 1911, 1913). Cannon (1911) conveys that ranchers in Arizona often locate wells based on the presence of mesquite. About the same time, Meinzer and Kelton (1913) were investigating the water resources of the Sulphur Spring Valley in Arizona. As part of that study, mesquite growth was measured in relation to measured or estimated depth to water table. Typically, the depth to water table ranged from 11 to 50 ft (3.3–15.2 m). The most vigorous mesquite growth was observed, however, in locations where depths to the water table were moderate, between 25 and 35 ft (7.6–10.6 m; Meinzer and Kelton 1913). The shallower depths resulted in less vigorous growth because of the accumulation of salts near the land surface by the evaporation of shallow groundwater. Transpiration rates between 2.8 and 3.5 mm/day have been reported (Interstate Technology Regulatory Council 2009).

It is interesting that observations of the relation between mesquite growth and depth to water table were documented as far back as the late 1880s. Harvard (1884) noted that a 60-ft (18 m) long taproot of mesquite could reach even a deep water table and observed that the shallower the water table, the better the growth (Harvard 1884). Others also noted the relation between the presence of mesquite and the relation to the water table (Coville and MacDougal 1903; Schwennesen 1918). G.E.P. Smith, who, as described in Chap. 1, first used automatic water-level recorders in wells installed in stands of trees to record the diurnal water-level fluctuation of trees transpiring groundwater, also noted in 1915

> *...that the trees [mesquite] send their roots down to the water table is easily proved, for the caving banks of rivers and arroyos reveal them. The mesquite, in particular, has deep, strong taproots, with a generous development of feeders.*
>
> (Smith 1915).

A short list of woody plants classified as phreatophytes is presented in Table 7.2. These plants are primarily observed

Table 7.2 Representative woody phreatophytes native to the western United States that may have phytoremediation potential (Modified from Robinson 1958).

Common name	Scientific name	Relation to groundwater	
		Depth to water table below land, ft (m)	Remarks
Boxelder	*Acer negundo*		Found along streams in mountains.
Alder	*Alnus*		Found near streams.
Hackberry	*Celtis reticulata*		Found near streams, and can reach 50-ft height.
Smoketree	*Dalea spinosa*		Found in gravel washes.
Sycamore	*Platanus wrightii*		Found near streams.
Cottonwood	*Populus spp.*		
Quaking Aspen	*Populus tremuloides aurea*		Found near streams and springs.
Mesquite	*Prosopis juliflora*		Extensive roots, up to 50–60 ft.
Live Oak	*Quercus agrifolia*	35 (10 m)	
Willow	*Salix*		
Elder	*Sambucus*		Found in moist areas.
Greasewood	*Sarcobatus vermiculatus*	60 (18.2 m)	
Sequoia	*Sequoia gigantean*		
Saltcedar	*Tamarix gallica*		
California palm	*Washington filifera*		Shallow roots, groundwater must be shallow.
Vanadium Bush	*Cowania stansburiana*		Used to indicate vanadium-uranium deposits.
Rabbitbrush	*Chrysothamnus spp.*		Grows in moderately alkaline soils; contains rubber (non-latex) up to 6%.

in arid areas of the western United States, although some also are present in the more humid East. One plant, the saltcedar, was not recognized in the early work of Meinzer (1927), because although present, it had not yet begun to invade the southwestern United States. Therefore, this supports conclusions that saltcedar probably was imported from the Mediterranean area into the United States sometime before 1927.

Another phreatophyte found in arid areas of western North America is the rabbitbrush (*Chrysothamnus spp.*). Although it can be found growing in areas where the depth to water table is deep, a thriving stand indicates a shallow water table between 8 and 12 ft (2.4–3.6 m). White (1932) first observed that it uses groundwater when he stated that wells installed in areas where rabbitbrush grew had a daily fluctuation in the groundwater level. Interestingly, these plants contain about a 6% non-latex rubber content of chrysil, which may serve defensive purposes or provide a way to store excess photosynthate for later use.

Other native phreatophytes that use groundwater but are more important in that they can indicate groundwater of poor quality include pickleweed. Pickleweed (*Allenrolfea occidentalis*) is a succulent shrub characterized by very small leaves and can grow where the salt content of soils and groundwater is high. For example, the soil where pickleweed grows usually contains about 1% salt. Plant-tissue samples from pickleweed growing near Malad Valley, ID, indicate that the sodium and chloride content of pickleweed can be as high as 26 mg/g (mg per gram; Robinson 1958). At Death Valley, CA, Robinson measured groundwater concentrations of sodium and chloride where pickleweed was growing and observed a conductivity of 31,600 μmho (micromhos) and a chloride content of 12,800 ppm (parts per million; Robinson 1958). While the presence of pickleweed may be used to indicate shallow depth to water table, the quality of the groundwater may render it unfit for irrigation or potable use.

The common rose also uses groundwater in its native habitats. Meinzer (1927) observed several roses growing at a spring mound near Big Smoky Valley, an otherwise arid area. Also, the 190-year-old Lady Banksia rose growing in Tombstone, AZ, must have as its source of water deep groundwater (Fig. 7.2).

Native palm trees also tend to be phreatophytes. The Washington Palm, for example, the only palm native to the United States, has a shallow root system but can survive in otherwise desert conditions because of their location near springs. This relation between palms and groundwater is the reason behind desert oases as described in Chap. 1. Unlike other arid groundwater plants, these palms can reach heights greater than 50 ft (15.2 m) and are used by weary travelers to guide them to oases of shade and cool water. Trees that can be classified as phreatophytes and found in the more humid eastern United States include birch, such as river birch (*Betula nigra*), sycamore, alder, walnut, and various oaks (Table 7.3).

An evergreen plant that may be useful in phytoremediation applications because of its moderate growth rate, survival in moist soils, and evergreen habit is the Eastern red cedar (*Juniperus virginiana*). Contrary to its common name, the Eastern red cedar is not a true cedar, or *Cedrus*. This tree is found throughout most of the eastern

Fig. 7.2 This enormous 190-year-old Lady Banksia rose grows in a courtyard in Tombstone, AZ, and is as popular a tourist attraction as the nearby OK Corral. In this arid area, the longevity of this plant is attributed to its use of deep groundwater (Photograph by author).

Table 7.3 Representative woody phreatophytes native to the eastern United States that may have phytoremediation potential and range of transpiration rates (Modified from the Virginia Natural Heritage Program and Interstate Technology Regulatory Council (2009)).

Common name	Scientific name	Range of transpiration (mm/day)
Canada service berry	*Amelanchier Canadensis*	–
Silky dogwood	*Cornus amomum*	–
Sweetbay magnolia	*Magnolia virginiana*	–
Redbay	*Persea borbonia*	–
Willow	*Salix spp*	2.0–50
Red maple	*Acer rubrum*	–
River birch	*Betula nigra*	9.1–15
Atlantic white cedar	*Chamaecyparis thyoides*	–
Sweet gum	*Liquidambar styraciflua*	–
Water tupelo	*Nyssa aquatic*	–
Pond pine	*Pinus serotina*	–
Loblolly pine	*Pinus taeda*	6.5–12
Longleaf pine	*Pinus palustris*	–
Sycamore	*Plantus occidentalis*	–
Swamp white oak	*Quercus bicolor*	–
Swamp laurel oak	*Quercus laurifolia*	–
Swamp chestnut oak	*Quercus michauxii*	–
Pin oak	*Quercus palustris*	–
Willow oak	*Quercus phellos*	–
Baldcypress	*Taxodium distichum*	11–18
White cedar	*Thuja occidentalis*	–
Poplar	*Populus heterophylla*	13–200
	Populus deltoides	

– none reported

Note: Caution is advised with the application of some of these trees at potential phytoremediation sites, especially those trees that produce acorns or other seeds that may be used by wildlife or human populations

United States, from Canada to Florida. In fact, Cedar Key, Florida is named for the abundance of Eastern red cedar that used to grow there. There are no cedars there currently because a pencil factory had been constructed there, depleted the eastern red cedar, and had to shut down after all the trees had been harvested. It also explains why the Eastern red cedar is sometimes referred to as the pencil tree. This evergreen tree prefers moist, well-drained soils and is often found growing near springs in limestone terrain.

Because of its small size and use of deeper sources of soil moisture and groundwater, the Eastern red cedar may be a perfect candidate for installation at sites to address concerns of the lack of groundwater uptake by plants during the winter when poplars and willows go dormant; this issue of dormancy with respect to phytoremediation and alternatives are discussed in Chaps. 10 and 16. The Eastern red cedar has a drawback for phytoremediation projects, however, in that it has a slow to moderate growth rate, such that it may take a considerable amount of time to produce any benefit for groundwater control. As a testament to its growth habit, it has been called the graveyard tree because if it is planted when a person is born, by the time it has reached its full height of 40–50 ft (12–15.2 m) the person usually has reached full maturity and may be nearing death.

7.1.3 Natural Plant and Groundwater Interactions as an Analogy to Plant Selection

The relation of plant distribution to the occurrence and depth of groundwater in the arid United States was introduced in Chap. 1. For example, cacti are more likely to be found in well-drained, nutrient-poor soils; weeping willows are more likely to be found near lakes, streams, and rivers than at the tops of mountains; and conifers are likely to be found in cooler, drier areas. These observations between plant distribution and water source were later confirmed to be true in humid areas using geochemical techniques, such as the stable isotopes of water.

Based on the process of natural selection, the distribution of plants is the result of a particular plant species being best suited to inhabit the environmental conditions provided by that niche; in other words, the prevailing plant has properties that provide it with a selective advantage against competing plants. This is especially true for plant survival and reproduction under natural, non-agricultural conditions, where plant distribution primarily is a function of seed-germination characteristics.

7.1.4 Plant Succession

The environment is always in a state of change. Some changes are more obvious than others. For example, the seasonal change in color of deciduous trees in the northern hemisphere occurs because of shorter days, lower light levels, decreased water availability, and cooler temperatures. The period over which these changes occur is variable and is a function of tree species; typically, however, change occurs over a few weeks.

Other changes take much longer to observe. One such example is called succession. If an area of land becomes barren by fire or clear cutting and is left alone, a predictable pattern of colonization by plants can be observed (Ricklefs 1979). The plants that typically invade such barren or disturbed areas are annual herbaceous plants from seeds that entered the soil after deposition by wind, water, or animal action. These plants tend to become established first because of rapid germination, fast growth, and high seed production and are classified by ecologists as r-specialists. The source of water to such plants typically is precipitation, which is taken up by their shallow fibrous root systems. Soon, slower growing plants become established, either herbaceous or woody. The deeper root systems of these plants permit them to grow taller and shade the annual grasses. This provides the opportunity for pine species to become established. Hardwood trees then become established, and outcompete the slower growing pines for resources, including water and groundwater. Ecologists call these plants k-specialists, and their arrival proclaims the climax community, which is the end of plant succession, at least until the site is disturbed again.

What does plant succession have to do with phytoremediation of contaminated groundwater? Many sites characterized by groundwater contamination are located in abandoned lands, often cleared of surficial evidence of past waste-generating structures or virgin forests. These sites when visited often are characterized by annual grasses and are in the first stages of community succession. If a phytoremediation plan is implemented at such a site, the natural succession of plant communities, as described above, is short circuited, with a goal of establishing a climax community as quickly as possible. If cuttings or whips of woody plants are used to establish the phytoremediation system, forest conditions can take 3–5 years to be realized. Therefore, the presence of a phytoremediation site can benefit local ecology by providing a climax community in a much shorter time frame than if such abused lands were not planted and plant succession proceeded at natural and slower rates. This benefit often will make phytoremediation a more acceptable remediation strategy.

7.2 Site Preparation, Design, and Plant Installation

A civilization flourishes when people plant trees under whose shade they will never sit.

Greek Proverb

Although phytoremediation and conventional remediation engineering represent different approaches to the remediation of contaminated groundwater, in many cases, similar data are used and both require monitoring of performance to ensure the efficient hydrologic control of the site. In the case of a classical mechanical engineering approach to the remediation of groundwater contamination, such as a pump-and-treat system, the number and size of wells to be pumped need to be rated to match the specific yield of the contaminated aquifer. This needs to be done to ensure that the wells do not continually pump dry and that the time needed to remove the required aquifer pore volume is met. Also, in placing well screens for each well the spanning of discrete redox zones must be taken into account, because oxic and anoxic groundwater will mix during pumping and may result in clogged filters that require routine maintenance to stay open. Alternatively, if a thermal heating design were to be used to degrade contaminants, engineers would have to calculate the amount of heat needed to volatilize a specific contaminant mass over a particular area and match this with the appropriate network of vapor-extraction wells to collect the vapor. Essentially, the desired data drive the design and engineering of such remediation strategies.

The design of a phytoremediation planting should be approached in a manner similar to the design of remedial actions that involve mechanical and civil engineering, with the additional consideration that the technology is based on living organisms. The rooting depth of plants is a key factor in determining the potential for plant and groundwater interaction at phytoremediation sites. Many other factors also need to be considered before, during, and after planting. For example, the parameters of temperature, light intensity, water availability, and gas exchange are needed as input to determine the most appropriate design.

Perhaps an appropriate analogy of this approach in terms of maximizing plant health at a phytoremediation site is a greenhouse. A greenhouse permits the regulation of the parameters essential to the stewardship of plant growth. Trying to design a phytoremediation site similar to conditions found at a greenhouse, however, would be prohibitively expensive even at small sites less than an acre in size. However, a given set of parameters at each site can be controlled by the phytoremediation designer.

Part of the design of a phytoremediation site in the future will be the increased use or application of molecular biology techniques. For example, some fruit and vegetable crops are genetically modified to enable them to grow in climates far from their original habitats. It may also be possible to use such genetically controlled plants for phytoremediation applications, such as those that transpire larger amounts of water than even the current best hybrids. This should be an area of exciting research.

7.2.1 Site Preparation

The addition of plants at a site for phytoremediation to achieve hydrologic goals is preceded, in most cases, by site-preparation activities. This is true of most endeavors that involve or have involved plants. For example, early North American Indians burned forests to create open spaces for planting. Colonial farmers in New England had to prepare their fields by removing stones and large boulders. In contrast, farmers in the wooded mid-Atlantic to Ohio Valley felled trees to prepare their land. Site-preparation activities also occur in other areas of the world to remove existing plants to plant agricultural crops, as is occurring in the South American rain forest region.

Such pre-planting activities are no different at phytoremediation sites where vegetation exists. At a minimum, surface obstacles in the areas to be planted will have to be removed or incorporated into the planting design. The soil may also have to be prepared. The existing soils can be ripped with a chisel plow or disked, especially if the soil has been naturally or mechanically compacted. If fill materials are brought to the site, they should be sampled for the concentrations of appropriate nutrients, amended if necessary, and checked for the presence of contaminants that may be detrimental to plant growth. If the native soil at the site is to be used, assuming that it has been tested and is not contaminated or categorized as a hazardous waste, it may need to be graded, supplemented with nutrient amendments, or at least aerated. It then may require tilling (remember J. Tull from Chap. 1) to mix the amendments into the soil.

The experience of the SRWC industry can be useful when trying to determine the nutritional needs of hybrid poplar trees used at a phytoremediation site. In general, their experience indicates that nitrogen can be added at a rate of 50 lbs/acre as inorganic fertilizer or an equivalent if an organic fertilizer is used during the growing season. Care must be exercised to avoid adding too much nitrogen, however, because the National Primary Drinking Water Standard (NPDWS) maximum contaminant level (MCL) for nitrate in groundwater is 10 mg/L.

The chemical and physical properties of the soil should be assessed prior to plant installation. Many plants that are useful for phytoremediation purposes, such as hybrid poplar, require a neutral or slightly subneutral soil pH. Many surface soils exposed to naturally acidic precipitation, which can range between 4 and 6 pH units, do not contain minerals that can buffer this acidic input. In such cases, an artificial buffer, such as lime (calcium oxide, CaO, or the hydrated form, $Ca(OH)_2$) could be added to increase buffering. Hybrid poplars, along with most plants, require soils in which the organic-matter content is between 3% and 8%. The organic matter provides a source of recycled nutrients and an *in-situ* microbial population. Peat moss is a good source of organic matter.

Terrestrial plants such as those chosen for phytoremediation thrive if soil has sufficient air spaces that permit oxygen diffusion to support growth and respiration. Factors that control the entry of oxygen into soil, or aeration, are the soil porosity; soil moisture; depth to the water table; degree of water-table fluctuation; and precipitation amounts. In general, clay soils tend to hold water and refuse oxygen diffusion and sandy soils drain water easily and are highly aerated. Amendments can be added to either soil types to control water content and make the soil more bioavailable, such as by adding organic matter.

The potential need for artificial irrigation also must be addressed on a site-by-site basis. Some sites require irrigation because of soil-moisture limitations, climatic factors, or simply to rapidly establish growth or to protect the investment of plant installation at large sites. In low-permeability soil conditions, it is possible for even young poplars to tolerate wet soils or flooded conditions, but only after an extensive root system has developed. Most plants used for phytoremediation for hydrologic control are riparian plants that grow in areas that receive pulses of high water levels from flooding. Irrigation, if required, can lead to shallow root systems dependent on irrigation water rather than deep root systems that tap groundwater. In most cases, water application rates that amount to an in./m is sufficient during early plant installation of cuttings or whips once the roots have started to grow. One inch of water over 1 acre is about 27,150 gal (102,642 L). If irrigation is required during the first year after plant installation, the rule of thumb for water volume to be added is at the rate of 0.5 gal/ft^2 (5.3 L/ m^2) of root area each week as necessary, unless at least 1 in. (2.5 cm) of precipitation occurs, which is the equivalent amount of water.

During site-characterization activities as outlined in Chap. 6, soil samples can be collected to the water table and analyzed in the laboratory for soil moisture. A soil sample is weighed, oven dried, and reweighed to calculate the amount, or percentage, of moisture lost. The borehole or hand-auger methods previously described can provide qualitative visual data about soil moisture conditions. For example, the presence of clay in otherwise sandy sediments provides a source of moisture to newly installed cuttings; these clay lenses, if laterally continuous, also may be locations of perched water-table conditions that occur after precipitation events. Being on site during any planned drilling activities ensures that such important observations are not overlooked.

If substantial site-preparation activities are required, a natural colonization of the site by a variety of plants likely will occur. This happens even if control measures, such as mulch or hay, are used. In many cases, the appearance of r-specialist plants, or weeds, can be appreciated, because they help stabilize the ground surface while trees are becoming established and can decrease the amount of precipitation infiltration, which will ensure that the trees planted for phytoremediation seek groundwater. Conversely, the weeds compete for limited site resources and may affect the growth of the phytoremediation plantings. If necessary, the growth of weeds can be controlled with appropriate chemicals or alternative farming practices, such as intra-row tilling. The application of herbicides should be used with caution, however, because many plants, such as poplars, can be injured by herbicides. (An alternative to synthetic chemicals for weed control is to spray a solution of white vinegar and water, at 15% concentration, onto the foliage of weed plants.) Finally, simple and inexpensive tree mats can be used around individual installed plants to suppress weed growth. This approach, however, may be problematic at sites where thousands of trees need to be planted.

A few liabilities are inherent in the installation of a phytoremediation system for the hydrologic control of contaminated groundwater. The biggest one is the amount of work that must be done below grade where many utilities are known, or thought, to be located. A utility survey is essential before any work begins that involves movement of the subsurface soil. A utility survey usually is a free service. Local ordinances regarding the maximum height of plants located near roads, intersections, or flight paths also must be followed. Many of these liabilities can be addressed, however, even before the start of planting as part of site assessment and characterization.

7.2.2 Site Plants and Planting

Because the installation of plants at a phytoremediation site leads to a grove, orchard, or plantation, it is worthwhile to consider how the planting relates to future plant-care duties. On the one hand, plants can be viewed as tools to solve an environmental problem and can be considered strictly from a mechanistic standpoint. Alternatively, each plant can be viewed as a living entity that has to survive on its own ration of air, water, and soil nutrients. In this view, plants are installed with a full awareness of the soil properties (as previously discussed) prior to the installation of the first plant. In the first viewpoint, the planting design can be constrained purely by limitations not necessarily related to the health of an individual tree or its interaction with groundwater. For example, the decision to plant along a particular spacing interval between trees may be based on the width of the tractor axle needed to maintain the area between rows. In most cases, these two extremes of planting mindset should be avoided and a compromise selected.

Conventional wisdom is that in order to establish a more ecologically efficient system of plants, a monoculture should be avoided. This wisdom is based on experience that a mixed planting that consists of different species and even genera is more resistant to infestation threats. This is partly true, because if one plant in a monoculture is attacked by a pest, few limits prevent all from being attacked. However, even if multiple genera and species are planted, most often the plants used are vegetative cuttings of a common parent and, therefore, are derived from vegetative reproduction and tend to lose resistance to infestation over time relative to plants derived from sexual reproduction. This liability can be avoided by using plants that have resulted from sexual reproduction or by using transgenic plants to which other genes have been added to proffer protection against disease or infestation.

The appropriate perspective must be kept in mind, however, during planting. Phytoremediation plants are not a food crop, but rather a tool to remove groundwater or contaminants at sites. Some losses to infestation are to be expected but can be easily remedied. The SRWC industry has demonstrated that even cutting trees to the ground can result in extensive new growth, and removal of damaged parts of trees stimulates new growth.

The footprint of the site boundaries controls the maximum size of the planting. As could be expected, if a constant transpiration rate can be assumed for a given genus, the more plants that are installed per site the more water that will be transpired from the site. George et al. (1999) provide data for conditions in Australia that indicate that for every 10% increase in planted area, the water table decreased 1.3 ft (0.4 m). These results indicate that plants that are installed as closely as possible to each other at a site maximizes biomass and, therefore, potential measurable changes in the water budget.

However, the goal of a phytoremediation system is not necessarily to establish biomass but that the biomass interacts with groundwater. As such, Meinzer et al. (1996; not O.E. Meinzer of the USGS!) reported that the transpiration from a stand of koa plants (*Acacia koa*) was higher at a larger spacing of 8 ft by 8 ft (2.5 m by 2.5 m) relative to a closer spacing of 3 ft by 3 ft (1 m by 1 m). This was in contrast to closer spaced plants that had higher biomass. In general, transpiration and stomatal conductance increased to a greater extent with a wider-spaced planting in response to increased *VPD*, solar radiation, and wind speed.

7.2.2.1 Seedlings

In many cases, selective pressure during evolution has resulted in plants that can reproduce themselves by using more than one method. This includes sexual reproduction and asexual reproduction.

Plant sexual reproduction occurs when the pollen produced from the stamen of a male plant enters the pistil of the female plant of the same species (or different in the case of a hybrid), fuses with the female egg, and forms a zygote. In the ovary, this ultimately becomes the embryo of a new plant, as a seed, encased in fruit in angiosperms but exposed in gymnosperms. Under natural conditions, the seed is dispersed, germinates, and grows into another plant. Under nursery conditions, these plants, or seedlings, are cared for in individual pots until they are at least 1-year old. Seeds from hybridized plants often result in less desirable characteristics than the parent, which is why most are sterile and are propagated by using vegetative methods as discussed below.

Phytoremediation sites can be planted with seedlings. Seedlings in their first year are little more than a shoot with some roots. They are often available free of charge from local forestry departments, where they are grown for reforestation purposes. Seedlings usually are established in a funnel-shaped container of soil, and immediate planting is not necessary as long as water and light are provided. Seedlings typically are planted by using manual methods, which include a dibble bar or shovel.

7.2.2.2 Vegetative stock

Because of the time needed to establish plants from seedlings, other techniques can be used, such as plant cuttings. A cutting from a plant, be it leaf, stem, root, etc., can form an entirely new plant without having to go through sexual reproduction, and is a form of asexual propagation. Asexual reproduction, or vegetative reproduction, does not involve the fusion of separate sex cells from separate plants. Rather, it is the propagation of a new plant from a piece of the existing adult plant. The produced plant is not the offspring of the parent but instead is a younger identical version of the adult plant. For example,

> *Romulus, once in a trial of his strength, cast hither from the Aventine Hill a spear, the shaft of which was made of cornelwood [cornelian cherry]; the head of the spear sank deep into the ground, and no one had the strength to pull it up, though many tried, but the earth, which was fertile, cherished the wooden shaft, and sent up shoots from it, and produced a cornel trunk of good size.*
>
> Plutarch (1914; translation)

There are downsides to generational advances with cuttings, because plants become more susceptible to disease, as the genetic mixing and dilution of sexual reproduction does not occur. As an alternative, the cloning of trees with desirable traits can stop future genetic dilution and instead capture the necessary genetic makeup to assist with environmental restoration.

A cutting can be taken from different parts of the plant during different times of the growing season. If a stem is used to produce a cutting, it can be from the growing tip, called a softwood cutting, or from the part of a stem that contains bark, called a semi ripe to hardwood cutting. Soft (herbaceous), firm (semi lignified), or hard (woody) is the way that Garner (2003) described making cuttings. In most cases cuttings made from the growing tips are best, as these contain the highest concentrations of natural growth hormones that will stimulate shoot and root growth, as discussed in Chap. 3. However, internode cuttings can be taken successfully even if many are made from the same cutting. This is because even though the tip contains the highest levels of auxin, the nodes contain higher levels than the internodes.

Hardwood cuttings supplied by most third parties typically are not available throughout the entire calendar year and, therefore, their availability will control when a phytoremediation planting occurs. Such cuttings typically are made in the fall when the tree is dormant, usually after the leaves have fallen and the nights are longer. These cuttings are stored in coolers until the next season and typically are available from January to July. They are shipped coming out of dormancy and soon begin to acclimate to warmer air temperatures; this in combination with rising ground temperature breaks the cuttings' induced dormancy. Thus, the ease of plant acquisition often will determine the timing of planting. For example, planting in the fall may require the purchase of older unsold stock that has remained in a cooler for more than 6 m. Similar circumstances can be corrected if plants have been placed in specially designed grow bags while in storage.

Cuttings also can be made from trees that are on site or have been planted previously. Cuttings acquired this way can be used to replace those previously planted trees that have been lost for a variety of reasons. The cuttings have to be done correctly, however, to ensure survival of both the donor tree and the cutting. Also, because these cuttings may be from cuttings themselves, there is an increased risk of potentially devastating effects if the plants are susceptible to pests or diseases.

Cuttings of softwood, or the new, green, unsuberized growth, are best when cut early in the morning when water supplies are highest prior to transpiration. These stems are actively growing. Cuttings should be no more than 12 in. (30 cm) long, and placed in plastic bags in the dark to prevent drying. Leaves should be removed in most cases, or the largest one at least cut in half. Growth hormone can be added to the cut.

Cuttings made from hardwood (seasoned old growth that is dormant) can be taken after the leaves have fallen. These can be planted as soon after they are collected and planted leaving only the top bud above the ground. If some time will elapse before the cuttings can be installed, the cuttings can be placed into a plastic bag wrapped with a wet towel and this placed in another plastic bag to decrease water loss. The bundle can then be placed in a cooler or refrigerator. Prior to planting, the cutting can be placed in a root hormone powder, such as indolbutyric acid (IBA). Cuttings made during dormancy should occur only after a period of colder temperatures.

After a cutting is taken, the plant hormone auxin travels from the cutting tip to the site of the cut, and begins to stimulate root growth in the separated cutting. This transfer of auxin must be an adaptive response to tree damage, perhaps by ice or wind, such that these natural cuttings would have a chance to grow on their own. At the site of the cutting where auxin travels, root cells begin to grow and form a callous, lumpy white mass that takes the form of a slender root. These roots have to form to take up water into the cutting before too much aboveground shoot growth occurs, because bud break and leaf growth are driven more by cell elongation through turgor (caused by water uptake) than by increased cell division. For poplars and willows, basal parts of 1-year-old plants perhaps are best. The top and bottom are cut, close to nodes; this is why cuttings made from these plants often are shipped from suppliers without tips. Some cuttings arrive straight and are 1-year-old cuttings; and some arrive with a heel, which is part of a 2-year-old plant.

The time from taking a cutting and removal to planting and root formation is characterized by vulnerability; essentially, the cutting is on its own. Cuttings have no roots, but do have many lateral buds and root initials. One may be tempted to use a commercially available rooting hormone or synthetic auxin to encourage the growth of roots at the bottom of the cutting, especially since one cannot observe root growth after installation. Too much rooting compound, however, actually decreases root formation.

Cuttings should be placed in an organic-rich soil that is well aerated. This is so that the early nitrogen needs of the cutting can be met and ensures that adequate water infiltration occurs to supply the expansion and growth of root hairs. Such organic-rich soil of high porosity also permits oxygen diffusion to support root respiration; soils of low porosity result in anoxic conditions and root death. Still, close contact should remain between the cutting and the soil, because the root primordia need access to the soil as soon as possible.

The cuttings can be installed as whips or as bare-root stock. Whips are large cuttings of complete branches, usually taken during the dormant season, and are stored under cool, moist conditions. Typically, nursery's or other third-party suppliers have a field of growing adult plants, and whip cuttings are taken after the plants go into dormancy following leaf drop and cooler temperatures. Cuttings are the most commonly available and least expensive form of plant propagation and, therefore, the most widely planted type of woody plant useful for phytoremediation.

Bare-root plants are cuttings that have a suberized root system. They start out as cuttings by the grower, are placed in soil, and then removed from the ground and all the soil is shaken off. This typically is done in the fall during dormancy, and the plants are kept cool and moist while stored. Advantages of installing bare-root cuttings over cuttings or balled-and-burlapped trees include the conveyance of extra stored carbohydrate in the roots to support early growth needs, and the decreased potential for soil pests to be transplanted in that little soil is present. On the other hand, the process of removing the soil from the roots removes the root hairs, which are the primary entry points for water absorption, as well as some of the microbial community in the rhizosphere. The roots of bare-root cuttings are suberized, so bare-root cuttings still need to develop new root growth and root hairs after reinstallation at a phytoremediation site.

Cuttings or bare-root cuttings are living organisms; this fact can be easy to forget because they can be shipped like common freight. When the stock arrives or is picked up and before planting begins, some steps can be taken to ensure survival of the cuttings. The first is to place the cuttings, tips up, in about 3 in. of water. This prevents air from entering the vascular system and causing embolisms and encourages the development of adventitious roots. Dead or broken parts should be removed. The plants should not be removed from either of these treatments until ready for planting. In the field, the cuttings should be kept away from wind and direct sunlight. Basically, they should not be allowed to dry out. After planting, the top terminal bud, if present, can be removed to induce growth, although most cuttings do not come with a terminal bud.

In general, plant installation should occur within 1–4 days after the cuttings arrive from the supplier. If a longer period of time is required before planting or if other logistical problems arise that prevent planting, the cuttings should be placed in cold storage, such as a refrigerated unit. Keep in mind that this will decrease the humidity of the air around the cuttings and increase the chance for the cuttings to lose

Fig. 7.3 Hybrid poplar cuttings placed in a bucket of water prior to planting (the plastic wrap has been removed to expose the cuttings for viewing). Note the profuse appearance of adventitious roots along the sides of many of the cuttings (Photograph by author).

stored water, so plastic wrap should be used and the cuttings set in buckets of water during storage (Fig. 7.3).

In general, the type of plant installation method will be based on the type of plant, or its stage of growth, to be installed. As stated previously, perhaps the most commonly installed plant at phytoremediation sites is the hybrid poplar tree. It can be obtained as a cutting of various sizes from 6 in. (15 cm) to 10 ft (3 m) with no roots, a rooted cutting, a bagged-and-wrapped root ball, or as a full-grown specimen in a multi-gallon container. It should not be surprising, therefore, that there are many different ways to install even the same type of plant. Regardless of plant size, the larger the hole that a plant is introduced into, the better as far as the life of the plant is concerned.

Common to all planting methods, planting can commence any time after the soil temperature has reached at least 50°F and the threat of frost has passed. This is to ensure that the soil temperatures can support root respiration and growth. In general, at least one-third of the bare-root cutting should be installed below ground (2 ft (0.6 m) for a 6-ft (1.8 m) cutting, for example), and for whips at least one-half, if not more, below ground. At least one lateral bud should remain above ground and in the right direction (facing upwards!). This is to ensure that one main stem grows upward, rather than having multiple, but weaker, stems.

If the stock is wide enough in diameter (0.5–2 in. [1.2–5 cm]) and at least 2–3 ft (0.6–0.9 m) long, the cuttings can be driven into the ground with a rubber mallet if the soil is soft enough and the soil chemistry is conducive to growth without amendment. For shorter cuttings, say less than 24 in. (60 cm) for example, site installation can be as simple as using a short piece of reinforcing bar (rebar) and a sledge hammer to create a small-diameter hole in which the cutting can be inserted. This technique works well in loose sandy soils or topsoil fill material but may not ensure cutting survival if the sediments are tight (clays and silt) or contaminated, as new and adventitious roots will be in contact immediately with the contaminated soil. If preliminary data indicate contamination in the soils to be planted and it is cost prohibitive to remove this and add fresh backfill, then a large-diameter auger hole filled with fresh uncontaminated loam would be better (Cook et al. 2010).

Large vegetative cuttings that exceed 6.5 ft (2 m), called whips or poles, can be planted in boreholes drilled or trenches installed with the appropriate equipment. This approach of using longer stock is used at sites where the depth to water table is greater than 15 ft. Holes also can be made manually with a post-hole digger or a 1- to 2-person rotary auger. The latter is a more rapid method to install a relatively low number of cuttings at smaller sites. Larger sites where even small cuttings will be planted do better with a more automated hole-creation method. A tractor with the appropriate front-loader or 3-point attachment can be used to rapidly create holes for planting using a rotary auger (Cook et al. 2010: Fig. 7.4).

As cuttings emerge from dormancy, leaves develop from the bud scales. This 'growth' actually is enlargement of the previously developed leaves with water. Water should be

Fig. 7.4 Drilling boreholes using rotary augers prior to planting hybrid poplar cuttings, Elizabeth City, NC, United States Coast Guard Support Center (Photograph by author; see Cook et al. 2010 for further explanation).

applied at this stage if no precipitation occurs. Adventitious roots develop below ground from stem nodes or root initials. Once the leaves are established, auxin present in the new leaves and apical meristem is transported to roots to encourage root cell growth and elongation.

If the drainage of the site is poor and precipitation is high, such that the potential exists for oxygen to be excluded from the subsurface, then the drainage needs to be improved before the plants are placed in the ground. To increase drainage around the roots, a large, deep hole can be dug and coarse-grained media can be added prior to soil backfill. Other techniques that can be used as a conduit for oxygen to reach the root zone include the installation near the roots of pipe material that contains holes, or well-screen material, such that air can enter the pipe and diffuse into the soils around the roots (Ferro et al. 2001; Quinn et al. 2001; Tsao 2003). The rate of oxygen diffusion will depend on the soil tortuosity and other factors, such as sediment, chemical, and biological oxygen demands, which may limit the actual amount of oxygen supplied to the roots. Also, these holes may become clogged over time with roots. In order to maintain air flow and prevent stagnation in the pipes, a U-shaped pipe with an influent and effluent stovepipe above ground can be used. At any rate, the installation of such pipes may help to bring air in contact with the subsurface to support root respiration. There are a few liabilities using such pipes with respect to contaminant fate, which are discussed in Chap. 15.

Alternatively, a trencher can be used to create a long trench where the plants can be added with backfill. This method, however, greatly disturbs the soil, and can result in increased permeability and, therefore, increased infiltration and recharge.

7.2.2.3 Time of Plant Installation

Timing of plant installation is as much an art as it is a science. In short, plants can be added to soil at most sites any time during the year; even in frozen soils. A random planting approach without regard to the relation between plants and season, temperature, light, and moisture will not result in the most successful planting, however. On the other hand, there is no single perfect time to plant in a particular area. In many cases, the perfect time is based on individual experience and preference or plant availability rather than plant ecological considerations.

In general, for most areas and most types of plantings, it is advised to plant in the fall, on a cool and overcast day, because constraints on plant-available water are decreased and below-ground soil temperatures resist seasonal fluctuations in air temperature. Warm soil encourages roots to continue growth, even while shoots are dormant. Planting during the fall is best for larger plants that already have extensive root and shoot biomass, assuming that the plants have been removed from the nursery using the correct methods. Dormant cuttings or whips can be planted anytime from immediately after harvest to before bud break the following spring. Fall planting is favored in the warmer Southern United States where winters are milder than in the Northeast. Bare-root plants should be planted right after leaf drop if fall planting is to be performed.

Conversely, spring planting also has been successful in some areas. Planting in the spring after the last frost, however, tends to induce shoot growth relative to root growth. This may cause excessive wilting because the roots cannot provide the top growth with sufficient water to meet increasing transpiration demands. The negative effects of this condition can be remedied, however, by pruning. For bare-root plants that must be planted in the spring, they should be installed before they break dormancy.

Regardless of planting in the fall or spring, an important but often overlooked criterion is the soil temperature. Generally, planting can be accomplished successfully when soil temperatures reach a stable 50°F. Warm air temperatures may indicate planting, but if the soil temperature is too low, seeds will not germinate even if water is available, and root growth from cuttings will be slowed. It is advantageous for tree growers to sell poplars from January through July, a good time for such dormant cuttings to be installed in most parts of the United States or northern hemisphere. The soil temperature also is an important variable in determining the type of plant for a particular area to achieve biomass at a phytoremediation site. An evergreen tree adapted to cold air and soil temperatures if planted in a more temperate area will die of dehydration, because warmer air temperatures result in increased transpiration while the cool soil temperatures decrease water absorption by the root hairs.

Acquisition of plants also deserves attention. As any florist knows, the length of time that a cut flower remains sellable is determined by the conditions in which it is stored. Cut flowers must be stored in cold, dark, and humid storage containers and wrapped in plastic with very little solid media, such as soil. The treatment of cuttings for phytoremediation purposes follows the same practice. Cuttings need to warm up slowly from being stored at cooler temperatures. This can be done by removing them from storage and placing them in a bucket of water in indirect sunlight, as stated previously. Cuttings or bare-root cuttings should be placed in bags to maintain high humidity levels or in buckets of water until planted so they do not dry out. As much as 1–2 weeks of warming time should precede planting the cuttings. Soil moisture should be high enough to reduce water stress in newly planted cuttings but not so high as to induce anoxic conditions, which leads to death.

In some areas, the use of birches generally is not recommended. They can be prone to disease caused by the bronze birch borer, which enters the top of trees and then feeds on the sap contained in the phloem. To guard against such infestation, the European white birch (*Betula alba* or *Betula pendula*) can be used only in areas where temperatures go down to 20°F. Such problems with birch borers are less likely with the *Betula nigra*, which is native to the United States as it is more resistant to attack from the borers than the birch imported from Europe.

7.2.3 Recharge-Reduction Design

Plants can be installed with the goal of reducing groundwater recharge, which can be accomplished in two ways. First, plants can be used to intercept precipitation and remove it by evaporation before it becomes infiltration. Second, plant roots can be used to remove soil moisture and infiltrating water by transpiration before recharge can occur. Moreover, the root systems of plants selected to decrease recharge do not necessarily have to reach the water table; the plants used can be facultative phreatophytes or even drought-tolerant grasses. These processes behave in a manner similar to vegetative caps that have been successfully installed at older landfill sites.

Whereas plants have been added to decrease recharge at a number of older, closed landfill sites, there are fewer examples of this practice for contaminated groundwater. A study where a primary goal was the planting of trees over a source area to reduce recharge to the water table occurred near Milwaukee, WI (McLinn et al. 2001). The site was a former fuel-tank farm adjacent to the Menomonee River. Because of the geologic history of this area of the United States, the shallow aquifer at the site is composed of low-permeability materials, called till, left behind after the last glacial retreat, to a depth of about 18 ft (5.4 m). Because of the site activities, the soil and groundwater were contaminated by petroleum hydrocarbons, including free-phase product, with some product trapped in the source-area sediments above the water table. A phytoremediation system was designed for the site to decrease recharge to the water table by planting trees to sequester infiltration. In 2000, following extensive site preparation using some of the approaches described here, 485 hybrid poplar trees (Imperial Carolina *deltoides x nigra*) were planted at the site. Of the 485 trees planted, 290 were planted in a row adjacent to a river near the downgradient boundary of the site. These 290 trees were installed using a hollow-stem auger to a depth of 9 ft (2.7 m) to have the roots as close as possible to the water table. In order to ensure that oxygen in the unsaturated zone was not limiting for root respiration, an air-injection aeration system was installed during planting; the air-injection system was discontinued in 2003, however. Additionally, up to 195 trees were planted in the source area where the fuel tanks previously were located. These trees were not planted as deeply, only about 4 ft (1.2 m). As of August 2006, this site was being monitored (Van Epps 2006). Unfortunately, little specific hydrologic data is available regarding the effect of the planted trees on recharge reduction. However, data do exist regarding the effect of the trees on groundwater contamination and are described in Chap. 13.

Another instance of using transpiration to reduce recharge through a contaminated area was investigated at an industrial-waste site in South Africa (Duthe et al. 2005). The site is located inland from the Indian Ocean. Most of the annual precipitation occurs during the winter months, with drier conditions prevailing during the summer. Potential evapotranspiration also is high during the winter and lower during the summer and greater than precipitation. An area greater than 30 ha has been contaminated by chlorinated solvents and total dissolved solids–materials disposed there between 1958 and 1994. To decrease potential leachate formation and additional groundwater contamination, hydrologic control of the source area was proposed through a variety of methods, such as mechanical pumping of groundwater to dewater the contaminated area, control of surface-water runoff through a drainage network, and reduction in recharge by plant transpiration. Simulations made by using the finite element numerical model HYDRUS-1D (Simunek et al. 2005) indicate that the addition of up to 1,600 plants in the contaminated area would result in limiting recharge to less than 0.3% of annual precipitation (Duthe et al. 2005).

A typical planting layout to achieve similar recharge reduction at a fuel-contaminated shallow aquifer is depicted in Fig. 7.5 (Cook et al. 2010). The trees installed were four types of hybrid poplar and willow. The trees were planted as cuttings on 5-ft (1.5 m) centers. No artificial irrigation was used. Tree mortality was high across the site, and the mortality caused primarily by installing cuttings in contaminated areas without the addition of clean backfill. Subsequent reinstallation of cuttings in contaminated areas using clean backfill decreased mortality. About 1-year after planting, the cuttings had grown greater than 10-ft (Fig. 7.6). The study concluded that large boreholes backfilled with clean soil and planting prior to the summer increased the success of installation.

7.2.4 Hydrologic-Barrier Design

At some sites it may be impracticable to control groundwater hydrology by using plants to reduce recharge because the source area may be beneath a building or in an area that is not suitable for planting. In these cases, plants can be used to control the flow of groundwater that may already be

Fig. 7.5 A typical planting distribution to achieve a goal of recharge reduction at a site in coastal North Carolina characterized by the release of jet fuel from an above-ground storage tank. The site had been regraded and planted with grass decades before these hybrid poplar cuttings were installed in the spring of 2006. This image shows the cuttings after 3 months of growth. Brad Atkinson (North Carolina Department of Environment and Natural Resources-Division of Waste Management) is shown for scale (Photograph by author; see Cook et al. 2010 for further explanation).

Fig. 7.6 Growth of hybrid poplar trees from cuttings after 1-year at the site in coastal North Carolina. Dr. Elizabeth Guthrie-Nichols (North Carolina State University) is shown for scale (Photograph by author; see Cook et al. 2010 for further explanation).

contaminated and near property boundaries, in much the way that many pump-and-treat systems often are installed near property boundaries. If this is done, the installed plants will need to have some root contact with the capillary fringe or water table to be successful.

Table 7.1 includes a variety of grasses, shrubs, and trees that possess the ability to have roots near and into groundwater. Although most are native to the western United States, some also can be found in the more humid eastern United States. This quote says it best:

> *The distribution of the native vegetation throughout the eastern part of the United States is influenced by the water table to a much greater extent than is realized by most persons who have not given much attention to the occurrence of groundwater. In some places in the woodlands near Washington, D.C., ferns have been observed adhering about as closely to tracts of shallow groundwater as the well-recognized groundwater plants of the West, and many similar examples of the relation of plants of a particular species to the water table can readily be found almost anywhere in the East.*
>
> O.E. Meinzer (1927)

What we know about such obligate and facultative phreatophytes is derived from a variety of sources, as discussed in Chap. 1. These include (1) historical recordings of plants associated with usable quantities and quality of groundwater in arid areas; (2) the use of phreatophytes to prospect for minerals in the western United States; (3) wilderness survival tactics, such as taught by the armed forces, to ensure survival; (4) the experience, research, and development of the pulpwood industry; and (5) the practice of dewatering land to support agriculture or other purposes. Each of these five sources of information has an obvious economic interest. More importantly, all can provide useful information to help guide plant selection for use at phytoremediation sites where groundwater is contaminated.

The relation of the roots to the water table is the determining factor in deciding whether or not a particular plant will be useful in achieving hydrologic control through a barrier design. Most phreatophytes establish roots to a depth that represents the capillary fringe directly above the water because oxygen is available for root respiration. On the other hand, the depth to water table, and by definition, capillary fringe, is not a fixed location but can change over time. Even for phreatophytes, if the water table and capillary fringe rise because of infiltration, there will be a period when previously exposed roots will be submerged. There are, of course, some phreatophytes, such as cottonwood, alfalfa, and mesquite, which can have at least some roots below the water table continually. This can occur if the osmotic pressure is higher in the roots (i.e., a lower water content in the roots caused by concentrated salts) than that in the water table. Moreover, this occurs even if salts are not present because the roots can remove groundwater more easily at pressures equal to or greater than 1 atm than water under tension in the capillary fringe.

Perhaps the second most important aspect of successful phytoremediation as a hydrologic barrier is the correct selection of plants to interact with the water table. The selection of the appropriate plants must meet a few criteria, such as whether (1) the plants will grow under the climatic conditions of the site, (2) the plants will interact with the water table, and (3) the plant interaction with the groundwater system will be to the extent that remedial goals can be met in a reasonable amount of time in a cost-effective manner.

A first step in plant selection for a hydrologic-barrier design is to determine what types of native plants are doing well at or near the site under ambient conditions. The presence of grasses may indicate that adequate conditions of soil nutrients and moisture are present. The presence of larger shrubs and trees may confirm what the grasses indicate but also that deeper sources of water are available. An advantage to such observation of native plants is the knowledge that they already are adapted to the climate of the area. The disadvantage, however, is that the presence of vegetation at a site does not directly indicate that the plants are using groundwater through uptake or decreasing recharge. These questions must be answered by field studies and extensive monitoring.

A second step is to look at the use of genetically engineered plants that have been specifically designed to perform a particular function. All characteristics are part of specific genes. More recently, specific genes linked to a desirable trait, such as the ability to detoxify a contaminant, have been identified in certain plants, isolated, and added to other plants to confer this trait; this topic is discussed in Part III. Other plants that have genes that result in longer or faster root growth also could be used to facilitate hydrologic control through the barrier design. In either case, the regulatory approval of the use of such genetically modified plants remains the biggest hurdle to their potential use at phytoremediation sites.

Plants have been added to affect groundwater flow at some sites where the results are published. For example, at a Superfund site in the southeastern United States characterized by the blending of pesticides between 1936 and 1987, pesticide-contaminated soil was removed, treated, and replaced. Pesticides, such as toxaphene and benzene hexachloride (lindane), were detected in groundwater (Leavitt et al. 2001). A pump-and-treat system was proposed to contain and treat hot spots of contamination that could not be excavated. Because the concentrations of pesticides were less than 50 μg/L and various metals were present, which would complicate treatment, it was determined through pilot tests that such a system would not be efficient. Therefore, a phytoremediation-based treatment of the contaminated groundwater using a barrier design was investigated.

The hydrogeology of the contaminated site consists of sands and clays. Groundwater flows in the surficial aquifer, which is about 20-ft (6 m) thick, from the about 28 acre (113,316 m^2) site, to an adjacent lake. In 1998, about 2,500 hybrid poplar trees were installed within a 2.2-acre (8,903 m^2) area. The trees were planted to depths between 2 and 12 ft (0.6 and 3.6 m), and extensive monitoring was initiated to determine the amount of groundwater being used by the trees. Over time following plant installation, the removal of water by the stand of trees went from less than 1 gal/min to more than 8 gal/min as the leaf area of the phytoremediation system increased from less than 21,527 ft^2 (2,000 m^2) to between 64,583 and 193,750 ft^2 (6,000 and 18,000 m^2). Leavitt et al. (2001) reported that the phytoremediation system removed 7.7 Mgal (29 million liters) of water between early 1998 and the summer of 2000. However, it is important to note that the authors did not discriminate the source of water being removed by the trees.

7.3 Environmental Factors That Affect Plant Growth and Groundwater Use

Even casual observation reveals that plants can grow in some of the most unexpected places. These unlikely environments range from large hardwood trees growing in cracks in igneous rocks to tiny herbaceous weeds growing in the cracks in sidewalks or parking lots. These plants are growing, but are they thriving? If plants are to be used as part of a phytoremediation system, simply getting them to grow is a first step, but it is far from the ideal scenario, where efficiency of plant growth and interaction with groundwater is important. There are many things that can be done to ensure that the plant growth at a particular site reaches its maximum in the shortest time to ensure interaction with groundwater.

Many environmental factors determine whether or not a plant simply survives or if it will thrive. The environmental factors that limit plant growth need to be identified and the ones that can be controlled removed as a limiting factor. Common environmental factors include the amount of light available; the physical and chemical composition of the soil, such as salinity and pH; the water availability over time, such as precipitation, soil moisture, surface or groundwater; and the geochemical composition of the soil moisture or groundwater. It also must be recognized that the notion, if a little works, more will be better, does not necessarily hold true with plants, such as nitrate application, because nitrate levels are regulated in groundwater. At a minimum, this approach will lead to wasted resources in terms of capital and potential environmental liability. Not all limiting factors can be modified, however. Perhaps the most obvious factor is light. Also, the genetic capability of plants, for example, cannot be affected other than during initial plant selection.

7.3.1 Groundwater–Soil–Plant–Atmosphere Continuum

As discussed in Chap. 4, water seeks its own level: water at higher elevations will flow toward lower elevations. The elevation of groundwater, or head, has a pressure equal to or greater than the atmosphere. This explains why groundwater will seep into dug holes or wells (Holzer 2010) because pressures are greater than 1 atm and, therefore, greater than the exposed hole in the ground. Water flow stops when the pressures reach equilibrium. In the unsaturated zone and capillary fringe where water is present under tension, however, movement is determined by negative pressure gradients, or water (matric) potentials. These concepts typically have been discussed separately, especially in terms of the water source to plants—for example, even in this book, the information contained in Chaps. 3 and 4 is separated. For the purposes here, it is more useful to integrate these ideas into one theme of a continuum of groundwater–soil–plant–atmosphere. This is because it is important from a phytoremediation perspective to understand the source of the water used by a plant.

For example, differences in the relative uptake of water from various sources along flood plains in the southwestern United States, such as surface water, soil water, and groundwater, was investigated by Busch et al. (1992). They sampled water from all potential sources to flood plain trees along the Bill Williams and lower Colorado Rivers in Arizona and analyzed the stable H and O composition of each source. Tissues from trees growing in the flood plain that had access to surface water, soil water, groundwater, and precipitation also were sampled for these stable isotopes and compared to the stable isotope composition of the various sources. The types of trees sampled included *Populus fremontii*, *Salix gooddingii*, and *Tamarix ramosissima*. Busch et al. (1992) also sampled groundwater in the flood plain, and adjacent trees were cored with an incremental borer to obtain xylem samples for stable isotope analysis. Branches and leaves also were collected at the time of core collection from the sampled trees. Soil samples also were collected near the trees.

For the Bill Williams River, the stable isotopes of H measured in tree cores were more similar to the stable isotope composition of groundwater than to soil water. One of the observations made by Busch et al. (1992) was that there is little difference between stable isotopes run on a branch sample relative to that of core material collected with a borer. This indicates that it would be more beneficial to the trees being investigated to remove branch samples compared to core samples, and the application of this approach to understanding groundwater contaminant behavior is discussed in Chap. 15. The similarity of results between branch and core samples should be confirmed in the field for each site, however.

7.3.2 Soil Physical Composition

In general terms, soil is the byproduct of rock weathering. This production of soil can be caused by abiotic processes, such as infiltrating or running water, freeze-and-thaw cycles, earthquakes, or by biological processes, such as microbial-nutrient acquisition, lichen, and root penetration. These processes make the minerals and elements that are contained in the original rock more bioavailable. Once mobilized by interaction with water, these nutrients are used by plants growing in the soil. It is the porosity of weathered rock and soil that aids in plant establishment by providing stability and a reservoir to hold water. Soil also is used by plants to keep the roots protected from the sun's radiant energy and to keep them from drying out.

Soil formation can take many hundreds of thousands of years. As plant succession occurs, lichens are replaced by vascular plants, and the thickness of organic matter on top of the weathered rock increases. Because of this process, soils typically are discussed in terms of profiles, or the characteristics of each vertical layer from the surface downward. The uppermost layer, called the O layer, or horizon, is high in percent organic matter, such as humus and detritus of dead plants. Beneath the O layer is the A layer, which can contain some organic matter along with the sediments, and in which infiltration can remove soluble minerals from the sediment, called the E layer, and subsequent mineral deposition in the B layer. Below the B layer is the parent or host bedrock. Plant roots can be present in all soil layers as long as water, oxygen, and nutrients are not limiting. Roots may have a hard time penetrating through the B layer where precipitates of iron oxides, called hardpan, can accumulate.

Anyone who has planted a tree realizes that the physical structure of the soil in the upper layers has a direct effect on the installation and growth of the selected plant. In the Coastal Plain areas of the eastern United States, the surficial sediments range from unconsolidated sands to impenetrable clays. In piedmont areas of the same region, brittle clay-rich weathered bedrock, called regolith or saprolite, is present on top of parent bedrock though alluvial sediments have been deposited in some intervalley areas. Bedrock is available to plant roots only through fractures that create secondary porosity. Even in unconsolidated sediments, soils made tight by natural processes, such as weathering or cementation or compaction, can prohibit the entry of water and air that are necessary for plant survival. This only can be overcome through the introduction of events that lead to secondary porosity, such as fracturing.

The common denominator that helps determine the viability of a particular soil for plant growth is its porosity and bulk density, as these parameters are closely related to water movement and storage. Bulk density is the dry weight of a soil sample per unit volume. The bulk density of a soil can, therefore, increase if volume decreases, such as by compaction. Soil bulk density, typically reported in grams per cubic centimeter (g/cm^3), is indirectly related to porosity. The average specific gravity of soil is 2.65 g/cm^3, but because soils also contain organic matter, air, and water, the term bulk density is used to quantify the weight per unit volume of sediment. A bulk density lower than 2.65 g/cm^3 in an oven-dried sample indicates the presence of pores, which can be filled with water or air.

As soil bulk density increases, the potential for root penetration into the soil decreases. Knowledge of the distribution of soil bulk density with depth, therefore, can provide information about potential restrictions to the development of deep root systems and groundwater interaction (Liang et al. 1999). Bulk densities should be less than 1.4 g/cm^3, as these sediment types contain pores large enough to hold water for plant use. These pores are needed not only to transmit and store water but also to allow aeration of the soil. Drainage is necessary to move new water in and old water out. As we saw in Chap. 3, plants usually die not because they are overwatered and cannot eliminate the excess water but because the low solubility of oxygen in water leads to root death after cessation of root aerobic respiration.

The aeration capacity of a soil can be examined during core collection or digging a hole as part of site-assessment and characterization activities. In general, the red color characteristic of oxidized iron minerals in the soil profile indicates that oxygen has been able to reach that depth, at least at some recent time in the past. Conversely, a gray color indicates the absence of oxygen, and the transition between red and grey delineates the maximum depth of oxygen diffusion. This also tends to indicate the seasonal high-water table, especially in low-permeability soils. These techniques are used widely by those in state agencies who permit septic systems, where the depth to the water table is required to be a certain distance below a drain field to ensure that the groundwater will not become contaminated.

One indication of the depth of the water table that can be deduced at some sites where the water table is shallow is the presence of evidence of previous or ongoing oxidation-reduction cycles in the pore water and sediments. If the pore water in the unsaturated zone or groundwater is anoxic, any dissolved iron present in solution will precipitate from the water if exposed to oxygen from the unsaturated zone or to oxygen dissolved in infiltrating precipitation. The conversion of ferrous to ferric iron is an abiotic reaction that occurs spontaneously in the presence of oxygen, but iron-oxidizing bacteria also can produce ferric iron from ferrous iron. This typically occurs after a period of precipitation has ended, no precipitation occurs for some time, and the water table drops. Evapotranspiration that leads to a decline in the water table also causes this process to occur. Conversely, when the water table rises again, oxidized iron will redissolve back into the pore water or groundwater if these water sources are anoxic, and iron-reducing bacteria are present along with reduced labile organic matter. In some cases, alternating layers of red and grey soil will be found as soil is brought to the land surface with a hand auger. In other cases, at sites where this has occurred for a long time, as in wetland areas where groundwater is naturally anoxic, the precipitation of iron leads to the formation of a hardpan deposit, where a layer of the sediment is encountered in which precipitated iron has cemented the sediment together.

Another factor that can decrease the amount of oxygen available to roots and, therefore, affect plant health is the amount of organic matter in the soil and its bioavailability. Sandy soils that have at least 3% organic matter generally are considered to be the best to impart drainage, aeration, and moisture retention. Greater than 3% organic matter can lead to water repellency. This is because at a molecular level, organic matter has functional groups that are hydrophobic and repel the absorption of polar water molecules. Water droplets first must be broken up by the attraction of water molecules to soil particles, and if these attractions are inhibited by the presence of organic matter, the water droplet remains intact as a continuous film on top of the soil particles and does not infiltrate (Dekker 1998). Moreover, retardation of xenobiotics onto the soil organic-matter fraction tends to decrease contaminant bioavailability. In fact, residual weathered petroleum-hydrocarbon contamination has been shown over time to become integrated into the soil O horizon (Guthrie et al. 1999).

These factors of bulk density, porosity, and organic-matter content represent part of the soil physical composition. A variety of options are available to understand the soil physical composition at a site that is a candidate for phytoremediation and to determine if any pre-existing condition needs to be addressed. Hard copies or digital files of county soil maps can provide an initial generalized description of the surficial soils at a site. These may be less useful, however, in terrains where great heterogeneity exists over only a short vertical distance, such as in glaciated terrain. At the site, soil samples can be collected in various areas, mixed, and analyzed as a composite sample for properties or characteristics such as bulk density, porosity, organic matter, and grain-size analysis. To determine the ease at which water will infiltrate through these sediments, samples can be sent to the laboratory for falling-head permeameter tests. In the field, infiltrometer tests can be done to determine infiltration potential.

The presence of undesirable physical soil characteristics can be dealt with by using a variety of approaches. Poor soils can be removed and replaced by higher quality soils, or removed and amended with more desirable soils, or left in place for the plants to be installed with a more desirable, loamy material as backfill. Tight soils that cannot be removed can be tilled or ripped with mechanical equipment. An increase in soil permeability before planting is essential, even if deep-planting methods have to be used, to ensure contact with the water table at depth. Over time, however, plants naturally increase soil porosity and permeability even in tight soils, as roots grow, die-back, and slough off as they explore the subsurface. Moreover, this growth over time helps to establish a rhizosphere where perhaps none before existed.

In combination with soil porosity and bulk density, soil water potentials determine whether or not water is available for plants to remove. In clay-rich sediments, for example, there may be a higher amount of void space than in sand, and porosities can approach 50–70%. However, clay particles have a larger surface area than do sand grains of similar porosity and, therefore, hold water more tightly under higher tensions, so the water potential is more negative than that for sand. In other words, clay soils may contain more water than sandy soils, but it may not be bioavailable.

Individual root hairs seek water almost on the molecular level and, therefore, will follow the path of least resistance, i.e., through the most permeable sediments. As water enters the root hairs by diffusion and osmosis and is depleted in the soil, additional water must be acquired. The root hairs either grow toward a new source of water; or water will be supplied by infiltration, groundwater flow, or capillary movement; or the root hairs will die. This indicates the relation between the presence of roots and in zones of higher hydraulic conductivity. This further emphasizes the importance of the multidisciplinary approach to site assessment and characterization that was advocated in Chap. 6.

7.3.3 Soil Chemical Composition

Ashes to ashes, dust to dust.

The Book of Common Prayer (1979)

There was an old belief that in the embers of all things their primordial form exists, and cunning alchemists could re-create the rose with all its members from its own ashes…

H.W. Longfellow, Flower-de-Luce, Palingenesis (McClatchy 2000)

Nutrients that are locked up in the non-living inorganic material or previously living organic material of soils are made available to plants by soil microorganisms. Plants require almost 20 trace elements and nutrients to ensure successful growth. This coupling between soil and plant composition was revealed, in part, by the combustion of plant matter and the analysis of the material that remained or the gases emitted. The gases released reveal that the biomass we recognize as being part of plants, such as wood, bark, leaves, stems, etc., are almost entirely derived from the inorganic gas CO_2 (this is the missing information that led J.B. van Helmont, as described in Chap. 1, to state that only water was needed to make plant biomass). The gaseous composition makes sense, because photosynthesis requires CO_2 along with water—plants, after all, make their own food, so there should be little else that they require, correct? The chemical composition of the ash tells a different story, however, such that more than CO_2 and water are necessary to sustain plant life. And this includes plants that will be used for phytoremediation of contaminated groundwater.

As stated, the analysis of the ash leftover from plant combustion can reveal much about what else plants need to survive. At least 13 elements are considered essential to plant health; nitrogen (in the form of the oxidized anion NO_3^- or the reduced cation NH_4^+), phosphate (as the anions $H_2PO_4^-$, HPO_4^{2-}, or PO_4^{3-}), and potassium (K^+). These elements are the big 3 that compose most lawn and garden fertilizers. With the introduction of the Haber process in the early 1900s, nitrogen from air could be combined with hydrogen from coal oxidation to produce ammonia (NH_2). This process made access to this essential plant nutrient widespread and affordable and contributed to widespread use of such inorganic sources of nitrogen in commercial fertilizers rather than manure. Since then, most of the plant proteins, which contain nitrogen, that have been ingested by man have been derived from manmade ammonia. Plants also require calcium (Ca^{2+}), magnesium (Mg^{2+}), and sulfur (as the oxidized anion SO_4^{2-}). Because each of these compounds constitutes more than 1% of the dry organic weight of the plant, they are called *macronutrients*.

Each of these macronutrients is used for different processes in which the whole usually is greater than the sum of the parts. Nitrogen is used to synthesize proteins and co-enzymes. Phosphorus, typically taken up as a phosphate salt, is essential in conversion of ADP to ATP. Plants absorb phosphorus not as elemental phosphate but as inorganic phosphorus. Calcium is used to maintain cell membranes. Magnesium is in the heme group of chlorophyll *a*. Potassium helps maintain the process of osmosis discussed in Chap. 3 with regard to water uptake by root cells. Potassium in root-cell cytoplasm and vacuoles reduces the water potential, or concentration, in the cell and, thus, sets up conditions for the passive entry of water.

The elements present in plants at less than 1% of dry organic weight are called *micronutrients* and include iron (as Fe^{2+} or complexes of Fe^{3+}), manganese (Mn^{2+}), zinc (Zn^{2+}), boron (as BO_3^{3-} or $B_4O_7^{2-}$), copper (Cu^{+} or Cu^{2+}), and molybdenum (as MoO_4^{2-}). These micronutrients all are essential to plant life; zinc helps transfer phosphorus, boron helps during synthesis of the growth hormone auxin, and iron and manganese act as catalysts in many reactions, which change the rate of reactions without themselves being used up. Although iron is an essential micronutrient, too much iron can be toxic to plants, and amounts greater than 400 mg Fe/kg dry plant weight (equivalent to ppm) cause toxic effects. The role of some of these macro- and micro-nutrients is discussed in Chap. 11, with specific emphasis on their role in the phytoremediation of contaminated groundwater.

Even elements considered unessential and toxic to plants can be useful. For example, sodium and chloride can upset the osmotic gradient of the plant. But sodium and chloride are necessary for plant survival. In arid areas where recent recharge is essentially absent and the depth to the water table is on the order of hundreds of feet (tens of meters), groundwater can rise toward the land surface in response to lowered water potentials near the soil surface caused by evaporation and transpiration (Andraski et al. 2003). Much like the salt residue left in a pot after water has been boiled away, in arid parts of the United States, as the water near the land surface evaporates, salts do not evaporate and can form a crust on the surface soils at percent concentrations. This can be toxic to plants or to animals that eat the plants. For example, cows in California can become poisoned if allowed to eat grasses that contain high levels of selenium, a salt that is enriched in surface soils as a result of evapotranspiration of shallow groundwater. Saline soils have been operationally defined as having more than 4 decisiemens per meter (or 4 millisiemens per centimeter). On the other hand, some plants, such as *Spartina spp.* and tamarisk, can tolerate higher levels of salt, because they can keep the salt content of their cells higher than the salt content of water, even seawater; thus, the water content in the cells is lower than the water content in saltwater. This tolerance explains why these plants dominate the estuaries of the east coast and many riparian areas of the United States, respectively.

An interesting relation between root density of plants that are intolerant to salt and salinity levels in soils and subsurface water suggests that, in some cases, the salinity profile can change over time in response to root growth. At a site in western Australia, researchers removed cores alongside eucalyptus trees that were planted to lower a high water table composed of saline water (Rural Industries Research and Development Corporation 2000; Fig. 7.7). From observation of the cores, root density was greater near the surface (20–50 m/dm^3), decreased to less than 5 m/dm^3 3 ft (1 m) below land surface but increased between 6 and 21 ft (2 and 7 m). Chloride concentrations also were higher between 6 and 24 ft (2 and 8 m). These relations between depth, root density, and chloride were not constant but changed throughout the year in response to changes in soil moisture and the water-table elevation (Fig. 7.8).

The overall conclusion is that tree roots intersected the water table and took up groundwater, which tended to concentrate chloride in the area where the root density was the greatest. This emphasizes the advantage of transpiration by plants such that dilute elements are concentrated near their roots. Hence, the highest chloride concentrations in a vertical profile may indicate the recent (rather than past) location of water uptake by the trees. Moreover, live roots were observed to be growing below the water table, which the authors correctly assumed meant the groundwater contained dissolved oxygen which they, unfortunately, did not measure.

There are other processes that involve plants, groundwater, and soil salinization. The removal of forests during clearcutting for lumber or pulp production can result in an increase in net recharge to aquifers that previously had deep water tables as a result of removal by tree transpiration. Clearcutting of forests results in an increase in recharge, and the water table rises near land surface. Thus, mineralized groundwater can be influenced further by evaporation, which leaves behind salt in the soil profile. Subsequent irrigation of the salt-enriched soils can lead to soil salinization, especially if the source of the irrigation water is the mineralized groundwater, or if irrigation can lead to artificial recharge of the water table causing it to rise and be further

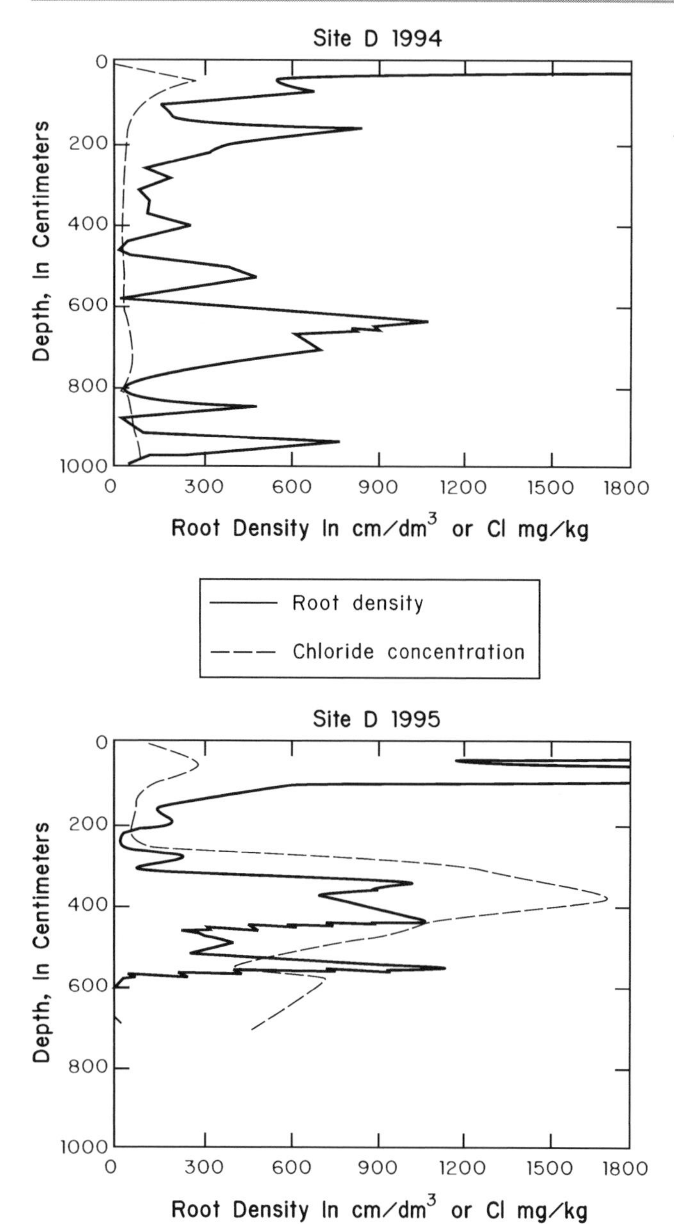

Fig. 7.7 Root density can change with depth over time and, in this case, affect the distribution of chloride in soil pore water (Modified from Rural Industries Research and Development Corporation 2000). One centimeter is equivalent to 0.39 in.

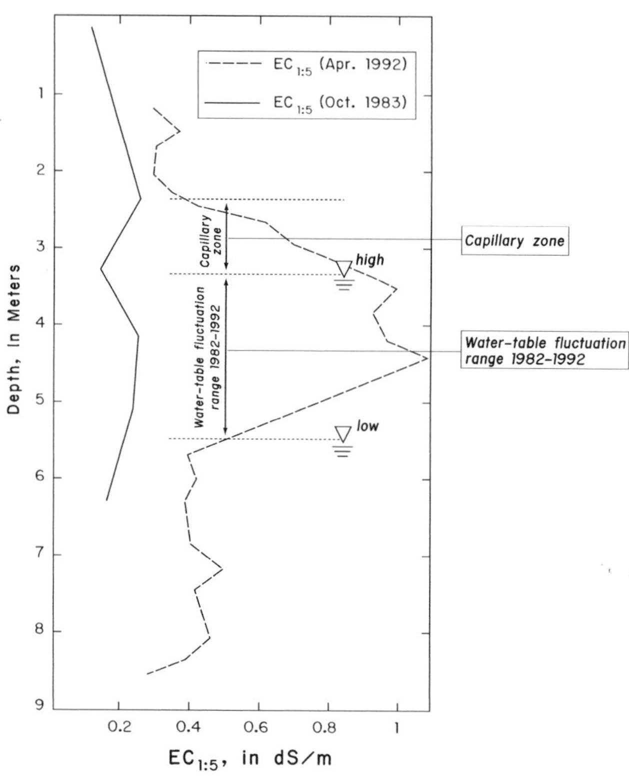

Fig. 7.8 The changing position of the capillary fringe and water table and effect on salinity, as reported as electrical conductivity (EC). When the water table was higher, salinity was lower (1983), but increased as the water table dropped (1992) (Modified from Rural Industries Research and Development Corporation 2000). One meter is equivalent to 3.2 ft.

affected by evaporation. In the United States, this process is called saline seep and is a common problem for western States like Montana (Jeffery Kuhn, Montana Department of Environmental Quality, written commun. 2008).

At a potential phytoremediation site, knowledge of the cation-exchange capacity of the soil, either under native or amended conditions, is important for plant health and should be measured. The cation-exchange capacity is the capacity for soil to contain an element of net positive charge, or cation, that can be reversibly replaced with a different element, also of net positive charge. This can occur in the upper soil layers. For example, hard water contains high concentrations of the cation calcium, which is considered undesirable to humans for aesthetic reasons. Hard water is rendered soft in many homes around the world by replacing the calcium with sodium. With reference to plants, J. Liebig stated in the early 1840s that the yield of any crop is limited by the minimum amount of any essential nutrient, which is called the law of the minimum (Miller 1938). Some information regarding the cation-exchange capacity of an area can be estimated from knowledge of the geologic history of the area. The geologic conditions of the region ultimately determine the presence or absence of these minerals.

The bioavailability of these minerals to plant roots depends on the balance of acids and bases in the soil profile, or pH, which is the concentration of the activity of the hydrogen ion. Most plants that can be used for phytoremediation purposes do not thrive in very acidic soils or very basic soils, even when cuttings or large plants are used. The ability of the soil to moderate changes in the acid or base content, such as occurs during the infiltration of low pH precipitation, is an indication of the buffering capacity of the soil. The soil pH can be improved by adding materials such as lime, a calcium oxide, CaO, formed when seashells or limestone, $CaCO_3$, are heated to 900°C. Lime addition acts as a soil buffer in that it can liberate hydroxyl ions as

$$CaO_{(s)} + H_2O_{(l)} \rightarrow Ca^{2+}{}_{(aq)} + 2OH^{-}{}_{(aq)} \quad (7.1)$$

Site-assessment and characterization activities should include a determination of the chemical composition of the soil profile before trees are planted. This soil chemistry can be evaluated by using field test kits or having soil samples analyzed in a laboratory. Local county or state extension offices generally provide an excellent starting point for this type of information. A good thing about site-assessment and characterization activities is that deficiencies in native soil can be overcome, although at some cost to the project.

Fertilizer also can be added to the soil during the initial planting or after the plants have become established. Addition prior to or during planting can occur either as an amendment to the soil used for backfill, or as the complete removal of old soil with replacement by an amended top dressing or fill. Addition to established trees is more difficult, but can be accomplished with a drop spreader of dry fertilizer or by injection of liquid fertilizer near roots. The fertilizer can be added at the beginning of the dormant season to stimulate root growth or at the end of the dormant season to encourage shoot growth. For cuttings the fertilizer is added to the backfill placed near the plant, but for an established tree the fertilizer is added away from the main trunk to near the drip edge of the plant's lower branch extent. Holes can be created with a length of steel rod, filled with fertilizer, and then water and soil. For small trees, such as 6 in. (15 cm) or less in diameter, the rate of application using this method is about 2–3 lb/in. of diameter. For larger trees, the dosage is increased to 3–5 lbs/in. of diameter. This rate is for a balanced fertilizer for gardens and trees, such as 10-10-10 (N-P-K), rather than the unbalanced lawn fertilizers that are high in nitrogen only and commonly applied to lawns and stimulate shoot growth at the expense of root growth. Moreover, some types of fertilizer can be injected directly into the tree, but as this wounds the plant, it is best done only as a last resort.

One of the indications of overall plant health is height relative to genetic potential. The relation between environmental factors and tree height has been examined for many years. The controlling factors that have been offered include resource availability, reduction in stress, and access to light. One of the more interesting factors is access to water and the ability to transport water to great heights above land surface where maximum light levels are also found. Koch et al. (2004) examined the water-transport characteristics of coast redwoods (*Sequoia sempervirens*) in relation to maximum height relative to physical constraints on water transport. They found that the maximum observed height of such trees of 410 ft (125 m) is a function of the maximum pull of water against gravity and the frictional resistance to additional flow provided by the xylem surfaces.

7.3.4 Soil Moisture Composition

The necessary reactants plants need to survive are in limited supply. This includes CO_2 and nitrogen above ground, and water and nitrogen below ground. In part, the limitations below ground are overcome by the movement of water through plants as part of the hydrologic cycle. The flow of water couples the plant's requirement for water with nutrient acquisition from dilute sources in the subsurface. Notice that the terms flow and movement were used in reference to subsurface water. This is because the presence of water in soils or sediments as soil moisture does not always mean that it is available to plants. Water held by soil particles under tensions higher than can be removed by plant root-hair cells results in the absence of water uptake.

This does not mean that soil moisture should not be measured as part of site-assessment and characterization. Soils that have too low ($<10\%$) or too high (100%) soil moisture can be detrimental to some plants. At sites where phytoremediation is being considered, low soil-moisture levels from little annual precipitation could detrimentally affect plants installed in that water may be limited prior to the ability of the roots to penetrate the water table. Under such circumstances, irrigation systems can be installed to alleviate the water limitations until root growth is established. Irrigation would be warranted if soil moisture fell below 80% of field capacity, or when the soil water reached tensions between -0.05 and -0.1 MPa. One way to determine the magnitude of low soil moisture and its effect on plant uptake is to use a nest of tensiometers to provide the necessary soil-water tension data for deciding whether or not to irrigate—some automated systems can be used to remotely control irrigation.

The irrigation system designed to overcome low soil-moisture conditions can be as simple as a water hookup and sprinkler or as complex as a built-in drip irrigation system. A drip irrigation system used in this context would provide a uniformly wet soil horizon rather than a means of simply limiting evaporation. Short, frequent irrigation could result in healthy growth but produce a shallow root system. Less frequent but longer irrigation periods, about 1 in./h, would be sufficient to deeply wet many surficial soils to encourage the roots to grow to deeper soil layers. Some researchers who installed irrigation systems observed that the roots of planted poplar trees tended to remain shallow and used the irrigation water rather than groundwater (Van Epps 2006). Some phytoremediation sites where poplar cuttings are installed in trenches also have irrigation pipes installed in the trench before backfilling.

An irrigation system similar to the type installed in many residential areas could be adequate for a phytoremediation system, with a water-supply line going through a manifold that then supplies individual drip-irrigation lines. The whole process can be automated such that a control box turns separate solenoids on and off in the manifold that controls the water flow to individual pipes. Filters also should be installed before the first series of drip-irrigation emitters, to prevent small particles from clogging the lines. Although this can help to reduce the chance that the emitters will become clogged, there is a possibility for the emitters to become clogged by algal growth or other bacterial growth, especially if concentrations of dissolved iron in the supply water are high. These problems can be dealt with by a periodic flushing with a low concentration of bleach, but this will need to be done for the life of the project or until the source of dissolved iron is removed.

Low soil-moisture conditions, however, do not always require the installation of an irrigation system to achieve positive results. A phytoremediation system of poplar cuttings planted in November 1998 in Charleston, South Carolina did not have an irrigation system, which was a concern as the planting coincided with the beginning of a 5-year drought in the study area. A deep root system was established, however, because the 6-ft (1.8 m) cuttings planted had been deeply installed near the water table at 3-4 ft (0.9–1.2 m) below land. Under natural conditions in a riparian area in Arizona, McQueen and Miller (1972) reported that saltcedar and willow thrived during drought conditions because the roots tapped the water table.

Soil moisture levels that are high are not always detrimental to plant health. Certain plants, albeit not ones used for phytoremediation such as cultivated rice, require continual flooded conditions. Other plants, such as those that characterize swampy areas or parts of flood plains that are inundated most of the season, also require water-saturated conditions. A key point is that the water remains flowing and does not become stagnant.

The type of soil present in the root zone of plants has a major influence on the relation between total soil moisture and total water potential, or the difference between the presence of water and its potential for movement into plant roots. For example, the wilting point, defined as a soil-water potential of -1.5 MPa, occurs in clays at a higher moisture content, about 15%, and at a lower moisture content of 2.5% for more permeable sands. This is because more air can penetrate sand than clay because of the higher degree of interconnected pores even though clay has a higher porosity than sand. The soil-moisture content also affects the viability of rhizosphere microbes (Cho et al. 2005), as these microorganisms also require water for survival.

Soil moisture often plays a critical role in the root density of many plants. Typically, as the depth below land surface increases, root density decreases. This holds not only for most plants but for obligate and facultative phreatophytes as well. Lower root density is compensated, however, by fewer deep roots characterized by higher root-hydraulic conductivities. This explains how water can be allocated to plants to support transpiration even when water potentials in shallow soils are more negative than the wilting point (Teuling et al. 2006).

7.3.5 Soil Topography

As described in Chap. 3, the topographic character of a site is a factor that can control soil-moisture conditions and should be evaluated as part of site-assessment and characterization activities. In general, water tends to collect at or near land surface in lower elevations. A site that has too steep of a land-surface gradient may not be a candidate for phytoremediation. This caution has less to do with plants not being able to grow but that either the depth to water table will be too great or the hydraulic gradient too steep and, therefore, rate of groundwater flow too high for measurable hydrologic control by plants.

7.3.6 Depth to Water Table

Not surprisingly, the depth to the water table is perhaps the most important factor in determining the *a priori* success of a phytoremediation system designed to control contaminated groundwater, as has been discussed. Although the term depth indicates some constant level where the water table is encountered, the position of the water table is not a constant but fluctuates around an average value (Holzer 2010). Many sites that need to be remediated are contaminated because a shallow depth to water table resulted in a lack of contaminant attenuation or because the source of the contamination was only slightly above or at the water table. For example, some USTs or pipelines are installed only slightly above the water table and are inundated during seasonal increases in the water table, thereby providing a direct conduit for contamination of groundwater if leaks are present.

The depth to water table and its effect on plant growth have been studied by agronomists for some time. Many shallow-rooted crops can be grown without irrigation or with supplemental irrigation all over the world in areas where the water table is shallow. Such crops include cotton, alfalfa, and barley. As the depth to the water table increases,

Table 7.4 Relation between plant type and depth to the water table (Modified from Meinzer 1927).

Plant	Depth to water table, in feet
Rushes and sedges	Water at surface or water table within a few feet.
Giant reed grass	Water at surface or within 1–8 ft.
Giant wild rye	Water near surface to 12 ft or more in arid areas, less relation in humid areas.
Salt grass	Near surface to 12 ft.
Arrow weed	Water at surface, or from 10 to 25 ft.
Willow	From surface to 12 or more ft.
Palm tree	Water within a few ft.
Greasewood	Water table from 3 to 40 ft.
Mesquite	From 10 to 50 ft.

however, the percentage of total plant transpiration from groundwater decreases (Ayers et al. 1999). In fact, this relation was stated by Grismer and Gates (1988) as

$$Q = G/ET = a - bD \tag{7.2}$$

where Q is described as the ratio of the amount of water supplied by groundwater, G, to total evapotranspiration, ET; a and b are constants that relate to the hydraulic properties of the soil and range from 0.7 to 0.36 and 0.20 to 0.17, respectively; and D is depth.

The depth to water table can vary many feet even in roughly the same region, so it is not surprising that the lengths of roots of certain plants that use groundwater also vary in relation to the depth to water table. For example, desert saltgrass (*Distichlis stricta*) can be found where the water table is no more than 12 ft (3.6 m) from land surface, whereas alfalfa (*Medicago sativa*), another member of the grass family, can have roots penetrate the subsurface to depths greater than 60 ft (18 m; Robinson 1958). It must be remembered that even deep-rooted plants, such as alfalfa, cottonwood, or willow, also can grow in areas where the water table is much shallower. However, in areas where only stands of alfalfa predominate at the expense of other plants that use groundwater, it is most probable that the explanation is the greater depth to groundwater.

In the early 1920s, O.E. Meinzer of the USGS made an intensive survey of the groundwater supplies of Sulphur Spring Valley, AZ, the Tularosa Basin, NM, and Big Smoky Valley, NV. As part of his investigations, he noted the relation between the depth to water table and the type of plant that predominated, as was introduced in Chap. 1. Although the reader is referred to Meinzer (1927) to see his entire list, he made some general relations as noted in Table 7.4 that have some applicability to phytoremediation of groundwater.

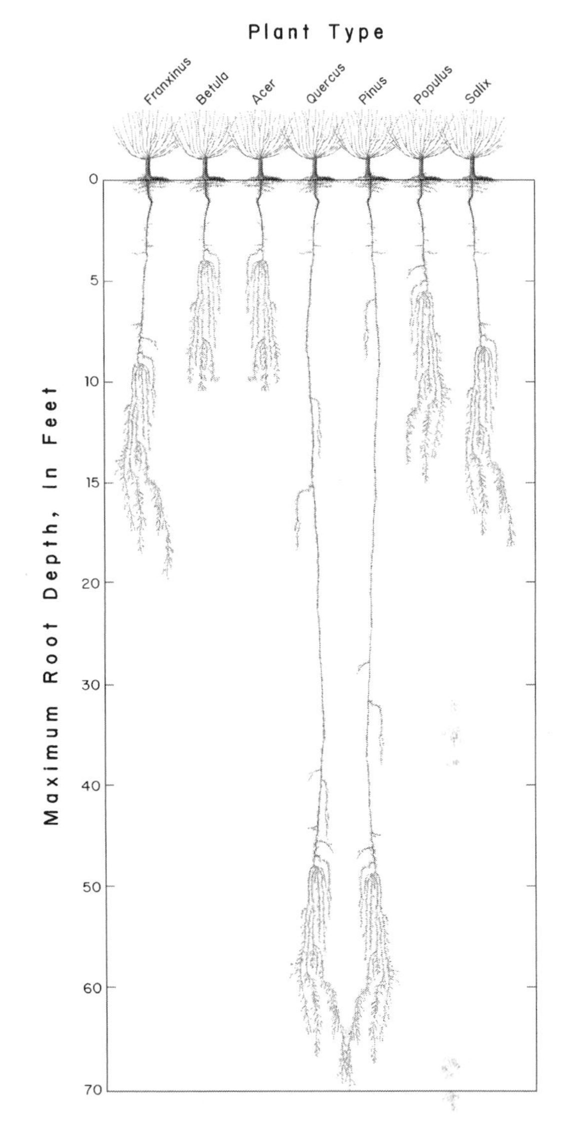

Fig. 7.9 A generalized comparison of the average maximum observed rooting depths, in feet, of phreatophytes commonly installed for phytoremediation, such as *Populus*, *Salix*, and *Betula*. These depths can be exceeded by native plants of *Quercus* and *Pinus* at some sites, but the total transpiration rate and, therefore, groundwater use is much lower than the shallow-rooted plants, which makes them less effective for use in phytoremediation of contaminated groundwater. One foot is equivalent to 0.304 m.

As stated previously, most sites where phytoremediation can be applied successfully typically have water tables between 5 and 15 ft (1.5–4.5 m) below land surface (Fig. 7.9). Water becomes more bioavailable closer to the water table, as water potential is the least negative.

That the depth to groundwater affects phreatophytes and the water budget with respect to groundwater can be observed in areas that are characterized by riparian ecosystems. In Chap. 5, native and invasive woody vegetation in the riparian zones of rivers in the American Southwest were shown to use groundwater. Many of the naturally flowing streams have been dammed or impounded in the last

100 years. This change in hydrology has affected riparian phreatophytes by changing the depth to groundwater. Horton et al. (2001) investigated the effect of various depths to groundwater on the physiology of riparian trees, such as poplars, willows, and tamarisk. One of their observations was an apparent linear relation between the decrease of groundwater use by the trees as the water table declines until it reaches a critical depth after which groundwater use per unit increase of water-table depth declines exponentially.

A water-based optimal root depth was determined for plants based on the carbon costs of deep roots relative to their use for water removal (Guswa 2010). In this investigation, deeper roots were related to areas where the precipitation amounts were roughly equal to ET_p. In areas where precipitation exceeds ET_p there is no benefit to a plant to support deeper roots and, conversely, in areas where precipitation is less than ET_p deep roots often are not found in all instances if groundwater is not available.

An interesting relation was observed between the depth of root penetration to the water table and how slowly the water table declined. Fenner et al. (1985) investigated the presence of root elongation in cottonwood trees in relation to groundwater levels that were regulated to some extent by regulated surface-water flows. They observed significant root elongation in the riparian cottonwoods if the water-table decline was on the order to 0.83 to 1.1 in./day (2 to 3 cm/day). Hence, fluctuations in the water table tended to lead to a more extensive penetration of deeper soil.

If the water table fluctuates, the rate of fluctuation may have a large effect on the survival of phreatophytes. If the rate of water-table decline is greater than the rate of terminal root growth and root-hair development, the plant may undergo water stress. This is particularly true in coarse-grained sediments that have a thinner capillary fringe relative to in fine-grained sediments. The highest density of root hairs of phreatophytes is concentrated in the capillary fringe. If the water table declines slowly, root-hair growth can occur at a similar rate, and the plant will not be water stressed. Conversely, if the rate of water-table decline is faster than the rate of root-hair growth, such as rapid drops that approach 3 ft (0.9 m), the tree will be water stressed and may die (Scott et al. 1999; Shafroth et al. 2000).

Conversely, a high water table can be detrimental or beneficial to a phytoremediation planting. It can be detrimental if the water has such a low flow rate or is in sediments that contain either organic or inorganic compounds that interact with and deplete oxygen to the point that root survival is diminished for most plants that are not adapted to such saturated conditions. If the water is oxygenated, however, the needs of water for transpiration, photosynthesis, and oxygen for root respiration are satisfied. If the water table is so high that anoxic conditions occur for long periods and the water becomes stagnant, the roots cannot respire and will die from a lack of energy. Plants, such as the baldcypress and tupelo present in swamps of the Southeastern United States, survive, in part, because the surface water in which they stand is not stagnant but is continually being exchanged by evaporation, transpiration, and groundwater discharge.

In studying the relation between transpiration by native cottonwood (*Populus deltoides*) and invasive saltcedar (*Tamarix chinensis*) along the Middle Rio Grande River, NM, Cleverly et al. (2006) related the depth to groundwater to transpiration, *LAI*, and groundwater levels during drought conditions. For the poplars, drought conditions did not affect evapotranspiration rates and the depth to water table remained fairly static at 9 ft (2.7 m) below land. For the saltcedar trees, however, evapotranspiration increased and the depth to water table decreased during the same period. The evapotranspiration increased from 0.23 to 0.35 in./day (6 to 9 mm/day) as the water table decreased about 0.2 in./day (7 mm/day), which indicates the decrease was caused by the increase in evapotranspiration. It is possible that the higher evapotranspiration rate measured for the saltcedar is driven to a greater extent by the atmospheric conditions than depth to water table.

Water tables that are within a few feet of land surface are more apt to receive recharge during precipitation events than deeper water tables that undergo the same amount of precipitation. This creates a unique scenario for a phytoremediation application. Even if the site is characterized by a relatively high ET_p that is near precipitation amounts, the water table will not trend toward ever lower depths, especially if precipitation and recharge are frequent. At sites with shallow water tables, a higher percentage of the precipitation becomes groundwater than at sites with deeper water tables, where less of precipitation reaches the water table. In the first case the water table actually may rise, as it is being replenished by recharge at a rate faster than it is taken up by trees. In the other case, the trees continually take up the groundwater and intercept infiltration in the unsaturated zone, and the groundwater level steadily declines.

The relation between the depth to water table and transpiration by phreatophytes was investigated by Gazal et al. (2006). In general, they found that as the depth to water table increased from 3.2 to 13 ft (1 to 4 m), the measured transpiration rate decreased from about 0.2 to 0.07 in./day (0.5 to 2 mm/day).

7.3.7 Semiconfined to Confined Groundwater Conditions

At some sites the groundwater to be assessed and characterized is under semiconfined to confined conditions.

Phytoremediation can occur under these conditions by using engineered means to have the roots reach the deeper confined aquifers. Boreholes can be advanced and planted with large cuttings, or poles, while the holes are backfilled with porous material. A case study where this process occurred is discussed in Chap. 13. For groundwater under semiconfined conditions, such a plant design and installation encourages upward flow to the roots by capillary action.

7.3.8 Groundwater Geochemistry

The ambient geochemistry of groundwater, in terms of the concentrations of dissolved solutes acquired along the flow path by interactions with the porous media, can influence the determination of the types of phreatophytes that grow in an area. This is important to consider during site-assessment and characterization, because simply determining the presence and depth of groundwater does not necessarily infer that plant growth can be sustained. For example, cottonwood and willow typically are not found in areas characterized by high concentrations of dissolved salts. However, certain hybrid poplar trees, such as the OP-367, can live in soils where the pore water is characterized by high salinities. Some native plants that can tolerate high salinity groundwater include greasewood, saltcedar, and pickleweed (*Allenrolfia ocidentalis*), although these plants rarely are used in phytoremediation applications.

Iron probably is just as important an element to measure in groundwater as are sodium and chloride. Although iron is a micronutrient for plants, at high concentrations it can be toxic. Iron concentrations in plant tissues can accumulate to between 400 and 1,000 mg Fe/kg plant tissue, but these higher concentrations can decrease plant health. Many aquifers that contain either naturally high concentrations of labile organic matter, such as aquifers that underlie swamps or peat lands, or aquifers that contain high concentrations of petroleum hydrocarbons from gasoline releases often have high dissolved iron concentrations from iron reduction under the prevailing anoxic subsurface conditions. Therefore, these indicators of higher iron concentrations should be assessed as part of site-assessment and characterization—many test kits that provide iron concentrations in the field are commercially available.

The classifications of plant type and groundwater geochemistry indicates that on some level interspecies competition for a limited resource occurs relative to subsurface moisture and groundwater, which produces a generalized differentiation of plants that use groundwater of different geochemistry. O.E. Meinzer's study of the three basins of the southwestern United States (described in Chap. 1) provided the opportunity to collect and analyze groundwater geochemical data from areas that had phreatophytes, existing wells, or boreholes that could be easily dug. Some of the conclusions he drew from the groundwater geochemical and plant-distribution data suggested that phreatophytes could exist if the mineral composition in the upper part of the water table ranged from low (TDS $<$ 1,000 ppm) to high (TDS $>$ 40,000 ppm) (Meinzer 1927).

7.3.9 Plant Ecological Conditions

Plants that rely on the water table for moisture are the ones that were successful in reaching the water table and in reproducing. These are the hardiest of each species, because before any growth from interacting with groundwater can occur, the seedlings must endure a time following deposition in which the water table is beyond the reach of the rootlet—this is natural selection at its most obvious. Also, because the air is dry in the arid western United States, the plants must not only reach the water table rapidly but also have the capability to use large quantities along the prevailing humidity gradient.

It is a common observation that poplar trees are hard to kill by even cutting the trunk at ground level. This is because within a few weeks during the growing season, numerous saplings will sprout up from the cambium all around the circumference of the cut tree. A unique feature of poplar trees is that they can produce new aboveground growth even after complete trunk removal to ground surface, or coppice. Upon reflection, this manner of survival makes sense for phreatophytes, because it allows the tree to continue life without having to expend energy to make and release seeds whose roots, following germination, may not reach the water table.

The relation of tree health to the ideal spacing interval between plants at a potential phytoremediation site raises many concerns. Close spacing typically is encouraged for the maximum removal of groundwater or prevention of recharge while maintaining individual tree health. The ultimate control variable on the minimal plant spacing to use should be the factor that limits plant transpiration. At most sites, this is the amount of solar energy input to an area that can be used to evaporate water. If water is not limiting, as can be assumed to be the case once trees reach the capillary fringe, then the factors that maximize evaporation should be enhanced. It has been shown that water use on a per-tree basis is higher in open forests relative to trees in a dense forest (Stewart 1984). Plants grown too close together have to compete for limited soil moisture and nutrients and often appear stunted. The upper Coastal Plain forests along the eastern seaboard of the United States provide an example of this scenario. The point is that larger trees have more leaves per a given area and, therefore, can potentially transpire more water and groundwater relative to areas of bare soil that are not planted.

Leaf size also is important in considering the closest tree spacing necessary to support hydrologic goals and protect plant health. Most trees have differences in leaves even on the same tree, with leaves grown in the shade near the ground being larger than leaves grown in direct sun near the top of trees. This is the result of smaller sized leaves at the top that can dissipate heat and larger leaves at the bottom that dissipate less heat but need to be large to capture fleeting levels of light.

The widely used *Populus* genera supply various organisms with a source of food to encourage their presence, which could jeopardize a phytoremediation planting. Pests are mostly the larval stages of various species of beetles (*Coleoptera*) and butterflies (*Lepidoptera*). For beetles, such as the cottonwood leaf beetle, the larval stage migrates from eggs that winter over in the leaf litter to the underside of leaves where they amass in groups that consume the leaf material. Rather than using the presence of these and other organisms as the criterion to enact control measures, it should be the extent of the infestation that is used to decide if foliar application of insecticide should be used. In all healthy ecosystems, there always will be some level of host and predation activity—this is to be expected.

Another interesting factor to consider when deciding which plants to use to interact with contaminated groundwater is the rate of growth. The adage of faster is better often can become the primary criterion for plant selection at a site. Choosing a plant that can grow faster relative to alternatives has the potential economic advantage of installing smaller, less expensive trees, and yet being able to demonstrate closed canopy conditions within a similar timeframe as if older and larger plants were installed. However, a fast growth rate does not necessarily indicate that more water will be moved through the plant and, hence, translate into successful phytoremediation for hydrologic control. For example, many pine trees can achieve fast growth rates similar to hardwoods, such as poplars and willows, but use less water per tree. Lower transpiration rates for pines that have similar growth rates to poplars can be explained by pines' lower *LAI*—plants can be fast growers but use less water because the total *LAI* is lower. Fast-growing plants that have lower water use due to lower *LAI* include Leyland Cypress (the naturally occurring hybrid *Cupressocyparis x Leylandii*); dawn redwood (*metasequoia glyptostroboides*); loblolly pine (*Pinus taeda*)—one of the few pines that actually prefer wetter soils such that its name is derived from moist depressions called loblollies; slash pine (*Pinus elliottii*); and pond pine (*Pinus serotina*).

7.3.10 Climate and Vapor Pressure Deficit

The water content of the air is an important control on the success of phytoremediation of contaminated groundwater. The transpiration gradient is set in motion at the leaf surface in response to the humidity difference in vapor pressure between the water vapor in the stomata relative to the air. Given all factors being equal, more water will be transpired by a plant during conditions of drier than more humid air. This is one reason why caution needs to be exercised when trying to apply results from one phytoremediation site to another, because the air humidity characteristics may be radically different between sites, even though the same plant may have been used and the depth to water table are equal.

Ambient air contains some level of moisture. This moisture, as water vapor, exerts a partial pressure on its surroundings. As the amount of water vapor increases, the partial pressure also increases. As this occurs, the partial pressures of the other atmospheric gases, such as CO_2, oxygen, and nitrogen, must decrease to conserve the total pressure, 760 mm Hg at sea level (defined hereafter as the North American Datum of 1983, NAD 83). For example, if the water vapor content of the air is 5%, the partial pressure of water vapor is 38 mm Hg; the balance of which (722 mm Hg) will be left to the other gases. Warm air requires more water vapor to reach saturation than cool air. The absolute humidity of the air is the partial pressure of water vapor, and saturated humidity is the total amount of water vapor that the air can hold at a specific temperature; therefore, relative humidity is the current percentage of total saturation.

Specifically, vapor pressure, VP, is the pressure that water vapor in air exerts on its surroundings. The maximum VP, or VP_m, is the vapor pressure of a molecule when it has come into equilibrium with its confining space. For example when water is heated, the VP_m occurs when water boils at the same rate that water condenses. This VP_m can be decreased, however, by solute addition. The ambient VP, VP_e, is the measured VP at conditions less than saturation. The vapor pressure deficit, VPD, between VP_m and VP_e represents the magnitude of the deficit between what VP is and what it can be at a given temperature. If VP_e is less than VP_m, a gradient is established so that water will move to balance the vapor pressures. Relative humidity is the more commonly used and recognized term for the ratio of the VP_e/VP_m (100).

The VPD of an area will affect plant transpiration and, therefore, phytoremediation potential. As a plant grows and has more and larger leaves and increasing LAI, the transpiration rate should also increase, especially if the VPD is high. This relation between LAI and VPD is depicted in Fig. 7.10. For two Eucalyptus species, when the LAI was highest during the summer (December in the southern

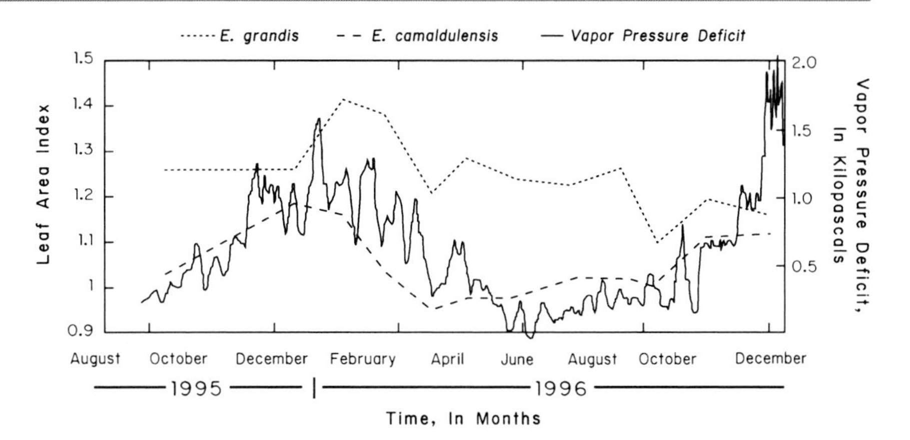

Fig. 7.10 The effect of leaf-area index, *LAI*, and vapor pressure deficit, *VPD*, (in kilopascals) on transpiration for two Eucalyptus species in Australia (Modified from Rural Industries Research and Development Center 2000).

hemisphere study site of Australia), the *VPD* also was high. This indicates that transpiration would be at a maximum. This also was during a period of low annual precipitation. Combined, these factors lead to plant and groundwater interaction that can be used for hydrologic control.

Wind also affects the rate of transpiration and, therefore, plant health by removing air in equilibrium with the leaf surface and replacing it with air out of equilibrium, thus continuing the transpiration process. In general, transpiration is higher on windy days than on calm days when all other factors are equal. Excessive wind speed can also damage the entire plant by toppling it during storms. This has a higher probability of occurring if high winds are preceded by precipitation that increased the soil moisture around the plant roots.

7.3.11 Site Operation and Maintenance, Pruning, and Fertilization

One of the reasons that phytoremediation is touted as an inexpensive remedial alternative is the notion that tree installation is inexpensive. While this may be true as far as the initial capital costs of trees relative to installing a pump-and-treat or air-sparge system, annual expenses are associated with a phytoremediation system that cannot be ignored. These annual expenses when run out over perhaps decades of the life cycle of a phytoremediation planting can result in costs that may approach those of more aggressive technologies. Even from the standpoint of those who promote the use of phytotechnologies, the short-term and long-term costs must be evaluated.

Much like a vineyard can be grown by using low-impact agricultural techniques but still remains a business that needs to produce a profit, the installation of a phytoremediation project also is part of an overall business model for those individuals doing the work. Costs that should be associated with a phytoremediation project include operation and maintenance costs. Pruning or grounds keeping is a major consideration. The idea of planting a site, walking away, and returning to it 10 years in the future to realize restored groundwater may be a reality in the future, but is wishful thinking for most projects.

As previously discussed, a phytoremediation planting is a microcosm of forest ecology, silviculture, and agriculture. What happens in forests happens in controlled phytoremediation plantings. In forests, the ground usually is littered with dead branches from lower parts of the tree that died from a lack of light. If such a branch were left to rot on the tree, it is possible that the rot would spread to the main trunk. Limb removal by storms is a form of wind pruning, after which the tree heals the wound to limit the spread of any microbial or fungal infestation. In a similar fashion, plants also undergo air pruning, in which limbs dry out and die, or new roots that no longer have access to moisture cease to grow. Some trees, however, such as poplars and willows, can self prune even during calm weather (Fig. 7.11). This form of pruning is not of dead plant parts but of live stems, and occurs at abscission zones.

Fig. 7.11 Evidence for the self-pruning habit of some willow trees growing near a pond in Blythewood, South Carolina. The pruned branches on the ground were not dead but had viable terminal and lateral buds (Photograph by author).

Manual pruning can be done at phytoremediation sites and is done to achieve the same effect, which is to protect the overall health of the plant. To understand what happens when a plant is pruned, it is necessary to understand how a plant grows at both the macroscopic tissue and microscopic cellular levels. Then and only then can the decision when and where to prune be made. Some of this information was covered in Chap. 3. Essentially, pruning is the selective removal of plant tissue to achieve a goal of a certain plant form or to remove older or diseased wood, such as stems and branches. The mere act of pruning, however, ironically enhances the formation of new growth, by causing dormant buds beneath the cut to become active and start to grow. One stem terminal bud, for example, if removed by pruning, will cause two new stems to take its place—this will increase the fullness of the plant.

The act of making a pruning cut is important not only for the shape of the plant but also its health. Only sharp blades should be used, and the cut made as close as possible to the main stem. Any amount left behind that projects from the main stem will die, and start to rot. This rot can then have an opportunity to invade the main plant. Also, the pruning cut should be made from the bottom up if using a scissor-type blade, but from the top down if a saw is needed to remove a large branch. In this manner, the weight of the branch helps keep the cut open and the saw from binding. However, an initial small cut from the bottom up close to the main stem will prevent the bark from tearing as the down-cut is finished.

Although pruning typically is associated with selective removal of plant biomass above ground, pruning also can be performed on root material. Root pruning can be accomplished to decrease root competition for resources at phytoremediation sites where the trees typically are planted within a few feet of each other. Some planting approaches also can decrease future root competition when cuttings or trees are installed in holes lined with impermeable materials, typically designed to limit lateral root growth and encourage deep root growth. The effect of root pruning on plant and water relations tends to be site and tree specific, however. Woodall and Ward (2002) reported that following root pruning of tree crops in Australia, sap flow in some trees was unchanged but ceased in others. They also reported that soil moisture increased in sediments after the roots of plants were removed, as the sink for the water had been eliminated. The perceived benefit of increased plant biomass, health, and water relations associated with root pruning may become more apparent at sites with relatively close tree spacing. Moreover, there may be genetic controls on root pruning that are plant specific and, therefore, the success of this approach may be hard to predict, or may result in varying degrees of success.

Similar to the question posed previously regarding the best time to plant, there are many answers to the question of the best time to prune. In general, pruning can be done anytime it is convenient for the person doing the pruning. There are positive and negative consequences, however, that should be considered before a cut is made based on convenience. Pruning can be accomplished during the winter when plants are mostly dormant, especially when removing dead branches. Winter pruning results in the production of wood at the expense of leaves or flowers or fruit. Pruning of live branches removes wood that contains many dormant buds. As a result, less leaves will be produced the following year because energy will be spent on new wood production, formed from adventitious buds, to replace the pruned wood. Fall pruning also can be done to remove old dead growth but will cause the remaining buds to be more numerous and grow faster in the spring. Late spring pruning should be avoided because the commencement of sap flow and higher ET_p lead to wounds that leak sap, which can become infected.

From this information, a general rule to be followed in most areas regarding pruning is that it should be done when the plant is dormant, typically between the fall and spring. This is because more food stored in the roots will be available for fewer parts of the shoots in the spring when top-growth resumes. This is counterintuitive, because pruning often is done in an effort to decrease the size and shape of a plant. For poplars, this is after leaf drop in more temperate areas of the Southeastern United States or in the spring in the cooler northern areas. During dormancy, most deciduous trees contain water stored in the xylem sapwood and even transferred to the heartwood prior to when the leaves dropped. Also, after the leaves drop, no new sugar compounds are formed by photosynthesis, so no additional sap is being formed. This is more of a problem for slow-growing trees that typically are not used in phytoremediation studies. All in all, pruning during dormancy produces cuts with very little fluid leakage. Very little pruning should be done the first year after planting a phytoremediation system, especially if whips or cuttings are used, other than removal of dead or damaged areas or removal of the terminal bud. For plants such as birches or maples, summer pruning is advised, because fall pruning causes too much leakage of downward moving sap. Pruning should be done when temperatures are above freezing, as the wood is too brittle when temperatures are at or below freezing.

With respect to pruning commonly used phytoremediation plants, eucalyptus trees require special care. This is because they tend to drop a lot of plant parts, called slash, that can include very large branches. These can be removed to limit the potentially negative consequences of such dropping, but keep in mind that reduction of a tree to its terminal growth will limit the *LAI* and keep transpiration below optimum levels.

The care for evergreens, especially the thin leafed or needle-type conifers, is slightly different than for deciduous trees. Evergreens drop their leaves like deciduous trees, but because of their efficient use of water and ability to grow in areas where water availability is limited, they drop their leaves every 2 to 3 years. Some evergreens, such as the baldcypress, that grow where water is not limited, drop their leaves annually, hence the adjective bald. One of the problems with pruning thin-leafed evergreens relative to deciduous trees is that rapid growth from the lateral buds does not occur. Hence, if too much pruning is done, there will not be enough leaf-surface area to make food for the plant, and it literally will starve to death. Therefore, it is not recommended to do heavy pruning of such trees; rather, light pruning should be accomplished throughout the year on a more frequent basis.

There always is some level of concern that the benefits of pruning come at the cost of essentially wounding the plant in the process. However, plants naturally lose limbs all the time, as can be seen the day following a major wind storm. In most cases, these limbs are older and partially dead anyway; or in the case of the lower shaded limbs of large conifers, the loss is a benefit to the plant because the limbs were a net drag on photosynthesis. Remember that the outer layer of all plant parts consists of a tough layer of epidermal cells, the first line of defense between the plant and the hostile external world. These cells can produce suberin in the cork cells beneath the bark. In most plant parts, naturally occurring tannins are produced and act as natural antimicrobial and antifungal compounds. After a wound or pruning cut is made, the exposed cork cells die, and the cutin and suberin are released and flow over the cut. Then parenchymal cells rapidly grow and cover the exposed area with a callus. To encourage a high rate of such coverage is the reason why pruning cuts are made as close as possible to the main trunk. Moreover, pruning is indeed not as harmful as sometimes envisioned and is best proven by the fact that the parts removed from the plant can often survive on their own, as occurs in vegetatively propagated plants.

To prevent organisms, such as fungal spores, from penetrating the entire plant by way of the phloem that may be exposed following damage by a storm, these cells when broken release callose, that essentially plugs the sieve tubes so that the exposed phloem cannot flow. In conifers, broken cells release resin produced by the duct cells of the aforementioned rays, which harden when exposed to air. Other examples of such wound healing and sealing compounds include gums, which are used to thicken some foods, and latex. Other compounds include the tannins, which are discussed in Chap. 11.

Poplar trees can be cut flush to the ground in the event of problems with top growth or damage by storms. This is purposefully done during coppice silviculture to produce and harvest trees for paper products or as an alternative energy source. That certain trees regrow after such cutting has been known for some time:

For there is hope for a tree,
if it be cut down, that it will sprout again,
and that its shoots will not cease.
Though its root grow old in the earth,
and its stump die in the soil,
yet at the scent of water it will bud,
and put out branches like a young plant.

Job 14:7–9 (RSV)

Why is all of this information about pruning important to those concerned with phytoremediation of contaminated groundwater? First, pruning will have to be done. Pruning makes more sense when the action is underscored by an understanding of plant physiology and the purpose behind the physiology, which is for the cells of a multicellular plant to essentially all work together to make and redistribute food. This is analogous to vineyard owners that purposely girdle grape plants below the branches that contain the grapes by removing the phloem after the fruit has formed. This ensures that all the food created by the leaves is routed to fruit production, size, and sugar content rather than to the roots.

As with any investment, there are concerns for the health of a phytoremediation crop after it has been installed. A farmer will worry about changes in the weather and the appearance of disease or infestation. It is no different for a grove of plants used for the phytoremediation of contaminated groundwater. The question is, should prophylactic control measures be used? Some plants can immunize themselves after mild exposure to certain pathogens. This is described more fully in Chap. 11. Chemical as well as biological agents can be used, but need to be accounted for in any assessment monitoring.

7.3.12 Growing Season Length and Effect on Acceptance of Phytoremediation

A widely held assumption regarding the relation between plants, water use, and groundwater is that deciduous trees, such as hybrid poplars, are relatively passive and inert with respect to water use during the dormant season (Interstate Technology Regulatory Council 2009). This assumption is valid and understandable for the most part, because the lack of leaves during the dormant season indicates that water loss from subsurface sources to the atmosphere does not occur through transpiration.

Fig. 7.12 A large mistletoe in a dormant deciduous tree. The illustration shows how a mistletoe obtains water from the xylem of a host plant even if the host plant is not actively transpiring.

But water is still present in plants even after leaf drop, as can be demonstrated by a few common examples. Wood cut during the winter is heavy on account of it containing water. This winter wood, called green wood, needs to be dried out, or seasoned, to remove this water before use the following winter. Tree branches that fall on homes during winter storms cause considerable damage to structures due to their water weight. Because there are no leaves and, therefore, very little water flow through dormant deciduous trees, the water present can be considered to be static. This scenario is similar to household plumbing, where the pipes contain water even though the faucets are turned off.

The presence of mistletoe in the tops of many deciduous trees indicates that there must be some water flow even during dormancy. Mistletoe produces its own food but is parasitic because it uses the host plant for anchorage and sends its roots into the host xylem to remove water (Fig. 7.12). Observations of multiple mistletoe plants present in many dormant trees suggest that some water flow is occurring during times when there are no leaves present to regulate water vapor losses. The author has hypothesized that this characteristic of mistletoe might make it a good candidate for site-assessment monitoring activities at potential phytoremediation sites.

To look rationally at this controversy of plant-water use and dormancy and its effect on the phytoremediation of contaminated groundwater, the principles behind the process of plant water losses through transpiration need to be remembered. The loss of water vapor through leaves is a consequence of the plant's need for the entry of CO_2. This loss can occur in any of the mesophyll cells that contain chlorophyll and chloroplasts. Even without leaves, the cambium beneath the bark of many plants also contains chlorophyll. The cambial cells are provided for gas exchange through lenticels; these opening also are a potential pathway for water loss. This water loss, although significantly less than loss by transpiration, will be replaced with water from the subsurface and provides a need for water movement in dormant plants.

The effect of a reduction in transpiration or its elimination is important in terms of the use of trees to control groundwater. The existence of a period of plant dormancy should not be the sole criterion used to determine if a phytoremediation system will be used at a particular site. The transpiration of a tree should be viewed in the context of the site's hydrogeologic characteristics. For example, even though some trees may transpire less during the dormant period for that species, these months also may be characterized by less precipitation and less recharge and, hence, impart little to no gradient to the water table.

Some aspects of dormancy can be used during the design of a phytoremediation project for contaminated groundwater. Matthews et al. (2003) performed numerical simulations of a phytoremediation system based on the removal of groundwater by *ET*. They reported an inverse relation between the length of growing season and plantation area. For example, a site with a shorter growing season required a larger area to be planted than a site with a longer growing season.

7.4 Summary

There are very few areas of the world where plants cannot grow. Even a small plot of unfertile soil proceeds through a rather predictable succession of plants, from invasive grasses to finally a climax community of hardwood trees. In a phytoremediation system, this natural succession is accelerated with the relatively rapid installation of phreatophytes.

Why is this information important to the phytoremediation of contaminated groundwater? The establishment of healthy plants should be the initial goal of all phytoremediation projects. However, the long-term success of phytoremediation requires careful management of the system over time to ensure the growth of plants and their interaction with groundwater.

8 Conceptual Frameworks for Phytoremediation to Achieve Hydrologic Goals

A logical application of the naturally occurring plant and groundwater interactions elucidated in Part I is at sites characterized by contaminated groundwater. The data collected during the site-assessment and characterization approaches described in Chap. 6 can be used also to assess whether a plant-based system can achieve the three major hydrologic goals at a given site. Because this assessment is being made in the context of state and Federal requirements to protect and restore groundwater resources, the use of objective approaches to evaluate the benefits of plant and groundwater interactions is warranted. Objective approaches, also called frameworks, that can be used are offered in this chapter.

The primary approach, or framework, to assess plant and groundwater interactions presented in this chapter is based on water budgets. At its most basic level, a water-budget framework states that if groundwater flow is fast relative to the removal of groundwater by transpiration, then hydrologic control will not be achieved. Conversely, if groundwater flow is slow relative to the removal of groundwater by transpiration, then some degree of hydrologic control may be achieved. In both cases, however, complete hydrologic control by plants may not be reasonable. Because a water-budget approach may not be useful at all sites characterized by contaminated groundwater, alternative frameworks also are discussed.

8.1 Initial Approaches to Assessment

For a given basin, the amount of water that enters through all compartments will equal the amount that exits through all compartments, assuming steady-state, or long-term, average conditions. This equality of water was discussed in Chap. 2 as part of the hydrologic cycle. At a site characterized by contaminated groundwater, the installation of plants that interact with groundwater will affect the site's water budget in at least two ways. First, the plants will decrease the amount of recharge by precipitation. Second, the plants will increase the amount of one of the outputs, such as discharge by transpiration from the subsurface. Both processes lead to a change in the amount of water stored or released from groundwater.

When a site is to be evaluated for phytoremediation to control or contain groundwater, various approaches can be used to estimate the influence of plants on the subsurface water resource. It is best to start with simple site-assessment approaches and proceed to more complicated approaches if necessary. Use of multiple approaches will decrease the level of uncertainty inherent to each approach. The following approaches that can be used for site assessment include the evaporation rate of water from free, exposed surfaces; the removal of water from plants caused by meteorological parameters; changes in groundwater levels caused by transpiration; and tank studies, as described below.

8.1.1 Free-Surface Water Evaporation

As described previously, the solar energy input to an area changes liquid water to vapor. This energy amount is fixed for a particular geographic area, although it varies within an area with the seasons. Between 540 and 590 cal are required to evaporate 1 g of water at 15°C. Because less than 400 cal/cm^2 are available on a clear sunny day, the maximum amount of water that can be evaporated from the surface of water is equivalent to about 6 mm/day (Kozlowski and Pallardy 1997). The point here is that the transpiration rate cannot exceed this evaporation rate, unless drier air is continuously advected over a planted area.

The outflow of water at a site by evaporation is, therefore, an excellent indicator of the maximum potential for water to be evaporated through transpiration and, therefore, at a phytoremediation site. In general, the free-surface water evaporation rate or index will be about 70% of the annual

J.E. Landmeyer, *Introduction to Phytoremediation of Contaminated Groundwater*,
DOI 10.1007/978-94-007-1957-6_8,

average precipitation for an area. For example, the free-water surface evaporation rate in arid southeastern California is between 60 and 85 in./year (152–216 cm/year), or 5–7 ft/year (1.5–2 m/year) (Lines and Bilhorn 1996).

In order to have an effect on groundwater the water removed needs to be from infiltration prior to recharge or from the water table. Even if the water table is deep, and the soil moisture levels low, the assessment of free-surface water evaporation will provide an indication of the potential for water to be removed by transpiration, a special case of plant-controlled evaporation. The free-surface water evaporation rate can be quantified at a site using indirect and direct methods, as described in Chap. 2.

8.1.2 Micrometeorological Data

Whereas the free-surface water evaporation rate will provide a general rate of evaporation of exposed water, the effect of various meteorological properties on how plants control evaporation to their advantage through transpiration can be assessed using site-specific meteorological data. The total removal of water by transpiration calculated from meteorological data is a part of potential evapotranspiration, ET_p. The ET_p of a site can be estimated using site atmospheric data, as described in Chap. 2. Monthly precipitation data can be collected onsite or from the nearest existing rain gage.

Various methods have been developed to estimate ET_p although perhaps the simplest are based on certain meteorological parameters. These methods include those offered by Thornthwaite and Holzman (1939); Penman (1948); Van Bavel (1966); Monteith (1965); and the Bowen ratio and Blaney-Criddle methods (Kramer 1983). These methods all are based primarily on an energy budget concept, in which inflows and outflows of energy are balanced. The newer method of growing degree days also can be used to estimate ET_p. As defined in Lorenz and Delin (2007), growing degree days (GDD) is the annual sum of the average temperature each day minus a base temperature for a particular area. The GDD method rather than ET_p was used by Lorenz and Delin (2007) to estimate the amount of net precipitation remaining per year that potentially may become recharge.

8.1.3 Transpiration Well

The above two approaches can be used to estimate the total amount of water that could be evaporated from a particular site. These approaches do not differentiate, however, between the sources of the water removed. This distinction is important when the goal of a phytoremediation application is recharge reduction or hydrologic control of groundwater flow. To differentiate between the amounts of groundwater removed by plants relative to other water sources, the transpiration-well approach can be used. Existing phreatophytes can be used or planted trees that have had the chance to grow and reach the water table can be used.

As described in Chap. 1, W.N. White took field observations of groundwater-level changes in wells installed in forests of phreatophytes and developed an empirical equation that could be used to determine the total amount of groundwater used by transpiration during the previous 24-h period. Equation 1.1 is re-listed here

$$Q = y(24r + s) \tag{8.1}$$

where Q is the depth to the groundwater; y is the specific yield of the soil; r is the hourly rise in water table (from midnight to 4 a.m., the time of assumed zero transpiration when groundwater levels recover by induced local convergent flow), and; s is the net fall, or rise, in groundwater during the same 24-h period. The variables r and s are derived from the water-table fluctuation data recorded in a monitoring well located near the plants. The nighttime rise in the groundwater level during periods of no precipitation is due to the movement of groundwater toward the plants in response to lowered water levels—the nighttime rise in groundwater level is not derived from the plant releasing water taken up during the day, because water in the plant is under tension and prohibits reverse flow by gravity.

This equation of White's provides a fundamental base that connects the hydrogeologic characteristics of the subsurface with plant use of groundwater and provides a metric to assess this interaction, i.e., the change in groundwater level. As roots take up groundwater from the capillary fringe, a gradient in water potential is established. This gradient causes groundwater to move upward to the capillary fringe. If the uptake of water by plants from the capillary fringe and replenishment by the upward movement of groundwater is faster than the rate of recharge of the aquifer by hydrostatic pressure, artesian flow, change in storage, or lateral flow from upgradient areas, then the water-table level will decline. When the plants are not removing water from the capillary fringe, for instance at night when transpiration is lower or ceases altogether, the water table rises to reach equilibrium. This is an important consideration to keep in mind when monitoring a phytoremediation site for hydrologic control, because the absence of groundwater-level fluctuation does not necessarily indicate a lack of plant and groundwater interaction. It could simply indicate that the groundwater-flow rate is faster than the plant uptake rate.

This approach should be performed with the appropriate calibration or null hypothesis controls. For example, groundwater-level fluctuations can be measured at the site before plants are added or measured in areas that are not planted after installation. As a result, the effect of the vegetation on groundwater can be observed.

8.1.4 Tank Experiments

The experiments of Charles Lee used tanks, or basins, where the quantity of water removed from tanks that contained plants could be compared to the water removed from tanks that did not contain plants. Lee's results were used by Robinson (1958) to determine the relation between evapotranspiration, *ET*, and depth to groundwater and temperature (see Fig. 2.9). Tank experiments to investigate *ET* from riparian trees also were conducted and results reported by Robinson (1970). Robinson (1970) reported *ET* as a volume per foliage amount or quantity of water per foliage amount, and this volume also was affected by depth to groundwater, length of growing season, and nutrient toxicity. Tank experiments are time and resource intensive, and the results may not be transferable to field conditions, however.

8.1.5 Foliage Volume

The concept of foliage volume was advanced by Bowie and Kam (1968) during their investigation into the consumptive use of groundwater by riparian vegetation in Arizona. Because leaves are the predominant location of water-vapor removal, an estimate of total leaf volume can provide an estimate of the amount of water lost by trees. Bowie and Kam (1968) considered a tree to be a cylinder, with tree height analogous to the cylinder height and the tree crown, or radial extent of leaves, analogous to the cylinder width. The volume of leaves contained within each cylinder is represented by an acre-foot, where 1 acre-ft (1,233 m^3) of foliage is a space 1 ft (0.3 m) deep and 1 acre (4,047 m^2) in area. The loss of water by *ET* from this cylinder of leaves should be related to the total foliage volume. This method was later used by Robinson (1970) to investigate *ET* by phreatophytes in Nevada.

8.2 Assessment of Potential Evapotranspiration, Recharge, and Groundwater Discharge for Phytoremediation Effectiveness

The various approaches and methods above provide either ET_p or groundwater-level changes, but not necessarily an analysis of the interaction between the two on a site wide scale. This analysis can be accomplished by comparing the ET_p to groundwater flux through a site, and the latter based on basic groundwater-level data. In general, control of site hydrology by phytoremediation begins with a decrease in water-table elevation; all else being held equal, decreases in water-table elevation can be attributed to the plant-facilitated reduction in recharge or to increased transpiration such that net recharge is low. Groundwater levels need to be measured in monitoring wells within and outside the planted area to ensure these changes are related to the plants and not to other site wide phenomena. A case study where these measurements are made and applied is given in Sect. 8.3.

8.2.1 Precipitation

For most sites being considered for phytoremediation of contaminated groundwater, precipitation can be measured at the site or obtained from an existing precipitation station nearby. Measurement of precipitation is necessary for the water-budget approach and provides the amount of water that enters the site. Precipitation varies annually at most locations across the United States. Even if the average annual precipitation is fairly constant, the frequency, duration, and amount of precipitation can vary significantly at one location over time.

Precipitation, one of the most readily quantifiable meteorological properties, can be measured by devices ranging from simple, inexpensive onsite precipitation gages to more sophisticated automatic monitoring devices such as tipping-bucket precipitation gages attached to a data logger. Most states have some network of raingages, either run by state, academic, or Federal programs that track precipitation amounts. Sources of precipitation information are many and include, but are not limited to, the following agencies: USGS; National Weather Service; National Oceanic and Atmospheric Administration; universities and state cooperative extension services.

8.2.2 Recharge

In most places in the United States less than 10% of annual precipitation becomes recharge. In South Carolina, for example, average annual precipitation is about 50 in./year (127 cm/year) with shallow groundwater recharge of about an average of 5 in./year (12.7 cm/year). In contrast, in Long Island, NY, where subsurface sediments consist of permeable glacial deposits, recharge is closer to about 50% of annual precipitation.

A water table that is within a few feet of land surface is more apt to be recharged during precipitation events than are deeper water tables. It also presents a challenge to using a groundwater-level change approach to monitor plant and groundwater interactions. This is because even though the shallow water table may be a source of water to plants, the water table can rise after precipitation events. This occurs because either the plants are facultative phreatophytes and are using infiltrating water or the groundwater is recharged at a rate faster than removal by trees.

Precipitation, recharge, and groundwater flow from other areas upgradient are not the only inputs that need to be considered as part of the water-balance approach to using plants for hydrologic control. For example, septic systems, storm drains, agricultural tile fields, or leaking water mains that may be located near the site often can contribute an unrecognized source of recharge to the water table. If this recharge flows through a source area that contains contaminants, it will create additional contaminated groundwater. Conversely, it may act to dilute the contaminant concentrations dissolved in groundwater at the site.

8.2.3 Potential Evapotranspiration

As was introduced in Chap. 2, the ET_p of a particular area is an estimation of the total amount of water from all sources that can be removed as vapor under given weather conditions. The inputs necessary to determine ET_p are the air temperature, air relative humidity, solar radiation, wind speed, and precipitation amount (van Bavel and van Bavel 1990). These data can be taken from monthly or annual averages, collected onsite using a weather station that has sensors to measure these variables, or from reference material. An amount for ET_p in depth (length) per time (such as millimeters of water per h or in. per month) will be calculated. The units of ET_p refer to the rate that a unit thickness of water in a unit area will exit the system as water vapor ET. Because ET_p usually is reported in units of length per time, multiplication of ET_p by the area of the site to be planted (in length squared) gives ET_p in units of volume (length cubed per time). This volume per unit time can be converted to gallons per unit time by the conversion factor of 1 ft^3 = 7.48 gal of water.

Knowledge of the magnitude of ET_p in relation to precipitation for a particular area or region is invaluable when phytoremediation to achieve hydrologic goals is being assessed. In general, if ET_p on a daily, monthly, or annual average is greater than precipitation, then the potential exists for a plant-based system to affect groundwater. If precipitation is greater than ET_p, the picture is not as clear—the answer is then dependent upon the depth to water table. If the water table is shallow, groundwater levels may rise even though water is being removed, and if the water table is deep, recharge will be decreased as plants take up infiltrating water.

As stated earlier, the magnitude of the ET_p at a site provides an estimate of the atmospheric demand for water, which can be supplied by surface water, soil moisture, capillary water, and groundwater. The fraction of ET_p that actually will be derived from the capillary fringe or water table will vary with each site and will be a function of the depth to water table, plant coverage and type, and hydrogeologic conditions. For most cases, a conservative estimate of groundwater contribution to ET_p can assumed to be no more than 25% for a site that consists of a water table of less than 10 ft in a sandy aquifer.

In general, if ET_p is less than precipitation within a common time period, be it weekly, seasonally, or annually, then net recharge of water to the shallow aquifer in the study area will occur. Conversely, if ET_p is greater than precipitation, then a net discharge of groundwater from the study area will occur, assuming that the source of the water being evaporated or transpired is groundwater. By strict definition, the successful implementation of plants at a site to achieve hydrologic containment or control would occur when conditions favor the *net* discharge of groundwater, by ET, from the site such that no water is stored, and ΔS in Eq. 2.4 is negative. This holds true when ET_p is derived from soil moisture, the capillary fringe, or from the saturated zone.

The balance between groundwater recharge from precipitation and groundwater discharge from ET_p is not a constant number but one that will change over time. The frequency of precipitation is not constant over time at a site, whereas ET_p tends to be more constant for a given climate. For example, in the humid southeastern United States, net recharge occurs in the winter and spring, even though precipitation amounts are higher in the summer. Groundwater recharge occurs during precipitation events but is quickly removed by transpiration. During winter, precipitation is lower, but ET is much lower than the summer. It is important to remember that for either case, groundwater recharge may not occur if a high percentage of impermeable surfaces are present. In these cases, water can still enter the groundwater system by groundwater flow from upgradient areas.

The energy parameters that affect the ET_p for an area can be considered to be a given property that cannot easily be modified. The hydrogeologic conditions at a particular contaminated site being evaluated as a candidate for phytoremediation also have an effect on the success of phytoremediation because of how the aquifer sediments determine the resistance to water bioavailability and flow to plants. The presence of a thriving grove of trees does not mean, however, that groundwater is bioavailable and being used by the plants or that the amount removed from the water table is sufficient enough in volume to affect site groundwater flow. Moreover, a lack of evidence to support hydrologic control may not become apparent at a planted site until the plant community is 3–4 years old.

8.2.4 Groundwater Discharge and Potential Evapotranspiration

In general, for hydrologic control to occur at a site where phytoremediation has been implemented, the volume of groundwater that flows through the planted area in a given time period, or flux, must be less than or equal to the flux of water lost by ET_p. If the groundwater flux exceeds that of ET_p, then a component of groundwater will be unaffected by trees.

Darcy's Law (Chap. 4) provides a fundamental framework to evaluate the groundwater flux that discharges through a given cross-sectional area of saturated media and forms the basis of the evaluation of hydrologic control by phreatophytes. In his laboratory, Darcy showed that the specific discharge, v, of water through sand-filled columns was directly proportional to the discharge, Q, and indirectly proportional to the cross-sectional area of the sand, A. The specific discharge also is proportional to the head gradient, Δh, but indirectly proportional to the distance over which the head gradient was affected, Δl. This reduces Darcy's Law into the equality $Q = kiA$.

In groundwater systems, the flow of water occurs only through the interconnected pores. As such, the interference of the soil matrix affects the volume and rate of flow, and this discharge of groundwater has to be normalized by dividing Q by the porosity, n, of the cross-sectional area of aquifer material to obtain the true seepage velocity of groundwater, as Q_v. Porosity and seepage velocity are inversely related. We saw above how to calculate the volume rate of groundwater flow as

$$Q = A\ i\,k/n \tag{8.2}$$

and that the Darcy velocity, Q_v, is ik/n, therefore

$$Q = A\ Q_v \tag{8.3}$$

The vertical flow of groundwater in a shallow water-table aquifer is dependent upon the resistance to vertical flow relative to horizontal flow, or anisotropy, as defined in Chap. 4. Many aquifers near the land surface that become contaminated consist of sediments that were deposited in horizontal, flat planes. Similarly, fining-upward sequences of sediments associated with aquifers composed of meandering alluvial sediments also impede vertical flow on account of the difference in hydraulic conductivity associated with the different sediments. If groundwater in such sediments becomes contaminated, the same equations can be used, but a retardation factor, f, must be considered to account for the slower movement of a contaminant relative to the bulk flow of groundwater on account of chemical, physical, and biological processes; this is discussed in Chap. 13.

In most cases, this evaluation of the removal of water by ET_p in relation to groundwater flux can occur assuming steady-state conditions, especially for newly installed phytoremediation systems characterized by small trees. As the plants grow and interact more directly with the water table, however, conditions are no longer at steady state but are transient, as groundwater levels decline. The effect of storage on the balance between removal of water by ET_p and addition by recharge no longer can be ignored.

Transient simulations presented in Matthews et al. (2003) indicate that in shallow water-table aquifers that have significant storage, a noticeable, long-term (not just one day) drawdown may take longer to be observed than in an aquifer with less storage. This finding has implications for the use of groundwater fluctuations as a master criterion to determine whether or not plant and groundwater interaction is occurring at a site. For example, in aquifers with high storage coefficients, the lack of a diurnal drawdown in a planted area does not necessarily mean that the trees are not removing groundwater.

The relation between ET_p and groundwater flux through a phytoremediation site ultimately will determine the size of a phytoremediation planting, for instance when the goal is to stop groundwater flow from occurring past a particular boundary. In general, the larger the volume of groundwater flux in the contaminated aquifer, the larger the planting area should be. The planting size of a phytoremediation system is linearly related to groundwater flux. Matthews et al. (2003) reported that for a sandy aquifer, an increase in anisotropy, as a ratio of flow in the horizontal relative to vertical direction from 0 to 200, resulted in an increase in planting area. Essentially as hydraulic conductivity, K, increased, groundwater flux increased and the planting size needed to be increased. For a less permeable aquifer sediment, the planted area also increased, but at a slower rate.

These and other fixed variables constrain the maximum groundwater uptake rate at all sites where phytoremediation is being evaluated. The ET_p is a measure of the maximum potential ET for a given area at a given location. Groundwater flux within this area also is not going to exceed the maximum amount of flow, if conditions are at steady state. The removal of groundwater by plant uptake also is relatively fixed for trees at different ages and will approach a maximum amount under closed-canopy conditions. To capture a given amount of groundwater flux within a fixed area in some cases will require planting older trees that each have a higher transpiration rate, because using smaller trees would require installing a greater number of trees, at closer spacing, or across a larger area than site property boundaries permit. Or, ground covers would need to be planted to reduce infiltration so that the trees could remove groundwater from upgradient areas only. Each of these options has limitations, such as the increased expense for installing older

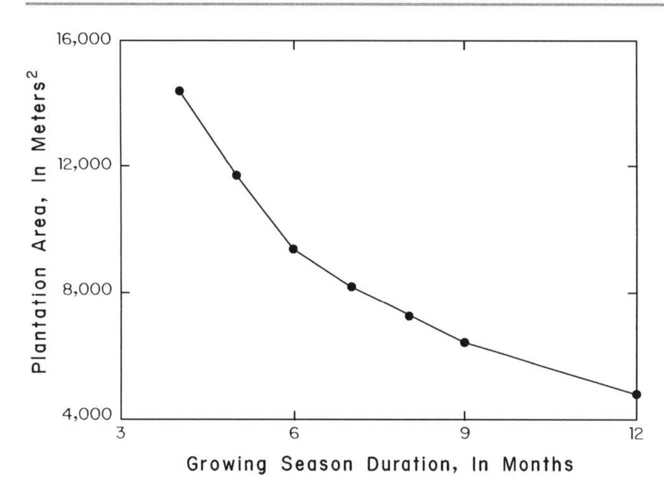

Fig. 8.1 Relation between the size of a planted area and time needed to achieve hydrologic capture, where the larger planted area achieves it faster than a smaller area (Modified from Matthews et al. 2003). One square meter is equivalent to 2.47×10^{-4} acres.

trees, the need for installation of an extra number of smaller trees, the need to purchase additional property, and the cost of increased time to successful phytoremediation (Fig. 8.1).

The ideal site conditions for phytoremediation to have an observed effect on the water table need to meet the following criteria (Table 8.1). Concerning the groundwater hydrogeology, the aquifer would be unconfined, with a shallow to moderate depth to a water table, primarily less than 15 ft (4.5 m) below land surface. Then, little effort would be required to plant even the small trees where roots would immediately be interacting with the water table. The groundwater quality would be fresh (low total dissolved solids) and not be influenced by saline surface-water bodies. The water-table surface would have little natural fluctuation, such that any fluctuation observed could be attributed readily to plant uptake. The aquifer would have no to low amounts of natural organic matter, contaminants present would be in the dissolved phase at concentrations much lower than solubility, and the aquifer would contain dissolved oxygen. The aquifer would be characterized by high porosity (greater than 30%) but with groundwater-flow rates slower than the water removal rate by transpiration. The aquifer would contain enough silt or fine sand to increase the thickness of the capillary fringe. The site would be located in an area of groundwater discharge where deeper groundwater flowlines converge upward toward land surface and plant roots. No surface water would be present to recharge the aquifer. Concerning the plants, the growing season would be year round, with abundant solar radiation. There would be little to no precipitation, and dry, low relative humidity conditions, with strong winds. No demographic obstructions would exist to prohibit the planting of the entire area above the aquifer, and deed restrictions would not convey. There would be no overhead restrictions such as powerlines, and there would be no insect pests, molds, diseases, beavers, deer, livestock, or voles. With respect to contamination, the source area would be completely removed, and any contamination remaining would be dissolved at concentrations less than solubility, and the physical and chemical properties of the contaminant would permit passive uptake into the transpiration stream.

Table 8.1 The parameters and range of values for an ideal phytoremediation site for hydrologic control.

Parameter	Range	Comment
Aquifer	Unconfined.	Depth to water table less than 15 ft (4.5 m) from land surface.
Water quality	Fresh.	Chloride less than 250 mg/L.
Water fluctuations	None.	Located away from tidal bodies.
Porosity, total	20–50%.	
Soil organic matter	None.	
Redox condition	Oxic.	Dissolved oxygen greater than 1.0 mg/L.
Flow path	Discharge area.	
Growing season length	12 months.	
Relative humidity	Less than 20%.	
Precipitation	Less than 4 in./year (10 cm/year).	
Wind speed	Greater than 5 mph (8 kph).	
Source area of contamination	Removed.	

As can be imagined, all these conditions are rarely found at one site. In fact, the conditions that make up the ideal site can be considered to be mutually exclusive, or at least not hydrologically defensible. For example, according to the above list, the meteorological conditions that would lead to the greatest amount of groundwater extraction by plants would be similar to those of the arid parts of the southwestern United States, in general, or of southeastern California in particular, but the hydrogeologic conditions there are characterized by a depth to water table far below land surface. The conditions that lead to the presence of a shallow water table more ideal for phytoremediation are typical of the humid parts of the eastern United States. There, however, abundant precipitation during the summer and fall can result in precipitation as the primary source of water to plants at a phytoremediation site. However, most of the above criteria are met in riparian zones adjacent to surface-water bodies characterized by shallow water-table conditions. These riparian zones are the ecological niches that support the phreatophytes found throughout the United States in arid and humid areas.

If a site has many or even a few of the parameters listed in Table 8.1 seeming to favor the implementation of phytoremediation to achieve hydrologic goals, other factors need to be considered. In shallow aquifers, if the groundwater flowlines are deflected deeper into the aquifer by recharge that did not pass through contaminated source material near the land surface, then downgradient discharge areas that will receive this contaminated, deeper groundwater, also must be planted. An example of this scenario is given in Landmeyer (2001) and Matthews et al. (2003). In this case, the full thickness of the shallow aquifer, which contains both clean and contaminated groundwater, can be affected by the plants as the removal of groundwater near the top of the water table will induce vertical flow to replace that removed from the capillary fringe.

On the other hand, if the shallow aquifer is relatively thin compared to its areal extent, then a two-dimensional approximation to groundwater flow can be used to determine the effect of a planting on groundwater. In this case, the trees would be removing groundwater across the full thickness of the aquifer, and many solutions to the capture zone of groundwater have been calculated for this condition, analogous to that of each tree behaving as a fully penetrating well. Ferro et al. (2003) noted that if parts of a contaminant plume of a given width needs to be captured by groundwater pumpage, the pump rate will move groundwater at two times the prepumped flow rate due to plant-induced changes in the hydrologic gradients.

Both the type and location of the contamination source areas and plumes need to be considered. For compounds that have specific gravity less than water and are released as a separate phase, they will tend to float on top of the water table. Once dissolved in groundwater, however, the molecules of even higher specific gravity will move with water. For example, a release of gasoline from an underground storage tank (UST) usually has a source area of free-phase compounds near the water table (many USTs are at shallow depths below land surface), as do aboveground storage tanks (ASTs) that have leaking pipes above the water table. In any case, as groundwater flows downgradient from these source areas, the shallower groundwater will be contaminated, whereas deeper groundwater will remain uncontaminated. In cases such as these, the plants need only control the shallow part of groundwater flow, not deeper flow (Fig. 8.2). In this case, to calculate the Darcy flow requires only the thickness of the aquifer that is contaminated.

For contaminants that have specific gravities greater than water, a free-phase release would displace groundwater until lower permeability sediment, such as a clay layer, is encountered. A dissolved-phase release of a contaminant enters groundwater according to contaminant solubility and the fraction of the contaminant that remains in the source material. After these processes occur, the dissolved contaminant could be transported with groundwater flow.

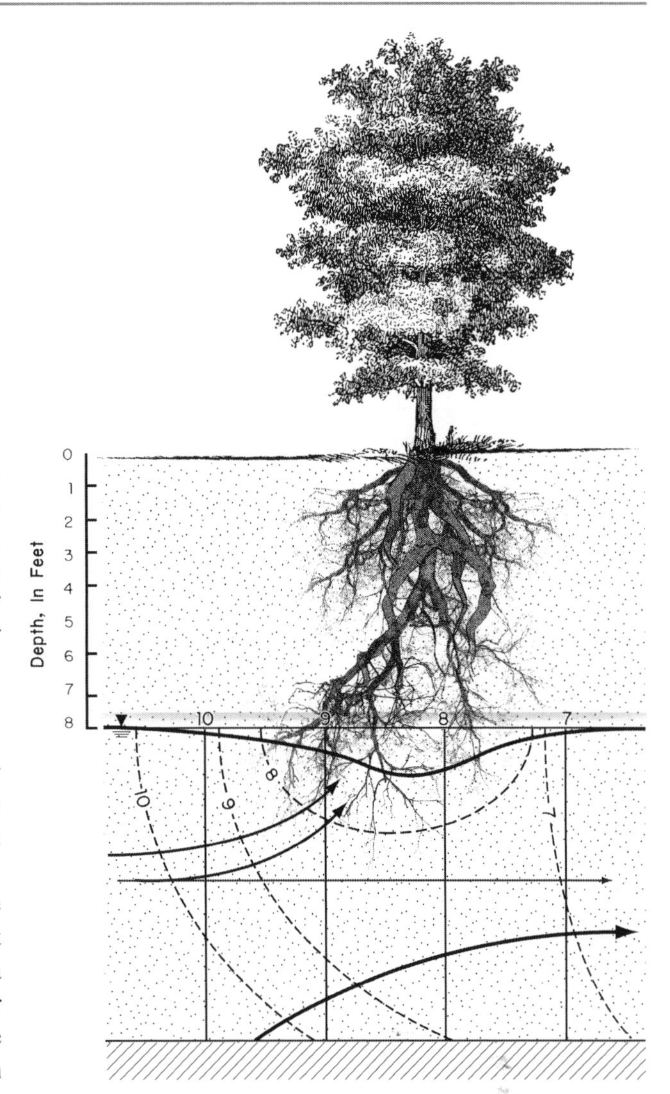

Fig. 8.2 Deep groundwater flow (shown as darker lines with arrowheads that cross dashed equipotential lines of 7–10 ft) can be deflected upwards beneath phreatophytes from an originally horizontal direction of groundwater flow (shown as lighter line with arrowhead that crosses solid equipotential lines of 7–10 ft). Some deep groundwater flow, including that from leakage from confining units (the thicker black line with arrowhead), may not be affected. One foot is equivalent to 0.304 m.

8.3 Case Study: Reduction in Groundwater Flow Across a Property Boundary, Charleston, South Carolina

The site of this case study is located in Charleston, South Carolina, on the eastern part of the Charleston Peninsula, and is the former location of an old manufactured gas plant (MGP), located on about 3 acres (12,141 m^2) of land. Site groundwater has been affected by residual coal tar materials disposed of around the site during site operations that extend as far back as 1850, as well as being affected by other adjacent manufacturing concerns related to materials rendered from coal tar.

The 3-acre study area is located in an area that has a history of various industrial land uses, a scenario typical of much of the coastline of the eastern United States. The study area was essentially under water throughout much of the eighteenth and early nineteenth centuries, until it and adjacent areas were filled with a variety of materials to accommodate the expansion of the City of Charleston. To help fuel this expansion, a MGP was located in the northwestern part of the current site in 1855, and permitted the City of Charleston to be the second city in early America, behind Philadelphia, to have gas-powered street lamps. The eastern edge of the former MGP was about 500 ft (152 m) west of the tidally influenced and saline Cooper River. The former MGP was in operation until 1957, as both a coal-gasification and water-gasification plant. Related industries, such as a coal tar and pine pitch refining plant, occupied adjacent properties during this time period. Between the former MGP and the Cooper River, a steam-generation facility was constructed in 1910 to support the MGP process. The present-day study area contains an operational electrical substation for the City of Charleston, the South Carolina Aquarium, the Fort Sumter National Park, and a parking garage.

The hydrogeology of the shallow water-table aquifer in the study area reflects this history of various site activities, as well its proximity to sea level. The water-table aquifer is composed primarily of fill material, with an average thickness of about 20 ft (6 m) at the site. The fill consists of sand, silt, wood, sawdust, concrete, bricks, and cinders and acts as an unconfined water-table aquifer with average depth to groundwater about 3 ft (0.9 m). The hydraulic conductivity as measured using slug tests of these heterogeneous fill materials varies widely, but an average value of 1 ft/day (0.3 m/day) suggests it is dominated by fine-grained material (Campbell et al. 1996). Beneath the fill aquifer are the more clay-rich depositional sediments of Quaternary age (Campbell et al. 1996).

The depth to groundwater in the surficial aquifer is between 2 and 4 ft (0.6–1.2 m). Groundwater flows to the south and east to discharge to an adjacent river, the Cooper River, located less than 2,000 ft (609 m) away. The river is the major hydrologic boundary to which onsite groundwater discharges. This hydrologic boundary does not, however, coincide with the site property boundaries, which are the locations of the regulatory points of compliance—it is common to have such nonhydrologic boundaries in areas characterized by dense populations.

Regarding the design of a phytoremediation project, the site has both positive and negative qualities, relative to those factors listed in Table 8.1. On the negative side, dense, nonaqueous phase liquid (DNAPL) exists at the bottom of the shallow water-table aquifer in some locations across the site, including in some areas of the phytoremediation planting; some saltwater encroaches from the adjacent saline river in the eastern part of the site; planting access is limited due to overhead powerlines of the current substation built near the footprint of the old MGP; seasonal hurricanes are a threat; the maximum daily relative humidity is high; the aquifer was rendered anoxic by the contamination; and, heterogeneous fill material is present in the shallow aquifer and this material consists of zones of high and low hydraulic conductivity. On the positive side, the depth to the water table is shallow (less than 4 ft [1.2 m]), the rate of groundwater flow is relatively slow; the growing season extends from April to November; strong winds come off the river which helps increase the *VPD* near the planting; and the site is in the U.S. Department of Agriculture Zone 8 (semi-tropical), meaning the location has ample solar radiation to sustain a phytoremediation planting.

A phytoremediation project was proposed for the site in 1998. Site constraints as far as places to plant dictated that 600 trees could be installed in a 18,000-ft^2 (1,672 m^2) area. In November 1998, about 600 hybrid poplar trees were installed. The hybrid poplar trees were planted at the site in two phases; phase one occurred in November 1998 and included the central to western part of the study area along the southern boundary of the site (Fig. 8.3), and phase two occurred in May 2000, in the remaining eastern part of the site.

Prior to the initiation of phase one, the existing surface soil material in the area to be planted was first excavated to a depth of 3 ft (0.9 m) below land surface, removed, and replaced with clean topsoil. It was later discovered that isolated lobes of DNAPL existed at some locations in the planted area at depths greater than 3 ft (0.9 m). The topsoil had a chemical composition that was slightly acidic, had a high cation-exchange capacity, with moderate amounts of organic matter (2%), and a chemistry equivalent to 80 lb/acre phosphorus, 18 lb/acre potassium, 77 lb/acre magnesium, 1,226 lb/acre calcium, and 11 mg/kg (milligram per kilogram, or part per million) sodium.

Following addition of the topsoil, about 600, 6-ft (1.8 m) bare-root cuttings were planted in November 1998, with 3-ft (0.9 m) of the cutting being placed below grade (Fig. 8.3). The cuttings consisted of a mixture of *P. deltoides x P. nigra* (DN Clone 34), referred to as the Imperial Carolina hybrid, and *P. charcowiensis x P. incrassata* (NE Clone 308), referred to as the 308 hybrid. These hybrids were selected because of their high transpiration rates and an effort to decrease plant mortality due to disease or infestation, as well as to be in accordance with the City of Charleston's historic district ordinances for landscape plants.

The cuttings were planted in five rows (Fig. 8.3) on 5-ft (1.5 m) centers, perpendicular to the dominant direction of

Fig. 8.3 Initial planting of five rows of hybrid poplar whips during phase one, November 1998, Charleston, South Carolina. Monitoring wells, such as shown in the foreground and far background, were installed within and outside the planted area to assess hydrologic control of groundwater by phytoremediation (Photograph by author).

groundwater flow from the former MGP site to beneath the planted area. Phase two consisted of installing 100, 6-ft (1.8 m) bare-root cuttings in March 2000 on the eastern half of the planted area. The cuttings installed were *P. deltoides x P. nigra* (OP-367). This hybrid, considered to be more salt tolerant, was selected because groundwater beneath the eastern part of the site contains higher amounts of salinity due to the proximity of the saline Cooper River.

For both planting phases, a mycorrhizal fungal inoculant was mixed with water and used to coat the cutting or bare roots before installation in the ground (MycorTree™ root dip, Plant Health Care, Pittsburgh, PA) (Fig. 8.4). To provide for erosion control during the establishment of a canopy of the cuttings, a ground cover consisting of a mixture of southern common alfalfa (*Medicago sativa*) and coastal Bermuda grass (*Cynodon dactylon*) was seeded at a rate equivalent to 20 lb/acre.

Fig. 8.4 Inoculation of bare root cutting hybrid poplars installed at the Charleston, SC, former MGP site, with a slurry of endo- and ecto-mycorrhizae prior to planting (Photograph by author).

A weather station was installed at the site in December 1998 after initial tree planting to collect data that would be used to evaluate which measured parameter, such as solar radiation, air temperature, and relative humidity, more strongly influenced the uptake of groundwater by the trees. The weather station consists of sensors that measure precipitation (Sutron Model 5600–04207), air temperature (R.M. Young Company), leaf temperature (Pessl Instruments), barometric pressure (R.M. Young Company), solar radiation (LI-COR Corporation), relative humidity (R.M. Young Company), and wind speed and wind direction (R.M. Young Company). The measurements were taken from all sensors at 15-m intervals, recorded on an adjacent data collection platform (Sutron 8210 Data Recorder), and transmitted by satellite to the USGS Water Science Center office in Columbia, SC. The data collection platform and all sensors were powered by 12-V marine batteries recharged by 40-Watt (W) solar panels (Solarex). The weather station was used to provide data to estimate ET_p rates based on the Penman equation (Penman 1948) as well as to compute changes in *VPD*.

8.3.1 Potential Evapotranspiration Relative to Groundwater Discharge

To evaluate the effect of the trees installed at the site in Charleston, SC on groundwater, comparison of estimated ET_p with estimated groundwater discharge was performed. The site parameters used as part of the ET_p assessment are presented in Table 8.2.

Table 8.2 Potential evapotranspiration (ET_p) for the Charleston, South Carolina, manufactured gas plant phytoremediation site.

Parameter	Value
Air temperature	70 degrees Fahrenheit (°F)
Relative humidity (RH)	72% high: 30% low
Solar radiation	0.633 KWh/m^2
Wind speed	5 mph (8 kph)
Precipitation	50 in (127 cm)
Results	
ET_p (at 72% RH)	0.518 in/day (0.548 mm/h)
	0.518 in/day/0.25 = 0.123 in/day (for 6 hours (h) of light)
ET_p (at 30% RH)	0.588 in./day (0.622 mm/h)
	0.588 in./day/0.25 = 0.147 in./day (for 6 h of light)
Average ET_p	0.135 in./day (0.342 cm/day)
Average ET_p/month (30 day)	4.05 in./month (10.2 cm/month)

Table 8.3 Hydrologic physical properties for the Charleston, South Carolina, former manufactured gas plant site.

Variable	Parameter	Value
Q	Groundwater discharge	Calculated
A	Area of saturated aquifer	600 ft by 10 ft aquifer = 6,000 ft^2 (557 m^2)
i	Hydraulic gradient	(3 ft/30 ft = 0.1)
K	Hydraulic conductivity	10 ft/day (0.48 m/day)
n_e	Effective porosity	0.35

The average ET_p of 4.05 in./month (10.2 cm/month) is an estimate of the total amount of water vapor that potentially could be removed from the site. The dimension of the area available for planting was 600 ft by 30 ft, or 18,000 ft^2 (1,672 m^2). If ET_p is applied across this area uniformly, a daily estimated ET_p would be about 1,514 gal/day.

If it is assumed that the planted area does not use precipitation or soil moisture in the unsaturated zone, and only would use groundwater, how does this daily demand for water vapor compare to the daily discharge of groundwater that enters the shallow aquifer? If the aquifer is no more than 10 ft (3 m) thick, and the effective porosity, n_e, is 35%, then the volume of voids can be estimated as (600 ft)(30 ft)(10 ft) (n_e) = 63,000 ft^3 (1,783 m^3), or about 471,240 gal of groundwater beneath the planted area. If the abstract rate due to ET_p is constant at 1,514 gal/day (5,722 L/day), and no precipitation or influx of groundwater from upgradient areas occurs, it would take about 311 day to dewater this aquifer thickness by evapotranspiration.

However, groundwater flows into the site from upgradient areas, precipitation does recharge the aquifer, especially during the wetter summer and fall months, and plant roots do not remove groundwater from the entire aquifer thickness. The first two processes result in an increase in head in the water table, as an increase in groundwater storage, because for that time, water inflow is greater than water outflow. As stated in Chap. 4, an increase in head differential will increase the groundwater-flow rate in that area, which can be estimated with the Darcy equation. The Darcy equation can be used as a tool to determine the specific discharge of groundwater through a unit cross section of the water-table aquifer. If the depth of the saturated thickness along a transect through the aquifer can be estimated, then the cross-sectional area, A, can be estimated. The difference in groundwater elevation measured in an upgradient, h_1, and downgradient, h_2, well separated by some distance, Δl, gives the head gradient, i, where $i = (h_1) - (h_2)/\Delta l$. The hydraulic conductivity values from field or laboratory methods can be used. These physical properties for the Charleston site are presented in Table 8.3.

Solving for Q yields a flow of groundwater through the site of given dimension of 44,880 gal/day (169,646 L/day). If multiplied by the effective porosity of 35%, then 15,708 gal (92,832 L) of groundwater flow through the 10-ft (3 m) thick aquifer each day. When compared to the calculated ET_p of 1,514 gal/day (5,722 L/day), there is more than enough groundwater entering the site to account for removal by ET_p, and the water table would not be affected.

If one assumes, however, that the predominant source of groundwater to plant roots will be in the upper 2-ft (0.6 m) section of the 10-ft (3 m) saturated thickness, we can recalculate the flow of groundwater to be 8,976 gal/day, (33,929 L/day) and when multiplied by 35% effective porosity, the result is 3,141 gal/day (11,872 L/day), relative to the 1,514 gal/day (5,722 L/day) that can be removed by ET_p. Hence, about 50% of the groundwater discharge through the area in the upper one-fifth of the aquifer could be removed by ET_p. If a mature poplar tree can remove at least 15 gal/day/tree (56 L/day/tree), then to achieve this removal rate would require between 100 and 200 trees in the 18,000 ft^2 (1,672 m^2) area at a minimum of 10-ft (3 m) on center spacing.

8.3.2 Groundwater-Level Fluctuation Monitoring

Groundwater levels have been measured in monitoring wells at the former MGP site in Charleston, SC, between 1994 (before planting occurred) up to 2010. Between 1994 and the installation of hybrid poplar trees in late 1998, groundwater levels across the site reflected a balance between the input and output of water to shallow groundwater. The input of water consisted of local recharge and lateral groundwater

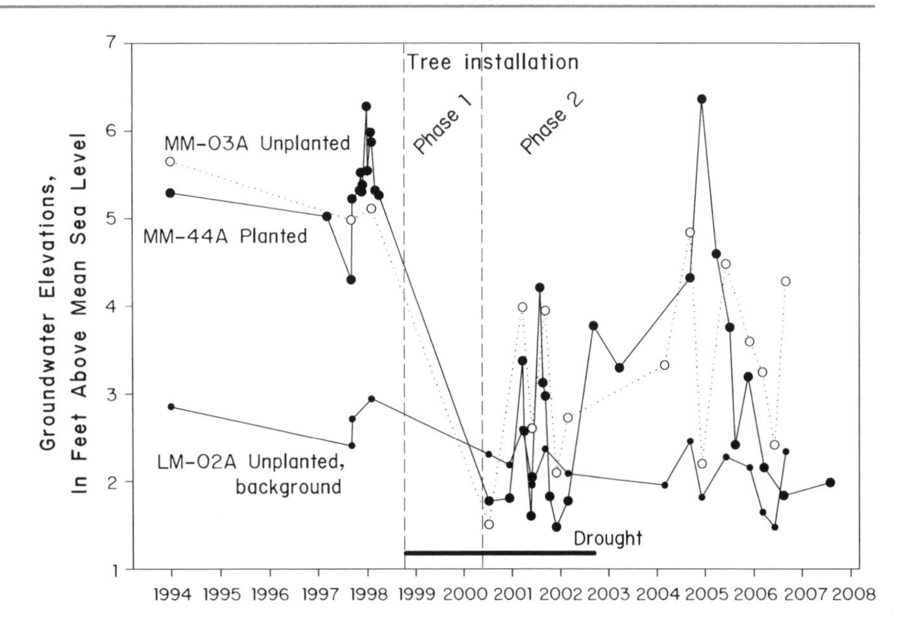

Fig. 8.5 Changes in groundwater levels measured in monitoring wells MM-03A (○) and LM-02A (•) in an unplanted area of the phytoremediation site and in monitoring well MM-44A (•) in a planted area, 1994–2007, at the manufactured gas plant site near Charleston, South Carolina. Between 1998 and 2003, the groundwater levels were lowered due to drought conditions. At the end of 2007, the groundwater levels measured in monitoring wells in the planted area were lower than the levels measured in wells in the unplanted area. One foot is equivalent to 0.304 m.

inflow of water recharged offsite. The removal of water included the discharge from the shallow aquifer by evapotranspiration of native grasses and oak trees and groundwater flow to the Cooper River (Campbell et al. 1996). Because groundwater levels in wells in the shallow aquifer were higher than groundwater levels measured in wells in the deeper aquifers, some groundwater was removed by vertical downward flow (Campbell et al. 1996).

Historically, in a monitoring well representative of the unplanted area, such as MM-03A, and in a monitoring well that represents the areas planted after 1998, such as MM-44A (Fig. 8.5), groundwater levels were uniformly high (greater than 5 ft (1.5 m) above mean sea level). However, precipitation amounts during the summer of 1998 were low, as this was the beginning of regional drought conditions across most of the southeastern United States.

After the poplar whips were established during phase one in late 1998, and up until late 2000 after phase two was completed, each year was characterized by conditions of precipitation more than 10 in. (25.7 cm) below normal amounts. As a result, groundwater levels monitored across the site decreased uniformly, in both the planted and unplanted areas (Fig. 8.5). Monitoring of water levels in wells during 2001, however, indicated that although drought conditions were affecting groundwater levels in both planted and unplanted areas, the decrease in groundwater level in wells in the planted area (characterized by trees now greater than 60 ft (18 m) tall) were greater than groundwater level changes measured in wells in the unplanted areas (Fig. 8.5). Between March and June 2001, the groundwater level declined 3.5 ft (1 m) and 2 ft (0.6 m) in wells MZ-55A (not shown) and MM-44A, respectively, compared to a decline of 1.5 ft (0.45 m) in background well MM-03A. The greater groundwater-level change observed in wells in the planted area appears to be related to the increase in transpiration demands associated with summertime highs in temperature and photosynthetically active radiation, *PAR*. Due to drought conditions, groundwater would be the most likely source of water to meet this increase in *ET* demand. Similar decreases in groundwater levels have been reported previously for pristine aquifers (Meyboom 1966) and contaminated groundwater systems (Eberts et al. 1999).

Groundwater levels in the site monitoring wells were observed to fluctuate directly with rainfall. Each rainfall event raised groundwater levels, and storage increased. During the drier winter months, the groundwater level averaged about 2.2 ft (0.6 m) above MSL. During the wetter summer months, groundwater levels were raised to about 4.6 ft (1.4 m) above MSL. However, during the winter months of 2005, unseasonably high rainfall amounts resulted in higher groundwater levels relative to the same time period during the 2003 and 2004 monitoring period. The increase in groundwater levels observed in wells following rainfall events provides direct evidence that recharge to the shallow water-table aquifer, and subsequent storage, occurs.

To determine the percentage of rainfall that becomes recharge, the ratio between the change in groundwater level and rainfall was determined for rainfall events for 2005. Because infiltrating rainfall fills only the voids in the aquifer, the change in groundwater level measured was normalized by aquifer porosity estimated to be between 0.20 and 0.30. The results are presented in Table 8.4.

The availability of rainfall and groundwater-level data at the site enabled the volume of recharge to be estimated. For 2005, the annual rainfall is about 40 in. (101 cm). If applied evenly across the 18,000 ft^2 (1,672 m^2) phytoremediation area, this equates to about 444,000 gal (1.6×10^6 L) of water input. The volume of rainfall that becomes recharge

Table 8.4 The percentage of rainfall that becomes recharge (%RC), for selected events during 2005, at the manufactured gas plant phytoremediation site, Charleston, South Carolina. The data shown are for a monitoring well located in the planted area. Greater than 100% recharge suggests recharge from rainfall was supplemented with additional soil moisture from previous rainfall events. Data given are in feet; convert to meters by multiplication by 0.3048; *n*, porosity; ΔGW, change in groundwater level.

2005	Rainfall (ft)	ΔGW Level (ft)	ΔGW (0.20)	ΔGW (0.30)	ΔGW Rain (n = .2)	ΔGW Rain (n = .3)	% RC (n = .2)	% RC (n = .3)
2/3	0.08	0.30	0.06	0.09	.06/.08	.09/.08	75	>100
4/2	0.08	0.50	0.10	0.15	.1/.08	.15/.08	>100	>100
5/16	0.13	0.25	0.05	0.07	.05/.13	.07/.13	38	53
5/17	0.08	0.50	0.10	0.15	.10/.08	.15/.08	>100	>100
5/30	0.15	0.40	0.08	0.12	.08/.15	.12/.15	53	80
9/28	0.20	0.60	0.12	0.18	.12/.20	.18/.20	60	90
Average							71	87

was calculated for all rainfall events during 2005, using measured groundwater-level changes and two estimates of aquifer porosity (Table 8.4). The calculated recharge volume is estimated to be about 326,000 gal (1.2×10^6 L) for 2005. This amount suggests that recharge was about 75% of annual rainfall (Table 8.4).

The calculated recharge to the water-table aquifer in the phytoremediation area is not constant but varies with time. The rather high percentage (71–87%) of precipitation that became recharge may be due to (1) that fact that the water table is near land surface in this low-lying area near the coast; (2) the presence of 3 ft (0.9 m) of topsoil that was added to the site prior to plant installation; (3) the potential of increased vertical permeability due to tree roots that have reached the water table, and; (4) the humid conditions of the coastal area that characterize the site.

8.4 Alternative Conceptual Frameworks for Groundwater Control

Because of the wide range of site-specific conditions encountered at sites characterized by groundwater contamination, it may not be practicable to rely on one single approach that could handle all site conditions. If such an approach were sought, by definition it would be too vague to provide much information that would be useful at a particular site, and it would be hindered by assumptions and have a high degree of uncertainty. It would be more appropriate to institute a multiple-line-of-evidence approach at sites to evaluate the potential for phytoremediation to achieve hydrologic goals. Evidence to support this conclusion is exemplified in the list of more than 140 phytoremediation sites and their phytotechnologies made available at the U.S. Environmental Protection Agency (USEPA) phytotechnology project profile at www.clu-in.org.

In addition to the framework of comparing ET_P to groundwater discharge just discussed with the case study for Charleston, SC, various alternative approaches exist. This section provides a few examples of these alternative conceptual frameworks. These examples were selected, in part, because information about these sites has been published and can be accessed more readily relative to other sites where only proprietary or confidential information has been generated. Also, the locations of these sites represent a cross section of climatic conditions from across the United States.

8.4.1 Water-Use Estimate Framework

One of the disadvantages of using ET_P as part of a framework to assess the potential effect of plants on groundwater is that it does not differentiate the source of the water removed. The fraction of ET_P used by phreatophytes can be estimated, however, as indicated with the previous example, in which the upper 20% of the full aquifer thickness was assumed to be the source of water removed by *ET*.

Another approach to estimate the fraction of groundwater that contributes to ET_P is given by Ferro et al. (2000). The uptake of groundwater by trees, V_t, can be estimated as being a fraction of the ET_P for the planted area. The total volume of water (V_t; in volume/time) used can be determined by

$$V_t = ET_P * \theta * LAI * A \tag{8.4}$$

where ET_P is the potential evapotranspiration (in./day or mm/day), θ is a water-use multiplier determined as a fraction of ET_P to represent that percentage of *ET* that is actually removed as actual evapotranspiration, ET_A, or crop coefficient, *LAI* is the leaf-area index, or leaf area per unit ground-surface area (dimensionless) that is discussed in Chap. 9, and *A* is the planted area (in feet or meters) assuming closed canopy is achieved (Fig. 8.6). Two examples are provided where this approach was applied.

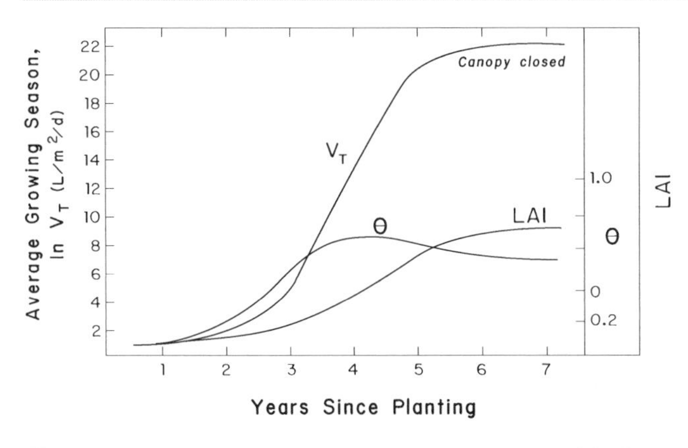

Fig. 8.6 Relation between the average growing season and leaf-area index, *LAI*, and time after planting as controls on plant transpiration of groundwater, V_t; θ is a water-use multiplier (Modified from Ferro et al. 2001). One meter is equivalent to 3.2 ft.

8.4.1.1 Case Study: Former Fuel Terminal, Utah

A phytoremediation system was installed at a petroleum-hydrocarbon contaminated site near Ogden, UT, in 1996. The 5-acre (20,235 m^2) site is a former light petroleum products terminal, used from the 1950s to 1989. The phytoremediation system was installed to control the migration of groundwater to offsite properties, the first hydrologic goal discussed in Chap. 6, as well as to enhance the degradation of contaminants. At this site, soil remediation of contaminants was addressed using grasses such as alfalfa (*Medicago sativa*) and fescue (*Festuca spp.*). The USEPA Superfund Innovative Technology Evaluation (SITE) Program, established by the USEPA Office of Solid Waste and Emergency Response and Office of Research and Development, to promote the evaluation of innovative technologies to remediate Superfund sites, was involved from the beginning.

The contaminated aquifer is a surficial water-table aquifer. It comprises silty sands, but no tests were performed to determine the hydraulic conductivity of the aquifer. Because it is a shallow system, depth to groundwater is about 6 ft (1.8 m) below land surface. Like many sites with shallow aquifers, however, this value is not constant, with high values near 2–3 ft (0.6–0.9 m) below land surface in the spring to lower values near 7–8 ft (2.1–2.4 m) below ground surface in the fall. The area is arid, with the ET_P often exceeding precipitation by as much as a factor of 2–10 (Ferro et al. 2001). For example, during the summer months, ET_P can approach 10 in./month (25.4 cm/month) and precipitation is less than 1 in./month (2.54 cm/month).

Poplar trees were planted in three rows, each 100 ft (30.4 m) long, in a location perpendicular to groundwater flow at the site and at the downgradient edge of a dissolved-phase plume of petroleum hydrocarbons. Forty hybrid poplar trees were planted (*Populus deltoides x Populus nigra* Imperial Carolina DN 34). Each of the 40 trees was installed in an 8-ft (2.4 m) deep, 10-in. (25.4 cm) diameter borehole created with a hollow-stem auger. In each borehole, a slotted 3/4-in. (1.9 cm) pipe was placed such that 1 ft (0.3 m) remained above ground, essentially to permit the entry of air to the tree roots. In each borehole, one 9-ft (2.7 m) long poplar whip, or pole, was planted such that 8-ft (2.4 m) was below grade in the sandy backfill amended with compost and slow release nutrients and 1 ft (0.3 m) was above ground level. No irrigation system was used. Five wells were installed along the direction of groundwater flow through the poplar tree rows, with two wells upgradient, one well within, and two downgradient from the rows. The wells were screened across the water table from 5 to 15 ft (1.5–4.5 m) to encompass the expected range of water-table fluctuations at the site. From 1998 to 1999, groundwater levels were collected using manual and automated methods.

The depth and distribution of roots was observed in 1998. Two intersecting trenches were dug almost 9 ft deep adjacent to a poplar tree. The exposed roots were counted along each exposed area. The soil in between the roots was washed away after pins were driven through the roots along one of the exposed faces. The *LAI* was estimated by Ferro et al. (2001) who, at the end of the growing season, removed and weighed all the leaves from trees that they had instrumented earlier with sap-flow sensors. Some leaves were then measured with a leaf-area meter, and the ratio of leaf area per unit leaf weight was determined. The leaf area of the tree so determined was divided by the ground area covered by the tree to get the *LAI*. These estimates were then compared to actual field measurements of these parameters to evaluate this alternative conceptual framework.

The uptake of groundwater by the trees, V_t, was estimated using Eq. 8.4. Some of Ferro et al. (2001) results are reproduced here. Calculations using Eq. 8.4 suggest that the V_t for an individual tree would be approximately 5 gal/day/tree (18.9 L/day/tree) during the 3rd year from planting to 18 gal/day/tree (68 L/day/tree) during the 5th year. Sap-flow measurements were made in the field to provide actual data for V_t to compare to the estimated V_t. The estimated ET_P was 5.5 in./month (140 mm/month), and the measured value was 4.04 in./month (102 mm/month). The estimated V_t was 3.8 gal/day/tree (14 L/day/tree), and the measured V_t was 2.8 gal/day/tree (10.5 L/day/tree).

One of the assumptions of this conceptual framework is that V_t is derived entirely from groundwater: that is, $V_t = V_{gw}$. During droughts this assumption may be valid. To account for uptake of any precipitation, however, the V_t can be normalized by the total precipitation measured, V_{ppt},

$$V_{gw} = V_t - V_{ppt} \tag{8.5}$$

Because precipitation, V_{ppt}, was estimated to be 65 gal/day (245 L/day) and the total water uptake, V_t, was estimated to be 510 gal/day (1,927 L/day), researchers concluded that up to 445 gal/day (1,682 L/day) of groundwater was used by the trees.

Much like the method described in the first part of this chapter, the total discharge of groundwater through a cross-sectional area of the aquifer was determined by the researchers using Darcy's Law to be 44 gal/day/ft of aquifer thickness (166 L/day/m). To account for the estimated 445 gal/day (1,682 L/day) of groundwater uptake by the trees, water from a 10-ft (3 m) thick section of the aquifer would need to be tapped to supply the estimated demand. The authors concluded that this demand of groundwater by the planted trees at the site should have resulted in a measurable water-table depression. None, however, was observed in the five monitoring wells. Ferro et al. (2001) suggest that the lack of water-table depression may be due to the fact that the groundwater flow rate was similar to the rate of groundwater uptake by the trees. Hence there would be no change in the water-table elevation, even though groundwater was being taken up.

8.4.1.2 Case Study: Superfund Site, Connecticut

The approach taken at this site was the same as described in 8.4.1.1. The site is located near Southington, Connecticut. The facility was used between 1955 and 1991 to recycle used industrial solvents. Depth to groundwater is between 4 and 5 ft (1.2–1.5 m) below land surface. Groundwater at this site was affected by volatile organic compounds (VOCs), such as chlorinated solvents, present in both the dissolved phase and DNAPL phase. In 1983, the site was listed as a Superfund site by the USEPA. The contamination is found throughout the aquifer thickness to bedrock at 30 ft (9.1 m) (Ferro et al. 2000).

To contain the DNAPL source and prevent additional downgradient flow of the dissolved-phase contaminants to offsite areas that include the Quinnipiac River, two traditional groundwater containment control structures had been used: a sheet-pile cutoff wall and a pump-and-treat system. The sheet-pile wall was constructed in a 700-ft (213 m) long section roughly perpendicular to groundwater flow, and placed up to 30 ft (9.1 m) deep through glacial overburden to bedrock. The pump-and-treat system consisted of 12 recovery wells, placed on the upgradient side of the sheet-pile wall. The average pumping rate for all 12 wells combined was less than 20 gal/min (75 L/min). This rate reflects the low hydraulic conductivity and low specific yield of the glacial overburden at the site. The pumped groundwater that contained VOCs was treated onsite by ultraviolet oxidation.

In order to determine if the amount of groundwater pumped by the wells could be enhanced or replaced by groundwater removed by the trees, a series of pilot plantings was performed in a 1.2-acre (4,856 m^2) part of the site. This initial phytoremediation system was installed in early 1998. Trenches were dug at least 3 ft below grade to near the top of the seasonal high water table, filled with peat and sand, and nearly 1,000 hybrid poplar trees (*P. deltoides x P. nigra* cuttings) were planted on 6-ft (1.8 m) centers across 0.8 acre of the 1.2-acre area of interest. By May 1998, only 60% of the planted trees had survived (Ferro et al. 2000); no explanation for the high mortality was provided. In early 1999, about 400 white willow trees (*S. alba*) were planted in the trenched area but in separate boreholes created with an auger, and only 5% mortality was observed. Other tree species were subsequently added to replace dead hybrid poplars. In the spring of 2002, the remaining hybrid poplars had to be removed due to a canker infestation by *Cryptodiaporthe populea*. Other native facultative phreatophytes were used, including pin oak, sweet gum, silver maple, river birch, tulip poplar, and eastern red bud.

Monitoring of the pilot phytoremediation system by Ferro et al. (2000) focused on the flow of groundwater through the stand of trees and sap-flow rates. Sap flow was determined between 5 and 7 times during May and September 2000 and 2003 using the heat-balance method. Sap-flow rates were normalized by the cross-sectional area of the sapwood of each tree measured and ranged from 6 to 16 gal/day/tree (26.6–71.9 L/day/tree). The highest sap-flow rate was measured in 2001 when precipitation was the lowest during the sampling period. The fact that ET_P exceeded precipitation at that time suggests that much of the sap flow consisted of groundwater. However, in 2003, the highest precipitation was recorded during the sampling period at the site, near 31 in. (80 cm), and exceeded ET_P such that recharge occurred, and sap-flow measurements probably reflect a mix of water from all sources. When these values were used by Ferro et al. (2000) to extend individual sap-flow rates to the entire planted area, the mean water use ranged from 2.1 to 8.0 gal/min (7.9 to 30 L/min).

Estimates were made using Eq. 8.4 to determine how much groundwater would be used by the trees. Estimates ranged from less than 1 gal/min (3.78 L/min) for the second year to almost 10 gal/min (37.8 gal/min) by the fifth year. If this rate is superimposed upon the average mechanical pump rate of 19 gal/min (71 L/min) attained by the pump-and-treat system, the pump-and-treat could be cut back to near 10 gal/min (37.8 gal/min) during the summer, as the removal by trees would account for the difference. Of course, Ferro et al. (2000) state that the pump rate would have to be increased during the winter to offset the reduction in transpiration by the dormant plants (Figs. 8.7 and 8.8).

Because groundwater-level data were not provided in Ferro et al. (2000), it is not clear if the water-table level decreased due to the mechanical or the planted system. This type of information would be useful, because it is unclear if well-and-plant interference would present a

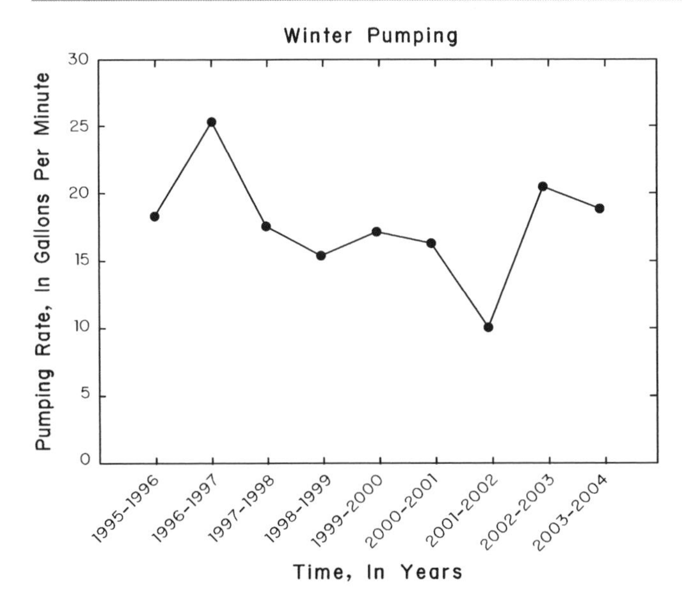

Fig. 8.7 Average mechanical pumping rate during winter, 19 gpm (gal/min) (71.8 L/min), for the pump-and-treat system with no groundwater removed by trees (Modified from Ferro et al. (2000)).

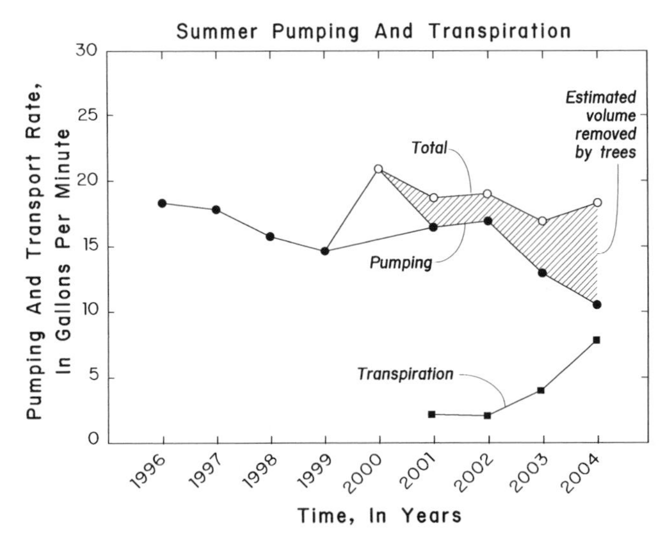

Fig. 8.8 Average mechanical pumping rate, in gpm (gal/min) could be cut back by almost 50% during the summer, when trees were removing groundwater (Modified from Ferro et al. (2000)).

complicating factor and decrease individual plant removal of groundwater.

8.4.2 Groundwater Flux Framework

This approach to determine the interaction between plants and groundwater is based on the water-balance equation. Instead of calculating the various components of the water balance and comparing them to each other, a flux approach is based on observed decreases in the groundwater flux through a cross-sectional area located downgradient from a planted site relative to that groundwater flux that enters the planted area in upgradient areas. The groundwater flux is *not* compared to ET_P, but to determine the change in groundwater flux caused by the plants. The expected decrease in groundwater flux would be observed as a decrease in groundwater level in monitoring wells located in the planted area. In this manner, the shortcomings of using ET_P are eliminated and are replaced by direct observations of changes in the groundwater system caused by trees. More on this method can be found in Eberts et al. (1999) and Landmeyer (2001).

8.4.2.1 Case Study: Air Force Plant 4, Texas

A phytoremediation project was initiated in 1996 at a site near Air Force Plant 4, located on the Naval Air Station west of Fort Worth, TX. The site has been used to manufacture aircraft since 1942, even before the United States entered into World War II. As part of aircraft construction, chlorinated solvents such as TCE were used and disposed of in landfills or fire training pits. As a result, TCE is present in the shallow aquifer downgradient from the aircraft manufacturing buildings. The facility is still used to produce aircraft such as the F-16 fighter jet. The plume of TCE-contaminated groundwater was first detected in 1982.

Phytoremediation was initiated to (1) reduce the mass flux of contaminated groundwater leaving the planted area, (2) alter the redox condition of the aquifer to one more favorable for the destruction of TCE, as is discussed in Chap. 13, and (3) remove groundwater and alter the groundwater-flow path from the site. The project was initiated by the U.S. Air Force, the Department of Defense Environmental Security Technology and Certification Program, and was a charter site for the U.S. Environmental Protection Agency SITE Program. Other researchers associated with the following agencies also have been involved at the site: Science Applications International Corporation; University of Georgia; U.S. Forest Service; USGS.

The contaminated shallow aquifer consists of alluvial sediments composed of particles sizes ranging from clays to sands and gravels with some limestone, and a porosity of about 25%. It is located in the Osage Plains of the Central Lowland Physiographic Province (Vose et al. 2000). Due to its alluvial progeny, the aquifer is thin, between 1.6 and 4.92 ft (0.5 and 1.5 m) thick, and depth to groundwater ranges from 2.5 to 13 ft (0.7–4 m) below grade. The hydraulic conductivity ranges from 3.2 to 98 ft/day (1–30 m/day), with an average groundwater flow velocity of 1.6 ft/day (0.5 m/day). The contaminated groundwater flows downgradient and discharges to a local creek. In this part of Texas, the subhumid climate provides precipitation of about 31 in. (78 cm) per year, with recharge of about 3 in. (7.6 cm) per year. Most precipitation occurs from May to October (Vose et al. 2000).

Hybrid poplar trees (*Populus deltoides*; poplars are native to this part of Texas) were installed in early 1996 in two areas using different approaches. Each planted area was rectangular in shape, 49 by 246 ft (15 by 75 m), and located perpendicular to the generalized direction of groundwater flow. The upgradient planting used 440 whips, or vegetative cuttings from mature trees of eastern cottonwood, and consisted of clones from local trees. The initial mean diameter of these whips at 3.9 in. (10 cm) above ground surface was 1.1 in. (2.8 cm) (Vose et al. 2000). The downgradient planting used 224, 1-year-old seedlings of eastern cottonwood grown at a local nursery. The initial mean diameter was 1.8 in. (4.6 cm). Rather than installing separate holes for each tree, linear trenches were dug to the depth of 3.2 ft (1 m), 7.8 ft (2.4 m) apart. This design facilitated the installation of a drip-irrigation system in the trenches. The drip irrigation system was used on alternating days following planting, due to drought conditions. It was determined that tree roots reached the water table during the end of the second growing season (1997).

The amount of water transpired by the trees was estimated using sap-flow measurements. Between May and October 1997, up to 14 trees were measured, including recently planted whips as well as the 1-year-old seedlings. Transpiration ranged from 0.42 to 2.4 gal/day/tree (1.6 to 9.2 L/day/tree) for the whips, and from 0.2 to 3.9 gal/day/tree (0.92 to 15 L/day/tree) in the 1-year-old trees (Vose et al. 2000).

Groundwater levels were monitored in all the wells installed at the site. The maximum decrease in groundwater level measured was 3.9 in. (10 cm), and this was in a well located between the two plantings. The change in groundwater flux due to the removal of groundwater by trees was estimated for each year after planting from 1996 to 1999. The change ranged from 2% to 12% less than when trees were just planted and no uptake of groundwater was assumed. As can be imagined from the reported decrease in groundwater flux of no greater than 12% from preplanted conditions, groundwater continued to discharge to the creek. Even in such a thin aquifer, not all groundwater flowlines were being diverted upward as a result of groundwater uptake by the trees.

8.4.3 Numerical Model Framework

This approach is based on numerical groundwater-flow models that simulate the various parts of the water budget at a contaminated site. For the model to function, a mass balance of water is calculated for each time step of the model. The simulation of plants and groundwater interactions in such groundwater-flow models is at best an approximation, however, as is discussed in Chap. 14.

8.4.3.1 Case Study: Areas 317 and 319, Argonne National Laboratory, Illinois

A phytoremediation project was installed at a contaminated site located at the Argonne National Laboratory, near Chicago, Illinois, in mid-1999. The site consists of two contaminated areas, the East Argonne Areas 317 and 319, totaling about 3.2 acres (12,950 m^2). The contamination resulted from the disposal of VOCs and tritiated water, respectively, in french drains used to dispose of the wastes. The french drains are no longer used, but the sites remain active facilities for waste processing and storage. The phytoremediation system designed for the sites was to (1) degrade the contaminants in groundwater (discussed in Chap. 13) and (2) provide hydrologic containment of the contaminant plumes such that groundwater flow across the downgradient property boundary could be decreased. The phytoremediation project was funded by the Department of Energy Accelerated Site Technology Development (ASTD) Program. After 1999, the U.S. Environmental Protection Agency SITE Program became involved. More information on the history of this project is given in Quinn et al. (2001).

The hydrogeologic setting of the disposal areas consists of multiple aquifers and confining units, reflecting the glacial history at the site. These layers of sediments are of widely different permeability and hydraulic conductivity, from low-permeability tills consisting of silts and clays to high-permeability sands and gravels. The aquifers systems are not of uniform thickness or of great lateral extent, and interconnection between separate sand layers occurs. Groundwater from these areas flows offsite to discharge in ravines in a downgradient forest preserve and ultimately to the Des Plaines River. Recharge to the aquifer system is by leakage from overlying units of high permeability or directly by precipitation. The hydraulic conductivity at the sites was determined from pump and slug tests, and averages 8.8 ft/day (2.6 m/day), but with considerable variation over relatively short distances.

Unlike the other case study sites discussed so far, because the contaminants were introduced to the subsurface through french drains dug into the overlying shallower aquifer and confining unit, the groundwater contamination was present in a confined aquifer at depths between 25 and 30 ft (7.6–9.1 m) below land surface. The contaminated confined aquifer of interest ranges from 3 to 10 ft (0.9–3 m) thick. Conventional planting approaches as have been previously described could not be used. Rather, a deep-rooting method called TreeWell™ Treatment System (or TreeMediation®, Applied Natural Sciences, Inc.) was employed to get the roots into the contaminated confined aquifer. In brief, a deep borehole is created to the appropriate depth through overlying aquifer and confining unit materials. A caisson is added to keep the hole from collapsing, as well as to limit or constrict lateral root formation, and materials having greater permeability than the surrounding aquifer are added to the borehole (Fig. 8.9). A tree is then planted in the borehole, and a pipe is installed from the root zone to the surface to increase air exchange. At the surface, the borehole is sealed such that no infiltration can occur.

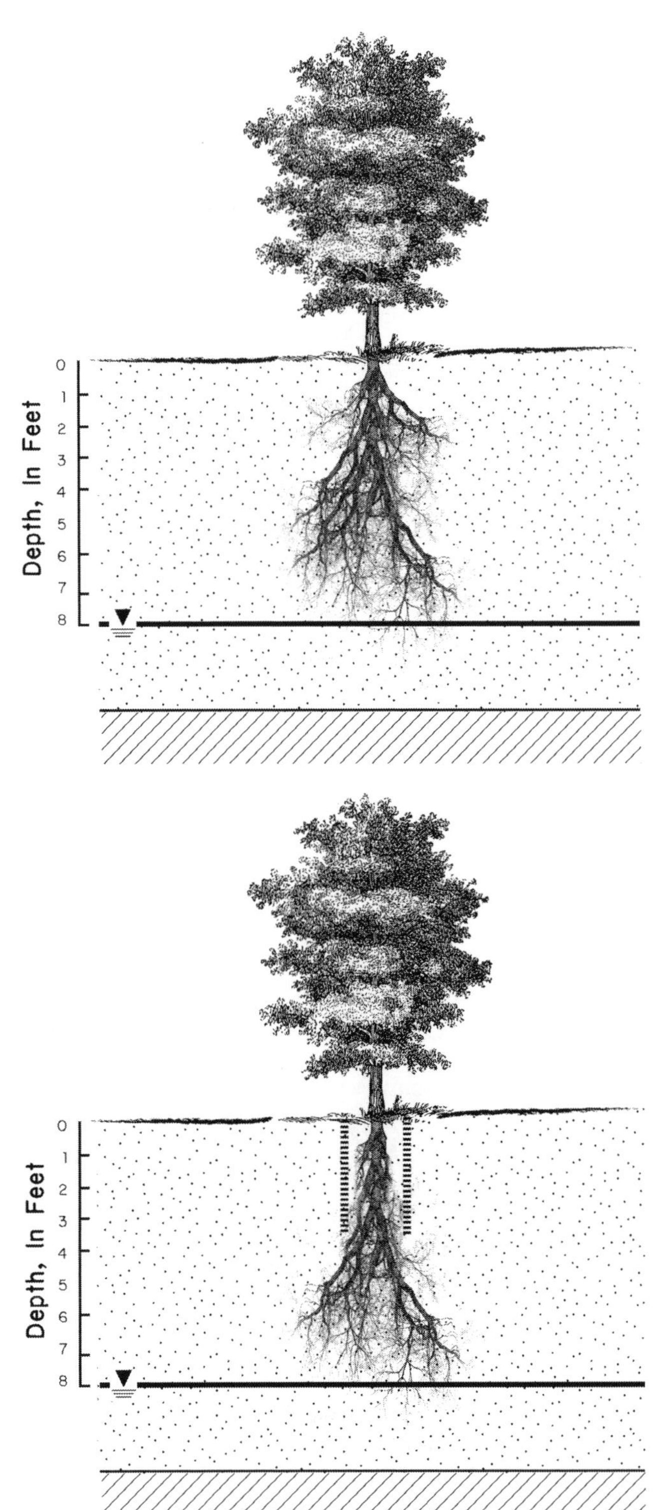

Fig. 8.9 The variation in root distribution with depth for hybrid poplar trees installed with and without shallow borehole constriction methods. One foot is equivalent to 0.304 m.

At the Argonne site, 420 poplar trees (hybrid HP 510 *Androscoggin poplar x* HP 308 *Charkowiensis incrassata*) were planted using this deep-rooting method. The trees were planted in 2-ft (0.6 m) diameter boreholes drilled to 30 ft (9.1 m), and lined with plastic. Each borehole was placed 16 ft (4.8 m) apart. The plastic lining is part of the TreeWell™ Treatment System to ensure that root growth is vertical and downward toward groundwater (Fig. 8.9). The backfill consisted of topsoil, sand, peat, and manure. Conventional planting methods were used to install 389 willows in shallow, contaminated soils.

A groundwater-flow model was used to estimate the future impact of the deep rooted trees on groundwater flow. The USGS model MODFLOW (McDonald and Harbaugh 1988) was used to simulate groundwater flow. A steady-state flow model was constructed and calibrated to groundwater levels collected over a 10-year period at the site prior to planting. The agreement between measured and simulated groundwater levels was very good, with a measure of spread, the root mean square error (RMSE) of 0.58 ft (0.1 m). Next, transient simulations were run to simulate the potential release of groundwater from storage in the system. Removal of groundwater by the trees was simulated using the ET module in MODFLOW. The ET module linearly relates the rate of *ET* to the depth of the water table; *ET* is at a maximum rate when the water table is high within a particular model cell and *ET* drops to zero at a predetermined lower water-table level called the extinction depth; more about this assumption is described in Chap. 14. This transient model was then used to predict the impact that the deep-rooted trees would have on groundwater flow at the site 6 years into the future from plant installation in 1999.

Simulation results suggested an impact on the groundwater-flow system as early as the summer of the second year, or 2001. Simulations indicate that in future years, the plants will exert a seasonal effect on the confined groundwater system, with lowered groundwater levels during the 6-month growing season at the site, from April to September, and higher levels during the other months. The effect was not cumulative, because the cycle of higher and lower head with seasonal changes in groundwater use by the plants was consistent for all simulated future years.

The rate of conservative solute movement with particles of groundwater from source areas to discharge areas also was simulated using the USGS model MODPATH (Pollock 1994). The particle-tracking simulations performed were used to determine if groundwater particles could be captured by the plants. The simulations indicated that hydrologic control and groundwater containment were reached. The simulations indicate that most of the particles released in upgradient source areas in the confined aquifer were captured by the deep-rooted trees. However, sap-flow measurements made on 1-year-old trees in 2000 indicated that tree-water flow was no higher than an average of 1.1 gal/day/tree (4.3 L/day/tree), an order of magnitude lower than the values used in the model simulations of 12.5 gal/day/tree (47.3 L/day/tree).

8.4.4 Water-Budget Framework

This approach is based on performing a site water budget, similar to that introduced in Chap. 2. In this case, the sources of water inflow and outflow to the site are enumerated and quantified, including the transpiration of water by planted trees. In essence, the site is viewed as a microcosm of the hydrologic cycle.

8.4.4.1 Case Study: Aberdeen Proving Ground, J-Field Superfund Site, Maryland

A phytoremediation demonstration project was implemented during 1996 at the J-Field Site at the Aberdeen Proving Ground in Maryland along the Chesapeake Bay. The J-Field Site consists of two parallel trenches used to burn a variety of chemicals and wastes between 1940 to the 1970s (U.S. Environmental Protection Agency 2000b). As a result, a legacy of contaminants, primarily chlorinated solvents, has remained in the soils, sediments, and groundwater at the site. The application of phytoremediation at this site was designed to (1) remediate the contamination (described in Chap. 13) and (2) alter the groundwater levels at the site to stop the movement of groundwater to nearby surface-water resources. The project was conducted by the U.S. Army, the USEPA Region III, and the USEPA Environmental Response Team Center. More information on this site can be found in Hirsh et al. (2003).

The contaminated shallow aquifer consists of low-permeability sediments, such as fine sands and clays such that the hydraulic conductivity varies between 0.3 and 8 ft/day (0.1–2.4 m/day). Aquifer tests performed at the site indicate that the specific yield of the shallow aquifer is no greater than 1 gal/min (3.7 L/min). Groundwater flows radially from the test pits in higher elevations after recharge and ultimately discharges in low-lying areas adjacent to freshwater marshes and a tidally influenced estuary.

Field measurements were made to evaluate the site water budget using sap-flow meters and a weather station. The researchers used the meteorological information from the weather station to determine the overall ET_P for the site. The climate is considered temperate, and precipitation averages 45 in./year (114 cm/year) distributed evenly throughout the year (Hughes 1995). The ET_P was then compared to the water flow though the trees, as determined by the sap-flow gage studies. One well was added in an unplanted area to represent groundwater conditions not impacted by the trees. Lysimeters also were installed at the site above the water table near monitoring wells, but were reported to be problematic (U.S. Environmental Protection Agency 2000b). Between 2000 and 2001, groundwater level measurements were made in 21 wells on a continuous basis with downhole pressure transducers. Because the site was located near a tidally influenced surface-water body, tidal-fluctuation data were collected in order to remove their effect on groundwater levels so that any tree influence on water levels could be observed. This method is described in Chap. 9.

In 1996, about 180 hybrid poplar trees (*Populus deltoides x Populus trichocarpa*) were planted in a 1-acre (4,047 m^2), U-shaped area that surrounds the downgradient part of the pits and is within the contaminant plume. The trees were planted at a spacing of 10 ft (3 m) from each other. The trees were planted in excavated holes to which was added plastic sleeves to encourage downward root growth rather than lateral root growth. Some trees died within the first few years, perhaps as a result of the beginning of a 5-year drought in the eastern United States that started in 1998. As such, 65 additional trees were planted in 1998, including hardwood species native to the area, such as tulip trees (*Liriodendron tulipifera*) and silver maple (*Acer saccharinum*).

Following the toppling of some trees during storms, it was concluded that the plastic sleeves placed in the boreholes before planting were reducing the support strength of the trees provided by lateral roots, and future use of the sleeves was not encouraged. Excavation of some trees indicated root depths of at least 7 ft (2 m) by 1998, only 2-years after planting. Tubing from the surface to the roots was placed in the backfill in an attempt to introduce atmospheric oxygen into the subsurface and thus enhance root growth by removing oxygen as a limiting factor for root respiration. A drainage system to remove surface-water overland flow from the areas was also constructed to reduce the amount of infiltration to the roots from aboveground sources.

The water budget was calculated for the site using meteorological data collected between 2000 and 2001 and using sap-flow data. The equation used was a modification of that presented as Eq. 2.4

$$P = SR + E + SW + T + GR + S' \quad (8.6)$$

where P is precipitation, SR is surface runoff, E is evaporation, SW is soil water, T is transpiration, GR is groundwater recharge, and S' is the change in soil water and aquifer storage. The weather data were used to estimate the ET, in a manner similar to the approach described earlier. Transpiration rates were determined by sap-flow measurements made in the field. Sap-flow data indicate that, on average, individual trees were moving water, some of which was groundwater, at a rate of 6.8 gal/day/tree (26 L/day/tree) during the growing season. This rate was used to estimate that the entire planted area may be using 1.058 gal/day (4,000 L/day) in 1999. Because closed-canopy conditions were not yet reached at this site, and these conditions may take longer given the 10-ft (3 m) spacing between trees, it may take as long as 30 years before the transpiration rates increase enough to double the total water flow through trees of 2,010 gal/day (7,600 L/day). Even so, these higher rates of groundwater removal would be about one fifth of the total groundwater flow through the area that discharges to adjacent surface-water systems.

Site researchers state that transpiration by the plants is a significant sink for groundwater during the summer growing season (Hirsh et al. 2003). In 1999, groundwater-level monitoring indicated a depression in the water-table surface beneath the planted area. The decrease in groundwater levels was observed to be about 5.4 in. (12 cm) during the summer growing season. The groundwater-level data were normalized for nonplant-induced changes, by correcting for tidal and barometric efficiencies. Daily fluctuations in the groundwater table in wells in the planted areas were measured to be near 1.8 in. (4.5 cm). During the summer, groundwater levels in the planted area become lower than even the adjacent marsh surface-water levels, and surface water would then have the potential to recharge the aquifer. This situation would reverse during the winter, when precipitation would be greater and *ET* would decrease. Interestingly, the authors suggest that groundwater flow is induced upward to the trees during the summer, but, unfortunately, data to support this assertion are not provided.

8.4.5 Plant-and-Monitor Framework

This approach involves the installation of plants and monitoring of wells for groundwater levels and contaminants in a manner similar to that of other remedial strategies.

8.4.5.1 Case Study: Landfill, Washington

A phytoremediation system was installed as part of overall remedial actions in early 1999 at a former landfill located at the Naval Undersea Warfare Center (NUWC), Division Keyport, Washington, about 11 water miles (17.6 km) from Seattle in central Puget Sound. The total landfill area is about 9 acres (36,423 m^2), and it is about 10 ft (3 m) above sea level. It used to be marshland connected to tidal flats but was filled in by the U.S. Navy. The landfill was operated between 1930 and 1973. It accepted domestic and industrial wastes generated by Naval activities. The landfill is unlined, but part of the landfill is covered by asphalt. The predominant groundwater contaminants are the chlorinated solvents PCE and TCE and their degradation byproducts.

The hydrogeology in the area comprises unconsolidated Pleistocene glacial deposits. These deposits range from sands to gravels, and silts to clays. Although the landfill proper comprises fill and debris, the native geology also is heterogeneous, with interbedded sands, gravels, and alluvial and fluvial deposits. The depth to groundwater at the site is about 10–15 ft (3–4.5 m), which approaches the maximum limit for efficient groundwater interaction. There is an unconfined aquifer, a discontinuous confining unit, and an underlying intermediate aquifer. The water table is affected by tidal fluctuations, which need to be monitored in order to determine the effect, if any, that the trees may have on shallow groundwater. The groundwater-flow rate in the shallow contaminated aquifer is about 29–83 ft/year (8.8–25 m/year). The site has wet winters and dry summers. Although the Pacific Northwest is known for its precipitation, the site actually receives a modest amount of about 30 in./year (76 cm/year), with most falling between October and March.

The shallow groundwater beneath the landfill and in the direction of local groundwater flow to adjacent surface water in Dogfish Bay has been documented. The site is being evaluated by members of the Naval Facilities Engineering Field Activity Northwest (EFA NW), the Washington Department of Ecology, USEPA Region 10, the USGS, the Suquamish Tribe, the local Regional Advisory Board, and the faculty and staff of the Universities of Washington and South Carolina. There are domestic drinking-water wells downgradient from the landfill, although in deeper aquifers. The goals of the phytoremediation planting are to affect the site hydrology and to decrease contaminant levels. Phytoremediation is stipulated as the remedial action in the Record of Decision (ROD) for the site.

The plantings occurred in two separate areas of the former landfill (Rohrer et al. 2000). Prior to plant installation, intensive activity was required to prepare the site for trees. The asphalt cap was removed. Landfill material was exposed during site preparation, of which some was left behind, but a large amount (24 tons of debris) was disposed of offsite. Clean fill was added to replace that removed to depths of between 1 and 2 ft (0.3–0.6 m) at the north planting and between 2 and 3 ft (0.6–0.9 m) at the south planting, for a total of about 3,100 cubic yards. To compensate for fill soil chemistry, lime and urea were added and turned into the fill with a chisel plow.

After soil preparation was completed, hybrid poplars were installed as 8-in. (20 cm) hardwood cuttings in April 1999. Irrigation was necessary after installation due to the depth to water table being near the maximum extent of efficient phytoremediation. The irrigation system was a drip type installed at 2-ft (0.6 m) intervals, with a maximum flow rate of 10 gal/min (27.8 L/min). The irrigation was derived from shallow contaminated groundwater by a shallow pumping well at each planted area. Apparently no previous calculations to estimate the effect of the trees on the site hydrology were reported. Dinicola et al. (2002) state that the trees had no immediate effect on groundwater levels or of contaminant concentrations.

8.4.5.2 Case Study: Portsmouth Plant, Ohio

The Portsmouth site is the location of a Department of Energy (DOE) uranium enrichment facility. It is located in central Ohio and was constructed in the early 1950s to enrich uranium. These activities stopped in 2001. Processes performed at the former oil-handling facility at the site's Gaseous Diffusion Plant left a legacy of chlorinated solvents in the water table and bedrock aquifers; at least five separate plumes were detected beneath the site in the early 1990s. At two of the plumes, concentrations of TCE ranged from 5 to 10,000 μg/L. These plumes were selected for remediation by phytoremediation as part of corrective action alternative studies done in the late 1990s. The depth to the surficial water table is about 10 ft (3 m). The depth to the contaminated bedrock aquifer is about 25 ft (7.6 m).

A phytoremediation system was installed at one plume in 1999 (Ferro et al. 2000). Trees were planted in the source area as well as downgradient from the plume. The trees were installed using two approaches: (1) 2-ft (0.6 m) wide trenches were dug to a depth of 10 ft (3 m), or to the top of the water table, and (2) 2-ft (0.6 m) diameter boreholes were made 10 ft (3 m) deep. Each was backfilled with fine sand. As such, 240 cuttings were installed in the trenches, and 526 trees were installed in the 10-ft (3 m) deep borings. Also at the site, 8-in. (20 cm) diameter borings were drilled 30 ft (9 m) deep to the local aquifer contact with bedrock. These boreholes were not planted but were filled with sand. The fine sand was used to physically wick-up water by capillary action to bring the contaminated groundwater in the deeper bedrock up into the area where the tree roots were growing. This method is essentially using a modification of shallow-rooted planting methods to remediate deeper groundwater contamination without using a proprietary method.

At another plume, the source of the chlorinated solvents was believed to be either a landfill or an old paint shop. A phytoremediation system was installed in 2002. A similar planting method was used, such that numerous 2-ft (0.6 m) wide trenches were dug to the top of the water table between 10 and 15 ft (3–4.5 m) deep and were backfilled with fine sand. Up to 3,300 cuttings were planted.

8.4.6 Other Conceptual Frameworks

Most of the approaches, or frameworks, previously discussed are based on decreasing the flow of groundwater to offsite areas by redirecting the groundwater to plants. An alternative approach presented here is to decrease the flow of groundwater through a source area. This process is accomplished using a constant-head moat, called an *ET* moat. The *ET* moat is created by a trench dug at a depth of the mean water table and a closed-loop drain pipe is installed. The trench is then backfilled. Due to the presence of the drain pipe, the water-table elevation can rise no higher than the elevation of the drain pipe, as long as recharge rates do not exceed groundwater-flow rates. Trees are then planted within the moat area of lower water table in order to remove additional water by *ET*. The limitation for widespread application is the high cost for the trench and pipe installation, disposal cost for removed material, and potential for root clogging of the drain pipes, which would decrease the efficiency of the constant head over time.

8.5 Summary

Naturally occurring plant interactions with groundwater affect groundwater levels and flow direction. This interaction provides the basis for various approaches (or frameworks) to affect contaminated groundwater through a decrease in recharge and increase in transpiration.

Why is this information important to the phytoremediation of contaminated groundwater? The effects of plants on contaminated groundwater often are not readily observed because the interaction occurs at depth. The approaches described in this chapter also indicate that a water-budget method used on its own is limited by the inability to account for the amount of groundwater removed by trees in the presence of multiple sources of water. Fortunately, there are geochemical methods that can be used in conjunction with the water-budget method to determine the source of water to trees at phytoremediation sites and are discussed in Chap. 9.

Monitoring Plant and Groundwater Interactions 9

In general, the combined processes of evaporation and transpiration can remove about 70% of annual precipitation from a basin or, on a smaller scale, a phytoremediation site. But how much of this water is removed by transpiration? How much of this transpired water is derived from groundwater? Fortunately, geochemical methods can be used to elucidate the various sources of water, including groundwater, that comprise sap flow. These geochemical methods can be used in combination with the water-budget methods discussed previously to decipher plant and groundwater interactions at contaminated sites.

9.1 Plant Physiologic Monitoring Methods

Various methods based on the fundamentals of plant physiology presented in Chap. 3 can be used to monitor the interaction between plants and groundwater to meet the three hydrologic goals presented in Chap. 6. The common denominator emphasized in most of these methods is the water status of plants. The measurement of plant-water status, or water potential, is only part of the story, however. Plant-water status does not indicate whether groundwater is the only source of the water being assessed in various plant tissues, and this limitation is addressed later in this chapter.

9.1.1 Water Potential

A property that can be used to assess plant-water status with respect to groundwater interaction for hydrologic control is water potential, introduced in Chap. 3. The concept of water potential provides an advantage over soil-moisture content, because the amount of water in sediment relative to saturated conditions does not indicate whether (1) the water is bioavailable to plants, or (2) the direction that water will flow. Because water will flow from less negative water potentials to more negative water potentials, the measurement of water potentials of the capillary fringe, soil, root zone, plant, and leaves, can provide an indication of the potential for water to be transpired under given conditions.

The water potential of each component of the soil–plant–water–air continuum can be measured, although such water potentials are not commonly measured at most phytoremediation sites. Soil-water potentials can be readily assessed in the field, however, using instruments such as tensiometers or psychrometers, as described in Chap. 3 and briefly summarized here. Tensiometers can be placed directly in the field to measure the water potential of the water in the soil near plant roots. A tensiometer is essentially a porous ceramic cup attached to a tube filled with water and then sealed. If the soil in which the tensiometer is placed is drier than the water-filled ceramic cup, water will exit the cup. Because no air enters the water-filled tube to replace the water that exited from the cup, a negative pressure develops in the tube. This pressure change can be measured with a pressure gauge installed in the air space created in the tensiometer.

With the psychrometer, a piece of plant material of unknown water potential is placed in a sealed chamber that also contains a droplet of a solution of known water potential. If the plant material has a lower water potential than the droplet of reference solution, and hence a lower vapor pressure by way of a higher solute concentration, preferential evaporation from the droplet cools the surface of the water; this temperature difference is measured using a thermocouple. Conversely, if the droplet of reference solution has a lower water potential than the perhaps less concentrated sample of plant material that contains water, the sample's evaporation will warm the reference droplet. Hence, if a particular solution's water potential is known and it results in no net movement of water to cool or warm the droplet, then the sample of plant material containing water must have the same water potential as the reference sample. Because a change in temperature can also cause a change in water potential, where a change in 0.01°C = 0.1 MPa, the chamber must be kept at constant temperature. As such,

J.E. Landmeyer, *Introduction to Phytoremediation of Contaminated Groundwater*,
DOI 10.1007/978-94-007-1957-6_9, © Springer Science+Business Media B.V. 2012

psychrometers are often only used in a laboratory setting. The psychrometer method has been used extensively by Boyer and Knipling (1965).

Another method to measure water potential involves placing a piece of plant material into a chamber, called a pressure bomb. Because the Cohesion-Tension theory is based on having negative water potentials up to −1 MPa in the xylem, the water in plants will be under tension. This tissue is pressurized in the bomb to restore the distribution of water potential between living and nonliving xylem cells in the plant material. A typical instrument that can make these measurements is the Scholander Pressure Bomb (PMS Instruments, Corvallis, OR), named after the research of Scholander et al. (1965). Because the act of taking a biopsy of plant material will release any tension present when the water column in the xylem is broken, water initially flows into the living cells by osmosis or capillary action. Pressurization of the chamber, however, will reverse this flow of water back to the xylem. The advantage of this instrument over the psychrometer is that it can be used in the field. There are those who suggest, however, that the bomb technique produces inaccurate and more negative water potentials than expected (Zimmermann et al. 2004).

The psychrometer and pressure bomb instruments can be used to assess the water pressure of water present in plant tissues removed as a whole component of many different cells. The turgor pressure of individual cells, however, also can be directly assessed using an instrument called a pressure probe (Zimmermann 1989). An air-filled glass tube sealed at only one end can be inserted into a cell. The pressure in the cell compresses the gas in the glass tube, and using the Ideal Gas Law the pressure can be calculated. The hydrostatic pressure of individual cells also can be measured with a similar approach that uses a glass microtube but filled with an incompressible oil rather than air. This oil can be readily distinguished from the sap that flows into the tube, and this sap flow can be offset by depressing a plunger, which can indicate the hydrostatic pressure of water in the cell.

Given the different instruments that can be used to measure water potentials, there are various ways to present water potential data collected from the field or under laboratory conditions. Because different instruments often lead to the use of different units, confusion can ensue, which inhibits the comparison of data collected using different approaches from different sites. Table 9.1 shows the range of water potentials in drying sediment and the units of measure most commonly used.

Models to simulate the uptake of water in the root zone and based on water potential were first applied in the 1960s (Gardner 1960; Green et al. 2006). Gardner's approach used an analytical solution, whereas other more recent approaches are based on numerically solving the Darcy-Richards equation.

9.1.2 Root Hydraulic Conductance

As described in Chap. 3, the term hydraulic conductivity is used both in plant physiology and hydrogeology to describe the movement of fluids though various media. With respect to plants, hydraulic conductivity describes the diffusive movement of water through the symplast from one cell to another cell by way of the cell membranes and plasmadesmata. Because cells are separated from each other by a semipermeable membrane, this acts to deter the simple flow of water. For example, if a cell at a certain water potential is placed in contact with water at a higher potential, water will move into the cell because it has lower water potential. The central questions here are, what is the initial rate of the water movement, and what controls this rate?

The resistance to water movement caused by the cell membrane, a force that can be quantified, is referred to as root hydraulic conductivity, L_p. Root hydraulic conductivity provides a way to assign a value to the degree of resistance to the diffusional flow of water between cells, and has units of volume of water per unit area of membrane per unit time per unit driving force ($m^3/m^2/s/mPa$). The velocity of water transport, W_v, from one cell to another can be described as

$$W_v = L_p(\Delta\psi) \tag{9.1}$$

Table 9.1 Range of units typically used to report measurements of water potential.

Status	Megapascals (MPa)	Kilopascals (kPa)	Millipascal (mPa)	Bars	Pounds per square inch (Psi)	Osmolality[a]
Wet	−0.001	−1	-1.0×10^6	−0.01	−0.145	0.0004
Field capacity	−0.033	−33.0	-3.3×10^7	−0.33	−4.786	0.0135
Plant available water, lower limit (wilting point)	−1.5	−1,500	-1.5×10^9	−15	−217.5	0.6157
Air dry	−100	−100,000	-1.0×10^{11}	−1,000	−14,503	41.0494
Oven dry	−1,000	−1,000,000	-1.0×10^{12}	−10,000	−145,037	410.4939

[a]Osmolality, in milliosmoles/kg

Over time, water uptake will decrease as the water potential difference, $\Delta\psi$, decreases, until no water is diffused into adjacent cells.

As water in the soil pores enters the root hair cells by diffusion, the water can reach the xylem only after first passing through multiple cell walls through the symplastic pathway or through cell walls and spaces through the apoplastic pathway. As water moves by diffusion, it encounters cell membranes along its path, which provide a large resistance to flow. Once in the xylem, however, these resistances to water flow by diffusion are no longer present, because dead xylem cells no longer contain cytoplasm or cell membranes. In roots, these resistances are quantified as root hydraulic conductivity, L_p, and can be estimated by rearranging Eq. 9.1 as

$$L_p = W_v / \Delta\psi \quad (9.2)$$

Root hydraulic conductance can be measured using the pressure chamber approach, in which a piece of root material is placed in the chamber and pressurized to reverse water held in tension to determine the value of $\Delta\psi$ for use in Eq. 9.2.

The few taproots that characterize most plants, relative to more abundant roots in the shallow parts of the soil horizon, have much higher root hydraulic conductivities than shallow roots. It has been suggested that this is a result of a difference in physiology of these deeper roots, which tend to have long and continuous xylem (Le Maitre et al. 1999). The higher root hydraulic conductivities also reflect the fact that the potential for groundwater to be encountered increases with increasing root depth penetration, and water potentials become less negative with depth nearer the capillary fringe. As taproots age, however, they tend to become more suberized and have lower hydraulic conductivities than young taproots.

9.1.3 Sap Flow

For all the methods available to examine the water status of trees at a phytoremediation site, the measurement of sap flow is the only method available to directly determine the flow of water. The sap flow method is based on either a heat-balance or heat-pulse equipment. Measurement of sap flow does not, however, yield the fraction derived from groundwater—a commonly held assumption, perhaps only defensible under drought conditions. Techniques to determine the fraction of sap flow derived from groundwater are presented later in this chapter.

Measurement of sap flow uses heat as a tracer and a mass balance of added heat to determine sap velocity. In general, a constant, known heat source is applied to a segment of stem or trunk. This locally heats the transpiration stream present in the xylem by about 1–6°C. Under steady-state conditions, heat added must equal heat lost, and this heat loss is quantified in four directions: conduction up the stem; conduction down the stem; conduction out through the heating element; and convection in the transpiring water. The flux of heat is balanced using Fourier's Law (Vieweg and Ziegler 1960; Sakuratani 1981; Baker and Van Bavel 1987).

A commonly utilized heat-balance sap-flow meter is the Flow32-1K™ (Dynamax, Houston, TX), a portable, noninvasive sap-flow sensor that requires no calibration. The flow meter consists of a flexible heater enclosed in insulation that is wrapped around the trunk, stem, or branch to be examined and is covered by foil to reflect incident radiation (Steinberg et al. 1990a) (Fig. 9.1). This is limited to smaller trees or branches of larger trees.

Constant heat also can be applied and heat loss recorded on larger trees using thermocouple dissipation probes (TDPs) (Probe12; Dynamax, Houston, TX; Fig. 9.2). The heated thermocouple can be readily observed using infrared photography (Fig. 9.3). Note that the heated probe is cooled and the heat plume travels upward in the direction of sap flow.

Using the thermocouples estimates sap *velocity*, and to calculate sap flow the velocity is multiplied by the stemwood area. Typical sap-flow results, for the large poplar tree in Fig. 9.1, are shown in Figs. 9.4 and 9.5.

In the heat pulse-method, sap-flow velocity is defined as the time required for a heat input of known quantity to travel from the heat source, such as a heater, to a thermocouple located a known distance up the stem (Huber 1932). These results of sap velocity also need to be multiplied by the stemwood area to yield sap flow.

Regardless of which sap-flow approach is used, certain considerations are common to both. For example, transpiration can exceed sap flow before the local solar noon, and sap flow can exceed transpiration in the afternoon and evening, (Steinberg et al. 1989). This indicates that a lag time exists between a plant's water demand and the water supply, probably on account of water storage or capacitance (Dugas et al. 1992). In smaller trees such as whips or 1-year-old trees the sap flow tends to be equivalent to transpiration, as there is no lag time on account of a lack of stem-water storage (Wullschleger et al. 1998). Moreover, sap flow often ceases in mid-afternoon, in relation to stomatal closure in response to leaf temperatures (Dugas et al. 1992)—this can be seen in Fig 9.4.

In addition to the lag time between transpiration and measured sap flow, solar orientation is an important factor to consider when making sap-flow measurements. Some researchers suggest installing the heater device in the afternoon, when trunk diameter is the smallest due to

Fig. 9.1 A sap-flow heater installed on the branch of a dormant poplar tree at a phytoremediation site near Charleston, SC. The sap flow is measured along with changes in groundwater level, in a shallow monitoring well located at the base of the tree (Photograph by author).

Fig. 9.2 A thermocouple dissipation probe (TDP) installed in the trunk of a poplar tree at a phytoremediation site near Elizabeth City, NC (Photograph by author).

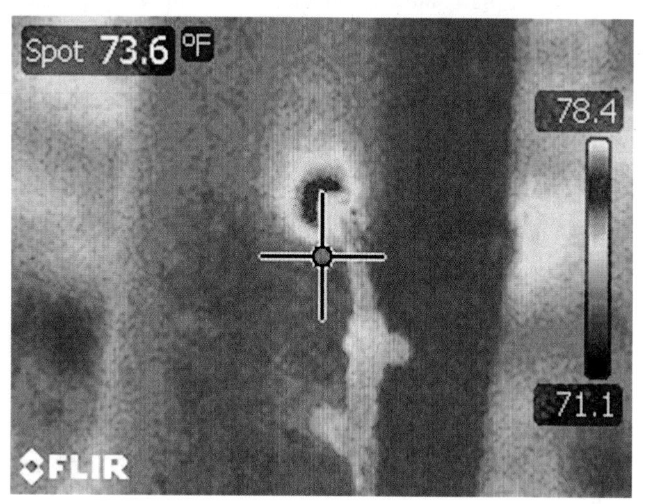

Fig. 9.3 The cooling of the heated thermocouple of a TDP sap-flow meter installed in the poplar tree shown in Fig. 9.2. The direction of sap flow is toward the top of the figure (Photograph by author).

transpirational water losses exceeding water supplies and the depletion of water storage (Kramer 1983) and tensional forces, with the diameter returning to normal at night when water loss is reduced (Hinckley and Bruckerhoff 1975). Steinberg et al. (1990b) noted that sap flow on the northern side of a tree was 41% less than flow on the southern exposure. Cermak et al. (1984) reported sap flow for shaded branches was less (by up to 79%) of sap flow measured on non-shaded branches. Multiple TDP probes can be placed around the circumference of a tree in order to determine the effect, if any, of solar radiation on sap-flow rates.

The measurement of sap flow usually is determined for an individual tree within a stand of multiple trees. These results can be applied, however, to estimate the sap flow for the entire stand of trees. Although perhaps not useful in situations of natural forests that cover thousands of acres, this approach is useful for phytoremediation purposes where most sites will contain less than a few thousand trees spread over only a few acres. The extension of sap flow and transpiration to full-scale transpiration is still as much an art as it is a science, however (Hinckley et al. 1994).

One method to apply individual tree-based sap-flow information to a larger stand is to multiply the individual tree's sap flow by an estimate of the total cross-sectional area of all the trees in the planted area. To do this accurately, plant diameter should be measured using a caliper.

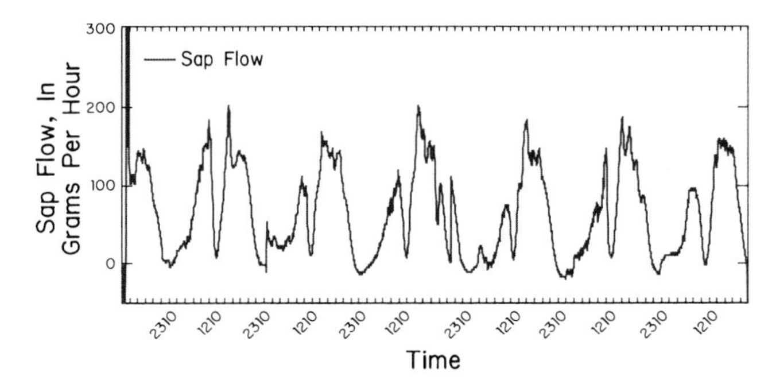

Fig. 9.4 Sap flow, in grams per hour (g/h) measured on a branch of the dormant hybrid poplar tree shown in Fig. 9.1 in Charleston, SC, for 6 days, where the depth to groundwater is between 2 and 4 ft (0.6–1.2 m) below land surface. Sap flow was low during the hot afternoons. This created a daily sap-flow curve of two peaks in high flow. This flow rate represents only part of the total flow of the tree (Landmeyer unpublished data, March 15, 2005).

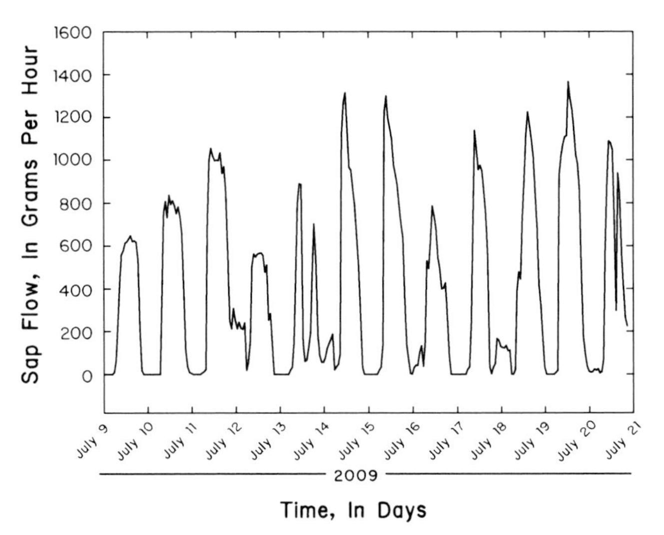

Fig. 9.5 Sap flow, in grams per hour (g/h) measured on a branch of the dormant hybrid poplar tree shown in Fig. 9.2 using the TDP method, Elizabeth City, NC. The lower sap flow from July 13 to July 14 was the result of relative humidity near 100% each day, whereas the other days did not exceed 75%.

Individual tree-based sap flow could be the mean daily sap-flow measurement. Sap-flow measurements could be taken on trees with different diameters or cross-sectional areas, and a regression line fit to explain how these two variables move with respect to each other. This equation could then be used to estimate the sap-flow rate at a particular site if the cross-sectional area is known (Rural Industries Research and Development Corporation 2000). Another method to apply sap-flow measurements made on individual trees to an entire stand is to relate the sap flow for a particular tree to its measured leaf area, as *LAI*, and is discussed later in this chapter.

Sap-flow measurements also can be calibrated using load cells (Ferro et al. 2001). In this method of computing sap

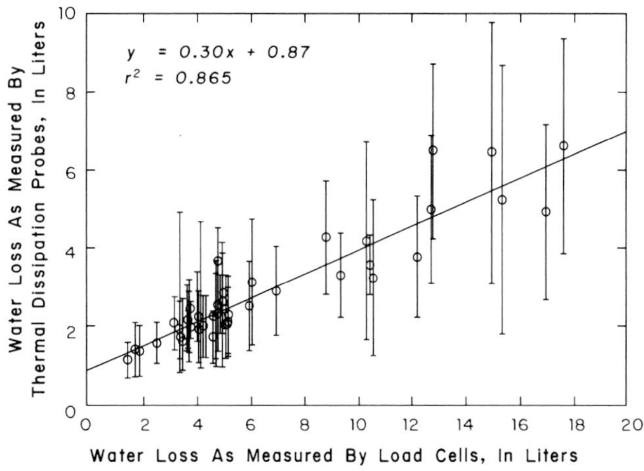

Fig. 9.6 Sap-flow volume, in liters (L), as determined by the heat-balance and load-cell methods (Modified from Ferro et al. 2001).

flow, individual trees grown in separate containers are placed on a scale, called a load cell. Over time, the loss of water from each tree is computed from the loss in weight of each tree. A control tree from which all branches have been removed also is placed on a scale to determine the water loss from the root zone. In Ferro et al. (2001), each tree also was measured for water flow using the TDP sap-flow method. In this manner, sap flow was calibrated to the water-loss method. Ferro et al. (2001) reported that the total water use of the plants by sap-flow monitoring was 45% of the total water use as determined by the load cell method (Fig. 9.6). It was unclear if this rather large difference could be explained by soil moisture losses from the soil not related to transpiration, however, which would not have been measured by the sap-flow method, or by the large standard deviations evident for ever increasing water losses.

Sap flow has been measured for phytoremediation systems at certain sites in the United States and can provide some information regarding the range of sap-flow rates to be expected at sites in similar areas. Sites where sap flow has been measured occurred when at least one of the researchers had a background in plant physiology or forestry and had some experience in making sap-flow measurements at uncontaminated sites. For example, at the site near Fort Worth, TX (Air Force Plant 4 described in Chap. 8), sap flow was measured in whips and 1-year-old plantings installed at the same time. These are the size of tree most commonly planted at phytoremediation sites. Sap flow for both whips and 1-year-old trees was higher during the summer, but the 1-year-old trees had higher sap-flow rates than the whips as measured on a per-tree basis, by almost a factor of 2, or 0.61 compared to 0.34 kg/h/tree (Vose et al. 2000) (Fig. 9.7). This difference can be explained by the fact that the 1-year-old plants had a larger average diameter, 2.9 in. (7.6 cm), compared to the 1.8 in. (4.7 cm) diameter of the

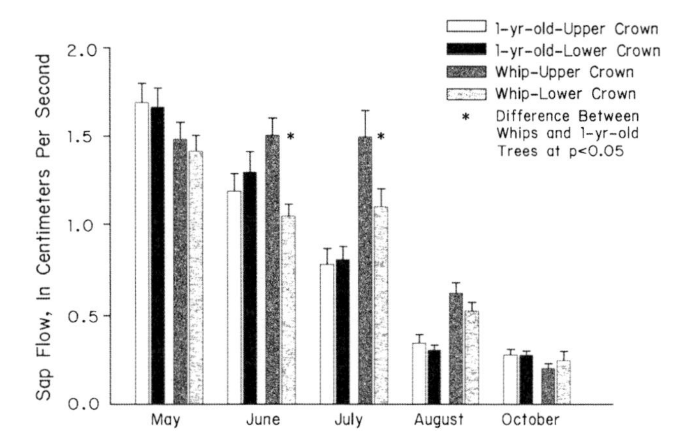

Fig. 9.7 Sap flow for both whips and 1-year-old trees was higher during the summer, but the 1-year-old trees had higher sap-flow rates than the whips as measured on a per-tree basis, by almost a factor of 2, or 0.61 compared to 0.34 kg/h/tree (Modified from Vose et al. 2000). Note that sap flow is not zero during the dormant season in October. One centimeter is equivalent to 0.39 in.

whips and had a higher leaf area index (Vose et al. 2000). As such, the measured sap-flow rates normalized by cross-sectional area of the plants, in kg/cm^2/h, indicated that transpiration was higher for the relatively smaller whips than for the 1-year-old plants (Fig. 9.7). This difference also may be attributable, however, to a shallower water table beneath the whips (Vose et al. 2000). In any case, the range of flow from 0.34 to 0.61 kg of water/h/tree is representative of the climatic conditions of this and perhaps similar areas.

A study by Schaeffer et al. (2000) in the San Pedro River basin, Arizona, provided measurements of sap flow using the heat-pulse approach. They measured transpiration rates of large, native cottonwoods in the riparian zone to be between 52 and 132 gal/day (200 and 500 L/day). Willow trees, which had smaller trunk diameters relative to the cottonwood trees, transpired less, from 8 to 26 gal/day (30 to 100 L/day). In addition to trunk diameter, the higher rate of sap flow in the cottonwoods may have related to the earlier bud break for the cottonwoods (March) relative to that of willows (April), a possible competitive selective advantage in terms of water acquisition. Moreover, Schaeffer et al. (2000) reported that the majority of the water flow measured in both tree species was derived from groundwater in the alluvial channel sediments.

Measurements of sap flow need to be examined in context of whether or not the sap (transpiration water) is moving in the plant or is static. Water movement would be expected to occur during the day when stomata are open, vapor pressure deficits, and a negative water-potential gradient exists from the roots to the leaves. Water movement could still occur during the night if relative humidity is low, however, even though the stomata are predominantly closed. Conditions of static flow often are encountered in the early morning, and water potentials are higher, or less negative, due to stored nighttime water. This difference in the water status in plants over a daily time interval is why sap-flow values should be measured frequently over a few days in order to account for any such lag time that occurs between the movements of water by transpiration during the day versus storage of water during the night.

A special method of using the heat-balance equation to determine sap flow was offered by Daum (1967). Rather than inserting two thin thermocouples into the xylem and heating one and measuring heat changes with the other (the TDP approach discussed previously), the method presented by Daum (1967) involved placing the thermocouples along with a metal plate beneath a flap of bark about 4.7 in. (12 cm) long, which was then resealed with caulking compound. When placed on the trunk and two equally sized stems above a crotch in a tree, Daum (1967) reported that sap flow increased at a sooner time and at a faster rate on the east stem relative to the west stem just after sunrise. Although a little-used approach, these results stress the importance of solar orientation and sap-flow instrumentation location when conducting sap-flow studies.

Mirck and Volk (2010) investigated the effect of seasonal changes on the sap flow measured on four varieties of willow trees in New York, U.S. Peak stand sap flow occurred during measurements made in the summer (June) although this peak sap flow (about 5 mm/day) was not measured during July and August. Sap-flow measurements made on the same trees during the winter (November) approached no-flow conditions (less than 1 mm/day).

9.1.3.1 Nocturnal Sap Flow

Nadezhdina (1999) measured sap flow during the night and suggested this nighttime flow replenished internal water storage lost during daytime *ET*. Transpiration at night can occur if the air is dry and water is unlimited. Benyon (1999) reported that 0.03 in. (0.8 mm) of water was used at night by a plantation of *Eucalyptus* trees. The main goal of the study was to evaluate the use of sap flow to estimate stand-level transpiration. Nighttime sap flow, however, was revealed during data interpretation. Interestingly, significant values of leaf conductance were measured as part of this investigation and suggested that the stomata remained open at night. Higher values of nighttime sap flow also correlated with high *VPD* and wind-flow conditions at the site. Overall, this nighttime release of water from the plantation amounted to about 5% of the total water transpired during the study. Such low sap flows often are not detected in routine sap-flow studies, because the heat-based method used is not sensitive enough to detect small changes in heat due to nighttime flow, or low-flow data are not recorded by sap-flow software.

Nighttime transpiration also was observed by Oren et al. (1999). Transpiration increased with increasing nighttime

VPD. It is thought that as long as water is available, there is no selection pressure against those plants that can keep their stomata open at night. There may be a slight increase in the potential for bacteria or viral entry, especially if moisture is present on the leaves, but this should not occur during times of high *VPD* unless artificial irrigation is used. Other researchers also have observed nighttime transpiration in willow (Iritz and Lindroth 1994) and *Populus spp.* (Hogg and Hurdle 1997), both plants commonly installed at phytoremediation sites. An important distinction to keep in mind is that nocturnal sap flow is a consequence of transpiration, not photosynthesis. For some C_3 and C_4 plants, however, stomata closure does not occur during the night (Caird et al. 2007).

When surface soils become dry, it is possible that the deeper water that is taken up by a plant from groundwater can be redistributed to the shallower soils during the night. This is driven by a gradient in water potential from the wetter xylem to the drier soil, so it is a passive process. This process is called nocturnal reverse flow or hydraulic redistribution (Hultline et al. 2003). Its occurrence is inversely related to the *VPD* during nighttime and does not occur after the surface soils are rewetted during precipitation. Hydraulic redistribution essentially is the allocation of water from wetter areas in contact with deeper roots to drier areas in contact with shallower roots from the same plant.

To summarize, the phenomenon of nighttime transpiration has implication for phytoremediation projects as far as the estimation of the total water budget. Even though this process of water loss is small, it still accounts for an effect on a site's water budget that may be beneficial to the hydrologic goals set for the site and should, therefore, not be ignored.

9.1.4 Stomatal Conductance

As was described in Chap. 3, the stomata represent a structural compromise between the need for the mesophyllic cells to be open to the diffusive entry of CO_2 into a wet boundary layer, while at the same time restricting the passive loss of water vapor to the atmosphere. Stomata help regulate the evaporation of water from a leaf such that transpiration is more complicated than evaporation of water from an exposed water surface. The stomata are regulated in turn by the water potential of the leaf cells. Hence, knowledge of the impact of stomatal conductance, or leaf resistance, to water-vapor flow to the atmosphere, is important in understanding the water dynamics in phytoremediation projects.

Most studies of stomatal (leaf) conductance have been performed on deciduous trees. Stomatal conductance ultimately determines the rates of the processes of transpiration, photosynthesis, and cellular respiration, as these processes are based on diffusion. In the past, plant physiologists have relied on a number of approaches to quantify stomatal size in order to relate it to both steady state and dynamic gas exchange, be it water, CO_2, or O_2. These approaches have ranged from simple observation of leaf surfaces in response to different conditions of *VPD* to photographic imaging systems (Weyers and Meidner 1990). Another method includes the addition of various fluids to a leaf surface to measure the amount of time for uptake, the uptake presumably occurring through open stomata. For example, the cobalt chloride method was a standard method used in the field prior to the 1960s (van Bavel et al. 1965).

The fluids in the above method were applied under normal pressure gradients but were under a pressure nonetheless. This progression led to the idea in the early 1900s of using a gas rather than a liquid under pressure to measure stomatal opening and stomatal conductance. Today, the accepted standard to measure leaf conductance is to use a gas-flow porometer. This device can be used to measure the effect of changes in humidity on sap flow by using a cup that contains a humidity sensor placed over a leaf for a short period of time. These measurements may have more relevance to trees such as poplars, which have stomata on both upper and lower surfaces of leaves, compared to other trees that possess only one stomatal surface. In these cases, some researchers have found no transpiration from the upper surface of a leaf but it is found on the lower surface (Zhang et al. 1999). However, porometer measurements on single leaves are not likely to be representative of the whole tree (McDermitt 1990), due to equipment-induced changes in the humidity and wind circulation patterns around sampled leaves. For example, attachment of equipment to poplar leaves prevents leaves from fluttering in response to wind that under natural conditions can remove large amounts of water vapor due to disruption of a humid boundary layer around the moving leaves.

Some standard pieces of equipment that are field portable include the LI-1600 steady-state porometer (LI-COR Biosciences™, Lincoln, NE) or a dynamic, transit-time porometer (AP4, Delta-T Devices, Cambridge, UK), or the SC-1 (Decagon Devices, Pullman, WA). These devices use a constant stream of gas (dry air) over the leaf surface to determine transpiration through the stomata. These devices calculate the rate of gas exchange by measuring the difference in humidity between the inside and outside of the leaf. For example, inside the leaf the humidity is assumed to be 100%, and outside the leaf the ambient humidity is measured by the sensors. In all cases, the results for stomatal conductance will be a function of the stomatal characteristics of the plant leaves, such as number, surface location, and degree of opening, as well as leaf temperature and atmospheric humidity. The units of measurement of stomatal conductance are expressed as a conductance (in millimoles per square meter

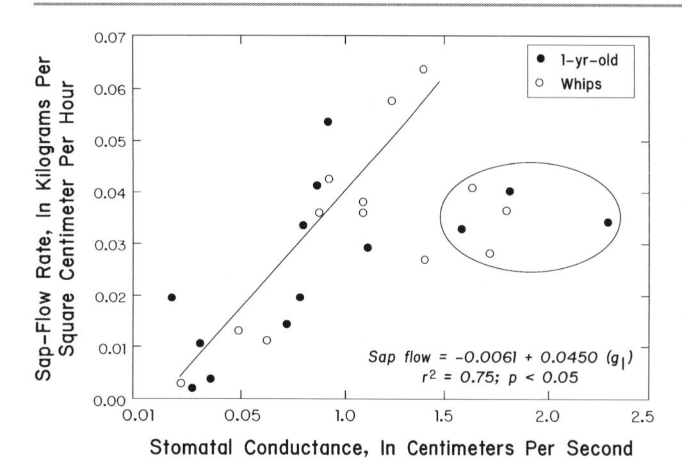

Fig. 9.8 Sap flow increases as stomatal conductance increases, and there is little difference between flow in young plants and 1-year-old plants (Modified from Vose et al. 2000). One centimeter is equivalent to 0.39 in.

per second; mmol/m^2/s) or as a resistance (meter squared per sec per mol; m^2/s/ mol).

Factors that affect stomatal conductance were examined at the phytoremediation site in Fort Worth, Texas (Air Force Plant 4 described in Chap. 8 and earlier in this chapter). Higher stomatal conductance was measured during the spring and declined throughout the season for both whips and 1-year-old trees (Vose et al. 2000). For the most part, the stomatal conductance of the whips was higher than the 1-year-old trees during the summer, but the average leaf conductance was relatively similar across the year at about 0.4 in/s (1 cm/s). As might be expected, the stomatal conductance for leaves near the top of the whips was higher than for leaves near the ground (Fig. 9.8). This relation did not hold for the 1-year-old plants, however (Vose et al. 2000). The authors speculate that the relation between stomatal conductance and height was not due to light competition, because in the 1-year-old plantation the *LAI* was relatively low (less than 2), but was due to leaf drop from lower branches in response to drought conditions that occurred during August (Vose et al. 2000). In addition, mean daily sap flow was directly related to mean daily leaf conductance (Fig. 9.8) (Vose et al. 2000).

9.1.5 Leaf Area Index

The total area of leaves has a significant effect on the amount of transpiration that occurs. Although transpiration is driven by meteorological factors such as solar radiation and *VPD*, the removal of water from plants is regulated at the leaf surface, such as by stomatal conductance. In order to estimate the effect of a phytoremediation planting on contaminated groundwater, measurements of the total influence that plants have on water uptake as related to leaf area, should be made. In fact, the component of the total leaf area, or Leaf Area Index, *LAI*, has been of interest to plant physiologists for a number of years. An estimate of the leaf area is important with respect to time, because it changes as the plant grows and changes seasonally in response to climate. Plant physiologists have been interested in *LAI* as a measure of forest productivity and overall forest ecology. For example, *LAI* is directly related to net primary production. From the viewpoint of a hydrogeologist concerned with implementing a potential phytoremediation project or monitoring an existing one, *LAI* can be used as a measure of the potential magnitude of plant-water loss and its implication for change in the site water budget.

The *LAI* is defined as the ratio of leaf surface area (L^2) of a plant or grove normalized to ground surface area (L^2) covered by the canopy. Measurements of *LAI* can be made using a Plant Canopy Analyzer (Li-Cor LAI-2000, LI-COR Biosciences™, Lincoln, NE), which measures the extent of light interception relative to a standard, which can be a separate system set up in an open area, if room is available. *LAI* can be viewed as the total one-sided green leaf area per unit ground-surface area. By convention, the *LAI* of a given site can vary from a high of 10 down to 1. *LAI* is one of many vegetation indices that are dimensionless but indicate relative abundance. The magnitude of *LAI* varies with such factors as the size and spacing of trees; *LAI* also tends to increase as trees age. For example, the *LAI* for young trees for a given area is near 1 and can approach 10 in dense stands of mature trees.

Because larger trees have more leaves than smaller trees, it may be appropriate to plant fewer trees per given area at a site that will grow larger on account of less competition for nutrients and water, as long as no decrease in *LAI* is reached by using fewer trees. A closer spacing of trees does not necessarily mean that a denser grove with higher *LAI* will be produced in a shorter amount of time, because closely packed plants often can become 'leggy' as they compete for position relative to each other to catch the sun's rays. At some well-established phytoremediation sites that are older than 5 years, the *LAI* typically is not constant across the planted area but varies depending upon the location of the *LAI* measurement in relation to the health of the trees, individual mortality, diseases, or increased leaf drop due to drought.

Measurements of *LAI* over time may be useful in quantifying the increases in *ET* during the development of a poplar grove at contaminated sites or in quantifying the effect of contaminant concentrations on tree health. For instance, if water becomes limited to plants, they drop older, yellowed leaves to decrease total transpiration water losses. Some plants will curl their leaves to reduce the surface area available to evaporation. In some sense, we have all experienced *LAI*; when we are hot and seeking relief from the sun, we go to sit in planted areas where the shade is the most complete (a *LAI* of 10) and avoid areas where some patches of sunlight are making it to the ground ($LAI < 10$).

Typically, more leaves equate to a higher rate of transpiration. More leaves can mean either a higher number of small leaves or a lower number of large leaves. For example, a deciduous forest on the eastern part of the United States may have a leaf surface area up to four times greater than the ground area the forest grows on. A maple tree's trunk covers 1 square yard, but the leaves, of which there may be as many as 100,000, have a combined surface area of more than 2,000 square yards, almost 0.5-acre (2,023 m^2).

In the mid-1990s, researchers in western Australia observed a relation over time between *LAI*, precipitation, and *VPD*. They were able to relate the seasonal increase in soil moisture with an increase in *LAI* (Rural Industries Research and Development Corporation 2000; Figs. 9.9 and 9.10). At the site they investigated, the climate is characterized by wet winters and dry summers.

The *LAI* also can be used to estimate transpiration rates. This can be done with the assumption that transpiration is directly related to leaf area by

$$T = sLAI/L \tag{9.3}$$

Where T is the transpiration rate (mm/day), s is the sap flow (kg/day), L is the leaf area of the stem used for sap flow, and *LAI* is the Leaf Area Index (Allen and Grime 1995).

9.2 Hydrogeologic Monitoring Methods

Although plants interact with groundwater as outlined in Part I, there is a paucity of data on how this interaction actually can affect groundwater levels at phytoremediation sites, and how this information can be used to assist water managers in making decisions about using plants to achieve hydrologic-based remedial goals at sites. This lack of information may simply reflect the fact that the water-table elevation history at a particular site often consists of one-time measurements made on the day of the sampling event, rather than continual measurements made over time. As such, only gross changes in groundwater levels and flow can be detected using this conventional way of monitoring groundwater levels. In fact, Quinn and Johnson (2005) suggest that the frequency of groundwater-level data collection may be as important as the total number of wells being monitored.

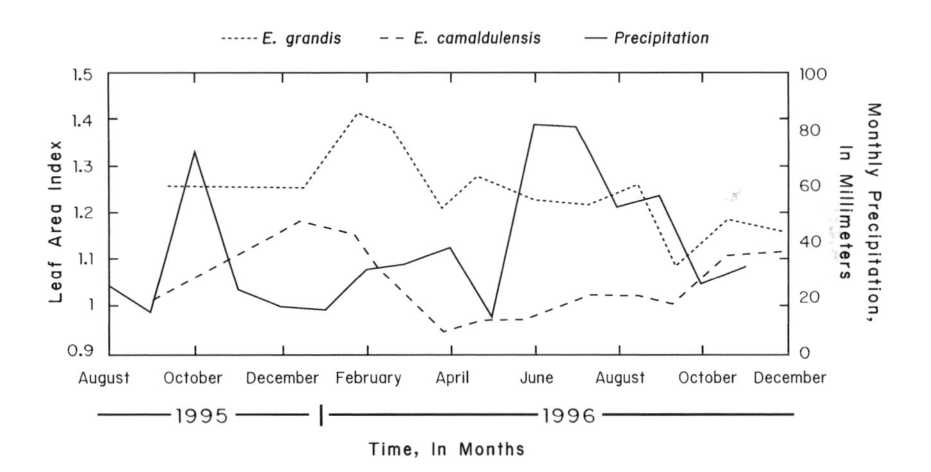

Fig. 9.9 Relation of *LAI* to precipitation for two species of Eucalyptus (Modified from Rural Industries Research and Development Corporation 2000). One millimeter is equivalent to 0.039 in.

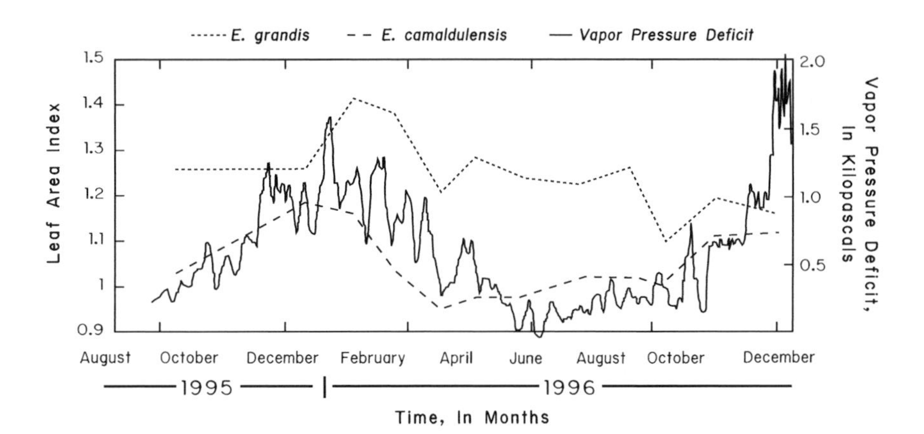

Fig. 9.10 Relation of *LAI* to *VPD* (in kilopascals) for two species of Eucalyptus (Modified from Rural Industries Research and Development Corporation 2000).

9.2.1 Groundwater Levels

A diurnal change in groundwater levels observed in areas where trees tap groundwater provides the most direct line of evidence of plant and groundwater interaction. Observation of plant-induced diurnal groundwater-level fluctuations goes back almost 100 years. As was introduced in Chap. 1, G.E.P. Smith of the University of Arizona presented a paper in 1922 that showed that daily fluctuations in water levels in wells placed in a grove of cottonwood and mesquite trees were due to groundwater withdrawal by these plants (White 1932). Five years later, O.E. Meinzer (1927) published the first data that showed the water table in areas that had plants known to rely on either groundwater or capillary water directly above often were characterized by fluctuations. White (1932) also stated that groundwater levels observed to decline during the day and not at night were caused by the plant-facilitated uptake of groundwater—the maximum daily drawdown observed was 0.13 ft (0.039 m). Later, Meyboom (1966) observed fluctuations in groundwater of about 0.10 ft (0.03 m) in wells near willow trees on the banks of small lakes. This groundwater level fluctuation was large enough in space and time to stop the discharge of groundwater to the lake during summer and cause lake water to recharge the aquifer (Meyboom 1966). In Davis and DeWiest (1966), a method to estimate groundwater use by trees was presented, which will be discussed in this section in more detail. As can be seen from these studies, the magnitude of the diurnal change in groundwater levels attributed to plants can be small, and therefore, inherently difficult to measure accurately.

Attributing such potentially small groundwater-level changes to trees often is problematic because other processes also can affect groundwater levels at a similar scale. These processes include barometric pressure changes, tidal loading, and pumpage—these processes may result in groundwater-level fluctuations that either mimic or obscure groundwater-level changes caused by plants.

In any case, measurement of groundwater levels, barometric pressure, and tides can be made to determine barometric and tidal efficiencies to show these factors affect the static groundwater level (Gonthier 2007) and how to remove these effects to be able to observe the influence of plant-water uptake. These changes we are hoping to observe are the barometric pressure-independent water-level changes and tidal-independent water-level changes. Groundwater-level changes caused by barometric pressures, however, are rarely observed in shallow, unconfined groundwater where planting has occurred. Barometric effects on groundwater in unconfined aquifers does not occur because the difference in atmospheric pressure transmitted to the free surface of a well and the atmospheric pressure transmitted to groundwater in adjacent aquifer material is negligible (Landmeyer 1996).

There can be a lag in time between the changes in barometric pressure in the atmosphere relative to a change in groundwater level, on account of the diffusivity of the unsaturated zone. Such a scenario assumes that the well screen is fully saturated, for no effect will be seen in a well if part of the well screen is above the water table. A lag time will not hold true, however, if the groundwater is characterized by a large fraction of dissolved or trapped separate-phase gas.

In order to observe the anticipated diurnal changes in groundwater levels due to plant uptake, all the potential changes that can affect groundwater levels need to be accounted for. For a change in groundwater level in a well, ΔW, is defined as the groundwater level at time $t + 1$ minus the water level at a previous time t, such that

$$\Delta W = W_{t+1} - W_t \tag{9.4}$$

The ΔW can be caused by any or all of the following processes, such as that caused by barometric pressure, ΔW_b, recharge, ΔW_r, pumping, ΔW_p, earth tides, ΔW_g, ocean tides, ΔW_m, evaporation, ΔW_e, surface-water level changes, ΔW_s, transpiration, ΔW_t, and all other possible processes, ΔW_o (Gonthier 2007). In sum, Eq. 9.4 becomes

$$\Delta W = \Delta W_b + \Delta W_r + \Delta W_p + \Delta W_g + \Delta W_m + \Delta W_e + \Delta W_s + \Delta W_t + \Delta W_o \tag{9.5}$$

In order to best examine the effect of plants on ΔW, groundwater levels should be measured between precipitation events, under conditions of relatively stable barometric pressure, in areas where no pumping from the site exists, and tides are under slack conditions, such that ΔW_r, ΔW_b, ΔW_p, and ΔW_m are near zero. However, because the time interval for changes in groundwater levels by plants is similar to the daily changes in barometric pressure experienced at most sites, these non-plant-induced changes will need to be determined and removed from the water-level measurements.

To do this, barometric and tidal efficiencies for each well will need to be calculated. At sites where barometric or tidal forces may affect groundwater level, measurements of barometric pressure and tidal height need to be monitored, or data need to be gathered from nearby existing monitoring stations. The barometric efficiency, α_b, is calculated as

$$\alpha_b = \Delta W / \Delta B \tag{9.6}$$

where ΔW is the groundwater level change and ΔB is the barometric change, in equivalent head units. In general, decreases in barometric pressure result in increases in groundwater levels. The tidal efficiency, α_t, is calculated as

$$\alpha_t = \Delta W / \Delta T \tag{9.7}$$

where ΔW is the groundwater-level change and ΔT is the amplitude of the tide across a 12-h period from high to low tide. In general, $\alpha_b + \alpha_t = 1$, in areas during conditions of no recharge or pumping.

To monitor anticipated small changes in groundwater levels, pressure transducers need to be used that can resolve up to a 0.01 ft (0.30 cm) or greater change in groundwater level. The manual tape downs often used at most sites during monitoring events are accurate to about 0.1 ft (3 cm) and are not useful in detecting typical plant drawdowns. Groundwater-level fluctuations also should be made over a period of at least a few days, to determine the presence or absence of a diurnal pattern. Many pressure transducers also meet this requirement of long-term deployment and come equipped with internal data loggers or can transmit the data to an external logger. Data also can be retrieved remotely to permit a near "real-time" examination of groundwater-level changes. The use of pressure transducers to measure groundwater levels is particularly applicable in gravel or sandy aquifers, because these highly permeable sediments often result in smaller groundwater-level fluctuations compared to changes observed in fine silt or clay. Moreover, pressure transducers that have been constructed and calibrated only to measure changes in barometric pressures need to be installed above ground at sites where pressure transducers are installed in wells.

An example of using the pressure-transducer method to monitor groundwater levels is given for the phytoremediation site at the former MGP sites near Charleston, South Carolina, discussed previously. There, groundwater-level fluctuations have been measured before trees were planted in 1998 in wells in the shallow silty aquifer. For example, groundwater-level changes were monitored between June and July 2000, 2 years after planting, using a pressure transducer (Hydrolab Minisonde™) placed in a temporary, 2-in. (5 cm) diameter well screened in the upper 0.5 ft (0.15 m) of the water-table surface. Precipitation data recorded at an on-site weather station revealed the monitoring period had both dry and wet conditions. Prior to precipitation, groundwater levels remained low with no diurnal fluctuations. After a 1-ft (0.3 m) rise in the groundwater level was observed in wells after an intense 4-h precipitation event, however, a maximum groundwater-level fluctuation of 0.07 ft (2.1 cm) was recorded and coincided with solar radiation also measured at the on-site weather station. This recharge event increased the elevation of the water table to be nearer the roots of the 2-year-old trees. Similar diurnal changes for cottonwood trees in Ohio have been reported by Gatliff (1994).

To determine the removal of groundwater by phreatophytes planted at a site, a method similar to that presented in Mower et al. (1964) can be used. Mower et al. (1964) used the Theis method for computing the effect of a pumping well on an adjacent stream. This holds because the removal of groundwater by phreatophytes can be envisioned as being analogous to that of a pumped well. This assumption is especially valid for riparian phreatophytes that produce a measurable decrease in river flow, as described in Chap. 5. Mower et al. (1964) also tried the transpiration-well method first developed by White (1932) to determine groundwater use by phreatophytes. As was previously discussed, this method is based on the assumption and observations by White (1932) that water from the capillary fringe is used by transpiring plants during the day, when the stomata are open. A decline is observed in nearby wells because the rate of depletion of the capillary water is greater than the rate that it can be resupplied by hydrostatic pressures. The reverse is true when the plants are not transpiring. It is important to note that since the advent of water-potential measurements that came after White, the capillary fringe can supply water to plant roots up to but not beyond tensions that represent the wilting point, or about -1.5 MPa.

Davis and DeWiest (1966) using basic groundwater-level fluctuation data presented a method to calculate the volume of groundwater transpired by plants. The amount of groundwater moving into a root zone per unit area, q, can be solved by

$$q = n_e V_H \tag{9.8}$$

where n_e is the effective porosity of the unit area and V_H is the velocity of the rise, or high, in groundwater level, also known as recovery, over time, normalized by any longer-term decreases in groundwater levels caused by changes in storage between consecutive daily lows. Changes in groundwater storage include that caused by recharge from precipitation infiltration or from nearby streams after a flood, or by discharge through springs. The solution to Eq. 9.8 is the volume of groundwater moving into area around the tree roots. This value can then be multiplied by the assumed area of root influence for either a single tree or a stand of trees in a phytoremediation application to obtain a volumetric rate of removal of groundwater per unit time.

Even though measurement of groundwater levels can provide the most direct evidence of plant and groundwater interaction, there are some precautions that need to be stated when examining the data. First, it is possible that roots may have grown into the well being monitored. In fact, root mass is typically found in many monitoring wells at many sites around the country even where phytoremediation plants have not been added (please refer to Chap. 13, Fig. 13.14). This condition is a consequence of hydrotropism, in which plant roots multiply quickly in areas where water supplies are the most readily available.

Second, the diurnal groundwater-level decrease being measured in a well often will be *lower* than that occurring in the adjacent aquifer, because a well has 100% porosity, whereas the aquifer material will be much less porous. A 1-in. (2.54 cm) water-level change in a well would be a 3-in. (7.6 cm) change in an aquifer of 30% porosity. The groundwater-level change measured in a well often is a minimum representation of that occurring in the adjacent aquifer sediments. This fact suggests that a lack of change does not necessarily mean that no groundwater uptake is occurring; it is simply an artifact of using wells as a metric. Also, if the efficiency of the well screen is lowered by clogging, a change in the groundwater level caused by plant uptake may not be measurable. Thus, the condition of the wells used to measure groundwater-level fluctuations is important. Because of the influence of well construction on the potential to discern groundwater-level changes caused by plants, care must be taken in well construction. For example, the screened interval should be kept to a minimum length, because longer screened intervals may provide a preferential flow path for plants to meet *ET* demands.

Groundwater-level fluctuations caused by phreatophytes were used to estimate the specific yield of a riparian aquifer in the Larned Research Site in the Arkansas River basin in Kansas by McKay et al. (2004). Wells were placed at increasing distances from the Arkansas River. The degree of water-table fluctuation was higher in the riparian ecosystem nearer the river and lower with increasing distance from the river. The water-table fluctuations were used in conjunction with measured soil-moisture levels measured by *in-situ* neutron probes. They used these data and the Skaggs method to determine aquifer specific yield (Skaggs et al. 1978). This method calculates the specific yield by using the difference in soil-moisture profiles at two different water-table positions. Using this method, McKay et al. (2004) calculated specific yield from the water-table decrease caused by phreatophytes and found that it was similar, or between 0.19 and 0.29 gal/min (0.7–1 L/min), to the values calculated by an on-site pumping test, that suggested specific yield was 0.16–0.31 gal/min (0.6–1.1 L/min).

This calculation of specific yield was based on an equation. Simple analytical groundwater models also can be used to evaluate pump-and-treat relative to phytoremediation as remedial strategies. By definition, for pump- and-treat to be effective, there either has to be hydrologic containment or the contaminated pore volume of groundwater has to be cleaned up. An aquifer pore volume (PV) is described by the USEPA as the volume of groundwater within a contaminated plume. Cleanup is defined when 10–100 pore volumes of groundwater are removed. The pore volume can be calculated if the dimensions of the height, width, and length of a particular aquifer plume can be estimated, as well as the porosity. The USEPA (U.S. Environmental Protection Agency 1996) reported that at 24 sites where pump-and-treat had been started and where the time needed to pump 20 pore volumes was calculated, the time ranged from 1 year at a small site (0.3 ha) to 3,015 years at a larger site (3,100 ha). The average time to achieve 20 pore volumes at all 24 sites was 274 years at an average site of 189 ha. There was not a direct relation between site size and time to remove 20 pore volumes, however, because the extraction rates at each site differed greatly.

The efficiency of pump-and-treat systems is a function of the removal, but this, of course, is a function of the aquifer hydrogeology. Expressed in another way, the removal or extraction rate, Q_e, is a function of the sustainable aquifer yield, Y_s. From what was presented in Chap. 4 regarding specific yield, S_y, the sustainable yield, Y_s, is dependent upon certain constraints that will ultimately impact Q_e, such as hydraulic conductivity, groundwater levels, and effective porosity. Fortunately, these physical properties can be assessed and measured and used in simple analytical models to determine whether or not pump-and-treat will be efficient. Moreover, as the removal efficiency approaches lower limits, a case can then be made that groundwater removal by plants would produce a closer and less expensive match to the sustainable aquifer yield.

Sustainable aquifer yield is related to the number of extraction wells used, the sites dimensions, and aquifer hydrogeology. The NRC (National Research Council 1994) reviewed 77 pump-and-treat sites and determined that the average number of extraction wells was 9, with the maximum number being 15. In addition to the total number of wells at each site, the distribution in terms of the distance between wells is important, because each pumped well will have its own cone-of-depression, or radius of influence, based on the pumped rate and the aquifer characteristics. If too few wells are used, it is possible that the cones of depression will not interact, and groundwater moving to the wells will pass uncaptured through the aquifer in the areas where the well's capture zones do not overlap. The efficiency of containment will decrease as either the distance between wells is increased or the pumping rate is decreased or both.

The maximum pumpage rate is ultimately controlled not so much by the mechanical properties of pumps but by the maximum sustainable yield of the aquifer. As the sustainable aquifer yield gets lower than 0.1 gal/min (0.3 L/min), then a closer well spacing and, hence, increased number of wells will be required. This scenario represents conditions under which trees planted on 5-ft (1.5 m) centers for phytoremediation become a scientifically defensible option for hydrologic capture.

The frequency of groundwater-level data collection at a phytoremediation site is an important consideration, as the data drive the type of approach used. If the goal is to understand diurnal impacts of plants on groundwater, then a short time interval is needed. If the goal is to determine a seasonal or annual site water budget, less frequent data collection will suffice. In many cases, however, more attention is paid to the location of monitoring wells rather than to monitoring frequency. Quinn and Johnson (2005) examined the benefits of using automatic groundwater-level sensors at their phytoremediation site near Chicago discussed in Chap. 8. In that study, continuous groundwater-level measurements were used to refine the site conceptual model to account for a source of subsurface seepage of stormwater runoff that previously had not been recognized.

9.2.2 Groundwater Flow

The uptake of groundwater by phreatophytes can affect not only the groundwater level in shallow aquifers, as described above, but also the horizontal direction of groundwater flow. For example, Fig. 8.2 in Chap. 8 depicts a hypothetical hydrologic scenario in a sandy, homogeneous, isotropic water-table aquifer. Under these idealized conditions, the vertical lines represent equal water potentials prior to the installation of poplar trees. Groundwater flow crosses these lines of equipotential at a right angle (Freeze and Cherry 1979; Fetter 1988), such that groundwater flow is from higher to lower water potentials, or from left to right in Fig. 8.2, and conditions are at steady state.

After roots reach the water table, groundwater levels can decline locally. The initial volume of groundwater removed to cause this decline is equivalent to that amount drained by gravity, or specific yield, S_y. This decrease in head can, theoretically, lead to the reversal of the downgradient flow of groundwater. More advanced analytical techniques can be used to determine the areal extent of this decline, or capture zone, using wells (Landmeyer 1994) or simulated trees (Gorelick et al. 1993).

9.2.3 Vertical Groundwater Flow

The plant-induced removal of groundwater also affects the originally uniform distribution of equipotential lines and can result in a vertical component of groundwater flow (Fig. 8.2) in an originally horizontal flow regime. This is not the same as plant-induced hydraulic lift, which affects the distribution of water *tension* in the unsaturated zone. The reduction in groundwater levels lowers water pressures beneath the trees throughout the *entire* saturated thickness, not just near the water-table surface. This induces a vertical flow component at depths greater than the depth of maximum root penetration (Fig. 8.2). The result is that planted areas that are not initially areas of groundwater discharge can become areas of groundwater discharge. The potential for vertical flow can be evaluated by installing vertically spaced, or nested, wells.

9.2.4 Groundwater Volume Removed by Plants

The amount of diurnal groundwater-level fluctuation caused by phreatophytes not only reveals the occurrence of plant and groundwater interaction but also the volume of groundwater removed by plants. This determination can be made because departure of the measured water level from static conditions represents the withdrawal of a finite volume of groundwater, initially S_y as described earlier, from the aquifer. For example, the maximum groundwater-level change, Δh, observed at the phytoremediation site near Charleston was 0.07 ft (0.02 m). This decline occurred over an assumed radius, r, of at least 1 ft away from a tree, and the effective porosity, n_e, was assumed to be 0.30, the groundwater volume removed can be estimated using $V = (B)(r^2)(\Delta h)(0.30)$. Therefore, a 0.07-ft (0.02 m) decline in the water table represents about 0.50 gal (1.8 L) of groundwater removed per day per tree. The 2-year-old trees actually transpired a larger volume of water per day of around 5 gal/day/tree (18.9 L/day/tree), the difference being supplied from soil moisture.

Another method was presented by White (1932). In his method, introduced in earlier chapters but repeated here because of its importance, the quantity of groundwater withdrawn by plants during a 24-h period is determined by:

$$GW = S_y(24r \pm s) \qquad (9.9)$$

where GW is the amount of groundwater transpired (L/T), S_y is the specific yield (% by volume; usually 2%, 22%, or 40% for clay, sand, or soil, respectively), r is the hourly rate of groundwater inflow within 24 h (L/T, in h), and s is the net rise (or fall) of the water table (equivalent to Δh) (see Freeze and Cherry 1979). The 0.07-ft (0.02 m) water-table fluctuation discussed above when evaluated using the White method indicates a removal of near 0.14 gal/day/tree (0.5 L/day/tree), similar to the result using the simple volumetric method described previously.

The clear advantage gained using either method is that an estimate of the volumetric uptake of groundwater by plants can be made without having to use intensive plant-physiologic methods. The White method is more applicable for plants using groundwater in coarse-grained aquifers relative to fine-grained aquifers. When the groundwater level drops in fine-grained sediments, the specific yield is released more slowly over time than for coarse-grained aquifers, especially for sites where the depth to water table is shallow. A unit drop in groundwater level for plants grown in each type of aquifer will yield different results for the amount of groundwater transpired.

9.2.5 Groundwater Discharge

Phreatophytes also can affect the discharge of groundwater in an aquifer. The change in groundwater discharge at contaminated sites after planting is a contaminant-*independent* process as it is a function of the local groundwater hydrologic conditions, as long as the concentration of the contaminant does not reach toxic levels. The discharge of a substance can be defined as that quantity flowing across a given area, perpendicular to flow direction, per unit time. In aquifers, Darcy's Law, or $Q = (A)(i)(K)$, can be used to examine the effect of plants on groundwater discharge. Prior to the addition of phreatophytes, groundwater entering the area, Q_{up}, would be equal to groundwater discharge from the area, Q_{down}, assuming no loss by evaporation or gain by recharge. This condition may exist after young trees are planted or during dormancy.

Groundwater discharge that exits the planted area can, over time, become less than the amount that enters the planted area, or $Q_{up} > Q_{down}$, for at least two reasons: first, groundwater discharge must be conserved, and second, groundwater levels will decline. It follows, therefore, that the term Q in the Darcy flow equation is decreased as the size of A decreases. Moreover, the rate of groundwater flow will increase from upgradient areas as the water table drops in the planted area (Eberts et al. 1999).

As indicated in the Darcy equation, few input parameters are required and can be easily obtained. For example, wells can be installed upgradient and downgradient of the planted area. The cross-sectional area, A, of groundwater flow can be estimated from well-construction data, such as depth, L, and width, W, from the scale of the site. These wells could be used to determine the hydraulic conductivity, K, of the aquifer, as described in Chap. 6, such as using the Hvorslev method (Fetter 1988). The hydraulic gradient can be calculated by dividing the vertical change in groundwater elevation between the two wells, Δh, by the lateral separation distance, $i = \Delta h / \Delta l$.

The groundwater discharge method can provide an objective criterion to determine if and when hydrologic containment or control occurs. Ideally, complete hydrologic containment or control would be implemented when $Q_{down} = 0$. This scenario, however, is not realistic, and could only be met if a trench were dug to the bottom of the aquifer and groundwater discharged by a pump at rates that exceeded the flow into the trench. At sites where this has occurred it typically has not been successful (see Widdowson et al. 2005a for site history). A more realistic objective criterion would be when the downgradient discharge of groundwater after planting is decreased by an agreed-upon percentage of the average, preplanting upgradient flux of groundwater. Some examples of this method exist at the Carswell Air Force Base site in Texas (Eberts et al. 1999).

9.2.6 Soil Moisture

Moisture present in the unsaturated zone above the water table provides an alternative source of water to plants, even occasionally to those plants considered to be obligate phreatophytes. When soil moisture levels decrease to tensions that are equal to or greater than the wilting point, deeper sources of water under less negative tension are needed to sustain plant survival. As such, the monitoring of soil moisture around plant roots above the water table can provide important information for what component of the total water transpired is derived from soil moisture during wet periods and for when plants are using groundwater during drier periods. Soil moisture does not indicate, however, the potential for water movement or flow direction.

Soil moisture can be determined using a variety of approaches. Electrical resistivity methods, such as the Beckman soil moisture meter, can be used. Soil moisture also can be measured by time-domain reflectometry (TDR; Campbell Scientific, Logan, Utah). Changes in soil moisture using TDR were measured at the Air Force Plant 4, Fort Worth, TX, phytoremediation site and related to input by precipitation and removal by transpiration (Vose et al. 2000). Soil moisture was highest during the spring when *ET* was lowest and immediately after precipitation events. At a given time, soil moisture was lower within the plantation relative to an open area populated by grasses and no trees.

Soil moisture also can be measured using a neutron meter or probes. Nnyamah and Black (1977) used neutron meters and tensiometers to investigate the depletion of soil moisture beneath thinned and unthinned forests that consisted of Douglas fir for a period of 4 weeks during which no precipitation occurred. Changes in soil moisture over the 4-week period agreed well with total *ET* measurements made using the energy-balance approach. During the drier period, water movement into the root zone increased from about 8–15% of the total water removed by *ET*. Conversely, less than 2% of *ET* was removed from trunk storage.

The sediment hydraulic conductivity not only affects infiltration of water downward but also the removal of water upward by evapotranspiration. Given the same conditions of sun, weather, and plants, *ET* will be high in the areas where the hydraulic conductivity is high and low where the hydraulic conductivity is low. As such, maximum rates of transpiration are as equally controlled by the unsaturated and saturated hydraulic conductivity, or soil water availability, than the demand for *ET* from *VPD* (Persson 1995).

9.3 Integrative Monitoring Methods

As stated earlier, a multiple-lines-of-evidence approach is encouraged regarding the monitoring of a phytoremediation system for plant-induced changes in site groundwater hydrology. This approach provides the confidence level necessary to make statements about processes that are occurring in the subsurface that often are not amenable to direct observation. This section describes approaches that contain aspects of both the plant and groundwater components of phytoremediation.

9.3.1 Geochemistry

The ambient geochemistry of groundwater, in terms of the concentration of dissolved solutes acquired along groundwater flowpaths through interactions with the porous media prior to a contaminant release, can play a role in the determination of the types of phreatophytes that grow in an area. For example, cottonwood and willow typically are not found in areas characterized by high concentrations of dissolved salts. However, certain hybrid poplar trees, such as the hybrid OP-367, have been developed that can live in soils where the pore water is characterized by high salinities. Some native plants that can tolerate high-salinity groundwater include greasewood, saltcedar, and bush pickleweed (*Allenrolfea occidentalis*), although these plants are rarely used in phytoremediation applications.

9.3.2 Stable Isotopes

Because plants can take up water from multiple sources, determination of the proportion due to groundwater is essential at a contaminated site. This evidence is especially convincing if diurnal fluctuations in groundwater levels are available. One of the more established methods to determine the source of water is to compare the stable isotopic composition of the plant water with that of the potential sources of water. Some examples will be given below.

Plants interact with many elements, but the primary ones are carbon, oxygen, and hydrogen. All these elements have stable isotopes. The stable isotopes of carbon were used in the early 1950s to show that the stable isotope value of plant carbon was about −27 per mil (also shown as ‰, per thousand, or a tenth of a percent), or contained less percentage of the heavy isotope ^{13}C, than the carbon in atmospheric CO_2, which has a stable carbon isotope signature of about −7 per mil (Ziegler 1995). This observed enrichment of the light isotope ^{12}C at the expense of the heavier isotope ^{13}C was explained to result from isotopic fractionation reactions that occur during the reduction of the CO_2, such that the kinetics of $^{12}CO_2$ reduction were more favorable, or faster, than the reduction of $^{13}CO_2$. Further investigation indicated that the enzyme rubisco, or ribulose bisphosphate carboxylate/oxidase, discriminates against $^{13}CO_2$ (Park and Epstein 1960).

Photosynthesis was discussed in Chap. 3, and the difference in how plants fix CO_2 gave rise to the classification of C_3 and C_4 plants (Fig. 9.11). This difference between how plants fix carbon was elucidated after archaeologists found that radioactive carbon, or ^{14}C, used to age-date samples of organic matter such as corn waste had *younger* radiocarbon age dates, in years before present (BP), than wood from the same site (Bender 1971). This was because corn and other monocots such as grasses turned out to have higher ^{14}C contents (and ^{13}C) than dicotyledon plants. This difference was explained in terms of the stable isotope value of the initial sugar produced during gas-phase diffusion of carbon fixation in different plants. For example, corn and other C_4 grasses fix CO_2 into a 4-carbon product that contains more ^{14}C and heavier ^{13}C, measuring about −14 per mil. This pre-fixation of carbon in C_4 plants is by the enzyme phosphoenolpyruvate (PEP) carboxylase, which occurs in the mesophyll cells of the leaves. Final fixation of carbon occurs by rubisco in the bundle sheaths of the leaf. Conversely, those plants that produce a 3-carbon sugar are the C_3 plants, and they produce isotopically lighter carbon, near −27 per mil.

The fixation of carbon is not limited to the C_3 and C_4 pathways. An additional pre-fixation pathway also discussed in Chap. 3 is the CAM pathway, or Crassulacean acid

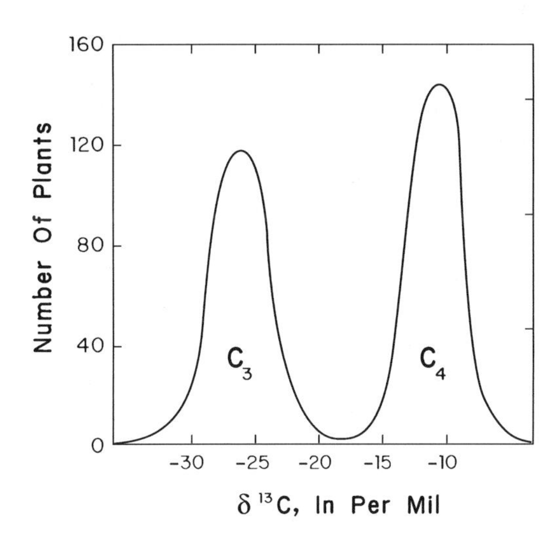

Fig. 9.11 Stable carbon isotope values and relative abundance for C_3 and C_4 plants.

metabolism. An example of a plant that fixes carbon this way is the epiphyte Spanish moss. The pathway is the same as for the C_4 plants, with a pre-fixation step, but differs in that the C_4 plants have the two steps separated by *location*, and the CAM plants have the two fixation steps separated by *time*. For example, the final fixation of carbon in CAM plants occurs in the dark of night by the same PEP enzyme, which produces malic acid which is then stored in cell vacuoles. Upon the return of light the next day, the malic acid is then decarboxylated to CO_2, which can then undergo fixation by the rubisco enzyme. This difference in biochemical reactions between C_4 and CAM plants produces different oxygen and hydrogen isotope ratios in plant water (Sternberg et al. 1986).

In addition to using the stable isotopes of carbon, the stable isotopes of water also can be used. The stable isotopes of water, as deuterium, as δ^2H, and oxygen, as $\delta^{18}O$, in tree tissues can be compared to values of potential sources of water (White et al. 1985; Dawson 1993; Scrimgeour 1995). All the potential sources of water for plants should be sampled and analyzed for these isotopes, including precipitation, soil moisture, runoff, surface water, fog, dew, and groundwater. Although all these sources are derived from precipitation, the various processes that affect the water in these components affect the stable isotope value and, therefore, for a specific area often will have diagnostically unique values. For example, evaporation of water will occur to surface waters to a greater extent than to groundwater, even if precipitation of the same stable isotopic composition is the source, and this evaporation affects the stable isotope composition of the water and water vapor, due to mass differences between H and D, as well as vapor pressures. Because of the wide variation in the stable isotopic composition of precipitation around the globe, most results will be site specific and cannot be readily comparable to other site results.

The reason that the stable isotope ratio of the various water sources can be used is that there is no fractionation of isotopes when water is taken up by those plants that cannot exclude salt, so that the stable isotope ratio of the source water(s) is preserved. The signatures are preserved because the entry of water into the root hairs is by mass diffusion, rather than molecular flow, so the water is not subject to kinetic fractionation. However, due to the transformation of liquid water to vapor in the leaves, the stable isotopic composition *will* be affected in these tissues, as well as in stems that are not completely suberized and, therefore, exchange gas with the atmosphere. The water in leaves will be isotopically enriched, or contain a greater percentage of the heavy isotope on account of discrimination during water vapor evaporation from the open stomata (Farquhar et al. 2007). Samples also can easily be obtained and routinely analyzed. The range of values that can be expected of various sources of water and, therefore, water in plant transpiration is depicted in Fig. 9.12.

One of the first studies that proved the usefulness of stable isotopes in elucidating the source of water used by trees was performed by White et al. (1985) (Fig. 9.13). In a swamp in Arkansas, the stable H isotope composition of precipitation was greater than −30 per mil and the groundwater was −12 per mil. The sap of baldcypress (*Taxodium distichum*) trees

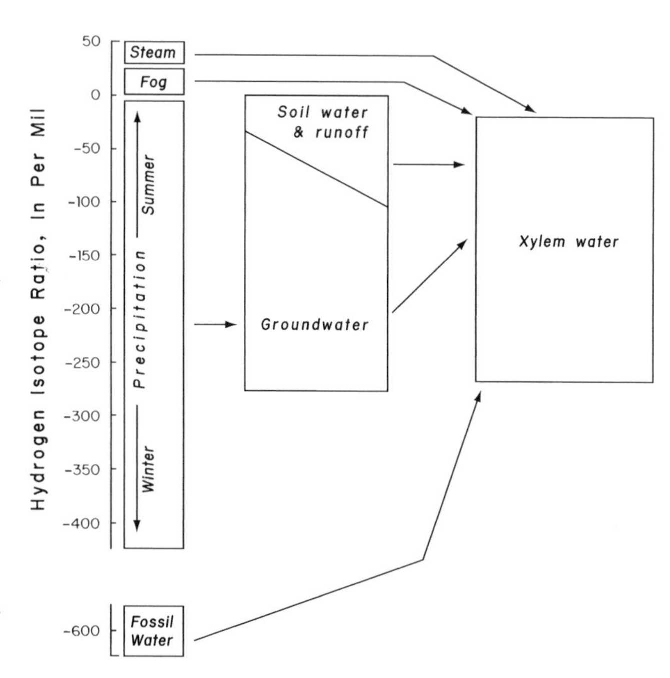

Fig. 9.12 Stable hydrogen isotopes of various water sources available to plants (Modified from Dawson 1993). Xylem is shown for woody plants.

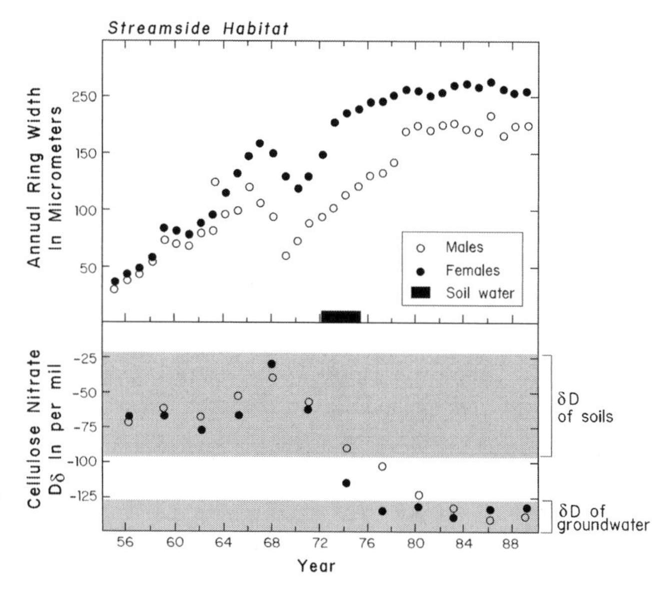

Fig. 9.13 Stable hydrogen isotopes (as δD, or deuterium) of precipitation, groundwater, and plant water (Modified from White et al. 1985).

was −12 per mil. As such, the authors concluded that the precipitation amount was too little to cause the trees to stop using groundwater. Conversely, pine trees (*Pinus strobus*) growing in a dry area in New York had a sap stable H composition similar to recent precipitation in the area, or −20 per mil, for at least 5 days following the precipitation event. After that time, the stable H isotope composition was assumed to be derived from water stored from past precipitation events. Moreover, for pine trees growing on wetter soils, the stable H isotope composition reflected a mixture of rainwater and groundwater but was composed predominately of groundwater around 5–6 days after the most recent precipitation. The precipitation stable H isotope can be seen in tree xylem between 15 and 36 h after the event for trees down to 40 min for certain monocots (Dawson 1993).

These results are useful to the evaluation of phytoremediation sites designed to interact with groundwater, because most sites where trees will be planted to control groundwater will experience precipitation events that deliver water to the root zone. The data that indicate that plants were alternatively using rainwater and groundwater needs to be considered when estimates are made regarding the total amount of groundwater that will be used by plants at other sites.

These data also stress the importance of recognizing that plants, from xerophytes to obligate phreatophytes, rarely rely on only one source of water. For example, at the Carswell Air Force Plant 4 phytoremediation site in Texas, plants that use groundwater during the dry season can use precipitation when available (Clinton et al. 2004). Researchers assessed the stable H and O values of groundwater, xylem water, and surface water, which was actually irrigation water used to simulate a precipitation event. After irrigation, sap flow increased by 61%, and the stable isotopes of H and O in the xylem rapidly became more enriched in the heavier isotope that characterized the irrigation water (Figs. 9.14–9.16).

The use of groundwater, soil moisture, or a mixture by trees can be estimated using an isotopic mass balance approach such as presented in Leenhouts et al. (2006). The fraction of *ET* that can be attributed to plants, *FT*, is given as

$$FT = (\delta_{ET} - \delta_E)/(\delta_T - \delta_E) \tag{9.10}$$

where δ_T is the isotopic value of water transpired by plants, δ_E is the isotopic value of water being evaporated from soil, and δ_{ET} is the isotopic value of water vapor within the vegetated area.

The implication from these results with respect to the use of phreatophytes for groundwater hydrologic control is that at least after precipitation events, the percentage of groundwater removed will be decreased as the plants use water from the recent precipitation. This use may decrease the

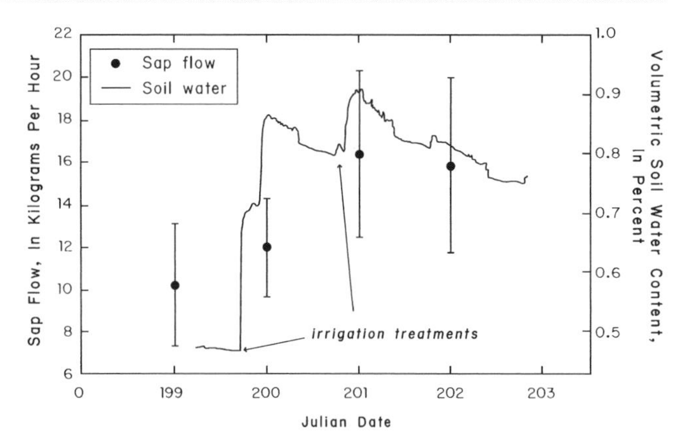

Fig. 9.14 Sap flow versus the soil moisture content before and after artificial irrigation (Modified from Clinton et al. 2004).

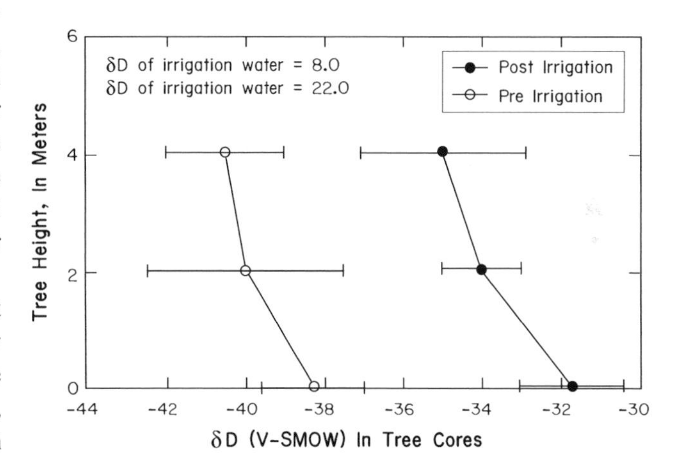

Fig. 9.15 Evidence that trees can use water from different sources is supplied by comparing the stable hydrogen isotope of the different source waters, such as irrigation waters with different stable isotope ratios, to that in tree tissue (Modified from Clinton et al. 2004). One meter is equivalent to 3.2 ft. SMOW is Vienna Standard Mean Ocean Water.

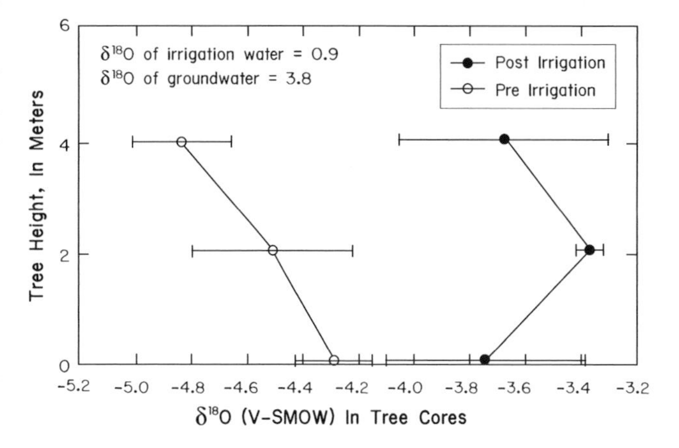

Fig. 9.16 Evidence of water source to plants by comparing the stable oxygen isotopes of two different source waters, groundwater and irrigation water obtained from surface water, to that in tree tissues (Modified from Clinton et al. 2004). One meter is equivalent to 3.2 ft. SMOW is Vienna Standard Mean Ocean Water.

efficiency of uptake of groundwater and the effectiveness of the phytoremediation system at sites that experience frequent precipitation, as suggested by Clinton et al. (2004). However, in areas with low-permeability sediments, the capture of precipitation in a wetting front suggest that little to no recharge of the water table will occur, a positive hydrologic goal at some sites. From a water-budget standpoint, the loss of efficiency when trees switch from a groundwater source to a precipitation source is less drastic.

The multiple sources of water used by plants can complicate the analysis of plant-water, stable-H data. To complicate matters even more, some plants such as cacti can store water from a common or various sources over time, such that the stable H isotope composition of xylem water will reflect a mixture of all water sources and will be an average over a period of time. Isotopic mixing models can be used to determine the relative contribution of the various water sources.

Because of the multiple sources of water that are available to plants, it might be expected that the most obvious source is the one preferred for use. For example, a long-held assumption is that trees growing near streams are using surface water. Many streams, however, consist predominantly of groundwater discharge, and is the primary source of water to streams in between precipitation events. Groundwater is continually in contact with the stream bottom and can provide a source of water to plants. Dawson and Ehleringer (1991) measured the stable H isotopes of stream water, groundwater, precipitation, soil moisture, and xylem water to show that large trees growing by streams, as well as larger trees growing some distance from the stream, do not rely on surface water to meet transpiration demands, as would be expected (Figs. 9.17 and 9.18). Rather, the isotopic composition of the xylem water in the large trees studied was identical to that of deep groundwater that discharged to the stream. Conversely, smaller trees contained water derived from soil water or stream water. A possible explanation for the lack of use of stream water by trees growing nearby is that groundwater supplying discharge to the stream is a more constant source of water to which mature root systems adapt.

As tree root hairs encounter water in the capillary fringe zone or the water table, transpiration can commence. Stable isotopes of water have been used to show that trees can deflect deep groundwater flowlines toward the surface and can in fact support the growth of nearby plants. This upward movement of groundwater to the unsaturated zone is called hydraulic lift (Richards and Caldwell 1987) (Fig. 9.19).

The physical basis behind hydraulic lift is best explained in terms of the various components of water potential in the root zone. In areas where transpiration and evaporation are removing soil moisture from shallow surface soils, the total water potentials will decrease, or become more negative.

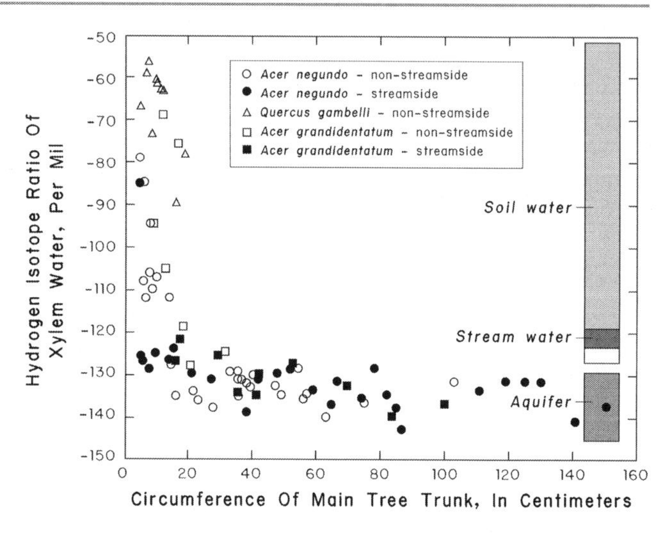

Fig. 9.17 Stable isotope values of different water sources (Modified from Dawson and Ehleringer 1991). One centimeter is equivalent to 0.39 in.

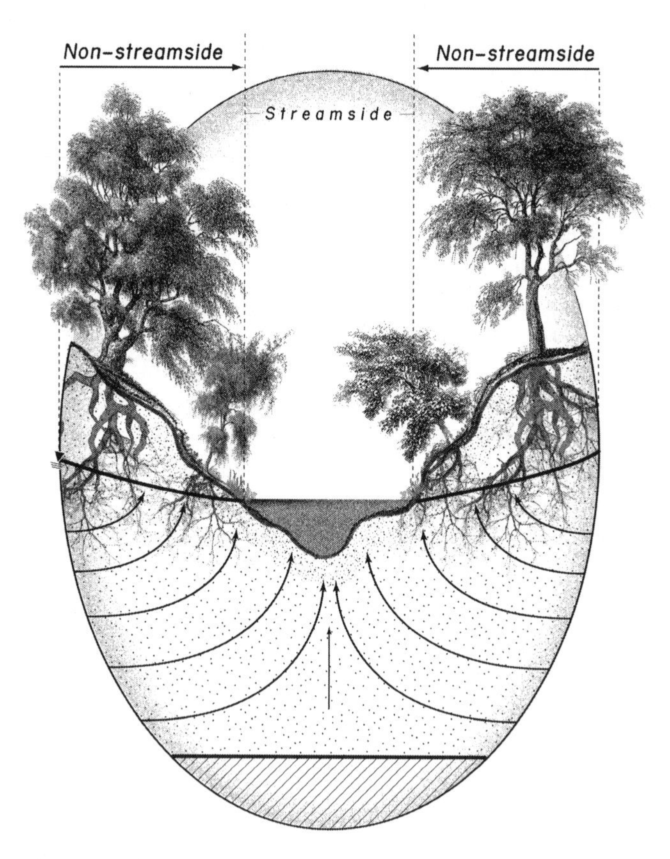

Fig. 9.18 Riparian trees often use groundwater rather than surface water (Modified from Dawson and Ehleringer 1991). Groundwater flow lines are deflected upward toward the roots and the balance to the surface water.

Deeper soil with higher, less negative water potentials will then move upward toward the surface to replenish this lost water. Much of this action occurs at night when transpiration and evaporation cease to remove water from the upper soils.

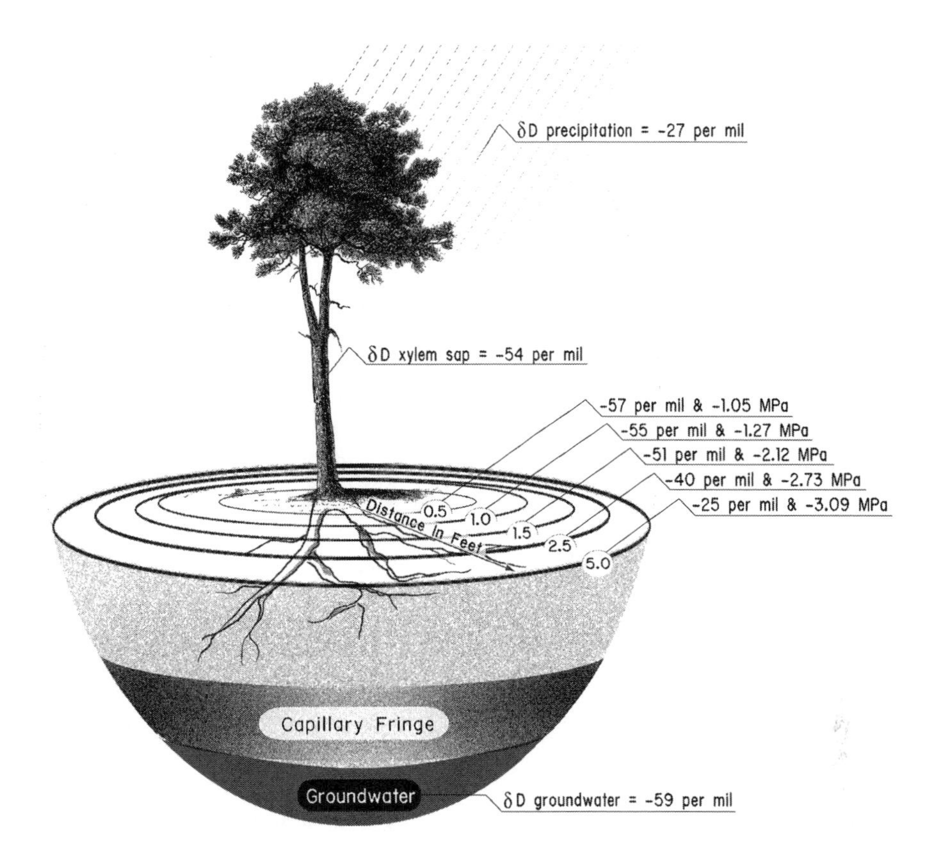

Fig. 9.19 The concept of hydraulic lift, in which water is distributed vertically from wetter sediments to drier shallower sediments, as shown using stable hydrogen isotopes. This is not direct groundwater uptake, as the water potentials are negative at the starting point of the root hairs (Modified from Richards and Caldwell 1987).

This upward flow of water from deeper zones to a drier unsaturated zone is a principle mechanism of unsaturated flow in arid regions (Andraski et al. 2005). Of course, this process will continue only as long as there are deeper sources of water in the capillary zone with less negative water potentials or groundwater at atmospheric pressures. If this source of water is beyond the root growth, or dries up, the plants will wilt and will not recover the following day.

Hultline et al. (2003, 2006) and Leenhouts et al. (2006) present evidence that the reverse of hydraulic lift can occur, that is, where water in roots in the shallow soils can move downward to roots in drier, deeper soils. This process may explain how phreatophytes whose roots, following the fluctuation of the water table, can be established initially after germination, and how roots can reach declining water tables over time.

Other studies that used stable isotopes support the notion of plant and groundwater interaction. Busch et al. (1992) reported that *Populus* and *Salix* used groundwater, whereas *Tamarix* used water from the unsaturated zone, based on differences in stable H and O isotopes of the different water sources. Snyder and Williams (2000) used the stable isotopes of water to determine the source of water to a riparian forest of poplar and willow, along with a mesquite understory. The stable H and O isotopes of the soil water, groundwater, and xylem water were compared. Willows had xylem stable isotope values similar to groundwater present at 12 ft (3.6 m), even when frequent precipitation of a different isotopic signature was available. In contrast, the source of water to poplars was between 26% and 33% from the soil layers after precipitation, in addition to groundwater. Cramer et al. (1999) used stable isotopes of water to investigate the interactions between deep-rooted phreatophytes and groundwater in Australia.

9.3.3 Meteorology and Plant Characteristics for Groundwater Uptake

The presence of phreatophytes provides a direct link between subsurface sources of water and the atmospheric demand for water. This is significant, because unlike surface water in lakes, ponds, or streams, groundwater is isolated from the atmosphere. While some groundwater can move upward from the water table toward the atmosphere by the physical process of capillary action, in most cases it has no direct connection with atmospheric processes. Groundwater does interact indirectly with the atmosphere with regards to recharge from precipitation and perhaps changes in water pressures due to changes in atmospheric pressures, but these are indirect interactions.

Various mathematical models have been created to estimate the relation between atmospheric conditions and plant transpiration of groundwater. They include energy balance, water balance, and direct measurement, as discussed in Chaps. 2 and 7. The classic method for estimating potential transpiration from a grove of plants using solely meteorological data is provided by the Penman-Monteith equation (Monteith 1965). As was stated previously, the Penman-Monteith equation is a physics-based model that couples the energy budget with aerodynamic parameters and essentially considers that the vegetation in a grove of trees behaves as one large, well-watered leaf with no moisture stress and no canopy or stomatal resistance to vapor transport. The equation is based on wind speed, relative humidity, PAR, precipitation, and air temperature, all of which affect sap flow and, therefore, transpiration. Water stress occurs during periods of low humidity and high air temperature and wind speed (Denmead and Shaw 1960). A modification to the Penman-Monteith equation was prepared by van Bavel and van Bavel (1988) into software that computes ET_P from weather station data.

In many areas of the northern hemisphere, the drier atmospheric conditions during the winter seasons, when plants are dormant, are actually *more* conducive to transpiration and, therefore, the uptake of groundwater than during the more humid and higher precipitation periods during summer, when plants are growing. During the winter, water loss can occur through bark, or green tissue such as stems, where photosynthesis can occur in deciduous plants that have lost their leaves.

In Texas, the effect of climatic variables on the water use of poplar trees was measured at the Air Force Plant 4 site. At a given air temperature, a higher *VPD* indicates a higher tendency for the air to accept water vapor and therefore be conducive to transpiration. As such, sap flow increased for both whips and 1-year-old trees as the *VPD* increased (Vose et al. 2000). For a given *VPD*, sap flow was lowest during August but highest during June. Sap flow did not increase with increasing *VPD* but reached a plateau after about 1.5 kPa (Vose et al. 2000). At that site, sap flow was linearly related to solar radiation for all the months observed (Vose et al. 2000). The highest sap flows were in June and the lowest in August for both whips and 1-year-old trees (Fig. 9.20).

The lower readings in August at the Texas study site would be considered contrary to what would be expected, save for the fact that August was considered to be under drought conditions, when little precipitation occurred and low soil moistures of less than 30% were recorded. The authors stated that severe water stress was avoided because of two reasons (1) the roots had reached the water table and (2) leaf fall of upwards of one-half the total leaf area occurred when *VPD* and air temperature increased as precipitation and soil moisture decreased. It is more likely, however, that the primary explanation is leaf drop, because no measurements of groundwater uptake were provided to assess the first hypothesis. If the plantation were older, and groundwater being used as a source of water, then perhaps the effect of the drought would have been less apparent, and water use measurements (sap flow, etc.) would have been higher.

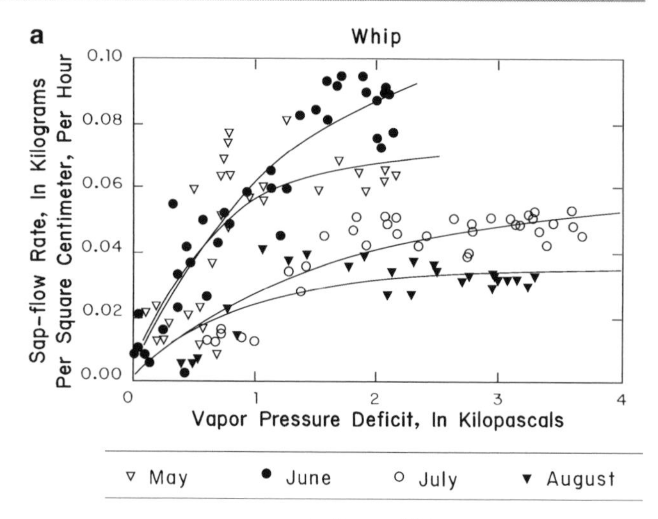

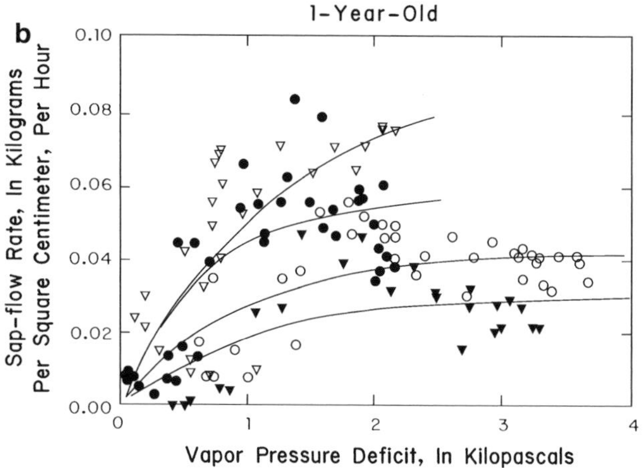

Fig. 9.20 Sap flow and *VPD* for (**a**) whips and (**b**) 1-year-old poplar trees (Modified from Vose et al. 2000). One centimeter is equivalent to 0.39 in.

The *VPD* is an important parameter for water flow through plants and can be determined readily at a potential phytoremediation site, as indicated in Chap. 3. To review, the *VPD* is not a single parameter measured directly, but is calculated from measurements of air temperature and relative humidity at the site, or looked up on a table that shows the relation between air temperature and *VPD* over a range of relative humidity. Basically, the *VPD* is the difference between the amounts of moisture in the air at a given time relative to how much moisture the air can hold at saturation conditions, or 100% relative humidity. In equation form, *VPD* is

$$VPD = VP_{(saturation)} - VP_{(air)} \tag{9.11}$$

The *VP*, or vapor pressure, is the amount of water vapor present, and a higher *VP* means more water vapor is present.

The $VP_{saturation}$ is the maximum amount of water the air can hold at a given temperature; any more water vapor added at this condition will condense from the air onto cooler surfaces, and this temperature is called the dew point. The $VP_{saturation}$ is directly related to air temperature; higher temperatures result in the air being able to hold more water vapor (Anderson 1936).

A similar variable that relates to the potential for air to receive more water vapor and a parameter that is readily obtained from meteorological studies is relative humidity, but *VPD* accounts for the effect of air temperature on the air's ability to hold or condense water vapor, whereas relative humidity is the ratio between the actual water vapor content to potential vapor content. As the relative humidity of the air increases every 20°F, the ability of the air to hold water will double. Hence, the same relative humidity can be stated for different air temperatures but will be of widely different moisture contents.

Anderson (1936) gives an excellent illustration of this point in his classic paper (1936) (Fig. 9.21). The amount of water vapor in the air does not correlate to the humidity or aridity often associated with various areas of the world. Death Valley, California, for example, is an arid area but has the same amount of water vapor in the air as a more humid area, such as Minnesota at a given time of year. Even the Moroccan desert has a relative humidity near 90% during the summer. As such, Anderson (1936) stated that *VPD* should be used rather than relative humidity when discussing water dynamics for biological systems. What is important in determining the degree of aridity is not the moisture content of the air itself but that content relative to the amount that the air could hold at a given temperature. This is similar to the solubility of gases in solution that are dependent upon the temperature of the solution, such as oxygen dissolved in water. The *VPD* is an *absolute* measure of the air's ability to hold water, rather than a *relative* measurement at a given temperature.

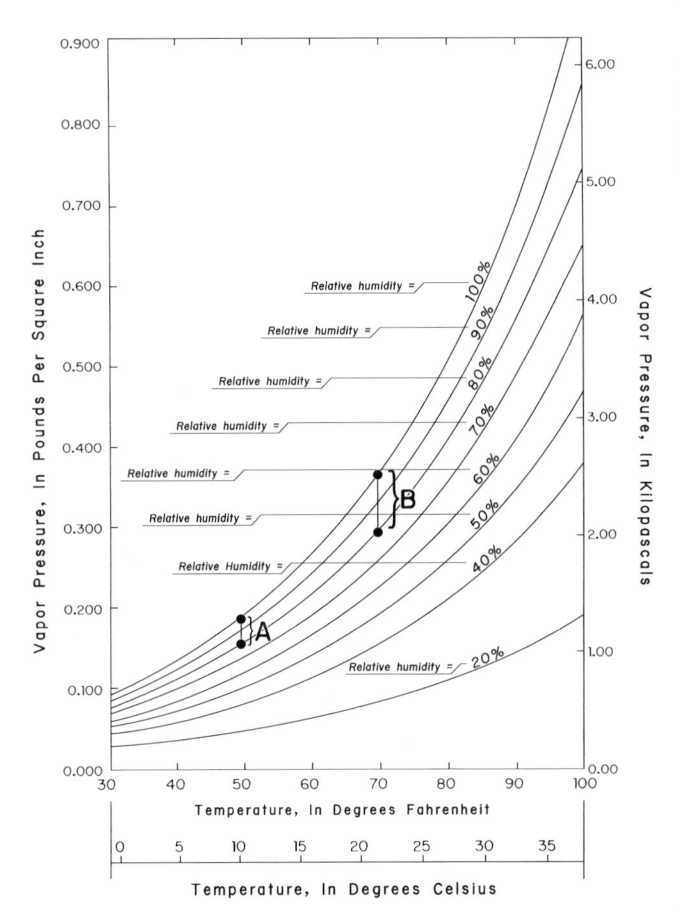

Fig. 9.21 Relation between air temperature (in °C and °F), vapor pressure (in pounds per square inch (PSI) and kilopascals (kPa)), and relative humidity (in percent) (Modified from Anderson 1936).

The relation between the air temperature and relative humidity also has a vertical gradient at most sites. The degree of the gradient is dependent upon location, plant type, and water status. An example of a vertical gradient in relative humidity and air temperature is shown in Fig. 9.22. Even though the air temperature was greater than 100°F near the grass, the relative humidity approached 50% there, almost five times that of the air only 1 ft above the grass.

As might be expected from Eq. 9.11, a lower *VPD* equates to being closer to condensation conditions, and higher *VPD* equates to more evaporation from an open surface, or more transpiration through plants. This has implications for groundwater use by plants in areas that

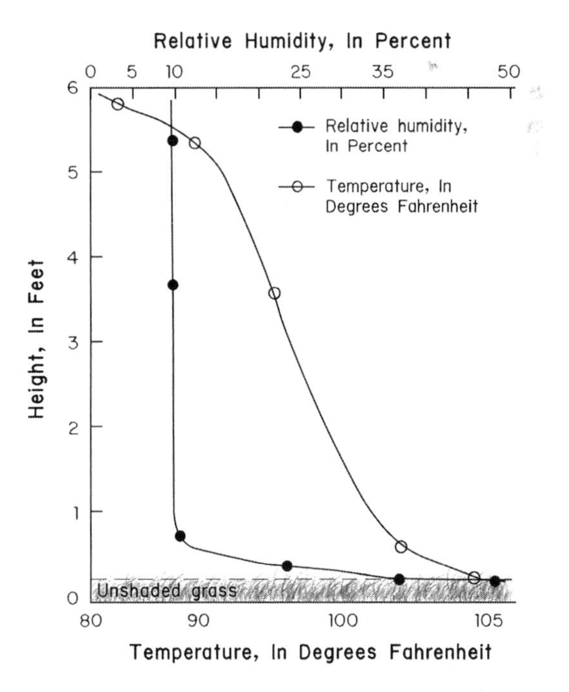

Fig. 9.22 Relation between air temperature and relative humidity at different heights above unshaded but well watered grass in Columbia, South Carolina, August 2007, during 10 consecutive days of record low relative humidity. One foot is equivalent to 0.305 m.

have humid growing seasons but dry winters. The *VPD* is lowest during the period of highest ET_P, solar radiation, and plant growth, but can be highest during the winter when ET_P is the lowest and plants are dormant.

Wind speed can affect the amount of water removed by evaporation and can affect the *VPD*. In areas with little wind, evaporation of water from open surfaces or leaves is controlled by diffusion gradients. As wind speeds increase, recently evaporated water close to these surfaces is removed, and evaporation is increased (Kucera 1954).

Eddy covariance was used by Scott et al. (2003) to examine the energy and water fluxes of riparian vegetation that used either shallow soil moisture or groundwater at sites in southern Arizona. They reported that the water-use characteristics of the deep-rooted plants under open canopy conditions was decoupled from atmospheric inputs such as precipitation, which more strongly controlled water use of shallower rooted plants. These results confirm the applicability of eddy covariance methods to investigate the linkage between plants and groundwater. Moreover, these results are particularly applicable to conditions at recently planted phytoremediation sites, where the deep-rooted trees have not reached closed canopy, and where interplanting of annual or perennial grasses will limit recharge.

9.3.4 Water-Balance Equation

For water distribution in a forest Spittlehouse and Black (1981) presented a form of the water-balance equation that required inputs of only daily solar radiation, precipitation, and the minimum and maximum air temperatures. Net radiation is determined by the solar radiation and air temperature fluctuation.

Another form of the water-balance equation that can be used to estimate the amount of groundwater removed by plants was provided in Leenhouts et al. (2006). It is based on measurements of *ET*, such that the groundwater taken up, *Q*, is given as

$$Q = ET - (P - \Delta S) \tag{9.12}$$

where *P* is precipitation and ΔS is the change in soil moisture in the shallow soil zone. Essentially, when *ET* is larger than precipitation and there is little change in soil moisture storage, groundwater uptake by plants will occur.

9.3.5 Remote Sensing

Remote sensing is the collection of information about an object without there being any direct contact with the object in question. Remote sensing technologies can be used to assess the interaction between plants and groundwater. For example, moisture levels in plants can be detected using thermal infrared (Hunt et al. 1987) and passive microwave sensing. These techniques have been shown to be able to delineate differences in tree types based on the unique spectral signature of trees. This can be used to identify possible groundwater use by plants. Becker (2006) provides an excellent review of the application of remote sensing to understand hydrogeologic parameters and plant interactions with soil moisture and groundwater. Remote sensing works on large tracts of land such as natural riparian systems, but its application to the smaller areas planted at phytoremediation sites is not well known, or at least widely documented.

Remote sensing technologies such as visible, microwave, and gravity sensors can indicate shallow groundwater levels, as well as fracture lineaments for preferential flow in bedrock aquifers. For example, Rodell and Famiglietti (2002) suggest that the Gravity Recovery and Climate Experiment (GRACE) satellite can be used to detect changes in groundwater storage and may indicate the extent of recharge, for unconfined aquifer systems. Other methods to detect the depth to water table include those based on heat capacity. Because the unsaturated zone contains water and air in pores, there is a lower heat capacity than when the pores are filled with water, because of the higher heat capacity of water. This heat can be detected with thermal infrared imagery.

Electrical capacitance was used by Preston et al. (2004) to determine the root mass of hybrid poplar trees. The advantage to such a method is that it is non-invasive. The authors were able to strongly correlate root electrical capacitance measurements with root dry mass and root wet mass.

If plants known to interact with groundwater can be used as indicators by remote sensing, then remote sensing will help detect groundwater, especially in discharge areas under natural conditions (Klijn and Witte 1999). The component of evapotranspiration derived from groundwater was examined using the USGS model MODFLOW and simple moisture transfer models, and York et al. (2002) concluded that in Kansas, between 5% and 20% of the *ET* was derived from groundwater. These techniques will further the collection of data that will provide more evidence of the interaction between plants and groundwater.

In another remote sensing technique, Raman Lidar can be used to measure the latent heat energy of the water flux from plant-water vapor. Cooper et al. (2000) used this technique to map *ET* from riparian trees in Arizona. The Lidar values from *ET* were calibrated to sap-flow measurements. Limitations of this technique include the requirement for uniform surface cover and flat terrain. Both the Lidar *ET* and sap-flow measurements agreed and indicated the *ET* was about 6 mm/day for the forest studied.

Multispectral remote sensing is the collection of reflected, emitted, or backscattered energy from objects across the spectrum of electromagnetic energy (Jenson 2000). Hyperspectral sensing can be used to indicate areas of groundwater discharge and possible consumption of this groundwater by phreatophytes. This technique can detect areas of constant wetness (Batelann et al. 2004).

9.3.6 Tracers

The introductions of dyes, or organic compounds that appear colored, have a long history of use in hydrogeologic studies. Dyes such as rhodamine are used often in surface-water studies to detect movement and dilution patterns. Dyes also have been used in groundwater studies to investigate groundwater flow in karst systems. Dyes also have been used in plant studies to determine the source of water being used by the plants. Robinson and Donaldson (1967) added the fluorescent dye pontacyl brilliant pink to the soil surrounding the roots of the woody phreatophytes willow and wildrose during the summer. Fluorometric analysis of the leaves indicated the presence of the dye in the leaves, roots, stems, and in gas bags that were used to capture transpiration. Moreover, the detection of the organic dye molecule in the leaves provides early evidence that these plants could take in a large organic molecule from the soil-water solution and translocate it throughout the plant's vascular system. Moreover, the use of radioactive isotopes, such as tritium and ^{32}P, as tracers can also provide information on the source of water used by trees.

9.3.7 ET and Groundwater Models

The effect of plants on groundwater levels can be evaluated using numerical models. The USGS model MODFLOW contains a module, called the ET Package (EVT), that simulates the interaction between groundwater and vegetation. EVT simulates *ET* based on the simulated water level, or head, in a model cell, where the head ranges from a maximum elevation (ET surface) to a minimum elevation (extinction depth). Maximum *ET* rates occur at or above the ET surface; at the extinction depth, *ET* is zero. Between these two extremes, the *ET* rate is variable and is linearly dependent on the depth of the head below the ET surface. These relations cannot be changed regardless of the differences in groundwater use by different phreatophytes, such as obligate or facultative, or wetland or transitional plants.

There are alternatives to this approach of a linear decrease in *ET* with increased depth of water table below the ET surface. One is provided by Banta (2000), in which the linear relation between *ET* and depth to water table below the ET surface is replaced with a non-linear curve that consists of segments of curves of different slopes defined by the user. This flexibility allows the modeler to more accurately simulate the higher zone of transpiration that occurs as the water table is within the root zone of phreatophytes, but it still assumes, like the original MODFLOW EVT, that the *ET* rate will decrease as the depth to water table increases below the ET surface. It also assumes, like the EVT, that the *ET* rate will increase as the depth to water table decreases. In fact, many plants will die and transpiration will stop if the roots are flooded by a high water table for a period of time such that oxygen levels are depressed.

Baird et al. (2005) presented a simple solution to these limitations. They developed two models that can take water-level data output from MODFLOW as input for a Riparian Evapotranspiration Package (RIP-ET); MODFLOW-2000 (Maddock and Baird 2003) and RIPGIS-NET (Ajami et al. 2011). The linear relation between *ET* and depth to water below the ET surface is replaced, in their models, by a set of plant-based physiological curves, or plant functional groups (PFG) that are representative and unique for different plant types, water tolerances, and root-depth ranges. This model decouples the *E* and the *T* in *ET*. For example, rather than a linear relation between *ET* and head decline, Baird et al. (2005) use a Gaussian distribution, or multiple, nonlinear segmented flux curves (Baird and Maddock 2005) such that at depths near extinction and at higher water tables, the *ET* rate is lower. As might be expected from the areas under these curves, the MODFLOW EVT method will typically overpredict *ET* relative to that estimated from the RIP-ET package. Maximum *ET* occurs when the water table is within the maximum root density, but with sufficient air supply to support respiration. Also, these models account for the opposite scenarios, where the water table increases and decreases the availability of oxygen to support root respiration.

A novel development of how to simulate *ET* in numerical models was reported by Shah et al. (2007). These authors not only suggest changes to how *ET* rates and the depth to water table are simulated but also recognize that plants will remove water from the unsaturated zone. They state that when the water table is within 1–2 ft (0.3–0.6 m) from the land surface most of the simulated *ET* is derived from groundwater. As the depth to water table increases, less *ET* water is derived from groundwater and more is derived from the unsaturated zone. Moreover, rather than using a linear relation to describe this relation between *ET* and depth to water table as is used in MODFLOW, an exponential decay function is more appropriate.

The Simultaneous Heat and Water (SHAW; Flerchinger 1991) model can be a useful tool to investigate the water budget of a site where plants will be installed. The SHAW model is a one-dimensional model developed to investigate the effect of snowmelt and soil freeze-thaw cycles on soil moisture levels. The model simulates heat, water, and solute transport in a profile from land surface to depth, and includes terms for plant and water interactions. The model is based on the Richards equation to describe water flow in the unsaturated zone. The plant interactions are simulated to specifically account for the effect on the water budget in the soil profile due to differences in rooting depth, plant size, and *LAI*. Transpiration is simulated as being controlled by the balance between stomatal conductance and leaf water potentials. Preston and McBride (2004) used the SHAW model to assess the impact of planting poplar trees over a decommissioned landfill in Ontario, Canada. Predictive simulations indicated that the poplars trees affect the site water budget by taking up precipitation and decreasing recharge.

Another plant–water–soil profile model is called UPFLOW (Raes and Deproost 2003; Raes 2004). This model can be used to estimate the steady-state amount of water that moves from the water table to the root zone under a variety of environmental conditions. The model is based on soil-water retention curves that represent the various soil types encountered in the unsaturated zone. The resultant profile is dependent upon the *ET* demand, the soil water content, depth to water table, plant root characteristics, soil properties, and salt content. The model will also calculate the depth when aeration levels may decrease to the point of anoxic conditions, which would jeopardize those roots continually submerged. The model assumes that higher plant-water uptake rates occur in the shallow soils associated with increased root density; this may not always be the case for obligate phreatophytes installed at sites where there may be lower root density with depth whose uptake will be offset by higher root hydraulic conductivities.

Hopmans and Van Immerzeel (1988) investigated the interconnection between groundwater movement to the capillary fringe under the influence of *ET* at a site in The Netherlands characterized by a shallow water table using the SWATRE model (Belmans et al. 1981). They used the model to reproduce field data and concluded that *ET* was controlled by the hydraulic conductivity of the soil profile through the capillary fringe. At locations where the hydraulic conductivity was lower, the *ET* demand could not be met by capillary rise.

The Hydrologic Evaluation of Landfill Performance (HELP, Schroeder et al. 1994) model has long been used to evaluate conventional and *ET* covers over decommissioned landfills. The model simulates the complex relation between hydrology, soil, plants, *ET*, and climate using a water-balance approach. Rather than being based on the Richards equation, such as SHAW, it is based on a water balance.

9.4 Summary

Water losses by evapotranspiration can approach nearly 70% of annual precipitation in most areas. Because phreatophytes effectively couple groundwater to the atmospheric demand for water, the hydrologic and meteorologic conditions of a site can affect the ability of plants, such as poplar trees, to remove soil moisture or groundwater. Conversely, trees also can affect groundwater levels, flow directions, discharge, and recharge.

Why is this information important to the phytoremediation of contaminated groundwater? The hydrologic changes that indicate plant and groundwater interactions can be observed and monitored by using both plant-based and hydrology-based methods. When combined with geochemical methods, such as the stable isotopes of water, plant and groundwater interactions at contaminated sites can be unequivocally demonstrated.

10 Economic and Regulatory Factors That Affect the Phytoremediation of Groundwater

"Money makes the world go around"
Money Song (Cabaret, the musical 1966)

Many of our natural resources, such as air and water, are considered common goods and, therefore, privately-controlled processes that affect these common goods, such as their extraction, production, and disposal, require state and federal regulation. Many common resources or goods are economically considered to be nonexclusive resources because of the general inability for one entity to claim the resource as being private property that can be excluded from use by other entities. Conversely, those commodities that are exclusive infer property rights, such that a price can be collected for an economic gain from the use of these goods by others. The scenario in which natural resources are essentially beyond the reach of commerce is more common in the United States than perhaps anywhere else in the world (Randall 1987). This, however, is changing, with recent trends in the privatization of water, as shown by the growth of the bottled water industry after the 2000s, or the privatization of agricultural plants by the agrichemical industry.

The regulation of the use of commonly held natural resources, also called rules of access by Randall (1987), can be imposed by a body of stakeholders tasked with the stewardship of that resource to guarantee the fair use of the resource by all that wish to use it. The stakeholders can be made up of an informal group motivated by self interest or by a central government of generally elected officials. For example, duck hunters who realized that overharvesting the duck population would result in the loss of their sport formed Ducks Unlimited to help regulate the number of ducks that individual hunters could kill, thus preserving the current duck population to ensure future hunting. For the more essential resources, those necessary for survival and commerce, the government usually delineates the rules of access to the resource. For example, such government regulation is required where multiple users of groundwater pump from an aquifer that crosses demographic and political boundaries. The regulation of such a common resource has as its goal a balance between economic competitiveness and the protection of the environment and human health from unintended harm caused by the use of the resource, and ensures that the resource will go to the highest use, such as drinking water.

These centralized governmental bodies also can require parties that access common resources not to degrade them beyond economically and toxicologically prescribed limits, and if exceedences occur, can require the responsible parties to restore the resource to pre-impacted conditions. To ensure compliance, regulations have the incentive that noncompliance will result in the assessment of fines; collection of those fines is termed enforcement. Enforcement provides the economic incentive for the responsible party to meet the regulatory requirements. Ironically, if the fine is low, it is often considered economically advantageous to continue resource degradation and pay the fine—in essence, pollution of a common resource becomes a cost-effective form of waste disposal. Hence, for a fine to be an incentive for a company to not pollute, the fine must be higher than the cost of waste-disposal.

The restoration of common resources needs to be achieved through cost effective methods; not too high to be prohibitive and not too low and be ineffectual. Where resource restoration for a particular site falls in between these two end members is the result of negotiations between the responsible parties and the regulators. A common complication that plagues such negotiations is the comparison of the value of money with the costs to achieve site remediation. As a result, any monetary values listed in this chapter, while reflective of relative values when written, are primarily offered for illustration purposes only.

J.E. Landmeyer, *Introduction to Phytoremediation of Contaminated Groundwater*,
DOI 10.1007/978-94-007-1957-6_10, © Springer Science+Business Media B.V. 2012

10.1 Economic Factors that Affect the Implementation of Phytoremediation

Many different economic drivers can determine which particular remedial strategy is selected for implementation at a site characterized by contaminated groundwater. Whereas a range of strategies may work to meet regulatory requirements, the technologies that often are preferred are those that also are cost effective.

The factors of economics cannot be ruled out of the selection of any remedial decision. For example, there may be a cost savings up front in the implementation of bioremediation or phytoremediation as a remedial strategy compared to a more engineering-intensive alternative such as pump-and-recovery. A lower initial cost may, however, be overshadowed by the costs associated with the long-term needs of phytoremediation, such as monitoring and operation and maintenance. As such, the total cost of these technologies may be equal to or exceed the costs of the alternative technologies.

Most sites under investigation for environmental contamination under RCRA or CERCLA are required to undergo a series of remedial investigation and feasibility studies (RI/FS). At each site, a range of remedial alternatives is evaluated to determine which strategy will provide the most environmental and human health protection and be cost effective enough so that it can be implemented as soon as possible. The alternatives are selected during negotiation between the responsible party and regulatory agencies. For the goal of hydrologic control of groundwater, alternatives to phytoremediation range from capping to reduce recharge, to digging trenches to intercept the water table, to full-scale pump-and-treat systems to remove groundwater from contaminated aquifers.

10.1.1 The Native Vegetation Versus Planted Vegetation Dilemma

Most groundwater contaminated sites tend to have at least some type of native vegetation present. The presence of native vegetation can be beneficial, as was discussed in Chap. 6 during site assessment and characterization activities, because the presence of such native plants indicates that, at some minimum level, conditions necessary to support plants exist and suggest a phytoremediation planting may be viable. In addition, if the plants are interacting with any contaminants at the site, then their growth indicates that concentrations of groundwater contaminants may be below toxic levels.

Such native vegetation often can be useful in planning a phytoremediation system. The application of existing vegetation to be part of a phytoremediation system can be called intrinsic phytoremediation. As we saw in Chap. 9, tissue samples of existing vegetation can be used to delineate the location of subsurface contamination, prior to the installation of initial or additional monitoring wells. If existing vegetation is being considered as part of the phytoremediation system, however, various factors that will contribute to the overall efficiency in groundwater containment or control should be considered. For example, the existing plants may not be phreatophytes and, therefore, will not directly take up groundwater. However, the water table may still be affected by the presence of native plants through recharge reduction, as was discussed in Chap. 5. Even so, this benefit on contaminated groundwater may be limited in areas where native plant density is sparse and, therefore plant-water uptake is less than the site recharge rate. Even if the plant density is thick, the existing plants, such as evergreens, grasses, or other drought tolerant plants, may not transpire much water.

Based on the limitations that affect the interaction of existing vegetation and groundwater, it often is prudent to artificially introduce plants to augment the background level of intrinsic phytoremediation at such sites—simply to increase plant density. Increased plant density brings the removal of groundwater by transpiration in line with the flux of groundwater at the site, as described in Chap. 8. Also, plants that are introduced can be those species that have a higher potential to degrade groundwater contaminants at rates faster than rates of native plants. We have seen for example that the readily available hybrid poplar trees have some of the highest transpiration rates (Pallardy and Kozlowski 1981; Interstate Technology and Regulatory Council 2009).

An advantage to using native vegetation at a site is a reduction in initial costs relative to the installation of phreatophytes. The cost associated with site preparation, installation, amendments, and labor, as introduced in Chap. 7, would not be necessary if existing vegetation were used. And as was discussed in Chap. 8, most sites have the very real need to minimize the size of the planting to maximize use of the land for other purposes, such as found at active gasoline stations or other industrial areas. If more land is needed to capture a plume of dissolved-phase groundwater contamination, for example, then additional lands may be needed. This acquisition may be cost prohibitive, even if native vegetation can be used.

A case study of the use of native existing vegetation to meet remedial goals was presented by Bankston et al. (2001). The site in question is a Superfund site in the Coastal Plain, or low country, of South Carolina. In this physiographic province, the topography is flat, the aquifers comprise fine silts and sands of alluvial deposition, and the depth to water table is shallow and in some areas is above land surface to support wetlands, swamps, springs, rivers, and creeks. Site groundwater was characterized by a wide range of contaminants, from chlorinated solvents to petroleum hydrocarbons. Groundwater flow was from the source area a relatively short distance of less than 200 ft (61 m) to a wetland swamp area that bordered the site. The groundwater-flow rate was estimated to be about 55 ft/year (16.7 m/year). The wetland is actually more characteristic of a swamp, because the flooded conditions support hardwood tree species such as tulip poplars, sweet gum, white oaks, and water oaks. Because the groundwater contaminant plume did not extend beyond the swamp, and because the hydraulic gradient from the source area to the swamp increased, the authors concluded that uptake of groundwater by the native swamp plants was partially responsible for hydrologic control of the groundwater contamination (Bankston et al. 2001).

This hydrologic control of the contaminant plume may not simply be due to the uptake of groundwater by the swamp plants, however. The presence of a wetland or swamp indicates, in most instances, that the area is a location of groundwater discharge for local groundwater flow or a place where surface runoff collects after precipitation events. As such, groundwater-flow paths would tend to terminate at such low wet spots even if trees were not present. This flow-path termination still represents hydrologic control, but it is not necessarily solely due to phytoremediation. It would have been interesting had Bankston et al. (2001) assessed the *source* of water in the transpiration stream of the plants to determine whether or not the plants were taking up groundwater and dissolved contaminants or if they were taking up the surface water in the swamp, which could have been a mixture of recent groundwater discharge that had not yet evaporated or transpired, and surface-water runoff. It is important in these types of investigations in which contaminated groundwater will interact with natural groundwater discharge areas to thoroughly document the influence of plants on groundwater.

Native plants may have an advantage over introduced phreatophytes when it comes to increased disease resistance. With clones, disease susceptibility can be a limiting factor relative to superior disease resistance in native plants. This is because under natural conditions of predominately sexual reproduction in native plants, there is a selective battle between plant health and insect infestation, and only the strongest survive to reproduce. With the clones that can be added to phytoremediation sites, plants are at a disadvantage in this battle, because each clone is an exact genetic copy of the parent only. The parent may have been selected for an original advantage, but without sexual reproduction and natural selection, clones are more susceptible to the ravages of new diseases or cyclical insect infestation. Moreover, such a scenario sets up the need for increased pesticide usage.

Another benefit of using native vegetation rather than clones is increased lifespan. The cause of reduced lifespan in clones could be a result of changes that occur in the chromosomes, particularly the teleomeric sequences, but insufficient data exist to support or refute this idea.

10.1.2 Hydrologic Control: Phytoremediation Compared to Pump-and-Treat and Trenching

In order to reach the specific remedial goal of hydrologic control at sites characterized by contaminated groundwater, the water table needs to be affected. Control typically has been accomplished using mechanical approaches, such as pump-and-treat systems or trenching. For a conventional pump-and-treat system, wells, pumps, electrical lines, pipes, and treatment all need to be installed before any groundwater is pumped. Trenching requires equipment, the installation of drainage materials in the trench, and disposal of the excavated sediments. The installation of trees is relatively less expensive in terms of capital investment and less mechanically intensive, assuming that an irrigation system, significant soil removal, or amendment, are not required. Strand et al. (1995), for example, showed that phytoremediation systems remove groundwater at 20% the cost of pump-and-treat systems. Once the pump-and-treat system is installed, annual costs of power and water treatment will continue, and water disposal also is needed for trenched projects. These concerns do not exist at phytoremediation sites, other than performance monitoring of the plant tissues and groundwater.

In reality, however, the perceived cost savings of phytoremediation, either during installation or operation and maintenance over the life of the project, may not necessarily be borne out. For example, if site assessment and characterization activities, which themselves represent a cost, indicate that the depth to the water table is greater than about 25 ft (4.5 m), more resource-intensive and proprietary deep-planting methods may need to be used—at a minimum, more expensive longer cuttings or poles will be required. If over time monitoring suggests that hydrologic control will require more acreage to be planted, the addition of more trees and expense of additional land, if available, also will increase costs. Losses of plants due to mortality will require new plants to be installed, with additional mobilization costs to replant. Moreover, because the use of phytoremediation to control groundwater hydrology is still a relatively new technology to many regulators, they may require the collection of performance data such as groundwater levels and samples on a more frequent basis than if a conventional alternative technology were chosen. Some regulators may even require that a backup containment system, such as pump-and-treat, be installed as a contingency to the phytoremediation system.

All remedial strategies, including phytoremediation, have associated costs. Some costs are borne up front when the project is started. Other costs include those future costs to be incurred over the life cycle of the project. The conventional approach used to account for such future costs includes experience, but mostly a life-cycle approach is used. A life-cycle approach sums up the total cost of the remedial strategy, from installation to projected costs of operation, maintenance, and monitoring for the anticipated duration of the treatment, with annual adjustments for inflation, salary costs, etc. This approach is one way to compare the cost effectiveness of one remedial strategy relative to an alternative remedial strategy.

To examine this comparison of cost effectiveness and phytoremediation more closely, the costs to treat and hydrologically control contaminated groundwater at a generic site will be compared using an example given by Tsao (2005). In that study, the cost to remediate a site by hydrologic control using an extraction system was compared to the cost of using plants to create a hydrologic barrier. The site was characterized by groundwater at 5–13 ft (1.5–3.9 m) below ground surface in an aquifer of low hydraulic conductivity, near 5×10^{-6} cm/s, and adjacent to a former petroleum refinery. A plume of gasoline that contained benzene, toluene, ethylbenzene, and xylenes (BTEX) had extended out beyond the refinery property boundaries into a neighborhood.

The initial cost comparison offered by Tsao (2005) used the life-cycle approach and is summarized in Table 10.1. In this example, implementation of the phytoremediation system would result in a cost savings of $1.29 million dollars, as the more expensive pump-and-treat alternative would not be deployed. Such a life-cycle approach is relatively straightforward but does not include the time value of money or the reduced power of a dollar due to inflation over the life of the project. Inclusion of this economic fact would tend to decrease the cost savings realized between the mechanical and plant-based hydrologic control systems.

Table 10.1 Life-cycle costs for pump-and-treat using horizontal extraction compared to phytoremediation.

Horizontal		Plant-based	
Extraction Item	Cost	Hydrologic barrier Item	Cost
R&D	$0	R&D	$110,000
Installation	$1,000,000	Installation	$200,000
Operation and Maintenance Monitoring	$750,000[a]	Operation and Maintenance Monitoring	$45,000 year 1 $25,000 year 2 $80,000 year 3–8
Total Life-Cycle			
Cost	$1,750,000		$460,000
Cost Savings	(−$460,000)		$1,290,000

[a]@$150,000/year for 5 years

To account for this devaluation of money over time, an approach based on a Net Present Value (NPV) cost comparison can be made. The NPV is the difference between the sum of the discounted cash flows, or net benefits. The NPV is used widely in the financial industry to assess the likelihood of an investment reaping a return above initial costs. The NPV approach also can be used for remediation projects such as phytoremediation to determine if the money spent will bring in a return on that investment or simply result in a cost savings as shown in the previous example.

To determine the NPV for a particular project being proposed for hydrologic control, or to decide which one of multiple projects should receive funding, the expected cash flow per year from the investment also is calculated. From this amount is subtracted the cost of capital to perform the project, such as all the costs needed to make a project happen, often done using an interest rate. From this amount is subtracted the initial investment costs, the balance being the NPV. A positive value for NPV indicates that a particular project would be economically advantageous. Moreover, the selected payback period has to be met, which is the time needed for the project costs to be recouped. If the research and development (R&D) has already occurred and the costs previously recovered, the payback period becomes moot.

Another factor that affects whether or not phytoremediation might be used at a site to control or contain contaminated groundwater is that the money that would go to set up a phytoremediation system, the capital investment, has to guarantee a rate of return on that investment. This is important for large companies that have multiple sites of contamination, because much of their expenditures are for activities that generate revenue. Although the use of phytoremediation at a site often does result in measurable cost savings, phytoremediation may not be selected because the rate of return is too low (Tsao 2005).

In the example presented by Tsao (2005), the NPV was determined using a 2.5% rate, as recalculated and shown in Table 10.2. The cost savings comparison still selects phytoremediation over pump-and-treat, but the savings realized is slightly lower.

There are other factors to consider when comparing which of the two technologies, pump-and-treat or phytoremediation, should be used to achieve hydrologic control. Regarding the above example, Tables 10.1 and 10.2 include tangible costs. Less tangible and less quantifiable factors also affect the costs of pump-and-treat relative to phytoremediation. These semi- to non-quantifiable costs include risk assessment and reporting costs, regulatory acceptance, and the favorability of the local community to each remedy. All of these factors can affect the NPV and, therefore, total cost of the remediation.

Table 10.2 Life-cycle costs for pump-and-treat compared to phytoremediation using a Net Present Value (NPV) cost comparison.

Horizontal		Plant-based	
Extraction Item	Cost	Hydrologic barrier Item	Cost
R&D	$0	R&D	$110,000
Installation	$1,000,000	Installation	$200,000
Operation and Maintenance Monitoring	$750,000[a]	Operation and Maintenance Monitoring	$45,000 year 1 $25,000 year 2 $80,000 year 3 to year 8
Total NPV			
Cost	$1,603,000		$416,000
Cost Savings	(−$416,000)		$1,187,000

[a]@$150,000/year for 5 year

Even when the NPV is used instead of the life-cycle approach, the final answer as to whether or not phytoremediation will be cost effective is not always clear. As pointed out by Linacre et al. (2005), the uncertainties inherent to a phytoremediation project can increase the projected costs. A phytoremediation planting in similar to an agricultural crop, subject to the whims of the weather. Some of this uncertainty is common to most groundwater remediation issues, due to the subsurface nature of these types of projects. Given the additional fact that phytoremediation systems are exposed at land surface, increased uncertainty may occur as a result of threats from storms, fire, and other forms of catastrophe.

One of the uncertainties of phytoremediation examined by Linacre et al. (2005) was the future property value of the land being remediated by either phytoremediation or by an alternative technology. In relatively non-urban areas, where the price of land is stable, the assumed lower price of such land (as opposed to the price of urban land) relates to a lower profit if the remediated land is sold, so the length of time to clean up is of little relevance. In urban areas, where the price of land may increase rapidly, the longer period of time attributable to phytoremediation may render it second choice relative to a more aggressive technology, even if preliminary NPV calculations indicate that the NPV would be positive with phytoremediation as the remedial strategy.

The USEPA, members of the Federal Remediation Technologies Roundtable (FRTR), and the Remediation Technologies Development Forum (RTDF) analyzed the costs for 32 pump-and-treat technologies at Superfund and RCRA sites (U.S. Environmental Protection Agency 2001). The FRTR includes members from the U.S. Departments of Defense, Energy, and Interior, as well as the U.S. Environmental Protection Agency, the Nuclear Regulatory Commission, National Aeronautics and Space Administration, the U.S. Coast Guard, and Tennessee Valley Authority. The RTDF includes members from government, academia, and industry. In their review, costs of pump-and-treat were considered averaged annual costs, and because they were ongoing projects, the NPV was not calculated. Rather, the authors chose to compare costs as costs per year or costs per 1,000 gal of pumped groundwater (U.S. Environmental Protection Agency 2001).

As stated previously throughout this chapter, costs vary and the remedial designs are often site specific. Of the 32 pump-and-treat sites evaluated, the total cost ranged from $1,700,000 to $5,900,000 (in 2001 U.S. dollars). Annual operating costs ranged from $180,000 to $770,000. Although the USEPA looked at treatment costs of the pumped water the life-cycle costs of each of the sites evaluated as part of the USEPA report were not calculated due to the need for site-specific information.

In some of the cases where pump-and-treat is occurring but inefficiently, phytoremediation may not be the appropriate substitute. A Superfund site at Fort Lewis, WA, provides such an example. The remediation is near the Fort Lewis Logistics Center. Extensive groundwater contamination by chlorinated solvents was found in the shallow aquifer beneath the property as well as in the shallow and deeper aquifers offsite about 1 mi (1.6 km) downgradient, beneath a town on the shores of American Lake (U.S. Geological Survey 1998). In 1995, a pump-and-treat system was completed in two areas of the site, and treated groundwater is added back to the shallow aquifer. One well field was installed in the source area, and one well field was installed near the property boundary downgradient of the source area. Because estimates of the amount of solvent spilled are near 110,000 lbs, groundwater pumpage is about 5 Mgal/day (204 million L/day). The contaminants removed from groundwater amounted to about 1,400 lbs, so it was estimated that at least 78 years would be needed to remediate the plumes (U.S. Geological Survey 1998).

In order to accelerate the remediation timeframe, alternative clean up technologies were offered, including phytoremediation. This would be an impracticable alternative to pump-and-treat at this site for many reasons, and these limitations are acknowledged (U.S. Geological Survey 1998). For example, up to 790 acres (3.1×10^6 m^2) of poplar trees would need to be planted if the goal was to replace the mechanical pump-and-treat system. Although in certain areas of the plume, especially in the shallow aquifer near the source area (where the water table is between 4 and 12 ft below land surface), installation of a phytoremediation system would not be detrimental to the pump-and-treat system. In fact, phytoremediation could assist in the removal of contaminated groundwater without the associated cost of pumping, after the installation costs were recovered. The other alternatives discussed were natural attenuation and a permeable reactive barrier (PRB).

An additional problem of using cost comparisons of any kind as a criterion to select a remedial option is the presence of many different definitions of what constitutes costs. For example, there is a need to define gross costs, which may come encumbered with various indirect charges, such as overhead or profit, versus net costs, or the actual unit cost per item. These definitions often are hard to decipher because their costs may be built into the total costs of the project.

10.1.3 Use of Phytoremediation as a Supplemental Remedy

The use of phytoremediation as the sole remedial strategy to contain groundwater is more of the exception than the rule. In most cases, phytoremediation will be used in conjunction with another remedial strategy, such as source removal. In some cases, performance verification monitoring data collected over time will provide data to support the use of phytoremediation as the sole remedial strategy, but the lag time may be on the order of tens of years. Some sites will have other engineered strategies implemented to control groundwater, such as trenches (Widdowson et al. 2005a).

Ferro et al. (2005) investigated the cost savings projected for a pump-and-treat system and phytoremediation system at a VOC-contaminated site. The goal was to decrease the number of wells in the existing pump-and-treat system in order to decrease treatment costs and replace the removed wells with plants. The total costs to set up the phytoremediation system were $282,600 (Ferro et al. 2005). The cost savings of using phytoremediation to supplement an equivalent reduction in the pump-and-treat system can be determined, assuming that the cost of the contaminated groundwater is known. In the case offered by Ferro et al. (2005), the cost was assumed to be $0.05 gal. If the system was treating groundwater at a rate of 19 gal/min (71 L/min), and this decreased to 10 gal/min (37 L/min), then a cost saving would be created. This 9 gal/min (34 L/min) is made up by the phytosystem, which cost $282,600. By 2010, the predicted reduction in pump-and-treat costs, since phytoremediation, would approach $470,000.

10.1.4 Operation, Maintenance, and Disposal Issues When Using Plants

One of the most comprehensive reports on the costs associated with using phytoremediation to hydrologically control contaminated groundwater was on work performed at the Air Force Plant 4, Fort Worth, Texas. The costs to propose, prepare, install, maintain, analyze, and report during a 3-yr period were all covered (see Chap. 8 for site specifics; Table 10.3). These costs, however, are site specific, and relate to hydrogeologic conditions, meteorological conditions, and plant hydroecological conditions that are specific to the Forth Worth area. As such, the authors concluded that these costs are correct within an order-of-magnitude (U.S. Environmental Protection Agency 2003). The assumption is that for up to 4 years after tree installation, a supplemental groundwater control technology, such as pump-and-treat or an interception trench, would be in place until the trees reached groundwater; these costs, however, were not included in their analysis.

As would be expected, a change in any one of the above factors would result in a different cost per item and, therefore, overall cost. More trees may not necessarily increase costs in a linear fashion because many vendors offer discounts on unit prices for large-volume orders. If the area to be planted increased in size, this change would increase the cost of trees, the cost of the irrigation network, and the time needed to sample wells.

Table 10.3 Phytoremediation item, cost, and percentage of total project cost (Modified from USEPA 2003).

Item	Cost	Percent of total cost (%)
Site preparation	$42,650	9.1
Characterization		
Planting		
Irrigation system		
Well installation		
Permitting and regulations	$55,000	11.8
Capital	$37,833	8.1
Sap flow meter		
Water-level recorders		
Weather station		
Groundwater sampling gear		
Fixed costs	$3,783	0.8
Consumables	$19,480	4.2
Soil fertilizers		
Trees		
Supplies		
Labor	$108,000	23.2
Maintenance		
Monitoring and sampling		
Utilities	$12,900	2.8
Irrigation water		
Treatment/disposal	0	0
Waste handling	$7,500	1.6
Analytical services	$172,855	37.1
Annual monitoring for 10 years		
Operation and Maintenance	$5,000	1.1
Demobilization	$1,050	0.2
Total	$466,051	

A consequence of phytoremediation is the production of biomass, the raw material similar to that produced by the short rotation wood culture industry. The ability to recover some of the initial costs of plant installation at a contaminated site makes phytoremediation an attractive remedial option compared to a more engineered solution. The revised interest in using biomass to produce fuels or as a source of carbon sequestration also makes phytoremediation more attractive as an exit strategy for when the goals of a phytoremediation system have been accomplished.

10.1.5 Laboratory and Greenhouse Studies

Due to the wide variety of terrain and soils across the United States, the question is raised about the transferability of results of poplar tree growth at one site relative to application at another site. The same tree may have a wide tolerance for changes in soil salinity but be very sensitive to air temperature changes. Before considerable expense is encountered, it is reasonable to examine the effect of such variables on plant growth and water use using controlled experiments in the laboratory or greenhouse.

10.2 Regulatory Factors That Affect Implementation of Phytoremediation

Contaminated sites require cleanup to maintain compliance with state or Federal laws and regulations. Although responsible parties and remediation professionals can propose a wide range of remedial activities for the restoration of a particular site, the remedial strategies used must be approved by state and Federal agencies. These guidelines for appropriate (legal) actions affect all aspects of site remediation, from cradle to grave, or from site characterization to treatment installation to waste disposal. Each component of site remediation is regulated under a separate law.

Most wastes released to the environment deemed to have the potential to cause harm to humans or wildlife are regulated under CERCLA or RCRA. CERCLA, enacted in 1980, is more commonly known as Superfund because of the magnitude of time and money that must be spent at many sites under its enforcement. The power of CERCLA was extended under amendments made in 1986 through the Superfund Amendments and Reauthorization Act (SARA). In sum, these regulations are designed to ensure that the most appropriate remedial activity (corrective action or removal) is performed at a site listed under Superfund oversite and that it will not endanger human health or the environment nor lead to the spread of contaminants to other areas. If removal is warranted, where materials are to be transferred to offsite areas for disposal, the procedure must follow Superfund regulations. During onsite corrective actions, Applicable or Relevant and Appropriate Requirements (ARAR) are determined to meet site-specific goals. Contaminated media is determined to be hazardous if it meets the criterion specified under RCRA legislation enacted in 1976. Once a material is deemed to be hazardous, it must be tracked from removal to disposal to limit the possibility of release to other areas of the environment. Specific information on various other regulations such as the Clean Air Act, Clean Water Act, Safe Drinking Water Act, and the Toxic Substances Control Act is beyond the scope of this book.

10.2.1 Time Required to Reach Hydrologic Control

The hybrid poplars widely used at sites characterized by contaminated groundwater are fast-growing phreatophytes, but they require time for roots to reach the water table and for the *LAI* to increase transpiration to affect groundwater uptake under closed-canopy conditions. Similarly, at most sites groundwater flow is slow, and flow rates are usually 100 ft/year (30 m/year) or less. In fact, it may take a few years to achieve demonstrable results in relation to the interaction of plants with groundwater. In general, this can take from 3 years (Eberts et al. 1999) to 4 years (Landmeyer 2001). Because the goal of groundwater cleanup or control is to reach the remedial goals in as short a time as possible, the use of phytoremediation may not be the first choice for all sites when viewed from the perspective of the regulatory community.

Computationally simple models exist to determine the time necessary for contaminants in groundwater to be remediated using phytoremediation, as outlined in Burken and Schnoor (1998) and Schnoor (1997). The benefit of the computational ease of these models, however, is at the cost of the simplifying assumptions that need to be made, such as constant groundwater contaminant concentrations, steady-state plume distribution, and lack of microbial biodegradation. To summarize, the uptake of organic contaminants dissolved in water by plants can be described by:

$$U = (TSCF)(T)(C) \quad (10.1)$$

where U is the uptake rate of the contaminant (M/T); *TSCF* is the transpiration stream concentration factor, an expression that accounts for the variable uptake efficiency by plants for different contaminant compounds (dimensionless) (Burken and Schnoor 1997) that is described in Chap. 12;

T is the transpiration rate (L^3/T); and C is the concentration of the dissolved-phase contaminant (M/L^3). Values of the $TSCF$ for contaminant compounds range from inefficient uptake (low $TSCF$) for low-solubility compounds, such as pentachlorophenol (0.07), to very efficient uptake (high $TSCF$) for compounds with higher solubility, such as toluene or TCE (0.74). From Eq. 10.1, the time required to phytoremediation processes to achieve remedial goals can be estimated, from first-order degradation kinetics:

$$k = U/M_o \tag{10.2}$$

where k is the first-order uptake rate constant (per unit time), U is the contaminant uptake rate from Eq. 10.1 (M/T), and M_o is the initial mass of contaminant present (M). At any time t during remediation, the mass remaining in the aquifer can be determined by:

$$M = M_o e^{-kt} \tag{10.3}$$

where M is the mass remaining and t is the time. Solving for time yields

$$t = -(1n M/M_o)/k \tag{10.4}$$

where t represents the time needed to reach a remedial action level (T), M is the mass allowed at time t(M), and M_o is the initial contaminant mass (M).

An example of using this approach to estimate the range of cleanup times possible was performed at the former manufactured gas plant site near Charleston, SC, discussed previously. The criterion assumed for descriptive purposes was for poplar trees to decrease dissolved benzene concentrations up to 10% of initially measured concentrations. Using Eq. 10.1 to Eq. 10.4 different amounts of dissolved benzene were calculated to be taken up (Landmeyer 2001). The uptake rate increased with increased water removal by trees, assuming that all transpirational demands were being met by groundwater. Proportionally more groundwater was removed by the older trees, resulting in a shorter amount of time necessary to reach the clean-up goal, in this example.

As shown by Matthews et al. (2003), the issue of time to hydrologic control can be investigated using numerical simulations run under transient rather than steady-state conditions. Hydrogeologic characteristics of the site, such as low specific yields, may require a long time before groundwater levels are lowered across the site to influence groundwater flow. If these factors are discovered during site assessment and characterization, however, it is possible to increase the size of the planting to reduce the time for hydrologic control to be reached.

10.2.2 Transgenics and Other Obstacles to Public Acceptance

The proposed use of transgenic plants as part of phytoremediation plantings is becoming increasingly common. This field of research, called plant breeding, or plant biotechnology, is based on the production of plants with desirable traits that did not have these traits originally, using recombinant DNA technology. Much confusion stems from fears of eating food produced by such methods, and public acceptance has been slow. Ironically, the genetic mixture of plants with different characteristics to produce a more desirable outcome occurs naturally, and the frequency of this interaction increased as man experimented with early agricultural crops—such benefits reaped from transgenic foods are not widely known (The Economist 2005).

Successful cultivation, or domestication, of our common cereal crops that were initially wild plants enabled the world's population to double 10 times since the last Ice Age, from 10,000,000 to 6,000,000,000. It started with the use of wild strains of corn and wheat, for example, and then the crossing of these varieties through time to produce a transgenic plant that had very large seeds. Other cereal plants that had properties found useful by hungry humans were, therefore, preserved and are the result of natural genetic mutations—the original plant bears little resemblance to the present-day plant. For example, corn kernels grown today are up to eight-times larger than those produced by wild corn plants. This process of selection of corn with larger kernels was started thousands of years ago by Native North Americans. As a result of such focused crop domestication, corn grown today cannot reproduce by natural pollination; a consequence is that abandoned cornfields do not repopulate on their own.

The production of food crops by artificial selection of natural varieties gave rise to the interest in making artificial mutations by exposure to processes designed to damage plant DNA. However, even this is a random process. It was the need to be more proactive in determining exact outcomes of these mutations that led to the development of transgenic processes.

In a similar manner, the pulp-wood and phytoremediation industries benefit from such manipulations of native plants to produce hybrids to supply their industry. The controversy that surrounds the use of transgenic plants not intended for the food supply is from potential implications that surround the escape of these transgenes into native plant populations.

The advent of recombinant DNA methods increased the rate of testing and production of transgenic plants. The methods used to introduce the desirable trait (the expression of genes, or genomes, which act as the instruction guide or recipe for each and every cell) take the gene from one plant and add it another plant, using plant plasmids to carry the new gene. Another technique physically adds the gene to microscopic pellets that are then forced into the host plant cell. Once this transfer occurs, the added gene, or transgene, will become part of the plant's reproductive cells and, therefore, will be passed onto the next generation. In 2001, up to 69% of the cotton planted in the United States was genetically modified, as well as up to 26% of corn planted.

10.3 Summary

Because 50% of the population of the United States relies on groundwater for drinking-water supplies, it is critical that the eventual remediation of contaminated groundwater be as effective as possible. The remediation also needs to be cost effective, as capital is a limited resource. At each contaminated site, a delicate balance will exist between those responsible for bearing the cleanup costs, be it a private or societal-based entity, and those citizens who rely on clean groundwater. In the middle, acting as the fulcrum, are those who initiate, moderate, and enforce the remediation of contaminated groundwater.

Why is this information important to the phytoremediation of contaminated groundwater? The cost of any effort spent toward remediation of contaminated groundwater always will be a factor in making decisions about remediation, including the implementation of a phytoremediation system, from its evaluation as a remedial alternative to its long-term effectiveness.

Part III

Contaminant Interaction, Partitioning, Uptake, Transformation, Metabolism, and Loss

. . .I do wonder whether there will come a time when we can no longer afford our wastefulness—chemical wastes in the rivers, metal wastes everywhere, and atomic wastes buried deep in the earth or sunk in the sea. When an Indian village became too deep in its own filth, the inhabitants moved. And we have no place to which to move.

Travels with Charley: In Search of America
(Steinbeck 1962)

11 Plant Interactions with Biogeochemical Environments

Evolution has shown that at any given moment out of all conceivable constructions a single one has always proved itself absolutely superior to the rest.

Albert Einstein

Plants are essentially chemical factories that naturally produce sugar by using the raw materials of CO_2, light energy, and hydrogen from the splitting of water. A waste product of this production of sugar, oxygen, and its release to the environment led to the oxidation of previously reduced metals, such as iron, that currently are used by man. This oxygen release resulted in the demise of many predominant forms of anaerobic life early in the earth's history or forced them into seclusion through burial in sediments. It also stimulated the development of aerobic respiration as a way to deal with the toxic oxygen gas—this scenario set the stage for all other aerobic forms of life, including us, to develop. Plants carried out these processes while constantly responding to changes in their environment from natural threats, such as fires, volcanic eruptions, radiation emitted from cooling rocks, methane releases, advancing glaciers, herbivory, and toxic concentrations of metal deposits. The plants that survived had selective advantages relative to those that could not cope with these threats.

Plants were not only capable of responding to these natural threats, but could themselves manufacture a wide range of secondary chemicals, or metabolites, that could be used for defensive or offensive purposes to ensure survival and reproduction. Defensive chemicals include those made to protect plants against threats from other plants or from insects or animals. Offensive chemicals include those made by plants to sequester limited resources or to inhibit the acquisition of these resources by other plants. The fact that these complex organic compounds are synthesized by plants from the simple reactants of water, oxygen, and CO_2 is the envy of many organic chemists backed by modern laboratory facilities.

Since the beginning of the industrialization of many societies, plants also have been exposed to a variety of compounds called xenobiotics, from the Greek *xenos*, meaning stranger. Such chemicals are derived synthetically, such as the chlorinated insecticide dichlorodiphenyltrichloroethane (DDT), or as a consequence of other processes such as the production of chloroform during water purification, and often have no natural source. Many of these xenobiotic compounds interact with plants because the physical and chemical properties of the compounds impart solubility in water and, therefore, the compounds move readily through the hydrologic cycle—for example, chloroform is the most frequently detected volatile organic compound in ambient groundwater (Zogorski et al. 2006).

To ensure that the interaction between plants and xenobiotic compounds can be applied to the phytoremediation of contaminated groundwater, scientifically defensible evidence needs to exist to document that plants can take up such contaminants dissolved in groundwater, and detoxify, immobilize, or volatilize these contaminant compounds into less harmful forms. Questions related to phytoremediation projects that commonly arise and are addressed in Part III include the following:

- What happens to dissolved groundwater contaminants that enter a plant?
- Do contaminants remain in the leaves after seasonal leaf drop?
- Do contaminants enter fruits or nuts?
- Should evergreens be planted at sites to address regulatory concerns about the possible lack of groundwater uptake and contaminant processing during periods of dormancy for more commonly used deciduous trees?
- Does groundwater uptake and translocation occur in deciduous plants during dormancy?

To answer these questions, fundamental information about the chemical properties and interactions of common groundwater contaminants is reviewed. This information provides a basis for the subsequent process of groundwater uptake and contaminant detoxification by plants, which must follow these fundamental physical laws and the results be reproducible for phytoremediation to be scientifically defensible.

J.E. Landmeyer, *Introduction to Phytoremediation of Contaminated Groundwater*,
DOI 10.1007/978-94-007-1957-6_11, © Springer Science+Business Media B.V. 2012

11.1 Plants, Ecology, and Biogeochemistry

The German biologist Ernst Haeckel (1834–1919), perhaps best known for his statement, "ontogeny recapitulates phylogeny," and who also was an accomplished artist (as can be seen in some of his publications), is credited with providing one of the earliest definitions of ecology (Kormondy 1976) such that

> *...By ecology we mean the body of knowledge concerning the economy of nature...ecology is the study of all the complex interrelations referred to by Darwin as the conditions of the struggle for survival.*

Haeckel's phrase "economy of nature" can be reduced, at a minimum, to an understanding of the flow of energy and cycling of nutrients through an ecosystem. Nutrients generally are discussed in terms of cycles that imply a system at steady state rather than in terms of a unidirectional flow, which is more characteristic of the pathway of energy flow. It probably is more correct to discuss nutrient fate in terms of flows, because although steady-state conditions of nutrients can be reached when the loss of a particular nutrient is balanced by input from another source, nutrients also can be removed for a length of time by burial through geological processes and, therefore, are more representative of unidirectional flow. Also, cycles or flows of nutrients all are driven by solar energy and water as it moves through the hydrologic cycle. For example, water is the medium in which life's reactions occur.

Plants must be efficient and tenacious to survive and reproduce. Even though plants are surrounded by all the resources they need, these resources often are in dilute supply. Above ground, CO_2 in the atmosphere is essentially at concentrations low enough to be considered an impurity, and below ground most of the essential and micronutrients to sustain healthy plant growth, such as nitrate and iron, are either diluted or not readily bioavailable. Also, acquiring some nutrients requires the plant to spend energy, for some nutrients are characterized by a net negative charge similar to plants root hairs. Plants need to maintain higher internal levels of some dilute nutrients, so energy is spent in bringing these into the cells against a concentration gradient.

The relation between plants and nutrient cycles or flows generally is envisioned to occur while the plants are alive and actively photosynthesizing. However, plants can affect nutrient cycles even after they die. Except when plants are removed from a stand for lumber or agriculture or are consumed by fire, the minerals and nutrients that entered the plant during growth are returned to the soil during decay or leaf fall. That this cycle has been important was known long ago, as indicated by

> *Death, however, does not destroy matter but only breaks up the union of its elements which are then recombined into other forms.*
>
> Democritus of Abdera (460–370 B.C.)
> From Browne (1978)

The following sections describe approximations of the flow of energy and nutrients related to plant survival and, therefore, application at phytoremediation sites. To simplify the presentation, the flow of each component essential to plant life is presented separately. All of these flows occur simultaneously, however, and water can be considered to be the common denominator. In fact, the discussion of the hydrologic cycle in Chap. 2 is itself an introduction to the biogeochemical cycles discussed here. The rate of cycling will not be discussed, as the focus here is the pathway taken. However, there are some interesting details about the cycling rate of elements that pass through plants. For example, as would be expected from the fact that between 70% and 80% of the precipitation in a given area is returned on an annual basis to the atmosphere, eventually all water in the hydrologic cycle will pass through plants as part of photosynthesis about one time every 2 million years. In contrast, the oxygen produced by plants cycles one time every 2,000 years, and the CO_2 respired by animals and plants cycles once every 300 years.

Unless plants are grown in the laboratory under hydroponic conditions, most plants used for phytoremediation of contaminated groundwater are grown in soil. As a result, plant roots interface with soil particles, soil gas, water, and microbes. Much of the early work done in the area of the effects of nutrients and micronutrients on plant growth was based on observations of plant growth in the absence of a particular element. These simultaneous interactions are no better illustrated that in how energy and chemicals move through the subsurface system, as will be briefly shown below. In the context of a phytoremediation planting, knowledge of how energy and chemicals move among plants, soil, and groundwater provides a framework within which the flow of groundwater contaminants can, therefore, be evaluated.

11.1.1 The Flow of Energy and Electrons

> *The light of the sun, and not the warmth, is the chief reason, if not the only one, which makes plants yield their dephlogisticated air (oxygen). A plant not capable of going in search of its food must find, within the space it occupies, everything which is wanted for itself. The tree spreads through the air those numberless fans, disposing of themselves to encumber each other as little as possible in pumping from the surrounding air all that they can absorb from it, and to present this substance to the direct rays of the sun, on purpose to receive the benefit which that great luminary can give it.*
>
> Jan Ingenhousz (1779)
> Experiments on Vegetables

The seminal reaction of photosynthesis described by Ingenhousz requires the simultaneous presence of chlorophyll, CO_2, and water. If these reactants are simply placed together, however, they do not spontaneously interact, but require a plant and the input of electromagnetic radiation from the sun. Although often ignored, it is humbling to remember that the sun's energy is transmitted through space as waves or electromagnetic energy a distance of at least 93 million miles (148 million kilometers), a trip of roughly 8 min. It is this electromagnetic energy that causes chlorophyll in chloroplasts to enter an excited state and start the whole process of chemical synthesis of plant food.

Radiation is only one form of energy transmission and is defined as the emission of electromagnetic energy transferred as photon packets or waves. Other forms of energy transmission include convection and conduction. Convection is the transfer of energy by a mass that carries heat, typically a fluid such as air or water. Finally, conduction is the transfer of heat by the movement of molecules.

The effect of the sun on all life forms on earth generally is overlooked. Like water, this energy is something that we cannot do without. The sun is a location of a thermonuclear reaction, where the fusion of hydrogen (H) to helium (He) occurs to release some of the mass of H as radiant energy. About 50% of the radiant energy is released in the form of the visible spectrum of wavelengths. Total input of solar energy to the earth is about 13×10^{23} cal/year (Kormondy 1976). And half of this is lost by reflection or absorption in the atmosphere. The energy that makes it to the earth does so at about 2 cal per square centimeter per minute (2 cal/cm^2/min), known as the solar constant, where 1 cal by definition is the amount of heat required to raise 1 g of water 1°C. Interestingly, the earth intercepts only about 1/12 billionths of the sun's total emitted radiant energy.

As can be concluded from the units used, this represents a flow of energy input to the earth. Most of this radiation consists of wavelengths between 0.3 and 3 μm. Based on changes in cloud cover or the elevation of land above sea level, the fraction of the solar constant that reaches the surface to become available for plants is decreased, to between 1.2 and 1.4 cal/cm^2/min in temperate areas and to 1.6 cal/cm^2/min in desert areas where cloud cover is scarce. Values higher than the solar constant sometimes are measured, however, in mountains where direct sunlight and reflected light can reach a plant at the same time.

This solar energy is used by plants to drive photosynthesis and store this kinetic energy as chemical bonds in sugars that later can be released to drive the process of plant life. This process is referred to as "primary production" by ecologists interested in the transfer of this energy through various ecological trophic levels. As we saw in Chap. 2, this input of radiant energy also drives the evaporation of water from the earth's surface and drives the redistribution of water around the globe in the form of local precipitation.

The input and flow of energy can be stated mathematically using a mass-balance equation similar to that shown in Chap. 2 for the water budget. Hillel (1998) presents a form of the heat-balance equation that is listed here

$$J_n = (J_s + J_a)(1 - \alpha) + J_{li} - J_{lo} \tag{11.1}$$

Where J_n is the net solar radiation, where 'net' is the balance that remains after the outgoing flux is subtracted from that incoming; J_s is the energy in the form of short-waves coming from the sun, or advective flow; J_a is the short-wave radiation coming from the sky, or diffusive flow; J_{li} is the incoming, long-wave energy from the sun; and J_{lo} is the outgoing, long-wave energy emitted back to the sky by the soil, which is especially obvious during the night. On a clear day, little radiation is reflected back to space, but clouds increase reflection to the extent that the amount of radiation that reaches the soil is considerably decreased. The term α is a coefficient that describes this reflectivity, or albedo, of the earth's surface.

Much like incoming radiation is felt as your body is warmed, this energy also is absorbed by the soil, transforms to heat, and thus, warms the soil surface. As such, the above equation can be rewritten as

$$J_n = S + A + LE \tag{11.2}$$

Where S is the soil heat flux; A is the heat flux transmitted to the air; LE is the evaporative heat flux; and E is the rate times the amount of water evaporated. The transfer of heat to evaporate water is one of the greatest losses of heat. This helps explain the importance of the hydrologic cycle and why in most areas water losses by ET can account for up to 70% of an area's water budget.

One aspect of heat transfer to evaporate water by plant transpiration is that plants use this evaporating water to reduce heat gain. Plants cannot regulate their temperatures like mammals that can change their basal metabolism. Therefore, some mechanism must help control the temperature of a plant, because the amount of energy input to, say, a leaf, could potentially raise its surface temperature to nearly 11,000°F! This does not occur for several reasons. First, about 30% of the energy passes through the leaf or is reflected from the leaf surface. Second, the air spaces in leaves absorb some of the heat. Third, more than 50% of the energy is used to evaporate water from the leaf mesophyll tissues.

With every gram of water evaporated from a leaf, 580 cal of energy are removed with the evaporating water molecule. The balance of heat gains and losses can present a dilemma for a plant, because slightly increased temperatures result in increased rates of photosynthesis, but greatly increased temperatures also may result in cell death. During the night, plants radiate the heat gained and stored during the day into the atmosphere. This process, and the fact that the respiring cells of plants release heat, are why blankets placed over plants prior to a nighttime frost can keep the plants at temperatures above freezing.

Also of interest in the heat-transfer equation is the amount of energy transformed by photosynthesis into chemical bond energy. The primary source of energy upon which all life on earth depends is relatively inefficiently used. For instance, of the total solar input of about 5,000 kcal/m^2/day, or 13×10^{23} cal/year, more than half (56%), or 2,780 kcal/m^2/day, are in wavelengths that plant pigments cannot absorb (Loomis and Williams 1963). Of the incoming solar radiation that *is* absorbed by plant pigments, 2,220 kcal/m^2/day, of this energy is lost by heat gain or evaporation of water. Thus, only between 2% and 10% of the incoming solar energy is available for storage in carbon–hydrogen bonds during photosynthesis.

What happens to the energy that is absorbed by plants? As a brief review of Chap. 3, the absorption of light energy by organic molecules, such as chlorophyll *a*, increases the electrons to an excited state of energy higher than the ground state prior to light impingement. The absorbed energy will decrease through transformation into vibration energy as heat, emission of the radiation as fluorescent radiant energy, or it can undergo a transformation to a different molecule, or be broken up into simpler molecules.

As stated above, light in the form of electromagnetic radiation travels from the sun as both particles and waves. The magnitude of the energy varies, as measured by differences in wavelengths, and results in different amounts of energy or radiation. The wavelength of the radiation varies from ultraviolet to infrared, from about 200 to 1,600 millimicrons (mμ). When such waves of energy interact with objects, light reacts as packets of energy, or photons. The energy can activate reactions, such as those discussed previously, where electrons are stimulated to an excited state, and work can be done as conditions return to the ground state, but only if all activation energies are overcome.

In some cases, the excitation occurs but no work is done, such as when light is used for vision by animals. In the case of plants, however, work is done through photosynthesis. The light energy is used to split water, the liberated hydrogen used to reduce CO_2, and oxygen is released. The reduction of 1 Mol of CO_2 by hydrogen takes 120 kcal of energy input. Because the primary pigment in plants is chlorophyll *a*, it absorbs light energy around 42 kcal/mole, in the wavelength of 680 nm, or 0.68 nm, seen as red light. Chlorophyll can absorb wavelengths of light in the visible range of the electromagnetic spectrum, between 0.35 and 0.70 nm. About three photons of light would be needed to reduce 1 Mol of CO_2. The light absorption spectra of chlorophyll *a* is that it is bimodal, with the highest absorption occurring between 400 and 500 nm and again at about 680 nm, when in fact this is at the end of the light spectrum. Chlorophyll *b* absorbs wavelengths between 0.45 and 0.65 nm. This also explains why plants appear to be green, because chlorophyll reflects the wavelength of light in this zone, from 500 to 650 nm.

The explanation behind the irony of poor absorption of wavelengths where sunlight is at a maximum may be a result of the porphyrin structure of the chlorophyll *a* molecule. Porphyrins are four pyrrole rings joined by =CH– groups in an aromatic structure as discussed in Chap. 3. The outer electrons are mobile and can travel along the aromatic structure of alternating single and double bonds. Hence, lower activation energies are present so that low levels of radiation can excite the electrons. It is no wonder that the heme- in the hemoglobin of mammalian blood has a similar porphyrin structure to that of chlorophyll, but the core contains Fe rather than Mg.

An artifact of the input of electromagnetic radiation to plants is not the amount of incident energy absorbed by the leaves or transmitted through the leaves but the energy reflected. This energy is in the range 0.35–3.0 nm and, therefore, can be detected by remote sensing equipment (Jenson 2000). The near-infrared wavelengths are reflected and not absorbed as this energy would increase the leaf temperature and perhaps denature plant proteins. The energy reflected and detectable is a function of leaf color, such as pigment and, therefore, the tendency for absorption and reflection; structure, such as thin or thick leaves, heavy cuticles, etc.; and water content.

It is the chlorophyll *a* and *b* absorption spectra that can be used for remote sensing. If photosynthesis occurs in a leaf, reflectance will be at the green spectra and absorbance at the blue (450 nm) and red (650 nm) regions. This characteristic absorbance and reflectance range changes over time as the leaf undergoes senescence and then death. More importantly, the spectra changes if the leaf and plant are undergoing environmental stress, and increased reflectance occurs. An approach to assess initial acute or long-term chronic contaminant toxicity in plants could be based on these spectral changes.

Because plants need light in order to carry out photosynthesis, it is reasonable to assume that chlorophyll controls a plant's orientation to light sources. Chlorophyll *a* is not responsible, however, for the tendency for plants to grow in the direction of light, a process called phototropism. As early as the nineteenth century, it was noted that plants did not grow toward windows if a bottle of red wine was placed on the window sill. Instead, plants were observed to grow in the direction of natural, or yellow, light. It is the carotenoid pigments that reflect light in the yellow wavelength of less than 550 nm. Even the vision of vertebrates, such as humans, can be linked to this plant pigment, which has to be ingested by eating plants since it cannot be synthesized by our metabolism.

In 1926, the scientist Edgar Transeau made an interesting observation that put the theoretical values of energy transfer into the perspective of a real-world scenario (Transeau 1926). He was curious about the fraction of solar energy input to a corn field that was actually captured in the form of reduced carbon, meaning the corn produced. He assumed that 10,000 corn plants per acre were harvested. Of the 100% of solar energy input to the corn field, only 1.6% ended up as primary production, or energy stored in the bonds of glucose. Because very little solar energy, about 0.4%, was used by the plant to support respiration, the process of making plants out of solar energy was almost 77% efficient. Measurement of CO_2 in the air over growing crops indicated that up to 5% of the daily net radiation is transferred to plants during photosynthesis.

The amount of solar energy used to maintain transpiration, however, accounted for almost 45% of the input energy (Transeau 1926), and 54% was unabsorbed and lost. A similar representation of inefficiency for water uptake and that fixed into plants is shown in Fig. 11.1. For example, if a forest or food crop takes up 2,000 tons of water in one growing season, only 3 tons (or roughly 1%) of the water remains behind in the form of carbohydrate; the rest is returned, or cycled, to the atmosphere.

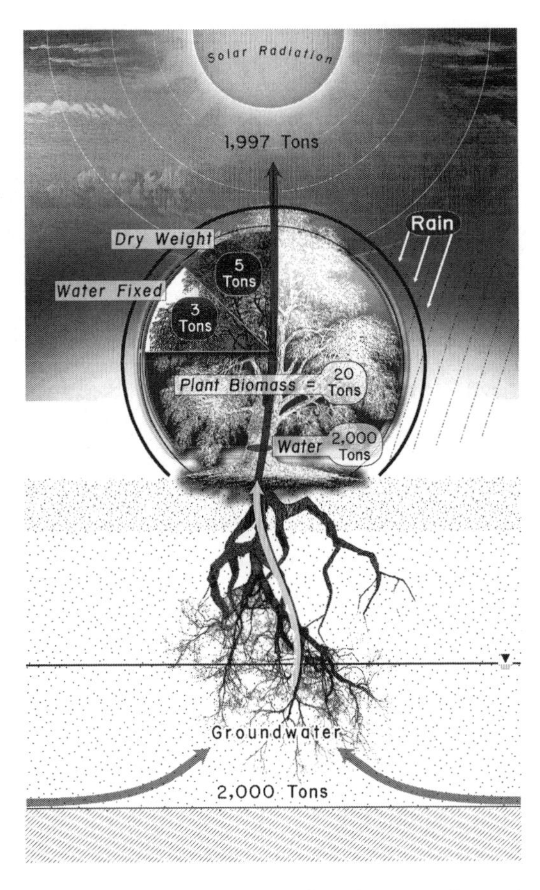

Fig. 11.1 The efficient flow of energy from the sun to plant carbon and the reverse inefficient flow of water through plants to the atmosphere.

The input of solar energy into the chlorophyll molecule involves electrons. The incident solar energy in the form of photons transfers their energy to the electrons of the chlorophyll molecule. The excited and energized chlorophyll molecule transfers this energy towards the production of ATP, a more usable and transferable form of the photon energy. The important component here is that unlike other forms of heterotrophic or chemolithic life that requires energy to be in the form of organic or inorganic species, the process of photosynthesis relies on essentially ambient solar radiation. The ATP so generated can then be used by a cell to synthesize the chemicals necessary for life.

The flow of energy is a one-way street. Much like the energy that is transferred to a light bulb is less than 10% efficient, with 90% of the input released as heat, the initial solar energy captured by plants in the form of reduced carbon compounds is lost as it transfers from producer to consumer. Decomposers, however, can release these elements, such as carbon, back to inorganic carbon that can be used again by plants. The flow of energy is somewhat cyclic as well, but a net loss also can occur, through burial by sedimentation.

These fundamental concepts of the flow and cycling of energy through plants provide a framework within which to evaluate the phytoremediation of common groundwater contaminants. Additional support for the use of phytoremediation of contaminated groundwater is derived from the fundamental interaction of plants within the natural biogeochemical cycles discussed below.

11.1.2 The Flow of Oxygen

Oxygen is a toxic gas. Oxygen has been present in the earth's atmosphere for less than half the age of the planet, which originally contained no oxygen and had an anoxic atmosphere. As was discussed in Chap. 3, the production of oxygen by plants as a waste product of photosynthesis gradually reversed this scenario. As the amount of oxygen increased in the atmosphere, the predominant anoxic life forms died, escaped oxygen's toxicity through burial during sedimentation, or adapted to oxygen by using it as part of their metabolism—anoxic life adapted by developing enzyme systems to deal with the oxidative effects of oxygen. In fact, enzyme-based antioxidant properties eventually evolved to be used as part of metabolism, which gave rise to truly aerobic respiration and aerobic-based life forms (Halliwell 2006). This change in atmospheric gas composition also produced vast amounts of oxidized minerals, such as iron, often seen as the iron-oxide rich deposits in many near-surface geologic deposits.

Although oxygen is a waste product of photosynthesis, plants are aerobic creatures that also require oxygen to release the energy stored in the food produced by photosynthesis for growth and metabolism (Fig. 11.2). Plant tissues above ground where oxygen usually is not limited and below ground where oxygen can be limited require oxygen to produce ATP. Oxygen is limited in some soils, the vadose zone, the capillary fringe, and groundwater because it is consumed rapidly by both biotic and abiotic reactions and has low solubility in water and low diffusivity in water relative to air. If an ample amount of oxygen is present, glucose formed during photosynthesis can be oxidized to 6 units of CO_2 and H_2O and produce 38 units of ATP. This is much higher than the production of two units of CO_2 and ethanol and only two units of ATP under anoxic conditions where fermentation occurs.

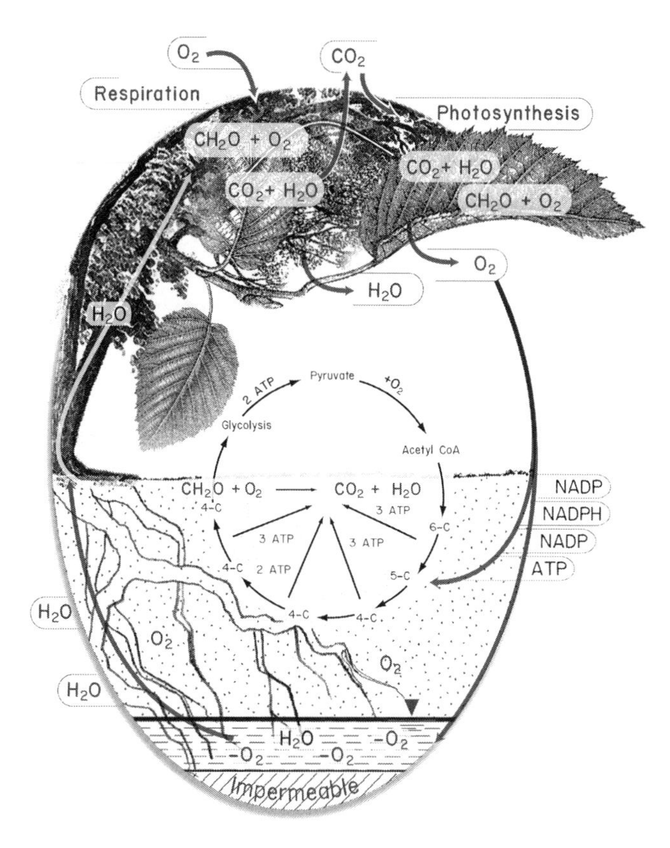

Fig. 11.2 A representation of the flows of oxygen and carbon during plant and groundwater interaction. Plants, like mammals, are aerobic life forms and respire, so plants consume as well as release oxygen. Roots can live in saturated soils or groundwater as long as oxygen is present, either from recharge or diffusion through the plant physiological structure or soil-pore spaces. Shown are water (H_2O); oxygen (O_2, $-O_2$ is anoxic); carbon dioxide (CO_2); carbohydrate (CH_2O); adenosine triphosphate (ATP); nicotinamide adenine dinucleotide phosphate (NADP); nictotinamide adenine dinucleotide phosphate, reduced (NADPH) (Glycolysis is discussed in the section on carbon flow).

The production of oxygen by photosynthesis and its consumption by aerobic metabolism is near a steady-state condition such that the excess in the atmosphere approaches 20% oxygen by volume. There are areas of the planet, however, where oxygen consumption exceeds oxygen production. In many surface-water systems, for example, photosynthetic organisms release excess oxygen to the water column on a daily basis, but at night this dissolved oxygen (DO) is depleted by aerobic respiration. In the sediments of wetlands and some aquifers a few inches to several feet below the water table, DO concentrations also are low because the rate of oxygen consumption is higher than input by recharge or gas exchange with the air in the unsaturated zone. Moreover, these oxygen dynamics are exacerbated by the low solubility of oxygen in water, no greater than about 9 mg/L at room temperature and atmospheric pressure.

It took centuries of scientific discovery to link the two processes of photosynthesis and aerobic respiration together. Until the 1700s, air was thought to be one element only, not a mixture of elements as we recognize today. Work by Stephan Hales (1677–1761), Joseph Black (1728–1799), Karl W. Scheele (1742–1786), Joseph Priestley (1733–1804), and John Mayow (1640–1679) and later by the chemist A. Lavoisier (1743–1794) began to show that air consisted of many different parts, and that one part was oxygen. The candle-and-mouse experiments performed earlier by Robert Boyle (1627–1691) had indicated that removal of air from a sealed jar resulted in extinguishing the candle and the life of the mouse, but the question remained whether or not one or more elements in the air was responsible. This question was solved by John Mayow when introduction of both candle and animal at the same time resulted in earlier cessation of flame and life than when they were added at different times.

Other observations of the burning candle used during this experiment indicated that as the flame was allowed to continue, the wax was used up. This led most scientists to the idea that oxygen in the air supported the flame, but the loss of mass of the candle to the air during burning was responsible for the loss of wax. The context of the time in the mid-1700s was called the "phlogiston theory" that stated when materials were burned a substance called phlogiston, from the Greek, meaning inflammable, was released. The heat and flame left the item undergoing combustion, or being detached from the ashes left behind after a wood fire, for example.

The experiments of Lavoisier refuted the presence of phlogiston by his use of quantitative approaches. He argued that the material undergoing combustion was supported by the removal of oxygen from the air. In fact, he was able to demonstrate that some materials undergoing combustion *gained* rather than *lost* weight. He was able to make this statement because he methodically collected measurements of weights during his experiments, and thusly contributed in making the study of chemicals, and the field of chemistry, more quantitative. Indeed, the chemicals during combustion were undergoing a combination with oxygen from the air.

Combustion, alas, was not a *release* of something to the air but an *absorption* of oxygen from the air. This led Lavoisier to develop the idea that respiration in animals and plants was the interaction of atmospheric oxygen with organic matter, and that both water and CO_2 were released. He also made the connection that the process of the burning candle was analogous to breathing by humans and animals, a process that was driven by the blood. Indeed, our lives are "burning" as much as is the wax of a candle. As such, he provided the data necessary to extinguish the phlogiston theory of combustion. Unfortunately, Lavoisier's own life was extinguished in 1794 when he was executed by guillotine upon being suspected of profiteering from association with a tax-collecting agency.

The interface between plants and contaminated groundwater brings to light a few concepts with regard to the oxygen cycle (Fig. 11.2). As we will see in the next few sections, minerals and nutrients rarely are cycled by themselves. Rather, they are associated with oxygen and, therefore, become soluble in water. Examples include nitrate and iron oxyhydroxides. In addition, in pristine systems, the unsaturated zone provides a source of atmospheric oxygen to plant roots. Under steady-state conditions, the input of oxygen is balanced by uptake during aerobic respiration. Even as trees grow larger and the demand for oxygen increases, the plant will survive as long as oxygen can reach the roots. If the oxygen source is removed, however, by soil compaction or if the water table rises and permanently floods the air-filled pore spaces, oxygen consumption demands will exceed oxygen supply and respiration will cease and roots will die.

Some plants, however, can handle oxygen limitations by shunting oxygen to the roots. Plants still need to respire under these anoxic conditions, and those that can transport oxygen by diffusion from the atmosphere to the roots have a selective advantage in environments where water is not limiting but oxygen is. The cortex and aerenchymal tissues that have interconnected pore spaces permit the diffusion of oxygen to the rhizosphere to assist with the rate of respiration necessary for cell growth. Again, as long as the delivery of oxygen is at a rate equal to respiration, the plant will remain alive. These anoxic conditions and the effect on plant survival must be considered when planting a site where groundwater is known to be anoxic, as is generally the case when petroleum hydrocarbons have been spilled or released to the subsurface.

11.1.3 The Flow of Carbon

Most carbon is stored, rather than flows, in the earth where up to 6.6×10^7 g/m^2 of carbon is in rocks such as limestone. These rocks also contain almost 90% of the near-surface store of oxygen in a form no longer available for respiration. Carbon is contained in previously synthesized organic matter that was buried over time, such as fossil fuels (around 8×10^6 g/m^2). From a biological standpoint, therefore, the atmosphere and hydrosphere are the major sources, 4×10^3 and 270×10^3 g/m^2, respectively, and sinks for CO_2. In each of these two compartments, carbon can take many forms, such as calcium carbonate, $CaCO_3$, in limestone, crude oil, CO_2 in the atmosphere, or as bicarbonate in neutral pH water. Unlike the hydrogen and oxygen in water, carbon primarily is present as a gas.

Plants need inorganic carbon to synthesize organic molecules, but this is only part of the story. These organic molecules are needed to capture the solar energy for later conversion into a source of energy to support plant metabolism and growth. As stated previously, plants take in simple CO_2 with H_2O for use in synthesizing all the complex organic compounds necessary for plant survival and reproduction. These reactions consist of a series of successive reductions designed to generate the compound adenosine triphosphate (ATP) a more useable and transferable source of energy. When plants or the consumers of plants die and are not rapidly buried, organic matter is mineralized to CO_2 by decomposition, and what is not decomposed is buried and the carbon thereby stored, especially under anoxic conditions. This cycle between carbon production and consumption or sequestration by the oceans is roughly in balance, because the rate of cycling is fast, as is evident in the relatively low concentration of CO_2 in air (only 0.032% or 320 ppm).

The reaction of photosynthesis, as part of the carbon cycle, was introduced briefly in Chap. 3. In this chapter the complexities of photosynthesis will be explored, as will the process of plant respiration. These details are important for those interested in implementing phytoremediation at sites characterized by contaminated groundwater, because photosynthesis supports plant growth, water use, and plant-detoxification reactions. Factors that affect photosynthesis and water use, therefore, affect the efficiency and effectiveness of a phytoremediation planting. This is similar to the implementation of an air-sparge system at some site with contaminated groundwater, in which air is added to the groundwater to remove and trap volatile contaminants. Such a system would never be permitted to be implemented unless data had been collected to support the soil's ability to transfer gases to and from the contaminated zone.

Photosynthesis requires light energy that is then transformed into chemical energy. Light required as the input of energy to split water is called the light reaction

$$\begin{aligned} &H_2O + ADP + P_i + NADP \\ &\quad \rightarrow O_2 + H + ATP + NADPH \end{aligned} \tag{11.3}$$

such that the energy-containing compounds ATP and NADPH are formed (Fig. 11.2). The light-harvesting compounds include proteins called phytochromes, but these compounds do not act the same way as chlorophyll. Attached to these proteins is a molecule of chromophore. Together, they absorb photons of light energy. This absorption is translated to other parts of the plant, which can lead to longer stems in shaded areas and shorter stems when afforded ample light. Other such compounds include phototropins, and their role is to help the plant position its leaves to follow the sun.

The chemical energy stored in ATP and NADPH from the light reaction is used to drive the reduction of CO_2 and production of sugar, and this reduction is called the dark reaction

$$CO_2 + ATP + NADPH + H \rightarrow CH_2O + ADP + P_i + NADP \quad (11.4)$$

The enzyme that causes this reduction is ribulose 1,5-biphosphate carboxylase/oxygenase, or rubisco, and the process is termed the Calvin Cycle.

Now that sugars are formed the plant cells have an available, but still untapped, source of energy. To release this energy to do work or for growth, the sugars are oxidized, through respiration (Fig. 11.2), back to CO_2 and H_2O as

$$C_6H_{12}O_6 + 8O_2 \rightarrow 6CO_2 + 12H_2O \text{ and } 686 \text{ kcal} \quad (11.5)$$

Not all of the 686 kcal released is available to do work, because some is lost as heat, much like that removed by a radiator from a running engine or the heat felt rising from a light bulb. The remaining energy is either used directly to perform work or is captured and stored as an intermediate compound such as ADP.

The first major step of respiration is glycolysis (Fig. 11.2). This step leads to the conversion of the 6-carbon sugar molecules formed during the dark reaction into two molecules of the 3-carbon pyruvic acid, or pyruvate. These reactions occur in the *absence* of oxygen in the cell cytoplasm, or cytosol, of respiring cells. The sugar is first reacted with phosphate (PO_4) using energy from ATP to create glucose-6-phosphate by

$$\text{Glucose} + 2\,PO_4 + 2ADP + 2NAD \rightarrow 2 \text{ pyruvate} + 2NADH + 2CO_2 + 2\,ATP \quad (11.6)$$

This continues to fructose 6-phosphate, then by energy from ATP to fructose 1,6-biphosphate, then to two 3-carbon fragments called dihydroxyacetone phosphate and glyceraldehyde 3-phosphate. These are further broken down into two molecules of pyruvate per molecule of sugar undergoing respiration. The most common form of glycolysis is called the Embden-Meyerhof-Parnas (EMP) pathway.

The production of two molecules of pyruvate, however, represents an incomplete oxidation of the original glucose synthesized by the plant. Aerobic organisms can obtain additional energy by coupling the additional oxidation of pyruvate to the reduction of the terminal electron acceptor of molecular oxygen. Pyruvic acid then is converted in the cell mitochondria in the *presence* of oxygen into acetyl-CoA (Fig. 11.2) and CO_2 by most aerobic organisms as part of the tricarboxylic acid (TCA) or Krebs cycle. The acetyl-CoA is then completely oxidized to CO_2 and water.

In the TCA cycle, acetyl-CoA is cycled through four oxidative steps where electrons are released to do work as they are passed down a redox gradient. Each of the four oxidative steps starts with isocitrate, alpha-ketoglutarate, succinate, and malate, and each oxidation results in the production of two electrons and two hydrogen atoms (Fig. 11.2). The electrons are passed to oxygen, almost as water is transported down a hill but through a series of waterfalls where work is done (Fig. 11.3). It is this electron transfer that drives the production of ATP from ADP and P, which in essence is the reverse, or uphill, process.

What is the source of oxygen in the TCA cycle (Figs. 11.2 and 11.4)? Is the oxygen in gas or dissolved form? For

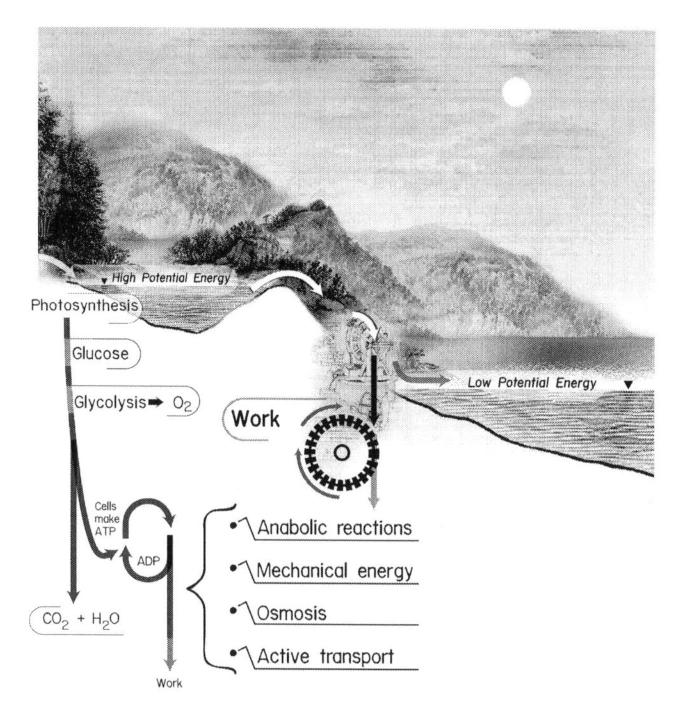

Fig. 11.3 A representation of the similarity between the flow of energy through a plant to do work and the flow of water harnessed to do work. Once the reduction of CO_2 in the plant has occurred, the potential energy of this CH_2O is high, much as water at the top of a water fall has high potential energy because of its elevation. This is only true if the water is allowed to flow over the fall rather than remain static. Once over the fall, the flowing water can be used to do work. In a similar manner, the potential energy stored by plants in chemical bonds in glucose also can do work for the plants during a chemical reaction.

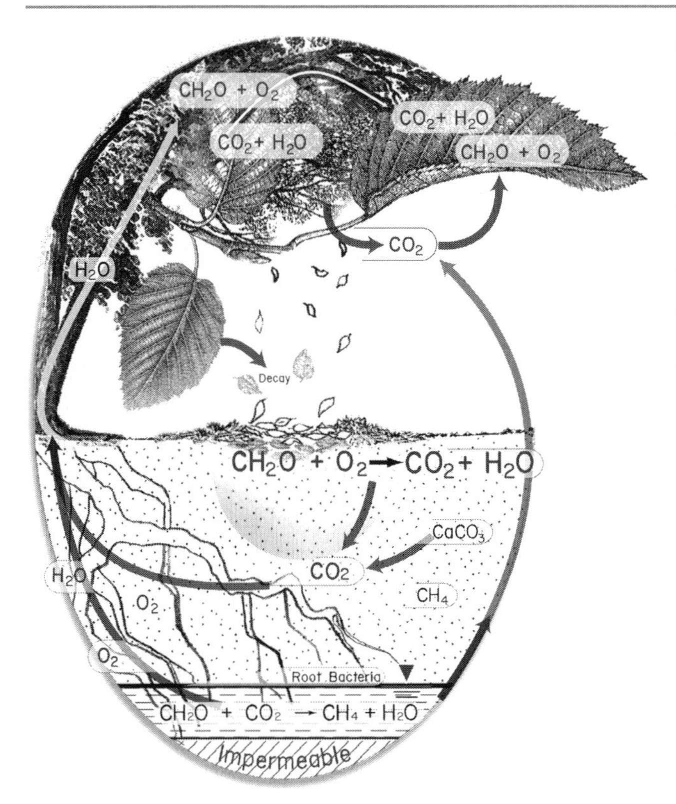

Fig. 11.4 A representation of the carbon cycle. Not only do plants take up carbon dioxide but also release it during respiration. The net release of carbon dioxide is an important metric when it comes to looking at carbon dynamics in plants. Shown are water (H_2O); oxygen (O_2); carbon dioxide (CO_2); carbohydrate (CH_2O); calcium carbonate ($CaCO_3$), and; methane (CH_4).

terrestrial plants, oxygen enters plants as a gas through the stomata and lenticels. Oxygen entry into plants is by diffusion along a concentration gradient from higher concentrations in the atmosphere (about 20%) to lower concentrations (or partial pressures) in the subsurface. Oxygen transport through a plant to roots is through interconnected air-filled spaces that exist between loosely packed cells collectively called the cortex. Oxygen is consumed during respiration, and oxygen also can diffuse into the rhizosphere, at a rate controlled by the oxygen concentration gradient present and potential abiotic and biotic sinks for oxygen.

The electron acceptors in the TCA cycle are as integral as the electron donors. The compound NAD (nicotinamide dinucleotide) is the electron acceptor for three of the four oxidation steps. The oxidized version of the compound, NAD+, (NAD is written as NAD^+, much like NH_4^+ to NH_3) is reduced by two electrons to form NADH. Dinucleotides are simply two mononucleotides linked by the phosphate groups. Electrons are passed between the reduced organic carbon and NAD to form NADH, much as a baton is passed between relay runners; NADH has reducing power. Energy is stored by ATP synthesis and released during electron transport to an acceptor, such as oxygen (Fig. 11.2).

The rate of chemical reactions in these flows is important, because common knowledge suggests, for example, that sugar stored in the cupboard tends not to disappear, even though ample supplies of oxygen are present. Sugar does not spontaneously react with oxygen to form CO_2 and H_2O because for this reaction to proceed requires the input of energy. This energy input, or activation energy, typically is derived from catalysts or enzymes. The role of enzymes in biochemical reactions was first investigated when yeast was ground up with sand, which resulted in non-living matter that was still able to cause the grape sugar to ferment because the disrupted cells released enzymes (Büchner 1897).

The interface between plants and groundwater brings to light a few concepts with regard to the general carbon cycle (Fig. 11.4). A plant needs CO_2, H_2O, and light to make its own food. It also needs the essential and trace nutrients to synthesize its macromolecules, such as proteins and fats. Terrestrial plants, therefore, depend on the atmospheric part of the carbon cycle for CO_2 as a gas, and do not need input of dissolved carbon, such as bicarbonate. Even aquatic plants take in free, dissolved CO_2 in the water column for fixation, and if dissolved CO_2 is limiting, will take up the CO_2 in the form of bicarbonate ions, as HCO_3^-. There is no analogy for plants to uptake other forms of reduced carbon, such as fuels, or oxidized forms of carbon, such as chlorinated solvents, to support photosynthesis or growth or macromolecule production.

11.1.4 The Flow of Nitrogen

Similar to animals, including man, plants need an external source of nitrogen in order to make proteins, such as amino acids, NH_2, or enzymes. With respect to the total dry weight of most plants, compounds that contain nitrogen approach 1–3%, and more than half of this total amount is stored in the leaves. Nitrogen also is used in the manufacture of non-protein compounds, such as hormones, and defensive chemicals, such as alkaloids, which are discussed in the next section of this chapter.

Unlike the direct uptake of CO_2 that exists in the atmosphere at low, part per million concentrations, nitrogen in the atmosphere is unavailable for use by plants even though it is present at levels near 80%, or is not bioavailable in the form of insoluble organic matter in soils or rocks. Most of the nitrogen in soil is in the form of organic matter, such as leaf litter and, therefore, available to plants only after microbial and fungal decay and release.

For nitrogen to become bioavailable to plants, certain processes must occur. For example, plants have adapted and developed the ability to take in and use nitrogen in both the oxidized form as nitrate, NO_3^-, and the reduced form as ammonium, NH_4^+. Ammonium is preferred by plants as it is already in the reduced state, but nitrate has higher water solubility and is sorbed onto soils less.

There are four main processes by which atmospheric N_2 gas can become available to plants for protein synthesis (Fig. 11.5):(1) N_2 can become 'fixed' by lightning during electrochemical reactions, to produce nitrate ions, NO_3^-; (2) N_2 can be artificially fixed during the man-made version of this natural reaction, called the Haber-Bosch process; (3) N_2 can be fixed into ammonia, NH_3, in the root zone, and (4) NH_3 can be converted into NO_3^- in the root zone through nitrification. Some plants do not need to directly rely on such sources of nitrogen, however, such as the various pitcher plants (*Nepenthes spp.*) and Venus fly traps that use modified leaves to entrap nitrogen-rich insects as their nitrogen source. These plants must rely on external nitrogen sources because the soil that they live in is nitrogen poor.

Above ground, fixation of molecular N_2 takes the energy of the lightning to split N_2 to $N + N + O_2$ to yield nitrate ions. This nitrate can be taken up foliarly by plants during precipitation or by the roots after precipitation. The Haber-Bosch process, named after Fritz Haber and Carl Bosch, revolutionized the agricultural industry in 1909 by taking nitrogen in air in the presence of H_2 from coal to make ammonia, NH_3. This process was used to replace the previous source of nitrogen from guano, or bird droppings, which was rapidly mined away by the late 1800s. That the process is important is indicated by the fact that one-half of the nitrogen atoms in human proteins have been produced by an ammonia factory (The Economist 2005). This process does essentially what nitrogen-fixing organisms do but requires high temperatures (400°C) and pressure (200 atm), whereas the biological process uses nitrogenase and hydrogenase enzymes to perform the same reaction at the lower temperatures and pressures of the soil zone. Finally, internal combustion engines with high pressures and temperatures can oxidize N_2 in fuel to nitrogen oxides (NO_x), although this is reduced in catalytic converters prior to exhaust.

Below ground, elemental gaseous N_2 must be fixed, or reduced, much in the same way CO_2 must be reduced for use by plants. The fixation of nitrogen is the rate limiting step of the flow or cycling of nitrogen as is evident by the large

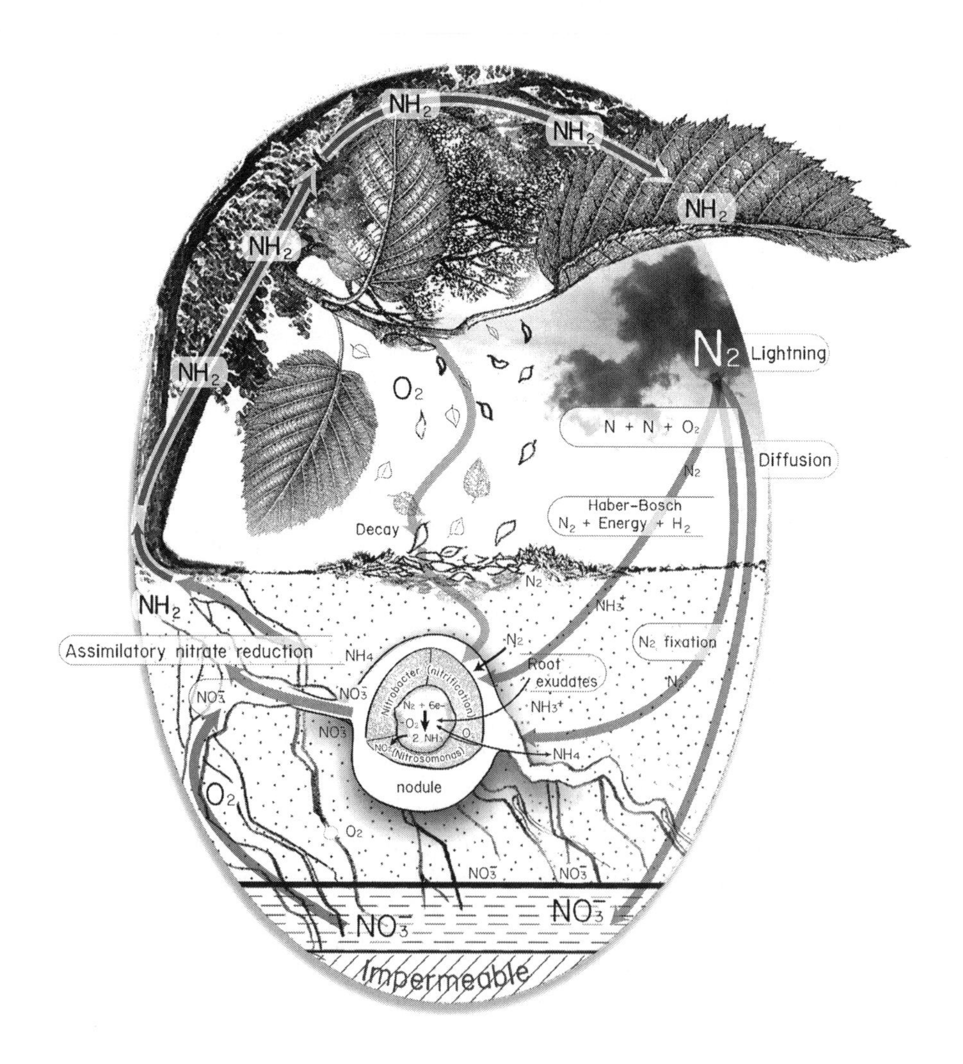

Fig. 11.5 A representation of the flow of nitrogen. Plants are surrounded by nitrogen in the atmosphere, but unlike their ability to take up carbon dioxide, they cannot directly take up nitrogen. This need is met by bacteria living in the roots. The energy needed by bacteria to perform nitrogen uptake is provided by plants, however, in the form of carbon compounds released in the root zone. Shown are water (H_2O); oxygen (O_2); nitrogen (N_2); nitrate (NO_3^-), ammonia (NH_3) and; ammonium (NH_4).

source of N_2 that remains in the atmosphere. No plant has been found that is capable of N_2 fixation alone. Plants require the help of bacteria to accomplish this fixation, and a source of energy such as electrons or hydrogen or ATP to initiate the process. This symbiosis between plant and bacteria is so well established that nitrogen-fixing bacteria cannot do their job if the plant is not present.

Nitrogen fixation is the keystone to all of the water-soluble amino acids in plants and animals. This starts with N_2 being split. Much like the splitting of water to create reducing power during photosynthesis requires the input of an external energy source in the form of solar radiation, the splitting of N_2 also requires the input of energy to the amount of 160 kcal/mole. Since this energy cannot always come from lightning, the energy needed to drive this reaction can be supplied in the form of reduced carbon compounds released by plant roots that are then used by nitrogen-fixing microbes in the soil to generate ATP. These microbes include those mentioned in Chap. 3 regarding the rhizosphere, such as *Azotobacter*, *Rhizobium*, and *Cyanobacteria*. Once the N_2 is split, the free N combines with H to form ammonia, NH_3, which requires the input of 16 ATP from the bacterial metabolism of reduced organic matter, such as that associated in the plant root zone. This interaction between root-zone organic matter, nitrogen-fixing bacteria, and N_2 reduction provides the linkage between rhizospheric microbes and plant roots—these microbes need the external input of energy in the form of organic matter and ATP to fix N_2, and the plants need the N_2 to make proteins.

The fixation of N_2 into NH_3 requires the enzyme nitrogenase. Nitrogenase contains two coproteins—an iron protein, called ferredoxin, and a molybdenum-iron protein. The reaction is $N_2 + 6H \rightarrow 2NH_3$, with the hydrogen coming from hydrolysis of water. This reaction cannot occur, however, in the presence of oxygen, because it destroys the nitrogenase upon exposure, by binding to oxygen better than to nitrogen, such that no nitrogen fixation occurs. Plant roots, however, also require oxygen to sustain root respiration, which would inhibit the activity of nitrogenase. This dilemma is solved by the root-colonizing bacteria, such as *Rhizobium*. These bacteria are found in nodules that form on many plant roots. At the center of these nodules, a protein called leghemoglobin scavenges oxygen from the air much like our own hemoglobin scavenges oxygen from the air in our lungs. The central orientation is crucial because it allows this reaction to occur away from oxygen. The physical separation and sacrificial uptake of oxygen by the leghemoglobin permits plants to fix N_2 under predominantly oxic subsurface conditions in the root zone.

These reactions do not occur, however, if nitrogen in the form of NO_3^- is provided artificially. Nitrogen fixation occurs in soil bacteria not associated with plant roots, but at much slower rates. These *Rhizobium* bacteria are found in the root hairs of certain plants, such as clover and beans, or legumes, as well as the roots of the evergreen hardwood wax myrtle. Nitrogen fixation leads to lower soil pH near the root zone, because as the roots take up N_2, which is electrically neutral, the intake of cations exceeds the intake of anions. When this occurs the plant roots release H^+. Farmers who recognize this process apply lime (CaOH) after growing and harvesting of legumes in order to stabilize the soil pH.

Some plants can take up ammonia after it has been reduced to ammonium (NH_4^+). The advantage to NH_4^+ uptake over NH_3 or NO_3^- is that NH_4^+ has lower water solubility and is not as rapidly leached from the root zone as is NO_3^-. Moreover, plants must reduce the NO_3^- to NH_4^+ following uptake. Most plants, however, still take up nitrogen in the oxidized form rather than the reduced form. The process of sequential oxidation is called nitrification, where NH_3 is oxidized to NO_2^{2-} (nitrite) and then NO_3^-; both reactions release, rather than require, energy. Specialized bacteria, such as *Nitrobacter*, can perform this oxidation and are aerobic reactions, as was discovered by the Russian scientist S.N. Winogradsky in the late 1890s. Since these reactions release energy, these bacteria use it and are considered autotrophs. Nitrate is taken up by the cells against concentration gradients similar to potassium uptake, as detailed in Chap. 3. Energy in the form of ATP is spent in pumping ions out of the cell to allow nitrate to enter.

Some nitrogen can be fixed by anaerobic bacteria in sediments or sediment and water slurries that are void of oxygen that would inhibit the effect of the nitrogenase. *Azobacter* and *Azospirillum* can do this, for example. After nitrogen is fixed to ammonia, further reduction to the ammonium ion (NH_4^+) can occur.

Plants often installed at phytoremediation sites are non-leguminous trees, such as alder or poplars. How do these plants meet their nitrogen needs? The nitrogen can be derived from free-living nitrogen fixing soil bacteria. In the case of alder, the non-filamentous bacterium *Frankia* is a source of nitrogen. Excess nitrogen from alder *Frankia* can be used by poplars in a mixed, densely planted setting (Côté and Camiré 1985).

Another potential source of nitrogen for non-leguminous plants or trees that do not contain *Frankia* is nitrate in groundwater. Groundwater can supply a relatively constant source of nitrogen, and phreatophytes, such as poplars, are able to utilize this source more competitively than shallower rooted plants (Arndt et al. 2004). In most cases, the source of nitrogen in shallow oxic groundwater is nitrate, as bacterial denitrification is inhibited in the presence of oxygen.

The return of molecular N_2 to the atmosphere where it started is accomplished by the process of denitrification, which occurs in the absence of oxygen (Fig. 11.5). This can include *assimilatory* denitrification to NH_2 as amino acids in plants, which can proceed in the presence of oxygen, or *dissimilatory* denitrification by nitrate reduction as an alternative electron acceptor during the oxidation of organic matter in the absence of oxygen, such as

$$CH_{2n}O + NO_3 \rightarrow CO_2 + H_2O + N_2. \quad (11.7)$$

The presence of denitrification in anaerobic soils that are characterized by high organic content and wet conditions, such as wetlands or swamps, reduces the fertility of the area for plant growth, as it produces nitrogen in a form that plants cannot directly use, and the nitrogen returns to the atmosphere. This is why carnivorous plants, such as pitcher plants or the Venus fly traps found in these environments use animals and insects as a nitrogen source.

The effect of endocrine-disrupting (ED) compounds on nitrogen fixation was investigated by Fox et al. (2001). They found that nitrogen fixation by alfalfa (*Medicago sativa*) was altered by the presence of EDs such as the pesticide DDT and the herbicide 2,4-D. Apparently, the presence of these chemicals altered the signal between the plant and the rhizospheric bacteria.

One of the reasons behind the habit of many gardeners and farmers to plow plant materials back into the soil is to stimulate the nitrogen cycle for a localized area. This process puts plant material into the ground for microbes to break down into ammonia and increases oxygen diffusion in soil to be used to oxidize ammonium to nitrate. In fact, this technique of plowing to accelerate the nitrogen cycle was used by the French forces during the Napoleonic wars to overcome shortages of nitrate used to manufacture gunpowder—soil and manure were mixed and frequently turned to accelerate the conversion of ammonia to nitrate, called 'nitrate gardens.'

The interaction between plants and the flow of nitrogen is perhaps best seen in what happens to local nitrogen dynamics when trees are removed from an area. This is best done with analysis of the differences in nitrogen-stable isotopes in the annual growth rings of trees. The lighter ^{14}N atoms react faster than the heavier ^{15}N atoms. The difference, or $\delta^{15}N$ isotopic signature, of stable nitrogen isotopes can be used, therefore, to delineate the predominant source of nitrogen in plants or to represent changes in nitrogen sources over time. This approach was used by Bukata and Kyser (2005) to show that the $\delta^{15}N$ in the annual rings of trees measured gets heavier by 1.5–2.5 per mil with tree-ring age. The authors attributed this shift in heavier $\delta^{15}N$ isotope in tree rings to increased bacterial nitrification of NH_4^+ to NO_3^- and subsequent leaching from the soil.

11.1.5 The Flow of Phosphorus

Similar to the plant need for nitrogen is the plant need for phosphorus. Plant cell membranes contain phospholipid bilayers. Phosphates also are used by the plant to buffer against rapid changes in pH. But the most important need for phosphorus is for the synthesis of ATP; without phosphorus, plants cannot synthesize ATP. Unlike carbon and nitrogen, which have the atmosphere as their source, phosphorus does not have a gaseous phase at environmentally relevant temperatures and pressures and is primarily found in soils derived from marine sedimentary rocks that contain phosphorus-rich organic matter, rocks, or minerals. Thus, the flow of phosphorus through an ecosystem starts with the weathering of rocks and the entrance of phosphorus into the hydrologic cycle. Phosphorus also can enter the hydrologic cycle by runoff.

Much as plants require inorganic nitrogen, so plants require inorganic phosphorus, or oxidized phosphorus as the orthophosphate ion. Phosphorus needs of plants are about one-tenth that of N_2. Too little phosphorus results in stunted plant growth, but too much causes excessive growth, especially in aquatic plants that grow in surface waters that receive phosphorus-rich wastes—this scenario is one of the challenges facing managers of the Florida Everglades.

Unlike the flows of carbon or nitrogen, the flow of phosphorus is limited because much of phosphorus is locked up in the anoxic layers of buried geologic material (Fig. 11.6). Plants take up the phosphate ion (PO_4^{3-}). Immobile phosphate tends to result from binding with sodium, calcium,

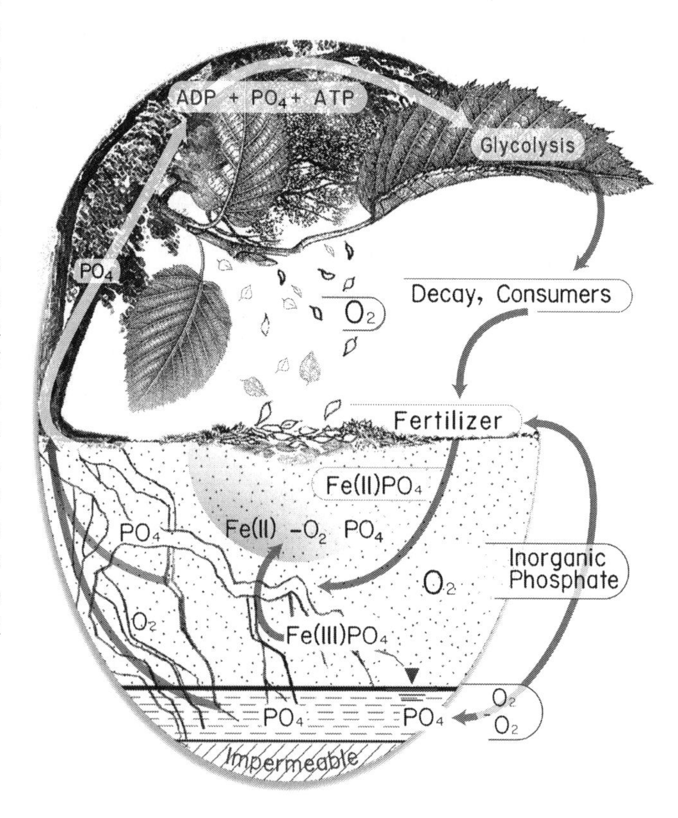

Fig. 11.6 A representation of the flow of phosphorus. Shown are oxygen (O_2); phosphate (PO_4); ferrous phosphate ($Fe(II)PO_4$).

magnesium, or iron, especially in the presence of oxygen, which renders the phosphate unavailable for plant uptake. Alkaline conditions also immobilize phosphate, whereas acidic conditions tend to mobilize phosphate. Phosphate also can be mobilized by microbes. Converse to the immobile phosphate salt, plant fertilizers that have mobile forms of phosphate are derived from phosphate-rich rocks, such as apatite, which are chemically treated to a form in which only one of the phosphate molecules is linked to a metal.

11.1.6 The Flow of Iron

As discussed in Chap. 8, terrestrial plants must harvest from the soil through the solvent action of water the necessary essential elements, trace elements, and micronutrients needed for survival. This is more difficult for terrestrial plants compared to aquatic plants that already live in a watery solution of such elements in the dissolved form. From a biological and biomechanical standpoint, the process of concentrating dilute trace elements, either by aquatic or terrestrial plants, is similar to how a filter can concentrate material, and this process is the ultimate source of these nutrients in all other animals in the food chain. This is one reason why root systems are characterized by a large surface area. This process of nutrient concentration and uptake is facilitated when the needed element is already in solution, such as bicarbonate ion, dissolved oxygen, or nitrate.

What happens, however, when the source of the element is a solid phase, such as for iron (Fig. 11.7)? Although it has long been recognized that iron is not part of the chlorophyll molecule itself (Willstätter and Stoll 1913), iron is required by the enzyme ferredoxin to make chlorophyll. It also is important to both plants and animals in electron transport to derive energy for growth from the oxidation of organic matter. In this case, iron is used to transport electrons through reversible redox reactions. Iron is part of one of the enzymes (FAD) needed during the conversion of acetyl-CoA into ATP (Fig. 11.2), or succinate dehydrogenase (FADH to FAD).

Iron is present naturally in most near-surface soils where plant roots grow and is necessary for plant growth and human health. Plant iron concentrations range from 10 to 100 μM (micromolar) as total Fe (Bauer and Hell 2006; Kim and Guerinot 2007). Iron in the soil of the root zone, however, is predominately in its oxidized, solid phase and is not bioavailable for uptake by plant roots, especially in the concentration range needed by plants. This situation is analogous to the comment made by the poet Samuel T. Coleridge in that a sailor is surrounded by water, but without a drop to drink. Plants are similarly surrounded by iron present in the earth's crust (4.5% by weight and is the fourth most abundant element) but cannot directly use it in most cases. Soil

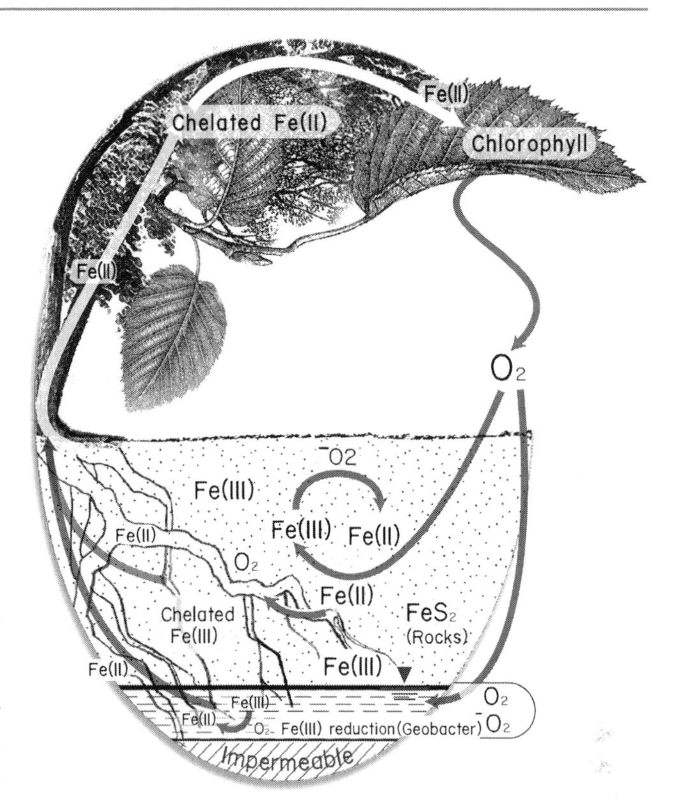

Fig. 11.7 A representation of the flow of iron. Shown are oxygen (O_2); ferric iron (Fe(III)); ferrous iron (Fe(II)), and; iron sulfide (FeS_2).

often contains up to 10,000 times the amount of iron found in plants growing in the soil. On the other hand, too much iron accumulation can be detrimental to plant survival. Strong oxidants like peroxides generated by the TCA cycle interact with iron and form free radicals. These radicals can damage cellular components, including DNA. This is why plants regulate total iron uptake.

As a result of this limited iron bioavailability, many food crops do not provide a sufficient source of iron for humans, and iron must be added or the food "enriched" with iron supplements in order to meet the nutritional needs of humans. This is especially true for foods consumed by infants during the first years of life, and for pre-natal nutritional needs. In developing parts of the world, reliance on iron-deficient crops and its effect on local populations are termed "hidden hunger." Some vegetables contain higher amounts of iron than cereal grains, but even so, iron from plants is much less (2%) absorbed by the human body than iron from animal meat (20%).

Because of the low solubility of oxidized, ferric forms of iron and the need for iron to be in the reduced state of dissolved Fe(II) for plant uptake, plants that can increase the solubility of iron have a selective advantage. Plants do this mainly in two ways, referred to as Strategy I and Strategy II. As energy is expended by plants to facilitate iron uptake using these strategies, it is tightly controlled.

In Strategy I, the goal of the plant is to solubilize the Fe(III). The Fe(III) is solubilized by either reduction with enzymatic reductases or the release of protons to reduce the soil pH and increase Fe(II) solubility; the Fe(II) produced is then free to enter the cell. Iron uptake typically follows this pathway; the Fe(III) is reduced to Fe(II) by a plasma membrane redox reaction (ferrireductase), and the Fe(II) is taken up by a transporter (Guerinot and Ying 1994). The reduction of ferric iron is done by dicotyledonous plants. In fact, low iron availability induces the production of ferric reductase. The reduction occurs in the plasma membrane of the epidermal cells of roots. By comparison, soil bacteria that encounter other metals, such as mercury (Hg), quickly route the toxic element through the cell by conjugation with organic functional groups, such as $-CH_4$, to increase Hg solubility and *decrease* uptake by increased elimination. The Strategy I system for iron works in reverse; organic compounds are released to increase solubility to *increase* cell uptake.

In some cases, plants can release enough organic matter to decrease the content of dissolved oxygen near the oxidized iron and render ferric iron to ferrous iron by microbially mediated iron-reduction reactions. This process also can occur in the uptake of ferrous iron by plants that grow in periodically flooded soils or high water tables where concentrations of dissolved oxygen are depressed. If the organic matter released is acidic, the pH is lowered and, with the presence of chelators, enhances Fe(II) uptake. Ironically, it is plants that ultimately are to be blamed the inaccessibility of iron, because iron was initially oxidized by early photosynthesizers that released oxygen.

In Strategy II, plants release organic acids into the root zone that then complex with the Fe(III) into a form, called a chelate, that can be taken up. Strategy II is more common for monocotyledonous plants. These plants synthesize, secrete, and then take up phytosiderophores that are used to chelate the Fe(III) (Römheld and Marschner 1986). Siderophores are nonprotein amino acids (Graham and Stangoulis 2003), such as mugineic acid (Kawai and Alam 2006) and rhizoferrin. These phytosiderophores are released primarily in the root tips and by newer roots. The chelation of iron then permits uptake to occur by the apoplastic pathway rather than the symplastic pathway. The Fe(III) complexes then can enter cells.

Strategy II is similar to methods used by bacteria that also release siderophores, such as ferrichrome. However, plant siderophores, or phytosiderophores, result in much faster iron uptake than bacterial siderophores (Römheld and Marschner 1986). One molecule of chelate will chelate one molecule of a particular metal, such as Fe(II) or Fe(III). The Fe(III)-chelate complex is extremely stable under environmental conditions. Researchers also have found that the weed *Arabidopsis thaliana* contains the enzyme ferric reductase, which directly can reduce ferric iron to ferrous iron to facilitate uptake (Robinson et al. 1999). The enzyme is the bridge needed to connect the roots with the ferric iron in the soil. This bridge is used to funnel electrons from reduced organic matter in the plant roots to ferric iron to complete the reduction and mobilization and subsequent uptake by specific transport systems. In some cases, these same plants can release excess protons into the pore water to decrease the pH and increase iron solubility.

The uptake of bioavailable forms of iron by plant roots is by diffusion along concentration gradients and advection within the transpiration stream. For diffusion, natural chelators produced by plants increase the extent of diffusion by increasing the iron concentration outside of the root zone. This increased iron solubility is achieved even in the presence of dissolved oxygen (Oborn 1962). The production of siderophores by rhizospheric fungi and bacteria may help explain their presence on most plant roots—each is competing for iron. Once inside the plants, iron is less mobile, as evidenced by the production of new growth of yellow or chloritic leaves after iron supplies have been depleted, whereas older leaves remain green. This yellowing can be observed in both old and new leaves, however, because chlorophyll is constantly breaking down and needs to be synthesized continually.

Additional insight into the interaction among plant roots, iron, and iron uptake has been provided by studies using the stable isotopes of iron, which exist as ^{54}Fe and ^{56}Fe. Guelke and Von Blanckenburg (2007) investigated the isotopic fractionation of iron stable isotopes during the uptake of iron by the two pathways described above. During Strategy I iron uptake, the reduction of Fe(III) results in the uptake of Fe(II), which is isotopically lighter, or depleted, in percent heavy isotope by 1.6 per mil relative to that remaining in the soil iron pool. During Strategy II iron that is associated with siderophores results in the uptake of iron that is isotopically heavier, or enriched, in percent heavy isotope by 0.2 per mil, than the soils.

Entry of iron into the roots and transport within the plant occurs by two methods. Passive uptake of iron occurs by diffusion along concentration gradients through the apoplast, or the cell walls and spaces between cells. The endodermis interrupts this transport to the inner part of the root (stele), however, by way of the Casparian strip, which is made of hydrophobic suberin. Still, solute does enter through this barrier. Alternatively, active uptake of iron occurs from cell-to-cell in the symplast, which requires selective transport from cell membrane to cell membrane. With metals, such as iron, the free ions can be taken up, and the metal ion-chelate complexes can be taken up by the apoplastic pathway along concentration gradients. The chelate EDTA and its metal complexes have been detected in plant xylem sap (Nowack et al. 2006). Once in the plant, chelates are then absorbed by shoots.

Phreatophytes that may have part of their root systems in the water table or capillary fringe may already encounter ferrous iron in anoxic pore or groundwater, especially at contaminated sites. This can cause the pore-water concentration of Fe(II) to increase (Jones and Etherington 1970). In this case, a plant can receive all of the iron it needs without having to resort to Strategy I or II. If the water-logged condition is too long, however, Fe(II) concentrations can reach toxic levels. Although this increases iron availability, root respiration is limited by a lack of oxygen. In swamps, this lack of oxygen is overcome, however, by plant development of extensive aerenchymal tissues that act as "pipes" to permit the diffusion of atmospheric oxygen into the plants that then moves by diffusion into the subsurface, as stated earlier. The presence of this additional oxygen, however, tends to, ironically, oxidize the reduced iron, rendering the originally available Fe(II) unavailable.

The entry of oxygen not only supplies the respiration requirement but also may be a process that controls ferrous iron uptake to levels that are not toxic. It also is possible that excessive ferrous iron is dealt with by differential compartmentalization to various plant tissues after uptake, with the goal of not interfering with the manufacture of chlorophyll.

Comparatively less information is known about the translocation of iron from root to shoot after uptake, although iron does make its way to the xylem by way of the symplastic pathway and its barrier the Casparian strip. Iron tends to accumulate in plant leaves rather than in roots. Chloroplasts contain 80% iron, which is required to synthesize chlorophyll and act in proteins in the electron-transport pathway. Iron is believed to be translocated in plants in the form of ferric citrate. Tannic acid can inhibit iron absorption through binding. This may be because Fe(II) if reacted with H_2O_2 can produce the highly toxic hydroxyl radical (OH•). This reaction, discovered by Fenton in 1876 (Wardsman and Candeias 1996), is called the Fenton reaction, and is widely used in the groundwater remediation industry to rapidly oxidize dissolved-phase organic contaminants.

Finally, Guerinot and Salt (2001) made the observation that the same process of enhanced iron uptake could be beneficial not only for phytoremediation but for food crops as well. The common denominator is that each process is the result of metal uptake by plants. The redox status of soil is determined in part by its water content, and this affects plant health and survival. High concentrations of dissolved iron generally, as we will see in Chap. 13, are found in groundwater at sites where reduced organic matter, such as gasoline or jet fuel, has been released. Conversely, high Fe(II) concentrations can be found near swamps and forests with copious amounts of natural organic matter in flood plains. If present in groundwater, dissolved iron itself can become a contaminant if levels exceed the National Secondary Drinking Water Standard (NSDWS) MCL for iron.

11.1.7 The Flow of Sulfur

Like iron, elemental sulfur is an important element for plant growth (Ernst 2004). Sulfur is necessary for the synthesis of plant proteins and coenzymes. The sources of most sulfur used by plants are sulfur-containing rocks and soils, and the ocean. It is transferred to plants through the atmosphere and hydrologic cycle. Elemental sulfur cannot directly be used by plants, however, unless it is first oxidized to sulfate (SO_4^{2-}). Plants that contain higher amounts of sulfur as sulfate tend to also have higher protein contents. In its reduced form, however, sulfur can be toxic to plants. Elemental sulfur also can be reduced under anoxic conditions to H_2S by bacteria such as *Desulfuromonas acetoxidans*.

Sulfur in its dissolved and reduced form as sulfide can inhibit plant growth. For example, Bradley and Dunn (1989) reported that the biomass and height of *Spartina alterniflora* were inhibited by hydrogen sulfide concentrations at 1 mM. The authors clearly point out, however, that this concentration of dissolved sulfide was one of many possible factors that can affect such growth in the field. It is important to note that H_2S can support chemolithotrophic microbes that oxidize the H_2S back to elemental S, such as is done by *Beggiatoa*. This S can be oxidized further to SO_4 by *Thiobacillus*.

11.1.8 The Flow of Potassium

Elemental potassium is used by plants to control the intracellular movement of water. Like phosphorus, potassium is present in large amounts in soil, but typically is present in the unavailable form, being sorbed onto clay particles in the soil or aquifer sediments.

11.2 Plants and Natural Chemical Compounds

The physician Paracelsus (1493–1541) stated that "all substances are poisons; there is none which is not a poison, and it is the dosage that differentiates a poison from a remedy." That a physician made this statement so long ago may at first seem surprising but is actually a logical source for such a comment. Early medicinal practices recognized that plant-based herbal and chemical approaches to health restoration were based on providing *small* amounts of the same compound that had been observed to have lethal consequences in larger doses. The same goes with other compounds and their effects on living organisms. Chemicals can have negative effects on living organisms; it is the dose, however, that determines the exposure risk.

Biologically, the primary goal of all life forms is to stay alive. Organisms must feed themselves, find water, and avoid predation; if successful, they can pass on their genetic material to their offspring. Plants are exposed to sunlight, and the process of photosynthesis provides food, but the control of water supply and avoidance of predation have led to various adaptive processes by which to achieve these goals. From root depth to defensive chemicals, plants are anything but passive sessile organisms at the mercy of nature. Allelopathy is perhaps the best example of this.

11.2.1 Allelopathy and Plant-Chemical Warfare

Upon first glance, most plants seem to be fairly passive and defenseless and unable to control their environment, because plants are relatively fixed in place. This notion of passivity, however, is far from the truth. Plants are living creatures that respond to or alter a variety of things in their immediate environment, usually to their competitive advantage.

Certain plants have been successful because of deterrents to predation that have evolved over time (Fig. 11.8). For example, the presence of visually attractive flowers is an adaptation to sexual reproduction that uses insects and birds for gene transfer. On the other hand, plants synthesize chemicals not for use as a source of energy but solely to keep the plant or parts of a plant from being destroyed by herbivores. The manufacture of such protective compounds has evolved because the loss of leaves to predation does not provide the plant an advantage, and plants lose fewer leaves if they can produce foul-tasting or toxic compounds. These few examples indicate how complex "sessile" plants really are, and how they can take simple substrates, such as CO_2 and water, and synthesize incredibly complex and toxic compounds for both defensive and offensive chemicals.

Fig. 11.8 As shown by this honey locust (*Gleditsia triacanthos*), thorns are an obvious indication that plants are not passive when it comes to defensive approaches to stay alive. In this case, the development of thorns may have occurred during the Cretaceous Period for protection against damage by dinosaurs (Photograph by author).

Plants appear to be not only immobile but seemingly unprotected and, therefore, easy prey. Long before the manufacture and application of pesticides, however, plants became naturally toxic to other plants and potential herbivores by producing their own defensive chemicals. It has long been known that plants can manufacture compounds that are harmless to themselves but toxic to other organisms. On the simplest, single-cell level, penicillin is produced by a mold to ward off attacks by invading harmful microbes. Perhaps closer to individual experience, dermal contact with poison ivy can result in a localized allergic reaction. The causative agent is a volatile organic compound (VOC) in the leaf that contains allergens designed to decrease successful herbivory.

The ability for multicellular organisms, such as plants, to control their local environment by reducing competition through a kind of low-grade chemical warfare is called allelopathy—it literally means "mutual suffering." Evidence exists from Theophrastus and his writings during 300 BC about the negative effect that chickpeas had on the growth of other weeds. Pliny the Elder also commented that chickpea and barley made the cultivated land inhospitable to other crops. Allelopathy can be considered as an extension of simple physical competition for resources (Whittaker and Feeny 1971). Trees growing close together have root competition to acquire limited resources, although this seems to be less true in groves of aspen or bamboo clones, where evenly spaced plants may result from resource optimization. A selective advantage is provided a plant to gain resources if it can essentially contaminate the surrounding soil by the release of allelopathic chemicals.

Allelopathic compounds can be produced by a wide range of plant parts. The leaves, or needles, of pine trees, for example, release organic acids upon decomposition around the root zone at sufficient concentrations to decrease the soil pH and render it inhospitable for plant seeds, other than pine, to germinate (Fig. 11.9). Other allelopathic compounds are produced by plant shoots and leaves and enter the soil zone either by washout from precipitation or after these parts have fallen to the ground and release their chemicals or are volatilized from the plant leaves. Moreover, some of the most toxic poisons across both the animal and plant world are synthesized by fungi associated with tree roots.

Fig. 11.9 This monoculture of short-leaf pines (*Pinus echinata*) in South Carolina is caused by the root production and release of allelopathic chemicals (Photograph by author).

Allelopathic compounds are produced primarily by woody plants and perennials in arid environments where competition for scarce water is fierce, and allelopathic compounds have a better chance of accumulating in soils that have low moisture content. Examples include plants from the Genera *Salvia* and *Artemesia*. In more humid areas, black walnut (*Juglans nigra*) produces glycoside compounds that are inhibitory to other organisms after transformation by exposure to oxygen into the active chemical juglone. Juglone is 5-hydroxy-1,4-naphthoquinone, and it acts to inhibit respiration in plant cells. It has low solubility in water and, hence, stays in the soil and resists washing from precipitation. It was long been known by farmers that it is next to impossible to grow crops next to walnut trees. All parts of the plant, including leaves, stems, fruit hulls, the bark and roots release the precursor to juglone, which affects other plant growth even after leaf fall.

Some plant-produced chemicals not only affect the growth of other plants, but also affect the growth of plant pests. One of the first insecticides for use on insect pests on plants was derived from a widely used plant in the late 1800s. The compound nicotine sulphate is a contact insecticide and attacks the central nervous system upon contact or inhalation, if used as a fumigant. Nicotine is synthesized in the roots, and is transported in the xylem in response to insect damage to shoots and leaves. An extract of leaves in water were used as insecticides as early as the 1690s. It is found in the tobacco plant (*Nicotiana rustica, N. tobacum*), which contains between 6% and 8% nicotine. This plant derives its name from John Nicot, French ambassador to Portugal, who sent specimens back from North America to Europe. Native tobacco plants actually contained so much nicotine that they had to be hybridized with tobacco plants that contained lower nicotine concentrations for subsequent use as a legal stimulant so as not to be too toxic to humans. Nicotine sulphate (40%), however, is toxic to bees, birds, and fish, and extracts also have been used as a pesticide. The formation of the alkaloid nicotine is a way for plants to store excess nitrogen as nitrogen-containing organic compounds.

Other plant-derived insecticides include the pyrethrum esters, or pyrethrins, derived from *Chrysanthemum cineraiifolium*, and rotenome, derived from tropical plants such as cube barbasco and timbo and *Tephrasis virginiana* (Devil's shoestring) in the southeastern United States. The recent marketing of such "natural" insecticides is not new; rotenone from the root of *Derris eliptica* and pyrethrum from *Chrysanthemum cinerariaefolium* were used as insecticides in the early 1800s. Results of recent research indicate, however, that the concentration of many natural compounds in household cleaning may adversely affect indoor air quality; these compounds, such as oils, react with atmospheric ozone to produce formaldehyde.

Plants release other chemicals derived from their fixation of carbon dioxide. They produce oils, called "essential oils," from glucose as an alternative way to store food for future use. Some examples include well-known herbs, such as peppermint, catnip, and rosemary. The oils are collected in the cell vacuole and transported to the trichome, or modified root hairs, on the epidermal leaf surface. Possible reasons for the use of photosynthate to synthesis these essential oils are as volatile pheromones used to attract pollinators or as offensive release as protection from predators. The medicinal smell of the Eucalyptus tree is derived from the high oil content in the leaves that reach the soil either during precipitation or leaf fall and render the soil infertile for the seeds of other plants to germinate.

Another type of plant oils are the terpenoids, which also are used in many natural insecticides. Terpenes include chemicals such as cineole and camphor (Fig. 11.10). Terpenes produced by pine trees and released to the atmosphere are the primary reason that the Smoky Mountains of the southeastern United States are smoky—the volatilized terpenes appear as a hydrocarbon haze. In fact, the characteristic odor of freshly mown grass is the plant's response to wounding, regardless of whether by insects, cows, or lawnmowers, in that the grass is essentially releasing mono- and sesquiterpenes and other compounds. In some cases, these VOCs are thought to be released as a signal to potential insect predators to solicit increased predation of insect herbivores.

One of the interesting things about plants and allelopathic potential is the role that a lack of pesticides or herbicides has on the plants production of allelopathic compounds. This has implications in the production of organic foods, because organic agricultural practices are based, in part, on not adding synthetic chemicals to the crops. Organically grown plants, however, if besieged by pests, can produce their own equally toxic compounds.

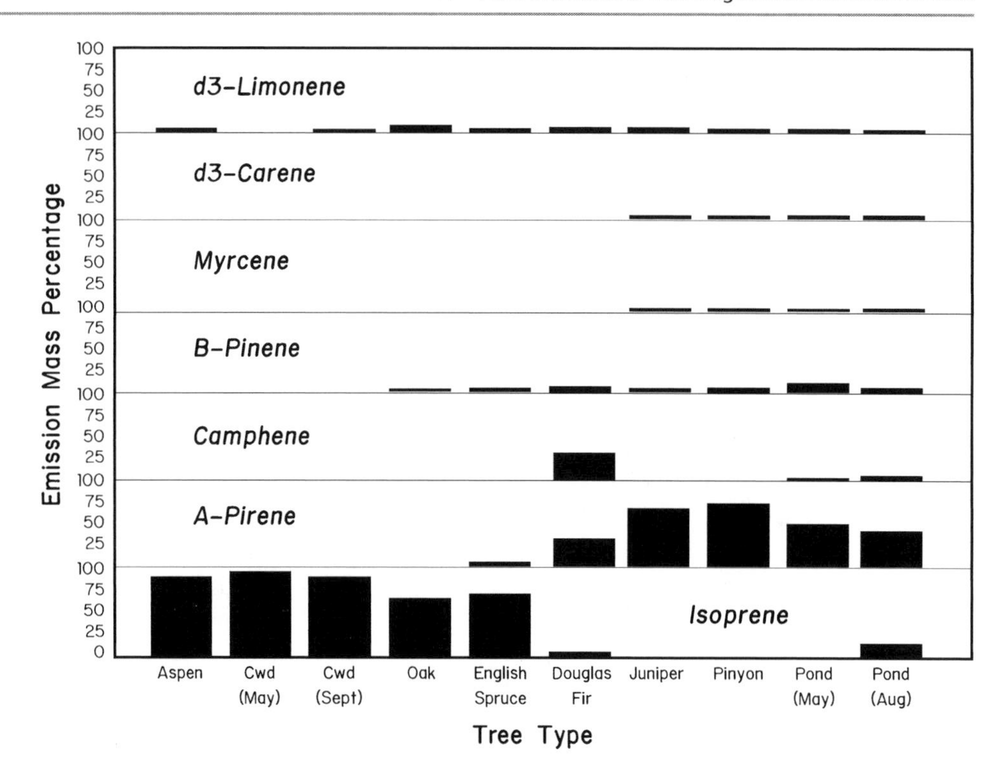

Fig. 11.10 A variety of tree genera and species release natural volatile organic compounds, such as camphene and *d3*-limonene. (Cwd; cottonwood.)

Another example of plant self-defense can be found in oak trees, which produce tannins to inhibit predation by herbivores. As oak leaves age, the concentration of tannins increases inside each cell, and tannins are stored in cell vacuoles. These tannins bind rapidly with proteins and, therefore, can interfere with enzymes that control cell metabolism. Tannins are useful for plant tissues that contain openings and are, therefore, exposed to potential invasion by viruses, bacteria, and fungal spores, such as the roots (root hairs), leaves (stomata), bark (lenticels and cork), heartwood and many seeds. During herbivory, the vacuoles are broken, and the released tannin combines with plant proteins and renders the plants nutritionally unavailable to the herbivore. However, because younger leaves have less tannin, they can be consumed without adverse effect on the population of predators, and some leaf-eating beetles can devour a leaf by avoiding the vacuoles altogether, which is the reason for the appearance of "laced" leaves.

Heartwood contains a higher concentration of such tannins and is, therefore, naturally protected from invasion, which explains the desirability of heartwood as a source of building supplies. Tannins from bark are used to render animal skins stable against microbial degradation; after soaking in tannic-acid solution, the skin becomes protected from microbial attack. Oaks that grow in infertile soils tend to contain higher concentrations of tannins. As winter approaches, the green, red, orange, and yellow pigments fade and give way to brown, the color of tannic acid, which remains because it is less degradable than the other colors.

Rivers that drain the broad, sandy land of the coastal plain along the eastern seaboard of the United States typically are dark colored even though they contain very little suspended sediment. The dark color, called "blackwater," is derived from the tannic acids that leach from the trees that grow along the banks, and was noted as early as the eighteenth century by William Bartram, a naturalist who traveled throughout the southeastern colonies between 1773 and 1778 (Van Doren 1955). These are the same tannins that are used to protect the various parts of the tree from fungal attack. Because of the antimicrobial property of tannins, black water actually is suitable as a source of drinking water. In fact, the crews of early sailing ships along coastal North America would fill barrels with blackwater because the tannins in the water and its lower pH would act as a biocide to inhibit microbial growth during extended sea travel, especially after supplies of beer ran out. Such blackwater was an important source of relatively clean drinking water on land in the time before refrigeration and was even believed to possess medicinal properties.

Many plants contain antimicrobial compounds for protection (Broekaert et al. 1997). Plants and animals are continuously exposed to potentially pathogenic organisms above ground and in the soil. Whereas animals have developed immune systems that are immunoglobulin-based, plants have developed chemicals to ward off or remove infestation. Of the many chemicals plants produce, peptides are one of the most common. Peptides consist of a number of amino acids. Plant peptides contain cysteine as a building block. For example, thionins are peptides that contain cysteine interconnected with disulfide (Broekaert et al. 1997). In plants thionins inhibit bacterial and fungal growth. These antimicrobial compounds tend to be found in the cells of the plant periphery.

The presence of antimicrobial compounds in the root zone would seem to contraindicate the data of increased microbial populations in the rhizosphere. However, there must be a selective screening process that removes the harmful organisms and keeps the beneficial ones. For example, the nitrogen–fixing bacteria that the plant needs to acquire nitrogen must not be adversely affected by plant antimicrobial compounds. The key may be the properties of the mycorrhizal bacteria themselves and the establishment of a positive feedback loop. These bacteria produce growth hormones for the plant as well as antimicrobial compounds to suppress plant pathogens.

The leaves usually are the primary site of defensive processes because the leaves are the site of food production for the plant, and loss of leaves by grazing or insect damage can be trouble for a plant. Physical modifications of leaf veins into spines occur at margins of leaves, such as in hollies. Chemical modifications of leaves, such as the production of compounds that impart a bitter taste; crystals of calcium oxalate, present in the common houseplant *Dieffenbachia*, which damage the mouthparts of insects, and; poison, such as in the poison hemlock, *Conium maculatum*. In the latter case, the poisonous compound is the alkaloid conine, and was the last thing that many early condemned prisoners, including Socrates, tasted. The plant chemicals found in leaves can be used for cooking and for medicinal purposes. The essential oils are used for flavoring, whereas the alkaloids are used for recuperative purposes.

Assessments of the shallow groundwater resources of the United States by the USGS as part of the National Water-Quality Assessment Program (NAWQA) have revealed that the most commonly detected volatile organic compound in shallow, ambient groundwater is chloroform, also known as trichloromethane (Zogorski et al. 2006). Chloroform detection in groundwater is attributed to the recharge of previously chlorinated drinking or waste water. There also are natural sources of chloroform to groundwater. These natural sources are derived from plants, either green plants or fungi in terrestrial ecosystems, or from the upper layers of the ocean from phytoplankton. Laturnus et al. (2002) reported that in forested areas where shallow groundwater did not contain chloroform, concentrations of chloroform increased in shallow soils. Moreover, the concentration of chloroform in the upper soil layers exhibited a seasonal trend with lower concentrations in the winter and higher concentrations in the summer. These compounds can persist in aerobic soil layers because of the stability of oxidized organic material under these conditions. The plant production of chloroform may be a selective advantage as a defense mechanism against infection, similar to alkaloids.

Other plant allelopathic compounds include glycosides, such as the juglone from walnut trees described previously. When in contact with water, glycosides yields sapogenins, compounds that essentially dissolve lipids, such as those found in cell walls. Cyanogenic glycosides can produce hydrogen cyanide. Glycoside concentrations in plants increase in those exposed to xenobiotics, and glycosides play an important role in the detoxification of xenobiotics in groundwater after uptake; this is discussed in Chap. 12.

Some plant defense chemicals are not synthesized by the plant but by other organisms associated with the plant. Some fungal endophytes of grasses impart protection against herbivory and damage from insects by producing alkaloids that are retained in the grass vacuoles. The alkaloid lysergic acid diethylamide (LSD) is produced by such grass fungal endophytes. The beans of the castor-oil plant (*Ricinus communis*) contain the protein ricin, which is 10,000 times more toxic than rattlesnake venom. Ricin is a lectin that has two polypeptide chains that are connected by a disulfide bridge. When ingested, ricin enters the cell cytoplasm and inhibits protein synthesis.

Allelopathy has many connections to the use of plants to remediate contaminated groundwater. Disease susceptibility can be a limiting factor in choosing clones over native plants. This is because under natural conditions of predominant sexual reproduction, there is a selective battle between plant health and insect infestation, and the strongest survive to reproduce. With clonal selection, this battle is not fought as effectively, because each clone is an exact genetic copy of the parent. Such a scenario promotes increased pest infestation. Many examples of the negative results of this can be found, but perhaps the best example is the potato blight in Ireland in the 1840s.

There often are concerns raised by stakeholders who oppose the implementation of phytoremediation at contaminated sites because of the potential for translocation of subsurface contamination to the above-ground parts of leaves. It is instructive to note by comparison, however, that many common plants that surround us are, in effect, toxic (Westbrooks and Preacher 1986). Table 11.1 provides an incomplete list of commonly recognized popular ornamental plants sold in garden centers or planted in gardens for human consumption. Also, many of these plants are specifically planted by state or municipalities interested in preserving the aesthetics of a particular area. Anecdotally, most of these plants seem to thrive with little or no damage by plant pests, a testament to their evolutionary patience—it may also explain why these plants are so widely sold as houseplants.

The notion that plants are not simply at the mercy of the environment and have, in fact, demonstrated the ability to produce chemicals not only to sustain their growth but to protect them from predators and to diminish the effect of resource competition is in direct contrast to the notions

Table 11.1 Common plants and their toxic compounds.

Plant name	Genus species	Toxin
Poison ivy	*Toxicodendron* spp.	Phenolics
Azalea	*Rhododendron* spp.	Grayanotoxin
Oleander	*Nerium oleander* L.	Cardiac glycosides
Philodendron	*Philodendron* spp.	Oxalate
Foxglove	*Digitalis purpurea* L.	Cardiac glycosides
Asparagus	*Asparagus officinalis*	Various toxins
Fig	*Ficus carica* L.	Furocoumarins
Tomato	*Lycopersicon* spp.	Alkaloids
Apple	*Malus sylvestris* Mill.	Glycoside
Dumbcane	*Dieffenbachia* spp.	Oxalate
Privet	*Ligustrum vulgare* L.	Glycosides
Lantana	*Lantana camara* L.	Lantanine
Sago palm	*Cycas circinalis* L.	Glycoside
Ginkgo	*Ginkgo biloba* L.	Alkyl resorcinol
Yews	*Taxus* spp.	Alkaloid
Caladium	*Caladium* spp.	Oxalate
Century plant	*Agave* spp.	Oxalic acid
Daffodil	*Narcissus* spp.	Alkaloids
Hydrangea	*Hydrangea* spp.	Cyanogenic glycoside
Wisteria	*Wisteria* spp.	Glycoside
Boxwood	*Buxus sempervirens* L.	Alkaloid
Holly	*Ilex* spp.	Illicin
Eucalyptus	*Eucalyptus* spp.	Glycosides and oils
English ivy	*Hedera helix* L.	Saponin
Queen Anne Lace	*Daucus carota* L.	Furocoumarins
Carolina Jessamine	*Gelsemium simper.* L.	Alkaloids
Chili pepper	*Capsicum frutescens* L.	Alkaloids
Tobacco	*Nicotiana tabacum* L.	Alkaloids

spelled out at the beginning of the environmental age in the early 1970s that can be traced to the publication of books such as "Silent Spring" by Rachel Carson (1962). In that book, Carson portrayed a scenario where plants were the unlucky recipients of the chemists' laboratory concoctions when, in fact, the plants could produce their own toxic compounds. Moreover, it is the very encounter of plants with allelopathic compounds that can be considered as the first natural analogy to the response of plants to exposure to released xenobiotic contaminant compounds, especially in groundwater. These reactions are discussed in Chap. 13.

11.2.2 Plants as Environmental Indicators

Coal miners used to carry canaries in cages as they descended into mine shafts, because the bird would die upon exposure to the odorless gas methane or to a lack of oxygen, which would alert the miners to go no farther. Today, the same task is carried out by portable gas detectors that help miners determine when methane reaches levels high enough to cause a risk of explosion or asphyxiation. In this case, the canaries acted as a biological surrogate, or bioindicator, of risk to human health. Plants also can be used as bioindicators, of the effects of either acute or chronic exposure to compounds. For example, in some water-quality studies, the bioindicator has been algae, where blooms of algae are associated with the release of excessive nutrients. One area of research that has received attention is terrestrial plants used as bioindicators of heavy-metal exposure. This is because certain plants can accumulate high concentrations of some heavy metals.

The application of using plants to indicate risk exposure to decrease contaminant levels and, therefore, reduce risk is an interesting story in itself. Initially, plant exposure to chemicals was considered to be a direct route to wildlife and human populations through ingestion. Plants also were grown at contaminated sites to determine the extent of contaminant transfer, accumulation, and release of the specific classes of contaminants found at the site. In fact, there are many sites where plants have been added for the sole purpose of determining the degree of plant exposure to the chemicals at the site.

In many cases, whole plants are not necessary to determine the interaction between plants and contaminants. The use of plant-tissue cultures to study the effect of plants on chemicals is widespread. This is because tissue-culture tests can be made with relative ease with respect to whole-plant studies, and only small amounts of the chemical in question need be used. Although algal cells typically have been the focus of tissue-culture studies, the use of terrestrial plants that make up phytoremediation systems have been investigated (Wickliff and Fletcher 1991). As part of their investigation, Wickliff and Fletcher (1991) used tissue cultures of *Rosa cultivar* (Paul's Scarlet) to which they added a surrogate for a xenobiotic, 1,3-dinitrobenzene (DNB). The toxicity of DNB on plant-tissue culture was examined, as well as the potential transformation of DNB by plant tissues, by tracking the fate of radiolabel ^{14}C-1,3-DNB. The plant growth rate, measured as dried cell weight, was unaffected by DNB concentrations up to 1 mg/L. As the concentration of DNB increased to 10 mg/L, however, the dried cell weight decreased. Part of the explanation for the lack of a deleterious effect on growth can be found in the mineralization study results. Radiolabeled DNB was transformed up to 90% by the plant-tissue cultures, possibly into a gaseous fraction that did not affect the plant (Wickliff and Fletcher 1991).

The production of CH_4 at manufactured gas plants produce wastes that contain cyanide (CN), which interacts readily with iron. Most vascular plants naturally produce cyanide as a byproduct during the plant-synthesis of ethylene. Therefore, plants possess the potential to tolerate CN if it is released into the environment as a result of industrial use of cyanide (Larsen and Trapp 2006). Poplar trees (*Populus trichocarpa*) have been grown in the lab in a solution up to

1,000 mg/L ferrocyanide, with toxic effects noticed only at 2,500 mg/L (Trapp and Christiansen 2003).

Since the early 1990s, interest in using vascular plants to phytoremediate contaminated groundwater is the result of a major shift in thinking concerning the interaction between plants and contaminants. The uptake and translocation of herbicides, pesticides, petroleum hydrocarbons, and chlorinated solvents by plants has primarily been viewed in terms of plant uptake being a vector for increased risk to wildlife and human populations. Research into the interaction between plants and herbicides increased but primarily to determine the effectiveness of the mode of action of the herbicide rather than on risk exposure, and led to the promulgation in 1982 by the USEPA of the Federal Insecticide, Fungicide, and Rodenticide Act (FIFRA).

A shift in the view that plants were vectors for contaminant exposure to a view that plants could be used to decrease environmental risk followed only after multiple laboratory and field studies were conducted to determine the fate and transport of contaminant compounds in plants. For example, observations made in the 1980s that pesticides persisted longer in unplanted areas compared to planted areas was attributed to the presence of pesticide-degrading microbes in the root zone and, therefore, were not present in unplanted areas (see Walton and Anderson 1992). During the 1990s, native plants at sites contaminated with polychlorinated biphenyls (PCBs), polyaromatic hydrocarbons (PAHs), halogenated benzenes, and munitions (Schnoor et al. 1995) were known to take up these contaminants, and metabolize them into less harmful byproducts (McFarlane et al. 1990).

11.2.3 Plants and Toxicity Assessment

Plants have been used since at least the 1970s to help not only indicate the presence of a particular compound but to assess the hazard level posed by certain chemicals on ecosystems, including man. Plants often are used in tests to determine phytotoxicity and act as sentinels to guard against wider ecosystem contamination, as discussed previously. In phytotoxicity tests, plants are exposed to chemicals, and various factors thought to be affected by the chemicals, such as plant growth and health, are observed. Sentinel tests are similar, in that the plants are used as early-warning systems to detect degradation of ambient environmental conditions. In many studies, freshwater algae are used to assess the potential toxicity of a chemical, because although algae are structurally simple, they are easy to handle in the laboratory, have high turnover rates, are ubiquitous, and are similar to more complex vascular plants in terms of photosynthesis.

According to USEPA and Food and Drug Administration (USFDA) regulations, phytotoxicity testing is required of all new chemicals. The USFDA, established in 1906 during the Woodrow Wilson administration, has regulatory authority over the testing and approval of chemicals used in the manufacture and production of food additives, food processing, and the chemicals used in processing products from animals. A substance is considered toxic by the USFDA if the maximum environmental concentration exceeds the concentration of that chemical that is found to cause adverse effects in test species, or exceeds 1% of the LC_{50} (Harrass et al. 1991). The LC_{50} is an acute lethality test, in which the magnitude of dead test organisms is the toxic amount and the concentration that causes lethality in half (50%) of the organisms. For the most part, the effect of chemicals is observed as changes in seed germination, root elongation, and seedling growth, although other factors could be tested, such as an enzyme assay, tissue-culture growth, and life-cycle changes (Fletcher 1991).

The bioassay tests approved by the USEPA are the Seed Germination/Root Elongation Toxicity Test (EG-12) and the Early Seedling Growth Test (EG-13). Life-cycle changes include changes during the manufacture, use, and disposal of a regulated product. These tests of the effects of chemicals on plants typically are referred to as bioassays. The USEPA has used bioassays that involve the exposure of algae to chemicals under laboratory conditions.

The potential toxicity of a chemical to a plant can be related to the ability of a chemical to enter a plant's vascular system. In general, the properties that render a chemical more hazardous to the environment also increase the potential for uptake by plants. For example, chemicals that have greater solubility in water tend to be more of a risk to water quality and also are more apt to be taken up by plants. More information on this relation between the physical properties of a chemical and its interaction with plants is discussed in Chap. 12. The use of vascular plants to test the toxicity of chemicals historically has been limited to aquatic rather than terrestrial plants. This is based on the assumption that the chemicals are more apt to be released to aquatic environments and will be more mobile in water. For example, recent research has shown that even treated municipal wastewater can contain a variety of chemicals used by man that are not degraded during the wastewater-treatment process and, therefore, enter the aquatic environment after release to streams (Kolpin et al. 2002).

The vascular plant most often used in such tests is the common duckweed plant (*Lemna*), although *Elodea* also has been used. For algae-based toxicity tests, the effect of a particular compound is measured as the difference between the number of cells at the beginning of exposure relative to the number of dead cells at the end of exposure, whereas multicellular plants are evaluated with respect to acute toxicity, root elongation, and seedling growth. Other factors that are evaluated for a negative response to plant exposure include measurement of differences in CO_2 uptake (photosynthesis), and the commensurate production of O_2. A reduction in CO_2 uptake or of O_2 production could be used to indicate a toxic effect of a chemical, as long as the production of CO_2 and consumption of O_2 by plant-cellular respiration are accounted for.

The results of toxicity testing using plants can be used to determine the need for further actions. If no adverse exposure or toxicity effects are observed with respect to seed germination, root elongation, and seedling growth, then the effect of the particular chemical on the environment can be deemed low. Conversely, if an effect is detected additional steps can be taken, such as preparing an environmental impact study (EIS). Finally, the effects of organic chemicals on terrestrial plants are recorded in the internet-based database PHYTOTOX, which is composed of a bibliography and dose-response information (Royce et al. 1984).

11.2.4 Geobotanical Prospecting and Phytomining

Plants are what they take up, so to speak—they accumulate and redistribute chemicals, such as iron for example, into the heartwood to increase structural support. Analysis of plants can be performed, therefore, to determine the presence of elements in the soil, such as iron. Cannon (1971) reported the relation between above-ground plant distribution and growth and below-ground variables that may be used to indicate the presence or absence of water, minerals, or geologic processes. This report confirms previous investigations (Meinzer 1927) that plants can indicate the depth to and sometimes the quality of groundwater. The presence of certain geologic strata and the resultant weathered soil has control on plant distribution, either from the standpoint of mineralogy or permeability to air and water.

The detection of ores or mineral deposits is connected to plants in two ways. First, the analysis of plant matter for particular minerals and the assumption that these minerals are present in the soil are referred to as biogeochemical prospecting. On the other hand, the use of the distribution or appearance of plants and their health are known as geobotanical prospecting. The application of geobotanical prospecting and the relation between plants and local hydrogeology for remediation purposes also are referred to as phytomining. The interaction between plants and heavy metals has long been studied. Plants are protected from negative effects of the heavy metals by the mycorrhizae that exclude the entry of the metal into the plant. A review of this topic can be found in van der Lelie et al. (2001).

In the simplest application, the presence of halophytes, such as mangrove, and *Spartina* indicates the presence of sodium chloride. Other plants indicate the presence of zinc, such as yellow violet (*Violet lutea*), or selenium (*Astragalus pattersoni*). The vanadium bush (*Cowania stansburniana*) found in the southwestern United States is used by prospectors to indicate the location of vanadium and uranium deposits. These plants absorb the uranium dissolved in groundwater that contains dissolved oxygen, under which the oxidized uranium is mobile. All these are examples of natural selection, in which the locations of plant species are determined by competitive exclusion.

11.3 Plants and Extreme Natural Environments as an Analogy for Groundwater Contamination

Plants exist in environments of many extremes. Plants are exposed to the electromagnetic energy contained in incident solar light. They use the light energy to split water to process sugar from gaseous CO_2. Plants also need to protect themselves, however, from damaging solar energy. Similar to the use of sunscreen compounds to protect human skin from ultraviolet A and B, plants produce a compound called zeaxanthin to protect themselves from solar radiation (Fleming and Niyogi 2005). This process of protection is called feedback de-excitation. The compound zeaxanthin, a carotenoid similar to the non-photosynthetic structures that give rise to different colors in plant leaves and fruits, apparently permits overheated chlorophyll to disperse its heat.

Plants are exposed to extremes in a variety of factors even in the same location. For example, temperatures on this planet can range from over 100°F to below −100°F. Moreover, the same location may have a wide fluctuation in temperature. The length of solar radiation per day changes over time. Short-term moisture levels can fluctuate rapidly if precipitation is infrequent. Insects and animals, including man, ingest various parts of plants for their own sustenance. Some plants can tolerate high concentrations of metal deposits that would kill other plants that do not possess such metal-detoxifying processes.

Below are additional examples in which plants have thrived in less than ideal geochemical or physical conditions, which indicates that the interaction between plants and contaminated environments, including contaminated groundwater sites, is the rule not the exception.

11.3.1 *Spartina* and Mangrove Monocultures

The interaction of plants with natural metals provides an interesting story about how harsh environments can actually be niches for some plants to the exclusion of other plants. For example, the saline marshes and estuaries of the eastern coastline of the United States have what would at first seem to be the most inhospitable conditions for plant growth, such as organic-rich, anoxic sediments that are high in concentrations of the toxic gas hydrogen sulfide; high concentrations of sodium chloride in the sediment pore water; and a daily cycle between inundation by saline water and exposure to dry conditions in areas affected by tides. For most plants the presence of salt acts to upset the entry of water into cells. In areas with high salt concentrations or high osmotic pressure and low water concentration, plants can have higher water concentrations and actually *lose* water to the surrounding area; as a result, the plants die from a lack of water (Gough et al. 1979).

Such conditions are so harsh to most plants that the eastern coastline is often dominated by only *Spartina alterniflora* and *Spartina patens*. Sodium is a trace element in plants, but an essential element for most animals. These plants deal with the deleterious effects of salt by increasing the content of sodium chloride in their root tissues above that of the surrounding saltwater. This lowers the water concentration in the plant cells below that of the seawater, thereby allowing the osmotic entry of water into the plant (Flowers et al. 1977). *Spartina* is acclimated to high saline conditions, but growth decreases as salinities increase above 40 g/dm^3 (Bradley and Morris 1991).

As the *Spartina* plants transpire, salt-free water exits the stomata because the excess NaCl has been excluded from entry into the plant beyond the root membranes. This ion exclusion by the roots was reported to be 91% of the theoretical concentration that would have been taken up based on transpiration rates (Bradley and Morris 1991). The accumulated salt near the roots in the rhizosphere is washed away by each tidal event. Any incidental salt that enters the transpiration stream is excreted along the plant leaves by special ducts or glands, after which the salt crystals are washed away by the tides. For example, Bradley and Morris (1991) reported that about 50% of the ions that entered the transpiration stream were removed from the plant tissue by leaf excretion. Moreover, even though ions were excluded, concentrations in the roots of the plants remained high enough to reverse the osmotic gradient so that water entered the roots.

In such marsh conditions, salinity is not the only environmental factor that results in the selection of a monoculture. The concentration of oxygen in the pore water of the sediments in these saline marshes and estuaries is low. This is because the oxygen that enters the upper layers of sediment after each tide is rapidly consumed by aerobic organisms present in a thin layer of sediment and satisfies the abiological oxygen demand exerted by the presence of reduced mineral species. Plant roots are living tissue and require oxygen, however, in order to respire. For *Spartina*, the need for oxygen in an oxygen-poor system is accomplished by diffusion. Oxygen levels in the air near 20% diffuses from the stomata through the plant toward lower oxygen concentrations in the roots. Moreover, as a result of root respiration, CO_2 levels in the root zone are higher than the CO_2 levels in the atmosphere, and the higher partial pressure of CO_2 in the root zone establishes a diffusion gradient from soil to air. To ensure that the diffusion rate of these gases is at a maximum, these vessels remain dry in *Spartina* and are not filled with water, which would slow the rate of O_2 and CO_2 transport and result in plant death. Leakage of excess O_2 into the immediate area around the root zone also is responsible for the rhizosphere and its effect on plant health and contaminant remediation, as discussed in Chaps. 12 and 13.

Saline conditions also can be tolerated by woody plants. For example, mangroves trees are found along the southernmost coast of Florida. Like *Spartina*, they have been able to cope with high salinity by excluding it after uptake of the highly saline water into their xylem. As with *Spartina*, uptake and exclusion leaves behind a more concentrated solution in the root-zone pore water. Measurements of pore-water salinities near red mangrove (*Rhizophora mangle*) roots in Tampa Bay, FL, indicated that in the 0.9–2.2 ft (0.3–0.7 m) deep root zone, sediment salinity concentrations were two to three times those in the surface water (Greenwood et al. 2006; Fig. 11.11).

Fass et al. (2007) reported that the ability of mangrove trees to create zones of high-chloride pore water also has led to increased salinity in groundwater in areas of Australia. Although such enrichment of the salinity of shallow groundwater near mangroves would be expected, Fass et al. (2007) reported the occurrence of saline shallow groundwater many miles inland. These pockets of high-salinity groundwater represent locations of when sea levels were higher some 4,000–6,000 years ago than during recent times. The high-salinity groundwater that formed at that time sank into deeper aquifer units over time.

Mangroves have adapted to the low-oxygen content of the sediment pore water in which they grow. In this case, form follows function. The characteristic structural features of the mangrove *Rhizophora* are the stilt-like roots that exit the trunk above the mean high-water line and below the lowest leaves. These roots are always exposed to the atmosphere. These plants probably are derived from those that can be found today along the west coast of Africa. Their aquatic habitat, like many riparian species, consists of a seed that can float and be transported great distances on open water and prevailing currents, much as with palms and coconut seeds. Although mangroves are always found near water, they are derived from predominantly land plants (spermatophytes) that produce seed.

In the previous examples, surface water is the source of the salinity excluded by the plants. Groundwater also can have high salinity values and effect trees. The effect of high salt concentrations on tree species used for phytoremediation

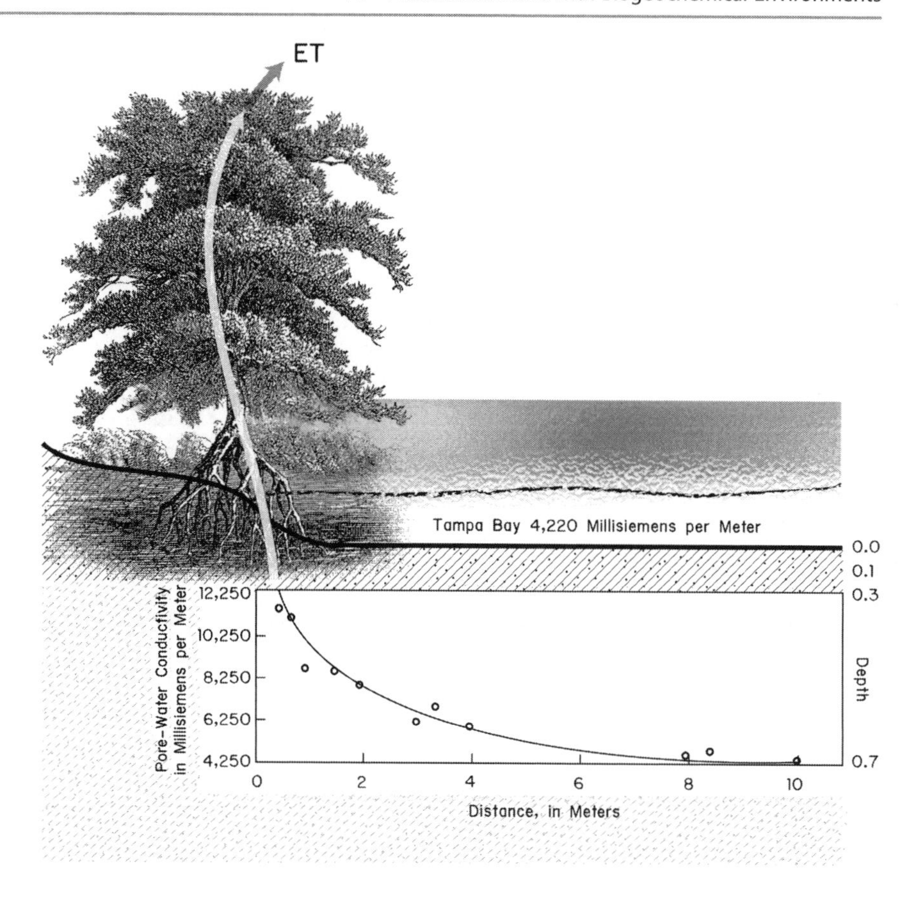

Fig. 11.11 The effect of the exclusion of salts on the salinity of pore water in the sediments in the root zone as a mangrove tree uses water. Modified from data presented in Greenwood et al. (2006). One meter is equivalent to 3.2 ft.

was studied by Shannon et al. (1999). Cuttings of eight poplar species, including DN-34 and OP-367, were exposed to salt between 3.3 and 7.6 dS/m were observed to have reduced growth rates and shed leaves prematurely. The authors concluded that the poplars studied were not very tolerant to salt, especially in relation to salt-tolerant eucalyptus trees (*Eucalyptus spp.*). The study by Shannon et al. (1999) supports the observation of higher salt tolerance in DN-34, with no defoliation at 5.53 dS/m—this salinity is equivalent to about 3,000 mg/L of total dissolved salt. Even for the more salt-tolerant poplar species, however, the negative effects of salinity on growth are enhanced when transpiration is higher and evaporation rates are higher.

Such tolerance to salt by phreatophytes also may be an adaptation to acquiring water from deeper within the soil zone than can be reached by shallower plant roots. Nilsen et al. (1983) reported that Honey Mesquite (*Prosopis glandulosa*) growing in the Sonoran Desert of California used groundwater from between 4 and 6 m below land surface. Near-surface, pore-water soil moisture was characterized by increased salinity relative to deeper depths because of soil-water evaporation but also hydrologic lift. Consequently, soil-water potentials in the surface soils were lower, between −4.0 and −5.0 MPa relative to deeper soils that had −0.2 MPa. This gradient caused deeper groundwater to flow vertically upward. The authors report that these data suggest a mechanism by which water stress could be avoided.

11.3.2 Extreme Temperatures

Prevailing temperatures affect the types of plants that can grow in a particular area. Because plants cannot respond to changes in temperature that occur in a location on a seasonal basis like migratory animals, they have had to adapt to these changes. In fact, most plants in temperate areas actually need for temperatures to change to promote sexual reproduction. For example, certain flowers that grow from bulbs, such as daffodils, have to undergo a period of cooler weather before they will grow, produce flowers, and can be pollinated. Deciduous trees also have a similar response to changes in temperature. Close inspection of some flowering trees during early dormancy reveals buds that do not open unless they are exposed to a period of cooler temperatures.

Temperatures affect plants not only changing over time but also over space. Temperature changes with a change in elevation, and this produces a gradient of plants. An example is the lack of plant growth beyond a certain elevation and temperature decrease in mountains, which is called the timberline. In such colder conditions or even climates, the few plants that do grow consist of mosses and grasses.

That the growth of the above-ground parts of plants is related to temperature is obvious. Below ground, however, temperature changes are not as drastic. Root elongation also is related to temperature, with maximum rates of growth near 30°C that decrease with additional increases in temperature. There are exceptions of course for plants in desert climates, where growth can proceed at temperatures above 60°C. One of the influences of temperature is that even though above-ground air temperatures may decrease and depress shoot growth, below-ground temperatures are more stable. Roots can elongate even during winter months when trees appear to be senescent. If roots are growing, respiration is occurring, and this requires the uptake of water. This issue of dormancy and its effect on phytoremediation is discussed in Chap. 16.

11.4 Plant Interactions with Contaminated Soil and Water

We have seen how plants have been used as indicators of environmental contamination and for toxicity assessment. Plants also have been used as a means to decrease contaminant levels in the soil and aquatic environment. This has been achieved mainly through constructed wetland systems or landfarming approaches, both practices which have a long history in the United States and even longer in Europe. In fact, the application of plants for phytoremediation of contaminated groundwater can trace its inception to these earlier plant-based contaminant or waste-reduction practices. This scenario is strikingly similar to the linkage between the application of *in situ* subsurface microbes for monitored natural attenuation and the long history of using microbes as the main part of municipal wastewater treatment. Essentially, plants have been and are being used to provide cleaner water.

11.4.1 Natural and Constructed Wetlands

Wetlands are more complex than simply shallow land areas that contain water and plants most of the year. To be defined by water managers as possessing characteristics of a true wetland, the land has to be inundated with water a specific amount of time each year. But how deep does the water have to be? Does it have to be flowing or stagnant? Is the water source groundwater or surface water? What types of herbaceous and woody plants are to be found? The answers vary, but the point is that most wetland systems provide a unique physical location where groundwater is essentially exposed at land surface and provides the opportunity to evaluate the interaction between plants and water. Moreover, these phenomena can be used to understand how phreatophytes installed at contaminated groundwater sites interact with the unseen water table some distance beneath the land surface.

In general terms, a wetland is a zone of transition of plant-water availability that ranges from aquatic constraints to terrestrial conditions. In legal terms, the U.S. Army Corps of Engineers, the Federal agency charged with wetland regulation, defines a wetland as having shallow water at least some length of time, such as during the growing season, having anoxic sediments, and vegetation. In the 1989 Federal Wetland Delineation Manual, a wetland was defined as land where the soil is saturated within 18 in. of the surface for at least seven consecutive days during the growing season. This was changed in 1991 to reflect that the land must completely be under water for 15 consecutive days or saturated to the surface for 21 consecutive days.

In biological terms, wetlands are ectones—areas that are not always wet or always dry—and are found between predominantly dry and predominantly wet land. With respect to groundwater, wetlands are areas were uplands interface with lowlands and the water table is near or at the land surface. Wetlands are transitional places between aquatic and terrestrial ecosystems. But they also are only a temporary part of the continuum from exposed land to wetland to swamp and back to terrestrial land as the wetlands fill in.

The water level is not the only variable that can be used to describe wetlands. Vegetation also can be used. Wetlands contain a variety of plants. As we have seen, many of the plants that possess characteristics of phreatophytes, such as poplars and willows, are found in wetland or near-wetland environments. Submerged aquatic macrophytes also are present. Wetland plants tend to be herbaceous.

Other types of wetlands exist. Swamps are wetlands characterized by having more woody plants. Marshes tend to have emergent macrophytes, and bogs and fens have grasses, mosses, and some trees. Some states along the eastern coast of the United States are characterized by elliptical-shaped landforms of unknown origin called Carolina Bays that often contain wetlands. Other wetlands are called sloughs and bogs. Wetlands also can be classified by their water quality. Some wetlands are dominated by freshwater and some by saltwater. For saltwater wetlands found along the coasts, fresh surface water draining the uplands mixes with saline water from the oceans in estuaries. Also along rivers, wetlands can be found as oxbow lakes and in periodically inundated flood plains.

The science behind the use of modern-day, westernized wastewater treatment can be traced back to the use of natural wetlands to "treat" waste. For example, wetlands in Europe have long been referred to as "wastes" (Horne 2000). Using such natural systems to cleanse wastes probably can be considered the beginning of bioremediation, monitored natural attenuation, and even, by connection with the rhizosphere, phytoremediation. Wetlands naturally improve water quality by the release of oxygen by plants and the removal of suspended particles and turbidity in these low-energy environments. Plants help keep the turbidity low by anchoring the sediment and preventing re-suspension. Their presence increases the number of microbes present in the submerged rhizosphere of the plants, and reduced organic carbon compounds can be oxidized by these microbes.

A potential liability of using wetlands to cleanup wastes is that the system primarily is anoxic. Under oxic conditions, the transformation of reduced organic contaminants, such as oils and fuels to carbon dioxide, occurs more rapidly than under anoxic conditions. Anoxic conditions promote denitrification, however and, therefore, facilitate the biological removal of elevated nitrate, a common nutrient in runoff and wastewater. Plant roots and leaves can take up nitrate and phosphate dissolved in the water column.

The linkage between wetlands and municipal wastewater treatment at first seems like a non-sequitur. In the days prior to environmental laws that regulated pollution levels in water in the United States, raw sewage was dumped directly untreated into rivers. This practice occurred with little negative consequence as long as the population of the area was constrained and great distances separated those dumping wastes upstream and the downstream users of the water for potable purposes. It was not until the typhoid epidemics occurred in the nineteenth century that the idea of using chemicals to treat drinking water was discussed and finally implemented in 1948 in the United States and amendments to the Federal Water Pollution Control Act (FWPCA) in 1970 took this idea further by requiring the treatment of raw wastewater prior to discharge. In the developing world, however, raw sewage often is not treated and water for drinking is often obtained from surface water contaminated by sewage.

The FWPCA was amended in the early 1970s and gave rise to the Clean Water Act (CWA). The Clean Water Act addresses water pollution through encouraging reductions in point and non-point sources of contaminant release. The National Pollutant Discharge Elimination System (NPDES) specifically addresses the amount of pollutant that can safely be added to a surface-water body. One of the sources is runoff from highways and parking lots, which has been associated with increased bacterial levels in waterways, and with the increase in concentrations of PAH particles in stream-bed sediments (Mahler et al. 2005).

Stormwater discharge from roadways is considered under the NPDES program to be a municipal separate storm-sewer system, or MS4. To reduce the load of pollutants that could reach a regulated surface-water body under MS4, various best management practices (BMPs) can be used. Ironically, one of the BMPs is the construction of a surface-water body designed to receive runoff prior to release to the regulated surface-water system. These appear as holding ponds at the edge of mall parking lots, usually surrounded by chain-link fencing (Fig. 11.12). Contaminants that enter the holding ponds during runoff events are reportedly removed by sedimentation, biodegradation, and plant-nutrient uptake.

Fig. 11.12 An artificial wetland to treat runoff from a commercial parking lot in Florida. The plantings primarily consist of cattails and sedges, as seen in the middle of the water, with willows along the banks (Photograph by author).

Natural wetlands are not necessarily the most efficient way to treat wastes or contaminants. This lack of efficiency primarily is a result of preferential water flow through only a limited part of the wetland. A constructed wetland removes this liability to pollutant removal by increasing the residence time and space for water flow through the wetland. More control over the water flow also permits control over the water level, which in turn helps to control the distribution of plants. Constructed wetlands enhance water-flow contact through these sediments by promoting a loose rather than compacted bottom layer. The hydraulic conductivity in the bottom sediment of a constructed wetland will be higher than that of a natural wetland (Horne 2000). Beneath these more permeable sediments, however, is placed a lower-permeability liner, such as clay.

Constructed wetlands are more prevalent outside the United States than inside, for there are roughly 300 in North America compared to 500 in Great Britain alone. The preferred plants for constructed wetlands include cattail (*Typha spp.*) and bulrush (*Scripus spp.*). However, many others are available for use and are being evaluated (Lewis and Wang 1997). The purpose of constructed wetlands is to maintain the surface-water level at a condition at or near the land surface to satisfy the needs of the plants that grow there.

An additional benefit of constructed wetlands is the maximization of areas where redox gradients are present. Zones of anoxic and oxic conditions within a relatively short distance can harbor a wide diversity of microbes that contain high pollutant degrading potential. This has been observed at gasoline-contaminated sites where anoxic groundwater that contains BTEX and MTBE discharges to surface water (Landmeyer et al. 2001). For example, the oxic and anoxic interface in the one foot-thick stream hyporheic zone led to the formation of a microbial system that mineralized MTBE in groundwater prior to discharge. The contaminant does not enter the wetland by overland flow, but rather the wetland processes occur vertically as groundwater discharge brings contaminants to the wetland redox interface. This also has been shown to be the case for chlorinated solvents (Lorah and Olsen 1999; Dinicola et al. 2002).

A constructed wetland can emphasize redox interfaces in areas where organic matter accumulates at the bottom of the surface-water column. This accumulation of organic matter promotes anoxia and denitrification. This process of nitrate removal is preferred over the uptake of nitrate by the wetland plants, because the removal of nitrate by plants would lead to huge increases in biomass that would have to be managed. Phosphorus, on the other hand, is taken up by plants, but it can be released back into the system when the plant dies.

Relative to the use of plants at a phytoremediation site to control a subsurface plume in groundwater, the rate of surface-water movement through a constructed wetland is fast. However, wetlands are a surface expression, almost like *ex-situ* treatment, and the risks associated to exposure potential are higher than with the same contaminants in groundwater. For example, flood conditions may increase the risk of exposure to downstream areas.

A wetland system was used to examine the biological effect of plants and associated rhizospheric microbes on concentrations of nitrate and the xenobiotic perchlorate. Perchlorate is a chlorinated hydrocarbon used as an explosive, propellant, and pyrotechnic. It also has natural sources. The rocket fuel propulsion industry has been implicated as the source for perchlorate being detected in public water-supply wells in California. Even though no MCL currently exists for perchlorate, the potential health effects have promoted research for remediation tools to remove perchlorate from contaminated groundwater.

Krauter (2001) investigated the use of a bioreactor that contained wetland plants to remove not only perchlorate but the high concentrations of nitrate associated with perchlorate detection. Each of four bioreactors built were filled with coarse aquarium gravel, presumably to simulate the flow of groundwater through a shallow aquifer. Wetland plants placed in the bioreactors included bulrush (*Scripus spp.*), sedges (*Cyperus spp.*), and cattails (*Typha spp.*) collected from areas in California near Livermore. To deliver perchlorate- and nitrate-contaminated groundwater to each bioreactor, a 55-gal (207 L) drum was filled with groundwater obtained from a well known to be contaminated with perchlorate (4.5 μg/L) and nitrate (68 mg/L); it was allowed to flow by gravity through the attached bioreactors. The flow was used to inoculate the system and create microbial biofilms acclimated to perchlorate prior to adding clean water to which a known concentration of perchlorate was added.

Nitrate concentrations decreased from 80 to 4 mg/L within the first day of the test. Perchlorate decreased from 44 to less than 4 μg/L within 4 days (Krauter 2001). This follows the usage of preferred electron acceptors; oxygen, nitrate, and then perchlorate. In the absence of oxygen, nitrate undergoes assimilative and dissimilative nitrate reduction to nitrogen gas. Chlorate can be transformed to chlorite in the presence of nitrate reduction. Microbes can reduce the oxidized perchlorate to chlorate ion and then to chlorite. The wetland plants are key to this process, because they release organic matter to stimulate nitrate reduction, and can maintain these reactions over time.

One of the first manmade wetlands constructed specifically to treat stormwater runoff was in the early 1980s in Florida. In 1983, the Northwest Florida Water Management District, and the USEPA and the Florida Department of Environmental Regulation, were concerned that Lake Jackson, located near Tallahassee, FL, was being negatively affected by the accumulation of stormwater runoff resulting from rapid urbanization. As with many lakes that consist of deep water and shallow coves, the coves are the first to experience changes in water quality as they are nearer the source of runoff from the banks and have less water volume for dilution. At Lake Jackson, one of these coves is called Megginnis Arm, and it was here that the manmade treatment system was constructed. It consisted of a retention pond to capture the volume of flow generated by storms. Next, the water flowed through a sand filter and then to the 9-acre marsh. The marsh was constructed to contain three treatment areas; one containing primarily sawgrass, one containing primarily rushes, and the third containing broadleaf perennial herbs. The most efficient removal of nutrients from the stormwater runoff occurred in these three treatment areas.

In the southeastern United States, there is a large constructed wetland that has been used to treat wastewater prior to discharge since 1997. This wetland treatment area is north of Augusta, SC, and is an extension of the naturally occurring Phinizy swamp. Prior to 1968, untreated wastewater from Augusta was discharged to a local ditch that made its way to a local creek that then discharged to the Savannah River. In 1968, a wastewater-treatment plant was built. In 1997, the Phinizy swamp was extended by the addition of 12 constructed "cells" used to slow the path of the treated wastewater as it makes its 2-day trip through the swamp. The plants in these cells, which essentially are defined by berms, contain cattails and rushes, which can handle the high nitrate and phosphate concentrations in the wastewaters. The design of the constructed wetlands attempts to mimic the flow of flood waters through flood-plain vegetation sometimes observed in natural flood-plain riverine ecosystems.

11.4.2 Plants and Biosolids

Much like the hydrologic cycle described in Chap. 2 consists of a continual, natural cycle of evapotranspiration and precipitation, man's use of water has added an artificial subcycle to include consumption, storage, and discharge of water. For example, water can be taken from a surface-water source, treated, delivered to homes, used, and then sent to groundwater through the tile field of a septic system. This pathway truncates the hydrologic cycle until the groundwater discharges to a surface-water body and once again can evaporate.

Any residual material left behind after the physical, chemical, and biological components of wastewater treatment is called sewage sludge, or biosolid. The term biosolid has come to replace sewage sludge to reflect more positively the potential market for its use. This material can be solid, semisolid, or liquid. This material can be stabilized to reduce its volume and concentration of pathogenic organisms through a number of processes. Digestion uses microbes under aerobic or anaerobic conditions to breakdown the material into simpler forms. Stabilization also can occur if the pH of the material is increased by adding chemicals, such as lime. This also reduces odor-causing compounds and reduces the number of pathogenic organisms, such as bacteria, viruses, and protozoa. The solids also can be dried by being spread on paved surfaces or sand beds. Additional stabilization can occur through composting, heat drying (pelletization), and chemical fixation. This material has become a resource for use as fertilizer, after meeting regulatory acceptable levels of contaminants and pathogens, that can be applied to crops. Similar to the natural decomposition of dead organic matter that makes stored nutrients and minerals available to living plants, or manure or mulch applied around plants, the use of biosolids as a way of releasing nutrients back to plants is a natural part of the global cycling of nutrients.

One of the concerns with the use of biosolids for incorporation into the plant base used for consumption is that the wastewater-treatment plants treat industrial wastes as well as municipal wastes. Although most wastewater-treatment plants do not accept industrial wastes unless the industry performs a pre-treatment, as defined under 40 CFR Part 403, unpermitted wastes can enter wastewater-treatment plants and can sometimes lead to disastrous results. An example is provided by the release of tributyltin (TBT) to a surface-water system after it was sent (unpermitted) to a wastewater treatment plant near Columbia, SC (Landmeyer et al. 2004). This event led to the permanent closure of the wastewater treatment plant and many private ponds.

The potential for such contamination events to occur and enter the food chain through biosolid application to farmland fortunately is small. This is because less than 1% of the United States farmland acreage is approved for the application of biosolids. Moreover, biosolids destined for use as fertilizer must conform to chemical limits set in 40 CFR Part 503 Standards for the Use and Disposal of Sewage Sludge (United States Environmental Protection Agency 1995). This regulation only applies to the levels of metals in the biosolids, not organic chemicals. However, the Milwaukee Metropolitan Sewerage District, a large municipal treatment plant in Milwaukee, WI, has approval to package its biosolids as a commercial lawn fertilizer, which can be purchased in home centers around the United States. This biosolid purportedly contains dried microbes (fecal coliforms); trace minerals essential to plant growth, such as iron and calcium; and other heavy metals, such as cadmium, lead, and selenium. It meets 40 CFR Part 503 Class A "Exceptional Quality" requirements. This means that the biosolids in question have to meet the limits established for the three criteria for presence of pollutants, pathogens, and attractiveness to vectors (rodents, mosquitoes; United States Environmental Protection Agency 1994).

In sum, waste treatment by plants is probably one of the earliest artificial interactions between plants and man's wastes. Wastes are applied directly to plants for the benefit of the plant and to decrease the biosolid volume. In the case of the phytoremediation of contaminated groundwater, the plants are directly applied at contaminated site to decrease the waste concentration.

11.4.3 Natural Attenuation

As we saw in the section on cycling of nutrients, it should become apparent that these systems possess the ability to deal with permutations to normal conditions. For example, if an acid is added to a buffered system, the acid will not affect the pH of the total solution until after the buffer is depleted. The same process of buffering, or assimilative capacity, in the environment occurs for a wide range of pollutants. Many civilizations depend on the assimilative capacity of surface-water systems to receive untreated effluent.

The process of contaminant reduction as part of the assimilative capacity is driven by both abiotic and biotic processes (National Research Council 2000). In groundwater remediation, intrinsic bioremediation refers to the *in situ* capacity for groundwater to clean up contaminants. Because groundwater is below ground removed from the atmosphere and direct precipitation, and because of oxygen's low solubility in water, aquifers tend to be oxygen limited. These limitations can be overcome by engineering ways to deliver oxygen into the aquifer, which is called engineered bioremediation.

An additional term regarding the assimilative capacity of groundwater systems is called natural attenuation, and it is analogous to assimilative capacity used in surface waters. It describes all of the naturally occurring processes, both biologic and abiotic, that act to decrease contaminant levels. For example, contaminant concentrations can decrease to acceptable levels by dilution. This can occur when clean water is added to contaminated groundwater, such as occurs during recharge through uncontaminated unsaturated-zone sediments and lateral groundwater inflow from upgradient areas or adjacent aquifers. Additional important processes include volatilization of contaminants from the water table to the unsaturated zone, sorption onto natural organic matter, or trapping in pore water of clays. These processes, however, may act to transfer aqueous-phase contaminants to the soil or atmosphere, and do not solve the overall problem or contaminant removal. Attenuation processes may, however, lower the overall risk posed by contaminants by lowering the concentrations to acceptable levels at points of exposure.

A brief history of natural attenuation in relation to groundwater remediation is warranted. In the early 1970s, around the time of the Love Canal contamination investigation, contaminant remediation consisted of engineered approaches, such as dig-and-haul for sediments, and pump-and-treat for groundwater. In the early 1990s, evidence mounted that particularly for pump-and-treat approaches that initially decreased contaminant levels in the aquifer, the concentrations rarely approached regulatory levels and never approached zero but, rather, leveled off at some higher value following an asymptotic pattern (National Research Council 1994). As a consequence, many researchers began to investigate the hypothesis that rather than contaminants accumulating in the groundwater environment, as was the implication for inefficient contaminant removal exhibited by pump-and-treat systems, contaminants could be decreased by in situ processes.

Initial evidence that natural processes were occurring at contaminated sites came in the form of monitoring data. The best example occurred at the Borden site in Canada. At that site, a plume of BTEX in groundwater was expected to continue to migrate downgradient from the source area and undergo attenuation by dilution and sorption. The monitoring data indicated that the measured plumes were smaller than would be predicted by sorption and dilution alone. It was theorized that subsurface heterotrophic microorganisms were oxidizing the contaminants.

That subsurface microbes would play a role in contaminant degradation was a radical idea at the time. This is because knowledge of the presence and distribution of microbes in the subsurface below the O layer was still in its infancy the late 1980s and early 1990s. Up until then, microbial numbers were known to decrease with depth in the soil column as concentrations of organic matter and oxygen decreased. Deeper sediment-water systems were considered to be as oligotrophic as the deep oceans. That mindset changed, however, when deep subsurface cores collected and were analyzed for the presence of bacteria that had been assumed to be there based on geochemical evidence (Chapelle 1993).

Most bacteria that inhabit groundwater are a group of fungi or simple plants that lack chlorophyll and, therefore, rely on external sources of energy for growth. They are either parasites, heterotrophs, or saprophytes. Some species reduce CO_2 to make organic compounds, such as methanogenic bacteria, much like plants reduce CO_2 to make sugars and starches. Some bacteria, however, are autotrophic and can make organic compounds from simple inorganic compounds, such as CO_2, H_2S, Fe(II), or H_2.

The application of microbes to restore contaminated soils, water, and groundwater has a rich history. First, it could start with wetlands where, as we noted earlier, plant-based remediation is important, but the majority of remediation is due to microbes. Microbial-based contaminant degradation processes rely on external sources of energy, in which organic compounds are broken down to release energy to drive growth. This is opposite of plants, which use the energy of the sun to make organic compounds that are used within the plant to support growth. Hence, there is a need for many more enzymes in heterotrophic bacteria to deal with external organic compounds that are not required by plants. Whereas bacteria can derive energy and growth from these compounds, plants do not. More on this difference is discussed in Chap. 12.

An interesting result of a study performed by Schnoor et al. (1995) was that plant enzymes were detected in contaminated soils that had been characterized as undergoing attenuation. This may be due to the increased role of microbial processes in the rhizosphere or the direct effects of the release of enzymes, as is discussed in Chap. 12.

There is a relation between bioremediation and phytoremediation of contaminated groundwater. They should not be confused, however. The presence of plants tends to stimulate the number of microbes relative to unplanted areas. Plants can interact with the atmosphere, and change redox by allowing oxygen to diffuse into potentially anoxic soils. Phytoremediation processes are directly plant-oriented processes, not just plant-stimulated processes. Much work has been done examining the effect of aquifer microorganisms on contaminated groundwater. Examination of the effect of microbes in relation to plants on groundwater contamination has intensified, particularly into soil remediation, but the extension to groundwater restoration is less well developed. Plants have many advantages over microbes when it comes to biodegradation. Whereas microbes need to derive energy from the contaminants, plant roots obtain their energy from respiration of the food they make and, therefore, have more energy available to remediate contaminants.

Plants can indirectly affect natural attenuation even if not used in concert for clean up. This occurs for at least two reasons—soil-moisture fluctuations and introduction of root organic matter. The first reason is because the removal of soil moisture by plants can affect the soil bulk density by decreasing the water content and increasing the air content. As soils become more oxic they also have decreased water content, but water is necessary to support microbial metabolism. Secondly, the flow of carbon, nitrogen, and phosphorus as described in this chapter are affected by the availability of microbes and moisture.

11.4.4 Plants and Riparian Buffers

The area above ground or below ground connected with a surface-water body generally is termed the riparian zone. The surface water can be moving water, such as creeks, streams, or rivers, regardless of whether they flow continually or not. The surface water also can be quiescent, such as lakes or ponds. The extent away from the interface between land and surface-water level, in either direction, is further categorized based on ecological criteria.

The riparian zone also has a legal definition. An area adjacent to the surface-water body can be used as a buffer in which no environmental degradation is permitted to occur, and if it occurs, it must be remediated. This not only protects the riparian buffer as a resource, it is designed to protect the surface water from degradation by contaminants, wastes, or decreases in DO. The exact width of the zone is dependent on the regulatory agency, intended use, the regulatory requirements, and the source of the impacts to the water.

The processes occurring in the riparian zone that act to protect the surface-water system include physical, chemical, and biological processes. The presence of plants and their root systems decreases the probability of sediment erosion and decreases the velocity of surface-water runoff by displacing the suspended-sediment load (and contaminants) before entering the surface water. These sources of problems fall into the category of nonpoint sources. Sediment input can endanger fish survival and reproduction; carry attached contaminants, such as bacteria and other fecal problems; reduce clarity; and increase the cost of treatment for drinking water downstream. Contaminated groundwater, however, can act as point sources to surface-water systems. The same can be said for plumes of nutrient-contaminated groundwater, such as high-nitrate levels in groundwater that flows beneath agricultural or livestock lands. This input of excess nutrients can cause eutrophication, or blooms of algae and other plants. Buffers that contain tall, woody plants with associated high *LAI* control the amount of light that reaches streams and rivers; this, in turn, controls the amount of primary production by algae, which decreases the BOD on available DO in the water column.

An interesting study of the effect of trees, such as the phreatophytes used to remediate contaminated groundwater, on water flow and quality was reported by Kollin (2006). A fire in San Diego, CA, in late 2003 (the Cedar Fire) affected about 13% of the city boundary, or about 28,000 acres (113×10^6 m^2). Up to 50% of the tree canopy in this area was lost. As a result, it was calculated that runoff during storm events increased about 12×10^6 ft^3 (3.3×10^5 m^3). This same loss in canopy and stormwater retention can occur without fire, however, as other factors can remove tree canopy, such as rapid or mismanaged development.

Engineered riparian buffers have similar characteristics to natural systems (Mayer et al. 2005). A gradational change occurs in plant types related to water use and source from the riparian zone to more upland plants that derive most water from recent or stored precipitation. Flood water is retained during times of high flow for subsequent slower recharge to the soils and aquifers and movement through the rhizosphere of the riparian vegetation. As long as the water and contaminants can contact the plant rhizosphere, roots, or be up taken, contaminant transformation by plants, the processes which are the focus of Chap. 12, can occur.

11.5 Summary

Plants trap solar energy and convert it to ATP to form glucose, cellulose and lignins, allelopathic chemicals, and to support respiration. Many xenobiotic compounds interact with plants because they have physical and chemical properties that make them readily dissolve in water. The interaction between plants and their natural, even seemingly inhospitable, geochemical environment provides the fundamental basis that phytoremediation can be used remediate contaminated groundwater—this is not the case for purely physical or chemical processes of groundwater remediation such as air stripping or Fenton's oxidation.

Why is this information important to the phytoremediation of contaminated groundwater? The notion that plants are not simply at the mercy of the environment and have, in fact, demonstrated the ability to produce chemicals not only to sustain their growth but to protect them from predators and to diminish the effect of resource competition is in direct contrast to the notions spelled out at the beginning of the environmental age in the early 1970s. This encounter of plants with allelopathic compounds and the early use to treat wastewater are perhaps the best evidence to support the application of plants for the remediation of xenobiotics in contaminated groundwater.

Chemical and Physical Properties That Affect the Interaction Between Plants and Contaminated Groundwater

12

Hidden from view, 30 rusty underground storage tanks were slowly releasing the soluble fuel compound methyl tertiary-butyl ether (MTBE) to the shallow water-table aquifer above the wellfields of the City of Santa Monica, California. The city's water utility was unaware of this invisible threat, and continued to pump fresh, clean groundwater from deep below the increasingly contaminated shallower aquifer. In the fall of 1995, however, the first molecules of MTBE were sucked up by the city's pumps and were detected in the drinking water sent to the city's permanent and tourist customers. In early 1996, the situation worsened as increasing concentrations of MTBE were being measured in the water supply. Eventually, the city had no choice but to shut down its wellfields, and was forced to purchase surface water from nearby Los Angeles. Water utility managers do not expect to be able to use the previously reliable, deep aquifer system again for many years.

This short summary of a true story received national media attention in January 2000 on the TV show "60 min." Although this specific incident occurred in California, the scenario could have occurred at any of the 400,000 leaky USTs across the United States in which gasoline that contained MTBE may have been stored. A leaky UST that contains gasoline enhanced with the fuel oxygenate MTBE can be accidentally released to the subsurface through corrosion in the joints of underground piping. After escape from the UST, gasoline as pure free product can migrate through the pore spaces of the unsaturated zone under the influence of gravity until it encounters the water-table surface. There, the lighter specific gravity of the gasoline will cause it to float on the water table as a separate phase. MTBE, and other gasoline compounds such as benzene, will then partition between the gasoline source itself and the air present in the unsaturated zone above the free product, the water present in the unsaturated zone as well as the water table, and the organic matter present in the soil.

To understand the fate of a particular contaminant such as MTBE, benzene, or chlorinated solvents in the subsurface, the potential interactions between the contaminant and the various physical and chemical components present in the subsurface need to be elucidated. Typically, contaminant fate is described in terms of the degree of partitioning that can occur between the contaminant and the major phases of the subsurface, such as soil organic matter, lipids, water, and air. This interaction is a function of the physical-chemical properties of the contaminant, such as water solubility, lipid solubility, vapor pressure, etc. Next, a determination needs to be made regarding how the properties of the chemical will interact with the system in question.

For the purposes of the phytoremediation of contaminated groundwater, the plant is an additional pathway of contaminant interaction. Plants can be considered an extension of the other phases that will interact with a particular contaminant, because plants offer phases that include solid, organic matter, organic lipid, and water, as well as gas phases. In addition, the fact that plants interact with groundwater adds the additional components of life, from the rhizosphere to the living tissues of the plants themselves. As such, the total interactions between contaminant and subsurface environment have to include both purely physical as well as biological interactions, and these can be defined by contaminant partitioning processes.

12.1 Contaminant Partitioning in the Subsurface and Plant Uptake

In general, plants consist of three phases: organic lipids and solids, water, and air. The unsaturated zone and most groundwater also can consist of these three phases. The additional interaction between groundwater, plants, and xenobiotics can be described in terms of basic processes such as advection, diffusion, sorption, and transformation through metabolism that result from interactions between these phases. These interactions often can be examined

J.E. Landmeyer, *Introduction to Phytoremediation of Contaminated Groundwater*,
DOI 10.1007/978-94-007-1957-6_12, © Springer Science+Business Media B.V. 2012

from the perspective of the physical and chemical characteristics of the compounds and the phase(s) in question. This relation can be quantified using the concept of the degree of partitioning of a particular species between its original phase and other phases present. The extent by which this interaction occurs can be quantified with a partition coefficient.

The interaction of xenobiotic solutes in groundwater will be discussed in terms of plant and groundwater interactions. For example, what is the degree of partitioning between soil-water, roots-water, wood-water, and leaves-air for various types of xenobiotics? Answers to these questions and others can be examined in the framework of previous work done on chemical partitioning, such as the partitioning of a chemical between a nonpolar organic solvent and polar water, between the dissolved phase and a gaseous phase, or between the dissolved phase and the solid phase.

In general, processes that result in a particular species being partitioned into at least two phases can be described in terms of the following ratio

$$K_d = C_a/C_b \tag{12.1}$$

where K_d is the partition coefficient (or constant) for a compound, C, present in two phases, C_a and C_b. A compound will have many coefficients to describe the chemical's partitioning from its original phase into solution, into the gas phase, into a lipid phase, and so on. Most partition coefficients typically are determined experimentally under controlled laboratory conditions. As with any simplification of a complex interaction, such as in using a partition coefficient, the concept of equilibrium needs to be fulfilled by assumptions that may not always be defensible under field situations.

Perhaps the partition coefficient that has received the most attention, in terms of scientific and layperson recognition, is that related to xenobiotic bioaccumulation. Since the 1970s, contaminants that had the potential to bioaccumulate, or strongly partition, into the food chain were studied. Today, the bioaccumulation of certain chemicals, such as heavy metals, by non food crop plants is specifically engineered, and the interaction between organic chemicals and plants that can take up but not bioaccumulate these chemicals also is engineered. As we will see later in this chapter, plants possess the ability to decrease the threat of chemical bioaccumulation in their tissues similar to the manner in which the mammalian liver provides chemical detoxification.

For these processes to occur, the chemicals must first enter the plant from the surrounding environmental media. The various partitioning that occurs in the subsurface to control this uptake are described first.

12.1.1 Water–Soil–Contaminant and K_{ow}

The potential that an organic contaminant solute will partition into soil organic matter can be described in terms of a partition coefficient. This coefficient is determined experimentally and can be approximated by the extent that a particular solute is hydrophobic and will tend to dissolve into a surrogate lipid-phase relative to being hydrophilic and, therefore, not dissolve into an organic phase. To determine this partition coefficient, the organic solvent widely used as the surrogate for nonionic (or net nonpolar) organic matter, such as lipids found in plants, is the hydrocarbon *n*-octanol, also known as 1-octanol or simply octanol. These values are determined experimentally in the laboratory under controlled conditions over a range of solute concentrations.

Most unsaturated subsurface sediments consist predominately of about 50% mineral matter, about 25% water and air, and less than 2% organic matter, by weight. An organic contaminant released to the subsurface will partition into the water, air, mineral, or organic phases that are present. The resulting partition coefficient is an estimate of the tendency for the solute, or compound, C, to dissolve into the organic lipid, or solvent, C_o, or water, C_w

$$K_{ow} = C_o/C_w \tag{12.2}$$

where K_{ow} is the partition coefficient (or constant) that describes the magnitude of the chemical's affinity for partitioning into either the nonpolar octanol phase, as C_o is the solute concentration in the octanol phase (kg/m^3), or water phase, as C_w is the solute concentration in the water phase (kg/m^3). K_{ow} is a surrogate for the water solubility of a particular chemical; a higher value of K_{ow} implies that the solute is more likely to be dissolved in the octanol or lipid phase and, therefore, have a lower water solubility, and a lower value of K_{ow} indicates the solute is less likely to be in the octanol phase and more likely to be dissolved in the water phase.

The contaminant solubility in water is proportional to the contaminant's mass and can be approximated by the partition coefficient between the organic phase and water phase, K_{ow}. By definition, a K_{ow} of 1 indicates that a solute will partition equally between the octanol phase and the water phase. A K_{ow} greater than 1 indicates that the solute will partition more into octanol phase. For example, a K_{ow} of 50 indicates that the compound is 50 times more likely to partition into octanol than water. For contaminants released to the environment, the greater extent that a contaminant dissolves into a contaminant mixture, the higher the K_{ow} and the lower its solubility in water. How this partition coefficient can be used in the phytoremediation of contaminated groundwater is discussed in a following section.

The octanol-water partition coefficient is such a widely used concept in chemistry as well as geochemistry that a little history is warranted. In the 1970s, the correlation between a chemical's K_{ow} and its toxicity and potential for bioaccumulation in crop plants was studied and showed to correlate strongly with changes in biological activity if plotted against the log transform of K_{ow} (Leo et al. 1969; Hansch and Leo 1979)—K_{ow} can range up to values over 5,000, so the log transform lowers the values and makes it easier to compare contaminant-environment interactions. With respect to the uptake of chemicals by plants, the parameter of log K_{ow} is considered the "gold standard" for researchers tasked with contaminant fate studies in various aquatic and terrestrial ecosystems. As might be anticipated, much of this early work was done on pesticides.

The concept of the partitioning of an organic contaminant compound between the organic phase, the solid phase, and the water phase has been used to determine the extent of bioaccumulation. Bioaccumulation is a measure of the relative accumulation of the compound present in water into environmental media. The solid phase can be the organism in question, such as fish. In fact, much of what is known today about the interaction between various chemicals dissolved in water and organic matter is derived from the early work into the bioaccumulation potential of these compounds in fish.

The removal of a contaminant solute from the aqueous phase onto a solid immobile phase is called sorption. Sorption, the transfer of mass from a liquid phase to a solid phase, can occur by adsorption, absorption, and ion exchange, where adsorption is the interaction with the surface chemistry, absorption is the interaction with the bulk chemistry, and absorption that is exchangeable is called ion exchange. Sorption in colloidal material can result in a mobile phase, however.

The partition coefficient, or soil-water distribution coefficient, K_d, is where

$$K_d = C_s/C_w \qquad (12.3)$$

where C_s is the concentration in the soil and C_w is the concentration in the water. This partition coefficient, K_d, also is related to the amount of soil organic matter (as % SOM) by

$$K_d = K_{om}(\%\ \text{SOM}/100) \qquad (12.4)$$

where K_{om} is determined from the log K_{ow} of the particular contaminant because K_d tends to be proportional to the lipophilicity of the contaminant. The soil sorption constant, K_{oc}, is defined as

$$K_{oc} = (K_d/\%\ \text{organic carbon}) \times 100 \qquad (12.5)$$

12.1.2 Water–Soil–Air–Contaminant, and Henry's Law

The above partition coefficients are important to consider when the contaminant solute is present in the dissolved phase, such as occurs during many groundwater contamination events. Many organic contaminants also can exist in the vapor phase, and this section describes the partitioning for such solutes into a vapor phase.

The extent of this phase change is determined in part by the chemical's vapor pressure, the concentration gradient between water and soil and air that often is driven by diffusion, its water solubility, and its potential to be sorbed onto aquifer media. There will be an equilibrium established between the phases of contaminant-water-soil-air, such as is the case with the first three discussed above. The equilibrium concept is important, because it reveals that as the gases are evolving, they also can reenter the solution at an equal rate once equilibrium concentrations are reached.

Under conditions of equilibrium, the partitioning of a particular contaminant compound between itself in the liquid phase and gas phase can be thought of as

$$H = G/A \qquad (12.6)$$

where H is the Henry's Law partition coefficient (Pa m^3/mol), G is the contaminant concentration (or partial pressure) in the vapor phase, and A is the concentration, or solubility, of the contaminant in the aqueous phase. The Henry's Law partition coefficient is dimensionless. The air-water partition coefficient, K_{aw} is

$$K_{aw} = H/(RT) = C_a/C_w \qquad (12.7)$$

where C_a is the contaminant's equilibrium concentration in air, C_w is the equilibrium concentration in water, R is the universal gas constant (8.314 J/mol/K), and T is the temperature (K). Studies have shown that if the K_{aw} is greater than 10^{-4} (dimensionless), the chemical is more apt to be found in the gas phase rather than in the soil or water, K_{aw} between 10^{-4} and 10^{-6}, the chemical can be present both in the air and water, and for K_{aw} less than 10^{-6}, uptake would be by the water phase. The degree of volatilization is dependent upon the vapor pressure as well as the concentration in water.

Plant roots that reach the water table must do so through at least some thickness of unsaturated zone. The plant roots share the pore spaces with water and air. Because VOCs dissolved in water also can have high vapor pressures and, therefore, exist in the vapor phase in near-surface environments, the possibility exists that the vapor phase will be taken up by plant roots. This uptake could occur with the unsaturated zone or the water table being the source of the VOCs (Lahvis et al. 1999). We have already established that the necessary plant structures exist to carry atmospheric gases to the root zone. Even then, gaseous uptake into the cortex is one of two pathways for VOCs to partition, with the other being into the transpiration stream of the xylem. Moreover, some pesticides, such as the fumigants ethylene dibromide (EDB) and dibromochloropropane (DBCP), are designed to slowly volatilize over time after placement in the soil in order to most efficiently decrease the number of pests.

This partitioning also can be used to explain the transfer of organics from the leaf water surface to the atmosphere from inside the stomata after translocation within the plant. As was discussed earlier, gas exchange at the stomata occurs by diffusion and can occur in both directions depending upon the gas concentration gradients. The gain or loss of gases in the stomata is not simple, however, as the flow must first overcome resistances in gas conduction, as was discussed in Chap. 3.

12.1.3 Water–Soil–Air–Plant–Contaminant

The partition coefficients described above also can be applied to understanding the interaction of subsurface contaminants with plants. For example, the distribution of a solute between the water and lipid phases previously discussed provides a direct analogy with the potential fate of the water and lipids that comprise a plant, such as lipid-rich plant cell membranes or water in the xylem. Moreover, the movement of solutes in the transpiration stream also will partition onto plant membranes according to the passive equilibrium-based distribution of the solute between water and lipids.

12.1.3.1 Molecular Mass

The molecular mass of a contaminant compound also will influence its potential for diffusional uptake by plant roots. This is because entry to the xylem must be through membranes, either the cellular membrane and wall after symplastic uptake or the Casparian strip by way of apoplastic uptake. If an organic compound has a molecular mass less than 1,000, it can cross both of these boundaries, assuming an osmotic or diffusion gradient is present. Another chemical property that may be used to determine if a chemical could be taken up by plants is the molecular weight. Chemicals with a lower molecular weight tend to be taken up by plants at a greater rate than chemicals with a higher molecular weight.

12.1.3.2 Log K_{ow}

Because natural SOM can act like octanol in natural systems, the potential partitioning of organic contaminants onto SOM was derived by Chiou (2002) as

$$C_{pt} = kC_w(f_{pom}K_{pom} + f_{pw}) \tag{12.8}$$

where f_{pw} is the fraction of water in the plant, K_{pom} is the contaminant partition coefficient between plant organic matter and water, and $f_{pom} + f_{pw} = 1$, and f_{pom} is the fraction of organic matter (OM) in the plant. As such, if $k = 1$, then the chemical in the external solution will be in equilibrium with that in the plant, under conditions of passive uptake; if $k > 1$, then active uptake has to be assumed (Chiou 2002).

The contaminant concentration in water can be determined if the concentration of that released is known, through the relation

$$C_s = K_d C_w \tag{12.9}$$

where C_s equals the concentration of the contaminant in soil, C_w is the concentration in the water, and K_d is the partition factor of a compound between soil and water. But the concentration has to be normalized to the amount of SOM present by

$$C_{som} = C_s / f_{som} \tag{12.10}$$

where C_{som} is the SOM-normalized contaminant concentration in soil and f_{som} is the fraction of SOM in soil. The contaminant uptake by plants can now be reduced to the following

$$C_{pt} = k(C_{som}/K_{som})(f_{pom}K_{pom} + f_{pw}) \tag{12.11}$$

One observation of earlier researchers was that the results presented by Briggs et al. (1982) could not be reproduced accurately. This observation in part is explained by the relative difference in percent composition of the water, carbohydrates, and lipids in various plants and in different parts of the same plant. Li et al. (2005) set up a series of laboratory experiments to evaluate the effect of plant-lipid content on contaminant uptake.

In an extension of this uptake research, Briggs et al. (1982) found that the log K_{ow} was positively related to the uptake into and translocation throughout non-woody plants. They looked at the uptake of different classes of organic compounds that had different degrees of lipophilicity with respect to the test plant barley. A useful surrogate for lipids in water and plants is 1-octanol and is used to determine the potential for uptake relative to water. Essentially, a chemical that is hydrophobic will partition into the organic octanol. Chemicals that have low to intermediate log K_{ow} from between 1 and 3.5 are more likely to be taken up and translocated through a plant. Briggs et al. (1982) indicated that peak uptake would be for those compounds with log K_{ow} of 1.8. These chemicals tend to have moderate water solubility's, a trait that also makes them more likely to be placed on the USEPA's priority pollutant list.

Chemicals with a higher log K_{ow} and tend to be hydrophobic may enter the plant or be absorbed to the root tissues, but these chemicals may become bound in plant tissues, and lead either to bioaccumulation or to biotransformation, as will be discussed later in this chapter. Briggs et al. (1982) noted that this affinity for root uptake approached equilibrium rapidly, often being less than 48 h. Chemicals that have a lower log K_{ow} are anticipated to have low uptake into plants, because these compounds may be too soluble and therefore not be able to pass through the root lipids. This relation between the physical property of a contaminant and its potential for uptake and fate in plants is a predictive relation, not a deterministic one (Fig. 12.1).

In a classic study, it was observed that the organic chemical lindane was taken up by food crops such as carrots at higher amounts if grown in sandy soils relative to more organic-rich soils (Lichtenstein 1959). Essentially, uptake of lindane was decreased in the organic soils because of the increased partitioning of the organic lindane onto the organic matter present in soils, which decreased the concentration of lindane in the pore water available for uptake by the plant. This example illuminates that in a general sense, the difference in plant uptake is directly related to the physical and chemical properties of the contaminant, the physical and chemical properties of the soil it is growing in, and the chemical characteristics of the individual plant. The more a chemical is absorbed by organic matter in the soil, or onto roots in the soil, the less likely it will be taken up into the xylem by plants.

As would be expected from the results of Lichtenstein (1959), hydrophobic chemicals will be retained on soil organic matter and not be taken up readily by plants. This is especially true for those compounds whose log K_{ow} is greater than 3.5. As a result, such contaminants actually become part of the soil organic horizon and in this way become sequestered from the more labile cycling and flow of carbon. These contaminant carbon compounds also can be bound to the cell membranes in the apoplast or the Casparian strip of the endodermal tissue of plants. Hydrophilic compounds with log $K_{ow} < 1$ have a different fate—these compounds readily move through the plant apoplast, but not through cell membranes, so transport across the endodermis is limited. There are exceptions to this rule, such as is evident by the presence of MTBE in the transpiration stream of plants.

Some contaminants in groundwater sorb onto the immobile organics present in aquifers and, therefore, these contaminants are assumed to be less mobile relative to contaminants that do not sorb onto organics. In some cases, however, a portion of the organics present in aquifers are in the dissolved and more mobile phase—contaminants sorbed onto such dissolved organic matter (DOM) can be mobilized in groundwater, and the process is referred to as facilitated transport. The DOM consists of humic acids, which occur naturally as a result of production by both living and dead plants.

Groundwater contaminants, such as polycyclic aromatic hydrocarbons (PAHs), also can become more bioavailable (i.e., dissolved in water) for subsequent uptake by plants (Wilcke 2000). A laboratory method developed by Tao et al. (2006) utilizes accelerated solvent extraction with

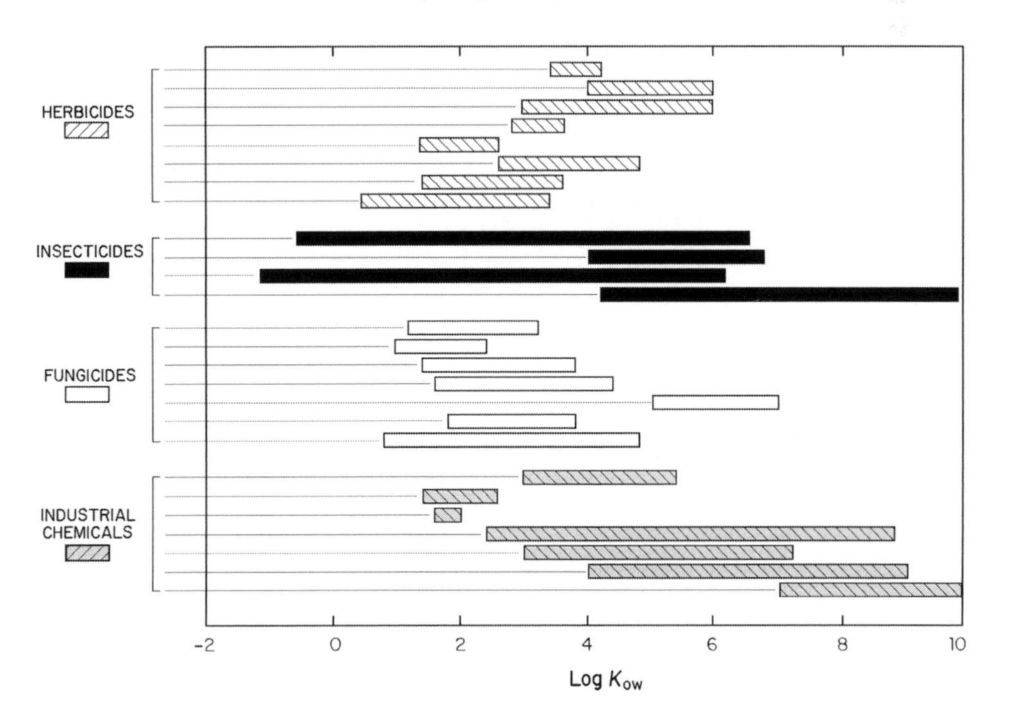

Fig. 12.1 Log K_{ow} for various organic compound classes that may be detected in groundwater.

water, *n*-hexane, dichloromethane, and acetone of soils contaminated with PAHs and can be used to relate the degree of extraction with potential for plant bioavailability and therefore uptake. These various preferential uptake or removal mechanisms will ultimately decrease the contaminant concentration at the root interface.

12.1.3.3 Root Concentration Factor

The efficiency of chemical sequestration into root tissues is called the root concentration factor, or *RCF*. This relates that entry of the contaminant into the roots through the vapor or aqueous phases:

$$RCF = C_{root}/C_w \tag{12.12}$$

where C_{root} is the solute concentration in the roots, and C_w is the solute concentration in water. *RCF* is essentially an experimentally determined bioaccumulation factor in which the concentration in the solution is related to the plant concentration. Because plant cell membranes are composed of a lipid bilayer, it acts to control the flow of substances on the basis of lipophilicity or hydrophilicity. Most compounds that are lipophilic can pass through the membranes more easily than the more highly water soluble compounds that are less lipophilic, as was experimentally demonstrated by Shone and Wood (1974). As might be expected, the *RCF* can be related to a compound's log K_{ow}, with higher *RCF* for compounds with higher log K_{ow}.

An equilibrium condition will be reached between the solute concentrations in the root with that in the solution. This condition will occur faster for root hairs and finer roots than for large tap roots because of higher surface-to-volume ratios for root hairs. Root interaction with contaminants can occur along the entire length of the root, but passive-diffusion-based uptake is limited to the unsuberized cell walls of the root hairs. For most nonionic organic solutes, this equilibrium will be controlled by diffusion. As this is occurring, the solute also will have a tendency to sorb onto the organic matter (lipids) present in the root itself. The extent to which this will occur is controlled by the log K_{ow} of the solute, with more absorption occurring with solutes of higher log K_{ow}. The relation between the magnitude of *RCF* and log K_{ow} was experimentally indicated by Briggs et al. (1982), as

$$RCF = 0.82 + 0.03{K_{ow}}^{0.77} \tag{12.13}$$

or

$$\text{Log }(RCF - 0.82) = 0.77 \log K_{ow} - 1.52 \tag{12.14}$$

In the case of MTBE (log $K_{ow} = 1.14$), the *RCF* is 1.04.

Specifically, in most cases initial uptake of a chemical can be described as following first-order, or concentration-dependent, kinetics. For example

$$dQ_{at}/dt = K_a(Q_m - Q_{at}) \tag{12.15}$$

where Q_{at} is the chemical amount retained on root surface at time t, Q_m is the chemical retained on root at maximum t, and K_a is the absorption rate constant. Integration of Eq. 12.15 gives

$$\text{Log } Q_m/Q_m - Q_{at} = K_a t \tag{12.16}$$

This equation can be used to estimate the potential for a chemical to be accumulated at concentrations in the plant above that present in solution. This also explains the derivation of the *RCF*. When the *RCF* is unity (1), the root concentration is equal to the external concentration. An $RCF > 1$ indicates that the plant has the ability for that compound to accumulate in it at a higher concentration than that external to the root.

The *RCF* can be considered as independent of concentration once equilibrium has been established between the contaminant in the subsurface and its concentration in the root. The uptake is not always dependent upon first-order kinetics, because it is possible for little uptake to occur if all sorption sites on the roots are filled (Collins et al. 2006), or if they are actively growing. Moreover, *RCFs* are given as constants for chemicals, but they may vary for different plants. Variation is expected because the lipid content of the epidermal cells differs between plants.

A potential shortcoming with comparing experimentally determined *RCFs* with field observations is the assumption that the log K_{ow} is constant, even if metabolites are formed in-situ. This assumption will lead to an underestimate of the *RCF* (Thompson et al. 1998). Also, when the *RCF* is determined, what part of the root is actually involved in the partitioning expression? The answer is not straightforward, because roots consist of organic material, such as lipids, but also of water and possibly gas in the cortex.

12.1.3.4 Transpiration Stream Concentration Factor

The efficiency of a solute to flow to shoots from the roots can be expressed in a manner similar to the *RCF*, called the transpiration stream concentration factor (*TSCF*) such that

$$TSCF = C_{shoot}/C_w \tag{12.17}$$

where C_{shoot} is the solute concentration in the plant, and C_w is the solute concentration in water. The *TSCF* is essentially an experimentally determined bioaccumulation factor in which the concentration in the solution is related to the plant concentration. A *TSCF* of 0 means no uptake, and 1 means 100% entry of the solute into the plant and describes conditions of essentially passive movement of a nonionic, chemically neutral contaminant, such as a conservative tracer with the flow of water in the plant. In most cases, the *TSCF* are less than 1 for chemicals that have lower log K_{ow}.

The *TSCF* is typically determined in hydroponic treatments in the laboratory. Most experiments measure *TSCF* by the proxy of the loss of solute in solution, rather than the gain of solute in the plant. The *TSCF* also can include some component of the *RCF*, so sufficient time should be allowed to allow equilibrium conditions to appear.

The *TSCF* is similar in concept to that of the bioconcentration factor (*BCF*), where

$$BCF = C_{plant}/C_w \tag{12.18}$$

This was investigated primarily because the translocation of chemicals taken up by the roots to the other parts of plants also can follow passive processes. Shone and Wood (1974), however, noticed that the concentration of the herbicide simazine inside the xylem of barley plants grown in solution was *less* than that in the stock concentration. They referred to this selection as a consequence of the *TSCF* (Shone and Wood 1974).

Studies performed by Briggs et al. (1982) extended this work to other pesticides and found that the *TSCF* was less than 1 for all compounds tested. They plotted a relation between *TSCF* and log K_{ow} and determined the maximum *TSCF* was when log K_{ow} was about 1.8, or intermediate lipophilicity (Fig. 12.2).

Burken and Schnoor (1998) state the *TSCF* for many contaminants taken up by poplar trees commonly planted at phytoremediation sites is

$$TSCF = 0.756[-(\log K_{ow} - 2.50) \times 2/2.58] \tag{12.19}$$

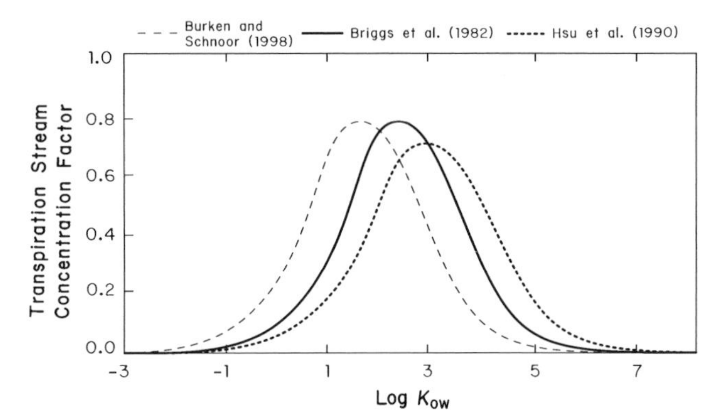

Fig. 12.2 Briggs relation between the chemical and physical properties of various xenobiotics (log K_{ow}) and influence on uptake into the transpiration stream (TSCF).

One possible explanation for the shape of the curve in Fig. 12.2, where the maximum *TSCF* is observed for compounds with intermediate lipophilicity, was given by Cunningham et al. (1996). Water can cross the cell membranes in the root stele easily before and after the Casparian strip by the symplastic pathway. The root tip has no Casparian strip, and water entry by way of the apoplastic pathway can occur. Compounds will move into the root with the bulk flow of water until these barriers are reached. At these barriers the compounds are sorbed into plant membranes, after which they must desorb for release into the xylem. It is the desorption step that causes more hydrophilic compounds with low to intermediate log K_{ow} values to be preferentially released into the xylem.

Another explanation is that the solute may not behave conservatively once in the transpiration stream. Many physical, chemical, and biological processes affect the solute concentration in the xylem, as described later in this chapter. Suffice it to say such losses can be accounted for by the following equation presented by Briggs et al. (1982) as

$$TSCF_{corr} = \left(TSCF_{appttk}\right)/\left[1 - \exp^{(-kt)}\right] \tag{12.20}$$

The *TSCF* calculated for given groundwater contaminants is provided in Table 12.1. However, for given contaminants and trees, the *TSCF* has been shown to vary (Struckhoff et al. 2005; Collins et al. 2006). For example, Davis et al. (1998) showed that for TCE, the *TSCF* can vary from 0.1 to 0.9 depending upon the plants exposed to TCE. These authors proposed that diffusion may be the cause behind these

Table 12.1 Log K_{ow} and transpiration stream concentration factor (TSCF) for common groundwater contaminants encountered at phytoremediation sites.

Contaminant	Log K_{ow}	*TSCF*
Benzene	2.13	0.71
Toluene	2.65	0.74
Ethylbenzene	3.13	0.63
m-Xylene	3.20	0.61
o-Xylene	2.95	0.70
MTBE	1.20	0.41
1,3,5-Trimethylbenzene	3.42	0.56
Naphthalene	3.37	0.56
Fluorene	1.98	4.18
Pyrene	5.32	0.03
Chrysene	5.61	0.02
Nitrobenzene		0.72
TCE	2.3	0.26–0.74
1,1,1-TCA	2.5	0.84
1,2,4-Trichlorobenzene	4.25	0.21
RDX	0.87	0.25
EDB	1.80	Unknown

different *TSCF* for the same contaminant and trees. They approached the *TSCF* issue by determining the partition coefficient for the compound TCE between the phases present: air, water, and wood. The dimensionless Henry's Law partition coefficient for TCE is 0.82, air–wood is 74 L/kg, and water-wood is 51 L/kg.

As was described in Chap. 3, water and solutes enter the transpiration stream after following plant entry through the symplastic or apoplastic pathways. Regardless of pathway, at the endodermis all water and solutes must pass over the cell membrane of the endodermal Casparian strip. It is here where the physical and chemical properties of the solute determine its potential for further entry into the plant's water system. This potential for entry has been predicted based on the log K_{ow}. In fact, the strength of the relation between log K_{ow} and plant-contaminant uptake explains the usefulness of *TSCF* as a master variable to describe contaminant entry into plants because it incorporates the differential entry of contaminants across the living cell membranes in the Casparian strip. Hence, the *TSCF* is a more defensible surrogate for membrane permeability than perhaps even log K_{ow}.

We saw in a previous section that solutes gain entry into plants by two main processes, passive uptake or active uptake. In Chap. 4, it was described how the scientist De Saussure in the early 1800s showed that even though solute uptake by roots is dependent upon the soil solution concentration, there was a selective membrane through which it must purchase entry into the plant's vascular system. Many chemicals are first taken up according to passive gradients controlled by diffusion. Additional uptake, especially in slowly transpiring plants, may be limited by diffusion, but assisted by plant metabolism. Passive uptake is suggested if the sorption (uptake) rate is a function of the solution concentration, as described by first-order kinetics. Active metabolism is indicated if oxygen becomes limited, or if the chemical concentration inside the plant exceeds that outside the plant. Active transport, on the other hand, involves the expenditure of plant resources to facilitate a reaction, usually against thermodynamic gradients. In sum, passive uptake processes occur along decreasing chemical potential gradients, whereas active uptake occurs against these chemical potential gradients.

Passive uptake, especially in the cortex cells outside of the endodermal cells of the Casparian strip, is related linearly to contaminant concentrations. If present in a system with no soil, just a water–solute solution, passive uptake can be written as

$$C_{pt} = kC_w \tag{12.21}$$

where C_{pt} is the concentration of the contaminant in the plant, C_w is the concentration of the contaminant in the external water, and k is the partition coefficient. Most groundwater contaminants are partitioned into the plant by passive processes, controlled by diffusion, water solubility, and the permeability of the endodermal cell membranes. However, the rate of uptake does not increase indefinitely, on account of saturation of available sorption sites within the plant. Ma and Burken (2002) showed that for TCE, there was a liner relation between the TCE concentration in the root solution and that measured in the transpiration stream of plant cuttings. They reported that when TCE concentrations ranged from 1 to 50 mg/L, the calculated *TSCF* was 0.26.

Just as there are limitations with the *RCF*, there are limitations with the *TSCF* that need to be kept in mind. The *TSCF* assumes equilibrium conditions and does not consider the impact of whether or not the contaminant is present in the subsurface as a dissolved or vapor phase. This is an artifact of initial development of *TSCF* for nonvolatile herbicides. Moreover, changes in concentration due to metabolism for rhizosphere degradation also are not considered.

Plant uptake from groundwater is directly related to the *TSCF* and the amount of water transpired:

$$\text{Plant uptake} = TSCF \times C \tag{12.22}$$

where *TSCF* is the transpiration stream concentration factor, plant uptake is the volume of transpired water (L), and C is the bulk pore-water concentration (mg/L). To account for the removal of contaminant mass in the root zone prior to uptake, this equation can be rewritten as

$$\text{Plant uptake} = C(1 - f)TSCF \tag{12.23}$$

where f is the fraction of contaminant degraded in the rhizosphere; f is 0 if no degradation is present and the resulting plant uptake is not decreased (Golpalakrishnan et al. 2007). Uptake of a contaminant that is dissolved in water in plants that are actively transpiring can be taken up by the advection of water.

The *TSCF* also assumes that the concentration of the compound remains unchanged as it moves through the xylem. This scenario most likely is the exception rather than the rule, given the ability for contaminant detoxification processes, as described later in this chapter.

12.1.3.5 $K_{water-wood}$

Xylem is a potential site for chemical and physical reactions to occur. The effect of these reactions on transpiration stream solute concentrations can be approximated by a water-wood partition coefficient, K_{lw}, where the l stands for lignin, as

$$K_{lw} = C_{dw}(\text{mg/g})/C_l(\text{mg/L}) \tag{12.24}$$

Where C_{dw} is the contaminant concentration in dry wood, and C_l is the contaminant concentration in the transpiration stream (Ma and Burken 2002). Similarly,

$$K_{wood} = C_{wood}/C_{water} \quad (12.25)$$

The partitioning of lipophilic compounds from water to wood is a function of the lignin content of the wood because lignin binds cellulose to provide structural integrity to wood, and is hydrophobic and, therefore, can attract lipophilic compounds as they move through the plant in the transpiration stream (Golpalakrishnan et al. 2009). The strength of this water-wood partitioning is directly related to the log K_{ow} of a compound and the lignin content of the plant in question. For example,

$$\text{Log } K_{wood} = -27 + 0.632 \log K_{ow}(\text{for oaks}) \quad (12.26)$$

and

$$\text{Log } K_{wood} = -28 + 0.668 \log K_{ow}(\text{for willows}) \quad (12.27)$$

An additional partitioning occurs as the water and solutes are moving through the transpiration stream: the presence of the large absorption potential of the wood itself. This includes the compounds of cellulose and lignin, as previously discussed, which attract those contaminants that are lipophilic. Trapp et al. (2003) calculated a K_{wood} partition coefficient

$$K_{wood} = C_{wood}/C_w \quad (12.28)$$

where C_{wood} is the chemical concentration in the wood, and C_w is the chemical concentration in the plant transpiration stream. As might be inferred, K_{wood} is similar to the K_{ow} in such a way that Trapp et al. (2001) reported a linear regression relation of

$$\text{Log } K_{wood} = -0.27 + 0.632 \log K_{ow} \quad (12.29)$$

12.1.4 Leaf and Tree Tissue Processes That Control Contaminant Loss

Leaves provide an interface between the water in the transpiration stream and moistures levels in the atmosphere. Leaves contain stomata, and these contain the invaginated wet layers of mesophyll cells where by diffusion CO_2 enters and O_2 and H_2O exit. Contaminants in the transpiration stream and water that supplies the cells in the stomatal opening may partition within the following compartments: the leaf lipids (internal–oils; external–waxy cuticle), water (external and internal–humidity, the xylem, cytoplasmic water, and water stored in vacuoles), or atmospheric gases–external and internal. For a volatile organic compound such as a groundwater contaminant, the leaf interaction with the atmosphere can drive the volatilization of a compound into the atmosphere. These partitioning processes will be discussed briefly here. The reader is referred to Riederer (1995) for a comprehensive treatment of this topic.

As has been discussed already, the partitioning of a chemical in solution to an atmosphere in contact with that solution can be described using the Henry's Law constant (Eq. 12.7). This transfer is based solely on the physical and chemical properties of the solute, independent of the fact that it is occurring in the living cells of the plant leaf. Another useful model for the concentration ratio of a contaminant between air and water at the leaf-air interface is

$$K_{aw} = C_{air}/C_{water} \quad (12.30)$$

Riederer (1995) presents an instructive model for relating the various physico-chemical properties of potential solutes to the potential for leaf concentration versus leaf loss, as

$$K_{la} = v_a + v_w/KAW + V_l K_{la} \quad (12.31)$$

where K_{la} is the gross leaf to air partition coefficient and V is the volume fractions of the compartments relative to the total plant leaf volume, where $v_i = V_i/V_t$ as a way of normalizing the magnitude of each process; leaf-cuticle partitioning is not included here. Representative leaf water-air partition coefficients, or log K_{aw}, range from -0.57 for toluene to -3.85 for methanol, where the log K_{ow} is 2.62 and -0.71, respectively (Riederer 1995). If there is a high lipid content of the leaf or low air volume in the stomata, the result will be an increased leaf concentration of the compound. K_{la} is inversely proportional to K_{aw} but proportional to K_{ow}. For a given K_{aw}, the K_{la} remains constant for compounds with log K_{ow}'s of less than 2.5. This fact is attributed to the dissolution of the chemical into the cellular water in the stomatal opening of mesophyl cells.

Plants' release of organic contaminants to the atmosphere as a vapor is an important mass-loss process and is based on diffusion. The extent of this diffusion of a particular contaminant after uptake will be controlled by the position of the stomata, the number and depth of the stomata, the contaminant concentration, and the diffusion coefficient of the compound to air through various tissues from the xylem-to-atmosphere pathway. These parameters can be used to calculate a stomatal conductance for that compound (Riederer 1995). Another approach is the following

$$K_{aw} = C_l(\mu g/g)/C_a(\mu g/L) \quad (12.32)$$

Where K_{aw} is the air-wood partition coefficient, C_l the concentration of organics in plants (in μg/g), and C_a the concentration in air (μg/L); this is similar to the Henry's law approach mentioned earlier in this chapter.

In order to examine the magnitude of diffusion of various groundwater contaminants from trees likely to be planted at phytoremediation sites, Baduru et al. (2008) provided the first measurements of the diffusional loss from excised tree tissue in the laboratory of common groundwater contaminants following advective transport upward. For each contaminant, the measured decrease in contaminant concentration from the xylem to atmosphere pathway was simulated using a 1-dimensional diffusion equation. The higher the molecular weight of the compound examined, the lower was its diffusivity (Baduru et al. 2008).

This relationship between contaminant fate and diffusivity can be used to calculate the potential for mass loss through tree tissues such as the trunk, stems, or branches. Baduru et al. (2008) suggest that the loss rate of a contaminant by volatilization can be estimated if the contaminant diffusivity, D (cm^2/s), diffusion pathway length, x, and the surface area, A, mass, M, and tree-tissue density, ρ, are related in the following expression

$$K_v = (A/M)(D/\Delta x)(\rho) \tag{12.33}$$

The use of this relationship is hindered by the range of values reported for the diffusivities of various compounds encountered at contaminated groundwater sites. For example, the reported diffusivities, in $D \times 10^{-7}$ cm^2/s, for TCE range from 0.01 to 25, for MTBE range from 1.78 to 8.00, and for benzene range from 0.80 to 2.98, and these were generated using hybrid poplar trees (Zhang et al. 1999; Ma and Burken 2002; Baduru et al. 2008).

The partitioning of an organic solute in water in the leaf also can partition onto the organic matter present in the leaf itself, such as lipids, membranes, among other polar and nonionic components of plant cells and tissues. The membranes of vacuoles are composed of lipids, as is most cell membranes. In sum, the resulting partition coefficient is an estimate of the tendency for the solute, or compound, C, to dissolve into the organic lipid, or solvent, O, or water, W, such that

$$K_{ow} = C_o/C_w \tag{12.34}$$

where K_{ow} is the octanol/water partition coefficient, C_o is the solute concentration in the octanol phase (kg/m^3), and C_w is the solute concentration in the water phase (kg/m^3). This can be related to the partitioning of solutes to plant lipids, K_{cl}, such that

$$K_{cl} = K_{ow} \tag{12.35}$$

12.1.5 Root Uptake of Contaminants in the Volatile Phase

Up until now, we have considered the entry of dissolved-phase contaminants into plants. Many priority pollutants encountered at sites characterized by groundwater contamination also will be present in the gaseous phase; this form of contamination also can enter plants and needs to be considered as part of the overall phytoremediation strategy.

Volatilization is the transfer of a contaminant from the liquid phase to a gas phase. We already saw how this process drives the hydrologic cycle as water changes phases from liquid to vapor. As was discussed in Chap. 2, the extent of a contaminant's ability to change state is a function of its vapor pressure, which essentially is a way to describe a compound's *gas* solubility, rather than the compound's *water* solubility.

An example perhaps better illustrates the potential for volatile groundwater contaminants to enter plants. Struckhoff et al. (2005) investigated the source of PCE detected in cores of tree tissues at a site near New Haven, MO. Both the subsurface soil and groundwater were contaminated with PCE. There was a stronger relation between PCE tree-core concentration and the soil PCE concentration than that of the groundwater PCE concentration. As such, it could be assumed that PCE probably entered the tree roots for subsequent partitioning into the water and then translocated by diffusion. At the contaminated site, the location of the trees cored was about 120 m (396 ft) from the Missouri River. Depth to groundwater was about 7 m (23.1 ft) but was higher if flooding occurred. The partitioning between water and wood for PCE was measured in the laboratory and found to be 49 L/kg. The partitioning between air and wood was 8.1 L/kg.

One of the confounding problems encountered in investigating the potential for volatile entry into plants is that the gas-phase exchange between the groundwater and the air from the unsaturated zone air is most likely not at equilibrium. This situation will be more prevalent at sites where the water table fluctuates in response to precipitation, *ET*, or river stage.

The uptake of gas solutes is primarily a passive process that depends on the movement of solute particles into plants imposed by concentrations gradients. This movement was related to the following equation by Fick in 1855:

$$D = -D_c \mathrm{d}C/\mathrm{d}x \tag{12.36}$$

where D is the net movement of particles across a unit area, D_c is the diffusion coefficient, and dC/dx is the concentration gradient. The D_c is proportional to temperature and inversely proportional to molecular weight. This diffusion can occur at the cellular levels as well as the tissue level. As pointed out by Trapp (1995), diffusion into roots can be approximated by diffusion into a cylinder (an 'ideal' root) such that

$$G_r A = 2L\pi D_e / \ln(R_2 - R_1/R_1) \quad (12.37)$$

Where G_r is the soil-root conductance (m/s), A is the root surface area, L is the root length, D_e is the effective diffusion coefficient, to account to a reduction in D_c caused by the tortuous pathways that must be taken before entry into a plant root on account of irregular-sized grains, R_1 is the root radius, and $R_2 - R_1$ is the length of diffusion between the root and soil or water matrix. D_e can be estimated from air- and water-filled pore volumes, according to

$$D_{we} = (\theta 10/3)/\varepsilon^2 D_w \quad (12.38)$$

$$D_{ae} = [(\varepsilon - \theta)10/3]/\varepsilon^2 D_g \quad (12.39)$$

where D_{we} and D_{ae} are the effective diffusion coefficients if the pores are filled with water or air, respectively, θ is the fraction of water-filled pores, $\varepsilon - \theta$ is the fraction of air-filled pores, ε is the total porosity, and D_w and D_a are the molecular diffusion coefficients in water and air. The amount of chemical available for diffusion also is constrained by the physico-chemical partitioning of the compounds in the medium.

The flux, D, across a membrane can be estimated as

$$D = -P(a_{\rm i} - a_{\rm o}) \quad (12.40)$$

where D is the diffusional flux (kg/m^2/s), P is the membrane permeability (m/s), and a is the activity inside, i, and outside, o, the membrane (Trapp 2003). As might be expected, compounds that have a higher lipophilicity tend to cross cell membranes with greater ease, so P is directly related to the K_{ow} of the contaminant.

Cho et al. (2005) investigated the ability for plants to take up and rapidly remove volatile organics such as TCE and PCE from contaminated soils above the water table by gas transport through the roots to the atmosphere. They compared under laboratory conditions the volatilization of such VOC in planted versus unplanted treatments. They reported that volatilization was faster and more complete in the unplanted treatments, and that the presence of plants (grasses) actually decreased contaminant volatilization. Other studies (Ma and Burken 2003) reported that the component of chlorinated solvents that were taken up by trees could be metabolized within the plant, rather than released to the atmosphere. In either case, the contaminant is removed from the contaminated subsurface media.

12.1.6 Uptake, Partitioning, and Transport Conceptual Models

This section brings together the physiological structures of plants in relation to water and vapor uptake outlined in Chap. 3, the chemical partitioning described earlier in this chapter, and the affect on the ultimate fate of the contaminants commonly found in groundwater.

Chemicals in the roots zone can enter the plant through the water or vapor phase. Simple uptake can be described as entry by diffusion. Plant roots exposed to water and soil air that contains a contaminant will take up the contaminant until equilibrium is established between phases. These interactions can be described by the various partition coefficients discussed above. The first process to consider from the plant perspective is the *RCF*. This in itself is related to the log K_{ow}. Keep in mind that roots are mostly water (85%) with few lipids (less than 1%). By now, it should be apparent that log K_{ow} is a master variable when it comes to contaminant fate in plants.

Chemicals can enter the plant through the apoplastic or the symplastic pathways. To enter the transpiration stream, however, the chemical has to be transferred into the symplast, i.e., through the Casparian strip of the endodermal layer of cells. As we saw earlier, the extent that this passage can occur is related to the chemical lipophilicity. The *TSCF*, therefore, is related linearly to the log K_{ow}. In a simple sense, the mass flow of solute, N_t (kg/s) is related directly to the flow of water in the plant, Q_w (m^3/s) and the concentration of the subject chemical C_w:

$$N_t = (Q_w)(C_w) \quad (12.41)$$

Because of the *TSCF* discrimination, the concentration of a chemical in the xylem C_x (kg/m^3) is

$$C_x = (TSCF)(C_w) \quad (12.42)$$

12.1.6.1 Mechanistic Models

Trapp et al. (1994) integrated most of the primary partition processes that occur between plants, air, water, soil, and contaminants into one model, called PlantX. It was reviewed as one of many models in a model comparison study by Collins and Fryer (2003). These discussions are for neutral contaminants only, not electrolytes or weak electrolytes. PlantX considers the diffusion of contaminants in water or air to roots using the partition coefficients of H and K_{ow} and the *RCF*, the movement of contaminants into the transpiration stream using the *TSCF*, translocation in the xylem, concentration changes caused by plant cell metabolism and growth, and the diffusion from the leaves into the air.

Trapp (2002) developed a simple dynamic model based on the advective rather than diffusive uptake of nonionic organics. The model is used to calculate the concentration of a particular compound if a transpiration rate and plant growth rate can be calculated or assumed. These determinations are based on a mass balance approach, similar to that presented for water budgets in Chap. 2. Essentially

$$Mass\,change(M) = Input(I) - Loss(m) \quad (12.43)$$

If the input, I, of a chemical is constant and the loss of mass, m, is proportional to the total mass, M, through a constant, k, then

$$\mathrm{d}m/\mathrm{d}t = I - k_m \quad (12.44)$$

The solution to this equation at any time t from initial conditions at time $t = 0$, m_o, is

$$M_t = m_o \mathrm{e}^{-kt} + I/k\left(1 - \mathrm{e}^{-kt}\right) \quad (12.45)$$

This model is dynamic up until steady-state conditions where $\mathrm{d}m/\mathrm{d}t = 0$.

Mass can be converted to chemical concentration by dividing the mass, m, by the total sample mass, M, or volume, V, such that

$$C = m/M \text{ or } C = m/V \quad (12.46)$$

Therefore, the mass-loss equation becomes a concentration-loss equation

$$\mathrm{d}C/\mathrm{d}t = I - km \quad (12.47)$$

12.1.6.2 Empirical Models

Alternative empirical-based models exist to evaluate plant uptake of neutral contaminants, including a simple regression-based model by Travis and Arms (1988) where

$$\mathrm{Log}\, B_v = 1.588 - (0.578 \times \log K_{ow}) \quad (12.48)$$

Where B_v is the bioconcentration factor for plants as the ratio of the chemical concentration in the shoots as dry weight divided by the concentration in the soil. The log K_{ow} of the organic compounds (mostly pesticides) used for the regression ranged from 1.15 to 9.35. The B_v was then converted to a BCF according to

$$BCF_{wet} = \left(B_{vdry}\right)(1 - W)\left(\rho_{wet}/\rho_{dry}\right) \quad (12.49)$$

where W is the plant water content and ρ the soil bulk density.

Another model is that of Paterson and Mackay (1995) based on the fugacity of a contaminant. They describe a system of models called SNAPS (Simulation Model Network Atmosphere-Plant-Soil). Both types of models are limited by their assumptions. Other factors that could affect the contaminant concentration are ignored in these models.

12.1.7 Plants and Contaminant Interactions

The uptake of contaminants released to groundwater such as organic solvents or petroleum hydrocarbons will tend to follow the passive uptake pathway of diffusion and osmosis and will be retarded due to the physical-chemical properties of the particular compound as it comes into contact with the organic parts of the plant. This retardation is important, as the bulk flow of water from soil to root hair to root xylem to xylem to leaf to atmosphere is continual as long as stomata are open. Passive uptake can be reduced to at least two phases: (1) the root cells come into equilibrium with the aqueous concentration in the pore water and (2) organic transfer to the cell walls of the root epidermis by sorption.

The uptake of water and contaminants was studied by Wild et al. (2005b). They point out that even though much has been revealed about the entry of extracellular water into the plant (see Chap. 3), little is known about how organic solutes are transported. Wild et al. (2005b) used two-photon excitation microscopy (TPEM) to observe the movement of solutes that exhibit auto-fluorescence inside plant cells. They observed the movement of anthracene through the epidermal cells wall through the cell membrane into the epidermal cytoplasm. The movement was diffusional, and rapid, with penetration into the cytoplasm of the epidermis observed within 72 h. Wild et al. (2005b) point out that their results challenge the dogma that organic compounds are limited to storage in lipophilic cell components and do not enter the transpiration stream and, therefore, cannot undergo transformation.

Plant leaves are exposed to the sun's radiation. The outer layer of leaf cells, such as the cuticle, acts to permit entry of particular wavelengths of light, much as the cell membrane excludes certain solutes but permits water to enter. Although harmful ultraviolet A (UVA) wavelengths are mostly attenuated by these cells through carotenoid compounds, some of this energy does penetrate deeper into leaf tissues. Wild et al. (2005a) showed that some of this UVA can be used by plants to photodegrade organic chemicals that enter the plant.

For the fate of contaminants in the subsurface, Wild et al. (2005a) looked at the entry of the PAHs anthracene and phenanthrene, into root cells, an extension of their work into the fate of foliar anthracene described above. Using TPEM techniques, they observed no PAHs in the root cap or apex but detected these compounds in the outer layer of root epidermal cells in these cells. In the root hair cells, the PAHs moved from the epidermal cells into the cortex. They report that the PAHs appeared to be concentrated into linear streams from the epidermis toward the vascular bundles. Some (5%) of the PAHs were observed in the vacuoles, but most was observed in the cell walls. No PAHs were detected in shoots. Wild et al. (2005a) also reported the presence of the degradation product of anthracene, anthraquinone, in the cortex, at levels up to 50% of the parent by the end of 56 days. They report that such degradation was observed in the older parts of the root cortex, not the growing apex. It would appear that only the more mature cells would contain the necessary enzymes to perform the degradation.

With respect to the entry of the PAHs in plant cells in relation to entry of water (discussed in Chap. 3), the studies of Wild et al. (2005a,b) are informative. They observed the uptake of anthracene and phenanthrene by root cells, and their results indicate the route of uptake was the apoplast, where movement is *between* cells through cell walls. The rate of uptake appeared to be related to the difference in compound solubility, where anthracene is 0.075 mg/L and phenanthrene is 1.65 mg/L. Uptake and movement of phenanthrene was faster than for anthracene. However, these compounds were not observed to reach the xylem. These results help explain the widely reported observation of decreased soil or groundwater concentrations, but little shoot collection of PAHs, with most collected in the root tissues.

As might be expected, there are very few cases in which the contaminant will interact only with a plant in the absence of soils. Hence, the pathway of a chemical from the unsaturated zone or groundwater solution to the vascular system of a plant has to include the interaction with the SOM, which can be explained as a series of partitions of the chemical in solution to the soil and plant water and organic tissues (Chiou et al. 2001; Chiou 2002). This is shown in the correlation between log K_{ow} and the *TSCF* reported in the Briggs et al. (1982) studies described earlier. Essentially, plant uptake is inverse to the soil organic-matter content, where high SOM equates to a lower uptake relative to lower levels of SOM. The initial driver that determines the extent of plant uptake of a contaminant is the interaction between the contaminant and water, and then the interaction between the contaminant and soil. Previously reported data exist to describe these interactions, or partitions, between water and soil and water.

Investigation of the interaction between plants and organic chemicals dissolved in water used for transpiration has proceeded like other questions regarding partitioning processes—the use of conceptual hypothesis and models followed by testing and documentation. The models look at the entry of compounds into root hairs, their transfer to the vascular system, their translocation to leaves, and the interaction with the atmosphere. Collins and Fryer (2003) point out that the conceptual models range from simple regression-based concepts that relate the concentration of a particular compound in a plant to the external concentration in the environment (deterministic, empirical model, based on lab data) to more complicated mechanistic models.

For an example of the first type, many studies and predictions of plant uptake are based on the physical property of the chemical such as log K_{ow}. In this case, the plant becomes the "octanol" for the contaminant dissolved in the water. Other properties include the relation between the molecular weight of a compound to its concentration in a plant after steady-state conditions have been reached. Collins and Fryer (2003) indicate that although these models are simple to understand conceptually, they are limited in that they do not consider all the other factors that might affect plant interactions with the contaminants. In fact, all of these models are based on equilibrium partitioning of the contaminant to the plant and the fact that this uptake is a passive process. We know that *pulses* of contamination often are the rule at contaminated sites rather than the exception and that steady-state conditions probably do not prevail in plants exposed to differing concentrations of contaminants.

Collins and Fryer (2003) provide evidence to suggest that the currently used empirical equilibrium approaches, such as *RCF* and *TSCF*, are inadequate relative to more dynamic-based models, because it is unlikely that the chemical in a plant will remain constant over time and that equilibrium is unlikely. However, these results are for short time periods, whereas most studies are based on longer term monitoring intervals that meet the assumption of equilibrium conditions for contaminant partitioning. This is particularly true when the source term of contaminants to the plant roots is rather constant, as often is the case for groundwater with residual contamination in source areas.

As water and contaminants travel up the transpiration stream to the leaves, they do so in the dead cells of the xylem. Some removal of contaminants can occur in this tissue following the log K_{ow} characteristics of the compound with increasing partitioning to the xylem with increased log K_{ow}. Keep in mind, however, that such high log K_{ow} contaminants may not enter the plant roots in the first place. The proximity of the xylem to the living cambium suggests that any contaminants removed in this chromatographic fashion may undergo transformation as these cells respire and carry out metabolism. These processes will be discussed later in this chapter.

Reactions that proceed from reactants to products that stop reacting once equilibrium has been reached are considered passive. In other words, the chemical potential established by a thermodynamic gradient of a particular compound is the same in each phase of a mixture, such as octanol and water. Processes based on concentration gradients only, however, will be satisfied once solubility limits are reached, and no further uptake will occur, and excess "free-phase" chemical will remain. No chemical energy is spent by a plant during passive uptake. Basically, the plant cannot control the entry of such compounds into its structure.

Computationally simple models exist to describe the fate of groundwater contaminants and plants (Schnoor 1997; Burken and Schnoor 1998). However, the simplifying assumptions necessary to use these models, such as constant contaminant concentrations, steady-state distribution, and no microbial biodegradation, can constrain their use to one of site conceptualization. Essentially, the uptake of organic contaminants dissolved in water by plants can be described by

$$U = (TSCF)(T)(C) \tag{12.50}$$

where U is the uptake rate of the contaminant (M/T), $TSCF$ is the transpiration stream concentration factor, T is the transpiration rate (L^3/T), and C is the concentration of the dissolved-phase contaminant (M/L^3). Values of the $TSCF$ range from inefficient uptake (low $TSCF$) for low-solubility compounds, such as pentachlorophenol (0.07), to efficient uptake (high $TSCF$) for high-solubility compounds, such as TCE (0.74).

The time required to reach remedial concentrations can be estimated, from first-order degradation kinetics, as

$$k = U/M_o \tag{12.51}$$

where k is the first-order uptake rate constant (per time, t), U is the contaminant uptake rate (M/T), and M_o is the initial mass of contaminant present (M). At any time, t, during remediation, the mass remaining in the aquifer can be determined by

$$M = M_o e^{-kt} \tag{12.52}$$

where M is the mass remaining (M) and t is the time (T). Solving for time yields

$$t = -(\ln M/M_o)/k \tag{12.53}$$

where t represents the time needed to reach a remedial action level (T), M is the mass allowed at time, t (M), and M_o is the initial contaminant mass (M).

12.2 Plant Rhizosphere Processes and Contaminant Fate

Water must often pass through the root zone as part of the hydrologic cycle. The dimensional extent of the root zone, as was discussed in Chap. 4, will vary considerably depending upon the type of plant and its location. Fibrous root systems typical of grasses will have a greater percentage of lateral roots, whereas taproot systems typical of woody plants will have a greater percentage of deeper roots. In sandy soils with naturally low amounts of soil organic matter, the mere presence of a root mass will increase the content of organic matter. As a plant grows and extends its below-ground space, relatively organic-poor soils become enriched by the root system over time. This enrichment is both a consequence of the presence of the root tissue itself as well as root exudates released by the roots as they grow and by the death of roots. For example, the root tip produces mucigel to aid its penetration into unrooted soil. Root cells slough off as roots elongate. Jordahl et al. (1997) reported that hybrid poplars grown in a lab incubator in sand released less than 0.30% of the total biomass as soluble exudates.

As plant roots move through soil, they tend to follow the path of least resistance, too, through the pore spaces that are most interconnected. This explains the relation between root cross-section and pore size. It is not surprising, then, that root-soil contact exceeds 60% (Kooistra et al. 1992). If water is a limiting factor and the water potential of that in roots decreases, then roots will shrink in diameter, and the extent of soil and root contact will also decrease.

The term "rhizosphere" reflects the confluence of the research in the nineteenth and twentieth centuries that described the interactions between microbes and the soil zone, as was introduced in Chap. 3. The extension of research into the increased numbers of microbes on plant roots in soil gave rise to the term rhizosphere by Lorenz Hiltner in 1904. Rhizospheric communities in roots are not limited to terrestrial plants. Aquatic plants have been shown to have increased numbers of microbial communities (Federle and Ventullo 1990). This increased number also is associated with the increased degradation by the microbes associated with cattail (*Typha latifolia*) roots relative to root-free sediments.

As we will see later this chapter, one of the selective reasons for the interaction of plant roots with bacteria and fungi is that their presence renders a greater degree of protection to the whole organism than by each alone. This dependency is similar to one in which the intestinal flora of animals can render imbibed toxicants harmless, to an amount that is equivalent to or greater than the amount the liver can process. This is because these microflora have similar enzyme systems as the mammalian liver.

Because contaminants also can be present in the root zone, degradation in the rhizosphere where contaminants are in the soils or unsaturated zone, has important implication to groundwater contamination, because groundwater is often contaminated by surface or subsurface releases to the soils above the water table. These soils become contaminated, and if not removed, this material becomes a long-term, residual source of addition groundwater contamination.

12.2.1 Microbial, Fungal, and Root-Zone Processes

It was observed that the root zone, or rhizosphere, is characterized by greater numbers and species diversity of microbes compared to areas that do not have plants (Paul and Clark 1989). A possible explanation for this difference is that the plant-root-released organic matter can support microbial metabolism, and this metabolism in turn may increase the bioavailability of essential minerals required by the plants. There are two general classifications of this observed association between plant roots and microorganisms. Microbes, in particular fungi, that are found to colonize the exterior parts of subsurface plant structures are called ectomycorrhizae. Fungi that colonize the interior of the plant roots within the cortical cells are called endomycorrhizae. Bacteria that colonize the interior of plants are called endophytes. The endomycorrhizae can be further divided into the arbuscular and ericoid mycorrhizae (Vosátka et al. 2006). The arbuscular mycorrhizae (AM) are present in more than 80% of vascular plants. They do not form fruiting bodies and, therefore, their presence is not always noticeable. The AM fungi have been show to be present in PAH- and PCB-contaminated soils. The ericoid mycorrhizae (ERM) are mostly ascomycetous fungi that inhabit the roots of *Ericaceae*, and are not widespread. Ectomycorrhizae (ECM) fungi can occur widely in trees. They consist of basidiomycetous and ascomycetous fungi; unlike the endomycorrhizae that produce invisible spores, these produce fruiting bodies that appear as mushrooms. These fungi appear to be able to use some fraction of lignin as well as cellulose as growth substrates. Therefore, they appear to be able to degrade more recalcitrant organic compounds, such as the multiple ringed (>2) PAHs.

The effect of these rhizospheric organisms on contaminants released to the subsurface is threefold. First, these rhizosphere microbes have an increased potential to degrade xenobiotic compounds (Fig. 12.3) because many plant exudates are structurally analogous to contaminant compounds (Fletcher and Hegde 1995; Reilley et al. 1996). Second, organic exudates from growing roots also stimulate cometabolic processes, which result in the fortuitous biotic degradation of chlorinated solvents. Jordahl et al. (1997) found that the numbers of benzene-, toluene-, and xylene-biodegrading microbes were higher in soil samples if poplar trees were present compared to adjacent soils (Fig. 12.3). Third, the presence of roots in a subsurface often devoid of sedimentary organic matter prior to planting will increase

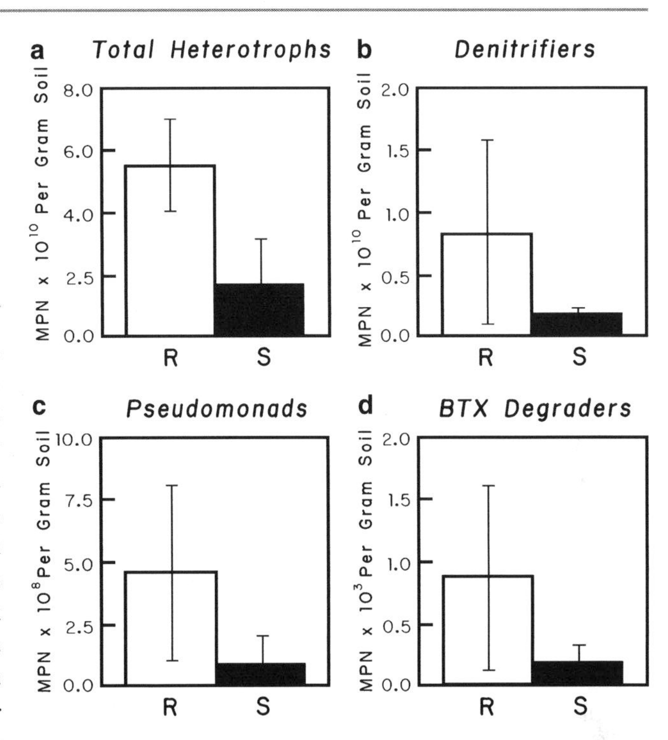

Fig. 12.3 Increased microbial activity in the rhizosphere of plants (R) relative to unplanted sediments (S). MPN is the Most Probable Number, an indication of the number of colony-forming units of bacteria growing in the culture medium. The 1-sigma standard deviation for the samples is shown.

the absorption potential of the soil (Brigmon et al. 1998). The increased number of microorganisms associated with plants does not necessarily mean a *de facto* increase in contaminant degradation potential, as the biodegradation of a particular class of xenobiotic compounds is ultimately determined by (1) contaminant bioavailability and (2) the appropriate enzyme expression for contaminant metabolism.

One of the first studies of the effect of enhanced degradation of chemicals associated with the rhizosphere was performed by Hsu and Bartha (1979). In their experimental design, rather than using a soil substrate for their test plants, which would confound their study on account of competing factors such as root sorption, they grew the plants in an aqueous medium (hydroponically). A study by Westover et al. (1997) suggested that with the herbaceous plants they studied (*caespitose*, *Festuca idahoensis*, and *Poa secunda*), plant composition may determine somewhat the degree of community structure of rhizospheric bacteria and fungi. The degree of this influence and its distance of regulation from the plant root will be examined below.

Along a similar line of reasoning, does the presence of contaminants induce in the rhizospheric bacterial populations an increase in the genetic capability to degrade these contaminants? The assumption is that plants that are exposed to contaminant compounds would have a selective advantage if the rhizospheric community increased its potential to detoxify or utilize these contaminant by gene expression of contaminant degradation potential; in other words, would the number of bacteria that contain enzymes to degrade contaminants increase? Rhizospheric bacteria may provide a selective advantage to those plants as far as the detoxification of deleterious compounds, such as allelopathic compounds. If these processes may also act to protect plants from xenobiotics, then the processes could be used for phytoremediation purposes.

A study by Siciliano et al. (2001) examined this issue further. As was stated in Chap. 11, plant encounter with allelopathic compounds could lead to an increase in those bacteria capable of degrading that particular compound, thus providing protection to the plant that had these bacteria. This may explain why studies with many compounds show an increased rate of mineralization and degradation in the rhizosphere that contains higher contaminant concentrations. Because this relation exists between plants for certain microbial communities, it is interesting to note the influence is stronger inside and near the root surface.

They (Siciliano et al. 2001) also examined the relative expression of catabolic genotypes for petroleum hydrocarbon degradation, such as *alk*B (alkane monooxygenase), *ndo*B (naphthalene dioxygenase), and for nitroaromatic degradation the genotypes *ntd*A*a* (2-nitrotoluene reductase) and *ntn*M (nitrotoluene monooxygenase) in unplanted soil, rhizosphere soil, and the endophytic part of roots grown at a contaminated site in California. They report that the *alk*B and *ndo*B genes were two to four times more prevalent in bacteria in the root itself than found in the soil and rhizosphere soil, respectively. In addition, they report that the *ntd*A*a* and *ntn*M genes were 7–14 times more prevalent in the root bacteria than in the soil and rhizosphere soil.

From the perspective of the phytoremediation of contaminated groundwater, these results present a dilemma of whether or not to inoculate plants with rhizospheric bacteria during plant installation. Studies have indicated, as mentioned previously, that contaminant remediation is increased in the soil zone in planted areas relative to unplanted areas. Will native bacteria colonize the installed plants, or should inoculants that contain the enzymes for specific contaminants be added to the soil during planting? A concern here is that funds will be used to add microbes but they may not be viable or may simply die off or be outcompeted.

Although it is generally accepted that there are more bacteria in planted soils versus unplanted soils, there are conflicting data about the effect of these bacteria on contaminant concentrations. Some research indicates that the rates of degradation in planted areas relative to unplanted controls are statistically significant, whereas others state the opposite, that the difference is not statistically significant. For example, Knaebel and Vestal (1994) reported that for plants in contact with agricultural chemicals, the degradation rates of these compounds was higher in planted versus unplanted soils, but the total amount of degradation was not statistically significant in comparison to unplanted controls. Brandt et al. (2006) reported that there was little experimental evidence to suggest that plants such as Vetiver (*Vetiveria zizanioides* (L.) Nash) added to petroleum-hydrocarbon contaminated sediments led to an expected enhancement of the root-associated biodegradation of these compounds present in contaminated soils. In fact, they report than both plant biomass and height were *decreased* in the presence of contaminated soils.

On the other hand, Nichols et al. (1997) presented evidence to support observations that the presence of plants leads not only to an increase in microbial numbers relative to unplanted areas, but the number of contaminant-degrading bacteria as well. They grew alfalfa (*Medicago sativa*) and alpine bluegrass (*Poa alpina*) in soil where one treatment was then contaminated with a mixture of organic chemicals and other treatment was not. The mixture consisted of hexadecane, 2,2-dimethylpropyl benzene, *cis*-decahydronaphthalene, benzoic acid, and pyrene, which are good analogies for BTEX and PAH groundwater contamination. After 9 weeks of testing, they determined that the number of organic-chemical degrading bacteria in the rhizosphere of the alfalfa-planted treatment was higher in the unplanted treatment that also was contaminated, or 4×10^7/g versus 6×10^6/g, and in the bluegrass-planted treatment there was 1×10^7 /g versus 1×10^6 /g in the contaminated versus uncontaminated soils, respectively.

In the early 1990s, intense investigation into the relation of bacteria and their enzymes systems and their effect on subsurface contaminants and remediation was in full swing. One of the first reports of the linkage between bacteria, contaminant degradation, and plants was an investigation by Donnelly et al. (1994). They reported that natural plant organic compounds, in their case flavonoids, also could support soil bacteria that can degrade PCBs.

An additional study designed to investigate the hypothesis that plant-exudates and plant roots can induce the expression of genes was performed with the contaminant naphthalene (Kamath et al. 2004). They reported the induction of the gene *nah*G by plant-released compounds such as salicylate, with no expression observed in the presence of root exudates. In fact, the converse occurred, where increased root extracts inhibited *nah*G expression. This suggests that the increased microbial mineralization of PAHs such as naphthalene in planted versus unplanted soils may be a consequence not of exposure to root exudates but of the exposure of the rhizospheric bacteria to the contaminant in the root zone.

Because the bacteria in the rhizosphere are associated with the roots of plants, the distribution of roots becomes an important factor in achieving remediation. Root density is different for different types of plants, and this knowledge can be used to add the correct plant to remediate different distributions of contaminants. For example, shallow contaminated sediment could best be remediated by grasses, which tend to have fibrous roots, where the roots are distributed in the upper parts of the soil column such as the O-horizon. Conversely, deeper sediment contamination or contamination of the water table or capillary fringe needs to come into contact with plant roots that are distributed more deeply, such as poplar trees and willows, or perhaps a deeply rooted prairie grass.

12.2.2 Rhizotron Methodology

Roots are inherently difficult to study directly because they are located underground. In many cases, to observe roots requires excavation or removal of dirt from around the roots *in situ*, both of which are invasive techniques. Noninvasive techniques, however, have been developed to enable root-growth patterns to be studied directly. One such method involves the observation of roots through an enclosure, or rhizotron, that consists of at least one clear panel. These macrocosms can be multiple feet in dimension, are necessarily expensive, and be used to monitor root growth for one or many plants.

Root-growth enclosures on a smaller scale more applicable to phytoremediation projects use smaller devices called minirhizotrons. These consist of tubes of clear material such as plastic, glass, acrylic, and butyrate, a few inches in diameter, that can be inserted into the root zone. A camera can be lowered down the tube to view the soil and roots in contact with the tube walls (Taylor et al. 1990).

The advantage of both methods is that the roots are not destroyed during observation and measurements at the same location over time can be made. One disadvantage of the minirhizotron method, that roots tend to grow in the space between the tube and soil, can be overcome using an inflatable minirhizotron as described by Gijsman et al. (1991).

12.2.3 Root-Zone Changes in Subsurface Sediment Chemistry

Roots can not only influence contaminants dissolved in groundwater but can also influence soil chemistry and, therefore, influence the fate and behavior of redox-sensitive solutes. For example, root respiration releases CO_2 as some percentage of gross primary production (GPP), the uptake of ions by roots alters the ion concentration remaining in solution, O_2 can enter and leave the subsurface, and roots can release organic compounds. A report by Sachs (1875) stated that a nutrient culture solution that contained anions such as nitrate became more basic as the nitrate was depleted. This would occur as the plant roots release bicarbonate ions (HCO_3^-) to maintain electrical neutrality at the root surface. In roots, this occurs according to the reaction

$$R-OH + NO_3^- + 8H \rightarrow R-NH_2 + 3H_2O + OH^- \quad (12.54)$$

As nitrate is reduced in the plant, hydroxide ion is released and must be excreted by the plant to maintain internal pH consistency. In shoots, it is possible for it to be excreted as an organic anion that may or may not be stored within the plant itself. The converse also can hold true for some plants that sequester more cations relative to anions, and these roots release hydrogen ions (H^+) to maintain electrical neutrality. This happens if nitrogen is made available as ammonium, such that as this cation is taken up, H^+ ion is released (Miller et al. 1970).

Plant root respiration releases CO_2 into the rhizosphere. This CO_2 can also lead to changes in soil pH, primarily when soils become flooded, because the removal of CO_2 would be limited to the solubility of CO_2 in water. In the unsaturated zone, CO_2 can migrate readily away from the root source by diffusion.

12.2.4 Release of Root Exudates and Increased Bioavailability

The release of organic substance by plant roots in the rhizosphere can be described as a passive or active process, depending upon when the organics are released. For example, the release of organic matter that follows plant death is a passive release. The active release of carbohydrates, proteins, sugars, tannins, mucigel, and ethylene occurs when bacteria exude the genes to promote the plant to release such organic material, such as during an infestation or allelopathic encounter. In this instance, the release of certain exudates may be a result of the plant needing the rhizosphere bacteria as protection from naturally occurring threats to the plant. The protection may be derived from the physical barrier the growth permits, or by actual secretions by the fungi, such as organic acids or chelators. Under natural conditions where plants are exposed to terpenes and alkaloid compounds as the result of allelopathic competition for resources, plant roots secrete pectins or lignitic compounds that act to sorb the toxicant prior to plant entry.

The amount of organic matter that is actively released by living roots is a subject of great controversy. On one hand, it is believed that plants actively release tens of percent of their GPP into the subsurface. However, the photosynthate used to create these exuded compounds is no longer available to support plant cell metabolism and growth. For this process to be a selective advantage to plants, there would have to be an equally important benefit to the plant to exude such high levels of organic compounds extracellularly. Using this argument, other researchers state that no more than 10% of photosynthate is excreted by roots (Gleba et al. 1999). The total amount of organic matter released will differ for different plants or similar plants under different environmental conditions, however.

As was discussed in Chap. 11, plants require the micronutrient iron for successful growth. This is in part because even though iron in not an element of the chlorophyll molecule, it is used in the synthesis of chlorophyll. Iron has to be in the soluble form for uptake, either as dissolved ferrous iron (Fe(II)) or as an Fe(III)-organic complex that is soluble.

For the plant-root released organic compounds to be able to complex with Fe(III), the organic acids have to contain oxygen, which acts as an anion (Luther et al. 1992). Both plants as well as plant-associated bacteria in the rhizosphere can produce organic chelates, such as siderophores. Luther III et al. (1992) reported that at a salt-marsh in Delaware, Fe (II) concentrations in shallow, anoxic pore water was highest during the summer (200 μM) and highest in the shallowest depths that coincided with the root zone of the *Spartina spp.* grown there. These results suggest that the increase in Fe(II) is a result of the increased loading of organics (ligands) from the marsh plants. The generation of Fe(II) can occur by nonreductive dissolution of Fe(III) or by reductive dissolution. In addition to root exudates, plants also release secondary plant metabolites into the soil (Singer 2006). These secondary plant metabolites typically contain isoprene, phenlypropene, alkaloid, or fatty acid structures.

12.2.5 Cometabolism

Metabolism refers to growth of an organism. It can be defined by the use of a substrate, such as carbon or nitrogen, as a source of energy or material to be added to the cell. In turn, the compounds are mineralized. These reactions all are enzyme specific. Conversely, substrate change can occur to compounds that are not enzyme specific and provide no energy or carbon source. This process is called cometabolism, a widely discussed but not completely resolved process. Essentially, cometabolism is the process assigned to observations that a contaminant can be degraded indirectly as the result of the microbial action on another compound that itself is used to support microbial growth or a source of energy. For example, bacteria produce an enzyme that is used to degrade compound A and also can degrade compound B—the microbe does not obtain energy or growth from this second reaction, however.

The most widely recognized process of cometabolism related to contaminated groundwater is done by methanotrophic bacteria that can oxidize methane using the methane monooxygenase (MMO) and reduce oxygen and simultaneously degrading TCE. This type of cometabolism also occurs for reduced organics, such as the fuel oxygenate MTBE, in which case alkanes are the initial compound oxidized. The relation of cometabolic processes to phytoremediation of contaminated groundwater is that plants and their assorted rhizospheric microbes keep conditions aerobic and through the plant-facilitated release of enzymes that induce the cometabolism of contaminants.

12.2.6 Endophytes

Plants not only have bacteria associated with their roots, but also in the tissues inside the plant as well. The application of these bacteria, called endophytic bacteria, to degrade contaminants in the transpiration stream has been investigated by Barac et al. (2004). If these results are reproducible, the role of endophytes in plants at groundwater contamination sites would emphasize the potential for plants to decrease the accumulation of bound contaminants or byproducts in plant tissue that occurs as part of the overall detoxification process.

12.3 Reduction-Oxidation Processes Controlled by Plants and Contaminant Fate

As was described in Chap. 11, a series of reduction-oxidation (redox) cycles occur in plants, such as during photosynthesis and respiration. Also, the electrons released during these reactions are then concentrated to form a potential gradient of hydrogen ions, within the cell, which helps reform ATP from ADP. The presence and absence of CO_2 and O_2, therefore, exert a major control on plant growth.

This same interaction of redox reactions and plants holds true for the fate of priority pollutants dissolved in contaminated groundwater. Some of these contaminants are present in the oxidized form, such as PCE, and some are present in the reduced form, such as benzene, and some are partially oxidized and reduced, such as DCE, TNT, and MTBE. While oxygen can enter contaminated groundwater through recharge events and diffusion, its low solubility in water and the presence of redox reactions rapidly depletes oxygen from the subsurface. This is why at many contaminated sites the groundwater chemistry is dominated by anoxic redox reactions, such as nitrate, sulfate, iron, or CO_2 reduction. Plants have adapted to this fluctuating availability of oxygen, regardless of whether their environment was rendered anoxic because of contaminant release or because of naturally high levels of BOD, such as in a swamp or wetland environment.

12.3.1 Influence of Oxygen, Carbon Dioxide, and Methane on Plants and Contaminated Groundwater

We saw from Chap. 3 that a plant will be adversely affected if too little water is available. Also, too much water can lead to the death of a plant where the rate of oxygen uptake by root respiration exceeds the rate of oxygen diffusion into water. It has been demonstrated that plants with up to 55% of their roots flooded by an artificially raised water table showed decreases in leaf growth within 4 days (Reicosky et al. 1985). Although this decrease in growth could be attributed entirely to increases in water potentials associated with the flooding event, an alternative explanation of the poor growth observed was the simultaneous decrease in ambient oxygen content due to the low solubility of oxygen in water.

12.3.1.1 Oxygen Diffusion into the Root Zone

Land-based plants, in general, have the majority of their roots located in the subsurface and, therefore, are surrounded by concentrations of oxygen that are equivalent to that in air (Conrad 1995). As a consequence of aerobic root respiration most plants, including phreatophytes, need to have a major percentage of their roots located in areas of the subsurface that are above the zone of constant water saturation. As the water table rises, or during flood conditions that affect riparian ecosystems, or in wetlands and swamps when the water level is high and oxygen concentrations lower, plant roots require oxygen by using other structures to remove this oxygen limitation. Moreover, the position of the water table and effect on oxygen conditions was found to reduce the growth of alfalfa (Bornstein et al. 1984).

Because we are discussing the flow of gases from the atmosphere to the subsurface though plants, some theory of gas flow is warranted. If the temperature of a gas is increased relative to that of the gas through a porous partition, the gas flow will occur from the warmer gas to the cooler gas; this is called thermoosmosis. This flow of gas was reported by Grosse (1989) for both wetland and riparian plants, such as alder (*Alnus glutinosa*).

The plant cortex works to overcome this oxygen limitation in the root zone as was described in Chap. 3 and is accomplished by oxygen diffusion from the atmosphere to the soil. The oxygen concentrations are kept low in the root zone by root respiration and consequently sustain diffusion-dominated oxygen transport from the atmosphere to the roots. Moreover, the production of CO_2 increases the concentration in the roots above that of the atmosphere, so CO_2 exits the root zone along a vertical diffusion gradient.

The best examples of this process and its implication for the phytoremediation of contaminated groundwater are found in the submerged macrophytes and trees, such as baldcypress, that transfer oxygen to the root zone as well as transfer methane, produced in the flooded, anoxic conditions near the roots, to the atmosphere (Armstrong and Armstrong 1987). The presence of this process in woody plants was investigated and reported by Armstrong (1968). Woody plants are found in swamps, and phreatophytes have roots that are in the water table at least some time of the year. In the case of the woody plants investigated, oxygen from the atmosphere was transported as a gas into the anoxic, waterlogged root zone, but the oxygen entered through the bark rather than the leaves. If the gas openings (lenticels) were covered with grease, then oxygen diffusion slowed down. In freshwater systems, the "knees" of baldcypress trees growing where the surface-water level fluctuates are extensions of lateral roots primarily located in the anoxic, bottom sediments, presumably to allow gas exchange to occur (Kramer et al. 1952), although this has not been proven unequivocally. The swamp cypress *Taxodium* has a series of protruding pillars of tissue that arise from the submerged root to above the water line. These are filled with air channels that are believed to transport oxygen to the other parts of the roots.

In the saline mudflats of Florida, for example, mangroves (*Avicennia*) have roots in anoxic mud but also have roots that grow off the main stems that project into the air called pneumatophores (Scholander et al. 1955). In such marsh conditions, the concentration of oxygen in the pore-water of the sediments is low. This is because any oxygen that enters the upper layers of sediment after each tide is rapidly consumed by a thin layer of aerobic organisms, as well as by abiotic oxygen consumption by reduced mineral oxidation. Mangroves transport oxygen to the anoxic subsurface in coastal areas along Florida, for example. The grey mangrove (*Avicennia marina*) found along coastal Australia accomplishes this transport through gas-filled pneumatophores that store oxygen during periods of low water level (tides) and transmit oxygen to the roots during periods of high water. In most cases, the transport of oxygen from the atmosphere into the pneumatophore through lenticels is a passive process driven by diffusion. Skelton and Allaway (1996) observed the time course of oxygen concentrations measured inside mangrove roots during unflooded and flooded conditions. They reported that during low water conditions oxygen entered the exposed roots. Following flooding during high tide, the oxygen concentration decreased but remained at levels for 10 h that could support aerobic respiration.

However, plants roots need a continual source of oxygen in order to live and grow. These limitations also affect

Spartina, as well as other plants whose roots are present in anoxic environments, such as freshwater marshes and swamps. As would be expected in such reducing sedimentary environments, the transport of excess oxygen away from the roots would oxidize any reduced species such as manganese and iron. The measurement of a plant's ability to transport oxygen into reduced sediments away from its roots has been called radial oxygen loss (ROL) (Michaud and Richardson 1989). Plants that have this trait to transport oxygen include cattail (*Typha latifolia*) and rush (*Juncus effusus*).

Wieβner et al. (2002) investigated the amount of oxygen released into solution in contact with the roots of several aquatic plants. All were found to release oxygen, at rates between 0.01 and 1.41 mg O_2/h, until fully oxidized *ex-situ* conditions were reached. Consumption of oxygen outside the roots is the main factor that determines the distance that oxygen will travel from the roots, as well as the concentration at a particular location. For microorganisms, the concentration of dissolved oxygen in pore water is the upper limit to define aerobic versus anaerobic pathways. The more reducing the solution the more oxygen is released. This process of oxygen release will affect the rhizosphere and, potentially, the redox status of shallow contaminated groundwater.

One of the factors that affects the presence of oxygen in the root zone is the level of the water table. This also can occur when surface-water levels are high, as during a flood, when the normally dry (and aerated) flood-plain vegetation becomes inundated. Kozlowski (1997) provides a review of the many processes that are affected when a plant experiences flooded conditions. These include a decrease in photosynthesis through stomatal closure, lack of mycorrhizae (which are aerobic, for the most part), and an increase in root decay. In flood-tolerant plants, however, photosynthesis and growth can remain unaffected, because such plants have adapted by the development of lenticels, thick aerenchymal tissues that allow atmospheric interaction between the roots and the air, and adventitious roots. The effects of flooding are more severe during the growing season relative to the dormant season.

12.3.1.2 Carbon Dioxide and Methane Diffusion from the Root Zone

Plants cannot only transport atmospheric oxygen to roots in anoxic environments but can transport reduced gases from the subsurface to the atmosphere. This has been previously demonstrated for aquatic macrophytes, such as rice, that have roots in anoxic mud. The methane produced by microbial methanogenesis, often seen as bubbles that rise when such sediments are disturbed, can also exit to the atmosphere through these plants. As such, these plants can act as gas conduits or shunts. In general, 80–90% of the methane generated in reduced wetland sediments exits the subsurface through plants through the aerenchymal tissues. Moreover, the release of methane is not constant throughout the day but has been reported to be higher in the early daylight hours than at other times for *Typha domingensis* and *Typha latifolia* in the Florida Everglades (Chanton et al. 1993).

If the oxygen enters through stomata or lenticels, how does the CH_4 exit? In general, CH_4 emission does not appear to be related to stomatal opening or closure (Whiting and Chanton 1996). Conventional thinking about methane movement and release to the atmosphere through aquatic plants is that it is driven by changes in soil and air temperatures. Peak emissions coincide with peak daily temperatures. These data suggest that diffusion is the driving force in gaseous transport from the anoxic sediment to the atmosphere through plants. Gaseous movement processes include diffusion (driven by concentration gradient) and convection (driving force of high gas pressure to low gas pressure). Peak gas transport emission did not correlate with peak stomatal conductance.

Two types of gas-flow exchange occur in plants that grow in anoxic sediments. The pressurized system relies on thermal transpiration, or the pressure difference between the higher oxygen pressures in the younger leaves in the atmosphere to the lower pressures in the roots underground and back to the atmosphere through older or dead leaves. Passive diffusion relies on molecular diffusion of gas molecules (a partial-pressure gradient).

Although CO_2 is produced by the roots during respiration, CO_2 also is produced by the other living cells in the plant, such as the trunk, stems, and cambium. At any given time throughout the year, both oxygen and CO_2 can be present in these tissues. During the winter for most trees, including conifers, these tissues contain lower concentrations of CO_2 but higher concentrations of O_2, as the lower temperatures and light levels decrease photosynthesis. Cores collected from decaying trees can emit gases that can be ignited when exposed to a flame. This phenomenon can occur in trees afflicted by the condition known as wetwood and is often found in cottonwood trees. It is an infection of the heartwood, usually by anaerobic soil bacteria that have entered the roots. These anoxic bacteria essentially ferment the wood and release fatty acids, which putrefy and are expressed to the bark under localized pressures (Hiratsuka 1987). These observations support the notion that other than the openings of lenticels, the bark of most trees is relatively impermeable to gaseous transport.

If roots reach groundwater rendered anoxic by the presence of reduced organic contaminants, CO_2 also will be present due to microbial oxidation of the contaminants. In either case, this CO_2 can be taken up by roots in the dissolved phase. This CO_2 can then be translocated in the xylem to other parts of the plant, and possibly be used as a source of carbon for reduction. The partial pressure of CO_2 in the subsurface unsaturated zone (soil pores) will be higher than that in the leaves as it is removed there to drive photosynthesis. In fact, Levy et al. (1999) reported partial pressures, as pCO_2, from 3,000 to 9,000 Pa within the woody stems of trees. They also reported that this transport of CO_2 in stems to leaves was equivalent to less than 7% of leaf fixation rates.

Wium-Anderson (1971) showed that CO_2 from sediment sources (as free CO_2 rather than HCO_3^-, the form used by higher plants) increased carbon-fixation up to five times more than CO_2 from the water in the hydrophyte *Lobelia dortmanna*. The absorbed CO_2, presumably from anaerobic microbial processes in the sediment, is transported inside the plant to the leaves, where it is fixed. In turn, the authors stated that a zone of oxidized iron in the bed sediments exists around the roots to a depth of 20 cm. This iron oxidation must not have detrimental impacts on iron uptake, since reduced iron is not likely to be limited in reduced sediments. Moreover, the presence of a large root surface area to above-sediment biomass in submerged aquatic macrophytes suggests the ability to take up dissolved CO_2 from the sediments.

An excellent example of this increased below-ground growth and CO_2 root absorption is expressed by, of all things, a terrestrial plant, *Stylites andicola* (Keeley et al. 1984). *S. andicola* is found in Peru in fens in higher ground adjacent to bogs in organic-rich peat areas. Two-thirds of the total biomass is underground. They lack stomata in the leaves, even though they produce an evergreen rosette. They have the ability to undergo crassulacean acid metabolism (CAM), with CO_2 uptake and conversion to organic acids at night, and with the reformation of CO_2 during the day for fixation as carbohydrate. This is similar to desert plants that during that day must tightly close their stomata. Under these conditions, CO_2 is taken in during the night, and fixed into simple organic acids, using CAM. The acids are stored in vacuoles until daylight, when the acids are decarboxylated into CO_2.

In one way, the acquisition of CO_2 by plants from subsurface sediment sources rather than atmospheric sources makes sense. The atmospheric concentration of CO_2 is low, no more than 0.030%. Conversely, it can be very high in anoxic sediments, where the degradation of organic matter leads to CO_2 production. Roots are present in this CO_2-rich media (Raven et al. 1987). This process is similar to the recycling of root-respired CO_2 back into organic carbon within the plant.

As stated previously, CO_2 is necessary for photosynthesis; terrestrial plants take up CO_2 from the atmosphere, and aquatic photosynthetic organisms take it from aqueous solution. Willows represent a phreatophyte that can be planted at a contaminated site where groundwater contains low DO and high CO_2. In these plants, even though the majority of the CO_2 is derived from the atmosphere, between 1% and 2% of the carbon taken in by leaves is taken in by the roots (Vuorinen et al. 1989). Essentially, the CO_2 is transported to the plants parts in a manner similar to that of other dissolved substances. An investigation by Brix (1990) also stated, using a ^{14}C-radiotracer study, that less than 1% of the total plant carbon-fixation, in *Phragmites australis*, was derived from sediment pore-water CO_2 in the root zone. These plants grow in anoxic mud and have extensive root systems. The internal concentration of CO_2 in such plants can exceed 8% of the total volume of air space.

Too much CO_2 in the soil gas of the unsaturated zone can be lethal to plants, however. At Mammoth Mountain, CA, volcanic activity has sent large plumes of CO_2 gas upward, and it has collected in the soil gas of the unsaturated zone. This was discovered when scientists were investigating the death of trees in an area of about 100 acres around the mountain (Sorey et al. 2000). Concentrations of CO_2 were present in the soil gas at concentrations greater than 20–95% of the soil-gas volume. Such high concentrations of gas also may be found in the unsaturated zone above groundwater plumes. The presence of CO_2 can be monitored with a soil-gas probe lowered into holes dug above the water table. This and other measurements are discussed in Chap. 15.

12.3.1.3 Plants, Gases, and Groundwater Contamination

What do these gas-transport processes in situ have to do with the phytoremediation of contaminated groundwater? The interest in gas transport in plants that remove groundwater from contaminated aquifers is twofold. First, most aquifers contaminated by petroleum hydrocarbons contain little dissolved oxygen. This is detrimental to plant root survival, as well as the efficiency of degradation of contaminants under aerobic processes. Thus, plant mechanisms to supply roots in anoxic environments with sufficient quantities of oxygen to support respiration may have positive consequences for contaminant degradation. Also, contaminated groundwater can become methanogenic, and the release of methane to the atmosphere can occur, and perhaps hydrogen sulfide, which can be toxic to plants at low concentrations. For aquifers contaminated by oxidized contaminants like PCE or TCE, oxygen limitations might not occur. In this case, the presence of plants and their release of root exudates could potentially render the groundwater anoxic and support reductive dechlorination (Eberts et al. 2005).

An elegant study by Wießner et al. (2002), having implications for phytoremediation of contaminated groundwater, was to determine the release of oxygen from the roots into the rhizosphere. Wießner et al. (2002) were able to vary the redox status of vials of plants growing in hydroponic solution by differential titration with titanium (III) citrate. Redox measurements in the hydroponic solution were correlated to dissolved oxygen concentrations in the solution by creating a calibration curve of different titanium (III) citrate concentrations relative to redox status measured as oxidation-reduction potential (ORP). The oxidation of titanium (III) citrate to titanium (IV) citrate was determined by adding different amounts of oxygen-saturated water to the media. They couldn't measure the oxygen zone directly but could do so indirectly by the measurement of titanium (III) citrate oxidation. Measurement of this was facilitated by mechanical stirring in the hydroponic solution, which displaced the oxygen produced near the roots into the bulk media.

Wießner et al. (2002) reported that under all initial redox conditions, oxygen was released by all the *Typha latifolia* plants studied. From initial reduced conditions of Eh near −400 millivolts (mV), within 6 h all plants studied had released enough oxygen to increase the Eh to millivolts above 0, or final near +300 mV, and dissolved oxygen concentrations as determined from titanium (III) citrate oxidation went from 0 to 0.9 mg/L in one plant, and to 0.5 and 0.3 mg/L in the other two plants. Under conditions of different initial redox status (from highly reduced to less reduced), the production of oxygen was higher when the redox was strongest (lowest Eh) and lowest when redox was lower (higher Eh). At the end of 24 h oxygen production ranged from 1 to near 8 mg/L. Moreover, the release of oxygen was continuous, even after oxygenated conditions had been established.

Because the zone of oxygenation near roots is so small, the impact on driving the aerobic microbial oxidation of reduced organic contaminants will also be small, although not insignificant. The main advantage for phytoremediation projects from the root-zone release of oxygen is the initial establishment of the young roots of plants often added to contaminated aquifer sediments that have no source of oxygen other than recharge. However, too much oxygen can increase the rate of all oxidative processes and, therefore, can lead to the depletion of cellular oxidative enzymes.

Many of these processes of diffusion gas transport have not been examined in poplar trees, however, and such studies would shed light on how poplar trees can use groundwater rendered anoxic due to high levels of contamination (Eberts et al. 2005).

12.3.2 Naturally Anoxic Aquatic Environments and Contaminant Fate

In the subsurface or in aquatic sediments, heterotrophic microbes can reduce electron acceptors such as nitrate, iron, manganese, sulfate, or CO_2 even after oxygen is depleted. These processes were facilitated by plants as they produced oxygen in the early earth and forced microbes deep into the sediments to avoid oxygen toxicity. Contaminants that are reduced can be oxidized using these various alternative electron acceptors, although the rates of mineralization are slower than if oxygen was used. Contaminants that are already in the oxidized state, such as chlorinated solvents, also can be used as electron acceptors in a manner similar to those above after oxygen and alternative electron acceptors have been depleted.

12.3.3 Plant-Induced Redox Changes at Contaminated Groundwater Sites

The interaction of plants and the redox reactions that occur in surface- and groundwater is perhaps best shown by an example. Take for instance dissolved iron. The NSDWS MCL for iron is 2,000 mg/L in surface-water systems. At a site near Cecil Field, FL, concentrations of iron were near the MCL in surface water fed by iron-rich groundwater. The presumed source of the observed high iron concentrations was the reduced conditions in the ambient shallow groundwater that discharged to the surface water, as the spring was downgradient of a landfill. A natural forest of oak trees, however, also was located between the landfill and the iron-contaminated spring. The trees probably contributed to the high concentrations of iron in the groundwater as a result of the release by the roots of high concentrations of organic matter that leached to the water table. The resultant high natural source of high BOD caused iron-reducing conditions in the shallow groundwater, the release of mobile ferrous iron, and transport to the spring. As such, the source of the iron in the surface water was from the natural input of organic matter from plants, rather than the landfill, the suspected anthropogenic source.

A similar process of plants rendering a shallow oxic aquifer to anoxic conditions, and its effect on TCE degradation, was reported by Eberts et al. (2005). They report that where cottonwood trees were planted over a shallow, aerobic aquifer contaminated with TCE, after 6 years of tree growth, the dissolved organic carbon (DOC) concentration in the shallow aquifer increased—this resulted in a lowering of dissolved oxygen concentrations and anoxic conditions. The anoxic conditions led to the reductive dechlorination of TCE following the establishment of iron- and sulfate-reducing conditions, and finally methanogenesis (Eberts et al. 2005).

Eberts et al. (2005) also report that as the DO content of the shallow aquifer decreased, the ratio of the mean concentrations of TCE/*cis*-DCE also decreased. Because *cis*-DCE is derived from the reductive dechlorination of TCE, a decreasing TCE/*cis*-DCE ratio indicates the conversion of TCE to *cis*-DCE. Increased bacterial numbers, determined by the most probable number (MPN) method, were observed for methanogenic bacteria capable of degrading

TCE to *cis*-DCE beneath an established cottonwood tree at the site. The tree was located in an area where the dissolved phase plume of TCE had been transported and oxygen conditions were lowered, presumably by the release of organic compounds by the tree (Godsy et al. 2003). The more oxic, uncontaminated aquifer by comparison was populated by aerobes, fermenters, and denitrifying bacteria, low numbers of iron and sulfate reducers, and no methanogens. DOC concentrations in the phytoremediation area ranged from 0.8 to 1.8 mg/L, enough to support methanogenesis, but not enough to decrease ambient nitrate and sulfate levels to levels below detection, suggesting that the aquifer is carbon limited, with respect to reduced carbon. Iron reduction also produced dissolved iron, which was detected in groundwater in the phytoremediation area.

At many sites, dissolved oxygen is rapidly depressed following the release of the high BOD characteristic of reduced petroleum hydrocarbons, such as BTEX and PAHs. Rentz et al. (2003) investigated the possibility of increasing the DO content around roots in high-BOD contaminated sediments and its effect on the growth of plants that might be planted at such sites to remediate the contaminated soils. This is significant, because many of the plants that could be planted at contaminated sites, such as poplars, may not have the gas-transport structures (aerenchyma) to transport the O_2 that aquatic macrophytes or other phreatophytes have. Additionally, as was shown at the site in Texas, poplar trees can decrease the DO by their release of labile organic matter. So, poplars can be a source *or* sink for DO in contaminated aquifers and vadose zones.

At their study site, Rentz et al. (2003) report the rapid disappearance of O_2 in soil gas with depth. Less than 1 m was required for complete removal, and the concomitant appearance of CO_2 and CH_4. As was reported above, CO_2 can be toxic to plants at high concentrations. The DO was depressed because the total petroleum hydrocarbon (TPH) concentration ranged from 820 to 11,000 mg/kg. Even at these concentrations, no toxic effects were observed in laboratory column incubation studies using poplar cuttings (*Populus deltoides* × *nigra* DN34). The researchers compared various inexpensive methods to deliver oxygen to the contaminated unsaturated zone in column studies in the laboratory, relative to unamended ambient conditions. These methods included aeration by insertion of a perforated tube into the smear zone, addition of high porosity gravel, use of proprietary oxygen releasing compounds in filters, as well as increased drainage.

The effect of these various methods of O_2 delivery was compared using net biomass—the biomass at the end of experiment relative to the biomass at the beginning. Net biomass was higher at the end of the experiment in the columns where a proprietary oxygen releasing compound (ORC®) was added in a filter in the smear zone, relative to the control and other treatments, for a 2.46-fold increase over the control. No nutrients were added during this experiment, so the researchers concluded that the observed increase in biomass was due to increased oxygen content in ORC® treatment. No oxygen content was measured, however. Also, ORC® contains high concentrations of PO_4, so the possibility exists that the release of this nutrient was partially responsible for the increased biomass in the control pots. Moreover, ORC® was originally manufactured in England to serve the needs of gardeners' attempts to increase the oxygen and phosphate contents of their characteristically clay-rich garden soils.

12.3.4 Water-Table Fluctuations and Contaminant Fate

The position of the water table affects and is affected by the local sources and sinks of water and also can affect contaminant fate. The water table can rarely be described by a single location in space or time. As was discussed in Chap. 4, it changes in response to the balance between the sources and sinks of water, such as recharge and *ET*.

Water-table fluctuation changes the composition of air and water in the soil pores of a sediment profile. Sediments located above the mean water table will contain air and water under tension. As the water table rises, the air will be displaced as the pore spaces fill with water. As the water table falls, some of the water drains from the pores by gravity, causing air to reenter. This change in gas composition will impact the fate of contaminant compounds in the soil, water, and gas phases near this water table and unsaturated zone interface. Contaminants such as benzene or MTBE in the dissolved phase during higher water-table elevations may diffuse into the air space between pores when the water table drops. This accelerates removal from groundwater and will increase the potential that these gas-phase contaminants will be biodegraded aerobically (Lahvis et al. 1999), volatilize to the surface, or be taken up by plant roots. If the water table is lowered by plant uptake of groundwater, this may increase biodegradation of contaminants such as naphthalene (Anderson et al. 2008), or BTEX (J. Burken, University of Missouri, oral communication, 2007).

12.4 Plant Biochemical Processes for Groundwater Contaminant Degradation and Detoxification

We saw in Chap. 11 that plants produce allelopathic compounds as an offensive act to survive in a competitive world or render themselves inhospitable to herbivory. Plants also must act defensively, however, in order to protect themselves from the allelopathic attacks from other plants or organisms. This kind of low-grade chemical warfare may explain the interaction between plants and the flora present in the rhizosphere, i.e., a plant can selectively "choose" which bacteria are present in the plant's rhizosphere based on exposure to deleterious chemicals, such that these bacteria will possess the necessary genes to code for degradation enzymes. Therefore, the interaction between plants and allelopathic compounds provides a natural analogy to the interaction between plants and xenobiotic chemicals found in groundwater.

Above ground, as plants respond to transpiration demands, water that enters the roots moves upward through the xylem to exit the leaves. During this transport, any compounds dissolved in the water have the potential to interact with the plant tissues of the vascular system according to the physical and chemical properties of the contaminant discussed earlier in this chapter. Some of these compounds can have deleterious effects on the living cells that surround the xylem. Cell culture experiments have indicated, however, that the cambium cells of plants can interact with xenobiotic compounds in the xylem and can transform them into less toxic compounds (Sandermann et al. 1977; Sandermann 1994). This is similar to how the blood in mammals must pass through the liver for detoxification.

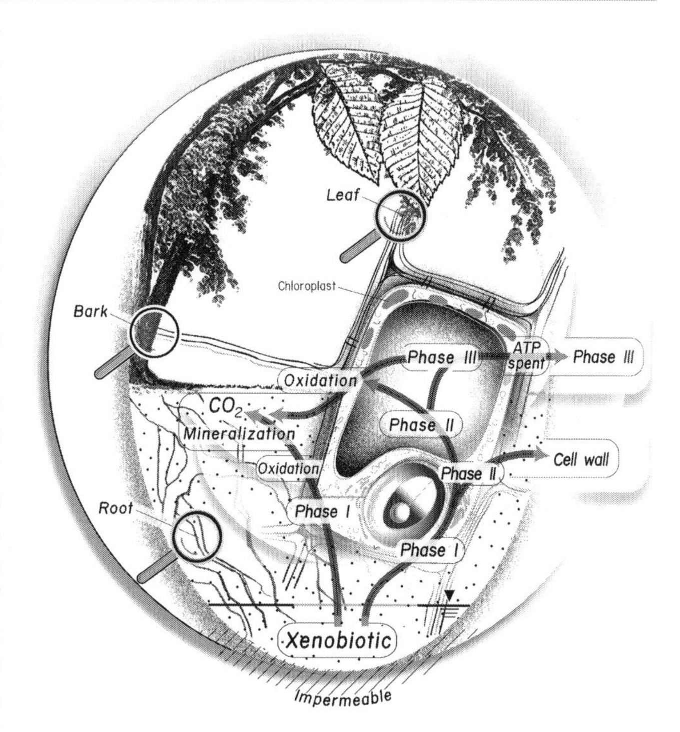

Fig. 12.4 Plant xenobiotic detoxification by Phase I, II, and III reactions after uptake from contaminated groundwater during phytoremediation.

In plants, those compounds with a higher lipophilicity that enter the transpiration stream will more quickly be absorbed to cells. For such lipophilic compounds to be eliminated from the plant, they must be transformed into more water-soluble compounds. In fact, this is the basis of most animal and plant detoxification mechanisms—increasing the water solubility of initially water insoluble xenobiotic compounds for removal from the organism. Much of this transformation is handled by enzymes. Obviously, these detoxification processes were not invented by plants to deal with groundwater priority pollutants; they instead evolved as part of a plant's evolutionary response to selection pressures derived from ex-situ and in-situ chemical bombardment.

A variety of processes in plants can act upon a wide range of potentially harmful compounds in order to render them into simpler, and less harmful, forms. Some processes are abiotic, whereas others involve biological processes where energy is extracted from the reaction. As these processes in plants are similar to those observed in the mammalian liver, the nomenclature to describe plant processes of contaminant degradation is taken from mammalian studies (for a review, see Burken 2003).

These detoxification processes include what are considered Phase I, II, and III reactions (Fig. 12.4). Phase I reactions, such as chemical activation, transformation, or functionalization reactions, involve oxidation and reduction reactions similar to those previously described, as well as oxidative metabolism, and hydrolysis reactions. Phase II reactions include detoxification reactions, such as conjugation reactions that irreversibly bind contaminants to plant tissues. Phase III reactions include compartmentalization or elimination reactions, where products of Phase I and II reactions are handled within the plant cellular organelles. Additional detoxification reactions include hydroxylation, dehalogenation, decarboxylation, and dealkylation.

These plant-facilitated, chemical detoxification reactions are described in detail below. An excellent review also can be found in various chapters in McCutcheon and Schnoor (2003). Knowledge of these processes is crucial to the application of phytoremediation at sites characterized by contaminated groundwater, because in order for the contaminants to be rendered harmless, detoxification by *ex-situ* mineralization in the root zone or detoxification in the plant after uptake must be demonstrated. For these reactions to be beneficial to solving groundwater contamination problems, the initial uptake into a plant has to occur. This does not mean, however, that all compounds taken up by plants can be detoxified, because some can move conservatively through the plant and be transpired unaffected through the leaves. Moreover, some reactions that lead to detoxification occur in the root zone without the plant taking the compound into the transpiration stream.

Much of the chemical detoxification information presented here was developed during the early investigation of the effect of herbicide/pesticide application on food crops, with little *a priori* concern of its implication for gasoline compounds and chlorinated solvents found in groundwater. Additional studies of the interaction between groundwater contaminants and plant detoxification reactions are warranted and offer a fruitful area for future research. Moreover, plants had developed these biochemical processes long before the release of manmade contaminants. Finally, even though these processes of chemical detoxification occur in plants, a controlling factor that will determine the success of this detoxification is how much contaminant source is present relative to contaminant sinks in the plant.

12.4.1 Phase I Reactions

Phase I reactions involve the transformation of potentially harmful compounds taken up by plants into more water-soluble byproducts or intermediates that undergo further detoxification by other processes. In general, Phase I reactions include oxidation, reduction, and hydrolysis, with the final product potentially being CO_2 (Fig. 12.4). Other Phase I reactions include hydroxylation, decarboxylation, and dealkylation. In these cases, organic molecule-based functional groups are either removed or added to the initial compound, called functionalization. This process renders a formerly hydrophobic contaminant to become less so after addition of a hydrophilic functional group, such as –hydroxyl, –amino, or –carboxyl groups, following enzymatic transformation by oxidation, reduction, or hydrolysis reactions.

Phase I reactions are initiated by enzymes produced in the endoplasmic reticulum of a cell's cytoplasm. In many cases, the exposure of plant cells to xenobiotics induces morphological changes in the cells, such that separate organelles are brought closer together, in an attempt to facilitate the transfer of electrons during the redox processes; this has been termed mitochondrial control (Kvesitadze et al. 2006). This is an interesting phenomenon, because the membranes that surround the endoplasmic reticulum are compounds primarily of lipids and, hence, act to attract the very lipophilic compounds that they act upon.

12.4.1.1 Oxidation and the "Green Liver"

Oxidation is the process where electrons are removed from a compound to form a negatively charged entity. The electron can be removed from a variety of areas on the compound, which are referred to using organic chemical nomenclature, such as the alpha (α), beta (β), and gamma (γ) positions.

In mammals, waste is continually produced as a consequence of life and metabolism. The *in-situ* production of potentially toxic metabolic byproducts, such as CO_2 or urea, are excreted to the blood stream and then the lungs by diffusion, or to the kidneys and then bladder. Mammalian livers purify the blood before its circulation through the body. Cells called hepatocytes act in concert with the rest of the immune system to destroy invading items, and this is accomplished through the release of oxidative chemicals that purposely cause inflammation. These "free radicals" produced by the body need to be neutralized, or they themselves will cause liver problems. How does the liver deal with this threat? The liver releases the compound glutathione, an antioxidant that can depress inflammation. This antioxidant also is found in plants, as is described below.

Potentially toxic compounds that enter the body from external sources are processed by the liver in a stepwise fashion. For example, the ingestion of alcoholic beverages involves the ingestion of ethanol, a known toxin. The ethanol is rendered harmless in the liver, however, by detoxification by oxidization to acetaldehyde, acetate, and then to CO_2 and H_2O. As you can see, the compound is broken down into its components that are then excreted through the process of exhalation and urination. The ingestion of such harmful chemicals induces the production of enzymes that facilitate this detoxification.

A major part of the Phase I oxidative detoxification system in mammals and plants is the production of microsomal cytochrome P-450 monooxygenases. These enzymes are present in most organisms, ranging from the *Archaea* to plants to mammalian livers. Cytochrome P-450 is not one compound, but rather a name that encompasses a large number, or group, of separate enzymes. Cytochrome P-450 works by adding functional groups such as hydroxyls to contaminant compounds. The P-450 monooxygenases initiate electron transfer by the NADPH reductase, with the electron grounding to the P-450 cytochromes. In animals, the P-450 enzymes reduce molecular oxygen to water and also render the contaminant compound more polar so that it can be excreted.

There are low levels of cytochrome P-450 enzymes in plants, but their role in detoxification is unclear, although its protective purpose proceeds undoubtedly by oxidative processes. Some oxygenases are present in the parts of the plant cells and in the apoplast as well as in the cell membranes, and some are present in the cytoplasm. Oxidation of xenobiotic compounds such as PAHs by these oxidative enzymes differs based on the source of the oxidation. For example, simple microbes such as the prokaryotes degrade PAH in the presence of oxygen using dioxygenase that contains two oxygen atoms, whereas eukaryotes use monooxygenases that have one oxygen atom.

Oxidative dehalogenation in plants by hydrolysis also can occur. Most dehalogenation reactions, however, are reductions carried out by reductases. Some evidence exists that poplar trees exposed to TCE in the transpiration stream can undergo oxidative dehalogenation to form trichloroacetic acid (TCAA) and dichloroacetic acid (DCAA), formed by the P-450 pathway (Shang et al. 2003). Interestingly, the higher dosage of contaminant exposed to a plant the higher the concentration of P-450 is measured, suggesting that its production is induced.

Other oxidative reactions in plants to detoxify threats involve the induction of plant peroxidases (POXs). These oxidases are used by plants to catalyze the transformation of many potentially harmful chemicals into less toxic forms. Xenobiotics, for example, can be polymerized into the soil humic fraction or root surface by POX and become essentially nonbioavailable. Peroxidases can decrease H_2O_2 concentrations in order to drive the oxidation of other substrates. This typically is a reduction of the H_2O_2 and the oxidation of other substrates. Because these POXs are prevalent in most plants, their role in detoxification is more well known than that of P-450. They are found in the cytosol of the cell.

Many different classes of organics can be oxidized by peroxidase. The putative enzyme peroxidase is a single peptide chain, with one heme group. Bacteria in the rhizosphere use a similar method to detoxify harmful substances but use a dioxygenase to accomplish this goal. Hence, these aerobic, heterotrophic bacteria can derive energy from this reaction as the toxicant is mineralized to CO_2 and H_2O. This full detoxification by rhizospheric bacteria is more common, however, when contaminant concentrations are low. When concentrations are higher, some gets mineralized but the majority is taken up into the plant and affected by Phase I reactions (Kvesitadze et al. 2006). Other oxidative enzymes released by microbes and fungi in the rhizosphere include cellulose, lignase, and protease, among others (Walton et al. 1994).

Another important detoxification enzyme in both plants and animals is glutathione-S-transferase (GST). Glutathione is a tripeptide that reacts with oxidizing agents to form a disulfide product. In this manner, GST acts sacrificially for the protection of more important proteins such as DNA, similar to how a zinc coating on various metal objects, such as nails and screws, acts as a sacrificial metal to prevent the underlying iron from oxidizing.

As can be inferred from the above comparison of Phase I detoxification reactions plants and animals, plants process xenobiotics using mechanisms that are fundamentally similar to how the mammalian liver detoxifies compounds. Although plants lack true excretory organs, they can store detoxification byproducts in vacuoles and in the lignin itself, and thus separate them from the rest of the plant. But this does not mean that plant exudates are not helpful in contaminant detoxification. Enzymes are released from plant roots and form part of the exudates that are associated with the rhizosphere. Compounds detected in the soil zone near roots include dehalogenases and nitroreductases (Schnoor et al. 1995). Peroxides have been found to polymerize contaminants onto the root surface or soil organic matter in the root zone. This action accelerates humification and renders the contaminant less bioavailable for plant uptake.

From the perspective of the phytoremediation of contaminated groundwater, plants have been shown to use such enzymes to break down contaminants such as chlorinated hydrocarbons. The C–Cl bond of these chlorinated compounds is attacked by monooxygenases, glutathione-S-transferases, and anti-auxin cell receptor binding, and the –Cl is replaced with –OH. Poplar trees were shown by Noctor et al. (1998) to contain high concentrations of glutathione. Komives et al. (2003) reported that, for poplar trees exposed to increasing concentrations of chlorinated herbicides, that increasing concentrations of glutathione were detected in poplar leaf cuttings.

12.4.1.2 Hydroxylation

Many xenobiotic compounds released to groundwater contain an aromatic or multiple-ring structure. The process of splitting these aromatic or heterocyclic rings and the subsequent addition of an –OH functional group is an oxidative process called hydroxylation. Hydroxylation reactions increase the contaminant reactivity in plant cells by increasing the compound's polarity and, therefore, hydrophilicity. Such "ring" cleavage occurs slowly in plants. Unlike bacteria which can render such rings all the way to CO_2, further degradation by plants after cleavage and –OH addition is limited. Instead of complete mineralization, these cleaved, hydroxylated compounds are incorporated into plant polymers as a bound residue. In many cases, because of the slow kinetics of hydroxylation of aromatic or multiple-ring compounds by plant cells, it is the rate-limiting step in contaminant detoxification (Kvesitadze et al. 2006).

Various contaminant compounds that can enter groundwater can undergo hydroxylation reactions after oxidation. Aromatic hydrocarbons that contain an organic functional group, such as the $-CH_3$ on toluene, can be oxidized by hydroxylation by –OH functional group addition in the *para* position. Organic compounds that contain a nitrogen, or –N, functional group undergo N-hydroxylation reactions, in which the –N is replaced by an –OH. Aliphatic hydrocarbons also can have an -OH functional group added, and proceed by the insertion of an oxygen atom between the C–H bond catalyzed by P-450 enzymes. Another oxidative reaction that undergoes hydroxylation is the addition of oxygen to a –C=C– carbon-carbon double bond, called an epoxidation reaction.

Specific examples of hydroxylation reactions by plants include those contaminants that are commonly detected in groundwater. The degradation of the aromatic compound benzene in gasoline is initiated by ring cleavage, and the formation of a hydroxyl intermediate such as muconic or formic acids. PAHs such as naphthalene also can undergo hydroxylation, and the intermediates then undergo conjugation as part of Phase II reactions. For chlorinated solvents, Newman et al. (1997) reported the presence of di- and trichloroacetic acids in plants exposed to the uptake of TCE. As such, these bound compounds are the hydroxylated byproducts of plant-cell TCE oxidation.

12.4.1.3 Decarboxylation

Decarboxylation is an oxidative reaction where a carboxyl functional group –COOH is replaced with hydrogen and CO_2 is released. Perhaps the most famous decarboxylation reaction is the conversion of pyruvate into acetyl-CoA at the beginning of the Kreb's cycle of aerobic respiration.

12.4.1.4 Dealkylation

Dealkylation is an oxidative reaction where compounds that contain –N, –O, or –S can have their alkyl functional group removed by addition of hydrogen atoms. These reactions are catalyzed by P-450 enzymes.

12.4.1.5 Antioxidants

Plants synthesize many compounds that are not used by the plant for energy or growth. This is a consequence of the plant evolutionary production of defensive processes for protection against both oxygen as a strong oxidant and the free radicals produced by ambient reactions, such as Fe(II) and H_2O_2. Oxidative stresses to plants are the result of the production of reactive oxygen species (ROS). These stresses are suppressed by the production of antioxidants compounds.

Free radicals, such as the hydroxyl radical (OH•), are harmful to living tissue because they contain at least one unpaired electron. Interaction of plant, or mammalian, tissues with free radicals is responsible for cellular damage, degeneration, and eventually cell death. The flow of electrons during oxidation is from the element being oxidized to the element accepting the electron. This reaction can occur spontaneously and abiotically, as with rusting metal, the chemical oxidation of Fe(II) to Fe(III) by atmospheric oxidation. The rate of this reaction can be enhanced by the presence of water and electrolytes.

Antioxidant compounds inhibit the oxidation process. As stated above, plants inhabit an essentially harsh environment, replete with ozone, UV radiation, oxidative chemicals, and free radicals, all which can destroy cells through damage to DNA. As well, a consequence of aerobic respiration in plants and humans, the Krebs cycle typically results in the reduction of free oxygen to water. However, if some oxygen is not reduced to water, this excess oxygen and can form superoxide anions (O_2^-). This can produce hydrogen peroxide and induce tissue damage, ultimately attacking DNA. To remove some of these threats, glutathione can be used as an antioxidant (Purvis 1997). Plants use the same glutathione and POX as antioxidants that humans use to combat oxidative stresses.

Another way that antioxidants work is by donating electrons to the oxidants, almost as surrogates, rather that having the cell DNA being oxidized. Such compounds include the vitamins C and E, as the carotenoids, beta-carotenoids, luteins, lycopenes, and terpenes found in many food crops. Limonene, found in oranges, stimulates the processes in Phase I and II, as does broccoli. Other antioxidant compounds include the flavonoids (cocoa, green teas), and catechins. These compounds are abundant in many fruits, such as pomegranates, grapes, blackberries, cranberries, and blueberries.

Because part of the plant's detoxification system in Phase I results in the generation of activated oxygen species, such as hydrogen peroxide, trees have various self-produced antioxidants to limit the damage. Such antioxidants include thiol-containing compounds such as the cysteine and glutathione previously discussed, and lipid-soluble compounds such as tocopherol (Vitamin E) and carotene.

The presence of antioxidant compounds in plant leaves and fruits can be explained by their interaction with the energy of the sun during photosynthesis. Singlet oxygen can be formed from oxygen in leaves by the input of too much solar radiation. Because this can kill cells by chloroplast oxidation, plants use these antioxidants to decrease the concentration of these oxygen singlets (Halliwell 2006). Others include those previously discussed in this section, such as the flavonoids and various vitamins. Flavonoids and bioflavonoids are plant pigments, and were called vitamin P by the discoverer of vitamin C, Albert Szent-Gyorgyi, in the 1930s.

Extracts of leaves and beans, such as tea and coffee, are used as beverages around the world. Leaves contain high concentrations of compounds such as flavonoids that act to protect the leaves from the damaging rays of the sun (since they can't put on sunscreen to absorb UV radiation like we can). These compounds are in highest concentration when the leaves are at their most vulnerable, when they first emerge from the winter buds. This is why green teas, the young leaves of the tea plants, are used for teas. These compounds protect the plant from the damaging rays of the sun, which can lead to the formation of free radicals that can cause cellular and genetic damage.

12.4.1.6 Reduction

As discussed above, the primary utilization of P-450 enzymes is as oxidative enzymes. These enzymes also can be used, however, to perform reductive reactions. Take, for example, the dehalogenases. In this capacity, rather than facilitating the addition of oxygen to a xenobiotic, they add hydrogen. These reactions happen best when concentrations of oxygen are low, such that there is no competition as an electron acceptor. For example, this reduction can include compounds that contain nitrogen, called azo reduction or aromatic nitro reduction, as well as the reductive dehalogenation of halogenated compounds (McCutcheon et al. 2003). Such reductions also can occur for the semi-oxidized contaminants such as TNT, where nitroreductases catalyze the reduction of the nitro groups on the central toluene structure.

The specific enzymatic capability to degrade chlorinated solvents by the dehalogenase enzyme exists in plants. The role played by nitroreductase in the roots of leguminous plants and other plants tissues was previously discussed in Chap. 11. For example, the kinetics of TCE transformation in leaf tissue was examined for trees growing above TCE-contaminated groundwater at the site in Fort Worth, Texas. All leaf samples collected showed dehalogenase activity. First-order rate constants of TCE degradation averaged about 0.049 h^{-1} for all plant species tested. This gives a more meaningful TCE half-life of about 14 h (U.S. Environmental Protection Agency 2003).

12.4.1.7 Hydrolysis

Hydrolysis is the process of splitting a molecule into two individual molecules. The most important hydrolysis reaction is the splitting of water into H and O during photosynthesis. Hydrolysis also results when functional groups interact with water, with –OH additions occurring most commonly. This occurs for organophosphates, carbamates, as well as esters and ethers, such as the conversion of MTBE to TBA in low-pH water. Because hydrolysis decreases the size of the parent compound, the smaller size of each individual compound renders it more susceptible to additional degradation, given the appropriate redox conditions.

The transfer of a glycosyl group to water occurs during hydrolysis. The glycosides are one of the largest classes of detoxification compounds in plants. The products of hydrolysis typically are further detoxified by Phase II conjugation reactions.

12.4.2 Phase II Reactions

Phase II reactions occur following oxidation reactions, are predominated by conjugation reactions, and require energy in the form of ATP to be expended by the cells (Fig. 12.4). The conjugation reactions increase the water solubility of the oxidized and functionalized compounds following Phase I reactions, or they produce water insoluble residues that are irreversibly bound into the plant tissue, such as occurs during lignification. Therefore, which pathway a particular contaminant will follow has important implications for the fate of xenobiotics released to groundwater. Conjugation reactions require enzymes such as GST. Less is known about the exact pathways of xenobiotic transformation, and rates of detoxification, though it is known that Phase II reactions occur at much slower rates in plants than in mammals.

12.4.2.1 Conjugation

Once a xenobiotic is taken up into a plant, the process of Phase I and Phase II detoxification can occur (Fig. 12.4). The initial step is usually interaction with cytochrome P-450 monooxygenases or POXs, as discussed above. After this oxidation, conjugation reactions occur where various sugars or amino acids interact with the activated xenobiotic to form glycoside compounds; these reactions are mediated by glycosyltransferases (Schröder and Collins 2002). The result is an inactivated xenobiotic. These Phase I and Phase II reactions act to protect the plant by removing the xenobiotic as quickly as possible by increasing the compound's polarity. Whereas in animals the end product is eliminated by excretion, in plants these byproducts are stored in vacuoles or in other organic matter in the plant, which is discussed below.

The interaction of the intermediate byproducts of Phase I reactions with a plant- or animal-produced compound is called a conjugate. Conjugates include plant protein, lignin, or organic acids. The resulting conjugates often are irreversibly bound to plant tissue. For example, these organic compounds cannot be extracted with chemical solvent extraction techniques. Up to 70% of contaminants that enter plants are rendered as conjugates (Kvesitadze et al. 2006). On the other hand, other conjugation reactions often result in the decrease in the toxic effect of the chemical through increased water solubility and intraplant mobility. In fact, the process of conjugation often is used in analytical chemistry to analyze water-insoluble compounds through derivatization to a more soluble conjugate.

The process of conjugation differs from bioaccumulation in that the parent compound taken up is changed into a less harmful form, and the process is regarded as beneficial in terms of risk reduction. Moreover, in the plant cells exposed to the contaminant, once the contaminant is conjugated, it no longer poses a threat to cell metabolism. However, these compounds are still present in the plant, as no mineralization occurred.

As was the case for most of the early investigation into the interaction between plants and xenobiotics, some of the first evidence of transformation reactions by conjugation was observed in plants exposed to pesticides. The plant enzyme GST was identified in the 1960s and 1970s to be present in both animals and plants. Transferases are enzymes that catalyze conjugation reactions, which lead to the interaction of the byproduct with endogenous plant cellular material. GSTs can facilitate the reaction between the contaminant functional group following Phase I reactions with the –SH group of the glutathione cysteine. In fact, evidence indicates significant DNA homologies of these enzymes from bacteria to mammals, including man. Essentially, many xenobiotics in the oxidized form tend to react with genetic materials, like DNA and RNA. Glutathione is hydrophilic, and conjugations of xenobiotic compounds that are more hydrophobic render these compounds more soluble in water. This is a protective mechanism that allows less exposure time of the xenobiotic to the animal or plant. These processes are not inducible, and remain in effect continuously.

It is possible for additional cell metabolism of the transformed xenobiotic to occur, or transfer to the external plant environment, such as the rhizosphere or atmosphere (Schröder and Collins 2002). In essence, in some cases the conjugated xenobiotic can be re-released, almost as an allelopathic agent. This fact of conjugated xenobiotic release has some interesting consequences. In the early summer of 2001, for example, in the horse country of Kentucky, more than 500 foals were stillborn or died after delivery, and many of the foals born alive had respiratory problems. Researchers initially thought that fungal spores in the grass were the causative agent. That year, however, also was a year of high numbers of eastern tent caterpillars, which just happen to like to forage on the leaves of local cherry trees (*Prunus spp.*). Cherry trees, like many other plants, contain toxic substances, including cyanide (CN) in their leaves. Cyanide is extremely toxic; between 50 and 70 mg (0.0025 oz) in *air* has a 50% chance of causing death to an average man, and it has no odor to warn of exposure. In humans, ingestion of 1 mg CN/kg/day will result in death. Cyanide poisoning occurs through blocking the binding of oxygen to hemoglobin in red blood cells during the initial point of electron transport. No electron transport means no ATP production and, therefore, no energy for growth.

The cyanide is not necessarily harmful to the plants, because it is conjugated with sugars to form a cyanogenic glycoside, compartmentalized in vacuoles or seeds, as amygdalin. In some plants, such as willows, death will occur only after exposure to 200 mg CN/kg/day. One of the explanations is that plants can take the CN ion and make asparagines from it. This can occur as long as the uptake rate is *less* than the rate of CN metabolism. Once the uptake rate is *greater* than the metabolism rate, however, accumulation of CN occurs and toxicity results. When these plant parts that contain the cyanogenic glycosides are eaten and burst open amygdalin will hydrolyze to hydrocyanic acid. This process effectively renders animals that eat such leaves, such as eastern tent caterpillars, a potent source of cyanide poisoning, and for the horses which came into contact when the leaves entered water troughs, etc., and became ingested by the pregnant horses.

The presence of natural toxins such as cyanide in plants is widespread, as was introduced in Chap. 11. For example, more than 1,000 plant species contain CN, including common plants, such as apples, whose seeds contain cyanide. These toxins probably developed as a selective advantage to predation by herbivores. The various tissues of cherry trees, as well as peach trees in the same genus, contain amygdalin.

Similarly, oaks (*Quercus spp.*) contain phenolic compounds, often collectively called tannins. Concentrations of these compounds are highest in green seeds (acorns) and young leaves. As discussed previously, black walnut (*Juglans nigra*) contains the phenol juglone in the bark, wood, nuts, and roots. The common landscape plant privet (*ligustrum spp.*) contains glycosides. Essentially, a rule of thumb is that if a plant doesn't seem to have any blemishes, such as holes or rough edges, it probably is a species that contains defensive toxic compounds, such as glycosides.

Although these compounds can be found in most parts of plants, the predominant location of storage in seeds seems to do more with inhibiting seed germination until conditions are right rather than thwarting ingestion. Other parts of the plants, once dead and fallen, also can release these toxins to inhibit the germination of other plants in an allelopathic manner.

The interaction of trees with cyanide also can occur when cyanide has been released to the environment as a contaminant. Sources of cyanide include manufacturing activities, such as electroplating. Blacksmiths, for example, use ferricyanide to harden iron. The surface soils and unsaturated zone sediments at many former MGP sites often contain cyanide. Trees have been shown to take up cyanide into their tissues. Following uptake, the cyanide is either stored or metabolized. At a former MGP site in South Carolina, hybrid poplar trees installed as part of a phytoremediation system and were growing over a plume of PAH-contaminated groundwater that also had CN were observed by the author to contain blue annual growth rings after being cut down, blue being associated with many compounds that contain CN.

12.4.2.2 Volatilization

Some compounds released to groundwater have the physical and chemical properties to move unattenuated though plants after uptake and be volatilized to the atmosphere. This movement of a contaminant unaffected through a plant is a form of phytoremediation, because the half-life of the contaminant once in the atmosphere will be considerably shortened by photooxidation, increased oxygen concentration, etc., relative to a longer half-life in anoxic groundwater. Volatilization of contaminant compounds from groundwater through plants is a logical extension of the rationale behind Phase I and II processes: production of a byproduct that has an increased solubility, in this case air, for elimination. In fact, this process of contaminant volatilization may be likened to a form of plant-based excretion. Burken and Schnoor (1999) investigated the potential for different organic contaminants to volatilize from leaves after uptake and translocation. The higher the vapor pressure of the contaminant, the more readily it volatilized from the leaves of hybrid poplar trees.

In order to understand the fate of a potentially volatilized compound on its route from the subsurface through plants to the atmosphere, the xenobiotic carbon tetrachloride (CCl_4) was examined following uptake into a tree (Ferrieri et al. 2006). They traced $^{11}CO_2$–labeled plant leaves after the plants were exposed to solutions of CCl_4. The tracer $^{11}CO_2$ is a short-lived ($t_{½} = 20.4$ min) radioactive isotope of C. The plants in question were OP–367 poplar clone cuttings grown hydroponically, to which radiolabeled $^{14}C–CCl_4$ was added, only after preparation from $^{11}CO_2$ and $^{11}CH_4$ and free Cl_2. They reported that following exposure to $^{14}C–CCl_4$ the plant isoprene emission to the atmosphere increased some two- to threefold relative to emissions prior to $^{14}C–CCl_4$ exposure. Also, the exposure of the plant to methyl jasmonate, a plant-defense signal transduction compound, decreased emission of $^{14}C–CCl_4$ and increased formation of nonvolatile trichloroacetic acid (TCAA) was observed.

12.4.3 Phase III Reactions

Because the accumulation of Phase II byproducts in plant cells might decrease the availability of enzymes needed to deal with future contaminant exposure, plants often compartmentalize these byproducts away from the cytoplasm (Fig. 12.4). Such reactions are the last step of detoxification and involve the stabilization of the insoluble conjugates to the plant tissues, such as lignin, or cell vacuoles, or even the cell apoplast or wall. It is analogous to mammalian excretion, in that the end products are no longer in direct contact with the cytoplasm of the cell.

12.4.3.1 Compartmentalization and Bound Residues

Compounds that are taken up or transformed can undergo covalent bonding to various plant tissues, in particular lignin, that cannot be chemically extracted. Such a compound is essentially nonbioavailable; the process of bound residues is considered a physical detoxification process. Such binding also can occur in the pectin, hemicellulose, and cellulose.

The common groundwater contaminant TCE was shown to be taken up at high aqueous concentrations by hybrid poplar trees and its fate tracked (Strand et al. 1995). TCE was found to be metabolized by the P-450 enzyme, since −OH intermediates are produced from the C−Cl bond. The fraction that was not metabolized was incorporated into the plant tissue itself; in fact, it was bound so strongly as to resist even solvent extraction. This process itself removes the contaminant from human exposure pathways, as long as the wood is not burned.

12.4.4 Other Processes of Contaminant Fate

Once in the xylem of a plant, groundwater and solutes are free to move throughout the plant under the water potential gradient induced by transpiration in response to atmospheric vapor pressure deficits. Along the length of xylem transport, the possibility exists that solutes will diffuse into adjacent tissues. This portioning is controlled by the log K_{ow} of the compound (Collins et al. 2006).

12.4.4.1 Fate in Shoots

Researchers have observed a decrease in contaminant concentration in tree-core material taken successively higher from land surface for trees growing above chlorinated solvent-contaminated groundwater (Vroblesky et al. 1999), as well as underground for roots (Ma and Burken 2003). Ma and Burken (2003) were able to demonstrate that at least some of this loss was due to diffusion from the root zone. Collins et al. (2006) suggest that loss of contaminants that were taken up by the tree could be explained by dilution in the growing tissue.

12.4.4.2 Endophytes

The interior of healthy, woody plants contains less living cells relative to dead cells. The phloem and cambium are alive, whereas the xylem cells die as soon as they form continuous vessels. Many of these dead cells fill with material over time to become heartwood, which is void of bacterial life. Bacteria can, however, enter the interior of a vascular plant in at least two ways. First, wounding or damage to the plant where bark is removed exposes the inner parts of trees to colonization by fungi and bacteria. If trees are "topped" during pruning, water can directly enter the dead xylem and set up conditions for rot.

Bacteria also can enter the plant below ground and become established in the cortex. It is there that *Pseudomonas* can colonize, rather than in the xylem, because it is external to the vascular system, where entry is controlled by the Casparian strip in the endodermis. This discrimination is essential to plant survival, because plants often live where water is not sterile, or is slowly moving and, therefore, a mechanism evolved by which plants restrict colonization and potential clogging of the vascular system. Currently, it appears that these bacteria are not pests, nor are they bacteria and fungi of decay, but instead they are probably symbiants, much like the bacterial flora of the guts of many mammals. Most endophytes are members of the common soil bacteria such as *Pseudomonas*, *Burkholderia*, *Bacillus*, and *Azospirillum*.

The potential effect of these endophytic microbes on the fate of contaminants in the transpiration stream of plants at phytoremediation sites is a topic of great interest. Although there is some ambient biotransformation of the solutes in the transpiration stream, it is envisioned that either naturally present endophytes with contaminant degradation characteristics or genetically modified bacteria with the capability to degrade specific contaminants will be added to plants in contaminated environmental systems. One such application would lead to the decrease of water-soluble contaminants, such as toluene or TCE, that are preferentially taken up and that have high enough vapor pressures that they exit the stomata into the atmosphere.

A potential advantage of this process would be to decrease the contaminants or byproducts that are essentially bound into plant tissue. For example, TCE is partially degraded in the xylem to the intermediate TCAA. If TCAA-degrading bacteria can be inoculated into plants growing in a TCE-contaminated system, TCE could be removed at the avoidance of the accumulation of TCAA in the plants. Van Aken et al. (2004b) reported that possibly the transformation of nitramine explosives in poplar trees could be due to a bacterial endophyte.

Siciliano et al. (2001) examined the question of the ability of plants to selectively enhance the number and types of endophytic bacteria that contain genes that code for enzymatic degradation of specific contaminants. They were able to document that genes that encode for the degradation of hydrocarbons, alkane monooxygenase (*alkB*) and naphthalene dioxygenase (*ndoB*), were two to four times more abundant in bacteria contained inside the plant roots relative to the bulk soil. Similarly at sites contaminated by nitroaromatics, the gene that encodes for nitrotoluene degradation, 2-nitrotoluene reductase (*ntdAa*) and nitrotoluene monooxygenase (*ntnM*), were 7–14 times more abundant in these intra-plant bacteria.

Taghavi et al. (2005) inoculated hybrid poplar trees with an endophytic bacteria *Burkholderia cepacia* VM1468 that contains the ability to code for toluene degradation. Following inoculation, the plants grew well in the presence of toluene and less toluene was released to the atmosphere by *ET*, relative to control poplar trees also exposed to toluene but not containing the bacterium. The gene that codes for toluene also was observed to be horizontally transferred to other endophytic bacteria. Such horizontal gene transfer can allow a microbial community to adapt to changes in environmental stresses. This area of the effect of endophytes on plants and contaminant cleanup will undoubtedly provide a wealth of research opportunities.

12.5 Summary

Organic compounds dissolved in groundwater enter the epidermal layer of root hairs only if the compound is not first absorbed by the soil or root itself. A useful parameter to predict the extent of plant uptake of groundwater contaminants is provided by the log transform of the partition coefficient K_{ow}. In general, compounds that have log K_{ow} between 1 and 3.5 will cross the Casparian strip, whereas compounds less than 1 will not enter as their solubility is too high, and compounds greater than 3.5 have low solubility and tend to partition onto soil or root surfaces.

Why is this information important to the phytoremediation of contaminated groundwater? Such a fundamental property of organic contaminant compounds provides a powerful approach to understand the potential for phytoremediation of groundwater contamination at a particular site based on the primary contaminant released. Once in plants, the fate of a particular organic compound will be governed by either diffusion through the bark, volatilization through the leaves, or *in situ* detoxification, where plants possess an arsenal of approaches to deal with the threat of chemicals. In some cases, these fundamental detoxification reactions can be applied through phytoremediation—the end result is a decrease in groundwater contamination.

Plant Control on the Fate of Common Groundwater Contaminants

13

When they came to Marah, they could not drink the water because it was bitter...
and the Lord showed him (Moses) a tree,
and he threw it into the water,
And the water became sweet.

Exodus *15: 23–25* (RSV)

In general, groundwater can become contaminated by two different processes; contaminant release from nonpoint and point sources. Nonpoint-source groundwater contamination reflects the widespread release of contaminants from sources that are dispersed throughout an area or cannot be attributed to an identifiable location. Such nonpoint sources include runoff or atmospheric deposition. Between 1985 and 2001, the USGS National Water-Quality Assessment (NAWQA) Program analyzed about 3,500 samples of groundwater collected throughout the United States. As part of that study, it was determined that almost 20% of the ambient groundwater samples contained 0.2 μg/L or greater of one or more of 55 volatile organic compounds (VOCs) analyzed (Zogorski et al. 2006). The most frequently detected VOCs were trihalomethanes (THMs), such as chloroform, and the chlorinated solvents PCE and TCE. In some specific areas of the United States, VOCs such as the fuel oxygenate MTBE, the fumigant and gasoline additive ethylene dibromide (EDB), and the soil fumigant dibromochloropropane (DBCP) also were detected. Encouragingly, however, 90% of the groundwater samples analyzed had VOCs concentrations less than 1 μg/L. Although some regulated compounds have MCLs near 1 μg/L, such as EDB, and vinyl chloride (VC), the low concentration detected for most compounds indicates either the lack of a constant contaminant source or the cleansing effects of natural attenuation processes.

In contrast to nonpoint source releases to groundwater, point-source releases have an identifiable, and often regulated, release location. Unregulated point sources occur at spills or accidental releases. For these sources, contaminants in groundwater often are found at percent-level concentrations. In the United States, the USGS Toxic Substances Hydrology Program specifically investigates the fate of a wide variety contaminants in groundwater from point sources. Such contaminants investigated include chlorinated solvents, petroleum hydrocarbons, tritium, and heavy metals.

The interaction between plants, groundwater, and contaminants from either type of source depends in most part on the relation between the abiological, physical processes that affect contaminant concentrations, and the rate of water flow through plants. For example, the uptake of water by root hairs is controlled by the physical properties of water, such as surface tension and capillarity. Uptake of dissolved solutes, such as contaminants, by root hairs also is controlled by the physical properties of the contaminant, such as water solubility and log K_{ow}.

Plants *can* control the movement of contaminant solutes through their structures, however, by affecting the rate of evaporation at the leaf surface and by the initial movement of water into root hairs in the subsurface. The only part of the interaction between plants and contaminants that is truly plant based, in the sense that it is not entirely related to physical or chemical phenomenon, is the cellular detoxification of xenobiotics discussed in Chap. 12. Even the root zone microbial degradation processes are predominantly *ex-situ* processes, more akin to bioremediation than phytoremediation.

A potential framework to address the interaction between plants and contaminated groundwater can be based on whether or not a particular contaminant fits into one of the following biogeochemical pathways:

- Flowing through one of the Phase I to III detoxification processes
- Flowing through the plant by transpiration and evaporation, and
- A physical partition into plant tissue

If a contaminant does not enter into at least one of the above pathways, then that particular contaminant may not be amenable to phytoremediation. A framework to determine

J.E. Landmeyer, *Introduction to Phytoremediation of Contaminated Groundwater*,
DOI 10.1007/978-94-007-1957-6_13, © Springer Science+Business Media B.V. 2012

the interaction between plants and groundwater contaminants based on these pathways is described in this chapter. Numerous examples and case studies from the laboratory and field are given to provide emphasis. Additional information on alternative frameworks regarding the interaction between plants and contaminated groundwater can be found in USEPA (2005a, b).

13.1 Early Evidence of Plant and Contaminant Interaction: Herbicides and Pesticides

The production of xenobiotic chemicals specifically used to remove insect and plant species considered to be pests increased dramatically after the 1940s following the end of the Second World War. Plant pests had been around long before that time, however, and the desire to eliminate them was not a new phenomenon. Various approaches had been used to deter faunal and floral pests, particularly for cash and agricultural crops. As was discussed in Chap. 11, an extract from tobacco leaves that contained the biochemical toxin nicotine was sprayed on leaves to render them inhospitable for ingestion by many insect pests. The use of plant-derived products to thwart pests also was extended to the planting of certain "toxic" plants near desirable plants in order to protect these from attack. Marigolds, for example, often are planted near cash crops to remove the threat caused by nematode worms that destroy plants by invading their roots. The marigolds release the allelopathic chemical pyrethrin that kills the nematodes.

This interaction between the production of naturally toxic compounds by plants for protection against plant pests was introduced in Chap. 11. It provides evidence that challenges the commonly held perception that the production of noxious chemicals was solely the responsibility of industrial chemists. In fact, chemists have often looked to plants in order to come up with ways to deal with plant pests. Even the idea for the development of widely used systemic insecticides that kill insects but not the plants they feed on was an extension of observations of natural plant–pest interactions seen in the field. For example, some wheat plants that grew naturally unmolested by aphids were found to contain selenium, a naturally occurring element. The wheat acquired the selenium in the form of sodium selenate from the soil and distributed it throughout the plant.

This natural systemic protection gave rise to the idea of using artificial chemicals applied to the foliage, roots, or trunk for systemic protection from predators. Over time, various chemicals were added to plants to make them 'toxic.' The uptake and distribution of these chemicals was widely studied, in order to ensure that the whole plant was protected. Such studies included those where the chemical, usually a systemic insecticide, was added to the plant by soil or trunk (cambium) injection using a Kioritz soil injector or Wedgle® Tip tree injection system, respectively (Gill et al. 1999). These studies examined the interaction between the chemical, its distribution in the plant, and the predator, rather than the interaction between the plant and the chemical, as is the focus of the phytoremediation of contaminated groundwater. Interestingly, the steady-state concentration of insecticide was achieved faster by a factor of 3–4 using trunk injection versus soil injection, probably because of the *RCF*.

As the number and volume of synthetic herbicides increased over time, the fate of these compounds was studied not only to determine the potential for these compounds to bioaccumulate in the environment but to increase the efficiency of their mode of action. Many compounds synthesized in the laboratory used to kill weed plants are similar to naturally occurring plant growth hormone compounds. In the laboratory, these compounds are slightly modified to increase their lethality. The herbicides based on plant growth hormones essentially act by making the plant grow itself to death by increasing the rate of respiration; the plant simply oxidizes more plant photosynthate than can be produced. The defoliant Agent Orange is a rapidly acting, growth-hormone-based herbicide. It is a mixture of the *n*-butyl esters of di- and tri-chlorophenoxynacetic acid (2,4-D and 2,4,5-T), which are plant hormones. As we saw in Chap. 3, plants use hormones for growth, protection, and reproduction.

Herbicides affect various aspects of plant growth. Herbicides can be classified in different ways, based on similarities in chemical structure, causative agent, or application schedule relative to the growth cycle of the target plants. The simplest herbicide is sodium chloride (NaCl), which works by upsetting osmosis. Some nitrogenous herbicides work by disrupting the light reactions of photosynthesis. Some disrupt respiration reactions through the use of various halogenated hydrocarbons. Some act as synthetic growth hormones that mimic the plant growth hormone auxin, such as carboxylic acids, and prevent cell division and protein synthesis.

More important to our understanding of the fate of xenobiotics in groundwater with respect to phytoremediation, herbicides also can be classified according to their mechanism of toxicity. The understanding of this interaction between plant and xenobiotics, such as plant uptake, detoxification, and fate, provides a fundamental basis to support a framework that can be applied to examination of the potential interactions between plants and common groundwater contaminants. Hsu and Bartha (1979) used hydroponic experimental methods to investigate the interaction between two commonly used organophosphate pesticides and the rhizospheric assemblages of test plants. The tests were done in glass flasks filled with water and air and a test plant to which radiolabeled pesticides were added and the fate tracked over time.

Perhaps the first interface between the interactions of plants used for phytoremediation, or hybrid poplars, with chemicals that were agricultural in use but are similar to those released to groundwater can be traced to a study by Burken and Schnoor (1997). They presented one of the first reports of the uptake and metabolism by plants of 2-chloro-4-(ethylamine)-6-(isopropylamine)-*s*-triazine, more commonly known as atrazine. Plant uptake of atrazine was hypothesized due to its log K_{ow} of 2.56.

What was interesting about their study, however, was the fate of the atrazine in the plants *after* uptake. To evaluate the fate of the translocated atrazine in the hybrid poplar trees, it was added as ^{14}C-atrazine. Atrazine was taken up near 30% over 80 days for poplar trees grown in soil, and 71% in 13 days for poplar trees grown in sand alone. The amount of atrazine radiolabel present as nonextractable, unbioavailable residue was 8.4% for cuttings in sand and near 16% for cuttings in soil. These bound residues are most likely less toxic than atrazine itself, but more importantly from the perspective of contaminant risk exposure, are rendered unbioavailable. Extraction and analysis of the poplar cuttings revealed that the balance of the atrazine underwent transformation to various metabolites from Phase I detoxification reactions such as dealkylation and hydrolysis.

13.1.1 Contaminant Half-Life Concept

A characteristic that can be used to describe the relative degree that a herbicide, or any other xenobiotic, may bioaccumulate in plants is the concept of a half-life. Contaminants present in plant tissues are exposed to the living processes of the plant and, therefore, can undergo the Phase I, II, and III detoxification reactions described in Chap. 12. Generally, these reactions result in a decreased contaminant concentration *in planta*, whose kinetics follow first-order, concentration-dependent kinetics, where the rate of detoxification is directly proportional to the contaminant concentration. Such first-order kinetics are analogous to the variable flow rate of water from a pipe stuck into the bottom of a water tank relative to the amount of water in the tank, where the rate of flow is faster when the water level is high and slower when the water level is low. These kinetics can be shown as

$$dC/dt = -\lambda C \qquad (13.1)$$

where dC/dt is the change of concentration over time, and λ is the reaction rate constant. The half-life, $t_{1/2}$, in general terms, is the time required for a compound added to a system to decrease to half its original amount. Mathematically, it is expressed as

$$t_{1/2} = \ln 2/\lambda \qquad (13.2)$$

The half-life of a compound is controlled by many factors. One factor is the ability for the contaminant to be degraded in the presence of various enzymes in the plant or to be excreted or stored in the plant. These enzymatic reactions are the most important aspect of the detoxification and degradation of a xenobiotic with respect to phytoremediation. These enzymes are usually not consumed in the reaction, but aid to catalyze the reaction to completion, usually at much lower temperatures and at higher rates than would be available without the enzyme. Degradation by these processes often results in the conversion of the contaminant back to the original photosynthetic reactants of CO_2 and water. In some cases, however, intermediate compounds are formed that are more resistant to further degradation than the parent compound.

The fact that some herbicides need to be applied to different parts of a plant implies that the uptake of xenobiotics occurs in different parts of the plant and suggests different contaminant detoxification pathways. Chemicals applied to the leaves are taken up by the plant through absorption and translocated within the plant by the phloem. Conversely, chemicals applied to the soil are taken up by the roots and translocated by the xylem. Others are applied to either area but are moved throughout the entire plant by the symplast, such as leaf-applied chemicals, or the apoplast, such as the soil-applied chemicals or the systemic herbicides. Herbicides that move through the plant through the phloem also can move to the roots, however, if applied on the leaves or shoots. On the other hand, herbicides that move primarily in the xylem move to the leaves if applied on the leaf, or if applied to the roots move throughout the plant after entry into the cortical tissues. These differences in the uptake and translocation of herbicides provide an important analogy into the potential for environmental pollutants released to groundwater to be taken up by plants during phytoremediation.

13.1.2 Contaminant Bioavailability

For most contaminants, once they are released to the environment, they are no longer considered to be in a stable system. In the case of petroleum hydrocarbons, an increase in time since release will result in a decrease in contaminant bioavailability. This is because the contaminants are exposed to various biotic and abiotic processes that act to remove the more soluble and volatile fractions from the source area. Some of these processes are the same that cause rock to become weathered into soils. In fact, in some cases the byproducts of contaminant weathering can become part of the soil organic horizon through the humification processes (Gregory et al. 2005), where they are irreversibly bound, adsorbed or absorbed, to sediment and soils.

In a related process, the presence of plants also affects the composition of soil horizons. This is caused by the release of plant root material either through exudation, root sloughing, or root turnover, which increase the sorptive capacity of the O-horizon over time. This may seem as counterproductive at phytoremediation sites, where trees are planted to *remove* contamination but over time will act to reduce the efficiency of uptake because contaminant uptake is decreased as contaminant bioavailability decreases and soil sorption increases. This decrease in bioavailability, however, actually helps to accomplish the goal of site remediation, even though the contaminants are not necessarily taken up inside the plant.

13.2 Plant Interactions with Aromatic Hydrocarbons: BTEX

All manmade petroleum hydrocarbons used for a wide range of purposes are derived from fossil fuels such as crude oil or coal deposits. These source materials are themselves composed of the remains of mostly plants and animals alive up to 450 MYa, when global temperatures were uniformly warmer and the continents more coalesced than today; after death, they were buried by fluvial sediments that removed them from oxygen and, therefore, slowed their decay back to CO_2 and water. These buried plant remains, after exposure to geologic time and pressure, have produced the resources of crude oil and coal that can be pumped or mined. These raw products have to be modified, and crude oil is refined or "cracked" to produce a wide range of products based on their boiling point as the oil is heated. It is ironic, or perhaps symmetrical, that phytoremediation can be used to clean up groundwater contaminated by products refined from the remains of ancient plants, many of which were phreatophytes themselves.

A common product of the refinery process is aromatic hydrocarbons composed of a ring of C–H bonds, with three double bonds. Benzene is the most carcinogenic of its homologues toluene, ethylbenzene, and the xylenes, but all can cause cancers such as leukemia. Collectively these compounds are called BTEX, short for benzene, toluene, ethylbenzene, and xylenes. Because these compounds are used in gasoline and it is widely distributed, it is a common groundwater contaminant and, therefore, its presence drives the need for remediation at many sites.

Some of the common aromatic groundwater contaminants and their physical and chemical properties related to plant uptake are described in Table 13.1.

Table 13.1 Physical and chemical properties of common aromatic groundwater contaminants with importance to plant bioavailability and phytoremediation.

Contaminant	Water solubility (mg/L)	Log K_{ow}	Log K_{oc}	Henry's constant (Pa m^3/mol)/dimensionless (H/RT)
Benzene	1,780	2.13	1.5	557/0.22
Toluene	520	2.69	1.75	673/0.24
Ethylbenzene	152	3.15	2.94	854/0.35
m-Xylene	160	3.18	2.20	700/0.31

13.2.1 Plant Interaction and Uptake Pathways

Laboratory studies have indicated that both herbaceous and woody plants can take up a variety of dissolved-phase petroleum hydrocarbons. Burken and Schnoor (1997) reported that the herbicide atrazine was taken up and subsequently metabolized by poplar trees (*Populus deltoides*). They extended that work to investigate the uptake, translocation, and volatilization of BTEX by poplar cuttings grown in hydroponic solutions. These compounds share the characteristic of many USEPA priority pollutants, that is, of an affinity for the dissolved phase, even though they have high vapor pressures. The log K_{ow} for these compounds is, in general, between 2 and 3.5 (Table 13.1).

Jordahl et al. (1997) investigated the influence of poplar tree roots on the fate of BTEX compounds as a function of increased microbial activity. As was discussed previously, root zones are sites of increased microbial numbers. For remedial purposes, however, the root zone needs to have bacteria that contain enzyme systems that will degrade the contaminant of interest. Jordahl et al. (1997) characterized the microbial populations in the rhizosphere of a mature hybrid poplar tree (*Populus deltoides* x *nigra* DN-34 "Imperial" Carolina) using the most probable number (MPN) technique on three soil samples taken in the root zone of poplar trees and compared to the same for an adjacent corn field with no tree roots. As would be expected from the "rhizosphere effect," there were more microbes associated with the roots of poplar trees than with the controls.

What is more important in terms of contaminant remediation, however, is not just the numbers of bacteria but the ability of these microbes to express genes to produce the enzymes needed to degrade contaminants such as benzene. More work in this area needs to be done to firmly establish a positive relation between plants, root microbial communities, and BTEX. It will then become possible to delineate more precisely those contaminant biodegradation processes that can be attributed to bioremediation or phytoremediation and to describe when bioremediation is enhanced by phytoremediation.

Aromatic hydrocarbons such as BTEX are found not only in gasoline but in jet fuel as well. Jet fuel is used by both commercial and military aircraft, and is derived from kerosene. Rather than stored in underground storage tanks like automotive gasoline, jet fuel, or "av gas," is stored in above ground storage tanks (AST). Releases from ASTs, therefore, tend to allow fuel to percolate down through the vadose zone sediments to the water table.

Karthikeyan et al. (2003a) investigated the fate of jet fuel, such as JP-8, in a laboratory study that measured the loss of JP-8 added to different columns that either contained vegetation or did not. The vegetation studied was alfalfa (*Medicago sativa*), horseradish (*Armoracia rusticana*), and fescue grass (*Festuca arundinacea*). At the end of a 3-, 8- and 12-month incubation of these plants with JP-8, the amount of fuel that remained in the vegetated treatments was only slightly less than in unvegetated treatments, where the difference between treatments was less than 1% (Figs. 13.1 and 13.2).

What is interesting about this study is not the specific results, which may differ depending on soil or plant selection, but that it raises the question of how exactly can phytoremediation be defined? For example, phytoremediation can be defined solely as a plant-based process strictly limited to plant–mediated processes (as defined in the introduction to Chap. 1) or it can be a plant-assisted process, such that the plant brings about conditions conducive to bioremediation, similar to co-metabolism. For example, in the above study in the vegetated treatment (Karthikeyan et al. 2003a) an upward flux of the JP-8, toward to the root zone was induced as a result of plant transpiration. This upward transport of the JP-8 contaminants increases the probability that oxic conditions in the subsurface will be encountered, as well as aerobic microbial processes that will lead to the oxidation of the jet fuel.

This upward advection of JP-8 was modeled using a 1-D approach (Karthikeyan et al. 2003b). Also, the upward movement of the contaminants might cause them to volatilize more rapidly than at deeper depths. At the end of their study, the loss of JP-8 was significant: up to 86% of that originally added was gone after 5 months. The authors concluded that the loss was due mostly to volatilization and biodegradation and facilitated by vegetation.

A study where a primary goal was the planting of trees over a source area to reduce recharge to the water table occurred near Milwaukee, WI (McLinn et al. 2001). The site was a former fuel tank farm adjacent to the Menomonee River. Due to the glacial history of this area of the United States, the shallow aquifer was composed of low-permeability till to a depth of 18 ft (5.4 m). Due to the activities at the site, the soil and groundwater were contaminated by

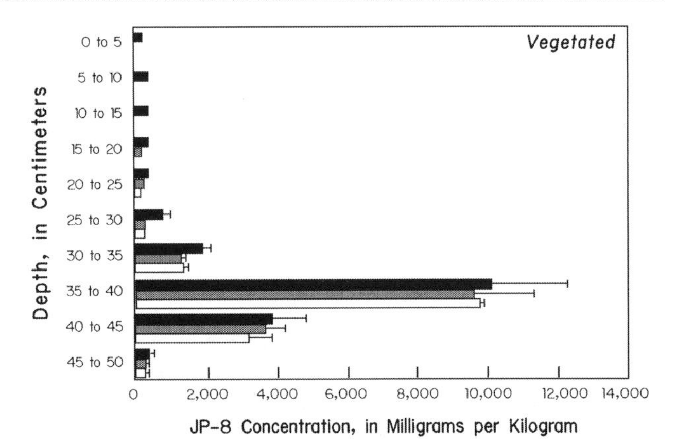

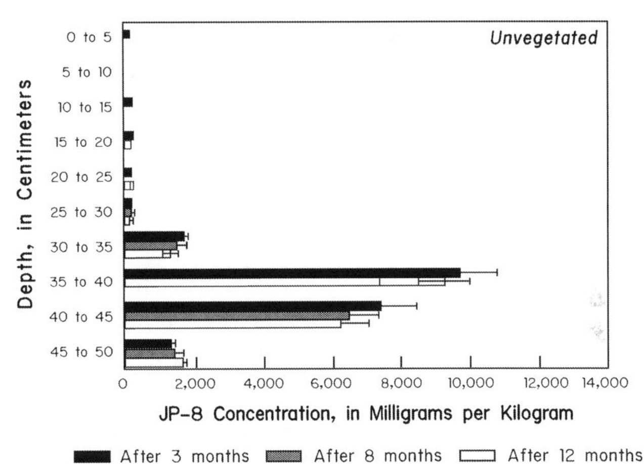

Fig. 13.1 The presence of plants did not greatly affect the removal of jet fuel from contaminated sediments (Modified from Karthikeyan et al. 2003a).

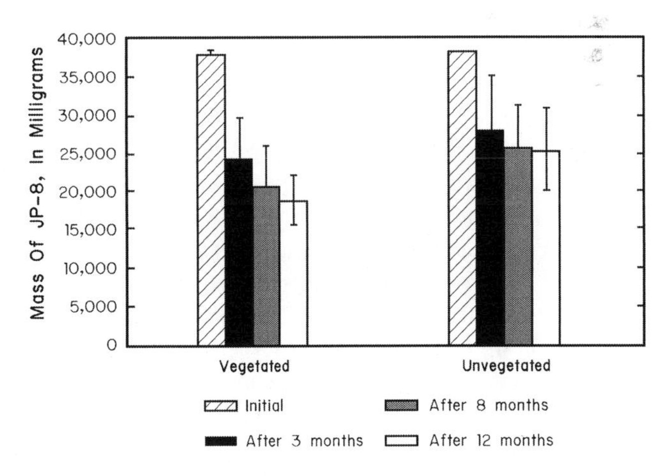

Fig. 13.2 The presence of plants did not greatly affect the removal of jet fuel from contaminated sediments (Modified from Karthikeyan et al. 2003a).

petroleum hydrocarbons, including free-phase product and residual product in the source area. The phytoremediation system designed for the site was driven primarily to decrease groundwater recharge by having the planted trees sequester infiltration in the source area.

In 2000, 485 hybrid poplar trees (*Populus deltoides* × *nigra*, DN34 "Imperial" Carolina) were planted after extensive site preparation. Of the 485 trees, 290 were planted in a row adjacent to the river at the downgradient edge of the site. These trees were installed using a hollow stem auger to a depth of 9 ft (2.7 m), to be as close to the water table as possible. In order to ensure that oxygen in the unsaturated zone was not a limiting factor for root respiration, an air-injection aeration system was installed during planting; it was discontinued, however, in 2003. In the source area, 195 trees were planted in the footprint where the former tanks were located. These trees were not planted as deeply, only about 4 ft (1.2 m). Phytoremediation at this site was monitored up until August 2006 (Van Den Bos 2002).

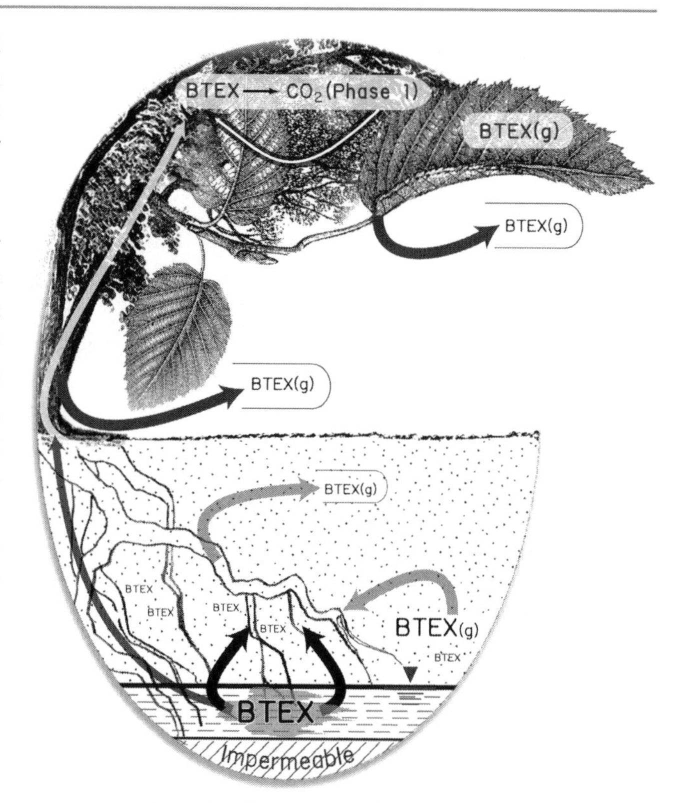

Fig. 13.3 Plant and groundwater interactions at a site characterized by BTEX-contaminated groundwater. BTEX(g) indicates the volatile phase.

13.2.2 Plant Transformation Reactions

Banks et al. (2003) investigated the fate of organic contaminants in treatment cells to which were added on-site soils contaminated by a diesel fuel release. Plant enzymes can oxidize benzene by Phase I hydroxylation reactions after ring cleavage to a variety of intermediates such as phenol, as well as to catechol *o*-quinone, muconic acid, and fumaric acid. These oxidized compounds can then potentially enter the TCA cycle for processing as an energy source by cellular respiration processes. Some pathways are shown in Fig. 13.3.

The concentration of BTEX, however, can affect plant health, and decrease these oxidative Phase I reactions. The toxicity of gasoline and diesel fuel to willow and poplar trees was investigated by Trapp et al. (2001b). Toluene was shown to decrease the growth of poplar cuttings at a concentration of 500 mg/L relative to no toluene exposure (Taghavi et al. 2005). The effect of benzene on plant cell structures was observed by Korte et al. (2000). Using ^{14}C-1,6-benzene, they noted pathological changes to the chloroplasts in some plants but noted that poplar, cypress, and ash were apparently resistant to the negative impacts of benzene. In part, it appeared that the potential negative effect of benzene on the plants was minimized by sequestration of the benzene and other soluble aromatic hydrocarbon contaminants into the cell vacuole, where it remained and, therefore, would not interfere with cell metabolism.

13.3 Plant Interactions with Polycyclic Aromatic Hydrocarbons

As the prefix "poly" suggests, polycyclic aromatic hydrocarbons (PAH) are compounds composed of fused aromatic rings. For example, coal can contain carbon up to 75% in the form of aromatics as high molecular weight polymers. Crude oil can contain up to 7% PAH, with gasoline containing lower amounts and diesel oils containing the highest amounts.

The largest source of PAHs is as a byproduct of the incomplete combustion, or rapid oxidation at temperatures near 700°C, of crude oil and coal. The combustion of coal during the coking process produces a coal-tar byproduct that can contain up to 50% by weight of PAHs. The PAHs present include naphthalene, phenanthrene, fluorine, fluoranthene, benzo[*e*]pyrene, benzo[*a*]pyrene, and perylene. The use of coal-tar-derived products, such as asphalt parking lot sealants, by homeowners and commercial businesses, was identified as a major source of PAHs detected in runoff; the asphalt-based sealants also released PAHs, but at lower concentrations (Mahler et al. 2005). The contamination detected was most likely caused by PAH-rich particle movement, such that mean concentrations reached 3,500 mg/kg, rather than dissolved-phase contamination.

PAHs also can be derived from natural sources. PAHs can be produced by fires and volcanic eruptions that occur at temperatures near 200°C, and by fungi, bacteria, and some plants, especially those that grow in coniferous forests and peat bogs. For example, the PAHs naphthoquinine and quinine are produced naturally in some plants, as was introduced in Chap. 11. Under anoxic subsurface conditions, these quinines are reduced to hydroquinones and then ultimately PAHs.

It is evident that the detection of PAHs in the environment could implicate many potential sources, both natural and industrial. Differing industrial sources have characteristic release patterns and, therefore, contaminant distributions. Spills tend to be characterized by higher concentrations in a smaller area or volume of environment, due to the physical chemical characteristics of many PAHs of low water solubility and high affinity for absorption onto soil organic matter.

13.3.1 Plant Interaction and Uptake Pathways

That portion of organic contaminants that are bioavailable will be affected by plant-based processes (Cunningham et al. 1995). As was discussed in Chap. 12, bioavailability is related to the physical and chemical properties of a contaminant such as log K_{ow}, pH, soil type, and degree of contaminant weathering. Plant interaction with PAHs can become lethal if the imbibed PAHs are exposed to UV radiation, because the energy the PAHs absorb can be transferred to very reactive singlet oxygen. The toxicity of a group of PAHs to willow trees was investigated by Thygesen and Trapp (2002).

A literature review of plant and PAH interaction was published in 1992 by the Electric Power Research Institute (Electric Power Research Institute 2002). The review presented information on the interaction of vegetation often found at former MGPs and was concerned not with the use of vegetation to remediate MGP contaminants but as a potential route of contaminant exposure; this mindset was common at a time prior to much knowledge about phytoremediation. The review concluded that the potential for uptake into plants was, as can be expected, related to the physical and chemical properties of the PAH. The log K_{ow} of most PAHs is high, on account of their low solubility and tendency to partition into organic matter. Because of the high log K_{ow}, plant interactions with PAHs would tend to favor initial absorption to root material rather than uptake by root hair cells, with uptake into the transpiration stream being restricted to those PAHs with lower solubilities. For example, PAHs that had five or more rings tended to undergo absorption onto roots and were not taken up in the transpiration stream by plants. On the other hand, PAHs that had from two to four rings could be taken up by plants, because of their increased, although still relatively low, solubility. These compounds include naphthalene, anthracene, benzo[*a*]pyrene, acenaphthene, fluorine, phenanthrene, anthracene, fluoranthene, pyrene, benzo[a]anthracene, and chrysene. It is not surprising, therefore, that most of our knowledge about the interaction of plants with PAHs is derived from research at former MGP (Anderson et al. 1997).

The PAH of concern at many contaminated groundwater sites is naphthalene, because it, like other lower ring PAHs, tends to be found in coal-tar-derived products as well as gasoline, and is the PAH with the highest solubility in water, being near 30 mg/L. High concentrations of PAHs are exposed to plants in soil with relatively lower SOM concentrations.

The *TSCF* of various PAHs was measured in a laboratory study where three types of plants were exposed to soils that contained phenanthrene, anthracene, fluoranthene, and pyrene (Mattina et al. 2006). The plants were all vegetable plants, and included zucchini, summer squash, and cucumber, all members of the *Cucurbitaceae*. They were grown in rhizotrons. The *TSCF* for the three-ringed PAHs of phenanthrene ranged from 2.8 to 11.6, and for anthracene ranged from 3.5 to 26.5. The *TSCF* for the four-ringed PAHs for fluoranthrene ranged from 1.1 to 5.17, and for pyrene ranged from 0.72 to 4.0.

The uptake of phenanthrene and chlorobenzene by black willow (*Salix nigra*) was investigated by Gomez-Hermosillo et al. (2006). Total uptake of radiolabeled contaminants in the laboratory was between 3.8% and 5.7% of the initial concentration of desorption-resistant contamination. This experiment was performed to test the assumption behind the conceptual models of *TSCF* and *RCF* that the water in the soil pores near the roots will contain contaminants that can be reversibly bound to the soil; that is, they are all bioavailable. In other words, does reversibly bound contamination enter plants? Most of the contaminant mass remained in the roots and was not translocated, such that translocation was between 0.38% and 0.47%. These results suggest that highly sorptive contaminants can be taken up by plants but at levels less than that predicted by log K_{ow}. This information should be useful for designing monitoring strategies at PAH-contaminated sites.

Groundwater beneath the former MGP site near Charleston, SC, described previously is characterized by dissolved-phase concentrations of monoaromatic petroleum hydrocarbons such as BTEX, and PAHs such as naphthalene, that are associated with the raw materials and wastes that were produced during the operation of the former MGP (Landmeyer et al. 1998a). The highest concentrations of the more soluble benzene and toluene compounds, up to 5 mg/L, and for the most soluble PAH compound naphthalene, up to 1,000 mg/L (Landmeyer et al. 1998a) are associated with groundwater in contact with the free-phase form of these wastes, known as DNAPL, or dense, non-aqueous phase liquid.

Due to the low groundwater-flow rate and occurrence of significant microbial biodegradation of these soluble compounds at the site (Landmeyer et al. 1998a), dissolved-phase plumes are limited in size to halos around discrete sources of DNAPL. Many of these discrete DNAPL sources have been identified and either have been removed by excavation or are being pumped near gravel-lined French drains installed at the site.

The contaminated groundwater is characterized by low, <1 mg/L, levels of dissolved oxygen, and the redox composition is dominated by sulfate reduction, with low sulfate and high sulfide concentrations, and methanogenic conditions (Landmeyer et al. 1998a). Although groundwater on the western, inland side of the site is relatively fresh, groundwater is more saline, being >1,000 microsiemens per centimeter (μS/cm), as it interacts with the more saline water from the Cooper River.

As stated in Chap. 8, hybrid poplar trees were planted at the site in two phases; phase one occurred in November 1998 and included the central to western part of the area along the southern boundary of the site, and phase two occurred in May 2000, in the remaining eastern part of the site. Poplar tree-tissue samples were obtained using an increment borer (Sunnuto Corporation) between 1999 and 2005 from trees growing in the area delineated by dissolved-phase groundwater contamination. Such tree coring methods have been used previously at other sites to determine the presence of a variety of chemicals in tree rings, such as chlorinated solvents (Vroblesky et al. 1999a), petroleum hydrocarbons related to fuel-oxygenated gasoline (Landmeyer et al. 2000), and metals (Forget and Zayed 1995). At the study site, cores were collected at a height of 1 ft (0.3 m) above ground on the southern side of each tree. Replicate cores taken about 2 in. (5 cm) apart were collected at each tree.

Results indicate the presence of benzene, toluene, and naphthalene (as well as other coal-tar related compounds, such as styrene) in various tree tissues sampled throughout the planted area in March 2002 (Table 13.2) that are also present in the underlying groundwater. These data suggests that the poplar trees are taking up contaminated groundwater as part of the transpiration stream during the time of sampling.

To relate changes in groundwater-level fluctuations with tree uptake of contaminated groundwater, the flow of water through representative hybrid poplar trees was estimated with a Dynagage sap-flow meter (Fig. 13.4; Flow32, Dynamax, Houston, TX). The Dynagage is used to heat the water in a tree using an imprinted circuit to which a constant low voltage (less than 5 mV) is applied; radial heat losses to the atmosphere and vertical heat losses are reduced with a Styrofoam wrap and heat from incident radiation is reduced by covering the gage with reflective tin foil. The sap-flow system was run for 1-week intervals during the summer months, powered by a marine battery. Data on the flow rate of water in the tree are reported in grams per hour (g/h), as induced by a change in temperature of the water in the tree, ΔT, between a heated and unheated reference area on the Dynagage.

Sap flow measurements were made on trees growing next to a series of monitoring wells. For example, a monitoring well (A3-T1) is located upgradient of the planted area on the eastern side of the phytoremediation plot. In this well, concentrations of benzene and naphthalene were greater than 6 mg/L, and less than 2 mg/L for toluene prior to 2001 (Fig. 13.5), suggesting a source of DNAPL some

Table 13.2 Detection and relative concentrations, in parts per billion by volume, of benzene, toluene, naphthalene, and other organic compounds in the headspace of vials containing tree materials from the phytoremediation site at a former MGP near Charleston, SC, March 2002.

Tree cored	Benzene	Toluene	Ethylbenzene	Xylenes	Naphthalene	Styrene
1	19.2	6.4	<20.0	21.1	<20.0	12.4
2	20.2	6.2	<20.0	23.2	<20.0	108
3	23.5	6.8	<20.0	29.9	<20.0	–
4	20.1	5.8	<20.0	23.7	<20.0	–
5	21.7	6	<20.0	24.3	<20.0	–
6	22.5	5.9	<20.0	14.3	<20.0	1,600
7	16.2	5.1	<20.0	20.3	<20.0	47.6
8	17.6	5.7	<20.0	22.7	<20.0	12.5
10	17.8	6.6	<20.0	22.3	<20.0	–
11	20.4	6.2	<20.0	26.6	<20.0	–
12	20.6	5.9	<20.0	22.9	<20.0	–
13	14.6	5.1	<20.0	17.5	<20.0	–

All results are in parts per billion by volume (ppbv), – not detected, < less than, Method detection levels were 10 ppbv (benzene); 5 ppbv (toluene); 20 ppbv (ethylbenzene); 10 ppbv (xylenes); 20 ppbv (naphthalene); 10 ppbv (styrene)

Fig. 13.4 Measurement of sap flow in a 7-year old hybrid poplar tree using the heat-balance method (Photograph by author).

distance upgradient of the well; DNAPL has not been detected in this well. Between 2000 and 2007 concentrations of benzene, toluene, and naphthalene decreased from their highest measured concentration by 80%, 99%, and 68%, respectively.

During this time period, concentrations of the compounds did not decrease steadily, but rather fluctuated from low to high back to low concentrations. An annual, seasonal variation in concentrations of these dissolved-phase contaminants was observed. For example, in 2001, 2003, and 2005, the lowest concentrations of benzene, toluene, and naphthalene were observed during the spring–summer sampling events, when transpiration increases to maximum and groundwater uptake potential increases to highest, and the highest concentrations were observed during the fall–winter sampling months.

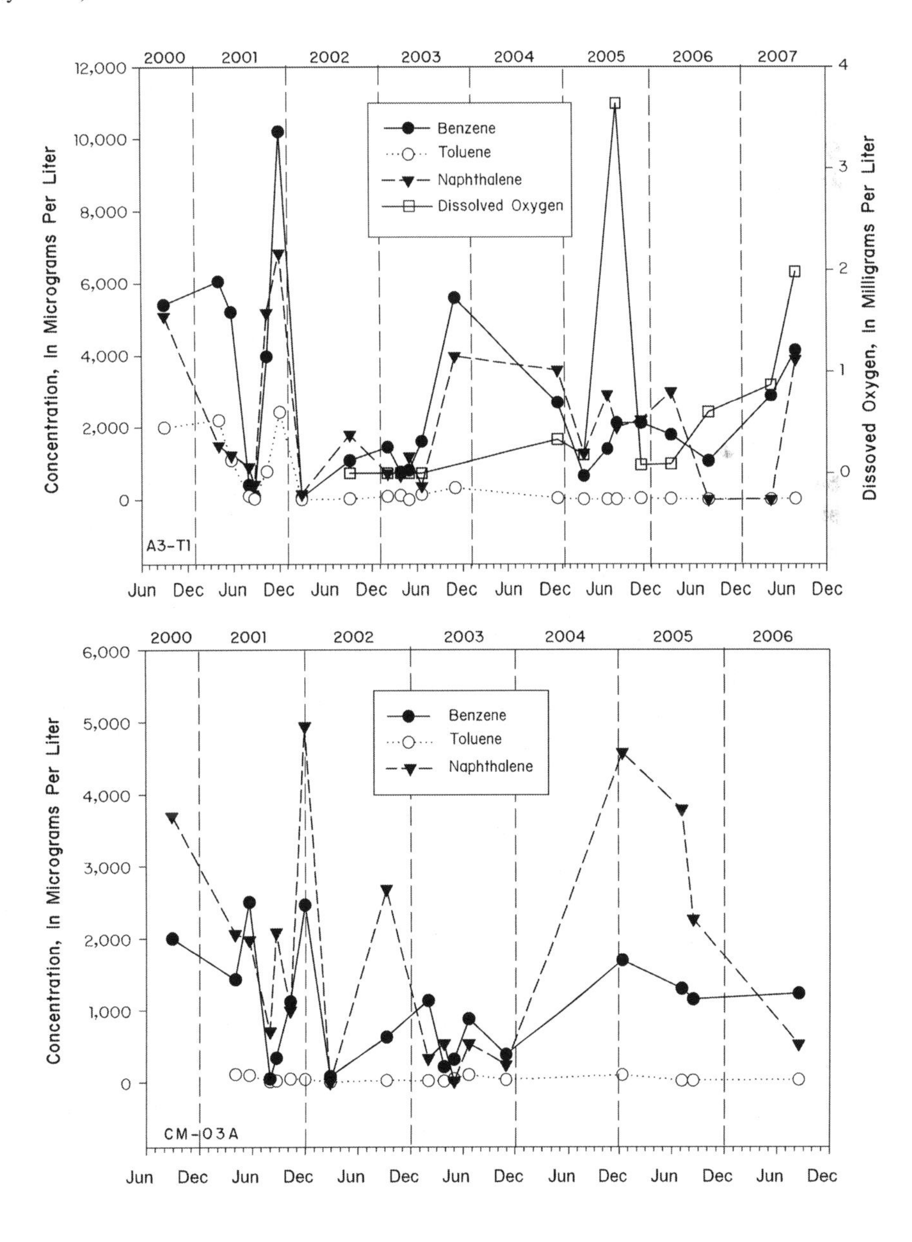

Fig. 13.5 Concentrations of benzene (●), toluene (○), naphthalene (▼), and dissolved oxygen (DO) (□) in monitoring wells A3-T1 (*top*) and CM-03A (*bottom*), 2000–2007, at the former MGP site, Charleston, SC. Vertical dashed lines represent calendar year divisions.

The elevated concentrations of benzene, toluene, and naphthalene measured during the fall–winter months did rebound over time but to progressively lower concentrations (Fig. 13.5). Similar information regarding the effect of phytoremediation on such plume control can be found in Van Den Bos (2002).

As the contaminant concentrations decreased in these wells, the level of dissolved oxygen in groundwater increased from 0 to 3.6 mg/L (Fig. 13.5). This increase may be due to (1) lowered demand on DO added to the aquifer during precipitation events, which had been shown at the site earlier to deliver up to 8 mg/L each event but decreasing rapidly thereafter; (2) an increase in DO from generation by poplar trees combined with the effects of (1); (3) a decrease in the generation of reduced species that consume DO, and; (4) an overall decrease in the amount of DNAPL upgradient of the well.

A few wells in the planted area were characterized by measurable levels of DNAPL at the base of the water-table aquifer adjacent to or below the well. For example in a monitoring well (CM-03A), where 1.12 ft (0.3 m) of DNAPL was detected in September 2005, benzene, toluene, and naphthalene decreased from their highest concentration by 54%, 80%, and 55%, respectively (Fig. 13.5). For benzene and naphthalene, this reduction was not as great as that observed for wells in the area were DNAPL was not present. Essentially, the levels of contaminant after 75 years and installation of hybrid poplar trees are not greatly changed from initial concentrations.

Even though the concentration decreases are not as great as observed in earlier years, a seasonal response in concentrations can still be seen, especially during 2001, indicating a positive influence of the trees on groundwater contaminant levels (Fig. 13.5). The influence of tree uptake is overshadowed, however, by the continual generation of dissolved-phase contaminants from the DNAPL.

In another monitoring well (USGS-A) located near the upgradient part of the western part of the planted area, contaminant concentrations are an order of magnitude lower than those seen in other wells in the phytoremediation area. Between 2000 and 2007, concentrations of benzene, toluene, and naphthalene in USGS-A decreased from their highest concentration measured by 99%, 100%, and 100%, respectively (Fig. 13.6). During this time concentrations of the compounds did not decrease steadily, but rather fluctuated from low to high back to low concentrations—an annual, seasonal variation in concentrations of these dissolved-phase contaminants was observed. For example, in 2001, 2002, and 2003, the lowest concentrations of benzene, toluene, and naphthalene were observed during the spring–summer sampling events and the highest concentrations were observed during the fall–winter sampling months (Fig. 13.6). The elevated concentrations of benzene, toluene, and naphthalene measured during the fall–winter months did rebound over time but to progressively lower concentrations (Fig. 13.6). As the contaminant concentrations decreased, the level of dissolved oxygen in groundwater increased over time from 0 mg/L to a high of 4.2 mg/L (Fig. 13.6).

In a monitoring well (USGS-C), installed near the downgradient area of the western part of the planted area, concentrations of benzene, toluene, and naphthalene decreased between 2000 and 2005 from their highest concentration measured by 75%, 70%, and 48%, respectively (Fig. 13.6). There also was a seasonal variation in concentrations of these dissolved-phase contaminants over time. For example, in 2001, 2002, 2003, and 2005, the lowest concentrations were observed during spring–summer sampling events and the highest concentrations observed during fall–winter sampling events. Concentrations of DO in groundwater increased from 0 mg/L to a high of 2.9 mg/L (Fig. 13.6).

Another site where PAH contamination provided the opportunity to investigate the interaction between poplar trees, groundwater hydrology, and contaminant fate was near Oneida, TN. The use of coal-tar to treat railroad ties to inhibit water uptake occurred between 1950 and 1973 at a site in the north-eastern part of Tennessee (Widdowson et al. 2005a). Even though it was stored in an AST and a holding pond, the viscous properties of most coal-tars enable them to be found in many locations at sites adjacent to where they were used. At the site in Tennessee, creosote contaminants were found to seep out at the bank of a local stream. The stream is hydrologically connected to the shallow aquifer comprising sands and clays to a depth of about 11.4 ft (3.5 m). As is the initial remedy at many sites where either DNAPL or LNAPL has been detected and is flowing toward a surface-water body, an interceptor trench was constructed in an attempt to cut off the movement of creosote to the stream.

At the bottom of the aquifer a dense shale of low permeability is encountered. This has acted to limit the downward vertical migration of coal tar DNAPL, and there exists a pool of DNAPL up to 11 in. (30 cm) thick. The rate of groundwater flow is about 0.06 ft/day (0.02 m/day). As is typical of many shallow aquifers, recharge is by precipitation, which averages about 59 in./year (152 cm/year). At the site, up to 1,026 hybrid poplar trees were planted in 1997; these trees were between 2 and 3 years old. An additional 120 trees were planted in 1998.

Monitoring wells screened at various depths within the shallow aquifer indicated that the saturated thickness decreased to less than 3.2 ft (1 m) during the summer months, following establishment of the trees. The researchers measured the depth of root penetration at some locations to be 6.5 ft (2 m) below land surface. For total PAH

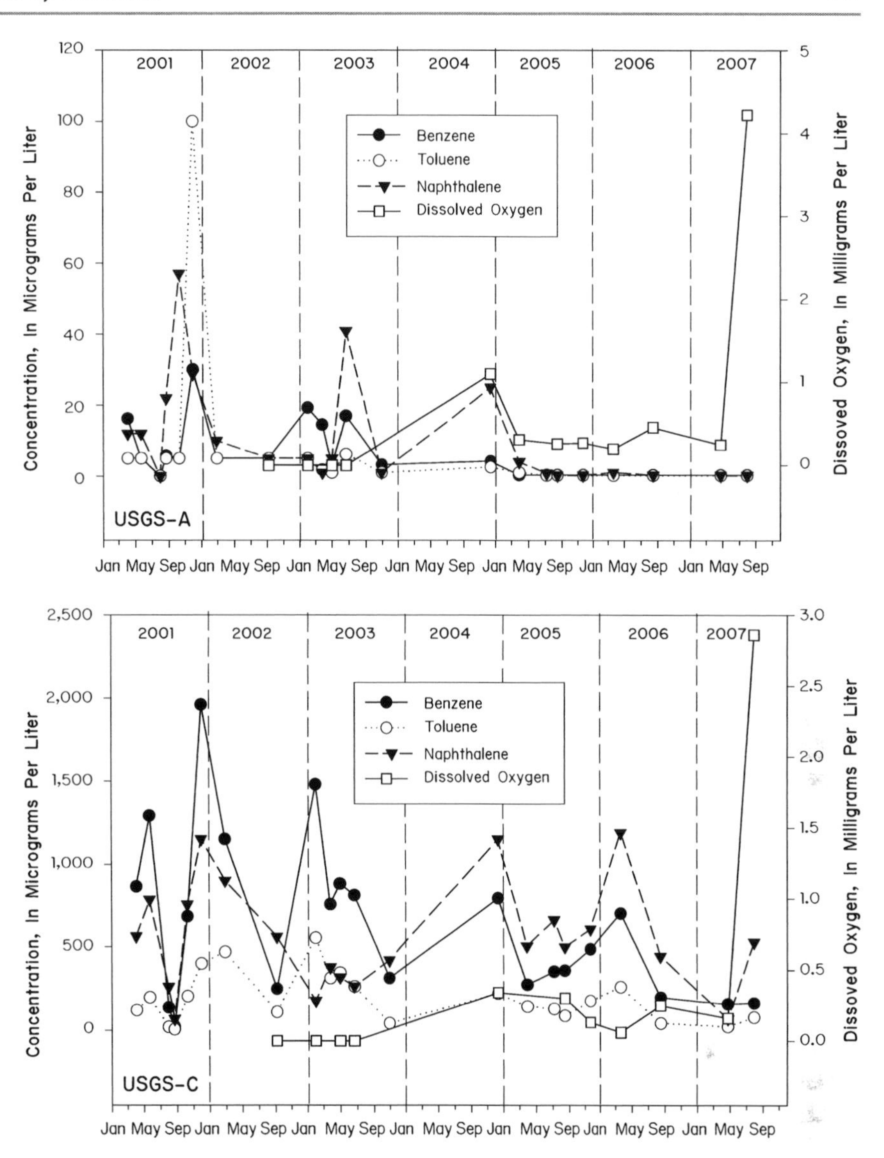

Fig. 13.6 Concentrations of benzene (●), toluene (○), naphthalene (▼), and dissolved oxygen (DO) (□) in wells USGS-A (*top*) and USGS-C (*bottom*), 2001–2007, at the former MGP site, Charleston, SC. Vertical dashed lines represent calendar year divisions.

concentrations collected in monitoring wells between 1997 and 2004, the concentrations remained high and stable; the average maximum was 20,000 μg/L. This elevated value is in part due to the presence of coal-tar DNAPL at depth below the trees. This caused the researchers to look at the PAH concentrations at various depths in the shallow aquifer, rather than total PAH, to determine the effect, if any, of the trees on PAH groundwater contamination.

Samples were collected between 1998 and 2003. Sample results indicate that after 2001, the concentrations of total PAHs near the top of the water table had decreased (Fig. 13.7). Conversely, PAH concentrations at greater depths remained unchanged. Similar decreases were observed in the ratio of naphthalene concentration to total PAH (as TPAH) concentration with depth (Fig. 13.8). Widdowson et al. (2005a) suggest that these concentration decreases correlate with the interaction of the poplar tree roots and the water table. Moreover, the installation of an interceptor trench at this site assists in decreasing the contaminant migration potential of the dissolved phase contaminants to the creek.

Marr et al. (2006) studied the relation of the phytoremediation plot at Oneida to the volatilization of naphthalene as a loss mechanism from the contaminated aquifer. The flux of naphthalene from the unsaturated zone to the atmosphere was measured over the year, both when the trees were actively removing groundwater and the water table was lowered and when the trees were dormant and the water table higher. Naphthalene flux was measured using a flux chamber. The highest flux of naphthalene to the atmosphere near 23 μg m^2/h was measured during the month of August 2004. The higher flux was attributed to a thicker unsaturated zone after the water table dropped about 1 m. The case could have been made stronger for the phytoremediation-induced increase in contaminant volatilization between the water table drop and plants, however, by

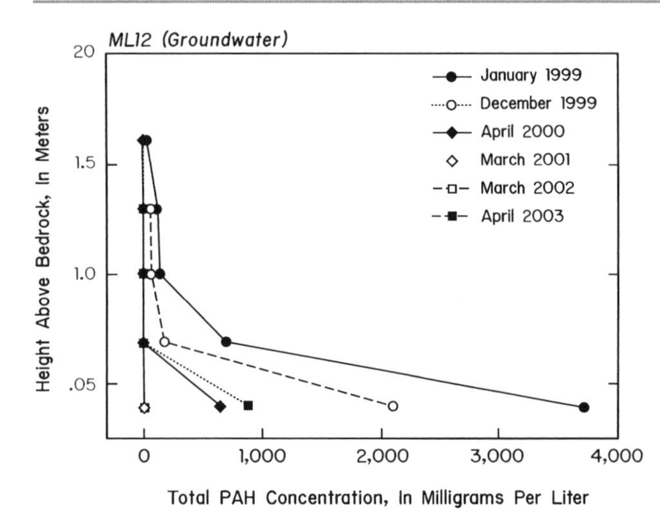

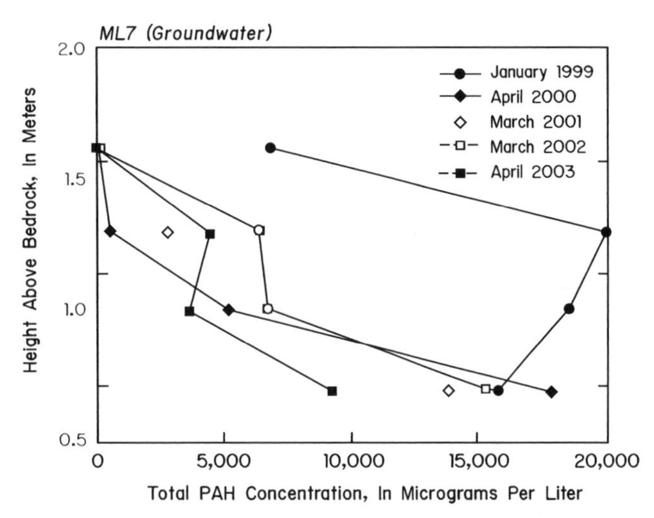

Fig. 13.7 The decrease in PAH concentrations following establishment of poplar trees at a site of PAH-contaminated groundwater (Modified from Widdowson et al. 2005a). One meter is equivalent to 3.2 ft.

including data for transpiration rates or estimates for *VPD* and contaminant loss.

13.3.2 Plant Transformation Reactions

One of the first studies to relate the root distribution of a tree to the fate of PAHs in the subsurface was by Olson and Fletcher (1999). They investigated some "volunteer plants," or those that grew naturally in a contaminated area, such as mulberry, at a site of a former waste disposal basin. The soil in which the plants were growing consisted of a soil–sludge that contained PAHs. The largest roots were found in the upper layer of soil–sludge to around 23 in. (60 cm), and the PAH concentrations were no more than 20% of that characteristic of the sludge in other areas. Between 23 and 39 in. (60 and 100 cm) deep, which was characterized by smaller roots, the PAH concentrations were similar to the unplanted areas. The authors hypothesized that the age of the roots had a direct impact on the overall PAH removal potential, such that older, more established shallower roots provided greater PAH degradation.

Ouvrard et al. (2006) investigated the production of organic-rich root exudates in the rhizosphere of plants exposed to PAH-contaminated soil and the relation of these exudates to the subsequent bioavailability of the PAHs for microbes or plant uptake. The organic compounds released by roots, such as malic acid, citric acid, and oxalic acid, can provide for the possible absorption of PAHs and other organic compounds. Ouvrard et al. (2006) determined that the presence of model root exudates such as malic and malonic acids was responsible for an increased removal of the PAH phenanthrene from solution following linear isotherms. They suggested that this enhanced soil sorption was a consequence of the organic acids modifying the soil aggregate structure to expose hidden sorption sites, rather than acting as sorption sites themselves. The problem with root exudates, however, is that they are readily metabolized substrates for the heterotrophic rhizospheric populations, and these exudates only become available if production is greater than consumption.

Many studies cited here on the effect of the enhanced degradation of PAHs in vegetated systems is due primarily to the enhanced microbial populations associated with the rhizosphere. However, other mechanisms associated with the presence of plants have been offered. Gregory et al. (2005) showed that organic matter released by plants can sequester PAHs and PAH metabolites and that these compounds become part of the natural cycle of humification in planted areas. Although the release of root exudates provides a labile substrate to support an active microbial community, they also act to help form the humic fraction of soils by decreasing larger fractions or increasing smaller fractions. The first process would tend to increase PAH degradation. The latter process would tend to decrease PAH attenuation because it leads to the production of sorption sites to render PAHs less bioavailable.

The effect of rhizosphere bacteria on PAH-contaminated soils in the unsaturated zone was investigated by Muratova et al. (2003). They exposed alfalfa (*Medicago sativa* L.) and a reed (*Phragmites australis* (Cav.) Trin. ex Steud.) to PAH-contaminated soils in pots in the laboratory over a 2-year period. Naphthalene was 8.71 mg/kg, and total PAHs were 79.80 mg/kg. Unplanted pots also were prepared. At the end of 2 years of interaction, the unplanted control PAH decreased from 79.80 to about 50 mg/kg PAHs. The planted control PAH concentration decreased to about 30 mg/kg. The authors attributed the increased loss of PAH to vegetation-induced microbial activity. Alfalfa plants had 1.3 times more total soil bacteria, although in the reed no increase was

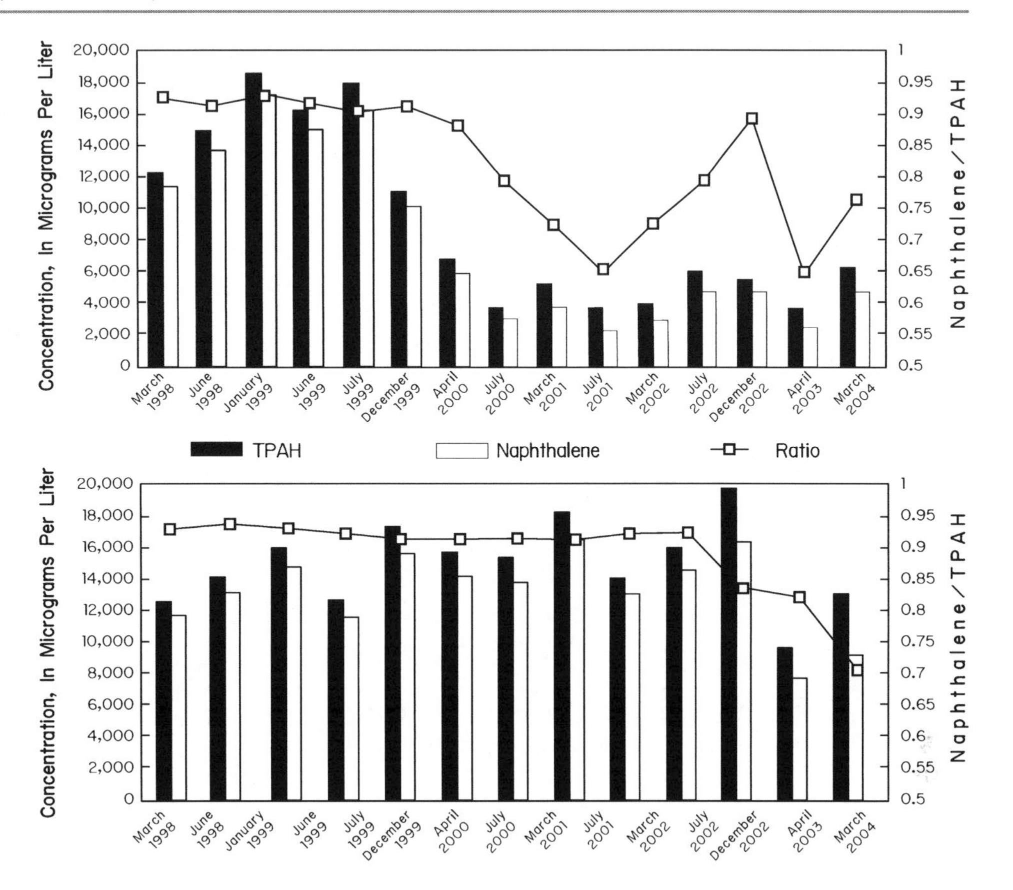

Fig. 13.8 The preferential loss of naphthalene from groundwater over time relative to total PAHs (as TPAH) present at different locations at the phytoremediation site in Tennessee (Modified from Widdowson et al. 2005a).

observed, even though both plants had similar higher PAH losses after the 2-year incubation. Nitrogen-cycle bacterial populations increased by two orders of magnitude in the planted versus unplanted pots. The number of PAH-degrading bacteria, or the phenanthrene-degrading bacteria used as surrogate, increased in the alfalfa-planted pots up to seven times more than as measured in the controls. Interestingly the number of PAH degraders present in the contaminated soil grown with the reeds actually decreased. The fact that PAH decreases were seen over time in these pots may be because reed can transport oxygen to the root zone, and this is sufficient alone to cause contaminant remediation.

Chen et al. (2003) investigated the fate of pyrene in the rhizosphere of tall fescue (*Festuca arundinacea*) and switchgrass (*Panicum virgatum*). They added 50 mg/kg of ^{14}C-pyrene and cold (nonlabeled) pyrene to uncontaminated soil which contained fescue and switchgrass. The fate of the ^{14}C-pyrene was traced in the soil, plants, and headspace. They reported that for tall fescue, 37% and 30% of the ^{14}C-pyrene was mineralized to $^{14}CO_2$ by the tall fescue and switchgrass, respectively, relative to 4% mineralization in the unplanted control. Radioactive ^{14}C that remained as plant biomass was 8% and 5% for tall fescue and switchgrass, respectively. Radioactive ^{14}C that remained in the soil unaffected by mineralization was 58% and 55% for tall fescue and switchgrass, respectively. In part, the high percentage of pyrene that remained in the soil can be explained by the bound residue, probably to soil humic and fulvic acids, that made it unavailable to plant uptake. The importance of this work is that pyrene concentrations decreased more in planted versus unplanted treatments and that greater than 30% of the loss was to CO_2.

An interesting outgrowth of this and other similar studies is the distribution of the pyrene in the bound soil residual organic matter and the implications for contaminant fate over time. Plants take up organic matter in the root zone but also release living and dead organic matter to the soil. Rhizospheric bacteria associated with this zone can help degrade these root exudates as well as contaminants. The contaminants also have absorption sites that can help immobilize the contaminant. This stabilization alone is important, even if such "aged" contaminant is less bioavailable to plant uptake or microbial mineralization. Guthrie et al. (1999) indicated that ^{13}C-pyrene added to soils remained in the soil humic fraction without undergoing mineralization.

Schwab and Banks (1994) also reported in a laboratory study the increased loss of PAHs in contaminated soil relative to unplanted soils. This loss was attributed to the 7–200 times greater number of microbes in the planted soils relative to unplanted soils. Most importantly, they added ^{14}C-pyrene to soils with and without surrogate plant exudates, such as formic and acetic acid. Although mineralization to $^{14}CO_2$ was low (<0.17%), mineralization was greater (by 36%) in the presence of the organic acids.

Reilley et al. (1996) reported that for anthracene and pyrene added to soils that were vegetated and unvegetated, 30–44% more degradation was observed when vegetation was present. The work was carried out in the laboratory, using fescue (*Festuca arundinacea*), alfalfa (*Medicago sativa*) sundangrass (*Sorghum vulgare*), and switchgrass (*Panicum virgatum*). Interestingly, the normal processes of natural attenuation, such as leaching, abiotic degradation, mineralization, and sorption were not significant factors in the dissipation of the added PAHs. Moreover, total accumulation of PAHs in the plant was less than 0.03% of that initially added to the soil.

After the first Gulf War, it was reported that crops in Kuwait could be grown in soil that had been contaminated with up to 10% crude oil and that over time and the presence of these plants the concentrations of the PAH decreased (Radwan et al. 1995).

Günther et al. (1996) investigated the effect of ryegrass (*Lolium perenne* L.) on the removal of hydrocarbons such as PAHs from contaminated soils under laboratory conditions. They reported that relative to contaminant decreases in unplanted controls, the planted treatments produced faster rates of contaminant disappearance, and increased microbial abundance. Because less than 0.1% of the total contaminants added were recovered from the plants themselves, it was assumed that the losses were attributed to soil microbial oxidation in the rhizosphere. A summary of the potential fate of PAHs in plants is shown in Fig. 13.9.

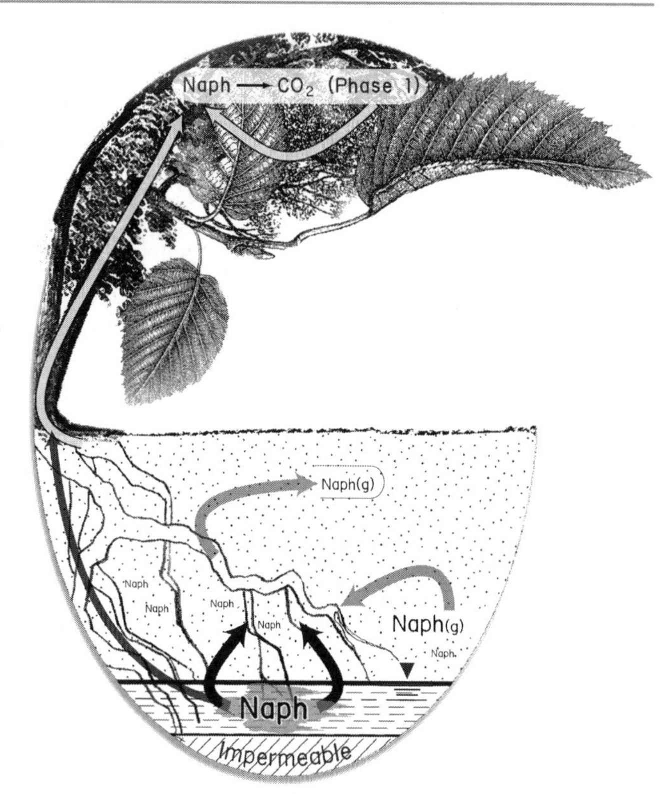

Fig. 13.9 Plant and groundwater interactions where PAH-contaminated (naphthalene, as Naph, shown as an example) groundwater exists. Naph(g) indicates the volatile phase.

13.4 Plant Interactions with Fuel Oxygenates and Additives

The term "fuel" can represent any reduced, hydrogen-saturated organic compound that when burned in the presence of oxygen releases its chemical bond energy, originally supplied by plant photosynthesis, and converted to mechanical energy. Essentially it is the reverse of photosynthesis. Fuels can be a solid, such as wood or coal, a liquid, such as gasoline, or a gas, such as methane, butane, or propane, or a mixture, such as liquefied petroleum gas (LPG).

The consumption of liquid fuels, such as gasoline, is of global scale and large volume. Considering that CO_2 and H_2O are supposed to be the only combustion byproducts, as indicated by the products of the respiration reaction, it may come as a surprise that this combustion has led to air-quality degradation. This is because the organic source of the fuel also contains nitrogen and sulfur compounds that, when oxidized in an internal combustion engine, produces oxides of nitrogen and sulfur, as NO_x and SO_x emissions, even though catalytic converters were mandated standard on car exhaust systems produced after the 1980s.

To improve combustion, additives that included lead-substituted organic compounds were mixed into gasoline stocks. Such leaded gasoline had high octane ratings, and the lead was deposited during detonation on the value guides and heads to protect the engine from excessive valve-train wear. Because lead built up on the valves over time, compounds were added to the gasoline to remove or scavenge these lead deposits. These lead scavengers included the halogenated organics EDB and 1,2-dichloroethane (1,2-DCA). The use of leaded gasoline was banned in the mid-1970s.

In an attempt to decrease the air-quality degradation resulting from the incomplete combustion of fuel in automobiles, the U.S. Congress mandated that in pollution-prone areas that fuel be supplemented with oxygenates, or petroleum compounds to which an oxygen molecule has been added, called reformulated gas (RFG). The oxygenates typically used are ethers or alcohols, with the form R–O–R or R–OH, respectively, where R denotes an organic compound. The RFG mandate called for a 2% by weight content of oxygen. The compound MTBE was selected to meet this oxygenate mandate. It could be sourced from waste stocks, and it mixed easily with existing product flow in pipelines,

Table 13.3 Reported physical and chemical properties of common groundwater contaminants that have importance to plant bioavailability.

[TAME, tertiary-amyl methyl ether]

Contaminant	Water solubility (mg/L)	Log K_{ow}	Log K_{oc}	Henry's constant (dimensionless, H/RT)
MTBE	>50,000	1.2	1.04	0.018
TBA	infinitely	0.37	1.57	4.8×10^{-4}
TAME	20,000	–	1.27	5.2×10^{-2}
Ethanol	infinitely	–0.16	1.21	2.09×10^{-4}
EDB	4,200	1.8	–	0.02

and was added to gasoline up to 15% by volume. However, as an ether, it was very soluble in water (50,000 mg/L). Also, USTs often leaked. It would turn out, by the end of the twentieth century, that this answer to protect air quality from degradation by gasoline engines had led, unexpectedly, to groundwater-quality degradation.

Some of the common groundwater contaminants and their physical and chemical properties related to plant uptake are described in Table 13.3. As this table indicates, the similarity of MTBE with alcohols suggests that its relatively low log K_{ow} and its relatively high solubility would render it less likely for plant interaction than, say, BTEX.

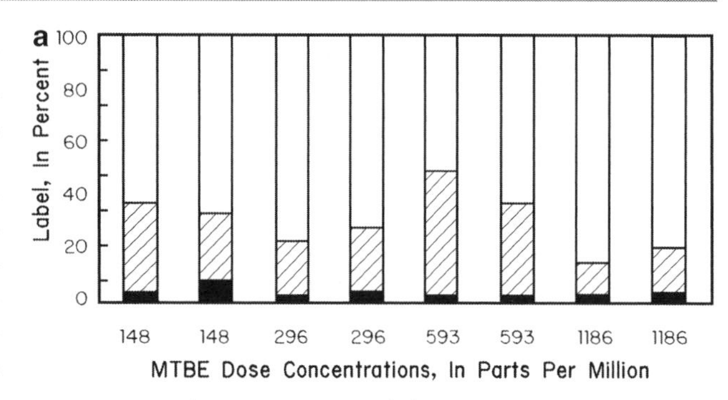

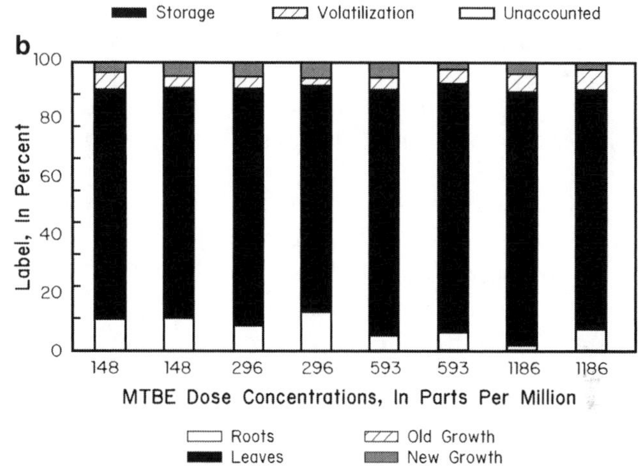

Fig. 13.10 The variable fate of ^{14}C-MTBE added to cuttings grown hydroponically (Modified from Hong et al. 2001).

13.4.1 Plant Interaction and Uptake Pathways

In the late 1990s, researchers at the Port Hueneme Naval Base investigating an extensive plume of gasoline that contained MTBE took samples of the transpirate that emanated from a Eucalyptus (*Eucalyptus spp.*) tree growing above the main axis of the plume and detected MTBE. Newman et al. (1999a) confirmed this observation in the laboratory with hybrid poplar and eucalyptus cuttings grown in solutions that contained MTBE.

The fate of MTBE in trees was investigated in the laboratory as part of a large modeling and field study into the effectiveness of using hybrid poplar trees to hydrologically contain and remediate a plume of MTBE that had been released at a site in Houston, TX (Hong et al. 2001). They added ^{14}C-MTBE to hydroponic solutions that contained 8-in.-long cuttings of hybrid poplar trees (*Populus deltoides* x *nigra* DN34). The experiment was set up with the necessary treatments to control for MTBE loss by biodegradation only (no cutting added), leaks (a glass rod was used in place of the cutting), and roots versus shoots (the cutting was cut at the location of the cap over the growth solution that contained ^{14}C-MTBE (activity of 1.48 Curies per millimole, Ci/mmol) and 4.32 mg of unlabeled MTBE. Samples of the various compartments were collected and analyzed for ^{14}C-MTBE.

At the end of the 10-d experiment, the mass balance of the ^{14}C-MTBE added to the growth solution indicated that for the 65% of ^{14}C-MTBE that was removed from the hydroponic solution, about 27% was lost through leaks in the apparatus, but almost 17% was lost by transpiration, which was indirectly calculated from the difference between the uncut and cut cuttings (Fig. 13.10). Up to 11% was lost by the stems that had been cut, which the authors indicate is significant in that at least some removal of MTBE would occur with dead or dormant trees. Very little MTBE (0.15%) was detected in the root zone. Negligible disappearance of ^{14}C-MTBE was observed in the control to test for microbial degradation. All of the label recovered was as ^{14}C-MTBE, as no intermediate compounds were detected, including as CO_2.

MTBE released to the atmosphere intact by transpiration would quickly be degraded by photo-oxidation or dilution. This is in contrast to its relatively long half-life of at least several years in oxic and anoxic groundwater. The researchers report that the *TSCF* for the ^{14}C-MTBE in the laboratory experiment was 0.5–0.8, within that predicted for preferential uptake. The high solubility of MTBE or its structure may allow its passage through the Casparian strip, where other hydrophilic compounds would be retarded.

The effect of the rhizosphere on MTBE was studied by Ramaswami et al. (2003). As was seen in previously referenced reports, the potential for MTBE to taken up and volatilized from shoots and leaves is a controlling factor in the fate of MTBE. Less is known about the fate in the rhizosphere, other than that MTBE fractions in some roots were high (Ma et al. 2004). While MTBE is recalcitrant under anoxic conditions, where it degrades to TBA (Bradley et al. 2002), MTBE has been shown to be less recalcitrant than previously thought, and can undergo complete oxidation to CO_2 under aerobic conditions (Bradley et al. 1999, 2001; Landmeyer et al. 2001). Hence, because the rhizosphere is essentially aerobic, it would be anticipated that MTBE would be degraded. The compromising factor is that water flow rates might be faster than contaminant biodegradation rates.

Ramaswami et al. (2003), however, did not observe MTBE degradation in aerobic treatments of MTBE and rhizospheric bacteria. They show results for DO concentrations in lab microcosms with no soil and for rhizosphere soils with MTBE-acclimated soils. Although they report no MTBE degradation in the rhizosphere treatment, the time course lasted only 48 h and the DO immediately decreased from 8 to 3 mg/L in 1 day; as such, the aerobic condition necessary to support aerobic MTBE metabolism was electron-acceptor limited. They increased the time course to 10 days and still reported little MTBE aerobic biodegradation. However, the DO concentrations are not shown, and the frequency of oxygen addition is not reported. It is likely that these microcosms did not suggest aerobic MTBE-degrading bacteria in the rhizosphere simply because of a lack of oxygen.

Rentz et al. (2003) investigated the possibility of increasing the DO content around roots in high-BOD contaminated sediments and its effect on the growth of plants that might be planted at such sites to remediate the contaminated soils. This is significant, because many of the plants that could be planted at contaminated sites may not have the gas-transport structures, or aerenchyma, to transport O_2. Additionally, as was shown at the site in Texas discussed in Chap. 12, poplar trees also can decrease the DO in shallow groundwater by the release of labile organic matter. Hence, poplar trees can be a source or sink for DO in contaminated aquifers and vadose zones.

Ma et al. (2004) investigated the fate of MTBE after exposure to hybrid poplar trees such as DN-34. In the laboratory, they determined the degree of partitioning of gas-phase MTBE when added to vials that contained samples of DN-34, such as cuttings, leafs, and roots. The experiment was then performed again but water replaced the air in the vials. They reported that MTBE partitioned to a greater extent to roots than to leaves or cuttings, and much more so than between water and cuttings. However, even these partition coefficients are low, due to MTBE's high water solubility near 50,000 mg/L.

They also grew cuttings in MTBE solutions after previous rooting and shoot growth in uncontaminated solution. They reported that growth of the cuttings was not inhibited during the 7-d exposure to MTBE in the solution. However, transpiration rates decreased significantly, between 36% and 59%, after MTBE was added compared to rates prior to MTBE addition. MTBE was detected in the diffusion traps in all the trees that grew in MTBE-spiked solutions (Fig. 13.11). The traps positioned at lower elevations contained more MTBE than traps located higher on the cuttings. They report that between 12% and 47% of the

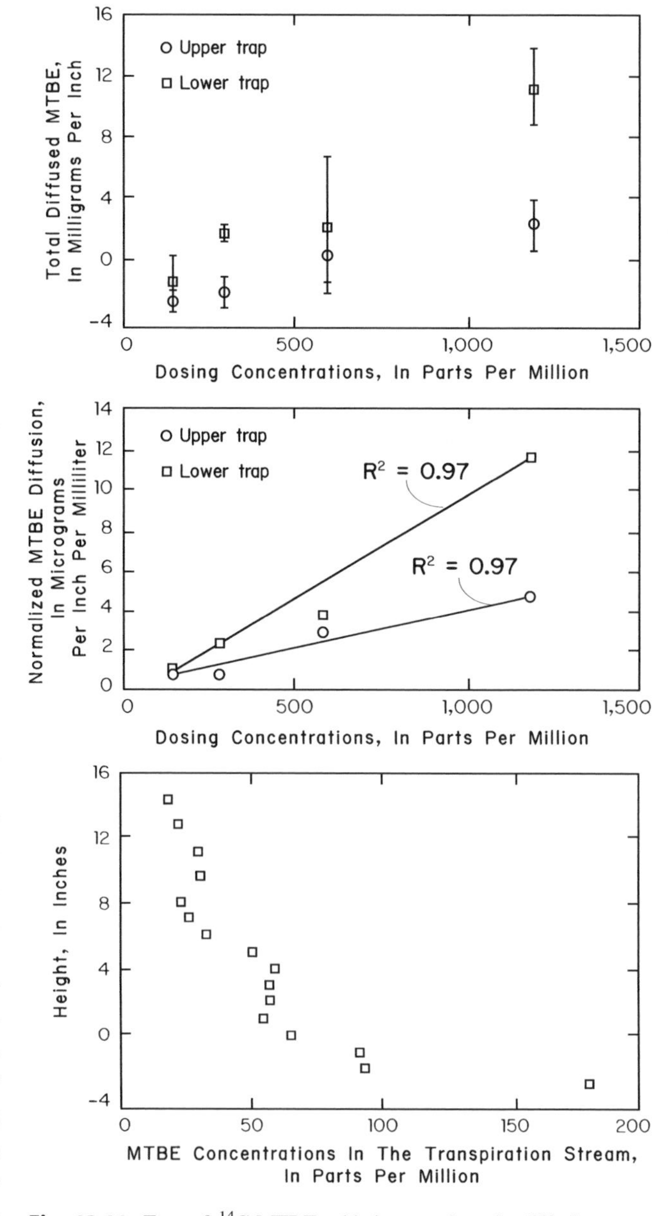

Fig. 13.11 Fate of ^{14}C-MTBE added to cuttings in diffusion traps (Modified from Ma et al. 2004). One inch is equivalent to 2.54 cm.

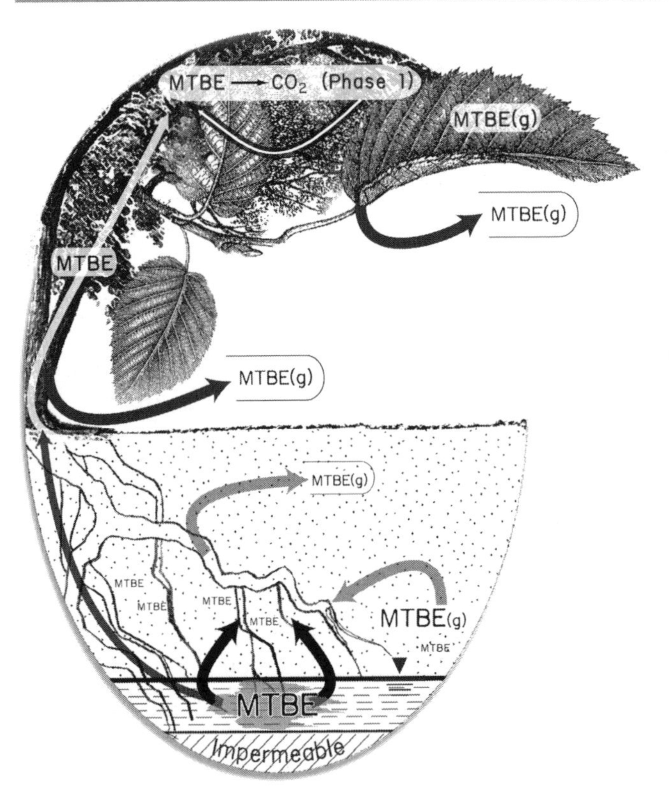

Fig. 13.12 The potential fate of MTBE in the dissolved and gaseous phases upon being taken up by a tree at a phytoremediation site. MTBE (g) is the volatile phase.

Fig. 13.13 At a UST release site in Beaufort, South Carolina, a dissolved-phase plume of BTEX and MTBE compounds extended downgradient in the surficial aquifer beneath a stand of 40-year-old live oaks (*Quercus virginiana*). Monitoring wells are evident in front and behind this particular tree (Photograph by author).

MTBE taken up by transpiration was released unattenuated. However, this measurement of MTBE loss is from stems only that were monitored with the diffusion traps; the effect of MTBE loss from leaves would increase this loss percentage. Much of the MTBE spiked into the solution that was not volatilized remained in the roots and lower stems of the cuttings. Such plant parts comprise the largest proportion of biomass. Very little was found to be incorporated into plant tissue, including the leaves.

Because the log K_{ow} of MTBE is about 1.2, it is not surprising that many researchers have shown in the lab that MTBE is transpired from solution with almost 100% recovery in the air (Rubin and Ramaswami 2001; Hong et al. 2001). This could imply either dissolved-phase or gas-phase uptake of MTBE from the subsurface (Fig. 13.12).

Further evidence to indicate that MTBE in groundwater is transpired to the atmosphere essentially intact and at near 100% efficiency was provided by Yu and Gu (2006). They exposed weeping willow (*Salix babylonica* L.) cuttings growing in hydroponic solutions in the laboratory to a range of MTBE concentrations. The effect of MTBE concentration on the plant itself was measured by changes in transpiration; only at MTBE concentrations near 400 mg/L was a decrease in transpiration observed. Such high concentrations would tend only to be found near source areas of a gasoline spill that contained 15% MTBE (in such a mixture, MTBE has about 2,500 mg/L maximum solubility in water). A mass balance indicated that most of the MTBE was removed from the solution and found in the air, and that MTBE detected in the plant had not been transformed (Yu and Gu 2006).

These studies support the conclusion that in most cases, MTBE is volatilized through the plant after uptake. Arnold et al. (2007) investigated MTBE loss by existing vegetation at a gasoline site in California characterized by MTBE- and TBA-contaminated groundwater. The existing vegetation consisted of mature conifers and was 33–43 ft (10–13 m) tall. The depth to groundwater was not reported, although it appeared to be between 5 and 15 ft (1.5–4.5 m) below ground. In order to determine if the MTBE- and TBA-contaminated groundwater beneath the conifers was being taken up by these plants, leaf gas was collected and the transpired moisture collected as a condensate in 125-mL airtight plastic bags. These samples were then transferred to standard 40-mL VOA vials and analyzed using standard method EPA 8260B.

At a UST release site in Beaufort, South Carolina, a dissolved-phase plume of BTEX and MTBE compounds extended downgradient in the surficial aquifer beneath a stand of live oak (*Quercus virginiana*) (Fig. 13.13). The depth to water table is on average about 13 ft (3.9 m) from land surface, and the soils and sediments comprise sand grains with very little (0.01%) natural organic matter (Landmeyer et al. 2000). Precipitation at the site approaches 60 in./year (152 cm/year), as it is located near the Atlantic Ocean. The live oak trees are at least 40 years old, and even though they tend to have a predominantly shallow and horizontal root structure, oak-tree roots were found at depth near the water table in some monitoring wells near

Fig. 13.14 Oak-tree roots on a water-level pressure transducer removed from a monitoring well at a phytoremediation site near Beaufort, South Carolina. These roots had not grown within the well casing and down to the water table; rather, the roots had grown in the unsaturated zone and entered the well through the well screen (Photograph by author).

trees. For example, Figure 13.14 shows root growth on an *in-situ* sampling device that was placed 1 ft (0.3 m) below the water table. These roots had not grown within the well casing and down to the water table; rather, the roots had grown in the unsaturated zone and entered the well through the well screen.

Tree core material was collected from trees located upgradient of the MTBE plume as well as from trees growing above the plume delineated by an extensive monitoring well network (Fig. 13.13, two monitoring wells are shown; Landmeyer et al. 1998a). Tree cores were collected about 1 ft (0.3 m) above ground surface on the northeast side of each tree sampled. The cores were placed in 40-mL VOA glass vials and sealed with a Teflon-lined cap. In the laboratory, the 40-mL vials were amended with 5 mL of pesticide-grade methanol and rolled on a heated hot-dog roller for 24 h to mix the methanol with the tree cores. The final volume in the vial was brought to 25 mL with addition of organic-free reagent water. A purge-and-trap method similar to EPA 8260B was then used to separate and identify the compounds.

MTBE, as well as other gasoline compounds detected in the groundwater such as BTEX, were not detected in the vial headspace in the trees that grew above uncontaminated groundwater upgradient of the source. Conversely, MTBE and BTEX were detected in the headspace of cores from trees that grew above the groundwater in the former source area and in areas downgradient of the spill above the dissolved-phase plume. Although concentrations of these compounds detected in the headspace of vials that contained tree cores were provided in Landmeyer et al. (2000), they are not repeated here because the relation between tree core headspace concentrations and aqueous groundwater concentrations is not well established, although Ma and Burken (2002) indicate a definitive relation. Finally, no MTBE (or BTEX) was detected in tree cores taken from a tree that was growing above the plume in the downgradient area, as the plume was below a lens of uncontaminated recharge in only that area.

The trimethylbenzene (TMB) isomers 1,3,5-TMB and 1,2,4-TMB also were detected in the headspace of tree cores that contained MTBE (Landmeyer et al. 2000). These compounds are relatively recalcitrant and, therefore, are used as conservative tracers to judge the extent of attenuation of other fuel compounds. These results of tree uptake of TMBs indicate that any study that proposes to use TMBs as conservative tracers may need to consider the effect of plants on these compounds.

Hong et al. (2001) describe a study where about 2 acres (8,094 m^2) of hybrid poplar trees (DN-34 and NE-19) were planted at a site in Houston, Texas, in 1998. MTBE was released to the water-table aquifer, where the water table is about 10 ft (3 m) below ground surface, and is composed of silty sands, thought to be an old river channel. Above these sands is a clay-rich sediment extending to land surface. The trees were planted in rows on 6-ft (1.8 m) centers, and the rows were separated by 8 ft (2.4 m). As for the water budget, precipitation is about 39 in./year (100 cm/year). The ET_P as determined using the Penman equation is about 59 in./year (150 cm/year), which suggests that if plants that reach the water table are used, ET_P will lead to a lower water table. The 10-ft (3 m) whips were planted in holes 1-ft (0.3 m) in diameter drilled into the clay sediments to the water table. Mulch (40%) and fertilizer-amended sand (60%) was used as backfill.

Sap-flow measurements on the 1-year-old trees at the time (1999) indicate that the water-uptake rate was about 4 gal/day/tree (15 L/day/tree) (Hong et al. 2001). Using this data and estimates of water input to the site from precipitation and irrigation, a total input of 293,000 gal (1.1×10^6 L) for 1998 and estimated output 734,000 gal (2.7×10^6 L) suggest that the trees potentially removed 441,000 gal (1.6×10^6 L) of water from the study site. However, because no groundwater-level data were shown, it is hard to determine if this increase in extraction was met by a water-table decline or by increased groundwater flow from upgradient areas or from deeper aquifers.

Since MTBE was banned in the United States as part of the 2006 Energy Policy Act that removed the oxygenate mandate for fuels, alternative additives such as ethanol have begun to replace the MTBE in fuels. One of the more popular choices is ethanol derived from the fermentation of plant material such as corn. Little information exists as to the interaction of ethanol with plants, but ethanol was found to be transpired by willows (*Salix babylonica*) under laboratory hydroponic conditions, as well as removed from solution after root accumulation (Corseuil and Moreno 2001).

Similarly, the interaction of EDB with plants is not well known. This is unfortunate, considering the extremely low maximum contaminant level for EDB in water of 0.05 μg/L. However, some of its physical and chemical characteristics suggest that a plant-based remediation strategy may be feasible (Table 13.3). A review paper by Davis and Erickson (2002) sheds some light on important plant and contaminant interactions that may occur at sites characterized by EDB-contaminated groundwater, either from use as a fuel additive or as a soil fumigant. One of the rate-limiting steps of EDB transformation is its slow diffusivity in water relative to air. Trees that can transpire groundwater or vadose zone water may accelerate EDB loss by increasing the percent air space in the contaminated soil or aquifer. However, the high solubility of EDB in water often rapidly removes it from near the water-table surface and, therefore, extensive plumes of EDB often are found in contaminated areas also characterized by permeable sediments, oxic conditions, and high recharge rates.

13.4.2 Plant Transformation Reactions

The fate of MTBE that is not transpired in plant cells is not entirely known (Newman and Arnold 2003). Typically, the transformation of ethers like MTBE would follow an oxidation pathway of Phase I detoxification. For example, MTBE can undergo *o*-dealkylation reactions and can be hydroxylated to *tert*-butoxymethanol and then *tert*-butyl alcohol and formaldehyde. This transformation is problematic because of the increased toxicity of these MTBE-transformation intermediates, also formed anaerobically by bacteria in groundwater (Bradley et al. 2001). Newman et al. (1998) reported that MTBE could be metabolized by cell cultures of hybrid poplars, albeit at a low percentage of total MTBE added.

13.5 Plant Interactions with Chlorinated Hydrocarbons and Solvents

Many of the petroleum-derived compounds are high-energy compounds—they are saturated hydrocarbons, in which carbon is reduced with hydrogen (–C–H). This molecular quality makes them excellent fuel stock. But compounds are needed for other purposes such as their solvent action or their heat-absorbing qualities. The synthesis of chlorinated hydrocarbons provided a chemical whose qualities make it useful for meeting these needs. The characteristics of chlorinated hydrocarbons are derived from its synthesis, from the substitution of Cl atoms for the H in the C–H of saturated hydrocarbons. This process partially oxidized the originally reduced compound, which renders it greater stability in the presence of oxygen. This is why chlorinated hydrocarbons are used for purposes where a low-flammability product is needed, such as the polychlorinated biphenyls (PCBs) used in electrical transformers and as fire suppressors.

The polychlorinated biphenyls contain chlorine atoms substituted on biphenyl rings in numerous combinations called congeners. Isomers of PCB exist when the same number of chlorine atoms exist on a ring but are located in different positions. Their synthesis in the early 1930s was heralded as a chemical breakthrough, because they resisted oxidation and could be used as heat removal solutions in electrical transformers and flame retardants. Their ability to resist combustion is due to the many chlorine atoms added to the organic molecule, as previously described. This substitution renders them already partially oxidized, so that they are more prevalent to be reduced and *accept* electrons rather than donate electrons through oxidative reactions. These compounds also were found to bioaccumulate and were later detected throughout the environment. In the lab, PCBs have been shown to cause mutations. In response, their use was banned in 1979 in the United States under the Toxic Substances Control Act (TSCA).

Another quality of chlorinated hydrocarbons is their ability to be used as a solvent and this is perhaps the most widely used means for these compounds, as degreasers and cleaners. The dry-cleaning industry uses these compounds rather than use soap and water to remove organic-based stains from materials. These compounds also are used by the tool-and-die, semiconductor, and commercial print industries to remove organic compounds from metal surfaces. Due to its widespread use and chemical stability in the presence of atmospheric levels of oxygen, it is not surprising that it is one of the most common pollutants of soil, groundwater, and surface water at hazardous waste sites.

Some of the most common chlorinated solvents found in the environment are perchloroethylene (PCE) and trichloroethylene (TCE; Moran et al. 2007).They may be readily detected in shallow groundwater because they are stable in the presence of oxygen, and dissolved oxygen concentrations are typically higher in shallower groundwater. Because they also have a specific gravity greater than water, for example TCE is 1.46 at 20°C, the release of pure-phase free product will tend to move through groundwater in response to gravity rather than groundwater flow. This is true only if the PCE and TCE are present in a free phase of pure product, however. Once these compounds are dissolved in groundwater at concentrations less than their solubility, for example at concentrations of TCE less than 1,100 mg/L, the dissolved compound will move with the prevailing flow of groundwater.

Perhaps some of the early resistance to using phytoremediation at chlorinated-solvent-contaminated sites resulted from statements made in Cunningham et al. (1996) that PCE and TCE would be hard to remediate using phytoremediation because they formed dense pools near the bottom of aquifers. While this statement is true when these compounds are released as pure products, in fact, these compounds can be routinely detected in tree-tissue samples where PCE and TCE have been released. The application of phytoremediation at sites characterized by chlorinated-solvent contaminated groundwater has become more common (U.S. Environmental Protection Agency 2006).

13.5.1 Plant Interaction and Uptake Pathways

A study by Brigmon et al. (1998) demonstrated that the presence of the rhizosphere in soils at a TCE-contaminated site in South Carolina increased the potential for natural attenuation even though very little attenuation was caused by the rhizospheric effect. Rather, the reduction in TCE concentrations observed in laboratory microcosms was caused by sorption to sediments. For example, up to 90% of the TCE added to laboratory microcosms was removed from solution within 7 days. Donnelly and Fletcher (1994) reported that some PCBs could be degraded in the root zone under similar conditions.

A study by Anderson and Walton (1995) demonstrated that TCE in the presence of planted soils had increased degradation relative to the unplanted soils. Their studies confirmed what other researchers have reported, that in soils planted with a legume (*Lespedeza cuneata*), loblolly pine (*Pinus taeda*), and goldenrod (*Solidago*), that mineralization of the ^{14}C-TCE to $^{14}CO_2$ accounted for greater than 26% of the total radiolabel added, relative to 15% in the unplanted soils. However, they report that the TCE taken up into the plant was between 1% and 21% of that added. In was detected in the leaves (needles for *Pinus*), stems, and roots. Moreover, the novel part of their study was the selection of plants used, because the root types ranged from fibrous to leguminous to tap. In addition, although the raw data they collected indicated a difference in uptake rate for ^{14}C-TCE between plant types, after correction for water uptake and use, the differences were no longer apparent, suggesting a linear relation between water use and TCE uptake; this result is not surprising in light of the fact that contaminant and water uptake are passive processes. These workers also looked at the effect of ^{14}C-TCE on a plant that had not been previously exposed to TCE. The soybean plant tested (*Glycine max*) did mineralize slightly more TCE than the controls, but growth was not inhibited.

The direct uptake of groundwater contaminants such as chlorinated solvents by poplar trees has been studied previously, and can occur by the aqueous pathway (McFarlane et al. 1990) or by the gaseous pathway (Bromilow and Chamberlain 1995; Neitch et al. 1999). Regardless of the physical state of the contaminant being taken up, however, the fate of the contaminant in the transpiration stream can be assessed using tree-core collection and analysis and is discussed in Chap. 15.

Fortunately, TCE has other chemical properties that make it amenable to remediation by phytoremediation. It has a relatively high vapor pressure of 80 mmHg at 20°C, a dimensionless Henry's Law constant of 0.38, and a log K_{ow} of 2.29. These properties indicate that TCE has the potential to be taken up in both the vapor and dissolved phases by roots (Schnoor et al. 1995). The process is more complicated, however, in subsurface environments that have considerable amounts of organic matter, which leads to more absorption of TCE onto the sediment surfaces. A report by Doucette et al. (1998) indicates that uptake of TCE vapors by plants was a major pathway of subsurface remediation of a chlorinated solvent plume located in Florida. This was an important pathway, because the majority of root mass was above the water table. This pathway of attenuation is often ignored in phytoremediation studies relative to the fate of the aqueous state and its importance should not be overlooked.

Struckhoff et al. (2005) confirmed the observation by Doucette et al. (1998) that the vapor phase of chlorinated solvents is an important avenue between their accumulation in the unsaturated zone and uptake by plant roots. They reported that the concentration of PCE in trees, from core material, more greatly reflected the concentration of PCE in the soil-gas concentration relative to the groundwater PCE concentration. The data were originally collected from a PCE-contaminated aquifer at New Haven, Missouri, adjacent to the Missouri River. Tree cores were collected as well as soil samples from above the water table, both for VOC analysis in the headspace. The correlation between tree-core PCE concentration and soil-gas PCE was higher than the correlation between tree-core PCE concentration and groundwater PCE concentration (Fig. 13.15).

To further examine this relation, the authors determined partition coefficients for PCE between plant-air and plant-water of 8.1 and 49 L/kg, respectively. The partition coefficients were determined by adding a known amount of PCE to cores collected from uncontaminated trees, and then sampled after 1 week.

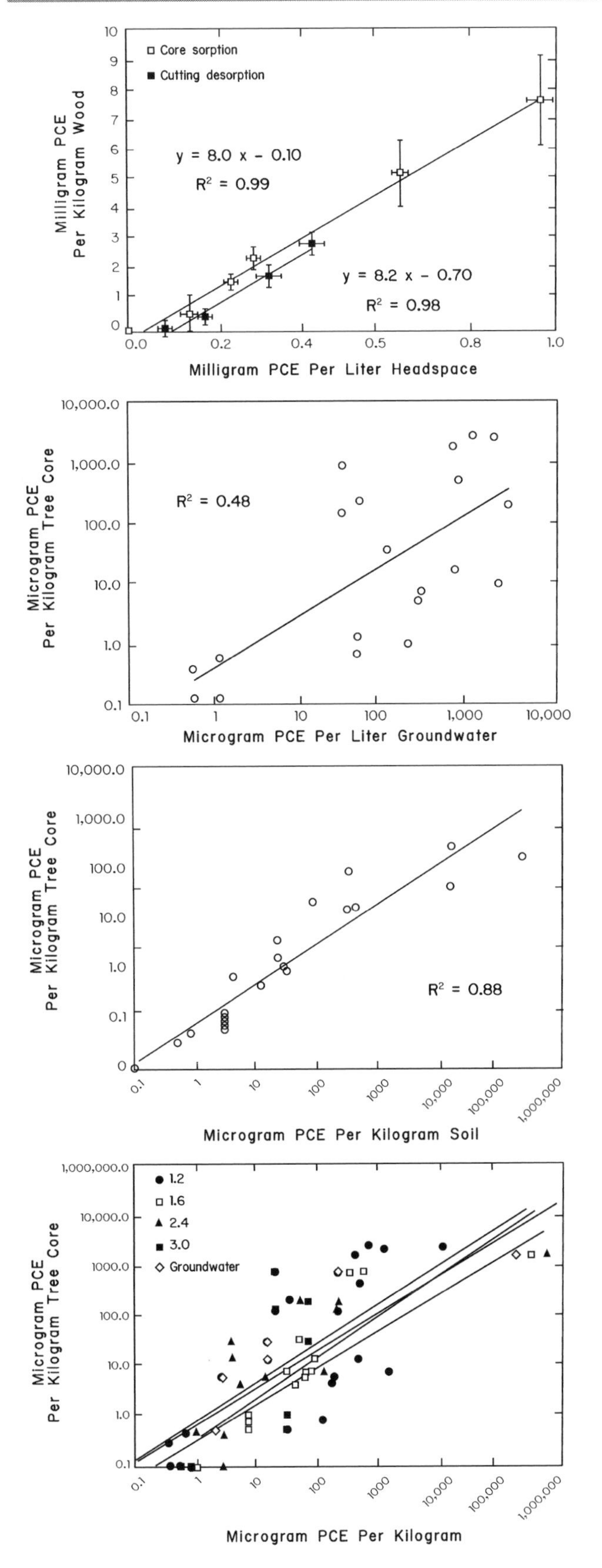

Fig. 13.15 PCE in headspace of cores collected from trees growing over PCE-contaminated soil and groundwater (Modified from Struckhoff et al. 2005).

For trees growing with most of their root mass above the water table or capillary fringe, which both contain dissolved contaminants such as TCE, the main interaction with the TCE would be the vapor phase. This was noted because soil-gas concentrations in clean fill were low even though the groundwater concentration was high.

Neitch et al. (1999) also compared the relative impact of TCE uptake into wetland trees such as baldcypress (*Taxodium distichum* (L) Rich) as a dissolved phase or gaseous phase. They observed that the uptake of TCE from the dissolved phase was controlled by transpiration by the physical process of osmosis, the TCE concentration, and the *TSCF* for TCE, all passive processes. The uptake of gaseous TCE was controlled by the Henry's Law partition coefficient for TCE, the TCE concentration, and the diffusive flux of TCE to the roots, as well as the percent of total root mass that consisted of air space, or aerenchymal cortex tissue, as would be expected for the investigated woody plants, which have to aerate anaerobic flooded soils.

Neitch et al. (1999) also reported their results of TCE uptake in baldcypress seedlings grown in the laboratory in containers (Fig. 13.16). Transpiration by the seedlings was measured as water loss from Marriotte bottles placed at a higher elevation than the water covering the seedlings in the bottles. TCE flux was determined using a static chamber technique (Neitch et al. 1999). The TCE flux through the seedlings after uptake from the dissolved phase was found to be higher during the day relative to night, or 80 and 50 μL TCE per hour, respectively, when transpiration was higher, and both values decreased from highs in August to lows in December. Interestingly, dead seedlings also showed the removal of TCE during the summer. Because water loss was still occurring due to capillary wicking effects through the dead plants, TCE was being removed although at lower rates. Therefore, during the growing season, the TCE concentration multiplied by the transpiration or *ET* rate can provide an estimate of TCE loss. A diffusive model can be an approximate model for TCE losses during the dormant period, where the total mass loss based on diffusion relative to transpiration will be small or no greater than 1% of total losses.

TCE loss flux was inversely correlated during the day with CO_2 plant uptake, but during the night, the production of CO_2 by plant respiration was not correlated with TCE loss flux (Neitch et al. 1999). This suggests that a controlling factor of TCE loss occurs through plant control of stomatal aperture.

Another mechanism that may be responsible for the phytoremediation of TCE is the plant-root formation and release into the rhizosphere of dehalogenase enzymes. These enzymes can be used to oxidize the strongly halogenated compounds present in the root zone. This production of such enzymes useful for chlorinated solvent

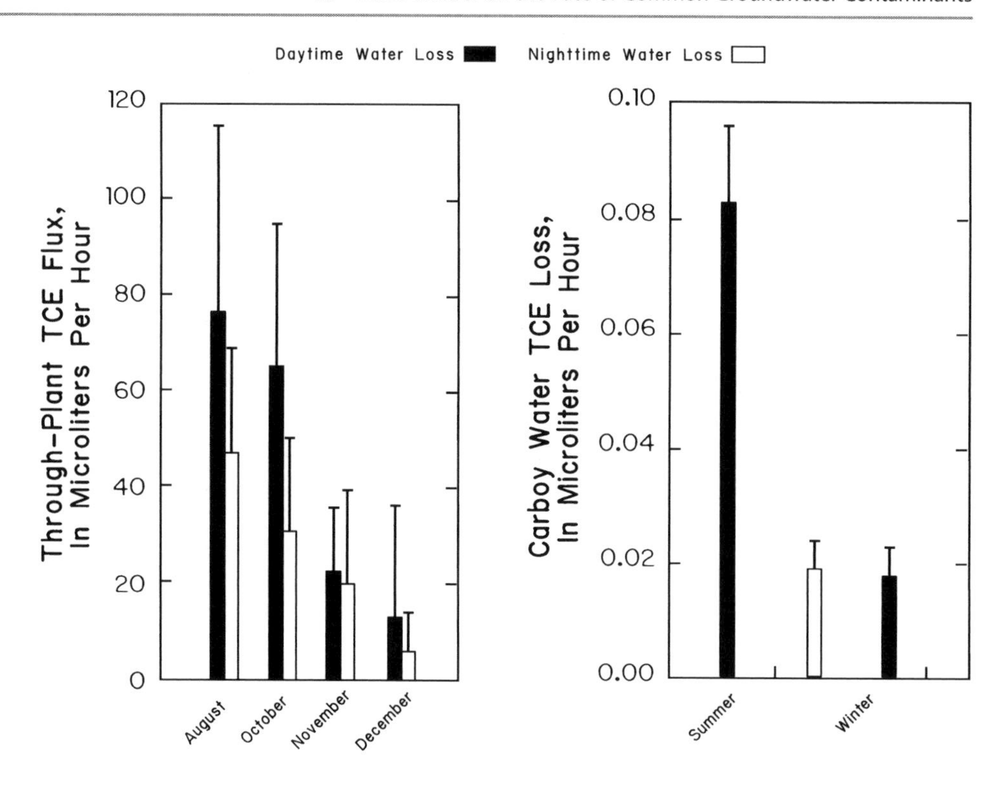

Fig. 13.16 The loss of TCE added to plants under laboratory conditions at different times of the season as measured through the plant and as loss from the water solution. Loss occurred even when transpiration was decreased during no-leaf dormant conditions.

remediation is an artifact of the evolutionary process that offered a defensive advantage to such plants, because as we saw earlier, some competing plants can release halogenated organics. Rather than a reductive dechlorination reaction, in which reduced organics are oxidized to provide a source of electrons to reduce chlorinated compounds, the dehalogenase directly oxidizes the TCE to CO_2 (Schnoor et al. 1995).

In oxic groundwater, TCE can resist degradation, as would be expected from its chemical structure. If plants release exudates that support microbial activity and the depression of DO levels, it may be possible for methanogenic bacteria to predominate and release methane. If oxic conditions exist in shallower parts of the aquifer, this CH_4 can be oxidized by methanotrophic bacteria, which, in turn, release methane monooxygenases (MMO), which also can enzymatically degrade TCE. This process gratuitously degrades TCE because the oxidation of methane requires a MMO (Wilson and Wilson 1985). Whether or not this can be considered phytoremediation is a matter for debate, however.

Brigmon et al. (1999) report the influence of the rhizosphere on the fate of TCE at a waste disposal site in South Carolina where TCE had been detected. The waste disposal activities stopped in 1974, and the surficial fill material was naturally populated over time by weeds to pine trees, specifically loblolly pines (*Pinaceae spp.*). Anderson and Walton (1995) had reported that the phenols released from pine trees supported TCE mineralization in the rhizosphere relative to unplanted soils. Whereas these studies showed that TCE could be mineralized to CO_2 by bacteria in the root zone, the question remained whether or not methanotrophic bacteria were influencing TCE concentrations. Since these bacteria can be found in the root zone, Brigmon et al. (1999) investigated the interaction of these plant bacteria on the TCE plume at the waste disposal site in South Carolina. They reported the presence of these bacteria on the roots and in the soil. However, at the site studied, up to 90% of the TCE released was absorbed to the soils.

The fate of TCE in the unsaturated zone in vapor form was studied by Narayanan et al. (1999). This pathway is important in light of the high vapor pressure of TCE, and because a fluctuating water-table level renders formerly saturated TCE-contaminated sediments to be exposed to air, and this will drive the gaseous diffusion of TCE into the void spaces that can interact with plant roots. To test this hypothesis of plant-root interaction with gaseous TCE, they created in the laboratory an artificial aquifer to which sediment, water, TCE, and plants (alfalfa) were added. Samples were collected periodically from this experimental water table, unsaturated zone, and plant tissue. A vertical-upward TCE gradient from the water table through the unsaturated zone was detected, driven by diffusion from the concentration gradient (Fig. 13.17). This is similar to that shown by Lahvis et al. (1999) for the fate of MTBE that volatilized to the vadose zone from gasoline-contaminated groundwater.

Vegetation lowered the transition zone between saturated and unsaturated conditions, and caused the diffusive flux of aqueous TCE in the water table to be upward into the unsaturated zone. This presents a scenario where the plants

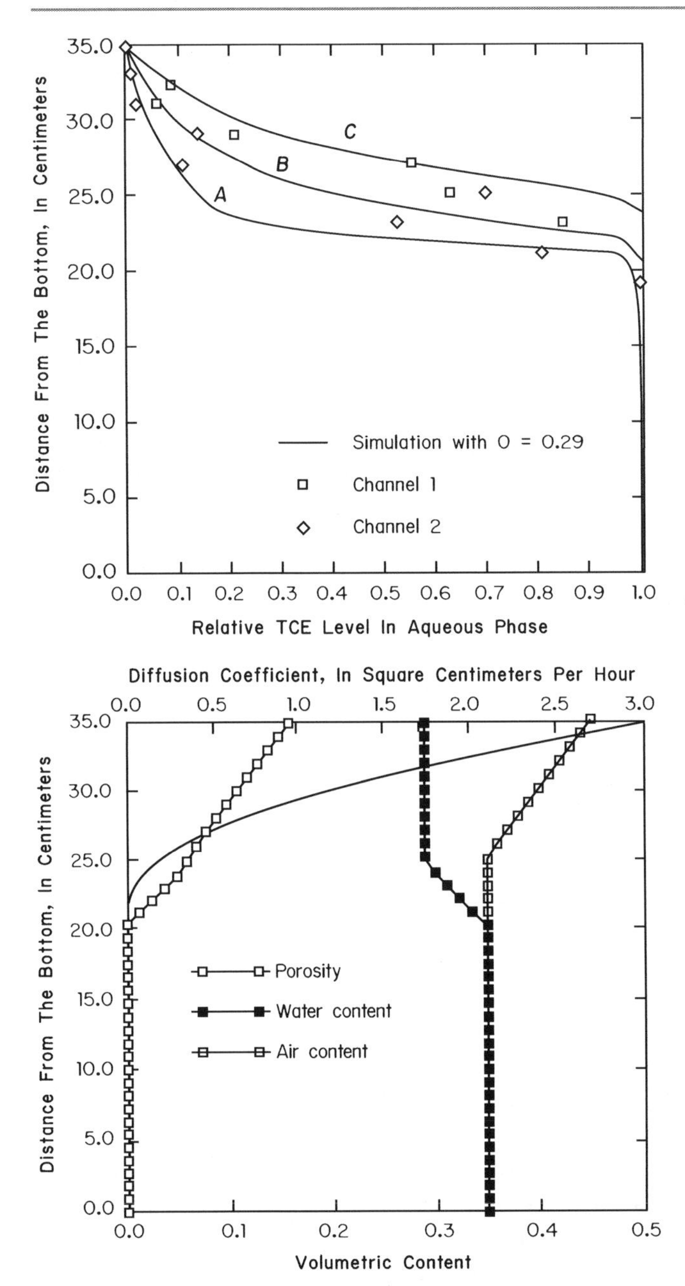

Fig. 13.17 Simulated diffusion of TCE from test trees as a function of porosity, water content, and air content (Modified from Narayanan et al. 1999). One centimeter is equivalent to 0.39 in.

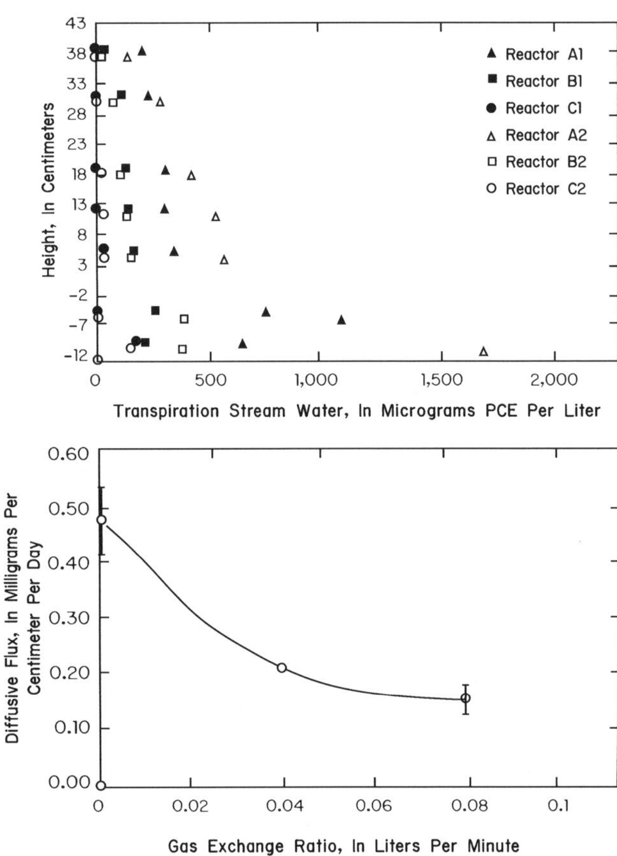

Fig. 13.18 Gas exchange rates for TCE (Modified from Ma and Burken 2003). One centimeter is equivalent to 0.39 in.

are taking up groundwater contaminants at lower, gaseous concentrations and are not being exposed to the higher aqueous concentrations. In doing so, they may be avoiding potential toxicity effects of the contaminants.

Ma and Burken (2003) used laboratory tests of poplar cuttings (DN-34) and diffusion traps placed around stems to determine the fate of TCE-laced irrigation water added to the plants. They also collected tree cores from the poplar trees that have been growing above a TCE-contaminated aquifer since 1996. The diffusion traps placed lower on the cuttings contained more TCE than the traps placed higher up the stem for the cuttings grown in soil. For the cuttings grown in hydroponic solution, the total mass of TCE in the diffusion traps was higher for the lower traps and lower for the higher traps. Moreover, TCE was detected in leaf tissue (Fig. 13.18).

TCE diffusion rates in the stems were directly related to transpiration rates (Ma and Burken 2003). The higher the uptake rates of water the higher the diffusion of TCE. In the tree center, where the xylem contained the TCE with decreasing concentrations up the tree (diffusion in vertical flow direction), TCE concentrations also decreased radially laterally from the center to the atmosphere, as a result of diffusion. On account of this vertical and radial diffusion to the atmosphere, it is not surprising that less than 0.05% of the TCE added to the soils was detected in the plants at the end of the experiment.

The lack of TCE detection in leaves may be a result of stem TCE diffusion rates being higher than the transport rate by diffusion of xylem TCE to the leaves. This is especially true for ring-porous trees that have most solution transport near the bark–atmosphere interface, resulting in a much shorter diffusion path than for diffusively porous trees (Ma and Burken 2003). These processes also happen below ground in long roots of small diameter. Whereas TCE was not detected in leaf material of poplars planted over a TCE-contaminated aquifer at the Carswell AFB site in Texas, some TCE vapor was detected emanating from the leaves of planted poplar trees at the APG site in Maryland.

Any dissolved-phase TCE in the transpiration stream that reaches the leaves may interact with the atmosphere in the stomata. There, the dissolved-phase TCE may transform into TCE vapor, similar to the evaporation of liquid water. Stomata that control the rate of transpiration also to some extent control the release of TCE; this does not include, however, the diffusive flux of contaminants such as TCE from the xylem out through the bark (Ma and Burken 2003) as was discussed above.

The concept of a unique *TSCF* for TCE, or other contaminant compounds for that matter, has not yet been proven. For example, the *TSCF* for TCE derived by theoretical methods is 0.62 using the method of Briggs et al. (1982), measured is 0.75 (Burken and Schnoor 1998), and in a laboratory study with ^{14}C-TCE was found to range from 0.02 to 0.22, and was a function of concentration (Orchard et al. 2000). These differences in *TSCF* are most likely due to differences in experimental setup, plant type, and use of a soil versus hydroponic solution.

The fate of TCE in groundwater that discharged to a wetland containing cattails and cottonwood trees was investigated by Bankston et al. (2002). Cattails can allow the diffusive transport of atmospheric oxygen to the root zone, and its diffusion into pore water that contains TCE could promote its co-metabolism by MMO-producing methanotrophic bacteria. Microcosms were built in the lab that contained site sediments and plants, sandy soil for the cottonwoods and organic soil for the cattails. To these microcosms was added ^{14}C-TCE and its fate monitored. Most of the recovered ^{14}C-TCE was in the volatilized form, being greater than 50%. In the cottonwoods and cattails, up to 33% and 39%, respectively, of the label was present in the plant tissues. Low amounts of $^{14}CO_2$ were present in the control and treatment microcosm, suggesting a minimum amount of oxidation by soil microbes.

Researchers at the University of Washington extended the investigation into the fate of TCE from the laboratory into the field (Newman et al. 1999b). Their test site was located near Fife, Washington. The test was not performed in a contaminated aquifer; rather, the test was performed in "cells" composed of 1.5-m by 3-m wide by 5.7-m long of 60 mil polyethylene filled with a sand layer on the bottom covered by a thicker layer of clay loam collected on site. This type of study is actually an intermediate level between laboratory testing and field application at a contaminated site. Fifteen hybrid poplar clone cuttings of *Populus trichocarpa* × *P. deltoides* were planted every 1 m in each cell in early 1995. To each cell was added a perforated pipe at the depth of the sand layer to which plain water or water laced with TCE mixed in 40% ethanol was added. TCE was added to three treatment cells over time, with one cell not receiving TCE acting as the control. The TCE concentration of the influent averaged about 0.38 millimolar (mM) between 1995 and 1997. The total amount of water added over the 3-year test varied with variations in plant transpiration. Various aspects of the fate of TCE were monitored, as briefly summarized below.

As was stated in Newman et al. (1997), the overall growth of the trees after exposure to TCE was not significantly different than the growth of the trees in the control cell. Mean tree height was 3 m after 1995, 7 m by 1996, and 11 m by 1997. Although a water budget was presented, the lack of a distinction between water lost by transpiration or evaporation from the cells precludes an accurate picture of the effect of the trees themselves. The amount of TCE and reductive dechlorination intermediates measured in the effluent of planted cells increased after leaf drop each year, indicating the effect of transpiration while in full leaf accounts for a large percentage of TCE loss, which is then decreased after leaf drop. Over the 3-year study, more of the original TCE added to the influent was collected as effluent in the control cell with no trees (67%) than in the cells with trees (1–2%).

The fate of TCE taken up by the trees was examined—TCE, and both dichloro- and trichloroacetic acid intermediates of TCE transformation, were detected in plant tissues, such as leaves, branches, roots, and the trunk. Some of the TCE removed from the influent was observed to be transpired by leaves, as shown by the measurement of 2×10^{-8} mol TCE/leaf/h. Scaling this individual leaf transpiration rate to the entire plant for the entire season leads to an estimate of about 0.30 mol TCE transpired per cell per year, or about 9% of the total amount of TCE lost per season. By the end of the experiment, 95% of the added TCE was removed in the planted cells.

A study of the interaction between TCE at a contaminated site with existing vegetation was performed by Doucette et al. (1998). The site was located on the east coast of Florida at Cape Canaveral. It had been exposed to releases of TCE from metal-cleaning activities for at least three decades. The shallow sediments that comprised the vadose and saturated zones to depths near 35 ft (10.5 m) from ground surface included coarse to fine-grained sands and shells. The average depth to the water table was between 4 and 7 ft across the site, and varied due to recharge and the level of adjacent surface-water bodies to which the groundwater discharged. Above a plume of TCE delineated by monitoring wells grew indigenous plants that included castor bean (*Ricinus communis*), live oak (*Quercus virginiana*), and saw palmetto (*Serenoa repens*). Samples of plant tissues, such as leaf, stem, and root, were collected from all three types of plants.

This occurred at the same time that groundwater, transpiration gas, and soil-gas samples were collected. The stem samples were cores from the live oak but whole stems for the smaller saw palmetto and castor bean. Plant samples were preserved in methanol in vials or jars sealed with Teflon-lined caps. The groundwater samples were collected from temporary well points installed with a hand auger, and with the well screen set at the water-table surface. Soil samples were obtained from the unsaturated zone above the water table during removal of material for well installation. Soil-gas samples were collected at 3 ft below the ground. All nonplant samples were analyzed for TCE using purge-and-trap gas chromatography with electron-capture detection, according to EPA method 8010B. Plant samples were analyzed using a similar method, with the addition of methanol exposure and shaking for 1 day, and the removal of a 250-μL sample that was injected into a purge-and-trap tube.

Soil-gas samples contained TCE, but the soil samples did not, indicating that TCE vapors were emanating from the contaminated water table. The highest soil gas was detected under saw palmetto, 3,400 parts per billion TCE by volume (ppbv), and ranged from 85 to 603 ppbv for the unsaturated zone beneath the live oak and saw palmetto, respectively. Up to 90% of all root mass was detected in soils no deeper than 2 ft, above the water table and source of contamination. The roots of the saw palmetto and castor bean contained higher concentrations, or 16.2 and 2.88 μg/kg plant fresh weight, respectively, than the other tissues from these plants, with no more than 2.8 and 2.14 μg/kg, in the stem and leaf, respectively. Conversely, the stems of the live oak contained the highest concentrations, near 3.68 μg/kg relative to the roots and leaves, that contained no more than 2.07 μg/kg. TCE in groundwater beneath the saw palmetto, where both root concentration and soil-gas concentrations were highest, was also the highest, at 64,500 μg/L. TCE in the groundwater beneath the live oak was 527 μg/L and beneath the castor bean plants was 3,741 μg/L.

The implication of these and other results where soil-gas contamination is being degraded by plants is that plants at phytoremediation sites can help decrease groundwater contaminants even when not taking up groundwater. The extent to which this occurs is a function of the vapor pressure of the contaminant, the resistance of gas flow in the unsaturated zone, and the type of plant-root system in place.

Clinton et al. (2004) reported that TCE concentrations decreased in tree cores collected at increasing heights above ground following artificial irrigation of the soil around the base of the tree with water that did not contain TCE. The concentration of TCE went down, as would be expected if the plant was now taking up a more dilute solution of water. It is possible, however, that TCE in the vadose zone was entrained in the water during irrigation.

PCB compounds also are common chlorinated hydrocarbons but are associated more with soil contamination rather than with groundwater contamination. This is because of the physical properties of PCB, and low solubility, which render them absorbed to soils. These factors vary, of course, with the amount of chlorine substitution. Because PCBs share similar properties with both the PAHs and chlorinated solvents, their fate in plants warrants discussion. For example, the white rot fungi, often present in the rhizosphere, have demonstrated the capability to degrade PCBs. Schnabel and White (2001) investigated the interaction of willow (*Salix alaxensis*) and balsam poplar (*Populus balsamifera*) that were exposed to a representative PCB in soils under laboratory conditions. The PCB was 3,3′,4,4′-tetrachlorobiphenyl (TCB), and was used in uniformly ^{14}C-radiolabeled form to trace the fate of the ^{14}C. In contrast to the fate of ^{14}C-TCE discussed above, for which the label was primarily in the shoots and leaves, up to 88.9% of the added PCB label remained in the soils in the root zone, and less than 1% in the roots themselves. This can be explained by the low solubility (near 0.175 mg/L) and high sorption of TCB, and its low bioavailability. Even so, Schnabel and White (2001) suggest that the presence of ^{14}C-TCB in the root tissue indicates that these compounds can enter the root cells. Moreover, the root cells had up to nine times higher concentrations of total radiolabel relative to the soil concentrations, indication of bioconcentration. TCB was not detected in any shoot material, so it can apparently enter the root hair cells but not pass the Casparian strip and, therefore, gain entry into the xylem.

PCB-degrading bacteria associated with the root zone of trees were found to degrade PCB-contaminated soil (Leigh et al. 2006). Higher concentrations of PCB-degrading bacteria, such as *Rhodococcus*, were associated with plant roots, at values between 2.7 and 56.7 times higher than in unplanted contaminated soils. The highest numbers of PCB-degrading bacteria were associated with the root zones of Austrian pine (*P. nigra*) and willow (*S. caprea*). These plants can produce in the root zone such compounds as terpenoids, tannins, phenols, and salicylic acid, which can serve as substrate for or induction by PCB-degrading bacteria.

Liu and Schnoor (2008) report that exposure of lesser-chlorinated PCB congeners to poplar plants resulted in the congeners being sorbed to the roots. Translocation of the PCBs beyond root to stems was for those congeners with lower sorption potential. Very little PCB was lost through the leaves and it was not clear if the congeners were metabolized. In Liu et al. (2009), poplars and switchgrass (*Panicum vigratum*, Alamo) were exposed to solutions that contained 3,3′,4,4′-tetrachlorobiphenyl and they observed plant-mediated metabolism of the PCB.

In field sites where phytoremediation is being considered, TCE often is not the sole component of groundwater contamination. Other chlorinated solvents may be present, as well as co-contaminants such as petroleum hydrocarbons. For many laboratory studies of the interaction between plants and groundwater contaminants, such as TCE, the contaminant interaction with the plant is studied by itself, rather than as a mixture, such as has been done to examine the *TSCF* of a particular contaminant (see Chap. 12). Sorption of VOCs such as TCE has been modeled as being a linear response. However, nonlinear responses have been observed in the laboratory for a mixture of TCE and TCA (Graber et al. 2007). TCE added to seedlings and to wood from a mature *Eucalyptus camaldulensis* was observed to follow a Langmuir isotherm. In all cases, less TCE was taken up into the tree tissue at a given TCE concentration if another contaminant was present. This study also pointed out that *TSCFs* derived from a single solute may not be representative of sites where a mixture of contaminants has been released (Graber et al. 2007).

For PCE, James et al. (2009) investigated the fate of PCE in water added to hybrid poplars growing in a controlled field-scale experiment. At the end of the treatment up to 99% of the added PCE was reduced and free chloride recovered in near an equal amount.

13.5.2 Plant Transformation Reactions

At the Aberdeen Proving Ground in Maryland described in Chap. 8, tree-tissue samples collected from trees growing above TCE-contaminated groundwater indicated the detection of trichloroacetic acid (TCAA), a breakdown product of TCE. TCAA also was detected in a study of axenic cell cultures dosed with TCE (Newman et al. 1997). Axenic cultures are sterilized cell cultures that do not contain bacteria and are the conventional way to observe the function of plant cells without the interference of bacterial cells. Nodule cell cultures, or spherical photosynthetic cell aggregates, also accomplish the same goal.

In that study, they took axenic cell cultures of a *Populus trichocarpa* × *P. deltoides* clone and added TCE. Cells were viable in the presence of TCE at 260 ppm TCE. The cell cultures were used to determine the fate of the TCE, and the intermediate byproducts of TCE mineralization, such as dichloroacetic acid, trichloroacetic acid, and trichloroethanol, were produced, similar to the fate of TCE in mammalian livers.

To determine if complete mineralization of TCE occurred, in addition to the production of the intermediate compounds, ^{14}C-TCE was added to the cell cultures. The production of $^{14}CO_2$ was monitored, and between 1% and 2% of the TCE was detected as CO_2 by the end of the experiment. Newman et al. (1997) reported that some of the label remained as an insoluble residue, which they suggested was due to abiotic binding to plant cell material. They reported that this TCE residue has important implications for the phytoremediation of TCE, because the bound residue might be perceived as being more desirable than gaseous TCE release to the atmosphere through leaves. This may not necessarily be the case, however, due to the rapid destruction reactions that can affect TCE in the atmosphere. Storage of TCE metabolites in plant tissue has a minimal effect on overall TCE taken up by plants, however. Shang et al. (2001) reported that the TCE-intermediate trichloroethanol is glycosylated in poplar trees, and if the TCE source is removed, the accumulation of trichloroethanol does not occur, suggestive of additional plant metabolic capability.

Strand et al. (1998) continued the investigation of the plant-mediated mineralization of TCE. They report that for axenic tissue cultures of poplar-tree cells exposed to TCE, carbon tetrachloride (CCl_4), and PCE, the complete mineralization of the contaminants to CO_2 occurred. At the field scale, TCE and CCl_4 were added to containers that held soil and poplar trees. The removal of TCE and CCl_4 approached 95% of that added relative to no loss from control containers with no plants.

Experiments to trace the fate of TCE also were performed using whole plants in PVC pipes grown in the greenhouse. Each pipe contained of a plant, soil at the top, and a sand layer at the bottom to which water was added through a small-diameter pipe. To some of the plant-pipe setups was added a solution of 50 ppm TCE in water. Poplar trees have been show to survive when grown in water that contained 50 ppm TCE (Gordon et al. 1997). The plants exposed to TCE did not suffer toxic effects but were somewhat affected morphologically. First, the plants exposed to TCE grew to 85% of the height of plants not exposed to TCE. Second, there was a demonstrable decrease in the amount of fine root material in the sand layer to which TCE-laden water was added relative to that in the plants that received water only. TCE was found in the stems to a greater extent than the leaves, and root material also contained TCE. The TCE concentration in leaf tissue was lower than that found in the stems, in part because TCE was released into the atmosphere near the leaves, as monitored using polyethylene bags that contained a charcoal filter placed around leaves. Amounts of TCE detected in the leaf-bag charcoal after desorption with pentane ranged from 0.053 to 0.811 μg TCE.

The importance of this paper in the study of phytoremediation of chlorinated solvents is that the experimental data show that the effect on TCE concentrations was caused by the plant *directly*, through the formation of TCE metabolites in the sterilized axenic cell cultures, not indirectly by microbial processes through a rhizospheric effect. Although poplar trees contain cytochrome P-450 enzymes under natural conditions, accelerated transformation of TCE

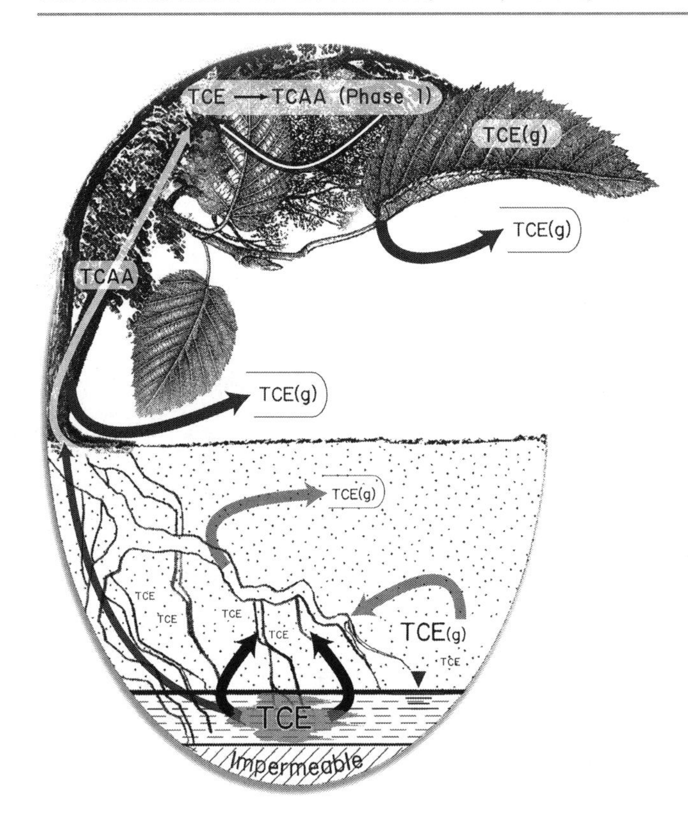

Fig. 13.19 Plant and groundwater interactions at a site characterized by TCE-contaminated groundwater. TCE(g) is the volatile phase.

could be achieved using transgenic poplar trees that contain mammalian genes that code for cytochrome P-450, as has been shown by researchers at the University of Washington. A summary of the fate of a representative chlorinated solvent TCE in plants is shown in Fig. 13.19.

13.6 Plant Interactions with Nitroaromatics, NDMA, Dioxane, Perchlorate, and Tritium

Whereas the aromatic ring structure of nitroaromatics is essentially the same reduced carbon-ring of aromatic hydrocarbons, the nitrogenous functional groups are partially oxidized. Therefore, these nitrogenous groups can be reduced, whereas the aromatic core can be oxidized. In some ways, these compounds reflect both major classes of degradation pathways, oxidation and reduction, acting both as a source of electrons and a sink for electrons.

Perhaps the most commonly known nitroaromatic to be released to groundwater is trinitrotoluene, or 2,4,6-trinitrotoluene, also known as TNT—TNT also is the most toxic of the nitroaromatics. TNT was first synthesized in 1863, for use as a red-colored dye. It is often found in surface soils at munitions manufacturing plants. This is particularly the case at military sites where TNT has been manufactured, stored, used, and disposed of, and these sites can be characterized with TNT-contaminated soils, sediments, and aquifers. In addition, the manufacture of TNT does not produce a 100% theoretical yield, and often TNT-contaminated sites contain the TNT impurities 2,4-dinitrotoluene (2,4-DNT) and 2,6-dinitrotoluene (2,6-DNT). All these compounds are priority pollutants as defined by the USEPA. Other explosives compounds that are based on nitroaromatics include hexahydro-1,3,5-trinitro-1,3,5-triazine (Royal Demolition Explosive, or RDX), and octahydro-1,3,5,7-tetranitro-1,3,5,7-tetrazocine (high-melting-point explosive, or HMX). The USEPA maximum contaminant level for RDX is 1.05 mg/L.

Nitroaromatic compounds can undergo a wide variety of biologically mediated transformation reactions in groundwater. Microbes in a TNT-contaminated aquifer near Weldon Spring, Missouri, have been shown to transform TNT, 2,4-DNT, and 2,6-DNT to amino-nitro intermediate compounds, as well as be completely mineralized to CO_2 (Bradley et al. 1994). That site is typical of many TNT-contaminated sites, where TNT is present in surface soils up to percent-level concentrations, as well as dissolved TNT concentrations in groundwater that receives recharge through these contaminated surface soils. 2,4-DNT and 2,6-DNT also are present in groundwater at such sites. For example, at Weldon Spring these concentrations were as high as 44 and 61.4 μg/L, respectively (Bradley et al. 1997).

The linkage between such microbial degradation of TNT with similar contaminant degrading capability of the microbes present in the rhizosphere around trees at phytoremediation sites was shown by Schnoor et al. (1995). They showed that the TNT-degrading nitroreductases originally believed to be related to soil bacteria were shown to be, in fact, derived from plant roots. TNT can induce plant toxicity, however, and was observed as a decrease in the transpiration of poplar trees exposed to TNT (Thompson et al. 1998b).

Another nitrogenous contaminant compound that has been detected in groundwater systems is N-nitrosodimethylamine (NDMA). NDMA is a probable human carcinogen. NDMA has both manmade and natural sources, being a byproduct of rocket-fuel production but also can be formed during the chlorination of water, respectively.

Perchlorate, as ammonium perchlorate, or NH_4ClO_4, also is associated with rocket fuel and also can be found in arid climates where it is formed naturally in evaporate deposits. If consumed, perchlorate inhibits the uptake of iodine by the thyroid; perchlorate is a probable human carcinogen. Perchlorate, however, has no federal drinking-water standard, but health advisories as low as 24.5 μg/L have been mandated. Perchlorate is soluble in water and has been detected in surface- and groundwater.

The compound 1,4-dioxane, $C_4H_8O_2$, also known as dioxane, is a peroxide. Dioxane is primarily used as a stabilizer for chlorinated solvents such as PCE and TCE but also is used in paints, lacquers, and varnishes. Dioxane detection in groundwater is problematic as dioxane is classified as a probable human carcinogen. Dioxane is miscible in water and not readily absorbed by sediments, and plumes of dioxane in groundwater can be extensive (Tillman 2009). The physical properties of dioxane indicate that it is conducive to phytoremediation, even with a log K_{ow} less than 0.

Tritium, ^3H, is the radioactive isotope of hydrogen, with a half-life of 12.32 years. Tritiated water is formed naturally in the upper atmosphere. Elevated levels of tritium are produced for thermonuclear purposes and biochemical tracers. Tritium enters the hydrologic cycle from these sources, including as leachate from landfills designed to contain low-level radioactive wastes in both humid (Vroblesky et al. 2009) and arid (Garcia et al. 2009) locations. The tritiated water can be present in liquid and vapor form at low-level radioactive waste landfills.

13.6.1 Plant Interaction and Uptake Pathways

Many studies of the fate of nitroaromatics have been performed; they involve investigations of both microbial and plant processes. The fate of nitroaromatic compounds is typically determined in soils, rather than groundwater. Early studies focused on the interaction between plants and TNT from the standpoint of human health exposure. Because the log K_{ow} of TNT is 2.0, the translocation of TNT was predicted. However, Burken and Schnoor (1998) did not find TNT in the shoots and leaves.

The fate of perchlorate exposed to woody plants was investigated by Nzengung et al. (1999). They report almost 100% removal of perchlorate by willows exposed to solution of 10–100 mg/L perchlorate. Loss from solution was attributed to uptake and transpiration and by rhizospheric degradation. Loss of perchlorate in the rhizosphere coincided with an increase in chloride concentration.

Yifru and Nzengung (2006) investigated the effect of woody plants such as black willow (*Salix nigra*) and hybrid poplar (*Populus deltoides* × *nigra*, DN34) on the fate of NDMA and perchlorate in water. They created microcosms of individual cuttings placed in 2-L bioreactors filled with a dilute Hoagland solution to which was added either 1 mg/L NDMA, 0.65 mg/L NDMA and 27 mg/L perchlorate, or just perchlorate. The fate of these two contaminants was monitored over time by sampling small aliquots from the Hoagland solution until non-detect levels were reached.

After 80 days, about 98% of the added NDMA had been removed from the growth solution by the cuttings during the summer, and 81% was removed during the winter. This large loss of NDMA was primarily by the removal of water by plant uptake, since the loss of NDMA varied linearly with the volume of water taken up over time. This is an important result if transferable to field studies of NDMA-contaminated groundwater, because measurements of transpiration can be used to estimate the potential for NDMA mass loss from contaminated groundwater. Moreover, these results challenge the commonly held belief that phytoremediation strategies are handicapped by a lack of contaminant removal during the winter; although these are laboratory studies, significant contaminant removal was observed during the winter. Yifru and Nzengung (2006) used the loss of NDMA over time, along with the loss of water as an indication of the transpiration rate, to calculate a *TSCF* of about 0.28.

If the NDMA and perchlorate were removed from solution during transpiration, did the chemicals simply volatilize from the leaves? In the experimental setup, only about 52% of the original ^{14}C-NDMA was recovered, the balance of which was presumed to be transported to the leaves and volatilized. Of this recovery, 46% was contained in plant tissues, and distributed between leaves (19%), main cutting stem (16%), branches (8%), and in the roots (4%). About 5% was recovered as being that volatilized from the total tissues above the growth solution. On the other hand, perchlorate was removed from the solution to levels below detection in 50 days, or 20 days longer if NDMA also was present. The distribution of the ^{14}C-perchlorate in the plant was as follows: leaves (98%), stem (1.3%), and roots and branches (less than 1%).

Aitchison et al. (2000) examined the fate of 1,4-dioxane in poplar cuttings (DN-34) about 10 in. (25.4 cm) long in both hydroponic and soil treatments. Within 9 days of exposure, more than 50% of the 1,4-dioxane was removed from the hydroponic solution. Most of that taken up, between 76% and 85%, was transpired. A *TSCF* of 0.76 was calculated from this mass balance. In soil treatments, cuttings were exposed to ^{14}C-1,4-dioxane. At the end of this experiment, only 19% of the label remained in the soil that contained the cutting, relative to 72% of the label that remained in sterilized soil controls.

Tritium, being essentially part of the water molecule, readily enters plants from the vadose zone (Garcia et al. 2009) and water table (Vroblesky et al. 2009). At the USGS Amargosa Desert Research Site (ADRS) in southern Nevada, the arid environment promotes the upward movement of any tritiated water in the subsurface following leakage from tritium waste-disposal facilities (Garcia et al. 2009). Although evaporation removed three times more tritiated water vapor from the contaminated subsurface at the ADRS, the presence of vegetation led to decreased infiltration of precipitation, which enhanced the upward movement of tritiated wastes. This study is perhaps the

Fig. 13.20 Simple apparatus developed by the author to collect tritiated water vapor from the riparian forest growing above tritium-contaminated groundwater. See Vroblesky et al. (2009) for more information (Photograph by author).

first to examine the effect of native plants on the phytoremediation of tritium-contaminated groundwater in an arid environment.

The phytoremediation of tritium-contaminated groundwater in a humid environment was examined by Vroblesky et al. (2009) near Barnwell, South Carolina. A low-level radioactive waste-disposal facility resulted in tritiated water in local groundwater and surface water in a down-gradient riparian forest that received tritium-contaminated groundwater discharge. Tree cores were collected and analyzed for tritium. The sap-flow rate was measured on some tritium-contaminated trees using the TDP approach. The results indicate that some trees may remove more than 17.1 million picocuries per day (pCi/day) when transpiration is high during the growing season. The tritiated water taken up by the roots is transpired from the leaves, but the tritium level in air samples, as collected near trees using a simple apparatus consisting of a frozen water bottle, funnel, and 40-mL VOA vial (Fig 13.20), did not measure high enough to create a tritiated-water vapor hazard.

13.6.2 Plant Transformation Reactions

Perhaps the first evidence that plants could take up and transform TNT was presented by Hughes et al. (1996). In their study, they hypothesized that the plants first reduced the TNT, as there was no evidence of mineralization. Plants can transform TNT by a Phase I reduction into nitroamine intermediate compounds such as 2-amino-4, 6-dinitrotoluene (Bhadra et al. 1999). These intermediates can undergo Phase II conjugation and be stored as unextractable organic matter in the plant tissues. This was the first study to report evidence suggesting that byproducts of TNT transformation could be conjugated in plants. Wayment et al. (1999) investigated the formation of conjugates in root tissue cultures of various plants exposed to TNT, including the aquatic plant *M. aquaticum*. They reported the detection of up to four different TNT conjugates, which were most likely formed by the addition of sugars to the amino group of the TNT taken up (Wayment et al. 1999). Burken et al. (2000) also indicated that the predominant first detoxification process for TNT can be an oxidative Phase I to mono-amino derivatives, with subsequent conjugation to more soluble compounds. These reactions are not as significant for removal of RDX in plants, however. To this end, researchers are focusing on engineering transgenic plants that express the bacterial genes to produce cytochrome P-450 and reductases that rapidly remove RDX and TNT from soils and leachate.

Chekol et al. (2002) investigated the fate of TNT applied to plants grown in soils of two different organic matter contents in a greenhouse study. The effect of organic content on TNT fate, by plants, is important. For the plant to interact with the TNT, the TNT has to be bioavailable rather than absorbed to the sediment. For the soil that contained 6.3% organic matter and 100 mg/kg TNT, plants such as the legumes alfalfa (*Medicago sativa*) and sericea lespedeza (*Lespedeza cuneata*) and various grass species, less than 1% of the added TNT was recovered in both treatment and controls. This indicates that the TNT was removed by sorption. Conversely, for the soils that contained 2.6% organic matter, higher amounts of TNT were recovered. Also in these soils, 85% of the initial TNT added was recovered from the unplanted controls, but only 23% was recovered from the planted soils. Moreover, the grass species characterized by a shallower, more dense, fibrous root system removed a higher percentage of added TNT relative the legume species, which are characterized by a deeper tap root. These results have implications as for what type of vegetation to install at a site where the contamination has been delineated as a function of depth.

The fate of explosives compounds in leaves after they fell was studied by researchers at the University of Iowa (Yoon et al. 2006). The nitroaromatic compound investigated was TNT, along with RDX and HMX. These compounds remain at many Department of Defense sites and other military sites as a legacy to munitions production since the 1940s (Bradley et al. 1994). The fate of these explosive compounds in the presence of plants has been investigated, which showed that

TNT and some of its intermediate breakdown products were observed to be taken up into plants and accumulate in the roots, whereas RDX and HMX were found primarily in the leaves of test plants (Groom et al. 2002).The fate of these compounds in leaves after they fell was studied by Yoon et al. (2006). Regulators are typically concerned about the potential risk exposure to contaminants by exposure to leaves that might contain contaminants taken up by the plant. The researchers added radiolabeled ^{14}C-TNT, ^{14}C-RDX, and ^{14}C-HMX to flasks that contained a solution of half-strength Hoagland solution, TNT mixtures, and a prerooted hybrid poplar cutting (*Populus deltoides* × *nigra* DN-34). Over 2 weeks, the removal of TNT from the solution was greater than the removal of RDX, with little removal of HMX. After uptake, the distribution of these compounds within the plant was depicted. Almost half of the ^{14}C-TNT taken up was detected in the roots after 30 days, whereas between one-fifth and one-half of ^{14}C-HMX and ^{14}C-RDX, respectively, were found in the leaves (Yoon et al. 2006). Due to the detection in leaves, dried leaves were exposed to deionized water to simulate exposure to precipitation after leaf drop and were resampled. Very little TNT was found in the leachate, but one-fourth and one-half of RDX and HMX was found in the water. Moreover, when dried roots were exposed to the deionized water, very little of any compound was detected in the leachate.

RDX exposure to poplar tree tissue cultures and leaf extracts resulted in the partial reduction of RDX to MNX and DNX and subsequent mineralization (Van Aken et al. 2004a). The authors present a model for the phytoremediation of RDX, such that RDX is taken up and translocated to the leaves where it is reduced to MNX and DNX during Phase I reactions, possibly by plant reductive enzymes of the "green liver," such as P-450. The heterocyclic rings are then broken and the metabolites mineralized to CO_2. It is possible that this CO_2 can then be used by the plant.

Thompson et al. (1998b) confirmed results of others that the interaction of plants with TNT-contaminated soil and water results in up to 75% of the TNT being bound to roots, with little translocation to the leaves. Of the fraction that made it into the plant tissues, it was transformed into the Phase II conjugate amino derivatives, such as 4-amino dinitrotoluene (4-ADNT) and 2-amino dinitrotoluene (2-ADNT), that remained bound in plant tissues, as well as other unidentified byproducts that were more polar than TNT. As with the other studies performed that looked at the fate of TNT in plants, little mineralization to CO_2 was detected. Thompson et al. (1998) calculated an experimental *RCF* and *TSCF* for TNT and compared it to previously calculated values. Their *RCF* was 49.0 and the *TSCF* was 0.46; the *RCF* is much higher than previously calculated by Briggs et al. (1982) or Burken (2003), whereas the *TSCF* was lower.

Van Aken et al. (2004b) isolated a bacterium believed to be a symbiotic endophyte often associated with plants, including the poplar trees used in the study, and could in pure culture degrade TNT, RDX, and HMX to reductive derivatives. The TNT could not be oxidized by the bacteria, although RDX and HMX were, and was identified to be a species of the genus *Methylobacterium*. TNT reduction produces ADNTs, which can conjugate with hemicellulose in the roots of willow trees (Shoenmuth and Pestemer 2004a, b).

Tanaka et al. (2007) have presented novel evidence that documents the importance of plant-detoxification processes on the fate of explosives after uptake by poplar trees (*Populus deltoides* × *nigra*, DN-34). In the laboratory, poplar trees were exposed to 50 mg/L of RDX under hydroponic conditions for 1 day. Following exposure, they observed amplification of genes related to the Phase I to Phase III detoxification processes, such as GST, cytochrome P-450, reductases, and peroxidases. This amplification was observed primarily in poplar leaf tissue, not so much in root tissue, although the majority of the added RDX remained in the roots following rapid uptake. Additional evidence as to the effect of this increased gene expression on the final fate of the RDX taken into the plant from hydroponic solution will be needed, however, to link these expressed genes to contaminant fate.

Tognetti et al. (2007) examined the influence of transgenic plants that express a bacterial flavodoxin to phytoremediate 2,4-dinitrotoluene (2,4-DNT). Flavodoxin is not naturally found in plant cells and is believed to shuttle electrons to the nitro group of 2,4-DNT for subsequent reduction. These transgenic plants could be installed at sites where levels of 2,4-DNT may be toxic.

Another nitroaromatic compound, although not an explosive, is nitrobenzene. Fletcher et al. (1990) grew soybean (*Glycine max*) in the presence of a range of concentrations of ^{14}C-UL-nitrobenzene, from 0.02 to 100 μg/mL as a mixture of labeled and unlabeled nitrobenzene. At the end of a 3-day incubation, the plants were analyzed for ^{14}C-nitrobenzene distribution. It is important to note that the authors examined the potential for contaminant effects on transpiration and photosynthesis from the exposure to nitrobenzene, and none was found. However, they did report that there was some visual evidence that the highest concentration used, 100 μg/mL of nitrobenzene, did indicate root growth reduction. In any case, up to 80% of the ^{14}C label remained in the roots, and 20% in the shoots.

Seyfferth and Parker (2007) conducted an experiment to determine the fate of dissolved perchlorate in the water transpired by lettuce (*Lactuca sativa* L.) grown under hydroponic conditions. Plants exposed to higher concentrations of perchlorate contained higher concentrations of perchlorate in leafy tissues, as microgram per kilogram (μg/kg) fresh weight. An increase in transpiration led to an increase in perchlorate concentrations in lettuce tissue, with accumulation in the leaves. However, these concentrations were less than would be predicted if the tissue concentration were only a function of the transpiration rate, suggesting exclusion, as stated by Seyfferth and Parker (2007). Alternative explanations include sorption of perchlorate onto root or xylem tissues, or Phase I to III degradation mechanism *in planta*.

The fate of perchlorate in aquatic plants also was investigated by Susarla et al. (2000). Perchlorate was taken up and chlorate, chloride, and chlorite were detected in the plants examined, such as sweet gum (*Liquidambar styraciflua*) and black willow (*Salix nigra*), with the most mass present in the leaves rather than roots. The fate of perchlorate in poplar trees also was investigated by Van Aken and Schnoor (2002). They exposed poplar trees (*Populus deltoides* × *nigra*) to ^{36}Cl labeled ClO_4^- (25 mg/L), and measured the concentration of perchlorate in the hydroponic solution over time. After 30 days, the original perchlorate concentration had decreased by 50%, with no apparent toxic effect on the plants. The perchlorate was taken up, translocated, and entered the leaves. There, perchlorate (as ClO_4^-) was found in abundance (26%), but it was transformed into various reduced species, such as chlorate (ClO_3^-; 4.8%), chlorite (ClO_2^-; 2.4%), and chloride (Cl^-; 1.6%). This is novel because the reduction of perchlorate had previously only been observed by anaerobic bacteria, in which perchlorate was used as a terminal electron acceptor. The plant-induced reductions that occurred in the poplar leaf tissues were related to the action of reductases and dismutases that are still present even though oxygen is available. This may explain the low yield of chloride from plant-perchlorate reduction.

Because perchlorate is highly oxidized, it can act as an electron acceptor under conditions of limited oxygen, such as anaerobic conditions, in groundwater. This is similar to the case with PCE and TCE. There needs to be an electron donor, however, to drive the reduction. This can be an abiotic or biotic process. At sites limited in electron donors, various organic compounds have been added, such as molasses or vegetable oils. Plants also add organic matter to the rhizosphere, both as a byproduct of living cells or as the shedding of dead cells. Shrout et al. (2006) investigated the effect of root organic matter as a source of electron donor to drive the microbial reduction of perchlorate in soil and water samples. They showed in microcosm studies that perchlorate-reducing bacteria used plant-root exudates as the sole source of carbon to drive the reduction. The exudates were prepared from hybrid poplar tree cuttings (DN-34) by taking the roots and homogenizing them with a blender device. The fact that root exudates can facilitate the bacterial reduction of perchlorate is encouraging, especially considering that perchlorate also has been shown to be taken up and partially reduced in plants (Van Aken and Schnoor 2002). The key to efficiency of this remedial process will be to get maximum interaction between plant roots and the perchlorate-contaminated media. This *ex planta* reduction of perchlorate by soil bacteria once again illustrates the need to be clear about the difference between phytoremediation, bioremediation, and rhizospheric processes, all of which interact to some extent.

The study by Aitchison et al. (2000) of the fate of 1,4-dioxane exposed to poplar cuttings indicated that the majority of dioxane taken up was transpired. For example, 50% of the 1,4-dioxane removed from the hydroponic solution and of this percentage, between 76% and 85% was transpired. The balance of dioxane not transpired was detected in the stem of the cuttings. The cuttings did not exhibit gross toxic effects to dioxane even at concentrations of 23 mg/L. The fate of the plant-volatilized dioxane was not examined.

Tritium that enters roots will be translocated to the leaves where near 100% of the tritiated water taken up will be transpired. While it is possible that some tritiated hydrogen may react to reduce CO_2 during photosynthesis, the amount that remains in plant carbohydrate is probably low because 99% of water taken up by plants is transpired.

13.7 Concerns About Plant and Contaminant Interactions

The exposure of plants to groundwater contaminants, by definition and historical precedent, should raise concerns for the ultimate fate of the contaminant. Questions include:

- Are the concentrations of the groundwater contaminant(s) toxic to plants?
- Can the groundwater contaminants enter the plants?
- Can these contaminants be degraded *in planta*?
- What is the contaminant half-life if exposed to plants?
- Where does the contaminant go once in the plant?

Many existing approaches can determine the effect of contaminant stresses on plants. They can be classified as either plant-level or molecular-level methods. Plant-level methods include gross observation of wilting in the presence of adequate water potential, change in leaf color, etc. Molecular-level methods include measurement of enzymes such as peroxidase, a compound found in most plant cells, and in largest amounts in the cell wall. In general, exposure of a plant to a variety of chemicals results in an increase in peroxidase (Byl and Klaine 1991). Other indicators used to monitor plant stress include chlorophyll *a* and dehydrogenase levels, as well as photosynthesis and respiration rates.

For example, the effect of toluene on representative plants was studied and results presented by Reporter et al. (1991). The study used plant cell cultures, rather than whole plants. In this way, the effect of toluene on plant cell growth could be observed readily. Although a wide range of plant species was studied, the results for alfalfa (*Medicago sativa* L. cv. Vernal) and black locust (*Robinia pseudoacacia* L.) only are summarized here because of their importance as a herbaceous and woody phreatophyte, respectively, and potential use at phytoremediation sites. Alfalfa and black locust cells were grown in the presence of 500 ppm toluene for 12 days, and controls did not contain toluene. At the end of 12 days, cell growth (as grams fresh weight) had increased in both the control and treatment.

One of the most obvious plant processes to be used to assess the toxicity of a particular chemical is the process of shoot growth and its related transpiration. Measurements of transpiration can be readily made. This is similar to lab tests with microbes in which respiration is measured. Although the gross characteristic of a dying or dead plant is easily detected, it is subjective and the results are not easily comparable among sites and different labs. A toxicity test devoid of this subjectivity was described by Trapp et al. (2000). Using their method, cuttings can be weighed and then placed in flasks filled with a nutrient solution. A stopper can be used to seal the space around the cutting and the flask neck, and the gaps sealed with silicone caulk. The flasks are then wrapped with various materials to prevent light from entering. The flask-cuttings can be placed in artificial or natural light to acclimate to conditions prior to addition of the test toxicant (no addition for the control). The cutting-solution-flask needs to be weighed every day; the loss of weight is a surrogate for transpiration. Transpiration of the treatment flasks is normalized by the control flask transpiration. This hydroponic test also can be accomplished using various solid media.

The results of various toxicity tests on plants suggest that it is prudent to assess the levels of soil and groundwater concentrations that the plants might encounter prior to implementing a phytoremediation planting. Contaminant delineation also is needed to determine *a priori* if the levels are too high and contraindicate plant survival.

13.8 Plant Selection for Specific Groundwater Contaminants

To implement the bioremediation of groundwater contamination, knowledge of the ecological niche of soil and groundwater microorganisms and metabolic pathways was shown to be a useful approach to determine the type of contaminants that can be transformed (Chapelle 1993). For instance, aerobic bacteria will use reduced organic contaminants as an energy and/or growth source if oxygen is present but cannot derive the same benefit from a highly oxidized compound such as PCE. Conversely, anaerobic bacteria will be able to reduce PCE and will only be able to transform reduced organic contaminants if alternative electron acceptors are available.

A similar approach can be used to assess if phytoremediation of groundwater contaminated by specific types of xenobiotics will occur. Essentially, knowledge of the biotransformation potential of different plants can be used to match the plant to the type of groundwater contaminant present, even if the goal is hydrologic control, rather than contaminant transformation, as discussed in Chap. 8. However, even this interaction between plants and contaminated groundwater will lead to contaminant transformation, which can be enhanced if the correct plants are used. Contaminant transformation can be enhanced by comparing the fate of groundwater contaminants with various plants, using the metric of transformation rate or transformation extent. Not all plants, however, possess the correct enzymes to carry out Phase I, II, and III detoxification reactions. Most plants, however, can offer an important microbial component to remediation through the establishment of root-zone microbes in the rhizosphere.

Some of the plant traits necessary for the efficient interaction with contaminated groundwater include a deep root system, a documented interaction with the water table or capillary fringe, the ability to interact with common groundwater contaminants at environmentally relevant concentrations, the ability to process these contaminants through *in planta* detoxification reaction, and robustness in the face of contaminants and low intensity agricultural practices (Table 13.4).

As can be seen, members of the family *Salicae*, such as willows and poplars, are useful for a range of contaminants that can be found in groundwater. Similar results were shown by Zalesny et al. (2005) for plant interaction with soils contaminated by petroleum hydrocarbons. They planted a wide range of hybrid poplars and willows at sites characterized by petroleum hydrocarbon contamination of soil and groundwater and monitored plant-growth parameters over time. The greatest survival rate was for the hybrid poplars relative to the willows, at 97% versus 56%, respectively. The larger cuttings of poplars had higher survival rates than did smaller cuttings. The smaller, less expensive cuttings experienced more growth in length and height, however, versus the larger cuttings. As such, sites that require a larger area to be planted or sites with a smaller budget need not be placed at a disadvantage. Moreover, their study confirmed that the widespread application of poplar and willow clones is not just due to their availability from the paper or nursery industries but also to their survival at contaminated sites.

Table 13.4 Relation between plant type and potential for interaction with particular groundwater contaminants. Where site conditions indicate a period of high water table such that the root zone would be saturated for more than 1 week, willows can be used instead of poplars.

Reduced organics
Petroleum hydrocarbons
Aromatics
Alfalfa (*Medicago sativa* L.)
Willow (*Salix spp.*)
Poplar (*Populus spp.*)
Pine (*Pinus sylvestris* L.)
Polycyclic aromatics
Poplar
Alfalfa
Buffalo grass
Oxidized organics (Kassel et al. 2002)
Chlorinated solvents
Populus spp.
Alfalfa (Narayanan et al. 1999)
Oxidized/reduced organics
TNT, HMX, RDX
Hybrid poplar
Willows (Shoenmuth and Pestemer 2004a, b)
Birch (*Betula pendula*)
Pine

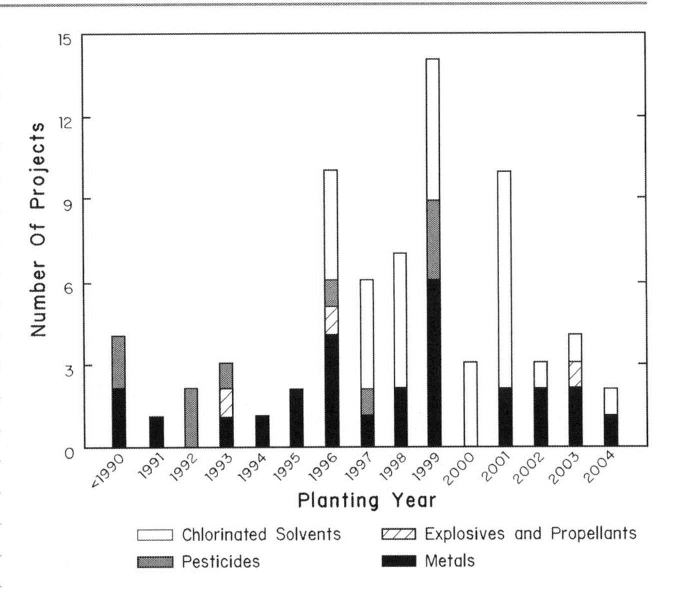

Fig. 13.21 Number of new phytoremediation sites and type of contaminant started per year (Modified from U.S. Environmental Protection Agency 2005c).

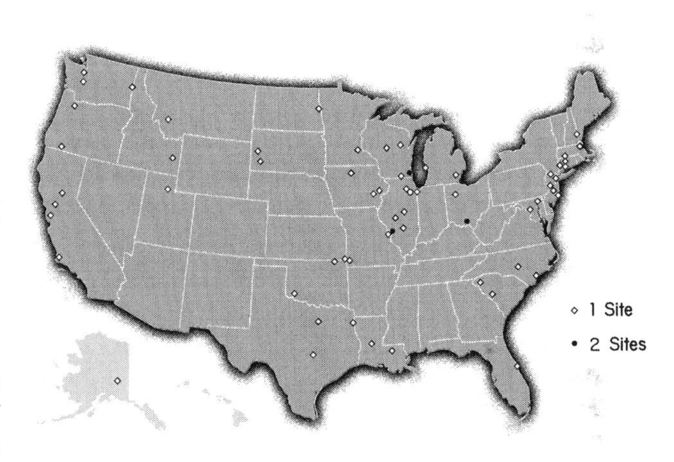

Fig. 13.22 Generalized distribution of phytoremediation sites in the United States (Modified from U.S. Environmental Protection Agency 2005b).

13.9 Historical Trends in the Initiation and Continuation of Phytoremediation at Groundwater Contamination Sites

Although the use of plants to restore contaminated water has been around for some time (see Chap. 11), the application of vegetation to remediate subsurface contamination such as groundwater was suggested only in the early 1990s (Cunningham and Berti 1993). The early 1990s had very few new projects, no more than five per year, where plants were specifically being used to remediate the subsurface contaminated by xenobiotics such as chlorinated solvents, pesticides, explosives, and metals (U.S. Environmental Protection Agency 2005b; Fig. 13.21). From 1996 on, however, there was a three-fold increase in the number of sites where plants were being applied as part of or were the sole remedial activity to address contaminated groundwater (Fig. 13.21). The highest number of sites (14) of phytoremediation application occurred in 1999 (U.S. Environmental Protection Agency 2005b).

The expansion of government-funded sites for innovative projects of high risk such as phytoremediation is typical of other such efforts that include government venture capital funding of the Interstate Highway System and national mail service. Of the 79 sites evaluated as part of the report (U.S. Environmental Protection Agency 2005b), they ranged across the United States in 31 States from the East Coast, Central Plains, and West Coast. The sites were located in areas that had cool seasonal temperatures and warm summers, and little to large amounts of precipitation, with great variations in climate and ET_P characteristics (Fig. 13.22). Although Cunningham and Ow (1996) stated that phytoremediation was still in its initial stages of research and development, this probably is a valid statement even today.

Not just the number of sites but the type of contaminant changed at phytoremediation sites since the 1990s (Fig. 13.23). Of the few sites where plants were being used prior to 1996, most were characterized by pesticide or metal contamination. This reflects the historical natural progression from the early studies of pesticide bioaccumulation and early experiments into metal-accumulating plants that was

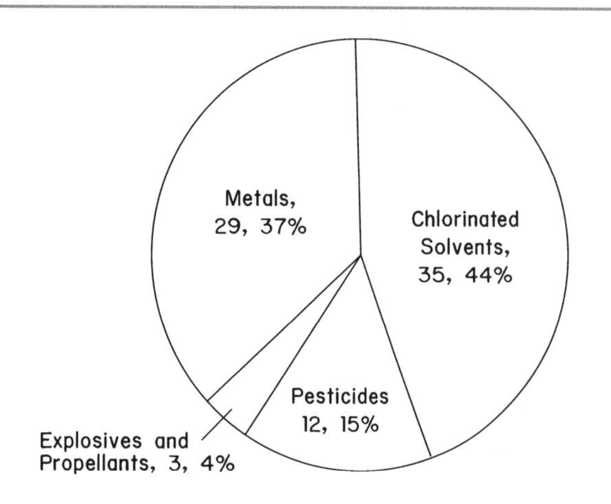

Fig. 13.23 The distribution of the types of contaminants being treated by phytoremediation (Number of sites, with percentage of total sites).

briefly depicted in Chap. 11. After 1995, however, the majority of sites where phytoremediation was attempted contained chlorinated solvents and petroleum hydrocarbons in groundwater and soils (U.S. Environmental Protection Agency 2005b). This reflects the priority pollutant nature of these compounds, their toxic effects at low concentrations, and the scientific communities' intensive research into these compounds with respect to natural attenuation processes. An additional class of compounds that received much interest was the petroleum hydrocarbons released at AST and UST sites (U.S. Environmental Protection Agency 2005b)—no data were collected, however, on these contaminants. Moreover, at some of the sites under Superfund authority, phytoremediation as a remedial strategy was written in the Record of Decision (ROD) for the site.

Since the late 1990s, the number of reported new sites has declined each year according to the USEPA survey (U.S. Environmental Protection Agency 2005b) (Fig. 13.20). There are many possible explanations of this trend. First, the lull could be the discontinuation of government-sponsored sites since 1999. It could indicate that responsible parties did not invest after government initiation of the project, or other remedial activities were selected or mandated by the regulatory agency involved in the project. A low interest in private investment into phytoremediation at some of these sites also could have occurred because of a real or perceived lack of return on capital. Also, at some sites there may be too much risk involved with environmental variables, such as droughts. The ever-changing political and environmental arena, however, may provide additional reasons to increase the involvement of research into new phytoremediation projects. These incentives include the use of vegetation to achieve carbon sequestration goals as well as remedial goals; the potential for cash flows from biomass generation; or as sources of carbon credit income. The limitation here is that in most cases phytoremediation sites will be less than 5 acres in size.

13.10 Summary

The plant-uptake of commonly detected groundwater contaminants is a reproducible fact based on a wide range of laboratory- and field-based studies. Easy-to-use conceptual models, such as *TSCF* and *RCF*, have been developed to guide our understanding of these interactions. At its most fundamental level, the degree of interaction between plants and groundwater contaminants is a function of the prevailing chemical and physical properties of the contaminants and the subsurface.

Why is this information important to the phytoremediation of contaminated groundwater? Plants *can* control the movement of contaminant solutes through their structures by affecting the rate of transpiration at the leaf surface and by the initial movement of water into root hairs in the subsurface. If a contaminant does not enter into at least one of the pathways discussed in this chapter, then that particular contaminant may not be amenable to phytoremediation.

14 Conceptual Frameworks for the Phytoremediation of Groundwater Contamination

In general, the log transform of K_{ow} is a useful physical property to prioritize which xenobiotics will interact with plants at a potential phytoremediation site. However, this physical property does not account for all factors that affect how groundwater contaminants will interact with plants, such as the groundwater-flow rate, the rate of transpiration, and the volume of groundwater contamination.

There are at least two different approaches, therefore, to evaluate the overall effectiveness of using phytoremediation to address contaminant remediation in groundwater. One is based on a mass-balance approach, where the flux of contaminants through a planted area is compared to the original contaminant mass. The other approach employs a method based on the one-dimensional solute-transport equation similar to that used to evaluate monitored natural attenuation, with the inclusion of terms to represent the effect of plants on contaminant fate. This chapter provides an introduction to both approaches and provides examples of their data requirements and implementation.

14.1 Contaminant Mass Reduction Framework

As shown in previous sections in this book, the uptake of groundwater by phreatophytic vegetation can affect groundwater levels, change horizontal and vertical groundwater-flow directions, and reduce the flux of groundwater to downgradient areas. As a result, such vegetation also can be used at contaminated sites to reduce the mass flux of aqueous-phase groundwater contaminants flowing through or contained beneath a planted area. These plants also can be used to decrease contaminant mass by the direct uptake and translocation of dissolved-phase contaminants, enhance biodegradation processes in the rhizosphere, and detoxify contaminants once in plant tissues.

14.1.1 The Conceptual Framework

To demonstrate the remediation of groundwater contaminated by xenobiotic compounds, the levels of the contaminant expressed as a concentration of mass per unit volume need to decrease over time and space. In other words, the concentrations must decrease and the plume must shrink. In most cases, site-characterization data will exist or can be collected in order to delineate the size of the contamination and the areas that contain dissolved-phase or free product, and the aquifer properties that determine the rate of groundwater flow.

To assess the potential for plants to hydrologically and geochemically remediate the contaminated groundwater, contaminant mass data prior to planting or upgradient of unplanted areas can be compared to contaminant mass levels that exits the site or at a specific location over time. If the contaminant mass decreases, then some level of phytoremediation will have occurred and the contaminant concentrations compared to general or site-specific remediation goals.

To accomplish this comparison, the measured, average contaminant mass flux, as determined in wells located upgradient of the planted area, M_{up}, as a product of the groundwater discharge, Q, in L^3/T, the area, A, through which the contaminated groundwater flows, and the concentration, C, of the contaminant from wells, $QACM_{up}$, in M/L^3 is compared to the measured average contaminant mass flux across the downgradient area, $QACM_{down}$ after mass losses, M_{loss}, due to biodegradation and volatilization are removed. This approach is an extension of the groundwater flux approach presented by Eberts et al. (1999).

Under conditions prior to the installation of vegetation, the average contaminant mass flux would be equal to the average contaminant mass flux leaving the area, or

$$QACM_{up} = QACM_{down} + M_{loss} \tag{14.1}$$

J.E. Landmeyer, *Introduction to Phytoremediation of Contaminated Groundwater*,
DOI 10.1007/978-94-007-1957-6_14, © Springer Science+Business Media B.V. 2012

assuming other processes such as sorption and dilution were constant in the contaminated aquifer beneath the area, with no changes in water storage. After plants are installed and take up groundwater that contains dissolved-phase contaminants, however, the average contaminant mass flux leaving the planted area would decrease, such that $QACM_{up} > QACM_{down}$. A maximum uptake rate could be expected during closed-canopy conditions, with the result being a larger gap between influent and effluent mass flux.

What about the fate of groundwater contaminants? This type of evaluation is essential if the actual mass loss of contaminants from a site need to be determined, rather than just a decrease in mass. A computationally simple model exists that describes the fate of sequestered contaminants, as outlined in Burken and Schnoor (1998) and Schnoor (1997) and presented in previous chapters. This model relates the contaminant fate to the transpiration rate, itself a flow that, when combined with a contaminant concentration, becomes a flux. The benefits derived from such computationally simple models, however, are at the cost of the simplifying assumptions that need to be made for their use. These assumptions include constant groundwater-contaminant concentrations, steady-state plume distribution, and no microbial biodegradation.

To summarize, the uptake of organic contaminants dissolved in water by plants can be described by Eq. 14.2

$$U = (TSCF)(T)(C) \tag{14.2}$$

where U is the uptake rate of the contaminant (M/T), $TSCF$ is the transpiration stream concentration factor discussed in Chap. 12 (dimensionless; Burken and Schnoor 1997), T is the transpiration rate (L^3/T), and C is the concentration of the dissolved-phase contaminant (M/L^3).

From Eq. 14.2, the time required for plants to take up a sufficient amount of groundwater to render the concentrations in the remaining groundwater at or under remedial concentrations, can be estimated from first-order degradation kinetics, such as

$$k = U/M_o \tag{14.3}$$

where k is the first-order uptake rate constant (per T), U is the contaminant uptake rate from Eq. 14.2, and M_o is the initial mass of contaminant present (M). At any time t during remediation, the mass remaining in the aquifer can be determined by

$$M = M_o e^{-kt} \tag{14.4}$$

where M is the mass remaining (M) at time t. Solving for t yields

$$t = -(\ln M/M_o)/k \tag{14.5}$$

where t represents the time needed to reach a remedial action level in the remaining groundwater, M is the mass allowed at t (M), and M_o is the initial contaminant mass (M).

The advantage of this simple contaminant mass-reduction approach is that it is essentially contaminant independent, unless, of course, the plants are negatively affected at toxic concentration levels of the contaminant. Because it requires the change in concentration over the length of a groundwater flowpath, however, it may not be useful at those sites characterized by slow groundwater-flow rates or where plants are installed over contaminant source areas. The following section offers an alternative approach that can be used as a framework under these conditions.

14.1.2 Case Study, Fort Worth, Texas

A phytoremediation project was initiated in 1996 at a site near Air Force Plant 4, Fort Worth, Texas (see Chap. 8 for more information on this site). At the site, hybrid poplar trees (*Populus deltoides*) were installed in early 1996 in two areas using different approaches. The groundwater flux and the contaminant concentrations were used to calculate the flux of TCE in the contaminated aquifer beneath the plantings. The volumetric flux of groundwater was calculated using Darcy's Law, and this product was multiplied by the average TCE concentration for the wells, located in a row, that define the cross-sectional areas up and downgradient of the planted area, perpendicular to groundwater flow. The parameters included the hydraulic gradient, $i = 2.25\%$, the cross-sectional area, $A = 807\ ft^2\ (75\ m^2)$, the aquifer thickness $b = 3.2$ ft (1 m), the aquifer width $= 246$ ft (75 m), the effective porosity, n_e, $= 23\%$, and the hydraulic conductivity of the saturated zone, $K = 19.6$ ft/day (6 m/day).

This calculation was done for various times of the year in order to reflect the seasonal differences in groundwater flux. The groundwater flux was calculated to be about 2,675 gal/day (10,125 L/day). This was multiplied by the average TCE concentration in a row of wells downgradient of the planted area. The researchers used this approach to estimate the change in calculated groundwater contaminant flux due to the removal of groundwater by trees for each year after planting until 1999. The decrease in calculated contaminant flux ranged from 2% to 12% of conditions before the trees were planted. Higher contaminant flux decreases were not observed, however, due to incomplete groundwater flow capture.

14.2 Framework That Accounts for Solute Transport and Plant Processes

The contaminant mass-reduction approach introduced above may be more useful at sites where the groundwater plumes have a discrete source area located upgradient of a phytoremediation application; i.e., the plants are downgradient of the contaminant source typically to stop the migration of a dissolved-phase plume to offsite areas. Although this scenario characterizes many sites, other sites may contain multiple source areas, have DNAPL, or are smaller sites where planting options are limited to the source area. These conditions require a different approach in order to evaluate the effectiveness of phytoremediation on contaminant fate. Fortunately, past study of natural attenuation processes produced solute-transport models that describe the chemical, physical, and biological processes that affect contaminant fate. These models can be applied for use at phytoremediation sites, with slight modification.

The factors that affect the concentration of a solute in groundwater include physical forces that drive the growth of plumes and the physical, chemical, and biological forces that resist plume growth (Chapelle et al. 2001). A meaningful expression of the interaction of these processes and their effect on solute transport in groundwater is given by the solute-transport equation as:

$$\begin{aligned}\text{Solute concentration} = (&\text{advection}) - (\text{dispersion} \\ &+ \text{diffusion} + \text{sorption} \\ &+ \text{volatilization} \\ &+ \text{biodegradation} \\ &+ \text{plant processes})\end{aligned} \tag{14.6}$$

and can be expressed mathematically as

$$\begin{aligned}\mathrm{d}C/\mathrm{d}t = {} & D\mathrm{d}^2C/\mathrm{d}x^2 - v\mathrm{d}C/\mathrm{d}x - P_bK_d/n_e\mathrm{d}C/\mathrm{d}x \\ & - kC - kV - kP\end{aligned} \tag{14.7}$$

where D is the coefficient of hydrodynamic dispersion (length squared per time), V is the velocity of groundwater (length per time), P_b is the soil bulk density, K_d is the linear sorption distribution coefficient, n_e is effective porosity, k is the first-order biodegradation coefficient, D is proportional to V and aquifer dispersivity (ft), kV is the first-order volatilization coefficient, and kP is the loss due to plant processes. Each of these processes is described below.

14.2.1 Advection and Dispersion

The movement of a solute dissolved in groundwater that moves through a porous media that is caused by the movement of the total solution is referred to as advection. Because the flow of groundwater can be described by the hydraulic conductivity as was discussed in Chap. 4, the higher the K value the higher the potential transport by advection of a solute, especially if the solute has a high solubility in water and low tendency for sorption. Because advection is related to the rate of movement of groundwater and therefore the aquifer hydraulic conductivity, K, Darcy's Law can be used to estimate the general extent of solute transport by advection for a contaminant that behaves as a conservative tracer.

The movement of a solute in groundwater does not behave as a "plug" of solute that moves uniformly through the subsurface. For example, the direct observation of the downgradient extent of solute transport often revealed that the solute was farther down the groundwater-flow path than predicted solely on advection using Darcy's Law. This can be explained by the fact that aquifers are not composed of homogeneous sediments and, therefore, flowpaths are not of equal length. This variation in K causes a variation in groundwater velocity. The difference in sediments encountered by the solute causes some solute particles to be retarded relative to the bulk movement of groundwater and some solute particles to move ahead of the bulk movement of groundwater.

These processes of dispersion are accounted for in the solute transport equation. Dispersion includes solute movement by mechanical mixing as well as the movement of solutes along concentration gradients, or diffusion. Dispersion, therefore, is mathematically described as

$$D = D_oT + \alpha V \tag{14.8}$$

Where D is dispersion, and D_o is the diffusion coefficient (L^2/T) that is specific for the solute, T is a factor used to describe the different tortuosity that the solute will encounter based on the heterogeneity of the sediments, α represents the dispersivity of the aquifer sediments (L) that is scale dependent (α increases as the flowpath increases) and sediment type dependent, and V is the groundwater velocity (L/T).

Dispersivity is often measured at the field scale using tracer tests with a conservative (nonreactive) solute. Under conditions of relatively fast groundwater-flow rates, say greater than 1-ft/day (0.3 m/day), then diffusive movement of a solute is considered negligible and Eq. 14.8 becomes

$$D = \alpha V \tag{14.9}$$

14.2.2 Diffusion

Diffusion is the movement of solute in response to a gradient in concentration of the solute over space. Our first introduction to diffusion was osmosis, a special case of diffusion through a membrane, such that water diffused into a plant cell that contained a higher solute concentration. With respect to solute diffusion in groundwater, the concentration gradient is that of the solute molecules themselves.

14.2.3 Sorption

Sorption of a solute to an immobile solid phase relative to groundwater flow removes the solute from the aqueous phase. The solid phase is the aquifer material in most cases. This solid phase also can consist of plant material. This removal is either onto the surface of the immobile solid phase or its interior. It can be irreversible or reversible. Sorption is an important property to simulate with respect to contaminant plume fate. Sorption does not determine the maximum extent of the plume at steady state; rather, it determines the time that it will take to reach steady-state dimensions, with longer times for aquifers with higher sorption capacity (Landmeyer et al. 1996b).

14.2.4 Volatilization

Many of the compounds released to groundwater have high vapor pressures and low to moderate water solubilities. As such, these xenobiotics have a tendency to be present as a vapor phase. If released into groundwater as a pure liquid or mixture, these compounds will establish equilibrium in the pore spaces of the unsaturated zone with respect to the pure-phase concentration. Because this equilibration with the air in the vadose zone decreases the concentration in the pure phase that can ultimately dissolve into groundwater, it can be considered mass removal and, therefore, a component of natural attenuation.

The equilibration is governed by diffusive flux along a concentration gradient. As was discussed in Chap. 12, volatilization permits uptake into plant roots. These volatilized compounds, once in the air of the pore spaces, also can undergo dispersion and diffusion, as well as undergo oxidization to support microbial growth (Lahvis et al. 1999).

14.2.5 Aerobic and Anaerobic Biodegradation Processes

The processes discussed above can all act to decrease the dissolved-phase concentration of a particular contaminant in groundwater by physical dilution with uncontaminated or less-contaminated groundwater, or by removal of a contaminant from the aqueous phase to a solid phase. In both cases, however, the total contaminant mass remains unchanged and remains in the subsurface, and the occurrence of reversible sorption can lead to the long-term release to groundwater.

In contrast, the mass of contaminants in groundwater can be decreased by biodegradation processes by the production of a less harmful intermediates or the complete mineralization to CO_2. The degree of this transfer is dependent upon the contaminant type and concentration, and the presence of the predominant electron acceptor in the contaminated groundwater (Chapelle 1993).

14.2.6 Plant Processes

The various physical components of plants can interact with contaminants above and below ground to the extent controlled by the chemical and physical properties of the particular contaminant. Some of the more important processes are described below.

14.2.6.1 Sorption

The presence of root material in a soil or sediment leads to an increase in the organic matter content. This is the process that leads to the formation of an O horizon in soil. The release of plant organic material, either in the form of exudates such as organic acids or as dead root material, acts to add labile and refractory carbon to previously mineral soils. This organic matter can act to absorb contaminants from solution according to the degree that the compound is attracted to organic matter, described by log K_{ow}. As was discussed in Chap. 12, the RCF is an indication of this capacity for chemicals to be sorbed onto organic root material from groundwater.

14.2.6.2 Rhizospheric Processes

The rhizosphere can affect contaminants by the presence of increased microbial and fungal populations, increased oxygen delivery from the atmosphere to the subsurface, the release of organic substrate, and release of contaminant-specific or non-specific degradative enzymes.

14.2.6.3 Uptake

The root systems of plants are the pathway for water uptake and delivery to the xylem but also act as the endpoint of the diffusive transport of atmospheric oxygen to the subsurface in the cortex of some plants. Hence, plants can take up solutes dissolved in groundwater as well as those that volatilize to the vadose zone.

14.3 Existing Conceptual Models

14.3.1 PLANTX

The model PLANTX (Trapp et al. 1994) describes the entrance of xenobiotics into plants. The compounds are simulated to enter the plant through soil, water, or the atmosphere through foliar uptake. Importantly from the perspective of phytoremediation, this model accounts for changes in the contaminant concentrations within the plant caused by the plant metabolism processes introduced in Chap. 11.

14.3.2 CTSPAC

Lindstrom et al. (1990) produced a one-dimensional analytical model called CTSPAC to simulate the transport of a contaminant from the soil to plant to atmosphere from a source in the vadose zone. Movement of water in the vadose zone is simulated by the Richards equation (see Tindall and Kunkel 1999 for equation derivation), and movement of solutes by convection, diffusion, dispersion, sorption, degradation, and plant uptake. The model simulates all aspects of plant physiology, including separate xylem and phloem compartments. This model was calibrated by Ouyang (2002) using the contaminant 1,4-dioxane as a model xenobiotic for which experimental properties that needed to be entered into the model were known experimentally after research by Aitchison et al. (2000).

14.4 Site-Characteristic Data Needed to Support the Framework That Accounts for Solute Transport and Plant Processes

Although the solute-transport equation in Eq. 14.7 contains multiple parameters, each is a physical process that can be measured at a site. If all the parameters in the equation are quantified using a combination of field and laboratory approaches, then the change in solute concentration over space and time can be solved, within the degree of parameter variability at the site. This section summarizes each parameter in the solute-transport equation.

Because advection is related to the rate of movement of groundwater and therefore the aquifer hydraulic conductivity, K, Darcy's Law can be used to determine the general extent of solute transport by advection. The hydraulic conductivity can be determined from single-well rising- or falling-head slug tests, a pumping test with some wells used to observe the time-dependent water-level drawdown, laboratory permeability tests done on aquifer material removed from the site, or look-up tables in reference books. More information on these field tests is presented in Chap. 4.

In order to calculate a Darcy velocity, the hydraulic gradient can be calculated from a water-table map prepared using synoptic groundwater-level measurements. Each depth to groundwater is then plotted on a map that shows the well distribution; the data can be recorded as "depth to water table" or as "elevation or altitude of water table above a common datum". The difference in water-table elevation, or head, Δh, divided by the distance, ΔL, between two wells, will provide the head gradient, i, that will cause groundwater to flow in response to gravity. The effective porosity can be determined by laboratory tests or from reference tables.

Movement of solute by diffusion can be estimated using different approaches. In the field, tracer tests can be performed using a conservative solute such as bromide or chloride. The diffusion coefficient can also be selected from reference tables.

Samples of aquifer material are collected and examined for sorption using standard methods. The absorption coefficient for a particular compound for site sediments can be determined in the laboratory by adding various concentrations of a compound to vials containing the sediment and determining the fraction that remains in the added solution. Once the samples are collected, a known amount, between 5 and 30 g, can be added to vials and then amended with a solution of the contaminant(s) of interest. This is allowed to equilibrate for a period of time on a shaker table. Samples of the liquid phase are then analyzed to determine the fraction of the contaminants that remain in the solution; the difference will represent the degree of sorption. A linear response can then be plotted as a function of different initial contaminant concentrations. The slope of this line can provide the partition coefficient, K_d, for that compound in that sediment. Contaminant volatilization can be determined using a field or laboratory approach, such as that described by Lahvis et al. (1999).

The microbial degradation of contaminant compounds is a special case of the microbial metabolism of substrates. For most field situations, this metabolism can be considered to follow first-order kinetics, such that the degradation of a substrate is not limited by the availability of the appropriate enzymes, or

$$V = kS \tag{14.10}$$

where V is the rate of substrate uptake (moles per time per g of cells), k is the rate constant (per time), and S is the substrate concentration.

This dependence of rate on substrate concentration is described in the Michaelis-Menten equation,

$$V = [v_{max}/(K_s + S)](SB) \tag{14.11}$$

where v_{max} is the maximum rate of substrate uptake, K_s is the substrate concentration at which $v = ½\ v_{max}$, S is the substrate concentration (moles per liter), and B is the amount of cells (g). If concentrations at a site are high, first-order kinetics may not apply.

Field data also can be used to determine biodegradation rates, and many examples of this approach exist in the literature (Chapelle et al. 1996; Wiedemeier et al. 1996; Landmeyer et al. 2009). Essentially, the difference in contaminant concentration between two wells can be used to determine loss rates assumed to be due to biodegradation. A third approach to determine biodegradation rates from field data is to use a method based on flux. This method overcomes some of the limiting assumptions of the flowpath method. It was used at a landfill site (Dinicola et al. 2002) and at sites of MTBE-contaminated groundwater discharge (Landmeyer et al. 2001) and MTBE-, TAME-, and TBA-contaminated groundwater discharge (Landmeyer et al. 2010).

14.5 Use of Existing Groundwater Flow and Solute-Transport Models That Incorporate Plant-Related Processes

The concept of a water budget was introduced in Chap. 2. Although deceptively simple, it provides a useful conceptual framework to assess the effect of different components of the water budget on a particular site's hydrology. Under simple site conditions or restrictive assumptions, it may be possible to answer some qualitative questions regarding the compartmentalization of water within a basin (see Lindgren 1903). Under most site conditions, however, contaminants are present and hydrogeologic conditions are more complex. Numerical models that represent solutions to calculus-based formulas that describe groundwater flow and solute transport are often the best tools to use.

Groundwater-flow models can be used to simulate the various processes that affect the movement of groundwater in aquifer media. Although such models do not account directly for the movement of solutes being carried in the water phase, groundwater-flow models can be used to determine the effect of one component of the proposed solute-transport equation, and that is the uptake of groundwater by plants. Because this uptake involves at least some water from the unsaturated zone, a model should be able to simulate flow in the unsaturated zone. The movement of water in the vadose zone often is simulated by the Richards equation.

14.5.1 Groundwater-Flow Models

The groundwater flow model MODFLOW (McDonald and Harbaugh 1988) is a widely used model to address groundwater flow issues. The most recent version is MODFLOW-2005 (Harbaugh 2005). It has a subroutine, or module, that describes the impact of *ET* on groundwater levels, and the reverse. If coupled with the model's particle tracking module called MODPATH, then it can be used to determine not only if plants are going to affect the direction of groundwater flow and the volume removed, but also if particles that represent dissolved substances will be captured by areas proposed to be or have been planted.

14.5.2 Case Study in Florida

The MODFLOW model was used to determine if existing vegetation could be used to capture contaminated groundwater before it discharged contaminants to a surface-water body (Halford 1998). At the site in Orlando, Florida, groundwater contaminated by chlorinated solvents from a dry-cleaning facility was discharging to Lake Druid about 500 ft (152 m) downgradient from the source area (Fig. 14.1). The shallow water-table aquifer consisted of fine-grained sand, which would increase the potential for any interaction between groundwater and trees if phytoremediation were used as a remedial strategy. The hydraulic conductivity of these sands was between 10 and 40 ft/day (3–12 m/day).

The conceptual model of the site depicts that water input (precipitation and irrigation) is about 57 in./year (144 cm) and that *ET* removes about 34 in./year (86 cm/year) (Fig. 14.2). The study area was developed in both the horizontal and vertical dimensions. Horizontally, it was discretized into a grid composed of cells where finite-difference approximations to the groundwater-flow equation could be solved. In order to reduce the potential for complications in the computed groundwater elevations that arise near model

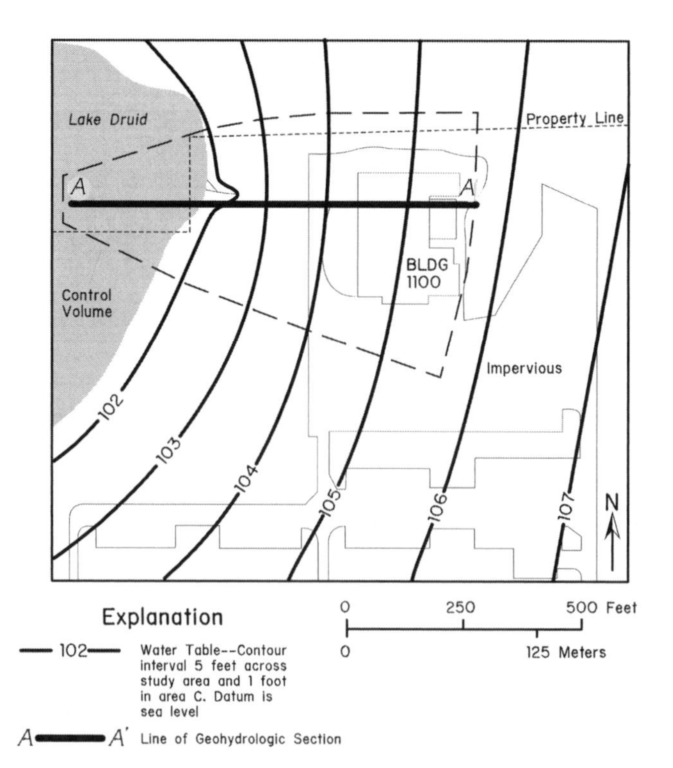

Fig. 14.1 Diagram representing Florida case study of effect of a potential phytoremediation planting on groundwater flow (Modified from Halford 1998). One foot is equivalent to 0.305 m.

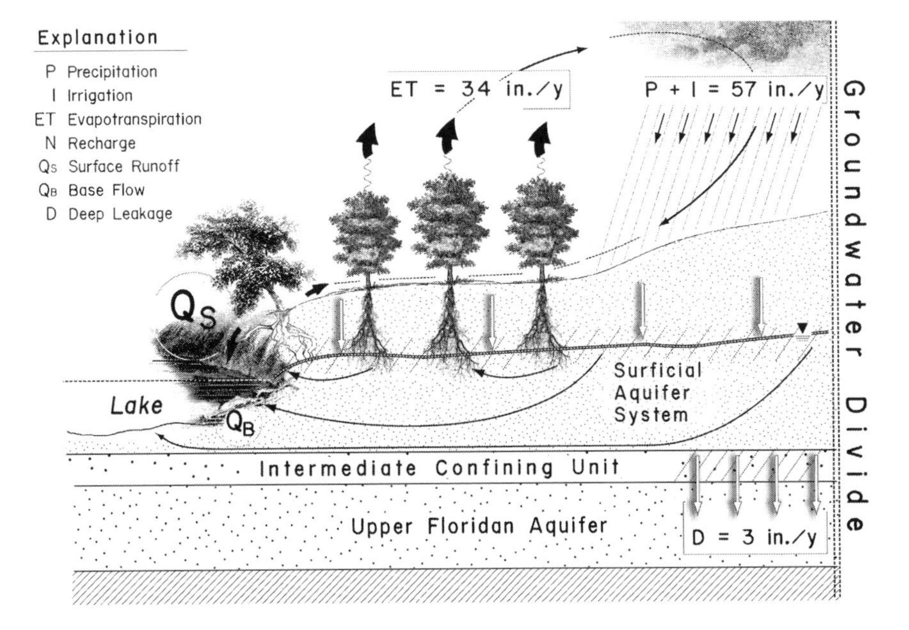

Fig. 14.2 The water budget for a contaminated groundwater site in Florida (Modified from Halford 1998).

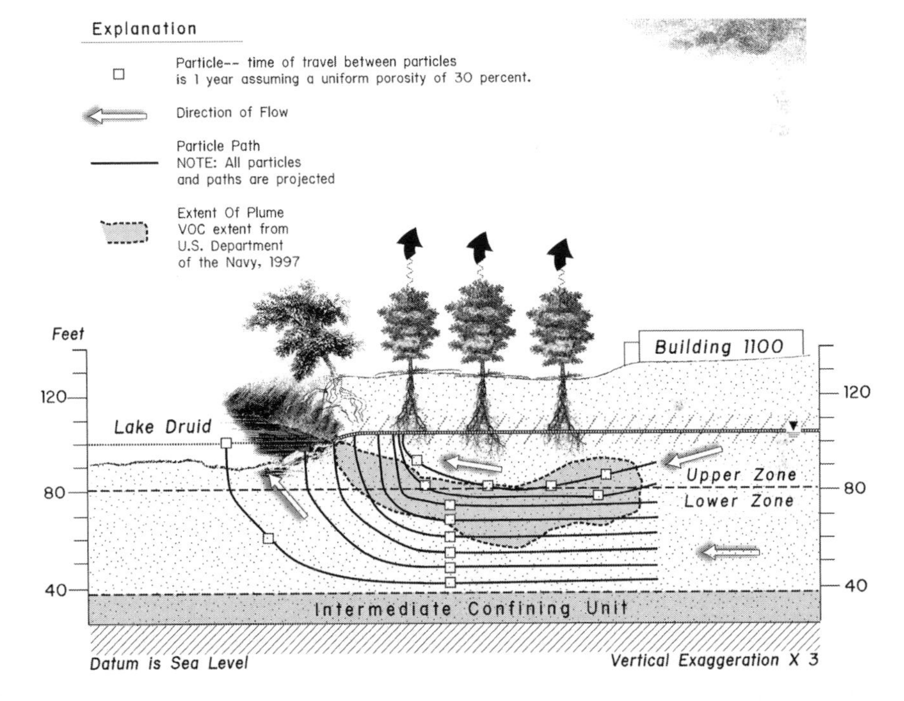

Fig. 14.3 The simulated effect of existing trees and transpiration (black arrows) on shallow and deep groundwater flowlines that discharge to Lake Druid, shown as the flow of simulated particles, at a VOC plume site in Florida (Modified from Halford 1998). One foot is equivalent to 0.305 m.

boundaries, called boundary effects, the model was extended beyond that of the site property boundaries. Vertically, the shallow aquifer was simulated as three layers, in order to account for groundwater and surface-water interactions. The model input for hydrogeologic characteristics was determined from an aquifer test conducted on site to determine hydraulic conductivity (average was 33 ft/day (10 m/day)). The discharge of groundwater to the lake also was simulated.

Even though MODFLOW is a groundwater-flow model, surface-water features can be simulated. The lake and streams at the site were simulated using the river package of MODFLOW. All surface-water features were simulated as gaining, indicating areas of groundwater discharge. Because the area between the source of contaminants and discharge was forested with native vegetation, MODFLOW was used to simulate the impact that the trees would have on groundwater flow. Evapotranspiration in this model was simulated not by using the *ET* package but by assuming that recharge was reduced in those areas that had trees relative to areas that received recharge but did not have trees.

The model results suggest that the existing vegetation removes only a small part of the overall groundwater flux from the source area to the surface-water body. Groundwater discharge to the lake was estimated to be about 20 gal/min (75 L/min) and groundwater uptake by *ET* was simulated to be less than 4 gal/min (15 L/min) (Fig. 14.3).

14.5.3 Groundwater Flow and Solute-Transport Models

14.5.3.1 FACT

The Flow and Contaminant Transport (FACT) model is a three-dimensional, variably saturated finite-element model of groundwater flow and solute transport (Hamm et al. 1997b). It is a potentially useful model for phytoremediation purposes because it includes the removal of water from the unsaturated zone by plant transpiration.

The FACT model was used to simulate plant and groundwater interactions at the Savannah River Site (SRS) in South Carolina. The area simulated was a pilot-scale test facility where wastewater that contained chlorinated solvents was discharged to unlined basins. Subsequent installation of monitoring wells in the area and downgradient toward the discharge area of the Savannah River indicated the chlorinated solvents were in the shallow groundwater downgradient and in the wetland flood plain of the Savannah River (Vroblesky et al. 1999b).

Researchers conducted a comprehensive investigation into the processes that affect the natural attenuation of the chlorinated solvents in groundwater. After the collection of the necessary groundwater-flow and solute-transport data, such as absorption and biodegradation data, as well as plant interaction data from laboratory experiments (see Neitch et al. 1999), the FACT code was used to develop a model to integrate the influence of these processes on contaminant movement (Hamm et al. 1997a). In the model, a TCE plume in groundwater was generated and allowed to migrate to the Savannah River. Different simulations were performed to account for the impact on the plume by sorption and biodegradation, although this was set to zero based on laboratory experiments with TCE and aquifer sediment. Another simulation was used to estimate the influence of *ET* on the groundwater and plume distribution. The extinction depth was set to 30 in. below the water table, and recharge was set at 47 in./year (119 cm/year). All simulations indicated that natural attenuation and *ET* processes did not affect the discharge of TCE to the river.

Some interesting observations were made regarding the influence of plants in the flood plain on groundwater levels as part of this modeling study. Continuous measurements of groundwater-level and surface-water-level fluctuations were made with pressure transducers and the data recorded on data loggers. Depth of shallow groundwater in the flood plain sediments extends to 8 ft (2.4 m) below land surface during periods of low groundwater level during the summer (June). During this time, diel groundwater-level fluctuations were observed in a well.

This observation is significant, in that the water-table elevation was at its lowest during this time, and a fluctuation was still observed. During periods of higher water table, this zone remains saturated. As such, the tree roots either follow the water table as it is lowered or remain suberized when the water table is higher. Capillarity may also explain this observation, where the roots are always above even the higher water table elevations but evaporative demands cause water to move upward by capillary action. Moreover, the increase in groundwater level in June was caused by recharge due to precipitation, and the removal of this recharge by the *ET* from the plants.

14.5.3.2 SUTRA

SUTRA is a code to solve for saturated-unsaturated variable density groundwater flow and solute transport (Voss 1984). It is a finite-element code. Although it does not directly simulate the uptake of water by plants, it does simulate the fate of density-dependent contaminants and, therefore, may have some application for phytoremediation.

14.5.3.3 SEAM3D with PUP

The numerical code SEAM3D was modified to account for solute transport coupled to plant processes such as sorption of solute to roots, translocation into plant vascular tissue, and *ET*. The extent of such interactions is based on the physical and chemical properties of the contaminant, such as K_{ow}, using the *RCF* and *TSCF*. The modification to SEAM3D entails the Plant-Uptake-Package, or SEAM3D-PUP (Widdowson et al. 2005b).

The model has been used to evaluate the effectiveness of phytoremediation on the proposed dimensions of the planting, the effect of plant density on maximum expected *ET*, the effect of plants on through-site contaminant flux, and the time of dissolved-phase contaminant mass removal after source material extraction. Simulation results indicated that (1) the width of a proposed phytoremediation system has a limited effect on solute-mass removal (2) having a higher density of planting near a source area has a greater impact on contaminant mass removal relative to a uniform planting over the entire plume area (3) if the *TSCF* of the contaminant is low, such that little is taken up by plants, plant roots will exclude uptake and concentrations in groundwater near the roots will increase (4) the dimensions of a plume under steady-state conditions are controlled by groundwater flow, which can be affected by plant-water uptake (5) splitting the phytoremediation system into two halves is less effective than one large mass planting, and (6) after source removal, the contaminant concentration in the groundwater near trees increased, but decreased in downgradient areas (Widdowson et al. 2005b).

14.5.4 Unsaturated-Zone Models

The fate of xenobiotics in the unsaturated zone is important to the success of phytoremediation of contaminated groundwater. Many contaminants that enter groundwater were released at the surface or in the unsaturated zone and will be transported to groundwater by density or infiltration and leaching. Residual contamination in the unsaturated zone represents a long-term continuous source of dissolved-phase contaminants to groundwater. The distribution of plant root systems, including for phreatophytes, is greater in the unsaturated zone and, therefore, plays a role in determining the concentration and structure of contaminants. The combination of enhanced plant processes and contaminant presence are why an understanding of such interactions is imperative to include in phytoremediation studies.

The models HYDRUS, RZWQM, and VS2D solve for water flow in the unsaturated zone using the Richards equation. Also, the uptake of water by plants is directly simulated. For solutes, the number of contaminants that can be simulated ranges from 1 to 5, and processes that affect these contaminants include sorption, volatilization (except for VS2D), dispersion, degradation, uptake by plants, and effect of various physical parameters on degradation such as soil moisture (Nolan et al. 2005).

HYDRUS (Simunek et al. 2005) is finite-element model that simulates the movement of water in the saturated zone as well as being used predominantly for simulating water movement in the unsaturated zone, under one and two dimensions (HYDRUS-2D). It solves for flow in the unsaturated zone using the Richards equation. It also solves for the uptake of water by plant roots, which can be selected from a database of values, and simulates the growth of roots using a logistic growth function. Also, *ET* values can be added and simulated. Contaminant fate of up to five solutes is simulated as being affected by sorption, volatilization, dispersion, degradation, and uptake by plants.

A model that can be used to examine the relation between plants, xenobiotics, and hydrology in the unsaturated zone is the Root Zone Water Quality Model (RZWQM). The RZWQM is a one-dimensional model developed by researchers with the U.S. Department of Agriculture (Ahuja et al. 2000). In an investigation of various such models, it was determined that RZWQM simulated pesticide fate and transport with the smallest error of all models examined (Nolan et al. 2005). Inputs required are extensive but similar to most unsaturated-zone models and include values for soil organic content, microbial populations, degradation-rate coefficients, among others; the effect of dispersivity on contaminant fate is not, however, simulated. Water flow is simulated using the Richards equation. Up to three contaminants can be simulated simultaneously.

VS2D stands for Variably Saturated 2-Dimensional Transport. It is a finite-difference code that simulates the flow and transport of solute in variably saturated porous media (Lapalla et al. 1983). Unlike HYDRUS, the root growth distribution can be set as a function of time. Unfortunately, only one solute can be tracked over time.

14.5.5 Guidelines for Model Evaluation

Groundwater-flow models are applications of numerical codes that approximate the flow of groundwater through porous media. The application of such codes to solve field problems is subject to user bias and the way that the conceptual model has been conceived. As such, in order to properly evaluate the application of any form of model to examine a problem, the objectives of the use of the model need to be stated clearly. Only then can it be determined if the model was used correctly or used within its set of limitations (Reilly and Harbaugh 2004).

14.6 Summary

Models are simulated approximations of physical phenomena, such as groundwater flow and solute transport. Simple models include those that describe the interaction of a chemical species with water or organic matter, such as the log transform of K_{ow} or the interaction between the sediment characteristics or an aquifer and groundwater flow as in Darcy's Law. Because the processes of groundwater flow and solute transport are complex, the complexity of models necessarily increases to more sophisticated analytical and numerical models. Some of the existing models incorporate plant and water relations, from either the vadose zone or the saturated aquifer. Recently, attempts have been made to simulate plant and water interactions along with plant-water-solute reactions. Such models will undoubtedly become more useful as they are applied to phytoremediation field trials before, during, and after installation.

Why is this information important to the phytoremediation of contaminated groundwater? Both simple and complex models provide the ability to test hypotheses at phytoremediation sites in a relatively rapid manner. Also, models can be (cautiously!) used in a predictive manner to estimate how long a phytoremediation system will need to remain active in order to reach remedial goals. In terms of the perceived cost effectiveness of phytoremediation, such information provided by models is invaluable.

15 Monitoring for Phytoremediation of Groundwater Contamination

In 1977, the Federal Water Pollution Control Act was amended as the Clean Water Act (CWA). In general, the CWA provides for the regulation of the release of contaminants to water. This regulation is monitored by the USEPA by effluent standards, and permits are required to ensure the discharge of acceptable levels of wastes. The CWA covers contaminant levels that affect aquatic life and recreational standards, although the CWA affects drinking-water quality by proxy. In 1974, the Safe Drinking Water Act (SDWA) was enacted to regulate the quality of drinking water, either existing or potential sources of surface or groundwater. This regulation also is monitored by the USEPA through water-quality standards, and municipalities are required by law to treat drinking water to these standards.

To track the manufacture, use, and disposal of potentially hazardous wastes that could affect water quality, the Resource Conservation and Recovery Act (RCRA) was enacted. It is managed by the USEPA, or states if so directed. It also addresses releases from underground storage tanks. The Comprehensive Environmental Response, Compensation, and Liability Act (CERCLA), or Superfund, was enacted to deal with those previously contaminated sites not covered by RCRA.

Plants can be installed at sites characterized by contaminated groundwater to achieve part or all of the site remedial goals required by these various regulatory programs. Although these goals will be site specific, they have in common the requirement of establishing plant growth. But it is not sufficient from a regulatory perspective to simply plant a phytoremediation system and then walk away. Part of most remedial programs requires monitoring of the groundwater or remedial apparatus to verify that remediation is occurring and to document its performance over time. This chapter provides some basic approaches that can be used to evaluate the success of phytoremediation implemented at sites characterized by contaminated groundwater.

15.1 Plant Physiologic Methods

A plant's survival depends on its interaction with the various components of soil, microbes, water, and air. These interactions provide the opportunity to evaluate the effect of plants when applied to remediate contaminated groundwater.

15.1.1 Rhizospheric Community Analysis

Rhizospheric processes are important to monitor and quantify at groundwater-contaminated sites because contaminants completely mineralized or absorbed in the rhizosphere are not available for plant uptake or potential transfer to other parts of the immediate environment.

Rhizospheric processes can be monitored using a combination of field and laboratory approaches. At the field scale, soil material representative of the root zone for most plants can be obtained with a stainless-steel hand auger. In the laboratory, the biodegradation potential of these sediment samples can be assessed using microcosm studies, where sediment samples are amended either with radiolabeled contaminants to track the appearance of biodegradation end products such as $^{14}CO_2$ or are amended with nonradiolabeled contaminants to track their disappearance. Microbial numbers can be quantified using the most probable number (MPN) technique (Jordahl et al. 1997), which is based on the appearance of growth in serial dilutions.

To further determine if this increase in growth translates to an increase in the bacteria needed to degrade specific contaminants, molecular approaches, such as fatty acid methyl ester analysis (FAME), phospholipid fatty acid analysis (PLFA), and denaturing gradient gel electrophoresis

J.E. Landmeyer, *Introduction to Phytoremediation of Contaminated Groundwater*,
DOI 10.1007/978-94-007-1957-6_15, © Springer Science+Business Media B.V. 2012

(DGGE), can be used. Brigmon et al. (1999) used a direct fluorescent antibody (DFA) technique to determine the presence of methanotrophic bacteria that produce MMO in order to understand the potential for TCE to be gratuitously degraded at a site, as was discussed in Chap. 13.

To resolve to a higher degree if the bacteria needed to degrade specific contaminants are present then molecular techniques are required. Molecular markers can be used to distinguish one species of bacteria from another. A type of marker widely used is called an antibody resistance marker, such as rifampicin and streptomycin. A very specific marker, such as beta-glucuronidase (*gus*A) gene, is called an introduced marker gene (Wilson 1995). This method can make visible, through a colored product, the plant-root colonization of strains that contain that gene. Another approach is the enzyme-linked immunosorbent assay (ELISA) method, where antigens on the cell surface are recognized by specific antisera. Molecular methods are useful because they are based on the genetic characteristics unique to specific microbes, such as those that inhabit the rhizosphere.

The DNA contained in cell nuclei is the key to understanding the inner and outer workings of both bacteria and plants. The DNA contains the genetic information to code for the production of various proteins that are used for different purposes. The first step in unlocking any question that may be related to DNA, therefore, is to isolate it from all the other materials that are contained in the cell or that the cell has come into contact with during preparation for DNA isolation.

DNA is not organized like other cell parts, that is, of carbon or hydrogen atoms, but rather of linear segments of DNA called genes. These are composed of a series of interconnections between four amino acid (protein) bases. This linear genetic code can be copied onto mRNA, which is then used to make linear proteins. Phospholipid fatty acids are present in all cells and can be extracted from cell mixtures easily. It is this characteristic that makes the use of PLFA a preferred method over use of MPN to determine microbial biomass in the subsurface. Gene probes can be used to measure the prevalence of specific genes at a site. The efficiency of their use as a provider of accurate numbers still remains in question, however.

15.1.2 Plant Tissue Samples

As was discussed in Chap. 12, the direct uptake of groundwater contaminants by phreatophytes such as poplar trees commonly installed at sites characterized by contaminated groundwater can occur by the aqueous pathway (McFarlane et al. 1990) or by the gaseous pathway (Bromilow and Chamberlain 1995; Neitch et al. 1999). Regardless of the physical state of the contaminant being taken up, however, once in the transpiration stream the fate of the contaminant can be assessed using similar approaches.

Fig. 15.1 An increment borer used to collect tree cores. The borer has been advanced but the inner sleeve not yet inserted (Photograph by author).

A simple approach that has received much attention is the collection of various tree tissues, such as bark, stems, leaves, as well as material cored from the trunk using an increment borer (Fig. 15.1). The collection of core material for analysis of tree health has been a standard technique of the forestry industry for many years (Grissino-Mayer 2003), and gained initial use for hydrologists as an indicator of hydrologic change (Phipps 1967; Helley and LaMarche 1973; Phipps et al. 1978; Zimmerman 1989). The method used may differ slightly among users but consists of inserting an increment borer into a tree, turning it to advance the borer, then inserting a steel sleeve, and then retrieving the core, as documented in user guides by Phipps (1985) and Grissino-Mayer (2003).

The relation between the presence of trace elements in the tree and their source from surface water was investigated using tree cores as early as the 1970s by Sheppard and Funk (1975), from soil (McClenahan et al. 1989), and from groundwater (Kalisz et al. 1988). The collection and analysis of other tree tissues also has a long history. Moreover, even some of these early investigators understood that the ultimate source of the minerals being investigated in the plant tissues was either from the soil or from the soil-water profile, including groundwater. Some plants can accumulate in them elements at higher concentrations than exist externally, such as selenium accumulation by *Astragalus spp.* (Anderson et al. 1961), or can contain lower concentrations than those measured outside the plant, such as is the case with sodium or chloride.

Due to the potential deleterious effect of such salts on the survival of plants due to disruption of the osmotic process of water uptake (Gough et al. 1979), the concentration of sodium and chloride in various plant tissues has been of interest since the mid 1900s. For example, the USGS geochemist Hem (1967) investigated the occurrence of Na and Cl in the leaves and stems of the riparian phreatophyte saltcedar (*Tamarix spp.*) growing above salinized groundwater along the banks of the Gila River, Arizona, and Rio Grande, New Mexico. The total content of chloride, as well as calcium, magnesium, and sulfate, in the leaves of saltcedar ranged from 5% to 15% of the leaf dry weight. Moreover, the leaves that contained the highest concentration of sodium and chloride grew above groundwater that also was characterized by high salinity, even though the saltcedar is not a halophyte. This study by Hem (1967) is perhaps the first published study that showed a link between groundwater quality and the uptake of a contaminant by a phreatophyte.

Hem (1967) also related the changes in mineral concentration in the leaves over time to the chemical composition of the groundwater, the time of year, and transpiration rate at the time of sample collection. In the field, about 25 g of plant tissue was collected and placed in a bag and then taken to the laboratory where the tissue was air dried; a smaller subsample was oven dried. These samples were placed in a beaker to which was added distilled water, and this extract was analyzed for the presence of ions. More than 30 trees were sampled and analyzed in this manner.

In general, the depth to groundwater had an effect on the mineral composition of the plant-leaf extract (Hem 1967). The lowest concentrations of calcium, magnesium, sulfate, and chloride were observed in samples from saltcedar trees that grew where the depth to water table was the deepest, and the converse also was true. Interestingly, the composition of the residue that remained on the leaves of saltcedar also was analyzed. Saltcedars can survive in salinized groundwater, even though they are not halophytes, by removing the excess salt to the outer part of the plant using special salt glands. These glands are locations of high salt concentrations. This and the guttation of salinized water may help these plants tolerate changes in the mineral content of groundwater. Other plants, such as succulents, can deal with high salinity by maintaining high internal water concentrations, and do not excrete salts.

Hem (1967) analyzed the mineral content of water shaken on leaves and related it to the chemical composition of groundwater pumped from a nearby well. Even though the author cautioned that little can be stated about the processes that led to the observation between solute leaf concentration and source water, the gross groundwater geochemistry appeared to be related to the chemistry of the leaf extracts (Hem 1967). For example, two saltcedar trees that had the highest chloride concentration grew above groundwater that was characterized by 3,000–4,000 mg/L chloride. At another location, shallow saline groundwater was related to the presence of high salinity in the leaves of saltcedar. Moreover, the sulfate/chloride ratio in leaves growing above saline groundwater was nearly the same as the groundwater sulfate/chloride ratio. Hem (1967) went on to state that lower concentrations of minerals were detected in young leaves and higher concentrations were found in older leaves.

The collection of cores of tree material to understand the presence or absence of the uptake of xenobiotics other than salt as shown by Hem (1967) from groundwater can be traced back to initial investigations on metals (Vroblesky and Yanosky 1990; Vroblesky et al. 1992). These reports indicated the tree cores contained not only native metals, such as those essential and trace elements necessary for tree metabolism, but also excess concentrations of metals related to the presence of higher than background concentrations of these contaminants in groundwater. The depth to the water table where the tree cores were collected ranged from 8.2 to 0.9 ft (2.5–0.3 m) adjacent to surface water. The groundwater may have been contaminated for some time, at least since the 1930s, and groundwater samples collected in 1987 contained between 19 and 88 mg/L iron and 52–2,150 mg/L chloride (Vroblesky and Yanosky 1990), relative to much lower concentrations of these elements in groundwater from uncontaminated areas, where these values were 0.1–4 mg/L, respectively. The concentrations of iron (Fig. 15.2) and chloride in tree rings were reported in rings formed since the 1930s, and this time period encompasses the time prior to and during the disposal activities that occurred upgradient of where the cores were collected.

Vroblesky and Yanosky (1990) did not use tree-ring chemical data for rings that were formed after 1980. This was because all trees, including those growing above contaminated and uncontaminated groundwater, showed elevated concentrations of iron in rings formed between 1980 and 1987 (Fig. 15.2), when the samples were collected. These elevated iron concentrations may reflect the flow of

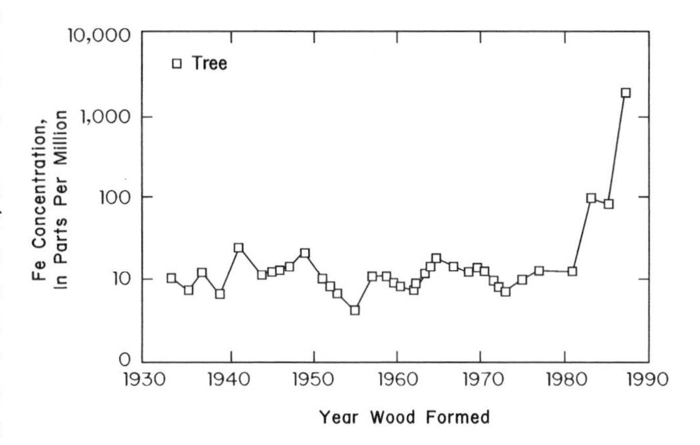

Fig. 15.2 The concentration of iron in individual tree rings between 1930 and 1990 (Modified from Vroblesky and Yanosky 1990).

water in the outer rings relative to less flow in the inner rings in ring-porous trees, although this explanation would not hold true for diffusely porous trees.

The technique used to measure the tree-ring metal concentrations was done so that the concentrations in each ring could be measured. In brief, after tree-core collection, the cores were dried in an oven, placed onto a plastic holder with cyanoacrylate glue, and then shaved with a surgical blade to a flat surface. The analysis of the prepared core was by proton-induced x-ray emission (PIXE) spectroscopy. Because most elements taken up by the xylem in a particular year are not translocated between rings formed during early or later years, the concentration detected in a particular ring reflects the source of water used by the plant at that time. On the other hand, some elements taken up such as potassium (K) were found to be translocated and concentrated in the heartwood of tulip trees (*Liriodendron tulipifera* L.) growing over K-contaminated groundwater after uptake (Vroblesky et al. 1992). The source of the potassium was potassium chlorate used in munitions manufacture. In that study, depth to groundwater was about 8.2 ft (2.5 m). Potassium concentrations in the contaminated groundwater from 1985 to 1987 averaged about 9.4 mg/L.

In this study by the same authors, for a few trees there was no outer ring enrichment of trace elements, as there was reported for iron, but other trees in which the K concentrations in groundwater were lower exhibited the outer ring enrichment. In fact, the potassium was higher in the heartwood relative to the sapwood in two trees growing over K-contaminated groundwater, opposite what was shown for iron and chloride. In trees growing over groundwater with lower levels of K, however, the trees showed higher K in the sapwood, relative to heartwood. The authors suggested that trees maintained potassium levels between 700 and 1,300 mg/L in the sapwood, when average groundwater potassium concentrations were no greater than 10 mg/L. If true, this would suggest the presence of an ion pump in trees similar to a sodium-potassium pump.

Nickel contamination of groundwater used by trees showed similar results of the occurrence of nickel in tree rings formed during periods of exposure to nickel with little translocation from sapwood to heartwood (Yanosky and Vroblesky 1992). Depth to water table was 0.9–8.2 ft (0.3–2.5 m), and dissolved nickel was as high as 0.2 mg/L in contaminated groundwater and 0.013 mg/L in uncontaminated groundwater. No attempt was made to relate groundwater trace-element concentration to what was measured in the tree rings. The tree-ring concentrations for the various elements were reported as part per million (ppm), similar to the groundwater concentration. However, the results of PIXE analysis are provided in mg/kg (or μg/g; equivalent to ppm) such that a direct comparison to groundwater concentration is tenuous at best, other than as a relative comparison. In a later publication (Yanosky and Vroblesky 1995), these results are reported as μg/g, rather than ppm.

Researchers also have shown that the tree-core technique for oaks and cypress trees has been a useful indicator of the presence of chlorinated solvents from groundwater contamination (Doucette et al. 1998; Vroblesky et al. 1999a) and in a variety of hardwood and softwood species (Vroblesky et al. 2004; Schumacher et al. 2004; Sorek et al. 2008), and petroleum hydrocarbons such as benzene, toluene, trimethylbenzene isomers, and methyl *tert*-butyl ether (MTBE) in oak trees (Landmeyer et al. 2000; Arnold et al. 2007). In those studies, cores of trunk material were collected using standard increment-borer techniques; core material was placed in glass vials that were capped with gas-tight seals, and compounds in the headspace were then identified using a field gas chromatograph with a photoionization detector (Vroblesky et al. 1999a). This analysis was done a day or a few days after collection in order to permit time for gas diffusion from the core to the vial headspace to occur.

The relative simplicity of this method has resulted in a variety of similarly-themed publications that confirm that tree cores are good surrogates for groundwater sampling from a qualitative point of view, especially for chlorinated solvents at contaminated sites or as a survey tool to find suspected contamination (e.g., Sorek et al. 2008; Larsen et al. 2008). To more rapidly determine the VOC concentrations, such as would be needed to direct field studies, Vroblesky et al. (1999a) demonstrated that heating the samples in a block heater or water bath for a few minutes provided similar analytical information about the tree-core VOC concentration.

In Vroblesky et al. (1999a), nearly 100 trees were cored, and the species included baldcypress, tupelo, sweet gum, oak, sycamore, and loblolly pine. The trees were located in a flood plain of the Savannah River, between South Carolina and Georgia, that received contaminated groundwater discharge. In Landmeyer et al. (2000), cores taken from the evergreen live oak (*Quercus virginiana*) and VOCs extracted with methanol, and the reduced carbon compounds such as BTEX and the fuel oxygenate MTBE were identified using gas chromatography/mass spectrometry. This detection of MTBE in trees at the field scale was later confirmed to occur in coniferous evergreens by Arnold et al. (2007).

The location on the tree where the core is collected is important for studies of trees growing above contaminated groundwater (Vroblesky et al. 1999a). In general, tree cores should be collected from near ground surface to breast height, in order to intersect as many annual growth increments (rings) as possible. This is because growth moves downward from the stem tips to the base, such that the higher up a tree a core is collected, the more recent are the annual rings intersected (Phipps 1985). It has been shown that

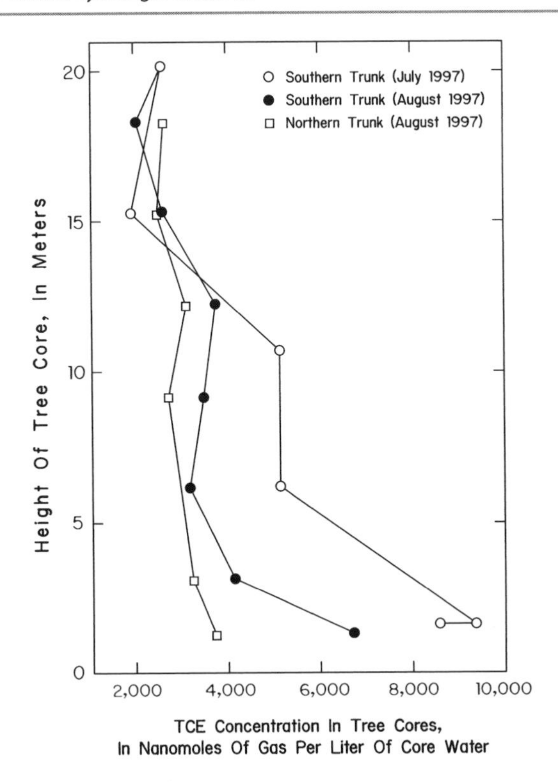

Fig. 15.3 The concentrations of TCE in tree-core samples decrease up a tree (Modified from Vroblesky et al. 1999a, b). One meter is equivalent to 3.2 ft.

concentrations of TCE in tree cores differed by 44–92% and for *cis*-DCE between 6% and 90% based on the core location around the tree at a given height (Vroblesky et al. 1999a) (Fig. 15.3). These results may be due to passive losses of VOCs from the xylem to the atmosphere through the bark; plant transformation of the parent compound; sorption; in-tree dilution as the sectoral ascent spreads out the root water to a more dilute ring ascent up the trunk or stem; or other processes.

A similar pattern of parent-to-daughter contaminant profiles was shown in trees located along a groundwater flowpath and downgradient from a PCE spill (Larsen et al. 2008). Cores collected from trees sampled at increasing distances from the spill had more daughter products, such as TCE and *cis*-DCE than PCE, reflecting the possible natural attenuation of PCE to these compounds in the contaminated aquifer. However, a full assessment of the use of tree-core samples in lieu of groundwater samples to assess natural attenuation was precluded by the lack of information provided on the redox status of the contaminated aquifer.

In most cases, core samples that should be representative of the flow of water through the plant can be collected on the south side of the tree. This is because the southern side of most plants is most often exposed to sunlight. Sap flow measured on the northern side of trees can be less than sap flow on the southern side (Steinberg et al. 1990b). For example, the southern side of maple trees is tapped by those wishing to collect the sap to make syrup. Also, the greatest amount of sap flow is derived from that part of the trunk beneath which is growing the largest branches above or roots below, which need more food to expand and lengthen. Schumacher et al. (2004) provided data for PCE in tree cores that indicated higher PCE concentrations for core collected from the southern side of the same tree relative to cores collected from the northern side.

It is good practice not to collect cores from the same general area over time, because the core hole created from a previous collection will be rendered into non-conducting (heartwood) tissue and, therefore, yield little, if any, water. As the tree expands in girth, this heartwood will become covered by newly developed sapwood. By extension, caution should be exercised when applying any plant-tissue approach to monitor the performance of a phytoremediation planting in which a large number of trees or cores from the same trees may need to be taken. A user guide prepared by the USGS (Vroblesky 2008) and is useful for phytoremediation of contaminated groundwater can be consulted prior to core collection.

Because the use of tree-core collection and analysis for phytoremediation projects is an application of a method historically used in forestry, there has been ample discussion about the effect of core collection on trees, especially over time in one tree (Grissino-Mayer 2003). One of the loudest and clearest voices in this discussion had been Dr. Alex Shigo, known as the Father of Arboriculture, who excelled at challenging the accepted practice of tree investigations during his life time. One of these was the practice of coring trees, which he unilaterally did not support. He cited that core collection exposed the tree to the entrance of fungal spores, bacteria, viruses, and insects.

Perhaps the best example for the apparent lack of a negative effect of tree coring on tree health, however, and perhaps actually an example of a stimulant to plant growth, can be taken from the maple sugar industry. Some maples have been tapped close to 100 years and still survive and produce sap the next season. Although boring does cause injury to trees, the injury is compartmentalized (sealed off to limit the spread of any infection) by the living tissue (Grissino-Mayer 2003).

One of the interesting data trends reported in Vroblesky et al. (1999a) was the difference in TCE concentration between different genera of trees growing over TCE-contaminated groundwater that had a uniform concentration. Tree cores collected from baldcypress and tupelo had similar TCE concentrations, but oak had lower TCE concentrations even though the groundwater concentration of TCE was the same. The difference between uptake by the different genera is probably due to the difference between xylem conductance of water in diffusely porous trees, such as baldcypress and tupelo versus ring porous trees, such as oak.

The effect of different tree tissue samples on contaminant concentration was investigated by Vroblesky et al. (2004). At the Ft. Worth, Texas site introduced in Chap. 8, tree-core samples contained higher TCE concentrations than samples of stems from the same tree. Collection of stems would be easier than collection of core material, but the rate of equilibrium for gas exchange to take place in a VOC vial is different for stem material than that of a tree core with no bark.

At a site in Colorado, higher TCE concentrations were detected in cores from trees that had shallower depths to the groundwater. In one case, TCE was detected in a cottonwood tree even though the depth to groundwater was 26 ft (8 m). This site was semiarid, with about 17 in. (44 cm) of precipitation per year, most of which was as snow. The trees present to sample were eastern cottonwood (*Populus deltoides* Bartr.) and were found growing along an adjacent stream.

At a site in Charleston, South Carolina, a TCE-contaminated aquifer was present beneath a 3-m thick clay unit. All trees sampled were growing above groundwater that contained TCE (Vroblesky et al. 2004). The cored trees consisted of loblolly pine (*Pinus taeda* L.) and oak (*Quercus* spp.). The TCE dissolved in groundwater was flowing underneath three sampled trees in which one had been growing above the plume since 2000 and two were growing at the leading edge of the downgradient part of the TCE plume. The TCE concentration in the tree cores collected from the tree growing over the plume in 2000 was a slightly more than 100 ppbv of headspace, whereas cores collected from the two downgradient trees were at 10 ppbv. From 2001 to 2003, however, these downgradient trees (trees SC2 and SC32) had increasing detections of TCE, as well as the upgradient tree (Fig. 15.4). This indicates the possibility that the two downgradient trees were acting as sentry wells and were indicating plume transport. These data also could suggest, however, that plants were accumulating TCE over time from a common fixed source. It also is not clear from the data if the TCE entered the plant tissues from the dissolved phase or as a vapor. In any case, these data indicate that tree-core collection and analysis does work if the goal is to detect the interaction between tree vascular systems and groundwater contaminants such as chlorinated solvents.

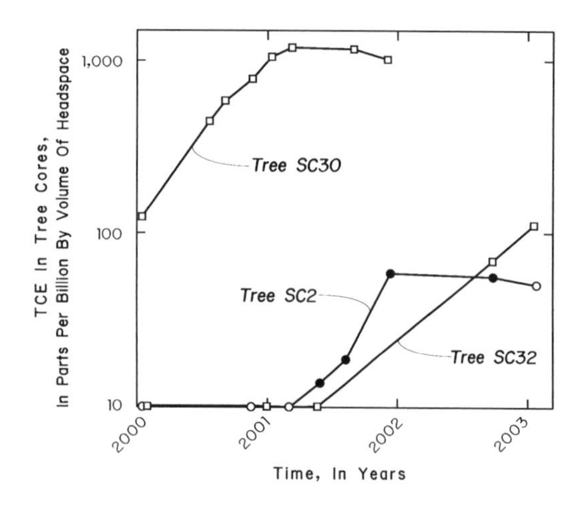

Fig. 15.4 The appearance of TCE in tree cores collected over time as TCE-contaminated groundwater moved beneath the tree (Modified from Vroblesky et al. 2004).

Tree-core collection and analysis to detect contaminants in groundwater has also revealed the usefulness for this approach to investigate VOC contamination present above the water table. Schumacher et al. (2004) found a stronger relation between tree-core results for PCE and the PCE present in subsurface soils than for the relation between tree-core PCE results and PCE in groundwater. In fact, the authors report a poor relation between groundwater PCE and tree-core PCE concentrations, which suggests that tree-core collection should not be the sole tool used to assess and delineate groundwater contamination at a site. The lack of relation between tree-core and groundwater contaminant level is more likely to occur when the depth to water table is near the maximum depth of root penetration, around 30 ft (9 m). As was stated in Chap. 12, volatile organic compounds can passively enter root hairs by diffusion in the vapor or dissolved phases. In either case, tree-core collection and analysis provides a relatively inexpensive way to delineate both saturated- and unsaturated-zone contamination by VOCs.

As was discussed previously, early research done to examine the interaction of plants and groundwater geochemistry was performed by Hem (1967), who used leaves and branches to investigate inorganics, and the various work done more recently by others looking at organics in tree cores. A combination of approaches was reported by Gopalakrishnan et al. (2007) that looked at organics in branch material at a site characterized by chlorinated-solvent-contaminated groundwater (see Chap. 7 for more details about the site). The authors recognized the inherent limitations to the collection of tree cores as proxies for groundwater contamination evidence, such as damage to trees and lack of suitability for smaller trees or younger plantings. They collected branches from willows and poplars that were planted to remediate chlorinated-solvent-contaminated soils and groundwater, respectively. The trees were 4 years old, and the diameters were about 1.9 in. (7.5 cm). In order to quantify the predictive ability of the tree branch approach, soil and groundwater samples near the sampled trees were collected and analyzed for PCE, TCE, and CCl_4.

The plant samples were collected by Gopalakrishnan et al. (2007) using pruning shears to remove the branch and leaves closest to the ground surface. Branch samples were cut to fit into 20-mL volatile organic analysis (VOA) vials. Leaves without petioles were placed in 20-mL VOA vials. In order to correlate these samples with tree-core samples, tree cores were collected at the same time and placed in 20-mL VOA vials. Prior to analyses of the head-space for VOCs, the tissue samples were first frozen for 12 h prior to heating for 4 h. Contaminants were correlated to soil and groundwater samples collected during this field work. The relation between the concentration of TCE in core and branch samples produced a regression coefficient of 0.70. This approach suggests the applicability of using branch samples rather than tree cores, at least in areas with little potential for atmospheric contamination sources for the contaminant in question.

TCE also was measured in core, branch, and leaf samples in areas of the site characterized by high and low levels of TCE-contamination of soil. The results were unique to each area. For example, in the area of high TCE-soil contamination, the concentrations of TCE in the samples increased from leaf to branch to core sample. The authors stated that this profile was controlled by TCE mass losses by radial diffusion (Gopalakrishnan et al. 2007). Such vertical loss along the transpiration stream also could be explained, however, by plant detoxification by Phase I and II reactions with the production of bound residues that were not analyzed. In the area characterized by low TCE-soil contamination, the exact opposite trend was observed: the TCE concentrations were higher in the leaves, then core, and then branch. This may be due to the presence of atmospheric contamination by TCE and foliar uptake. Moreover, these TCE profiles were similar to those observed by the authors for PCE and CCl_4.

The authors also presented a simple analytical model that depicted the relations between contaminant and tree processes that might control the fate of the contaminant in the tree tissue (Gopalakrishnan et al. 2007). They added equations that related the contaminant concentration in the plant to sorption onto the soil from the soil water; microbial degradation in the root zone; root uptake by aqueous phase; and sorption and diffusion from the xylem along the transpiration stream. The authors did not include changes to contaminant concentrations that might result from detoxification reactions or the role that the gaseous uptake of contaminants might play. The authors also assumed that the transpiration stream, Q, measured in the main trunk before it splits into two branches is the sum of the Q for each branch, Q_n.

A topic for further research is the effect of the high tensions in the xylem on core collection and, therefore, on the VOC concentrations measured. When a core is collected, the auger tip breaks through the bark, cortex, and phloem, which usually is under positive pressures, then the cambium, and finally the xylem, which usually is under negative pressures or tensions (Fig. 15.5). This causes the water column in the xylem to break, introduce gas, and induce loss of tension through cavitation. The effect of this cavitation on the measured versus *in-situ* concentration of VOCs is not known.

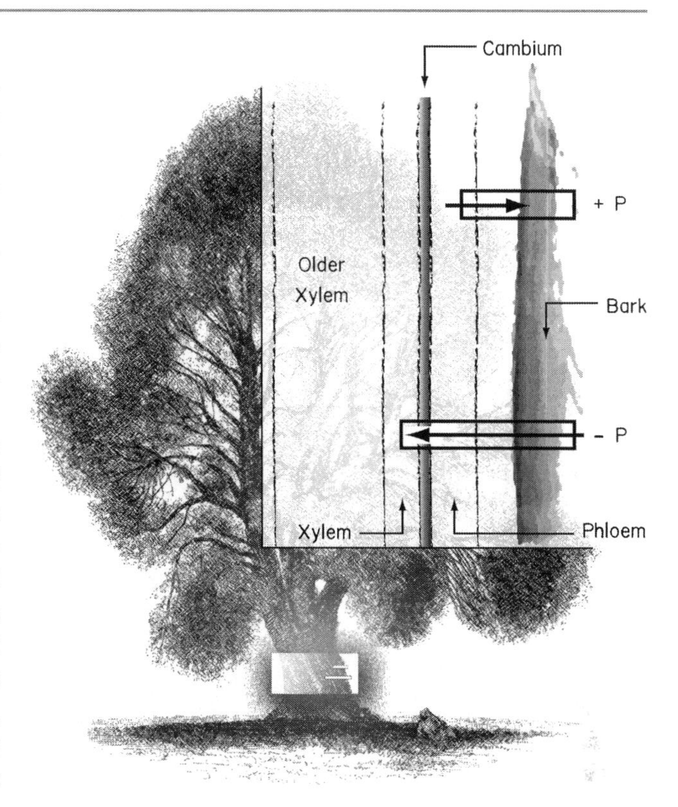

Fig. 15.5 The depth of core collection will determine the condition of pressure (*phloem*) or tension (*xylem*) that can be encountered during tree-core collection for VOC assessment (P is pressure.).

An additional problem with tree-core tissue collection is the assumption that the core location is one of growing tissue and, therefore, metabolically active. Cambial growth to produce phloem and xylem is not continual, even during the growing season for the tree being sampled. At certain times cambial growth will stop, if resources are limiting, for example. The cambium may be dead on a particular section and may not produce xylem. These scenarios will affect the result of cores taken from ring-porous trees and tend to underestimate the contaminant load (Figs. 15.6 and 15.7).

The xylem's anatomy also plays an important role in how contaminants are detected in ring-porous and diffuse-porous trees. Ring-porous trees have larger diameter xylem vessels produced in the spring, when moisture is readily available, and smaller diameter vessels as the season progresses. For diffuse-porous trees, the xylem vessels are smaller throughout the season. There is less communication between xylem vessels, therefore, in diffuse-porous trees.

The key to tree-core utilization is to match the level of analysis with the needs of the phytoremediation project. For example, if a rough delineation of groundwater or soil contamination prior to planting is needed, then tree cores can be used qualitatively to assign plume boundaries. On the other hand, more rigorous analytical work on the cores would be

Fig. 15.6 The vertical location of core collection up a tree's trunk will affect the results of any contaminant delineation survey, because the higher the level above ground, the younger the rings that will be sampled from the same corer. In the example above, ring 3 relates to growth in 1958, whereas this would equate to ring 1 higher in the tree.

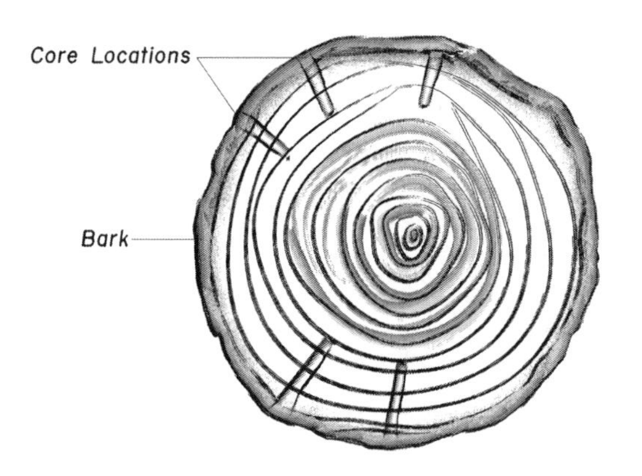

Fig. 15.7 The horizontal location, depth penetration, and the number of annual rings encountered that contain recent transpiration water need to be considered during the interpretation of contaminant delineation results.

necessary to attempt to correlate a tree-core contaminant concentration to the environmental concentration *in situ*. This can best be done for the contaminant of interest by first defining the partition coefficients for the contaminant of interest between the various compartments that the contaminant will encounter.

Ma and Burken (2002) did this for TCE, 1,1,2,2-tetrachloroethane, and CCl_4. They determined the partition coefficients for these compounds between the air, transpiration stream, and wood. As might be expected with passive processes, the partitioning was related to the physical and chemical properties of the contaminant. Ma and Burken (2002) were able to determine partition coefficients for these contaminants because they added known concentrations to vials that contained weighed cores.

Finally, the tree-core-tissue collection method may one day be replaced using passive *in-situ* methods. Rather than collecting multiple cores from the same tree over time at a site, a passive sampler could be installed in one core hole and then sampled repeatedly over time. This assumes that the installation of the sampler does not affect the movement of sap or contaminants through the tree and that the samples collected are representative. Such samplers could be based on semipermeable membrane devices (SPMD); time-weighted average solid-phase microextraction (TWA-SPME) methods (Burken and Ma 2006; J. Burken, pers. commun. 2009); or polyethylene devices (PEDs) (Adams et al. 2007). The ultimate method selected, however, will come down to the goals of each site project.

15.1.3 Diffusion Traps

The diffusion rate of TCE in stems was found to directly relate to the transpiration rate (Ma and Burken 2003). To determine this, they used diffusion traps, which consist of a short section of 1-in. (2.54 cm)-diameter glass tubing inserted over a cutting, and closed at each end by a stopper. The application of diffusion traps is most appropriate for contaminants that have high vapor pressures and high Henry's law constants. In the tree center, where the xylem contained the TCE with decreasing concentrations up the tree (diffusion in vertical flow direction), TCE concentrations also decreased radially laterally from the center to the atmosphere, caused by diffusion.

On account of this vertical and radial diffusion to the atmosphere, it is not surprising that less than 0.05% of the TCE added to the soils was detected in the plants at the end of the experiment (Ma and Burken 2003). The lack of TCE detection in leaves may be a result of stem TCE diffusion rates being higher than the diffusional transport of TCE in the xylem to the leaves. This is especially true for ring-porous trees that have most solution transport nearer the bark/atmosphere interface, resulting in a much shorter diffusion path than for diffusively-porous trees (Ma and Burken 2003). These processes also happen below ground along roots of considerable length and smaller diameter.

The diffusional loss of contaminants such as TCE can be approximated by Fick's diffusion equation. Mass losses from the transpiration stream to the air should be proportional to the concentration gradient between the tree and air, but inversely proportional to the length of diffusion of a molecule of contaminant. As a result, smaller diameter trees that have a greater surface area/volume ratio will have higher diffusional loss rates across a shorter length of stem than a larger tree.

Ma and Burken (2004) used the data presented in Ma and Burken (2003) to develop a model that accounts for the effects of various contaminant transport and fate processes in plants to explain the loss of contaminants from the tree tissue with ascending height. In both ring- and diffuse-porous trees, the water is conducted near the outer surface of the plant, closer to the bark. As such, the potential for diffusive loss of waterborne solutes increases.

15.1.4 Gas Bags

Various gases enter and exit plant leaves based on concentration gradients. The most important movement of gas is related to photosynthesis. Water vapor also exits plant leaves during photosynthesis. Couple this with the ability of plants to translocate and transpire VOCs and the study of gaseous exudates at the whole stand, whole plant, leaf, or bark surface is warranted.

Conditions that affect gas exchange are controlled at the leaf level. Knowledge of leaf-level gas exchange can often be used to scale up the results to the whole tree, but can be problematic. As an alternative, whole-canopy measurements typically involve infrared gas analyzers (IRGA) to measure CO_2 and H_2O vapor exchange. Open systems have been developed to amend these problems (Alterio et al. 2006).

Collection of gas emission of water vapor and other compounds can be performed while the leaves remain on the plant. For example, Tedlar® bags were used by Martin et al. (1999) to detect the release of various hydrocarbons from the branches and leaves of a wide variety of deciduous and coniferous plants (see Chap. 16 for more details). Ferrieri et al. (2006) used Teflon®-lined plastic bags sealed around individual plants grown hydroponically. The bags encased all the foliage and were secured around the base of the plant's main stem. To capture any emitted volatile contaminants from the leaves into the air, each bag contained a carbon-based molecule sieve cartridge of 500 mg of Tenax® GR placed in small-diameter tubes. In this study, CCl_4 and isoprene were measured.

A simple approach was presented by Andraski et al. (2003, 2005), who used gas bags to trap plant gas for analysis. Their goal was to investigate the fate of tritium contamination of the vadose zone at the USGS Amargosa Desert Research Site, near Death Valley National Park, NV. Andraski et al. (2003, 2005) developed a method where foliage from the native, deep-rooted creosote bush (*Larrea tridentate*) was manually stripped from plants, put into plastic bags, and then placed in the sun for evaporation and collection of water vapor. Tritiated water (^3HHO) and vapor behaves identically as non-tritiated water (HHO) and vapor. On average, roughly 170 g of plant matter produced about 23 mL of water. This approach is non-invasive to the subsurface, and has the advantage that the water transpiring from the leaves has entered in some part of the full volume of the root zone. The data indicated that plant-water tritium concentrations and soil-water tritium concentrations were directly related.

The measurement of contaminants in leaf samples directly or in gas bags many not necessarily indicate that the source of the detected compound was from the subsurface. This is because the atmosphere also can be a source of contaminants deposited on leaf surfaces. Atmospheric compounds can remain on the leaf surface or be taken into the leaf tissue, such as how the widely used herbicide glyphosphate enters target plants.

15.1.5 Infrared Analysis

The reaction of photosynthesis indicates that the measurement of CO_2 uptake by plants in a phytoremediation system over time could be used as an indicator of the impact that plants were having on contaminant remediation, or conversely, the impact that the contaminants were having on plant growth. Although laboratory methods to measure changes in CO_2 have been used widely in the plant sciences, these methods require the use of radiolabeled CO_2, as a gas for terrestrial studies, and bicarbonate for aquatic studies as a tracer of CO_2 uptake and photosynthesis. IRGA is a safer alternative to measuring photosynthesis and has been shown to be useful at the field scale.

15.1.6 Plant Fluorescence

Light energy captured by a plant in chloroplasts can be radiated back out of the plant at wavelengths between 650 and 800 nm. This characteristic of all photosynthetic plants is called fluorescence. Fluorescence begins when photons split water by oxidizing it into oxygen, hydrogen, and electrons. This process is 97% efficient, which means 3% of the energy is "lost" in the form of fluorescence. Dark-adapted leaves fluoresce when exposed to light. The degree of response is referred to as the Kautsky effect. The properties of bioluminescence and fluorescence can be considered as photosynthesis in reverse.

In many studies, the degree of fluorescence is used to determine the response of a plant to a variety of factors. Inhibitors of photosynthesis or stresses from water deficits, pest invasion, or contamination can affect photosynthesis and, therefore, fluorescence. Such fluorescence by chlorophyll *a* is directly related to the amount of chlorophyll *a*. In fact, chlorophyll *a* sensors that optically measure fluorescence are used in many surface-water monitoring programs.

15.1.7 Compound-Specific Isotope Analysis

Stable isotopes were used to determine that water was the source of the oxygen released by plants during photosynthesis, as described in Chap. 3. The carbon isotopic value of various carbon compounds derived from plants also varies, and can be traced back to the formation of the compound, or parent compound, by the fixation of carbon during photosynthesis. Most plants use the C_3 or C_4 fixation pathways.

Additional carbon isotope changes occur during the refinery process that converts crude oil to usable products. Most fuel compounds will have carbon isotope values near that of the plants, near −27 per mil, the fuel is derived from. BTEX compounds that biodegrade under oxic conditions produce CO_2 that is not fractionated from that of the compound being degraded and, therefore, also will be near −27 per mil. Under anoxic conditions, however, the lighter isotope, ^{12}C, reacts faster than ^{13} C, leaving the undegraded parent compound enriched with the heavier isotope. The application of carbon isotopic differences, in both stable as well as naturally radioactive and emplaced isotopes, for example, was used to determine the extent of local recharge of younger groundwater more recently in contact with the soil zone into older groundwater in a coastal area that has undergone much development and groundwater pumpage (Landmeyer and Stone 1995). In a contaminated aquifer, stable isotopes of carbon also were used to trace the pathway of biodegradation of contaminant compounds, such as BTEX, under oxic and anoxic conditions (Landmeyer et al. 1996a).

A concern with the above stable isotope approach at groundwater contamination sites that are candidates for phytoremediation is that CO_2 also can be produced by the degradation of non-contaminant compounds, which dilutes the stable-isotopic signature of the C in the CO_2 analyzed. Also, the approach is based on changes in the isotopic signal of the products of a reaction, rather than the reactants. Looking at changes in the reactants can be accomplished, however, by GC separation, followed by isotope ratio mass spectrometry (IRMS), and is called compound-specific isotope analysis (CSIA). This approach has been applied to groundwater contamination sites where fuels or solvents have been released (U.S. Environmental Protection Agency 2008).

15.1.8 Stomatal Conductance

In the past, plant physiologists have relied on a number of approaches to quantify stomatal size and how it relates to plant-gas exchange (see Chap. 9). These approaches have ranged from simple observation of leaf surfaces in response to difference conditions of *VPD* to photographic imaging systems (Weyers and Meidner 1990). Other methods included the addition of various fluids to a leaf surface and measuring the time of uptake, presuming that the uptake occurred through open stomata. Today, the accepted standard to measure stomatal conductance is to use a gas-flow porometer.

15.1.9 Leaf Area Index

The total leaf area exposed to the air has a significant effect on the volume of water lost by transpiration, and hence, potential for the uptake of groundwater. The leaf area index, *LAI*, is a measure of this potential and is defined as the ratio of leaf surface area of a plant or grove normalized to ground surface area covered by the canopy (see Chap. 9). *LAI* can vary from a high of 10 down to 0, and is dimensionless. The magnitude of *LAI* varies with such factors as the size and spacing of trees; *LAI* also tends to increase as trees age. For example, the *LAI* for young trees is near 0 and can approach 10 in dense stands of mature trees. Measurements of *LAI* over time may be useful in quantifying the increases in *ET* during the development of a poplar grove at contaminated sites or quantifying the effect of contaminant concentrations on tree health (Fig. 15.8).

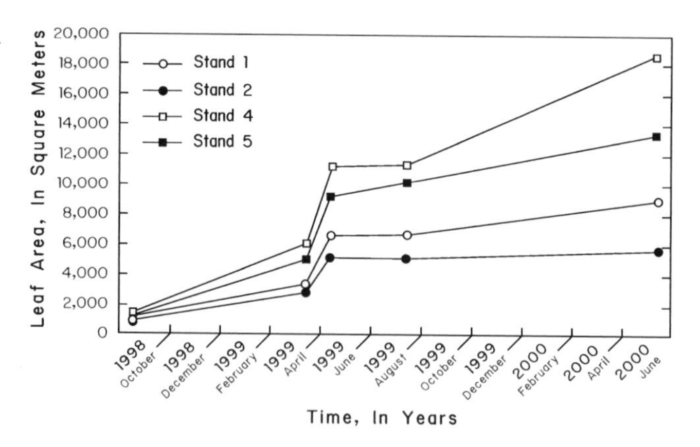

Fig. 15.8 The increase in leaf area index for multiple stands of trees over time at a phytoremediation planting.

15.2 Hydrogeologic Methods

Processes that occur in the subsurface often can be understood only after sufficient amounts of data are collected. The conventional approach to sampling groundwater is to use monitoring wells that are essentially the same design as those wells intended to remove groundwater for municipal or industrial use. Such wells consist of slotted screen placed in the saturated porous media. Data collected in the unsaturated zone cannot come from wells, as the flow that occurs there is in response to gradients in tension or water potential, not head. Water samples from the unsaturated zone still can be collected, however, using tensiometers and lysimeters which rely on vacuum to move water. These methods are often applied to phytoremediation purposes, but additional methods have been developed that are more specific to the root zone and capillary zone activities of plants, and will be discussed here.

15.2.1 Conventional Well-Sampling Methods

Conventional groundwater monitoring wells were first used in the early 1900s by O.E. Meinzer and his staff of USGS hydrogeologists that investigated the water resources of arid areas of the southwestern United States. They used water levels in wells to determine the relation between plant distribution and groundwater depth, among other tasks. Diurnal fluctuations in groundwater levels measured in wells, as done by G.E.P. Smith, were used to more accurately depict the interaction between plants and groundwater in natural environments. Wells were then used to relate the effect of plants on groundwater hydrology and geochemistry.

At phytoremediation sites characterized by contaminated groundwater, conventional monitoring wells also can be used to determine the effectiveness of the interaction between plants and groundwater. In order to determine the effect of individual trees on groundwater, wells should be placed as close to the trunk of the tree in question as is possible. On the other hand, if the effect of a mass planting of trees is to be examined, then groups of two or more wells screened at the same interval can be placed in pairs upgradient and downgradient of the planting, in order to determine the removal of groundwater from the aquifer caused by the plants, as would be evident from lower groundwater levels and, therefore, flux, in the downgradient pair of wells.

The sampling of monitoring wells to analyze groundwater for changes in contaminant compounds or other physical parameters can be accomplished using a wide variety of methods. These include those that require the removal of "stagnant" water that has presumably accumulated in the well bore since the last period of sampling. This can be accomplished using manual purging methods, such as a bailer, or automatic methods where a peristaltic or submersible pump is used (U.S. Geological Survey, variously dated).

15.2.2 Low-Flow Well-Sampling Methods

During low-flow groundwater sampling, it is recognized that purging a well of presumably stagnant groundwater prior to sample collection may not provide a realistic sample of the actual geochemistry in the aquifer adjacent to the well screen. The removal of groundwater from the well prior to sample collection, especially using a rapid manual method such as a bailer, can cause short-term changes in the hydraulic gradient between the water in the well and that in the aquifer. In many cases, the head is more rapidly lowered in the well than in the aquifer and groundwater will cascade into the well. This high flow rate can affect many of the physical properties and contaminants present in groundwater. For example, it could add dissolved oxygen to previously anoxic groundwater in the well. It also could preferentially volatilize VOCs.

With low-flow sample collection, peristaltic or submersible or check-valve-type positive-displacement pumps are run at low flow rates, less than 1 L/min, to minimize the negative effects of drawdown differences between groundwater in the well and in the aquifer. The pumped water is monitored using flow cells where the groundwater geochemistry can be monitored in real time to determine when physical properties such as DO, temperature, and specific conductance are stable (U.S. Geological Survey, variously dated). Then samples can be collected using standard methods with documented confidence that aquifer pore water is being sampled.

15.2.3 Diffusion and Dialysis Methods

There often are small-scale differences in contaminant concentrations across the vertical thickness of the unsaturated and saturated zones due to differences in contaminant sources and release histories, as well as differences in sediment characteristics such as hydraulic conductivity. This can result in sharp concentration gradients over small vertical and horizontal distances. The same situation occurs with root distribution within contaminated sediments. As the rhizosphere is a small zone around the roots limited to just a few millimeters, there has to be close contact of roots with contaminated water and sediments for degradation to be observed. Conventional groundwater wells and lysimeters may sample larger volumes of groundwater and soil than necessary to fully understand the interaction of roots and contaminants, especially with VOCs that are not amenable to accurate collection by vacuum-based lysimeter methods.

In order to sample such small zones, point-samplers based on diffusion are often used in groundwater investigations. These samplers, called dialysis samplers, can be used both within and above the water table (Hesslein 1976). This approach was used by Jackson et al. (2005) to investigate the fate of chlorinated solvents in the root zone of a plantation of poplar trees established at the Aberdeen Proving Ground in Maryland. They constructed a cylindrical dialysis sampler that consisted of taking a solid piece of Plexiglas about 5 cm wide and cutting out small cells from the pipe that could hold about 20 mL of deionized water added prior to installation. These holes were then covered with a semipermeable membrane. This membrane allowed contaminants in the pore water to diffuse from the contaminated media to the cells. The sampler was advanced inside the hollow rod of a direct-push rig. The samplers were installed in the root zone of a test tree in the planted area (about 1 ft away from the trunk) and left for 2 weeks to equilibrate.

Upon removal and analysis of the dialysis samplers, the depth profile of contaminant concentrations such as *cis*-1,2-DCE and TCE was variable in the planted area, both for parent compound and daughter-product formation, relative to a control sampler placed in contaminated sediment not planted with poplars. Whereas the control had more *cis*-1,2-DCE, the planted area had more TCE. Some dialysis cells had no water upon retrieval, which the authors suggest was a result of direct uptake through the membrane by adjacent roots. Redox- sensitive dissolved gases such as methane were lower in cells near the roots, suggesting that atmospheric oxygen may have diffused through the root cortex into the rhizosphere. Such dialysis sampling may be valid at sites where the majority of contamination resides in the vadose zone as a long-term source to groundwater.

15.3 Integrative Methods

The successful application of phytoremediation for groundwater restoration will warrant that a combination of methods be used to unequivocally determine that plant-groundwater-contaminant interactions are occurring at a particular site. This section discusses some additional approaches that can be used.

15.3.1 Root Zone Models

As was discussed in Chap. 12, much work was done on the fate of pesticides and herbicides applied to plants such as commercial crops. Models were developed to understand the fate of these compounds after application and include PRZM and PRZM-2, the Pesticide Root Zone Model (Carsel et al. 1984) available from the USEPA. Whether or not these models are applicable to those contaminants that commonly are encountered in groundwater is a subject for future investigation.

15.3.2 Push-Pull Tests

The fate of groundwater contaminants can be evaluated from laboratory and field tests. In many cases, the laboratory results often are not directly transferable to field situations. In the field, however, rates of degradation of contaminants can be calculated using a variety of methods. One of these methods is called a push-pull test.

Push-pull tests are field-scale controlled tests similar to that performed in the laboratory. A single monitoring well is used to inject, or push, a solution that contains a contaminant of interest that is expected to behave non-conservatively, along with a tracer that will behave conservatively. The same well is later used to extract, or pull, water out, while sampling for the presence of the tracer and the reactive solute. Comparison of the concentrations of these compounds during the push-and-pull cycles of the test are used to determine degradation rates. This approach may not be hydrogeologically feasible at sites that have aquifers with low hydraulic conductivity, or may be uneconomical at other sites due to disposal of extracted and possibly contaminated groundwater. However, these tests are useful in providing a gross loss of contaminants such as metal, chlorinated solvents, and petroleum hydrocarbons. The primary concern with any fluid-injection-based technology, however, is the degree to which mixing with the voids in the aquifer sediments occurs outside of the well bore.

Push-pull tests were attempted at a PAH-contaminated shallow aquifer in Oneida, Tennessee, where more than 1,100 poplar trees were planted in 1997, as described in Chap. 13. Widdowson et al. (2005a) reported the decrease in total PAH concentrations and naphthalene at shallow depths below the water table relative to deeper depths near DNAPL. They performed a series of push-pull tests to determine if there were any differences in biodegradation of naphthalene in the areas planted relative to unplanted contaminated areas (Pitterle et al. 2005). They injected the conservative tracer bromide at concentrations near 750 mg/L, dissolved oxygen, and naphthalene near 2 mg/L in the push-pull wells. The tests consisted of injecting 9.2 gal (35 L) of this solution at a rate of 0.1 gal/min (0.5 L/min). Hydrogeologic conditions at the site indicated that the radius of travel of the injectate from the well into the aquifer was about 11.8–15.7 in. (30–40 cm). Once injection was stopped, extraction was started and continued until at least three volumes of the injectate were recovered or after DO stabilized back to pretest conditions (Fig. 15.9).

Little can be said about the difference in push-pull tests between planted and unplanted areas because the tests were done at different times of the year; the planted test was done in June, whereas the unplanted test was done in February. Also, differences in temperature will affect the rate of aerobic microbial respiration, which may be the simplest process to explain the observed differences, such that the cooler groundwater temperatures would decrease mineralization, thus producing less DO consumption. It is possible that the biodegradation rates, k, presented from the push-pull tests

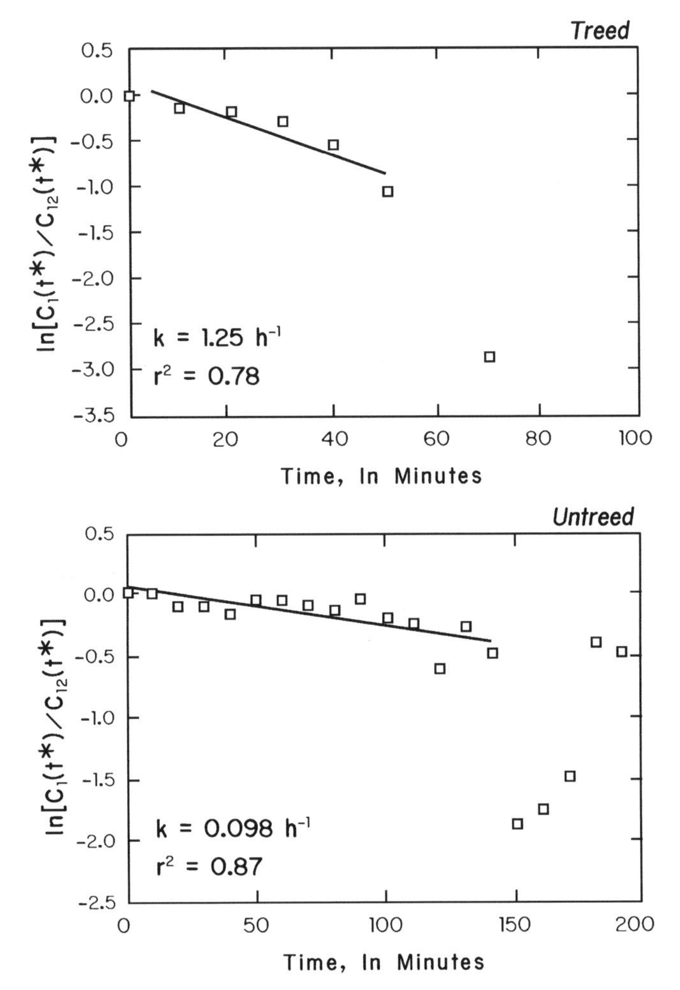

Fig. 15.9 Push-pull test results for the loss, presumably by biodegradation, of PAHs over time in planted (*treed*) and unplanted (*untreed*) areas of a phytoremediation site in Tennessee (Modified from Widdowson et al. 2005a, b).

more accurately describe aerobic microbial processes in the aquifer and may not be related to the effect of plants. This is because the wells that were used at the site were deep wells, on account of drought conditions having lowered the water table below the screened interval of the shallow wells located closer to the root zone of the planted poplar trees.

15.3.3 Stable and Radioactive Carbon Isotopes

Stable isotopic concepts related to plant sources of water of different isotopic composition were introduced in Chap. 9. To summarize, source waters for plants have different isotopic hydrogen and oxygen values to the extent that kinetic fractionation occurred during evaporation. If a sample of the xylem water is taken for isotopic composition evaluation, along with the compositions of the potential water sources, then the source(s) can be identified, as well as the extent of mixed sources.

The interaction of plant roots, the rhizosphere flora, and contaminants can be evaluated using stable isotopes as tracers. This approach, called stable isotope probing, is based on following a ^{13}C-labeled compound from the soil to microbial DNA. This approach has been used to understand which soil bacteria can degrade xenobiotics such as PCBs, and has been shown to validate the degradation of these compounds by plant-produced enzymes that are used by the plants naturally to degrade plant-derived aromatics.

15.3.4 Tree-Ring Chemistry and Aquifer Properties

The annual formation of tree rings is a record of the availability of environmental factors that affect growth, especially the availability of water. Studies described above also indicate that annual tree rings record the uptake of certain elements, such as chloride, originally dissolved in groundwater. Yanosky and Vroblesky (1995) suggested that if the chloride concentration of tree rings was measured in a single tree over time or at the same time but at multiple locations, this would provide an indirect method to determine the velocity of groundwater flow. If the hydraulic gradient is measured and effective porosity can be estimated, then the hydraulic conductivity, K, can be estimated using this approach. It may be most useful in those contaminated aquifers where pumping tests to determine hydraulic conductivity would not be feasible.

15.3.5 Lysimeters

Lysimeters are devices that can be used to collect water samples from the unsaturated zone for geochemical analysis. They were first used in the early 1960s. They are similar in design to the tensiometers used to measure soil water tension. For a lysimeter to collect water through its initially empty porous cup, however, a vacuum is drawn on the sample chamber, and the water sample is pumped to the test vial. The material of the porous cup should be selected based on the contaminant of interest at a site. Water samples collected from lysimeters installed near tree roots at phytoremediation sites provides perhaps the closest sample of the water quality that those roots are using.

15.3.6 Passive Soil-Gas Methods

The primary goal of phytoremediation of contaminated groundwater is to decrease the total mass of contaminants in groundwater or the vadose zone. If groundwater is contaminated with volatile or semivolatile organic compounds, the loss of contaminant mass over time caused by phytoremediation can be monitored using various passive methods that trap the contaminant vapors, or soil gas, for subsequent analysis. One simple soil-gas method is a passive soil-gas sampler that consists of various adsorbents surrounded by a thin tube of GORE-TEX® (W.L. Gore and Associates, Inc.). The sampler permits soil-gas vapors to enter but excludes water and other liquids. The samplers can be installed in the soil or in a well and are retrieved for subsequent analysis.

This method was used at a phytoremediation site installed near Elizabeth City, North Carolina; the installation of this site is discussed in Chap. 7 and additional information is contained in Cook et al. (2010). Soil-gas samplers were installed in a grid pattern on 100-ft-centers in 2006 prior to plant installation. Since plant installation, soil-gas samplers are installed and retrieved once each in the winter and summer. To date, the soil-gas sampling indicates that soil-gas masses of TPH, BTEX, and naphthalene have decreased following plant installation (Shaw et al. 2010).

15.4 Toxicity Testing

Phytoremediation involves putting plants in contaminated environments in an effort to restore these areas to precontamination conditions. Because many of the contaminants are carcinogenic, there are concerns that these chemicals will impart toxic effects on the plants. There is a relation between the type and concentration of chemicals and degree of plant toxicity—an extreme example is the use of herbicides to specifically induce death in plants. Some of the various tests and their usefulness to understand chemical toxicity affects on plants are discussed in Chap. 13.

15.4.1 Axenic and Nodule Analogs

Axenic cultures are sterilized cell cultures that do not contain bacteria. Axenic cultures are the conventional way to observe the function of plant cells without the interference of bacterial cells. Nodule cell cultures, or spherical photosynthetic cell aggregates, accomplish the same goal. These tests also can be used to understand chemical toxicity and plants.

15.4.2 Laboratory Approaches

Adam and Duncan (1999) investigated the effect of diesel fuel hydrocarbons on the growth of plants, considering that the level of diesel contaminants could be toxic to plants, especially if introduced to a site where seedlings are present. These authors reported that, for a wide range of grasses useful at sites for cleanup, the germination rate of seeds was inhibited at diesel fuel concentrations near 50 g/kg.

If seedlings are to be used, they will most likely be for grasses, as most deciduous trees installed at phytoremediation sites will be installed as cuttings or whips. Although the authors observed that the germinated plants had roots completely around diesel-contaminated sediments if uncontaminated sediment also was available, the roots would grow through diesel-contaminated sediment if no clean soil was present. Whether or not this colonization of diesel-contaminated soil was the result of rhizospheric microbes is unclear.

15.5 Summary

At this time, it is not sufficient to simply plant a phytoremediation system and then walk away. Most remediation efforts, including phytoremediation, require monitoring of the groundwater or remediation system to verify that, indeed, remediation is occurring and to document its performance over time such that human health and the environment are protected.

Why is this information important to the phytoremediation of contaminated groundwater? The basic approaches outlined in this chapter can be used to meet this need for the long-term evaluation of phytoremediation. The use of both plant- and hydrology-based approaches leads to a decrease in the uncertainty inherent to each method and provides a higher degree of confidence that phytoremediation is helping to achieve remedial goals at a site.

16 Economic and Regulatory Factors That Affect the Phytoremediation of Contaminated Groundwater

You're in charge of the last of the Truffula seeds. And Truffula Trees are what everyone needs. Plant a new Truffula. Treat it with care. Give it clean water. And feed it fresh air. Grow a forest. Protect it from axes that hack. Then the Lorax and all of his friends may come back.

The Lorax (Dr. Seuss 1971)

This passage from the end of the children's classic *The Lorax* provides a solution to the overuse of natural resources that was the central theme in the book, in this case, the removal of every last Truffula tree. But it also provides a message that some environmental problems can be solved, or at least left in a better condition, through the careful management of plants. In this case, phytoremediation may bring back cleaner groundwater, rather than the Lorax and all his friends.

Although planting trees does not require regulatory approval, their use as part of a remedial strategy to restore contaminated groundwater does. There are many factors that should be considered during planning a phytoremediation system that will affect its use and acceptance by the regulatory community. Many of these factors involve economic concerns as well. Some common economic and regulatory factors are discussed in this chapter.

16.1 Plant-Enhanced Contaminant Phase Transfer

During phytoremediation, plants are purposefully placed in contact with contaminated groundwater. Unlike the bioremediation or MNA of contaminated groundwater, there are realistic concerns with phytoremediation regarding the fate of contaminants in the treatment system itself—the plants. For example, it is often assumed that plants will act as conduits to bring subsurface contaminants to the surface and necessarily increase exposure rates. This can occur by contaminant volatilization or sorption to leaves that accumulate in the soil after leaf drop. This concern of contaminant transfer from groundwater to other media is valid, especially when the contaminants of interest are known or suspected carcinogens. Moreover, there are concerns that the contaminants will invoke a toxic response in the plants used for phytoremediation.

16.1.1 Natural Plant Toxic Compounds

When considering the legitimacy of stakeholder concerns about potential negative plant and groundwater interactions, particularly in determining whether or not a site is planted, some of these concerns may be alleviated through the introduction of stakeholders to the natural ecology of plants and chemical compounds.

As was discussed in Chap. 11, plants such as those implemented at phytoremediation sites are not helpless creatures at the mercy of groundwater contaminants. Plants have evolved external and internal defenses to prevent predation and ensure survival. The most common and obvious external defense is thorns or spines, essentially leaves that have been modified over time. As we saw earlier, some plants produce raphides in cells that render their leaves unpalatable to herbivores. Plants also can manufacture secondary metabolites that can be used as offensive and defensive chemicals. For example, alkaloids can be extracted from almost all parts of plants. Most alkaloids contain heterocyclic nitrogen compounds. Some of these plant chemicals include alkenes, which are characterized by a carbon-carbon double bond. These are all naturally produced chemicals that can be analyzed in the emissions of trees, as reported in the study by Martin et al. (1999). In this same manner, these naturally produced plant chemicals are part of the plant detoxification system that can be leveraged through the installation of plants at sites with contaminated groundwater.

J.E. Landmeyer, *Introduction to Phytoremediation of Contaminated Groundwater*,
DOI 10.1007/978-94-007-1957-6_16, © Springer Science+Business Media B.V. 2012

16.1.2 Plant Transfer of Subsurface Contaminants to the Air

The installation of many above-ground treatment technologies for groundwater contamination characterized by volatile organic compounds, such as BTEX, MTBE, or TCE, also requires that these chemicals be monitored for release to the atmosphere. If an air-strip, air-vapor extraction (AS/AVE) system is installed, for example, the ambient air near the extraction equipment is monitored for the potential for contaminant release to the atmosphere. In fact, such monitoring is required as part of corrective action plans (CAPs) or RODs to meet State or Federal air-quality mandates.

For phytoremediation of contaminated groundwater, there also is the concern for air-quality degradation from the release of unattenuated contaminants from the plants. This concern is well founded, given that chemicals such as MTBE and TCE have been documented to move through plants in the transpiration stream from groundwater to the air, as was discussed previously. This potential release to the atmosphere should be viewed, however, in the context of (1) the release of natural VOCs by plants in uncontaminated areas and (2) the ultimate fate of the VOCs released by plants at contaminated sites.

In the 1980s, former U.S. President Ronald Reagan was chastised about his comment that natural plants were responsible for 80% of measured atmospheric pollution in certain areas. In fact, his comment was meant to explain only one type of airborne pollutant, from the organic species called olefins, such as isoprenes and monoterpenes. Isoprenes are 5-carbon units that comprise 10-carbon units called terpenes, such as the commonly known turpentine. These turpenes are found in the sap of many coniferous trees. These olefins are naturally released in volatile form by many tree species. This synthesis and release is the cause of the haze that resides over most of the Appalachian Mountains, also called the Blue Ridge Mountains, because this haze of organics appears blue from a distance.

This phenomenon of the naturally occurring plant-release of VOCs is not limited to the eastern United States. Martin et al. (1999) reported that trees such as aspens (*Populus tremuloides*), cottonwoods (*Populus fremontii*), oaks (*Quercus gambelii*), fir (*Pseudotsuga menziesii*), spruce (*Picea engelmannii*), juniper (*Juniperus scopulorum*), and pine (*Pinus edulis* and *Pinus ponderosa*) growing in New Mexico all released to the atmosphere nonmethane hydrocarbons, monocarboxylic acids, and low-molecular-weight aldehyde and ketones. The pines emitted predominantly α-pinene, up to 100–10,000 nanograms per gram of dry weight per h (ng/g/h). Additional hydrocarbons such as isoprene, camphene, *d*-limonene, and β-pinene also were measured.

Even deciduous trees emitted isoprene at levels expected from the pines. Organic acids, such as the oxygenated aldehydes and ketones such as formaldehyde, were detected as an emission from predominantly deciduous trees. The gaseous emissions were detected in Tedlar bags placed over branches and sealed with tape or rubber bands. As discussed in Chap. 13, the amount of VOCs released by plants into the air will differ, in general, as a function of plant species and contaminant chemical and physical properties. Some volatile organic chemicals after translocation will diffuse to the atmosphere relatively unchanged.

On the other hand, release to the atmosphere of subsurface contaminants through plants actually will often enhance contaminant remediation. For example, for the fuel oxygenate MTBE, the half-life of this compound in groundwater where oxygen has been depleted is on the order of months to years, whereas if volatilized to the atmosphere after translocation through a plant, the half life can be on the order of minutes to hours, after attack by hydroxyl radicals in the atmosphere. Similarly, the chlorinated solvent PCE has a half life of near 2 years if not longer in oxic groundwater, but in the atmosphere it is lowered to between 1 h and 100 d. Benzene has been shown to volatilize from plants to the atmosphere (Collins et al. 2000) as well as TCE (Ma and Burken 2003) and become rapidly degraded. These scenarios may not be the case for every groundwater contaminant, however.

16.1.3 Fate of Contaminants in Leaf Litter

Some essential plant minerals or nutrients can have a gaseous phase and, therefore, if removed from the plant can readily be reassimilated. Other plants minerals such as phosphorus and iron, however, do not have a volatile phase, and as such, can be removed readily from plant access by leaching. Therefore, plants have evolved to retain certain easily leached minerals.

In a classic study, the loss of leachable versus sequestered minerals, such as calcium, was investigated using ^{134}Cs as a surrogate (Witherspoon 1964). This element was injected into the base of trees, and its fate monitored over time in the various compartments of a tree's environment. Up to 40% of the ^{134}Cs added ended up in the leaves, but the leaves returned more than half back to the trunk. The remainder, however, stayed with the leaves until leaf drop. Overall, no more than 20% of the ^{134}Cs left the tree by leaching. This study indicates that plants can store such mobile phases by sequestration into wood. The red heartwood of pine trees and oaks is a testament to the removal of these minerals to areas deep in the tree. This radial flow of nutrients from the phloem toward the center occurs by the rays.

What about the fate of VOCs in groundwater? Plants are composed primarily of water and both living and previously living organic compounds, much like the contaminated groundwater where plants are used for phytoremediation. Because even highly soluble organic contaminants have the potential to partition to organic matter, these compounds once inside the vascular transport system of a plant can partition into the plant tissue itself in a manner similar to the elements discussed above. However, the tendency for such compounds to partition to nonpolar lipids or to polar carbohydrate structures is low (Chiou 2002).

Newman et al. (1999a) reported the detection of TCE and its metabolites in various compartments of plants exposed to TCE during a 3-year highly controlled experiment in the field. TCE and its metabolites such as dichloro- and trichloroacetic acid were detected in all plant compartments measured. Weathered leaf litter measured after leaf drop the second year (fall 1996) revealed that while no TCE was detected in the leaf litter sampled, dichloro- and trichloroacetic acid and *cis*-1,2-DCE and *trans*-1, 2-dichloroethylene (*trans*-1,2-DCE) were detected. Because no units of measurement were provided, it is hard to comment on the risk that the detection of these compounds presents to human health and wildlife. Davis et al. (1996) reported that most contaminants are not water soluble or volatile enough, nor present at high enough concentrations, to present a significant risk through high concentrations in the atmosphere. Davis et al. (1996) reported preliminary calculations of potential maximum TCE transfer rates to the atmosphere, near 10 g TCE/m^2/d.

The fate of explosive compounds in the presence of plants has been investigated, and showed that for TNT and some of its intermediate breakdown products, these compounds were observed to be taken up into plants and to accumulate in the roots, whereas RDX and HMX were found primarily in the leaves of test plants (Groom et al. 2002) (Fig. 16.1).The fate of these compounds in leaves after they fall was studied by Yoon et al. (2006). Regulators are typically concerned about the potential risk exposure to contaminants by exposure to leaves that might contain contaminants taken up by the plant.

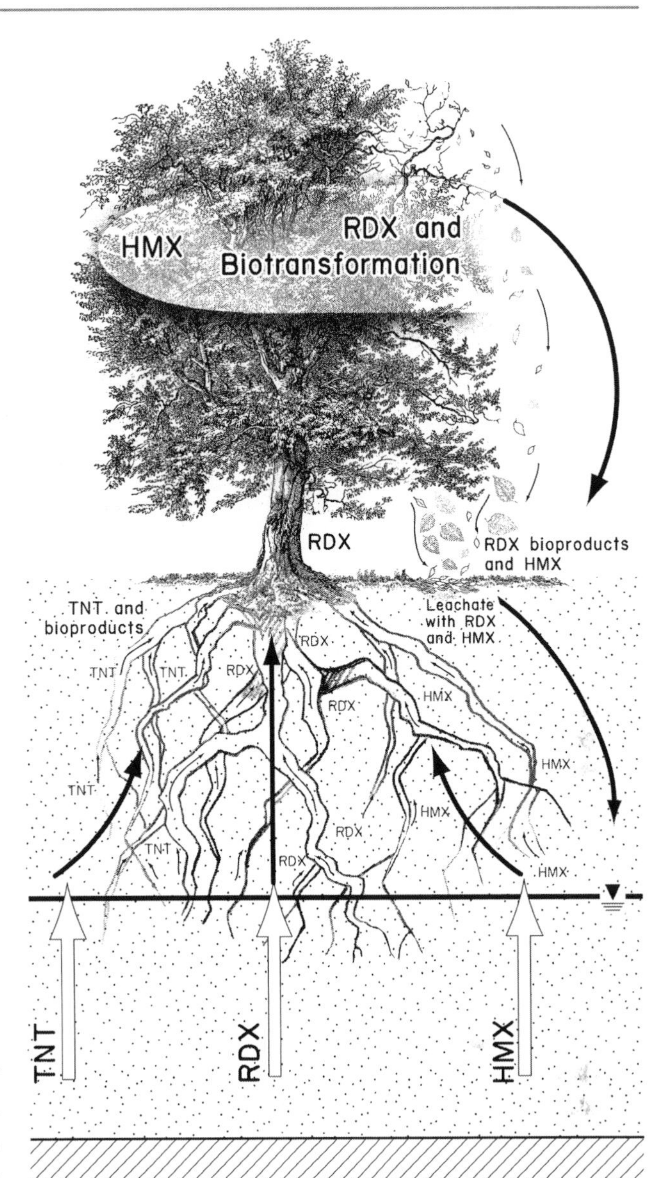

Fig. 16.1 Fate of TNT, RDX, and HMX in the phytoremediation of explosives-contaminated groundwater. The TNT is predominately degraded in the rhizosphere, and RDX and HMX are translocated to the leaves. These compounds are degraded in the soil after leaf fall (Modified from Yoon et al. 2006).

To address these concerns, Yoon et al. (2006) added radiolabeled ^{14}C-TNT, ^{14}C-RDX, and ^{14}C-HMX, to track the fate of the compound in different parts of the plant, to flasks that contained a solution of half-strength Hoagland solution, TNT mixtures, and a prerooted hybrid poplar cutting (*Populus deltoides* × *P. nigra* DN-34). Over 2 weeks, the removal of TNT from the solution was greater than the removal of RDX, with little removal of HMX (Fig. 16.2). After uptake, the distribution of these compounds within the plant was depicted. Almost half of the ^{14}C-TNT taken up was detected in the roots after 30 d, whereas between one-fifth and one-half of ^{14}C-HMX and ^{14}C-RDX, respectively, was found in the leaves (Yoon et al. 2006).

Due to the detection in leaves, dried leaves were exposed to deionized water to simulate exposure to precipitation after leaf drop and then resampled. Very little TNT was found in the leachate, but one-fourth to one-half of RDX and HMX was found in the simulated leachate. Moreover, when dried roots were exposed to deionized water, very little of any compound was detected in the leachate (Fig. 16.2 and 16.3).

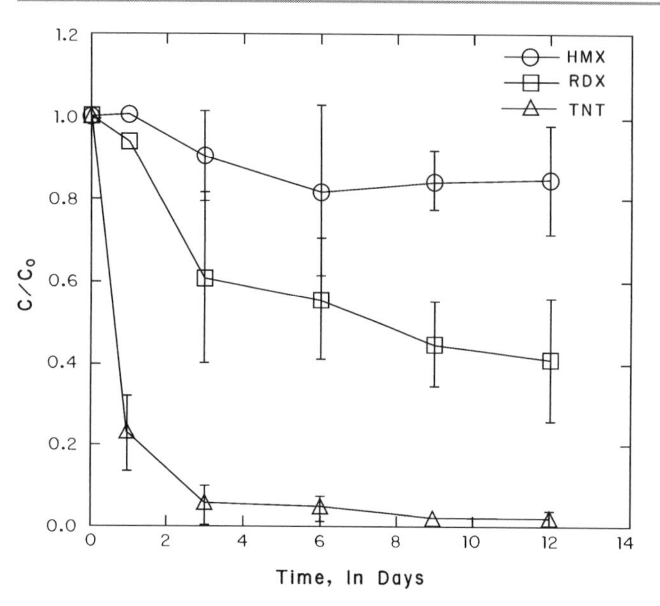

Fig. 16.2 The differential removal of TNT from solution relative to RDX and HMX after exposure to poplar cuttings in the laboratory. Much of the TNT removed entered the roots where it stayed, and the RDX and HMX taken up was translocated to the leaves (Modified from Yoon et al. 2006).

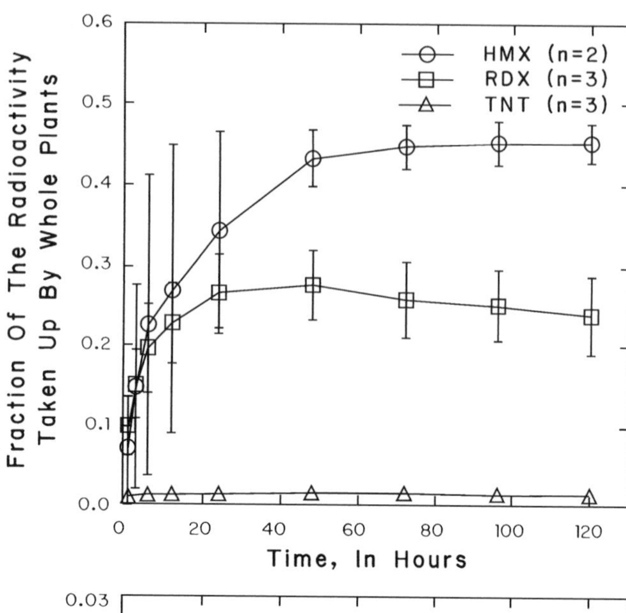

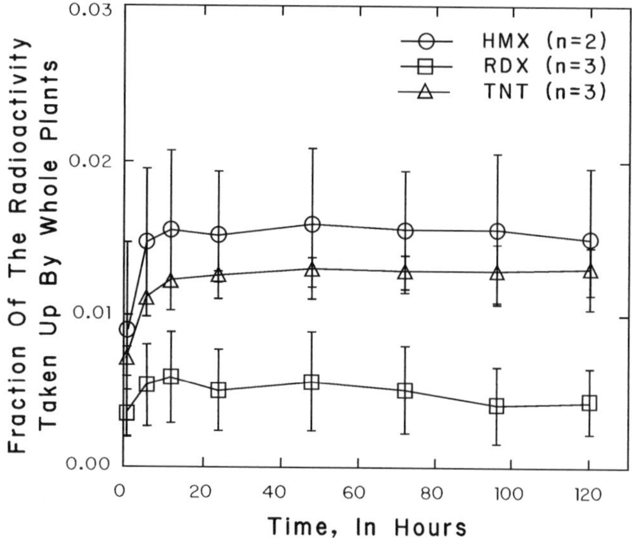

Fig. 16.3 HMX and RDX were preferentially taken up by plants relative to TNT in this laboratory experiment (Modified from Yoon et al. 2006).

16.1.4 Plant Detoxification Reactions

The products of Phase I reactions often further react with substances present in plant cells to form larger molecules which become essentially nonextractable. Conjugation of one chemical with another often results in the decrease of the toxic effect of the chemical. This process differs from bioaccumulation in that the parent compound taken up is changed into a less harmful form, and the process is regarded as beneficial. In plants, this occurs after the toxicant reacts with water-soluble cell components such as glutathione, amino acids, and sugars. This conversion renders the toxicant more water soluble and, therefore, reduces toxicity. Rather than be excreted in a way similar to that of animals, however, plants tend to store the compound, possibly in vacuoles.

Once a xenobiotic is taken up into a plant, the process of detoxification can occur. As was detailed in Chap. 11, the initial step is usually interaction with cytochrome P-450 monooxygenases or peroxidases (POX). After this oxidation, conjugation reactions occur where various sugars (glucose) or amino acids interact with the activated xenobiotic to glycoside compounds; these reactions are mediated by glycosyltransferases (Schröder and Collins 2002). The result is an inactivated xenobiotic of a glycoside. All these processes (Phase I and II) act to protect the plant by removing the xenobiotic as quickly as possible by increasing polarity and making it innocuous after storage in vacuoles or in other organic matter in the plant.

As detoxification reactions occur at a site over time, it is possible that the plants should be evaluated in terms of becoming a hazardous waste. To meet the definition of a hazardous waste as defined under 40 CFR Part 261.24, the medium in question, such as a soil sample or biomass sample, must meet guidelines established for toxicity characteristics. One of these is the Toxicity Characteristics Leaching Procedure (TCLP). In this test, environmental samples are evaluated in the laboratory for the potential for contaminant mobilization after exposure to leaching solutions of low pH. However, because plants tend to store detoxified contaminants in the form of nonextractable compounds, the potential for plant materials to fail a TCLP is remote. Such issues of plant toxicity may be encountered at the end of the life cycle of phytoremediation projects, and can be addressed using procedures such as the TCLP. These issues may need to be addressed in the near future, as many phytoremediation projects in the United States started in the 1990s and are about 15 years old.

16.1.5 Contaminant Fate in Food Crops

Even though fruit trees are not used for the phytoremediation of contaminated groundwater, the potential exists for contaminants to be transported to offsite areas where the contaminated groundwater may be unknowingly either taken up or applied through irrigation to crop plants. Although most fruits contain high amounts of water, the source of this water is predominantly that routed from the phloem rather than the transpiration stream, since this sap first must receive sugar from the leaves. For this reason, many studies have concluded that for hydrophobic contaminants such as PAHs, loading and movement of PAHs in the phloem to fruits is negligible (Trapp et al. 2007).

For xenobiotics that have the characteristics of a weak acid, however, these compounds may enter the phloem. Because groundwater contaminants primarily will be present in the xylem, to address this issue, samples from fruit trees and various vegetable plants grown in a residential area downgradient of a TCE-release from Hill AFB in Utah were collected to determine if TCE could be transferred from the TCE-contaminated groundwater to the fruits and vegetables. Depth to groundwater ranged from less than 10 ft (3 m) below ground surface to almost 20 ft (6 m) below land. Because this study site is in an arid area where *ET* is greater than precipitation at 45 in./year (114 cm/year) relative to 19.8 in./year (50 cm/year), respectively, groundwater is a potential source of water for plants. TCE was not detected in the headspace of samples of the fruits at levels above the detection limit, but cores collected from the tree trunk did have TCE (Doucette et al. 2007). Groundwater samples collected near sampled trees did not have TCE concentrations that could be related to plant tissue concentration.

Because of the variability in the data from the field, a greenhouse study was performed where a controlled amount of ^{14}C-TCE was added to representative apple and pear trees (Doucette et al. 2007). The level of ^{14}C in various tree tissues, including in some cases the fruit, was proportional to the level of TCE exposure, confirming other reports that suggest a diffusive uptake pathway as a function of contaminant concentration and transpiration. The highest concentration of ^{14}C was in the leaves. Because TCE was detected only in the roots and trunk, the ^{14}C in the other tissues such as the fruit was assumed to be TCE-transformation products (Doucette et al. 2007).

A greenhouse study investigated the translocation of a widely used chlorinated solvent to different parts of a plant including the fruit (Chard et al. 2006). These investigators were concerned with the question of whether or not fruit could be a sink for chlorinated solvents taken up in the root zone, because it raises a concern as to the risk exposure of phytoremediation to wildlife and humans. Unlike entry into leaves, where there is a loss mechanism by diffusion to the atmosphere through stomata and cell metabolism, only cell metabolism is present as a loss mechanism for fruits. This fact underscores the previous path of using sterile trees at phytoremediation sites rather than those that produce seeds, nuts, or berries. In this case, Chard et al. (2006) noticed that trees growing above a plume of TCE-contaminated groundwater contained TCE in concentrations proportional to the groundwater TCE concentration. Moreover, groundwater flow placed the plumes in residential areas where fruit trees were grown. Preliminary studies did indicate the presence of TCE in fruit from trees above the TCE plume.

To further address this potential incorporation of TCE into fruit, representative fruit trees were grown in a greenhouse setting. Two types of fruit trees were used: a dwarf apple (*Malus domestica* Borkh cv. 'Golden Delicious') and a 5-year-old peach tree (*Prunus persica* Batsch cv. 'Redhaven'). Irrigation water that contained no TCE was used as the control, and the treatments consisted of TCE concentrations of 5 or 500 μg/L ^{14}C-TCE and unlabeled TCE. Control trees were grown either separately or singly in each TCE treatment area to determine if atmospheric TCE uptake occurred. After 2 years of investigation, it was determined that even at the 500 μg/L TCE, no toxicity was observed. Fruit was produced both years by the peach trees but not the second year of the apple trees exposed to 5 μg/L TCE. Radiolabeled ^{14}C-TCE was found in all tissue samples analyzed in the treatment trees, but not in the control; this may indicate that the route of ^{14}C-TCE entry was through the roots and translocation, rather than foliar uptake of volatilized ^{14}C-TCE (Chard et al. 2006). Radiolabel was detected in amounts that decreased from leaves to branches to fruit. The radiolabel detected probably represented the byproducts of TCE degradation. Hence, although TCE was taken up by the trees, both in the field and laboratory study, no TCE was detected in the fruit.

This study expands on work done previously to investigate the fate of contaminants in sewage sludge spread for use in agricultural settings, a very common practice and highly regulated due to the use of organic-rich wastes as soil enhancements, as discussed in Chap. 11. Witte et al. (1988) studied the fate of PAHs in a greenhouse study where sewage sludge that contained milligram per kilogram concentrations of benzo[k]fluoroanthene and fluoroanthene was exposed to a variety of cereal and vegetable crops such as wheat, rye, carrots, and sugarbeet. At the end of the study, the leaves were found to contain more PAHs than the seeds and nuts. More interestingly, the concentration of PAHs in these tissues was not related to the concentration in the sewage sludge, as might be expected from the *TSCF* for these contaminants. Finally, because most sewage sludges are frequently monitored for contaminant concentrations, most PAH concentrations will be lower, less than 1 mg/kg and, therefore, would most likely not be taken up by plants.

In fact, some early studies indicate that the exposure of vegetable plants to the PAH naphthalene resulted in most of the compound staying in the root (Schwarz and Eisele 1984). For example, the plants were exposed to a nutrient solution that contained ^{14}C-naphthalene, and at the end of the experiment, for the pea plant (*Pisum sativum*), almost 60% of the ^{14}C radioactivity was detected in the roots, 37% in the stems, and 3% in leaves—also, less than 0.6% was detected in pea pods. The same experiment was done for onion (*Allium cepa*), and almost 95% was detected in the roots, and less than 3% in the bulb and leaves. For lettuce, more than 91% was detected in the roots, and no more than 4.5% in the stem and leaf parts. As was the same for even more recent studies of the fate of such compounds initially added, it is not known what form the ^{14}C was in as detected in the plants—^{14}C-naphthalene, ^{14}C-metabolite, etc. Most are probably metabolites and other nonextractable ^{14}C-containing compounds.

Garden plants such as carrots, spinach, and tomatoes were exposed to water that contained dissolved radiolabeled ^{14}C-TCE under laboratory conditions (Schnabel et al. 1997); it is unclear if it was uniformly labeled or not. The concentrations tested were in the range of that often found in contaminated groundwater, between 140 and 560 μg/L. At the end of the 106-d study, much of the added ^{14}C-label was detected in the headspace of the plant microcosms, and indicated that the ^{14}C-TCE had volatilized from the leaves after uptake by the plants from the contaminated water. A small amount (1–2%) of the label was detected in the plant itself, and was higher than that found in the soil, and was probably bound in the plant as a non-TCE transformation product(s) after oxidation by cytochrome P-450 or reduction and conjugation by glutathione, and of lower toxicity than TCE. Moreover, it was observed that the higher the dose of TCE, the higher the amount of TCE was in plant tissues, following the diffusion-based concept of the *TSCF*.

16.2 Potential for Transgenic and Mutagen Activation at Phytoremediation Sites

The natural and artificial mixture of different species to produce new ones has been going on since the beginning of the time when single-celled plants arose. The potential for these crosses is immense, but is limited to only those partners within species. Transgenics, however, changes this limitation barrier, as now genes of insects and animals can be added to plants, an extension of our desire to improve plant traits.

As was introduced in Chap. 10, transgenic plants are being used more frequently in phytoremediation plantings. Transgenic plants are plants that have been genetically modified such that desirable traits are induced using recombinant DNA technology in plants that did not have these traits originally, or undesirable traits have been removed. An example is the inclusion of genes taken from the soil bacterium *Bacillus thuringiensis* (or Bt) into plants. This bacterium has been available for use as a spray for foliar defense. The plants to which these genes are added did not evolve these adaptations through the force of natural selection; they were added in one fell swoop in the laboratory.

For trees used to remediate contaminated groundwater, such as hybrid poplars, researchers are attempting to add bacterial genes that encode for the production of enzymes that will decrease the negative effects of slower conjugation reactions that involve plant-produced glutathione. Other bacterial enzymes have been added to plants to help degrade explosives-related compounds and metabolites. It also is possible that traits such as metal accumulation or organic solvent degradation will be developed in plants and used to reduce risk at contaminated sites.

The development of transgenic plants also raises concerns previously addressed in Chap. 10, such as the widespread escape from cultivation with uncontrolled entry into the ecosystem (like happened with *Tamarix*). Also, the potential exists that a decrease in genetic variability will occur with transgenic plants used in phytoremediation applications.

The largest concern about transgenics centers around the transfer of genes from the transgenic plants to the neighboring plant community, particularly food crops or nuisance plants. For example, what would happen if herbicide resistant corn, modified genetically to be able to handle annual application of herbicides, was introduced to the weed population at large, the very plants that the herbicide was developed to eradicate? However, the perceived risk of introducing transgenics must be weighed against the benefit of environmental restoration using transgenics. Only experience in this arena will help determine the risks. Field trials of transgenics are regulated by the U.S. Department of Agriculture, Animal and Plant Health Inspection Service.

Linacre et al. (2003) acknowledged the need for continued use of transgenic plants in phytoremediation applications but stated that this use must come with the appropriate risk assessments and communication. The assumed risk is to wildlife and human health, but the consequence of inaction as far as remediation goes must also be quantified. For example, a dissolved-phase plume of TCE that reaches a domestic well may present a risk to those who ingest the TCE-contaminated well water. A phytoremediation effort to limit the spread of the dissolved phase plume using a transgenic plant may also have its risk of entry into the native population. A cost–benefit analysis at such sites will need to be done. Fortunately, at many phytoremediation sites transgenic usage occurs typically at great distances from food crops.

There are examples of the potential usefulness of transgenic engineering for increased biodegradation of common groundwater contaminants, such as TCE (see Chap. 13) with phytoremediation applications, as was shown in a study by Shim et al. (2000). TCE is stable in the presence of oxygen, on account of its oxidation during chlorine-atom substitution. TCE tends to only undergo extensive degradation if it serves as an electron acceptor in the absence of oxygen. TCE can, however, undergo degradation in the presence of oxygen during co-metabolism. It has been shown that the toluene *ortho*-monooxygenase (TOM) also oxidizes TCE completely to CO_2 and Cl^-. The genes that lead to the production of TOM can be added to the chromosomes of gram-negative bacteria, as was done to wheat root rhizospheric bacteria that demonstrated the ability to remove TCE from contaminated soils (Yee et al. 1998). In a separate study, James et al. (2007) reported that transgenic tobacco (*Nicotiana tabacum cv. Xanthii*) they developed expressed genes for cytochrome P-450 that could enhance degradation of compounds such as TCE, vinyl chloride, and benzene above that of native tobacco.

Another beneficial application of transgenic plants is the increased removal of TNT from explosives-contaminated soils. Travis et al. (2007) reported that transgenic tobacco that express bacterial nitroreductase genes can increase the rhizosphere degradation capacity of the transgenic plant relative to native tobacco. Moreover, the transgenic plants had more and deeper roots and a larger rhizosphere.

Natural rhizospheric bacteria collected from the roots of poplar trees were genetically engineered to have the TOM gene (Shim et al. 2000). Although recombinant bacteria from various plant rhizosphere added to poplar roots indicated a decrease in the recombinant TCE oxidizers over time to near 100% loss, the recombinant bacteria from tree colonizers were more competitive, with between 39% and 79% survival. The investigators concluded that this increased competitiveness was a function of the source of the bacteria from the rhizosphere of the host plant or similar surrogates, relative to unrelated plants or soils.

The opponents of transgenic plants for phytoremediation purposes also logically oppose the use of any plant that is not native to the contaminated area. This is because in some areas of the United States, or in upland contaminated areas of the humid east not supportive of riparian vegetation, the introduction of any non-native plant, even a hybrid, is not without some potential negative consequences. For example, what will happen to planted trees after a phytoremediation project is no longer funded? Should the plants be cut down? If so, they will grow back? Will they "escape" the site to populate other areas? *Populus* can form hybrids with other *Populus spp.* nearby.

If hybrids escape, there always is the potential for other areas to be overrun by the hybrid, and biodiversity will decrease. The best example of this potential escape is the gradual replacement of cottonwood-poplar riparian ecosystems by salt cedar in the southwestern United States. In order to assess the threat of introducing non-native hybrid plants into a contaminated part of a local ecosystem, Rotteveel et al. (2006) provide a decision tree to aid in risk evaluation. The key assesses the biological hazard, such as extent of invasiveness, through sexual propagation, or fast growth rates, and then helps to determine the extent of real risks, either at the planted site or offsite. Some of these concerns can be alleviated by planting sterile male plants as the hybrid sex of choice. Not only will there be no seeds formed by these trees within themselves, they also will not be able to sexually interact with native poplars.

In the same species gene flow is vertical. In different species, gene flow is horizontal. Natural gene flow occurs all the time. Gene flow can be defined as the incorporation of genes of one or more populations into the gene pool of other populations. It happens between domesticated crops and their native counterparts (Ellstrand et al. 1999)—the result is a hybrid. The hybrid can be fertile, and produce viable seeds itself, or sterile. As such, gene flow is a driver of evolution, sometimes of much larger impact than mutation or selection (Ellstrand et al. 1999).

Whether or not gene flow occurs in phytoremediation examples is not the question; rather, the question is, will the hybrid be undesirable? Gene containment such that phyto-plant genes do not leave the site is done by integrating the transgene in the plasmid genome. Also, much concern could be overcome using plant-generated genes that can produce the detoxification enzymes rather than relying on mammalian genes inserted into plants. There may also be less potential for transgenic-modified plants or plant-associated microbes to do harm if the work is done on endophytic bacteria rather than ectomycorrhizae.

It has been shown that some chemicals when added to plants themselves or their intermediate byproducts behave as plant activators. This describes a process where mutagenic compounds are activated from benign plant promutagens (Plewa and Wagner 1993). The potential exists for the activated mutagen to cause cancer in certain individuals.

The acceptance of transgenic plants for the use in the phytoremediation of contaminated groundwater has and will undoubtedly face the same challenges that faced the use of these techniques in the production of food or other commercial crops. Even after more than 27 years since the first genetically modified plant was created in 1983, no more than 69% of the cotton planted in the United States was genetically modified, as well as up to 26% of corn planted, and this is not for human consumption but for fodder. It may take as long for transgenes to become commonly used when designing a phytoremediation plan. As precise as the technology may be, there still is the potential for unintended consequences, unpredictability, and unanticipated outcomes.

16.3 Bioaccumulation Potential at Phytoremediation Sites

Plants are the basis of most food chains on earth. If plants become contaminated, the potential exists for the contaminant to move through the food chain. Even at contaminated sites where native vegetation exists, however, the plants present have not typically been assessed for contaminant levels to the extent that soil, water, and groundwater are required to be monitored by state and federal regulators.

Because of the interaction between plants and contaminants at contaminated sites, there exists the potential for bioaccumulation. Bioaccumulation, or bioconcentration, refers to the uptake and retention of a particular compound once it enters a living organism. As such, contaminant gains are greater than losses. The result is that an organism has a higher internal concentration of the compound relative to the concentration external to the organism (Schwarz and Jones 1997).

As was introduced in Chap. 12, a bioconcentration factor, or *BCF*, can be expressed as

$$BCF = C_{org}/C_{env} \tag{16.1}$$

Where C_{org} is the concentration of the chemical in the organism, such as a plant, and the C_{env} is the concentration in the environment, be it soil, water, or air.

The above equation can be enhanced after inclusion of a coefficient of bioconcentration, K_{bc}. This parameter reflects the bioavailability of the particular contaminant, as log K_{ow}, of the compound. Therefore,

$$C_{org} = K_{bc}C_{env} \tag{16.2}$$

There are forces within the plant that can act to decrease the potential for bioaccumulation to occur. These detoxification processes were discussed in Chap. 12 and include the uptake, oxidation, conjugation, and sequestration of metabolites into less bioavailable parts of the plant. The assumption is that these processes lead to a less toxic endpoint.

There are at least two considerations in terms of bioaccumulation that must be discussed when plants are exposed to contaminant compounds at phytoremediation sites. First, because plants essentially remove gaseous CO_2 from the atmosphere and convert it into solid biomass, it follows that this biomass itself acts as a sink for organic compounds. Chemicals that have lower solubility in water would tend to partition into the plant structure. However, the location of this sequestration would most likely be below ground in the root zone if the contaminant source is assumed to be groundwater. This distribution would change, however, if an atmospheric source were also present. The degree of bioaccumulation from this process is likely to be small, because even though most xylem cells from the previous year's growth are dead, some contaminant metabolism will occur in the living tissues in the cambium layer just below the bark. The cortex cells will permit the rapid volatilization of certain chemicals before concentrations are increased to high levels.

The second concern would be accumulation of contaminants in those parts of a plant that most likely might be a route of exposure, such as to humans or wildlife. Plants are primary producers, and at the base of most trophic levels. Chemicals can enter the plant from air, soil, or water pathways. Once in the plant, the chemical could stay localized at the point of entry, be translocated within certain parts of the plant, or be translocated unmediated to exit the plant by transpiration. In plants, the chemicals would reside in the fertilized ovaries of a female plant, the fruit, or in the sap of commercial plants, like the sugar maple. However, even though the xylem is located near the phloem, concentrations will not exceed those of the xylem on account that no physical process has yet been identified that shows that this occurs; see Marschner (1995) for a dissenting opinion. The implication for phytoremediation is that a sink for phloem-transported sugars such as fruits, will not be a sink for contaminants as long as they do not cross phloem membranes.

16.4 Technical Impracticability and the Role of Phytoremediation

The needs of consultants or responsible parties are often understandably at odds with the needs of regulators to carry out their jobs of environmental and human health protection. There is one area, however, that often both sides can agree on—the issue of technical impracticability (TI). This is a waiver for remediation at sites that are characterized by contaminated media that are not available for conventional restoration (U.S. Environmental Protection Agency 1993). A TI waiver does not mean that nothing is required to be done by the responsible party. At a minimum, such parties have to show limited site access or contaminant containment.

In these regards, phytoremediation can be used to achieve both of these requirements for the TI waiver, although control to the planted area, which could be considered an "attractive nuisance" may need to be implemented. If sources of contamination cannot be removed, for example due to site restrictions, then plants can be added to decrease the potential of leachate formation or extensive plume development. This has been accomplished at coal-tar contaminated sites in Tennessee (Widdowson et al. 2005a) and South Carolina (see Chap. 13).

16.5 Sustainability of Phytoremediation and When to Stop

Generation of wastes is a consequence of all biological systems. It also is common to all human cultures. The removal of wastes at rates that do not allow accumulation is unique to most civilized societies and often results in the re-use of once-contaminated lands. The full potential, in both non-economic and economic factors, of the contaminated subsurface may be actualized. The restoration of contaminated groundwater and its potential for re-use comprise part of the larger concept of resource sustainability.

In general, sustainability consists of two major parts: use of the resource while permitting no net degradation of the resource, over time. The concept of no-net degradation can apply to coastal aquifers under the threat of saltwater contamination as population centers grow, to contaminated groundwater in areas that may be developed, and for plant use of water to sustain ecological niches in arid areas with little surface water for drinking water or waste treatment. In many early studies of such interactions, the term safe yield was used, which is similar to the concept of sustainability.

It is interesting to note the general mindset that existed prior to the onset of environmental restoration, and how it fits in with recent conceptualizations of sustainability. For most of even recent history, the most pressing problems that humans had to solve were to acquire adequate water, food, shelter, and resources. When waste did accumulate, it was either burned or people moved to virgin land, as it was readily available. As population centers increased in density and space became a premium by the middle of the twentieth century, technology provided some solutions to the problems of food and shelter. This gave people the luxury of being able to become concerned about the quality of the water and food and shelter. The idea of moving to another place to use those less contaminated resources became less of an option as those areas had become inhabited by others perhaps wanting to do the same thing. Food was being produced in such abundant quantities on account of fertilizer application that any not taken up by the food crop unfortunately ended up in the surface and groundwater.

An example of plants and sustainability is given by the situation of groundwater use by humans and plants in the western United States as introduced in Chap. 1. Discharge of groundwater by phreatophytes in riparian corridors before it reaches streams is considered consumptive use in such areas and, therefore, had negative economic impacts on residents. In a paper that compares sustainability to safe yield, Alley and Leake (2004) describe a report by C.V. Theis that indicated that the capture of groundwater by wells rather than by phreatophytes would be an economic benefit. This was especially true for areas that wanted to attract new residents and growth. An equally valid stance once the growth has been achieved, however, is that natural greenspace or vegetated waterways have strong positive economic attributes for recreation, quality of life, ecological habitat, and current or future waste assimilation.

Alley and Leake (2004) also present an example of an approach to determine the sustainability of groundwater use in a basin in Nevada: drainage through Paradise Valley. In this arid area, the basin-fill aquifer prior to development received most of its recharge by surface-water leakage, which was balanced by natural discharge by evapotranspiration. A numerical model for the basin simulated the removal of half the natural recharge (44 out of 91 cubic hecta meters (hm^3/year)) for up to 300 years. As could be expected, a majority of this demand was met by a significant (72%) reduction in evapotranspiration. The resultant effect on the riparian vegetation was not simulated, but potential consequences could be lower humidity and more turbid surface water as the size of the riparian areas decreased.

At least 50% of the population of the United States relies on groundwater to meet its daily water needs. Because most people in the developed world need not fret about the quantity of water they can access, issues of the water quality are debated. The availability of water is a concern in arid areas where aquifers are being mined and coastal areas where saltwater contamination can render even productive aquifers unusable. These scenarios all go back to the central issue of sustainability; can we re-use water that becomes contaminated as long as it is restored by a natural or engineered process? Examples exist all over the United States at Brownfield sites, where previously industrialized sites are being remediated and developed. Groundwater can be used to decontaminate wastes, much like rivers and streams are allowed to while needing to meet, at the same time, recreational and aquatic criteria. In most cases, however, the time for such assimilation to take place in groundwater will be measured in years, rather than days.

Other aspects will affect the sustainability of phytoremediation over time relative to the costs associated with its operation and maintenance. Toxicity to contaminants, infestation, air-quality changes, demographic changes, and drought conditions all can negatively impact the future success of a phytoremediation plantation. The degree of reliability on hybrid clones whose health could be rapidly comprised by a single disease is unknown. History provides a ready example of the widespread, rapid use of a monoculture of one clonal plant and its inevitable downfall in the story of the potato in Ireland in the middle 1800s. Potatoes were a staple food for the poorer Irish, but the problem was exacerbated by the use of one type of potato, the Lumper. The parent plant had no resistance to the fungi *Phytophthora infestans* and, therefore, all clones had no resistance. The fungi *Phytophthora infestans* wiped out crops of potatoes in a few days and led to widespread famine. This scenario is to be contrasted to another monocultural use of the potato, by the Incas in Peru, but who used a wide variety of potato species. In that case, if a disease did arise, it was quickly contained and not disastrous (Pollan 2001).

Remediation projects have a life cycle that ends when remedial goals are met. As a given, even the use of transgenic plants with high rates of transpiration still in the end will have an upper bound as to how fast they can remove groundwater from a contaminated aquifer. It is just a matter of deciding what criteria should be used to determine when the end can be defined for a particular situation. For a phytoremediation-based project, the concept of sustainability can be used to help determine when the project should be stopped. The decision can be based on resource or economic criteria. In biological terms, a planting can be sustainable for a long period of time as long as solar radiation, water, and soil nutrients are not limited. It may be more likely that a planting will extend beyond the time period for which potential resources are allocated to the project.

Economically, this is where the use of the plants can directly restore the groundwater contamination and also affect the larger environment by increasing carbon sequestration and, potentially, providing a positive cash flow. This potential for the production of a marketable commodity may offset the negative consequence of the lengthy time needed to reach remedial goals at most sites (Robinson et al. 2003).

16.6 Climate Change, Carbon Sequestration, and the Role of Phytoremediation

In the 1970s and 1980s in the United States, the term acid rain was often discussed by scientific and laypersons alike. The blame for dying forests in the eastern United States was laid on western coal-fired power plants that, perhaps with some hubris, also met the power demands of this area. This was not the first use of the term to describe deforestation by acid precipitation, as the term acid rain was coined in 1872 by a British scientist named Robert Angus Smith. Lower-pH precipitation was seen as a new threat by many in the environmental field, and something had to be done to stop it. Currently, the greater threat of contamination from power plants is no longer from acid rain, due to technological abatement at the plants, but from the suspected release of mercury and CO_2.

Many compounds, including CO_2, CH_4, and H_2O vapor when released to the atmosphere at high enough concentrations, inhibit the release of reflected energy back from the earth to the atmosphere and are called greenhouse gases. The level of these gases has never been constant throughout the earth's 4.5 billion year history, but is rather is a continual state of flux. For instance, as the earth has cooled and before the onset of photosynthetic plant life more than one billion years ago, the level of CO_2 in the atmosphere was at least 10–15 times that measured today. As the invasion of land plants and the uptake of CO_2 to support life increased, the level of CO_2 dropped and that of O_2 increased. Over this period of geologic time, elevated atmospheric temperatures occurred throughout much of the ancient world after the continents coalesced into Laurentia and Gondwanaland. This very large range of elevated temperatures resulted in lush plant growth that, ironically, resulted in the production of the very coal and petroleum reservoirs being tapped today, the burning of which is being blamed for elevated CO_2 and global warming.

The gradual replacement of grasses by woody phreatophytes in arid areas has been shown to affect water use and carbon sequestration, because these plants are decoupled from local variations in precipitation (Scott et al. 2006). This increased productivity relative to shallow-rooted grasses leads to increased biomass. After leaf fall, however, much of this sequestered carbon may be returned to plants by decay by soil microbes. Although this process occurring at a phytoremediation site would not be significant, it does enhance the selection of phytoremediation over other groundwater remediation technologies.

16.7 Summary

In general, all groundwater supplies could be considered as a potential source of drinking water. In fact, some state regulatory agencies classify all groundwater, either fresh or saline, as a potential source of potable water and, therefore, must be remediated if contaminated. The phytoremediation of contaminated groundwater is part of environmental sustainability in that it permits a degraded resource to have the potential for reuse, either immediately or by future generations.

17 Epilogue

I realized, said Trout, "that God wasn't any conservationist, so for anybody else to be one was sacrilegious and a waste of time. You ever see one of His volcanoes or tornadoes or tidal waves...how about Dutch Elm disease? There's a nice conservation measure for you. That's God, not man. Just about the time we got our rivers cleaned up, he'd probably have the whole galaxy go up like a celluloid collar.

Breakfast of Champions (Kurt Vonnegut 1973)

Let's hope that Kilgore Trout's prediction won't come true, or at least won't be realized in full force. Nature may be powerful and can do great harm, but nature also can provide the means to heal and solve problems—from using plants for medicinal purposes to using plants to restore environmental contamination, including contaminated groundwater.

The phytoremediation of contaminated groundwater is a logical adaptation of the ecological development of a plant's naturally evolved detoxification system. Because uptake and detoxification is dependent upon the type of contaminant encountered in groundwater by plants, it also may be possible to remediate specific classes of contaminants with specific plants that possess appropriate detoxification pathways, or plants that do not possess the appropriate pathways could be transgenetically engineered to express these detoxification enzymes.

The future of phytoremediation of contaminated groundwater will be determined on the demonstration of its ability to have reproducible and verifiable evidence.

17.1 Phytoremediation Makes Evolutionary Sense

The phytoremediation of contaminated groundwater is part of an overall trend to use plants or plant products as a way to remove contaminants from people and from the environment. For households, there has been renewed interest in using plant-derived cleaners, especially if they are organically grown. Personal-care products derived from plant essential oils also are marketed as being a natural alternative to chemical products. Even the practice of aromatherapy, which may have as its source the use of burning various plant oils as a crude form of fumigation in order to expel the 'spirits' that were believed to cause sicknesses, has increased. The phreatophytic plants installed to clean contaminated groundwater were used by Native North Americans to cure various illnesses, and believed that plants found growing where the depth to water was shallow could be counted on to cure ailments of the fluid components of the human body such as blood or urine.

The rational explanation behind the past and current belief and faith that plants or plant-derived extracts will cure diseases is that plants, as well as animals, evolved in a competitive world. To survive, organisms had to develop chemical arsenals for defense against attacks from other life forms or threats from the inanimate environment, such as radiation or temperature. Because the stresses affected both plants and animals, they developed similar protective responses methods to ensure detoxification, such as the expression of oxidative enzymes in the P-450 group. Even though these chemicals are "natural," they are powerful. In fact, many plant-derived chemicals are more potent than some synthesized chemicals, such as the fruit of the manchineel tree in South America. As was discussed in Chap. 11, these plant-based detoxification systems that have evolved over the last 450 million years can be harnessed to restore contaminated groundwater where plants are installed.

The evolutionary progress of plants has been and will continue to be fueled by the ability of plants to continuously monitor the surface and subsurface environment for changes in the environmental aspects that affect their survival. This real-time monitoring can be hard to visualize because of the vast spans of time that has been necessary for evolution to affect these changes. That plants respond to their environment is seen as leaves change color and then drop off before some plants go dormant. In a similar fashion, certain plants also either naturally have or can be engineered to have the ability to detect other aspects of their environment, such as the presence of various xenobiotics.

J.E. Landmeyer, *Introduction to Phytoremediation of Contaminated Groundwater*,
DOI 10.1007/978-94-007-1957-6_17, © Springer Science+Business Media B.V. 2012

17.2 The Ideal Phytoremediation Plant

All plants use water, but not all plants can be used to restore contaminated groundwater. Removing contamination from groundwater involves access to the saturated zone, regardless of the remedial goal being pursued. At many sites where plants have been added to interact with groundwater, hybrid poplar trees are commonly selected. This selection reflects a wide variety of reasons, such as a long history of use by the paper industry, a rapid growth characteristic, good vegetative growth characteristics, high transpiration rates, a defined genome, the widespread distribution of parent poplar trees, and a deep rooting system.

But are hybrid poplar trees necessarily the best choice for a site characterized by contaminated groundwater? What is the ideal plant for phytoremediation of contaminated groundwater? Does such an ideal plant already exist? If not, can or should one be engineered? In all instances, such a plant would probably have to have some of the following characteristics:

- A deep root system where at least some highly transmissive roots reach the lowest level of a fluctuating water table
- A higher density of roots near the water table than in shallower soils
- Ability to release root exudates to encourage rhizospheric development
- A high transpiration rate
- A limited period of dormancy
- Perennial growth
- No toxicity responses to high levels of contaminants
- Well-developed cortex to promote gas exchange with the atmosphere
- Roots that can survive saturated conditions over time
- Roots that cause the water table to decrease during the day and recover at night, increasing the exchange of gas throughout the vadose zone to promote aerobic microbial contaminant biodegradation
- A fast growth rate and biomass production without the associated high reproductive rate
- Little need for irrigation or fertilization
- Classification as a C_3 plant
- The necessary plant detoxification enzymes to accomplish Phase I to III reactions, and
- Rhizosphere flora with appropriate enzymes to detoxify contaminants prior to entry into plants.

In many cases, those plants that meet most of the criteria above are those plants that have adapted to conditions characteristic of a riparian ecology and, therefore, come into contact with groundwater. These plants do not meet the other needs, however, such as not having a period of dormancy or a higher root distribution with depth. To overcome these limitations, hybrid poplar trees can be interplanted with coniferous plants that can transpire through winter months when the removal of groundwater and contaminants by the dormant poplars will be lessened.

Plants could be engineered to remediate specific types of contaminants found in groundwater. This bioengineering could be done either to the plant genetic material or to the genetic material of the rhizosphere flora. Specific genes, for example, that code for the production of enzymes known to facilitate the degradation of chlorinated solvents, could be added to native flora of plants that exist or are planted at such sites. Such bioengineering, however, needs to be highly regulated as to decrease potential negative impacts to the environment.

17.3 The Future of Phytoremediation of Contaminated Groundwater

The future of phytoremediation in solving problems of contaminated groundwater can be predicted, somewhat, in terms of the fundamentals of science and resource economics. The different perspectives of scientists involved in the implementation and oversight of phytoremediation projects needs to be validated, with the end result being a fuller understanding of each scientist's unique needs. In terms of economics, the future of phytoremediation will be determined by whether or not it is shown to be cost effective compared to other remedial strategies. As true for many early endevours, considerable investment regarding the effectiveness of phytoremediation was incurred mainly by governmental agencies because the risk return on investment required by private firms did not justify such large expenditure of capital. Currently, however, phytoremediation has become a part of the larger remediation industry, representing tens of billions of dollars annually around the world.

Also, the future of phytoremediation of contaminated groundwater will be determined by the answer to the question of "phytoremediation is a good theory, but does it work?" In some instances, phytoremediation can be viewed as being between science and mysticism—many proponents simply have faith that phytoremediation will always work at all sites. The future of phytoremediation of contaminated groundwater, however, has to be based on the systematic collection of quantitative data collected using calibrated approaches to provide reproducible conclusions over time. Those who rely on non-reproducible data in an attempt to promote phytoremediation will be dismissed as charlatans. Unfortunately, there are very few sites where all the necessary quantitative performance data are being collected, and there exists even a smaller number of sites that produce published data, a step necessary to encourage technology transfer.

The similarity between some phytoremediation processes and bioremediation processes also stresses the importance of establishing a firm foundation of observation and testing to parse which processes are plant related, plant caused, or not related to plants at all. This is because the field of natural attenuation includes both biotic and abiotic processes that can result in the decrease of contaminants in groundwater. It is typically thought that the biotic component is limited to aquifer microbes, but, as shown in this book, much evidence exists that plant-based reactions are an important but often unrecognized part of the natural attenuation process, especially when it comes to hydrologic control.

In any case, the similarity between microbial processes and phytoremediation may unnecessarily cause confusion. If a process of contaminant removal can be attributed to the plant-based process, then restoration can rightly be related to phytoremediation. On the other hand, if plants are fertilized with nitrates and phosphates, which stimulate contaminant-degrading bacteria, then it may be unclear if the degradation is entirely based on either phytoremediation or bioremediation, unless the appropriate control studies are performed. Perhaps the best way to separate the components of microbial versus plant restoration will be the use of axenic culture studies of plants cells with various contaminant compounds.

In order for phytoremediation to be most applicable to remediate contaminated groundwater, rather than be viewed by regulators or stakeholders as a less expensive method to escape a more expensive and perhaps more protective remediation scheme, the following steps should occur:

- Collect data that show the linkage between plants, groundwater, and contaminant uptake
- Publish phytoremediation failures as well as successes
- Evaluate phytoremediation as part of the overall site remedial strategy rather than as the only strategy
- Collect data to document the fate of chemicals taken up from groundwater by plants, and
- Generate reproducible results from laboratory and field studies.

This short list is not meant to promote the notion that consensus is required among phytoremediation practitioners in order for the science of phytoremediation of contaminated groundwater to be successful. In fact, such a consensus may actually harm the use of phytoremediation to restore contaminated groundwater—phytoremediation either will work or will not work, and not because some proponents declare that it will. In other words, consensus will not validate the application of phytoremediation of contaminated groundwater—the fundamentals of the technology will. Scientific facts are facts, and that which is not fact is not made so by consensus.

References

The original sources of much of the information presented in this book are in this reference list. These references should be considered an introduction to the phytoremediation of contaminated groundwater, and is far from being comprehensive because peer-reviewed research is being published on a daily basis. Every effort was made to include only those references that are publically accessible—unfortunately, this excludes much good work contained in a variety of unpublished reports, posters presentations, consulting reports, and the unfinished manuscripts found in the file cabinets of many scientists.

Adam, G., & Duncan, H. J. (1999). Effect of diesel fuel on growth of selected plant species. *Environmental Geochemistry and Health, 21*, 353–357.

Adams, R. G., Lomann, R., Fernandez, L. A., Macfarlane, J. K., & Gschwend, P. M. (2007). Polyethylene devices: Passive samplers for measuring dissolved hydrophobic organic compounds in aquatic environments. *Environmental Science & Technology, 41*, 1317–1323.

Ahuja, L. R., Rojas, K. W., Hanson, J. D., Shaffer, M. J., & Ma, L. (2000). *Root zone water quality model—Modelling management effects on water-quality and crop production*. Highlands Ranch: Water Resources Publications.

Aitchison, E. W., Kelley, S. L., & Schnoor, J. L. (2000). Phytoremediation of 1,4-dioxane by hybrid poplar trees. *Water Environment Research, 72*, 313–321.

Ajami, H., Maddock, T., III, Meixner, T., Hogan, J. F., & Guertin, D. P. (2011). RIPGIS-NET: A GIS tool for riparian groundwater evapotranspiration in MODFLOW. *Ground Water*, article first published online March 8, 2011.

Allen, S. J., & Grime, V. L. (1995). Measurements of transpiration from savannah shrubs using sap flow gauges. *Agricultural & Forestry Meteorology, 75*, 23–41.

Allen, S. J., Hall, R. L., & Rosier, P. T. W. (1999). Transpiration by two poplar varieties grown as coppice for biomass production. *Tree Physiology, 19*, 493–501.

Alley, W. M., & Leake, S. A. (2004). The journey from safe yield to sustainability. *Groundwater, 42*, 12–16.

Alley, W. M., Healy, R. W., LaBaugh, J. W., & Reilly, T. E. (2002). Flow and storage in groundwater systems. *Science, 296*, 1985–1990.

Alterio, G., Giorrio, P., & Sorrention, G. (2006). Open-system chamber for measurement of gas exchanges at plant level. *Environmental Science & Technology, 40*, 1950–1955.

American Society for Civil Engineers. (1989). In E. L. Johns (Ed.), *Water use by naturally occurring vegetation including an annotated bibliography* (216 p). New York: ASCE.

Anderson, D. B. (1936). Relative humidity or vapor pressure deficit. *Ecology, 17*, 277–282.

Anderson, J. E. (1982). Factors controlling transpiration and photosynthesis on *Tamarix chinensis* Lour. *Ecology, 63*, 48–56.

Anderson, T. A., & Walton, B. T. (1995). Comparative fate of [^{14}C] trichloroethylene in the root zone of plants from a former solvent disposal site. *Environmental Toxicology and Chemistry, 14*, 2041–2047.

Anderson, M. S., Lakin, H. W., Beeson, K. C., Smith, F. F., & Thacker, E. (1961). Selenium in agriculture. (U.S. Department of Agriculture Handbook 200 (65 p).). Washington, DC: U.S. Government printing office.

Anderson, T. A., Guthrie, E. A., & Walton, B. T. (1993). Bioremediation in the rhizosphere. *Environmental Science & Technology, 27*, 2630–2636.

Anderson, T. A., Hoylman, A. M., Edwards, N. T., & Walton, B. T. (1997). Uptake of polycyclic aromatic hydrocarbons by vegetation: A review of experimental methods. In W. Wang, J. W. Gorsuch, & J. S. Hughes (Eds.), *Plants for environmental studies* (pp. 451–480). Boca Raton: CRC Press, LLC.

Anderson, R. G., Booth, E. C., Marr, L. C., Widdowson, M. A., & Noval, J. T. (2008). Volatilzation and biodegradation of naphthalene in the vadose zone impacted by phytoremediation. *Environmental Science & Technology, 42*, 2575–2581.

Andraski, B. J., Sandstrom, M. W., Michel, R. L., Radyk, J. C., Stonestrom, D. A., Johnson, M. J., & Mayers, C. J. (2003). Simplified method for detecting tritium contamination in plants and soil. *Journal of Environmental Quality, 32*, 988–995.

Andraski, B. J., Stonestrom, D. A., Michel, R. L., Halford, K. J., & Radyk, J. C. (2005). Plant-based plume-scale mapping of tritium contamination in desert soils. *Vadose Zone Journal, 4*, 819–827.

Armstrong, W. (1968). Oxygen diffusion from the roots of woody species. *Physiologia Plantarum, 20*, 539–543.

Armstrong, J., & Armstrong, W. (1987). Phragmites australis-A preliminary study of soil-oxidizing sites and internal gas transport pathways. *New Phytologist, 108*, 373–382.

Arndt, S. K., Kahmen, A., Arampatsis, C., Popp, M., & Adams, M. (2004). Nitrogen fixation and metabolism by groundwater-dependent perennial plants in a hyperarid desert. *Oecologia, 141*, 385–394.

Arnold, C. W., Parfitt, D. G., & Kaltreider, M. (2007). Phytovolatilization of oxygenated gasoline-impacted groundwater at an underground storage tank site via conifers. *International Journal of Phytoremediation, 9*, 53–69.

Ayers, J. E., Hutmacher, R. B., Schoneman, R. A., Soppe, R. W. O., Vail, S. S., & Dale, F. (1999). Realizing the potential integrated irrigation and drainage water management for meeting crop water requirements in semi-arid and arid areas. *Irrigation and Drainage Systems, 13*, 321–347.

Baduru, K. K., Trapp, S., & Burken, J. G. (2008). Direct measurement of VOC diffusivities in tree tissues: Impacts on tree-based phytoremediation and plant contamination. *Environmental Science & Technology, 42*, 1268–1275.

Baird, J. B., & Maddock, T., III. (2005). Simulating riparian evapotranspiration: A new methodology and application for groundwater models. *Journal of Hydrology, 312*, 176–190.

Baird, K. J., Stromberg, J. C., & Maddock, T., III. (2005). Linking riparian dynamics and groundwater: An ecohydrologic approach to modeling groundwater and riparian vegetation. *Environmental Management, 36*, 551–564.

J.E. Landmeyer, *Introduction to Phytoremediation of Contaminated Groundwater*,
DOI 10.1007/978-94-007-1957-6, © Springer Science+Business Media B.V. 2012

Baker, J. M., & van Bavel, C. H. M. (1987). Measurement of mass flow of water in the stems of herbaceous plants. *Plant, Cell & Environment, 10*, 777–782.

Balling, A., & Zimmermann, U. (1990). Comparative measurements of the xylem pressure of Nicotiana plants by means of the pressure bomb and pressure probe. *Planta, 182*, 325–338.

Banks, M. K., Schwab, P., Liu, B., Kulakow, P. A., Smith, J. S., & Kim, R. (2003). The effect of plants on the degradation and toxicity of petroleum hydrocarbons in soil: A field assessment. *Advances in Biochemical Engineering/Biotechnology, 78*, 75–96.

Bankston, J. L., Duffey, J. T., & Bourquin, A. W. (2001). Innovative approaches for groundwater remediation using natural attenuation and phytoremediation. In A. Leeson, E. A. Foote, M. K. Banks, & V. S. Magar (Eds.), *Phytoremediation, wetlands, and sediments. Proceedings of the 6th International In Situ and On-Site Bioremediation Symposium,* San Diego, CA, June 4–7, 2001 (Vol. 6(5), pp. 33–40). Columbus: Battelle Press.

Bankston, J. L., Sola, D. L., Komor, A. T., & Dwyer, D. F. (2002). Degradation of trichloroethylene in wetland microcosms containing broad-leaved cattail and eastern cottonwood. *Water Research, 36*, 1539–1546.

Banta, E. R. (2000). *MODFLOW-2000, the U.S. Geological Survey modular groundwater model—Documentation of packages for simulating evapotranspiration with a segmented function (ETS1) and drains with return flow (DRT1)*(U.S. Geological Survey Open-File Report 00-466, 135 p).

Barac, T., Taghavi, S., Borremans, B., Provoost, A., Oeyen, L., Colpaert, J. V., Vangronsveld, J., & van der Lelie, D. (2004). Engineered endophytic bacteria improve phytoremediation of water soluble volatile organic pollutants. *Nature Biotechnology, 22*(5), 583–588.

Barlow, P. (2003). *Groundwater in freshwater-saltwater environments of the Atlantic Coast* (U.S. Geological Survey Circular 1262). Denver: U.S. Geological Survey.

Barrett-Lennard, E. G. (2002). Restoration of saline land through revegetation. *Agricultural Water Management, 53*, 213–226.

Batelaan, O., Hung, L. Q., Verbeiren, B., & DeSmedt, F. (2004). Mapping of wetness gradients by hyperspectral sensing of phreatophytes. *Geophysical Research Abstracts, 16*, 06459

Bauer, P., & Hell, R. (2006). Translocation of iron in plant tissues. In L. L. Barton & J. Abadía (Eds.), *Iron nutrition in plants and rhizospheric microorganisms* (pp. 279–288). The Netherlands: Springer.

Becker, M. W. (2006). Potential for satellite remote sensing of groundwater. *Groundwater, 44*, 306–318.

Belmans, C., Wesseling, J. G., & Feddes, R. A. (1981). *Simulation model of the water balance of a cropped soil providing different types of boundary conditions (SWATRE)* (p. 56). Wageningen: Institute for Land and Water Management Research.

Benayas, J. M., Bernáldez, F. G., Levassor, C., & Peco, B. (1990). Vegetation of groundwater discharge sites in the Douro Basin, Central Spain. *Journal of Vegetation Science, 1*, 461–466.

Ben-dov, Y. (1988). Manna scale, *Trabutina mannipara. Systematic Entomology, 13*, 387–392.

Bender, M. M. (1971). Variations in the 13C/12C ratios of plants in relation to the pathway of photosynthetic carbon dioxide fixation. *Phytochemistry, 10*, 1239–1244.

Bennison, E. W., & Bollenbach, W. M. (1947). *Groundwater—Its development, uses, and conservation* (509 p). St. Paul: Edward E. Johnson, Inc.

Benton, A. R., Jr., James, W. P., & Rouse, J. W., Jr. (1978). Evapotranspiration from water hyacinth (*Eichhornia crassipes* (Mart.) Solms) in Texas reservoirs. *Water Resources Bulletin, 14*, 919–930.

Benyon, R. G. (1999). Nighttime water use in an irrigated *Eucalyptus grandis* plantation. *Tree Physiology, 19*, 853–859.

Bernaldez, F. G., & Benayas, J. M. (1992). Geochemical relationships between groundwater and wetland soils and their effects on vegetation in central Spain. *Geoderma, 55*, 273–288.

Bhadra, R., Wayment, D. G., Hughes, J. B., & Shanks, J. V. (1999). Confirmation of conjugation processes during TNT metabolism by axenic roots. *Environmental Science & Technology, 33*, 446–452.

Blaney, H. F., Taylor, C. A., Nickle, H. G., & Young, A. A. (1933). *Water loses under natural conditions from wet areas in southern California (*Division Water Resources Bulletin 44, 176 p). Davis: Department of Public Works.

Blaney, H. F., Morin, K. V., & Criddle, W. D. (1942). *Consumptive water use and requirements—The Pecos River joint investigation reports of the participating agencies* (Natural Resource Planning Board, pp. 170–230). Washington, DC: U.S. Government Printing Office.

Bobeck, P. (2004). *English translation of Darcy's "The public fountains of the city of Dijon" 1856*. Dubuque: Kendall/Hunt Publishing Company.

Bond, B. (2003). Hydrology and ecology meet—And the meeting is good. *Hydrological Processes, 17*, 2087–2089.

Bond, B. J., Jones, J. A., Moore, G., Phillips, N., Post, D., & McDonnell, J. J. (2002). The zone of vegetation influence on baseflow revealed by diel patterns of streamflow and vegetation water use in a headwater basin. *Hydrological Processes, 16*, 1671–1677.

Book of common prayer. (1979). New York: Oxford University Press.

Boorstin, D. J. (1983). *The discoverers: A history of man's search to know his world and himself* (745 p). New York: Random House.

Bormann, F. H., & Likens, G. E. (1967). Nutrient cycling. *Science, 155*, 424–428.

Bornstein, J., Benoit, G. R., Scott, F. R., Hepler, P. R., & Hedstrom, W. E. (1984). Alfalfa growth and soil oxygen diffusion as influenced by depth to water table. *Soil Science Society of America, 48*, 1165–1169.

Bosch, D. D., Lowrance, R. R., Sheridan, J. M., & Williams, R. G. (2003). Groundwater storage effect on streamflow for a southeastern coastal plain watershed. *Groundwater–Watershed Issue, 41*, 903–912.

Bouwer, H. (1975). Predicting reduction of water losses from open channels by phreatophyte control. *Water Resources Research, 11*, 96–101.

Bouwer, H. (1978). *Groundwater hydrology* (480 p). New York: McGraw-Hill, Inc.

Bowen, G. D. (1984). Tree roots and the use of soil nutrients. In G. D. Bowen & E. K. S. Nambier (Eds.), *Nutrition of plantation forests* (pp. 147–179). London: Academic.

Bowie, J. E., & Kam, W. (1968). *Use of water by riparian vegetation, Cottonwood Wash, Arizona* (U.S. Geological Survey Water-Supply Paper 1858, 62 p). Washington, DC: U.S. Government Printing Office.

Bowser, C. W. (1957). *Introduction and spread of the undesirable tamarisk in the Pacific Southwest section of the United States and comments concerning the plant's influence upon indigenous vegetation.* Paper given at Pacific Southwest Regional Meeting, American Geophysical Union, February 15, 1957, 9 p.

Boyer, J. S., & Knipling, E. B. (1965). Isopiestic technique for measuring leaf water potentials with a thermocouple psychrometer. *Proceedings of the National Academy of Sciences of the United States of America, 54*, 1044–1051.

Bradley, P. M., & Dunn, E. L. (1989). Effects of sulfide on the growth of three salt marsh halophytes of the Southeastern United States. *American Journal of Botany, 76*, 1707–1713.

Bradley, P. M., & Morris, J. T. (1991). Relative importance of ion exclusion, secretion and accumulation in Spartina alterniflora Loisel. *Journal of Experimental Botany, 42*, 1525–1532.

Bradley, P. M., Chapelle, F. H., Landmeyer, J. E., & Schumacher, J. G. (1994). Microbial transformation of nitroaromatics in surface soils and aquifer materials. *Applied and Environmental Microbiology, 60*, 2170–2175.

Bradley, P. M., Chapelle, F. H., Landmeyer, J. E., & Schumacher, J. G. (1997). Potential for intrinsic bioremediation of a DNT-contaminated aquifer. *Groundwater, 35*, 12–17.

Bradley, P. M., Landmeyer, J. E., & Chapelle, F. H. (1999). Aerobic mineralization of MTBE and tert-butyl alcohol by stream-bed sediment microorganisms. *Environmental Science & Technology, 33*, 1877–1879.

Bradley, P. M., Landmeyer, J. E., & Chapelle, F. H. (2001). Wide-spread potential for microbial MTBE degradation in surface-water sediments. *Environmental Science & Technology, 35*, 658–662.

Bradley, P. M., Landmeyer, J. E., & Chapelle, F. H. (2002). TBA biodegradation in surface-water sediments under aerobic and anaerobic conditions. *Environmental Science & Technology, 36*, 4087–4090.

Brandt, R., Merkl, N., Schultze-Kraft, R., Infante, C., & Broll, G. (2006). Potential of Vetiver (*Vetiveria zizanioides* (L.) Nash) for phytoremediation of petroleum hydrocarbon-contaminated soils in Venezuela. *International Journal of Phytoremediation, 8*, 273–284.

Briggs, G. G., Bromilow, R. H., & Evans, A. A. (1982). Relationships between lipophilicity and root uptake and translocation of non-ionized chemicals by barley. *Pesticide Science, 13*, 495–504.

Brigmon, R. L., Bell, N. C., Freedman, D. L., & Berry, C. J. (1998). Natural attenuation of trichloroethylene in rhizosphere soils at the Savannah River Site. *Journal of Soil Contamination, 7*, 433–453.

Brigmon, R. L., Anderson, T. A., & Fliermans, C. B. (1999). Methanotrophic bacteria in the rhizosphere of trichloroethylene-degrading plants. *International Journal of Phytoremediation, 1*, 241–253.

Brix, H. (1990). Uptake and photosynthetic utilization of sediment-derived carbon by *Phragmites australis* (Cav.) Trin. Ex Steudal. *Aquatic Botany, 38*, 377–389.

Broekaert, W. F., Cammue, B. P. A., De Bolle, M. F. C., Thevissen, K., De Samblanx, G. W., & Osborn, R. W. (1997). Antimicrobial peptides from plants. *Critical Reviews in Plant Science, 16*, 297–323.

Bromilow, R. H., & Chamberlain, K. (1995). Principles governing uptake and transport of chemicals. In S. Trapp & J. C. McFarlane (Eds.), *Plant contamination, modeling and simulation of organic chemical processes* (pp. 37–68). Boca Raton: Lewis Publishers.

Brooks, J. R., Buchmann, N., Phillips, S., Ehleringer, B., Evans, R. D., Lott, M., Martinelli, L. A., Pockman, W. T., Sandquist, D., Sparks, J. P., Sperry, L., Williams, D., & Ehleringer, J. R. (2002). Heavy and light beers: A carbon isotope approach to detect C4 carbon in beers of different origins, styles, and prices. *Journal of Agricultural and Food Chemistry, 50*, 6413–6418.

Brooks, J. R., Schulte, P. J., Bond, B. J., Coulombe, R., Domec, J.-C., Hinkley, T. M., McDowell, N., & Phillips, N. (2003). Does foliage on the same branch compete for the same water? Experiments on Douglas-fir trees. *Trees, 17*, 101–108.

Brooks, J. R., numerous others. (2004). The cohesion-tension theory. *New Phtyologist,* 163, 451–452.

Brooks, J. R., Barnard, H. R., Coulombe, R., & McDonnell, J. J. (2009). Ecohydrologic separation of water between trees and streams in a Mediterranean climate. *Nature Geoscience, 3*, 100–104.

Brown, J. S. (1923). *The Salton Sea region, California: A geographic, geologic, and hydrologic reconnaissance with a guide to desert watering places* (U.S. Geological Survey Water-Supply Paper 497). Washington, DC: Government Printing Office.

Browne, C. A. (1978). *A source book of agricultural chemistry*. Manchester: Ayer Publishing.

Büchner, E. (1897). Alkoholische gärung ohne hefezellen (Vorläufige Mitteilung). *Berichte der Deutschen Chemischen Gesellschaft, 30*, 117–124.

Bukata, A. R., & Kyser, T. K. (2005). Response of the nitrogen isotopic composition of tree-rings following tree-clearing and land-use changes. *Environmental Science & Technology, 39*, 7777–7783.

Burken, J. G. (2003). Uptake and metabolism of organic compounds: Green-liver model. In S. C. McCutcheon & J. L. Schnoor (Eds.), *Phytoremediation: Transformation and control of contaminants* (pp. 59–84). Hoboken: Wiley Interscience, Inc.

Burken, J. G., & Ma, X. (2006). Phytoremediation of volatile organic compounds. In M. Mackova, D. N. Dowling, & T. Macek (Eds.), *Phytoremediation and rhizoremediation: Theoretical background* (pp. 199–216). The Netherlands: Springer.

Burken, J. G., & Schnoor, J. L. (1997). Uptake and metabolism of atrazine by poplar trees. *Environmental Science & Technology, 31*, 1399–1406.

Burken, J. G., & Schnoor, J. L. (1998). Predictive relationships for uptake of organic contaminants by hybrid poplar trees. *Environmental Science & Technology, 32*, 3379–3385.

Burken, J. G., & Schnoor, J. L. (1999). Distribution and volatilization of organic compounds following uptake by hybrid poplar trees. *International Journal of Phytoremediation, 1*, 139–151.

Burken, J. G., Shanks, J. V., & Thompson, P. L. (2000). Phytoremediation and plant metabolism of explosives and nitroaromatic compounds. In J. C. Spain, J. B. Hughes, & H. J. Knackmuss (Eds.), *Biodegradation of nitroaromatic compounds and explosives* (pp. 239–275). Boca Raton: CRC Press.

Busch, D. E., & Smith, S. D. (1995). Mechanisms associated with decline of woody species in riparian ecosystems of the Southwestern U.S. *Ecological Monographs, 65*, 347–370.

Busch, D. E., Ingraham, N. L., & Smith, S. D. (1992). Water uptake in woody riparian phreatophytes of the Southwestern United States: A stable isotope study. *Ecological Applications, 2*, 450–459.

Butler, J. J., Jr., Whittenmore, D. O., & Kluitenberg, G. J. (2005). *Groundwater assessment in association with salt cedar control—Report on year one activities* (Kansas Geological Survey Open-File Report 2005-19, 35 p).

Byl, T. D., & Klaine, S. J. (1991). Peroxidase activity as an indicator of sublethal stress in the aquatic plant *Hydrilla verticillata* (Royle). In J. W. Gorsuch, W. R. Lower, W. Wang, & M. A. Lewis (Eds.), *Plants for toxicity assessment: Second volume* (ASTM STP 1115, pp. 101–106). Philadelphia: American Society for Testing and Materials.

Cailloux, M. (1972). Metabolism and the absorption of water by root hairs. *Canadian Journal of Botany, 50*, 557–573.

Caird, M. A., Richards, J. H., & Donovan, L. A. (2007). Nighttime stomatal conductance and transpiration in C_3 and C_4 plants. *Plant Physiology, 143*, 4–10.

Campbell, B. G., & Landmeyer, J. E. (2003). Using MODFLOW to estimate tree-uptake rate for hydraulic control at a phytoremediation site in Charleston, South Carolina. *MODFLOW and more 2003: Understanding through modeling*. Denver: International Groundwater Modeling Center.

Campbell, B. G., Petkewich, M. D., Landmeyer, J. E., & Chapelle, F. H. (1996). *Geology, hydrogeology, and potential of intrinsic bioremediation at the NPS Dockside II site and adjacent areas, Charleston, SC, 1993–94* (U.S. Geological Survey Water-Resources Investigations Report 96-4170, 69 p).

Campbell, N. A., Reece, J. B., Taylor, M. R., & Simon, E. J. (2006). *Biology—Concepts and connections*. San Francisco: Pearson Education.

Cannon, W. A. (1911). *Root habits of desert plants* (Carnegie Institute Washington Publication No. 131, pp. 80–81). Tortugas: Carnegie Institute Washington.

Cannon, W. A. (1913). Some relations between root characters, groundwater, and species distribution. *Science*, new series 37, 420–423.

Cannon, W.A. (1923). The influence of the temperature of the soil on the relation of roots to oxygen. *Science*, new series 68, 331–332.

Cannon, H. L. (1971). The use of plant indicators in groundwater surveys, geologic mapping, and mineral prospecting. *Taxonomy, 20*, 227–256.
Canny, M. J. (1990). What becomes of the transpiration stream? Tansley Review No. 22. *The New Phytologist, 114*, 341–368.
Canny, M. J. (1997). Vessel contents during transpiration—Embolisms and refilling. *American Journal of Botany, 84*, 1223–1230.
Carsel, R. F., Smith, C. N., Mulkey, L. A., Dean, J. D., & Jowsie, P. (1984). *User's manual for the Pesticide Root Zone Model (PRZM)* (US EPA 600/3-84-109). Athens: U.S. Environmental Protection Agency, Environmental Research Laboratory.
Carson, R. (1962). *Silent spring*. Boston: Houghton Mifflin Company.
Cermak, J., Jenik, J., Kucera, J., & Zidek, V. (1984). Xylem water flow in a crack willow tree (*Salix fragilis* L.) in relation to diurnal changes of environment. *Oecologia, 64*, 145–151.
Chaney, R. L., Malik, M., Li, Y. M., Brown, S. L., Bewer, E. P., Angle, J. S., & Baker, A. J. M. (1997). Phytoremediation of soil metals. *Current Opinion in Biotechnology, 8*, 279–284.
Chanton, J. P., Whiting, G. J., Happell, J. D., & Gerard, G. (1993). Contrasting rates and diurnal patterns of methane emission from emergent aquatic macrophytes. *Aquatic Botany, 46*, 111–128.
Chapelle, F. H. (1993). *Groundwater microbiology and geochemistry* (424 p). New York: Wiley.
Chapelle, F. H., Roberston, J. F., Landmeyer, J. E., & Bradley, P.M. (2001). *Methodology for applying monitored natural attenuation to petroleum hydrocarbon-contaminated groundwater systems with examples from South Carolina* (U.S. Geological Survey Water-Resources Investigations Report 00-4161, 47 p).
Chapman, R. E. (1981). *Geology and water: An introduction to fluid mechanics for geologists* (228 p). The Hague: Martinus Nijhoff/Dr. W. Junk.
Chappell, J. (1997). *Phytoremediation of TCE using Populus* (Status report prepared for the USEPA Technology Innovation Office, 38 p).
Chapelle, F. H., Bradley, P. M., Lovley, D. R., & Vroblesky, D. A. (1996). Measuring rates of biodegradation in a contaminated aquifer using field and laboratory methods. *Ground Water, 34*, 691–698.
Chard, B. K., Doucette, W. J., Chard, J. K., Bugbee, B., & Gorder, K. (2006). Trichloroethylene uptake by apple and peach trees and transfer to fruit. *Environmental Science & Technology, 40*, 4788–4793.
Chase, J. S. (1920). *Our Araby: Palm springs and the garden of the sun*. Pasadena: Star-News Publishing Company.
Chekol, T., Vough, L. R., & Chaney, R. L. (2002). Plant-soil-contaminant specificity affects phytoremediation of organic contaminants. *International Journal of Phytoremediation, 4*, 17–26.
Chen, Y., Banks, M. K., & Schwab, A. P. (2003). Pyrene degradation in the rhizosphere of tall fescue (*Festuca arundinacea*) and switchgrass (*Panicum virgatum* L.). *Environmental Science & Technology, 37*, 5778–5782.
Chiou, C. T. (2002). *Partition and adsorption of organic contaminants in environmental systems* (257 p). Hoboken: Wiley.
Chiou, C. T., Sheng, G., & Manes, M. (2001). A partition-limited model for the plant uptake of organic contaminants from soil and water. *Environmental Science & Technology, 35*, 1437–1444.
Cho, C., Sung, K., Caopcioglu, M. Y., & Drew, M. (2005). Influence of water content and plants on the dissipation of chlorinated volatile organic compounds in soil. *Water, Air, and Soil Pollution, 167*, 259–271.
Clebsch, A. (1994). Selected contributions to groundwater hydrology by C.V. Theis, and a review of his life and work (U.S. Geological Survey Water-Supply Paper 2415, 70 p). Denver: U.S. Geological Survey.
Clegg, M. T., Gaut, B. S., Learn, G. H., Jr., & Morton, B. R. (1994). Rates and patterns of chloroplast DNA evolution. *Proceedings of the National Academy of Sciences of the United States of America, 91*, 6795–6801.
Cleverly, J. R., Dahm, C. N., Thibault, J. R., McDonnell, D. E., & Coonrod, J. E. A. (2006). Riparian ecohydrology: Regulation of the water flux from the ground to the atmosphere in the Middle Rio Grande, New Mexico. *Hydrological Processes, 2*, 3207–3225.
Clinton, B. D., Vose, J. M., Vroblesky, D. A., & Harvey, G. J. (2004). Determination of the relative uptake of ground vs. surface water by *Populus deltoidies* during phytoremediation. *International Journal of Phytoremediation, 6*, 239–252.
Collings, M. R., & Myrick, R. M. (1966). *Effects of juniper and pinyon eradication on streamflow from Corduroy Creek Basin, Arizona* (U.S. Geological Survey Professional Paper 491-B, 12 p). Washington, DC: U.S. Government Printing Office.
Collins, C. D., & Fryer, M. E. (2003). Model intercomparison for the uptake of organic chemicals by plants. *Environmental Science & Technology, 37*, 1617–1624.
Collins, C. D., Bell, J. N. B., & Crews, C. (2000). Benzene accumulation in horticultural crops. *Chemosphere, 40*, 109–114.
Collins, C. D., Fryer, M., & Grosso, A. (2006). Plant uptake of non-ionic organic chemicals. *Environmental Science & Technology, 40*, 45–52.
Conant, B., Jr. (2004). Delineating and quantifying groundwater discharge zones using streambed temperatures. *Groundwater, 42*, 243–257.
Conrad, R., 1995, Soil microbial processes and the cycling of atmospheric trace gases: Philosophical Transactions: Physical Sciences and Engineering (351):219–230.
Cook, R. L., Landmeyer, J. E., Atkinson, B., Messier, J. P., & Nichols, E. G. (2010). Field note: Successful establishment of a phytoremediation system at a petroleum hydrocarbon contaminated shallow aquifer: Trends, trials, and tribulations. *International Journal of Phytoremediation, 12*(7), 716–732.
Cooper, D. I., Eichinger, W. E., Kao, J., Hipps, L., Reisner, J., Smith, S., Schaeffer, S. M., & Williams, D. G. (2000). Spatial and temporal properties of water vapor and latent energy flux over a riparian canopy. *Agricultural and Forest Meteorology, 105*, 161–183.
Corseuil, H. X., & Moreno, F. N. (2001). Phytoremediation potential of willow trees for aquifers contaminated with ethanol-blended gasoline. *Water Research, 35*, 3013–3017.
Côté, B., & Camiré, C. (1985). Nitrogen cycling in dense plantings of hybrid poplar and black alder. *Plant and Soil, 87*, 195–208.
Coville, F. V., & MacDougal, D. T. (1903). *Desert Botanical Laboratory of the Carnegie Institution* (Carnegie Institution Washington Publication No. 6, pp. 5–8). Tortugas: Carnegie Institution Washington.
Cramer, V. A., Thorburn, P. J., & Fraser, G. W. (1999). Transpiration and groundwater uptake from farm forest plots of *Casuarina glauca* and *Eucalyptus camaldulensis* in saline areas of southeast Queensland, Australia. *Agricultural Water Management, 39*, 187–204.
Critchton, M. (1969). *The Andromeda strain* (304 p). New York: Alfred A. Knopf.
Culler, R. C., Hansom, R. L., Myrick, R. M., Turner, R. M., & Kipple, F. P. (1982). *Evapotranspiration before and after clearing phreatophytes, Gila River floodplain, Graham County, Arizona* (U.S. Geological Survey Professional Paper 655-P).Washington, DC: US Government Printing Office.
Cunningham, S. D., & Berti, W. R. (1993). Remediation of contaminated soil with green plants: An overview. *In Vitro Cellular and Developmental Biology, 20*, 207–212.
Cunningham, S. D., & Ow, D. W. (1996). Promises and prospects of phytoremediation. *Plant Physiology, 110*, 715–719.
Cunningham, S. D., Berti, W. R., & Huang, J. W. (1995). Phytoremediation of contaminated soils. *Trends in Biotechnology, 13*, 393–397.
Cunningham, S. D., Anderson, T. A., Schwab, A. P., & Hsu, F. C. (1996). Phytoremediation of soils contaminated with organic pollutants. *Advances in Agronomy, 56*, 55–114.

Curdy, E. M. (1923). *Leonardo da Vinci's note-books: Arranged and rendered into English with introductions*. New York: Empire State Book Company.

Curtis, H. (1983). Biology, Worth Publishers, Inc., New York, 1159 p.

Da Vinci, L., & McCurdy, E. (2002). *The notebooks of Leonardo da Vinci* (1184 p). Old Saybrook: Konecky & Konecky.

DaFarina, J. L. (1990). *Historical Atlas of crystallography* (p. vi). Dordrecht: Kluwer Academic Publishers.

Dalton, J. (1802). Experimental essays on the constitution of mixed gases; on the force of steam or vapor from water and other liquids in different temperatures, both in a Torricellain vacuum an in air; on evaporation; and on expansion of gases by heat. *Memoirs and Proceedings of the Manchester Literary and Philosophical Society, 5*, 536–602.

Darcy, H. P. G. (1856). *Les fountaines publiques de la Ville de Dijon*. Paris: Victon Dalmont.

Darwin, C. R. (1882). The action of carbonate of ammonia on the roots of certain plants. *Journal of the Linnean Society of London, Botany, 19*, 239–261.

Daum, C. R. (1967). A method for determining water transport in trees. *Ecology, 48*, 425–431.

Davis, S. N., & DeWiest, R. J. M. (1966). *Hydrogeology* (463 p). New York: Wiley.

Davis, L. C., & Erickson, L. E. (2002). Application of waste remediation technologies to agricultural contamination of water resources, Proceedings of meeting, July 30-August 1, 2002, Kansas City, Missouri

Davis, L. C., Banks, M. K., Schwab, A. P., Muralidharan, N., Erickson, L. E., & Tracy, J. C. (1996). Plant based bioremediation. In S. Sikdar & R. Irvine (Eds.), *Bioremediation*. Lancaster: Technomics Publishing Company.

Davis, L. C., Vanderhoof, S., Dana, J., Selk, K., Smith, K., Golpen, B., & Erickson, L. E. (1998). Movement of chlorinated solvents and other volatile organics through plants monitored by Fourier transform infrared (FT-IR) spectrometry. *Journal of Hazardous Waste Research, 1*, 1–26.

Dawson, T. E. (1993). Water sources of plants as determined from xylem-water isotopic composition: perspectives on plant competition, distribution, and water relations. In J. R. Ehleringer, A. E. Hall, & G. D. Farquhar (Eds.), *Stable isotopes and plant carbon-water relations* (pp. 465–496). San Diego: Academic.

Dawson, T. E., & Ehleringer, J. R. (1991). Streamside trees that do not use stream water. *Nature, 350*, 335–336.

de Jussieu, A. L. (1789). Genera Plantarum, Secundum Ordines Naturales Disposita, 555 p.

De Weist, R. J. M. (1965). *Geohydrology* (366 p). New York: Wiley.

de Wit, C. T. (1958). Transpiration and crop yields. *Versl Landbouwkd Onderzock, 64*, 1–88.

Decker, J. P., Gaylor, W. G., & Cole, F. D. (1962). Measuring transpiration of undisturbed Tamarisk shrubs. *Plant Physiology, 37*, 393–397.

Dekker, L. W. (1998). *Moisture variability resulting from water repellency in Dutch soils*. Published Doctoral thesis, Wageningen Agricultural University, the Netherlands, 240 pp.

Deming, D. (2005). Born to trouble: Bernard Palissy and the hydrologic cycle. *Groundwater, 43*, 969–972.

Denmead, O. T., & Shaw, R. H. (1960). Availability of soil water to plants as affected by soil moisture content and meteorological conditions. *Agronomy Journal, 54*, 385–389.

Detay, M. (1997). *Water wells: Implementation, maintenance, and restoration* (379 p). Chichester: Wiley.

Dhillion, S. S., & Zak, J. C. (1993). Microbial dynamics in arid ecosystems: Desertification and the potential role of mycorrhizas. *Revista Chilena de Historia Natural, 66*, 253–270.

Dickmann, D. I., & Stuart, K. W. (1983). *The culture of poplars in Eastern North America*. East Lansing: University Publications, Michigan State University.

Dinicola, R. S., Cox, S. E., Landmeyer, J. E., & Bradley P. M. (2002). *Natural attenuation of chlorinated volatile organic compounds in groundwater at Operable Unit 1, Naval Undersea Warfare Center, Division Keyport*, Washington (U.S. Geological Survey Water-Resources Investigations Report 02-4119).

Dixon, H. H., & Joly, J. (1895). On the ascent of sap. *Philosophical Transactions of the Royal Society, 186*, 563–576.

Donnelly, P. K., & Fletcher, J. S. (1994). Potential use of mycorrhizal fungi as bioremediation agents. In T. A. Anderson & J. R. Coats (Eds.), *Bioremediation through rhizosphere technology* (pp. 93–99). Washington: American Chemical Society.

Donnelly, P. K., Hedge, R. S., & Fletcher, J. S. (1994). Growth of PCB-degrading bacteria on compounds from photosynthetic plants. *Chemosphere, 28*, 981–988.

Doucette, W. J., Bugbee, B., Hayhurst, S., Plaehn, W. A., Downey, S. A., Taffinder, S. A., & Edwards, S. A. (1998). Phytoremediation of dissolved-phase trichloroethylene using mature vegetation. In G. B. Wickramanayake & R. E. Hinchee (Eds.), *Bioremediation and phytoremediation* (chlorinated and recalcitrant compounds, pp. 251–156). Columbus: Battelle Press.

Doucette, W. J., Chard, J. K., Fabrizius, H., Crouch, C., Petersen, M., & Gorder, K. (2007). Trichloroethylene uptake into fruits and vegetables: Three-year field monitoring study. *Environmental Science & Technology, 41*, 2505–2509.

Douglass, A. E. (1924). *Some aspects of the use of the annual rings of trees in climatic study* (Smithsonian Institution Report for 1922 (Publication 2731), pp. 223–239).

Dugas, W. A., Heuer, M. L., & Mayeux, H. S. (1992). Diurnal measurements of honey mesquite transpiration using stem flow gauges. *Journal of Range Management, 45*, 99–102.

Dunford, E. G., & Fletcher, P. W. (1947). Effect of removal of stream-bank vegetation upon water-yield. *American Geophysical Union Transactions, 28*, 105–110.

Dupuit, A. J. E. J. (1863). Theoretical and practical studies on the movement of water in open channels and through permeable terrains, 2nd edn, Dunod, Paris.

Durant, W. (1953). *The story of civilization, V: The Renaissance*. New York: Simon and Schuster.

Duthe, D., Lorentz, S., Cameron-Clarke, I., & Oliver, A. J. (2005). *Hydraulic containment, natural attenuation and phytoremediation as a combined remediation strategy for an industrial waste site in South Africa*. Presentation at the Third International Phytotechnologies Conference, Atlanta, GA, April 20–22, 2005. Accessed at January 1, 2006, clu-in.org.

Eberts, S. M., Schalk, C. W., Vose, J., & Harvey, G. J. (1999). Hydrologic effects of cottonwood trees on a shallow aquifer containing trichloroethene. *Hydrological Science and Technology, 15*, 115–121.

Eberts, S. M., Harvey, G. J., Jones, S. A., & Beckman, S. W. (2003). Multiple-process assessment for a chlorinated solvent plume. In S. C. McCutcheon & J. L. Schnoor (Eds.), *Phytoremediation: Transformation and control of contaminants* (pp. 589–633). Hoboken: Wiley.

Eberts, S. M., Jones, S. A., Braun, C. L., & Harvey, G. J. (2005). Long-term changes in groundwater chemistry at a phytoremediation demonstration site. *Groundwater, 43*, 178–186.

Eide, D., Broderius, M., Fett, J., & Guerinot, M. L. (1996). A novel iron-regulated metal transporter from plants identified by functional expression in yeast. *Proceedings of the National Academy of Sciences of the United States of America, 93*, 5624–5628.

Electric Power Research Institute (EPRI). (1992). *Uptake, translocation, and accumulation of polycyclic aromatic hydrocarbons in vegetation* (TR-101651, various pagination).

Ellis, A. J. (1917). *The divining rod: A history of water witching* (U.S. Geological Survey Water-Supply Paper 416). Washington, DC: U.S. Government Printing Office.

Ellstrand, N. C., Prentice, H. C., & Hancock, J. F. (1999). Gene flow and introgression from domesticated plants into their wild relatives. *Annual Review of Ecological Systems, 30*, 539–563.

Ernst, W. H. O. (2004). Sulfur metabolism in higher plants: Potential for phytoremediation. *Biodegradation, 9*, 311–318.

Fairchild, D. (1938). *The world was my garden; travels of a plant explorer* (494 p). New York: Charles Scribner's Sons.

Farquhar, G. D., Cernusak, L. A., & Barnes, B. (2007). Heavy water fractionation during transpiration. *Plant Physiology, 143*, 11–18.

Fass, T., Cook, P. G., Stieglitz, T., & Herczeg, A. L. (2007). Development of saline groundwater through transpiration of sea water. *Groundwater, 45*, 703–710.

Federle, T. W., & Ventullo, R. M. (1990). Mineralization of surfactants by the microbiota of submerged plant detritus. *Applied and Environmental Microbiology, 56*, 333–339.

Fenner, P., Brady, W. W., & Patton, D. R. (1985). Effects of regulated water flows on regeneration of Fremont cottonwood. *Journal of Range Management, 38*, 135–138.

Fernandez, M. D., Pieters, A., Donoso, C., Herrera, C., Tezara, W., Rengifo, E., & Herrera, A. (1999). Seasonal changes in photosynthesis of trees in the flooded forest of the Mapire River. *Tree Physiology, 19*, 79–85.

Ferrieri, A. P., Thorpe, M. R., & Ferrieri, R. A. (2006). Stimulating natural defenses in poplar clones (OP–367) increases plant metabolism of carbon tetrachloride. *International Journal of Phytoremediation, 8*, 233–243.

Ferris, J. G. (1949). Groundwater. In C. O. Wisler & E. F. Brater (Eds.), *Hydrology* (pp. 198–273). New York: Wiley.

Ferro, A. M., Chard, B., Gefell, M., Thompson, B., & Kjelgren, R. (2000). Phytoremediation of organic solvents in groundwater: Pilot study at a Superfund site. In G. B. Wickramanayake, A. R. Gavaskar, B. C. Alleman, & V. S. Maga (Eds.), *Bioremediation and phytoremediation of chlorinated and recalcitrant compounds* (Vol. C2–4, pp. 461–466). Columbus: Battelle Press.

Ferro, A. M., Chard, J., Kjelgren, R., Chard, B., Turner, D., & Montague, T. (2001). Groundwater capture using hybrid poplar trees: Evaluation of a system in Odgen, Utah. *International Journal of Phytoremediation, 3*, 87–104.

Ferro, A. M., Gefell, M., Kjelgren, R., Lipson, D. S., Zollinger, N., & Jackson, S. (2003). Maintaining hydraulic control using deep rooted tree systems. *Advances in Biochemical Engineering/Biotechnology, 78*, 125–156.

Ferro, A. M., Kennedy, J., Zollinger, N., & Thompson, B. (2005). *Groundwater phytoremediation system-performance at the SRSNE Superfund site*. Presentation at the Third International Phytotechnologies Conference, Atlanta, GA, April 20–22, 2005. Accessed at January 1, 2006, clu-in.org.

Fetter, C. W. (1988). *Applied Hydrogeology* (592 p). Columbus: Merrill Publishing Company.

Fleming, G. R., & Niyogi, K. K. (2005). Zeaxantin. *Science*, 307, 433.

Flerchinger, G. N. (1991). Sensitivity of soil freezing simulated by the SHAW model. *Transactions of the American Society of Agricultural Engineers, 34*, 2381–2389.

Fletcher, J. S. (1991). Keynote speech: A brief overview of plant toxicity testing. In: J. W. Gorsuch, W. R. Lower, W. Wang, & M. A. Lewis (Eds.), *Plants for toxicity assessment: Second volume* (ASTM STP 1115, pp. 5–11). Philadelphia: American Society for Testing and Materials.

Fletcher, J. S., & Hedge, R. S. (1995). Release of phenols by perennial plant roots and their potential importance in bioremediation. *Chemosphere, 31*, 3009–3016.

Fletcher, J. S., MacFarlane, J. C., Pfleeger, T., & Wickliff, C. (1990). Influence of root exposure concentration on the rate of nitrobenzene in soybean. *Chemosphere, 2*, 513–523.

Flowers, T. J., Troke, P. F., & Yeo, A. R. (1977). The mechanism of salt tolerance in halophytes. *Annual Review of Plant Physiology, 28*, 89–121.

Fogel, R. (1983). Root turnover and productivity of coniferous forests. *Plant Soil, 71*, 75–86.

Forchheimer, P. (1930). Groundwasserbenequxg in hydraulik. B.G. Turner, Leipzing.

Ford, B. J. (1985). *Single lens—The story of the simple microscope* (182 p). New York: Harper & Row Publishers.

Forget, E., & Zayed, J. (1995). Tree-ring analysis for monitoring pollution by metals. In T. E. Lewis (Ed.), *Tree rings as indicators of ecosystem health* (pp. 157–176). Boca Raton: CRC Press.

Fox, J. E., Starcevic, M., Kow, K. Y., Burow, M. E., & McLachlan, J. A. (2001). Nitrogen fixation: Endocrine disrupters and flavonoid signaling. *Nature, 413*, 128–129.

Franke, O. L., Reilly, T. E., Haefner, R. J., & Simmons, D. L. (1990). *Study guide for a beginning course in groundwater hydrology: Part I—course participants* (U.S. Geological Survey Open–File Report 90-183, 184 p).

Freely, J. (2004). *The western shores of Turkey: Discovering the Aegean and Mediterranean coasts* (148 p). London: Tauris Parke Paperbacks.

Freeze, R. A. (1994). Henry Darcy and the fountains of Dijon. *Groundwater, 32*, 23–30.

Freeze, R. A., & Cherry, J. A. (1979). *Groundwater* (604 p). Englewood Cliffs: Prentice-Hall.

Fritschen, L. J., Cox, L., & Kinerson, R. (1973). A 28-meter Douglas-fir in a weighing lysimeter. *Forestry Science, 9*, 256–261.

Fuller, M. L. (1906). *Undergroundwater papers, 1906* (U.S. Geological Survey Water-Supply Paper 160, 104 p).Washington, DC: US Government Printing Office.

Gage, E. A., & Cooper, D. J. (2004). Constraints on willow seedling survival in a Rocky Mountain montane floodplain. *Wetlands, 24*, 908–911.

Gager, C. S. (1934). *The plant world: Plant life of our earth* (136 p). New York: The University Society, Inc.

Galili, E., & Nir, Y. (1993). The submerged Pre-Pottery Neolithic water well of Atlit-Yam, northern Israel, and its paleoenvironmental implications. *The Holocene, 3*, 265–270.

Garcia, C. A., Andraski, B. J., Stonestrom, D. A., Cooper, C. A., Johnson, M. J., Michel, R. L., & Wheatcraft, S. W. (2009). Transport of tritium contamination to the atmosphere in an arid environment. *Vadose Zone Journal, 8*(2), 450–461.

Gardner, W. R. (1960). Dynamic aspects of water availability to plants. *Soil Science, 89*, 63–73.

Garner, R. J. (2003). *The Grafter's handbook* (p. 323). London: Cassell Illustrated.

Gatewood, J. S., Robinson, T. W., Colby, B. R., Hem, J. D., & Halpenny, L. C. (1950). *Use of water by bottom-land vegetation in lower Safford Valley Arizona* (U.S. Geological Survey Water-Supply Paper 1103, 210 p). Washington, DC: US Government Printing Office.

Gatliff, E. G. (1994, Summer). Vegetative remediation process offers advantages over traditional pump-and-treat technologies. *Remediation, 4*, 343–352.

Gay, L. W., & Fritschen, L. J. (1979). An energy budget analysis of water use by saltcedar. *Water Resources Research, 6*, 1589–1592.

Gazal, R. M., Scott, R. L., Goodrich, D. C., & Williams, D. G. (2006). Controls on transpiration in a semiarid riparian cottonwood forest. *Agricultural and Forest Meteorology, 137*, 56–67.

George, R. J., Nulsen, R. A., Ferdowsian, R., & Raper, G. P. (1999). Interactions between trees and groundwaters in recharge and discharge areas—A survey of Western Australian sites. *Agricultural Water Management, 39*, 91–113.

Gijsman, A. J., Floris, J., Noordwijk, M. V., & Brouwer, G. (1991). An inflatable minirhizotron system for root observations with improved soil/tube contact. *Plant and Soil, 134*, 261–269.

Gill, S., Jefferson, D. K., Reeser, R. M., & Raupp, M. J. (1999). Use of soil and trunk injection of systemic insecticides to control lace bug on hawthorn. *Journal of Arborculture, 25*, 38–41.

Ginzburg, C. (1967). Organization of the adventitious root apex in *Tamarix aphylla*. *American Journal of Botany, 54*, 4–8.

Glass, D. J. (1999). *U.S. and international markets for phytoremediation, 1999–2000*. Needham: D. Glass Associates, Inc.

Gleason, P. J., & Stone, P. A. (1994). Age, origin, and landscape evolution of the Everglades peatland. In S. M. Davis & J. C. Ogden (Eds.), *Everglades: The ecosystem and its restoration* (pp. 149–197). Delray Beach: St Lucie Press.

Gleba, D., Borisjuk, N. V., Borisjuk, L. G., Kneer, R., Poulev, A., Skarzhinskaya, M., Dushenkov, S., Logendra, S., Gleba, Y. Y., & Raskin, I. (1999). Use of plant roots for phytoremediation and molecular farming. *Proceedings of the National Academy of Science of the United States of America, 96*, 5973–5977.

Godsy, E. M., Warren, E., & Paganelli, V. V. (2003). The role of microbial reductive dechlorination of TCE at a phytoremediation site. *International Journal of Phytoremediation, 5*, 73–87.

Gomez-Hermosillo, C., Pardue, J. H., & Reible, D. D. (2006). Wetland plant uptake of desorption-resistant organic compounds from sediments. *Environmental Science & Technology, 40*, 3229–3236.

Gonthier, G. J. (2007). *A graphical method for estimation of barometric efficiency from continuous data—Concepts and application to a site in the Piedmont, Air Force Plant 6, Marietta, Georgia* (U.S. Geological Survey Scientific Investigations Report 2007-5111, 29 p).

Gopalakrishnan, G., Negri, M. C., Minsker, B. S., & Werth, C. J. (2007). Monitoring subsurface contamination using tree branches. *Groundwater Monitoring & Remediation, 27*, 65–74.

Gopalakrishnan, G., Burken, J. G., & Werth, C. J. (2009). Lignin and lipid impact on sorption and diffusion of trichloroethylene in tree branches for determining contaminant fate during plant sampling and phytoremediation. *Environmental Science & Technology, 43*, 5732–5738.

Gordon, M., Choe, N., Duffy, J., Ekuan, G., Heilman, P., Muiznieks, I., Newman, L., Raszaj, M., Shurtleff, B., Strand, S., & Wilmoth, J. (1997). *Phytoremediation of trichloroethylene with hybrid poplars: In phytoremediation of soil and water contaminants*. Washington, DC: American Chemical Society.

Gorelick, S. M., Freeze, R. A., Donohue, D., & Keely, J. F. (1993). *Groundwater contamination: Optimal capture and containment*. Boca Raton: Lewis Publishers.

Gough, L. P., Shacklette, H. T., & Case, A. A. (1979). *Element concentrations toxic to plants, animals, and man* (U.S. Geological Survey Bulletin 1466, 80 p). Washington, DC: U.S. Government Printing Office.

Graber, E. R., Sorek, A., Tsechansky, L., & Atzmon, N. (2007). Competitive uptake of trichloroethylene and 1,1,1-trichloroethane by *Eucalyptus camaldulensis* seedlings and wood. *Environmental Science & Technology, 41*, 6704–6710.

Graham, R. D., & Stangoulis, J. C. R. (2003). Trace element uptake and distribution in plants. *The Journal of Nutrition, 133*, supplement to 11th Meeting of the Trace Elements in Man and Animals (TEMA), Berkeley, CA, June 2–6, 2002, 1502S–1505S.

Gray, M. W. (1993). Origin and evolution of organelle genomes. *Current Opinions in Genetic Development, 3*, 884–890.

Gray, W. G., & Miller, C. T. (2004). Examination of Darcy's Law for flow in porous media with variable porosity. *Environmental Science & Technology, 38*, 5895–5901.

Green, C., & Hoffnagle, A. (2004). *Phytoremediation field studies database for chlorinated solvents, pesticides, explosives, and metals*. Washington, DC: U.S. Environmental Protection Agency Office of Superfund Remediation and Technology Innovation.

Green, S. R., Kirkham, M. B., & Clothier, B. E. (2006). Root uptake and transpiration-From measurements and models to sustainable irrigation. *Agricultural Water Management, 86*, 165–176.

Greenwood, W. J., Kruse, S., & Swarzenski, P. (2006). Extending electromagnetic methods to map coastal pore water salinities. *Groundwater, 44*, 292–299.

Gregory, S. T., Shea, D., & Guthrie-Nichols, E. (2005). Impact of vegetation on sedimentary organic matter composition and polycyclic aromatic hydrocarbon attenuation. *Environmental Science & Technology, 39*, 5285–5292.

Gribovszki, A., Szilágyi, J., & Kalicz, P. (2010). Diurnal fluctuations in shallow groundwater levels and streamflow rates and their interpretation—A review. *Journal of Hydrology, 385*, 371–383.

Grissino-Mayer, H. D. (2003). A manual and tutorial for the proper use of an increment borer. *Tree-Ring Research, 59*(2), 63–79.

Grismer, M. E., & Gates, T. K. (1988). Estimating saline water table contribution to crop water use. *California Agriculture, 42*, 23–24.

Groom, C. A., Halasz, A., Paquet, L., Morris, N., Oliver, L., Dubois, C., & Hawari, J. (2002). Accumulation of HMX (Octahydro-1,3,5,7-tetranitro-1,3,5,7-tetrazocine) in indigenous and agricultural plants grown in HMX-contaminated anti-tank firing range soil. *Environmental Science & Technology, 36*, 112–118.

Grosse, W. (1989). Thermoosmotic air transport in aquatic plants affecting growth activities and oxygen diffusion to wetland soils: In: D. Hammer (Ed.), *Constructed wetlands for wastewater treatment-municipal, industrial, and agricultural* (p. 469). Lewis Publishers, Chelsea.

Guelke, M., & Von Blanckenburg, F. (2007). Fractionation of stable iron isotopes in higher plants. *Environmental Science & Technology, 41*, 1896–1901.

Guerinot, M. L., & Salt, D. E. (2001). Fortified foods and phytoremediation: Two sides of the same coin. *Plant Physiology, 125*, 164–167.

Guerinot, M. L., & Ying, Y. (1994). Iron: Nutritious, noxious, and not readily available. *Plant Physiology, 104*, 815–820.

Guillen, M. (1995). *Five equations that changed the world: The power and poetry of mathematics*. New York: Hyperion.

Günther, T., Dornberger, U., & Fritsche, W. (1996). Effects of ryegrass on biodegradation of hydrocarbons in soil. *Chemosphere, 33*, 203–215.

Guswa, A. J. (2010). Effect of plant uptake strategy on the water-optimal root depth. *Water Resources Research, 46*, 1–5. Published online September 2, 2010.

Guthrie, E. A., Bortiatynski, J. M., van Heemst, J. D. H., Richman, J. E., Hardy, K. S., Kovach, E. M., & Hatcher, P. G. (1999). Determination of [^{13}C]pyrene sequestration in sediment microcosms using flash pyrolysis-GC-MS and ^{13}C NMR. *Environmental Science & Technology, 33*, 119–125.

Haitjema, H. M., & Mitchell-Bruker, S. (2005). Are water tables a subdued replica of the topography? *Groundwater, 43*, 781–786.

Halford, K. J. (1998). *Assessment of the potential effects of phytoremediation on groundwater flow around Area C at Orlando Naval Training Center, Florida* (U.S. Geological Survey Water-Resources Investigations Report 98-4110, 25 p).

Hales, S. (1969). *Vegetable Staticks (reprint)* (216 p). New York: American Elsevier, History of Science Library.

Halley, E. (1687). An estimate of the quantity of vapor raised out of the sea by warmth of the sun. *Philosophical Transactions of the Royal Society, 16*, 366–370.

Halliwell, B. (2006). Reactive species and antioxidants. Redox biology is a fundamental theme of aerobic life. *Plant Physiology, 141*, 312–322.

Hamm, L. L., Aleman, S. E., & Shadday, M. A. (1997a). *TCE contaminant transport modeling in the TNX area (U)*. Aiken: Westinghouse Savannah River Company Report WSRC-TR-97-0111, 139 p.

Hamm, L. L., Aleman, S. E., Flach, G. P., & Jones, W. F. (1997b). *FACT (Version 1.0) Subsurface flow and contaminant transport, documentation and user's guide* (Aiken: Savannah River Technology Center. Technical Report WSRC-TR-95-0223, 110 p).

Hansch, C., & Leo, A. (1979). *Substituent constant for correlation analysis in chemistry and biology* (339 p). New York: Wiley-Interscience.

Hanson, R. L., & Dawdy, D. R. (1976). *Accuracy of evapotranspiration rates determined by the water-budget method, Gila River flood plain, Southeastern Arizona* (U.S. Geological Survey Professional Paper 655-L, 35 p). Washington, DC: US Government Printing Office.

Harbaugh, A. W. (2005). *MODFLOW-2005 The U.S. Geological Survey modular ground-water model—the ground-water flow process* (U.S. Geological Survey Techniques and Methods 6-A16, variously paged).

Hardy, R. W., & Havelka, U. D. (1975). Nitrogen fixation research: A key to world food? *Science, 188*, 633–642.

Hari, P., Nygren, P., & Korpilahti, E. (1991). Internal circulation of carbon within a tree. *Canadian Journal of Forestry Research, 21*, 514–515.

Harr, J. (1995). *Civil action*. New York: Random House, Inc.

Harr, R. D., & Price, K. R. (1972). Evapotranspiration from a greasewood-cheatgrass community. *Water Resources Research, 8*, 1199–1203.

Harrass, M. C., Eirkson, C. E. III, & Nowell, L. H. (1991). Role of plant bioassays in FDA review: Scenario for terrestrial exposure. In J. W. Gorsuch, W. R. Lower, W. Wang, & M. A. Lewis (Eds.), *Plants for toxicity assessment: Second volume* (ASTM STP 1115, pp.12–28). Philadelphia: American Society for Testing and Materials.

Harris, W. F., Kinerson, R. S., Jr., & Edwards, N. T. (1977). *Comparison of below-ground biomass of natural deciduous forests and loblolly pine plantations* (Range Science Department Science Series). Fort Collins: Colorado State University.

Harvard, V. (1884). The mesquite. *The American Naturalist, 18*, 450–459.

Harvey, F. E., Swinehart, J. B., & Kurtz, T. M. (2007). Ground water sustenance of Nebraska's unique sand hills peatland fen ecosystems. *Ground Water, 45*(2), 218–234.

Healy, R. W., Winter, T. C., LaBaugh, J. W., & Franke, O. L. (2007). *Water budgets: Foundations for effective water-resources and environmental management* (U.S. Geological Survey Circular 1308, 90 p).

Heath, R. C. (1983). *Basic ground-water hydrology* (U.S. Geological Survey Water-Supply Paper 2220, 84 p). Washington, DC: US Government Printing Office.

Helfman, E. S. (1972). *Maypoles and wood demons—The meaning of trees*. New York: The Seabury Press.

Helley, E. J., & LaMarche Jr. V. (1973). *Historic flood information for northern California streams from geologic and botanical evidence* (U.S. Geological Survey Professional Paper 485-E, 16 p). Washington, DC: US Government Printing Office.

Hem, J. D. (1967). *Composition of saline residues on leaves and stems of saltcedar (Tamarix pentandra Pallas)* (U.S. Geological Survey Professional Paper 491-C, 9 p). Washington, DC: US Government Printing Office.

Hesslein, R. H. (1976). An in situ sampler for close interval pore water studies. *Limnology and Oceanography, 21*, 912–914.

Heuperman, A. (1999). Hydraulic gradient reversal by trees in shallow water table areas and repercussions for the sustainability of tree-growing systems. *Agricultural Water Management, 39*, 153–167.

Hewlett, J. D. (1982). *Principals of forest hydrology* (183 p). Athens: University of Georgia Press.

Heyward, F. (1933). The root system of longleaf pine on the deep sands of western Florida. *Ecology, 14*, 136–148.

Hillel, D. (1998). *Environmental soil physics* (771 p). San Diego: Academic.

Hinckley, T. M., & Bruckerhoff, D. F. (1975). The effects of drought on water relations and stem shrinkage of *Quercus alba*. *Canadian Journal of Botany, 53*, 62–72.

Hinckley, T. M., Brooks, J. R., Cermak, J., Ceulemans, R., Kucera, J., Meinzer, F. C., & Roberts, D. A. (1994). Water flux in a hybrid poplar stand. *Tree Physiology, 14*, 1005–1018.

Hiratsuka, Y. (1987). *Forest tree diseases of the prairie provinces*. Edmonton: Northern Forestry Centre.

Hirsh, S. R., Compton, H. R., Matey, D. H., Wrobel, J. G., & Schneider, W. H. (2003). Five-year pilot study: Aberdeen Proving Ground, Maryland. In S. C. McCutcheon & J. L. Schnoor (Eds.), *Phytoremediation:Transformation and control of contaminants* (pp. 635–659). Hoboken: Wiley.

Hitchcock, E. (1840). *Elementary geology* (416 p). New York: Ivison, Phinney, Blakeman & Co.

Hobhouse, P. (2004). *Plants in garden history* (336 p). London: Pavillion Books.

Hogg, E. H., & Hurdle, P. A. (1997). Sap flow in trembling aspen: Implications for stomatal responses to vapor pressure deficit. *Tree Physiology, 17*, 501–509.

Holzer, T. L. (2010). The water table. *Ground Water, 48*, 171–173.

Homer (1998). *The Iliad* Book 21, Penguin Classics, 704 p.

Hong, M. S., Farmayan, W. F., Dortch, I. J., Chiang, C. Y., McMillan, S. K., & Schnoor, J. L. (2001). Phytoremediation of MTBE from a groundwater plume. *Environmental Science & Technology, 35*, 1231–1239.

Hook, D. D., Brown, C. L., & Kormanik, P. P. (1970). Lenticel and water root development of swamp tupelo under various flooding conditions. *Botanical Gazette, 131*, 217–224.

Hooke, R. (1665). Micrographia, Octavo, CD-ROM edition, 1998, 292 p.

Hopmans, J. W., & Van Immerzeel, C. H. (1988). Variation in evapotranspiration and capillary rise with changing soil profile characteristics. *Agricultural Water Management, 13*, 297–305.

Horne, A. J. (2000). Phytoremediation by constructed wetlands. In N. Terry & G. Banuelos (Eds.), *Phytoremediation of contaminated soil and water* (pp. 13–39). Boca Raton: Lewis Publishers.

Horton, J. L., Kolb, T. E., & Hart, S. C. (2001). Physiological response to groundwater depth varies among species and with river flow regulation. *Ecological Applications, 11*, 1046–1059.

Hruska, J., Cermák, J., & Sustek, S. (1999). Mapping tree root systems with ground-penetrating radar. *Tree Physiology, 19*, 125–130.

Hsu, T. S., & Bartha, R. (1979). Accelerated mineralization of two organophosphate insecticides in the rhizosphere. *Applied Environmental Microbiology, 37*, 36–41.

Hubbard, E. (1916). *Little journeys to the homes of the great*. New York: Wm. H. Wise Co.

Huber, B. (1932). Beobachtung und Messung pflanzlicher Saftstrome. *Berichte der Deutschen Botanischen Gesellschaft, 50*, 89–109.

Hughes, W. B. (1995). *Groundwater flow and the possible effects of remedial actions at J-Field, Aberdeen Proving Ground, Maryland* (U.S. Geological Survey Water-Resources Investigations Report 95-4075, 39 p).

Hughes, J. B., Shanks, J., Vanderford, M., Lauritzen, J., & Bhadra, R. (1996). Transformation of TNT by aquatic plants and plant tissue cultures. *Environmental Science & Technology, 31*, 266–271.

Hultline, K. R., Williams, D. G., Burgess, S. S. O., & Keefer, T. O. (2003). Contrasting patterns of hydraulic redistribution in three desert phreatophytes. *Oecological, 135*, 167–175.

Hultline, K. R., Koepke, D. F., Pockman, W. T., Fravolini, A., Sperry, J. S., & Williams, D. G. (2006). Influence of soil texture on hydraulic properties and water relations of a dominant warm-desert phreatophyte. *Tree Physiology, 26*, 313–323.

Hunt, E. R., Rock, B. N., & Nobel, P. S. (1987). Measurement of leaf relative water content by infrared reflectance. *Remote Sensing of Environment, 22*, 4210–4435.

Interstate Technology Regulatory Council (ITRC). (1999). *Decision tree—Phytoremediation* (multiple pagination). Washington, DC: Interstate Technology & Regulatory Council.

Interstate Technology Regulatory Council (ITRC). (2001). Phytoremediation technical and regulatory guidance document (multiple pagination). Washington, DC: Interstate Technology & Regulatory Council.

Interstate Technology Regulatory Council (ITRC). (2003). Technical and regulatory guidance for design, installation, and monitoring of alternative final landfill covers (multiple pagination). Washington, DC: Interstate Technology & Regulatory Council.

Interstate Technology Regulatory Council (ITRC). (2009). Phytotechnology technical and regulatory guidance and decision trees, revised. Washington, DC: Interstate Technology & Regulatory Council.

Iritz, Z., & Lindroth, A. (1994). Night-time evaporation from a short-rotation willow stand. *Journal of Hydrology, 157*, 235–245.

Issar, A. S. (1990). *Water shall flow from the rock: Hydrogeology and climate in the lands of the Bible*. Berlin/Heidelberg: Springer.

Issar, A. S., Nativ, R., Karnieli, K, & Gat, J. R. (1984). *Isotopic evidence of the origin of groundwater in arid zones*. Vienna: IAEA.

Itoh, S., & Barber, S. A. (1983). Phosphorous uptake by six plant species as related to root hairs. *Agronomy Journal, 75*, 457–461.

Jackson, R. B., Moore, L. A., Hoffman, W. A., Pockman, W. T., & Linder, C. R. (1999). Ecosystem rooting depth determined with caves and DNA. *Proceedings of the National Academy of Sciences of the United States of America, 96*, 11387–11392.

Jackson, W. A., Martino, L., Hirsch, S., Wrobel, J., & Pardue, J. H. (2005). Application of a dialysis sampler to monitor phytoremediation processes. *Environmental Monitoring and Assessment, 107*, 155–171.

James, C. A., Xin, G., Doty, S. L., & Strand, S. E. (2007). Degradation of low molecular weight volatile organic compounds by plants genetically modified with mammalial cyctochrome P450 2E1. *Environmental Science & Technology,* 42(1), 289-293.

James, C. A., Xin, G., Doty, S. L., Muiznieks, I., Newman, L., & Strand, S. E. (2009). A mass balance study of the phytoremediation of perchloroethylene-contaminated groundwater. *Environmental Pollution, 157*, 2564–2569.

Jenson, J. R. (2000). *Remote sensing of the environment: An earth resource perspective:* (544 p). Upper Saddle River: Prentice-Hall, Inc.

Johnston, C. D. (1987). Preferred water flow and localised (*sic*) recharge in a variable regolith. *Journal of Hydrology, 94*, 129–142.

Jones, H. E., & Etherington, J. R. (1970). Comparative studies of plant growth and distribution in relation to waterlogging: The survival of *Erica cinerea* and *E. tetralix* L. and its apparent relationship to iron and manganese uptake in waterlogged soil. *Journal of Ecology, 58*, 487–496.

Jordahl, J. L., Foster, L., Schnoor, J. L., & Alvarez, P. J. J. (1997). Effect of hybrid poplar trees on microbial populations important to hazardous waste bioremediation. *Environmental Toxicology and Chemistry, 16*, 1318–1321.

Jordan, D. G., & Fisher, D. W. (1977). *Relation of bulk precipitation and evapotranspiration to water quality and water resources, St. Thomas, Virgin Islands* (U.S. Geological Survey Water-Supply Paper 1662-I, 30 p). Reston: US Geological Survey.

Kalisz, P. J., Stringer, J. W., Volpe, J. A., & Clark, D. T. (1988). Trees as monitors of tritium in soil water. *Journal of Environmental Quality, 17*, 62–70.

Kamath, R., Schnoor, J. L., & Alvarez, P. J. J. (2004). Effect of root-derived substrates on the expression of nah-lux genes in *Pseudomas fluorescens* HK44: Implications for PAH biodegradation in the rhizosphere. *Environmental Science & Technology, 38*, 1740–1745.

Karthikeyan, R., & Kulakow, P. A. (2003). Soil plant microbe interactions in phytoremediation. *Advances in Biochemical Engineering/Biotechnology, 78*, 1–50.

Karthikeyan, R., Mankin, K. R., Davis, L. C., & Erickson, L. E. (2003a). Technical note: Fate and transport of jet fuel (JP-8) in soils with selected plants. *International Journal of Phytoremediation, 5*, 281–292.

Karthikeyan, R., Mankin, K. R., Davis, L. C., & Erickson, L. E. (2003b). Modeling jet fuel (JP-8) fate and transport in soils with plants. *International Journal of Phytoremediation, 5*, 293–314.

Kassel, A. G., Ghoshal, D., & Goyal, A. (2002). Phytoremediation of trichloroethylene using hybrid poplar. *Physiology and Molecular Biology of Plants, 8*, 1–8.

Kawai, S., & Alam, S. (2006). Iron stress response and composition of xylem sap of strategy II plants. In L. L. Barton & J. Abadía (Eds.), *Iron nutrition in plants and rhizospheric microorganisms* (pp. 289–309). The Netherlands: Springer.

Keeley, J. E., Osmond, C. B., & Raven, J. A. (1984). *Stylites*, a vascular land plant without stomata absorbs CO_2 via its roots. *Nature, 310*, 694–695.

Kim, S. A., & Guerinot, M. L. (2007). Mining iron: Iron uptake and transport in plants. *FEBS Letters, 581*, 2273–2280.

Klijn, F., & Witte, J. P. M. (1999). Eco-hydrology: Groundwater flow and site factors in plant ecology. *Hydrogeology Journal, 7*, 65–77.

Kline, J. R., Reed, K. L., Waring, R. H., & Stewart, M. L. (1976). Field measurement of transpiration in Douglas Fir. *Journal of Applied Ecology, 13*, 273–283.

Kluitenberg, G. J., Butler, J. J. Jr., & Whittenmore, D. O. (2005). *A field investigation of major controls on phreatophyte-induced fluctuations in the water table*. ASA-CSSA-SSSA International Annual Meeting, November 6–10, 2005, Salt Lake City, UT.

Knaebel, D. B., & Vestal, J. R. (1994). Intact rhizosphere microbial communities used to study microbial biodegradation in agricultural and natural soils. In T. A. Anderson & J. R. Coats (Eds.), *Bioremediation through rhizosphere technology* (Chap. 5, ACS Symposium Series 563, pp. 56–69). Washington, DC: American Chemical Society.

Kobayashi, D., Kodama, Y., Ishii, Y., Tanaka, Y., & Suzuki, K. (1995). Diurnal variations in streamflow and water quality during the summer dry season. *Hydrological Processes, 9*, 833–841.

Koch, G. W., Sillett, S. C., Jennings, G. M., & Davis, S. D. (2004). The limit to tree height. *Nature, 428*, 851–854.

Kolb, K. J., & Davis, S. D. (1994). Drought tolerance and xylem embolism in co-occurring species of coastal sage and chaparral. *Ecology, 75*, 648–659.

Kollin, C. (2006). How green infrastructure measures up to structural stormwater services; quantifying the contributions of trees and vegetation. *Stormwater, 7*, 138–144.

Kolpin, D. W., Furlong, E. T., Meyer, M. T., Thurman, E. M., Zaugg, S. D., Barber, L. B., & Buxton, H. T. (2002). Pharmaceuticals, hormones, and other organic wastewater contaminants in U.S. streams, 1999–2000: A national reconnaissance. *Environmental Science & Technology, 36*, 1202–1211.

Komives, T., Gullner, G., Rennenberg, H., & Casida, J. E. (2003). Ability of poplar (*Populus* spp.) to detoxify chloroacetanilide herbicides. *Water, Air, and Soil Pollution: Focus, 3*, 277–283.

Kooistra, M. J., Schoonderbeek, D., Boone, F. R., & Veen, B. W. (1992). Root-soil contact of maize, as measured by a thin-section technique. II. Effects of soil compaction. *Plant Soil, 139*, 119–129.

Kormondy, E. J. (1976). *Concepts of ecology* (238 p). Englewood Cliffs: Prentice-Hall, Inc.

Korte, F., Kvesitadze, G., Ugrekhelidze, D., Gordeziani, M., Khatisashvili, G., Buadze, O., Zaalishvili, G., & Coulston, F.

(2000). Review: Organic toxicants and plants. *Ecotoxicology and Environmental Safety, 47*, 1–26.

Kovalick, W. W. Jr. (2005). *Phytotechnology: Current trends and prospects*. Presentation at the Third International Phytotechnologies Conference, Atlanta, GA, April 20–22, 2005. Accessed at January 1, 2006, clu-in.org.

Kozlowski, T. T. (1997). Responses of woody plants to flooding and salinity. *Tree Physiology Monograph, 1*, 1–29.

Kozlowski, T. T., & Pallardy, S. G. (1997). *Physiology of woody plants* (411 p). San Diego: Academic.

Kramer, P. J. (1983). *Water relations of plants*. New York: Academic.

Kramer, P. J., & Boyer, J. S. (1995). *Water relations of plants and soils* (495 p). San Diego: Academic.

Kramer, P. J., & Bullock, H. C. (1966). Seasonal variations in the proportions of suberized and unsuberized roots of trees in relation to the absorption of water. *American Journal of Botany, 53*, 200–204.

Kramer, P. J., & Kozlowski, T. T. (1960). *Physiology of trees*. New York: McGraw-Hill Book Company, Inc.

Kramer, P. J., Riley, W. S., & Bannister, T. T. (1952). Gas exchange of cypress knees. *Ecology, 33*, 117–121.

Krauter, P. W. (2001). Using a wetland bioreactor to remediate groundwater contaminated with nitrate (mg/L) and perchlorate (ug/L). *International Journal of Phytoremediation, 3*, 415–433.

Kubitzki, K. (1989). The ecogeographical differentiation of Amazonian inundation forests. *Plant Systematic Evolution, 162*, 285–304.

Kucera, C. L. (1954). Some relationships of evaporation rate to vapor pressure deficit and low wind velocity. *Ecology, 35*, 71–75.

Kvesitadze, G., Khatisashvili, G., Sadunishvili, T., & Ramsden, J. J. (2006). *Biochemical mechanisms of detoxification in higher plants*. Heidelberg: Springer.

Lahvis, M. A., Baehr, A. L., & Baker, R. J. (1999). Quantification of aerobic biodegradation and volatilization rates of gasoline hydrocarbons near the water table under natural attenuation conditions. *Water Resources Research, 35*, 753–765.

Lamarck, J. B. (1802). *Hydrogeologie* (Paris, 152 p). Urbana: University of Illinois Press

Landmeyer, J. E. (1994). *Description and application of capture zone delineation for a wellfield at Hilton Head Island, South Carolina* (U.S. Geological Survey Water-Resources Investigations Report 94-4012, 33 p).

Landmeyer, J. E. (1996). Aquifer response to record low barometric pressures in the southeastern United States. *Groundwater, 34*, 917–924.

Landmeyer, J. E. (2001). Monitoring the effect of poplar trees on petroleum-hydrocarbon and chlorinated solvent contaminated groundwater. *International Journal of Phytoremediation, 3*, 61–85.

Landmeyer, J. E., & Bradley, P. M. (2003). Effect of hydrologic and geochemical conditions on oxygen-based bioremediation of gasoline-contaminated groundwater. *Bioremediation Journal, 7*, 165–177.

Landmeyer, J. E., & Stone, P. A. (1995). Radiocarbon and $\delta^{13}C$ values related to groundwater recharge and mixing. *Groundwater, 33*, 227–234.

Landmeyer, J. E., Vroblesky, D. A., & Chapelle, F. H. (1996a). Stable carbon isotope evidence of biodegradation zonation in a shallow jet-fuel contaminated aquifer. *Environmental Science & Technology, 30*, 1120–1128.

Landmeyer, J. E., Chapelle, F. H., & Bradley, P. M. (1996b). *Assessment of intrinsic bioremediation of gasoline contamination in the shallow aquifer, Laurel Bay Exchange, Marine Corps Air Station Beaufort, South Carolina* (U.S. Geological Survey Water-Resources Investigations Report 96-4026, 50 p).

Landmeyer, J. E., Chapelle, F. H., Petkewich, M. D., & Bradley, P. M. (1996a). Assessment of natural attenuation of aromatic hydrocarbons in groundwater near a former manufactured gas plant, South Carolina, USA. *Environmental Geology, 34*, 279–292.

Landmeyer, J. E., Pankow, J. F., Chapelle, F. H., Bradley, P. M., Church, C. D., & Tratnyek, P. G. (1998b). Fate of MTBE relative to benzene in a gasoline-contaminated aquifer (1993–98). *Groundwater Monitoring and Remediation, 18*, 93–102.

Landmeyer, J. E., Vroblesky, D. A., & Bradley, P. M. (2000). MTBE and BTEX in trees above gasoline-contaminated groundwater. In G. B. Wickramanayake, A. R. Gavaskar, J. T. Gibbs, & J. L. Means (Eds.), *Case studies in the remediation of chlorinated and recalcitrant compounds*. The Second International Conference on Remediation of Chlorinated and Recalcitrant Compounds, Monterey, California, May 22–25, 2000 (pp. 17–24). Columbus: Battelle Press.

Landmeyer, J. E., Chapelle, P. M., Herlong, H. E., & Bradley, P. M. (2001). Methyl tert-butyl ether biodegradation by indigenous aquifer microorganisms under natural and artificial oxic conditions. *Environmental Science & Technology, 35*, 1118–1126.

Landmeyer, J. E., Tanner, T. L., & Watt, B. E. (2004). Biotransformation of tributyltin (TBT) to tin in freshwater river-bed sediments contaminated by an organotin release. *Environmental Science & Technology, 38*, 4106–4112.

Landmeyer, J. E., Bradley, P. M., Trego, D. A., Hale, K. G., & Haas, J. E. I. I. (2010). MTBE, TBA, and TAME attenuation in diverse hyporheic zones. *Ground Water, 48*, 30–41.

Lapalla, E. G., Healy, R. W., & Weeks, E. P. (1983). *Documentation of computer program VS2D to solve equations of fluid flow in variably saturated porous media* (U.S. Geological Survey Water-Resources Investigations Report 83-4099).

Larcher, W. (1983). *Physiological plant ecology*. Berlin: Springer.

Larsen, M., & Trapp, S. (2006). Uptake of iron cyanide complexes into willow trees. *Environmental Science & Technology, 40*, 1956–1961.

Larsen, M., Burken, J., Machackova, J., Karlson, U. G., & Trapp, S. (2008). Using tree core samples to monitor natural attenuation and plume distribution after a PCE spill. *Environmental Science & Technology, 42*(5), 1711–1717. ASAP Article Web Release Date 1/31/08.

Laturnus, F., Haselmann, K. F., Borch, T., & Grøn, C. (2002). Terrestrial natural sources of trichloromethane (chloroform, $CHCl_3$)—An overview. *Biogeochemistry, 60*, 121–139.

Le Maitre, D. C., Scott, D. F., & Colvin, C. (1999). A review of information on interactions between vegetation and groundwater. *Water SA, 25*, 137–152.

Leavitt, M., Thomas, P., Odle, B., & Hall, B. (2001). When pump and treat won't do: Lindane removal from groundwater in Phytoremediation, Wetlands, and Sediments. In A. Leeson, E. A. Foote, M. K. Banks, & V. S. Magar (Eds.), *Proceedings of the 6th International in situ and on-site bioremediation symposium, San Diego, CA, June 4–7, 2001* (Vol. 6, pp.189–198). Columbus: Battelle Press.

Lee, C. H. (1912). *An intensive study of the water resources of a part of Owens Valley, California* (U.S. Geological Survey Water-Supply Paper 294, 135 p). Washington, DC: U.S. Government Printing Office.

Lee, C. H. (1942). Transpiration and total evaporation. In O. E. Meinzer (Ed.), *Hydrology* (pp. 259–330). New York: Dover Publications, Inc.

Lee, D. R. (1977). A device for measuring seepage flux in lakes and estuaries. *Limnology and Oceanography, 22*, 140–147.

Leenhouts, J. M., Stromberg, J. C., & Scott, R. L. (Eds.). (2006). *Hydrologic requirements of and consumptive groundwater use by riparian vegetation along the San Pedro River, Arizona* (U.S. Geological Survey Scientific Investigations Report 2005-5163, 154 p).

Leicester, H. M., & Klickstein, H. S. (1952). *A source book in chemistry 1400–1900*. New York: McGraw-Hill.

Leigh, M. B., Prouzová, P., Macková, M., Macek, T., Nagle, D. P., & Fletcher, J. S. (2006). Polychlorinated biphenyl (PCB)-degrading bacteria associated with trees in a PCB-contaminated site. *Applied and Environmental Microbiology, 72*, 2331–2343.

Leo, A., Hansch, C., & Church, C. (1969). Comparison of parameters currently used in the study of structure-activity relationships. *Journal of Medical Chemistry, 12*, 766–771.

Levy, P. E., Meir, P., Allen, S. J., & Jarvis, P. G. (1999). The effect of aqueous transport of CO_2 in xylem sap on gas exchange in woody plants. *Tree Physiology, 19*, 53–58.

Lewis, M. A., & Wang, W. (1997). Water quality and aquatic plants. In W. Wang, J. W. Gorsuch, & J. S. Hughes (Eds.), *Plants for environmental studies*. Boca Raton: CRC Press.

Ley-Yadun, S., & Weinstein-Evron, M. (1994). Late Epipalaeolitic wood remains from el-Wad Cave, Mount Carmel, Israel. *New Phytology, 127*, 391–396.

Li, H., Sheng, G., Chiou, C. T., & Xu, O. (2005). Relation of organic contaminant equilibrium sorption and kinetic uptake in plants. *Environmental Science & Technology, 39*, 4864–4870.

Liang, J., Zhang, J., Chan, G. Y. S., & Wong, M. H. (1999). Can differences in root response to soil drying and compaction explain differences in performance of trees growing on landfill sites? *Tree Physiology, 19*, 619–624.

Lichtenstein, E. P. (1959). Absorption of some chlorinated hydrocarbon insecticides from soils into various crops. *Journal of Agriculture and Food Chemicals, 7*, 430–433.

Linacre, N. A., Whiting, S. N., Baker, A. J. M., Angle, J. S., & Ades, P. K. (2003). Transgenics and phytoremediation: The need for an integrated risk assessment, management, and communication strategy. *International Journal of Phytoremediation, 5*, 181–185.

Linacre, N. A., Whiting, S. N., & Angle, J. S. (2005). The impact of uncertainty on phytoremediation costs. *International Journal of Phytoremediation, 7*, 259–269.

Lindgren, W. (1903). *The water resources of Molokai, Hawaiian Islands* (U.S. Geological Survey Water-Supply Paper 77, 62 p). Washington, DC: U.S. Government Printing Office.

Lindstrom, F. T., Boersma, L., & Yingjajaval, S. (1990). *CTSPAC: Mathematical model for coupled transport of water, solutes, and heat in the soil-plant-atmosphere continuum. Mathematical theory and transport concepts* (Bulletin 676). Corvallis: Agricultural Experiment Station, Oregon State University.

Lines, G. C., & Bilhorn, T. W. (1996). *Riparian vegetation and its water use during 1995 along the Mojave River, Southern California* (U.S. Geological Survey Water-Resources Investigations Report 96-4241, 10 p).

Linnaeus, C. (1735). *System naturae* (12 p). Stockholm: Laurentii Salvii.

Linnaeus, C. (1753). *Species plantarum*. Stockholm: Laurentii Salvii.

Lite, S. J., & Stromberg, J. C. (2005). Surface water and groundwater thresholds for maintaining *Populus–Salix* forests, San Pedro River, Arizona. *Biological Diversity, 125*, 153–167.

Liu, J., & Schnoor, J. L. (2008). Uptake and translocation of lesser-chlorinated polychlorinated biphenyls (PCBs) in whole hybrid poplar plants after hydroponic exposure. *Chemosphere, 73*, 1608–1616.

Liu, J., Hu, D., Jiang, G., & Schnoor, J. L. (2009). In vivo biotransformation of 3,3',4,4'-tetrachlorobiphenyl by whole plants–poplars and switchgrass. *Environmental Science & Technology, 43*, 7503–7509.

Loeb Classical Library No. 70. (1916). *Enquiry into plants. Volume 1, Books 1–5* (572 p). New York: Loeb Classical Library.

Loeb Classical Library No. 330. (1938). *Natural history. Volume 1, Books 1–37*. New York: Loeb Classical Library.

Loeb Classical Library No. 474. (1990). *On the causes of plants. Volume 11, Books 3–4* (304 p). New York: Loeb Classical Library.

Loheide, S. P., Butler, J. J., & Gorelick, S. M. (2005). Estimation of groundwater consumption by phreatophytes using diurnal water table fluctuations: A saturated-unsaturated flow assessment. *Water Resources Research, 41*, 1–14. W07030R.

Loomis, R. S., & Williams, W. A. (1963). Maximum crop productivity: One estimate. *Crop Science, 3*, 67–72.

Lorenz, D. L., & Delin, G. N. (2007). A regression model to estimate regional ground-water recharge in Minnesota. *Ground Water, 45*(2), 196–208.

Lorah, M. M., & Olsen, L. D. (1999). Degradation of 1,1,2,2-tetrachloroethane in a freshwater tidal wetland: Field and laboratory evidence. *Environmental Science & Technology, 33*, 227–234.

Lowry, C. S., & Loheide, S. P. II (2010). Groundwater-dependent vegetation: Quantifying the groundwater subsidy. *Water Resources Research, 46*, 8 pp. Published online 17 June 2010.

Lubczynski, M. W. (2009). The hydrogeological role of trees in water-limited environments. *Hydrogeology Journal, 17*, 247–259.

Lucas, J. (1880). The hydrogeology of the Lower Greensands of Surrey and Hampshire: London, England. *Proceedings of the Institute of Civil Engineers, 61*, 200–227.

Luther, G. W., III, Kostka, J. E., Church, T. M., Sulzberger, B., & Stumm, W. (1992). Seasonal iron cycling in the salt-marsh sedimentary environment: the importance of ligand complexes with Fe (II) and Fe(III) in the dissolution of Fe(III) minerals and pyrite, respectively. *Marine Chemistry, 40*, 81–103.

Ma, X. M., & Burken, J. G. (2002). VOCs fate and partitioning in vegetation: Use of tree cores in groundwater analysis. *Environmental Science & Technology, 36*, 4663–4668.

Ma, X. M., & Burken, J. G. (2003). TCE diffusion to the atmosphere in phytoremediation applications. *Environmental Science & Technology, 37*, 2534–2539.

Ma, X. M., & Burken, J. G. (2004). Modeling of TCE diffusion to the atmosphere and distribution in plant stems. *Environmental Science & Technology, 38*, 4580–4586.

Ma, X., Richter, A. R., Albers, S., & Burken, J. G. (2004). Phytoremediation of MTBE with hybrid poplar trees. *International Journal of Phytoremediation, 6*, 157–167.

Mackay, A. L. (1991). *A dictionary of scientific quotations* (312 p). New York: Taylor & Francis.

MacFall, J. S. (1994). Effects of ectomycorrhizae on biogeochemistry and soil structure. In F. L. Pfleger & R. Linderman (Eds.), *Reappraisal of mycorrhizae and agriculture* (pp. 213–238). St. Paul: APS Press.

Maddock, T., III. & Baird, K. J. (2003). *A riparian evapotranspiration package for MODFLOW-96 and MODFLOW-2000* (HWR No. 02-03). Department of Hydrology and Water Resources, University of Arizona Research Laboratory for Riparian Studies. Tucson: University of Arizona.

Mahler, B. J., Van Metre, P. C., Bashara, T. J., Wilson, J. T., & Johns, D. A. (2005). Parking lot sealcoat: An unrecognized source of urban polycyclic aromatic hydrocarbons. *Environmental Science & Technology, 39*, 5560–5566.

Margulis, L., & Sagan, D. (2002). *Aquiring genomes, a theory of the origins of species* (240 p). New York: Basic Books.

Marion, W., & Wilcox, S. (1994). *Solar radiation data manual for flat-plate and concentrating collectors* (NREL/TP-463-5607). Golden: National Renewable Energy Laboratory (NREL).

Marr, L. C., Booth, E. C., Andersen, R. G., Widdowson, M. A., & Novak, J. T. (2006). Direct volatilization of naphthalene to the atmosphere at a phytoremediation site. *Environmental Science & Technology, 40*, 5560–5566.

Marsh, G. P. (1885). *The earth as modified by human action. A last revision of man and nature*. New York: Charles Scribner's & Sons.

Marshall, J. D., & Toffel, M. W. (2005). Framing the elusive concept of sustainability: A sustainability hierarchy. *Environmental Science & Technology, 39*, 673–682.

Marschner, H. (1995). *Mineral nutrition of higher plants*. New York: Academic Press.

Martin, R. S., Villanueva, I., Zhang, J., & Popp, C. J. (1999). Nonmethane hydrocarbon, monocarboxylic acid, and low molecular weight aldehyde and ketone emissions from vegetation in central New Mexico. *Environmental Science & Technology, 33*, 2186–2192.

Matthews, D. W., Massmann, J., & Strand, S. E. (2003). Influence of aquifer properties on phytoremediation effectiveness. *Ground Water, 41*, 41–47.

Mattina, M. I., Isleyen, M., Eitser, B. D., Iannucci-Berger, W., & White, J. C. (2006). Uptake by Cucurbitaceae of soil-borne contaminants depends upon plant genotype and pollutant properties. *Environmental Science & Technology, 40*, 1814–1821.

Mayer, P. M., Reynolds, S. K., Jr., Canfield, T. J., & McCutchen, M. D. (2005). Riparian buffer width, vegetative cover, and nitrogen removal effectiveness: A review of current science and regulations (EPA/600/R-05/118, 27 p). Washington, DC: U.S. Environmental Protection Agency.

McClatchy, J. D. (2000). *Henry Wadsworth Longfellow* (854 p). New York: Library of America.

McClenahen, J. R., Vimmerstedt, J. P., & Scherzer, A. J. (1989). Elemental concentrations in tree rings by PIXIE: Statistical variability, mobility, and effects of altered soil chemistry. *Canadian Journal of Forest Research, 19*, 880–888.

McCutcheon, S. C., & Schnoor, J. L. (eds.). (2003). *Phytoremediation—Transformation and control of contaminants* (987 p). Hoboken: Wiley-Interscience, Inc.

McCutcheon, S. C., Medina, V. F., & Larson, S. L. (2003). *Proof of phytoremediation for explosives in water and soil in Phytoremediation – Transformation and control of contaminants* (pp. 429–480). Hoboken: Wiley-Interscience, Inc.

McDermitt, D. K. (1990). Sources of error in the estimation of stomatal conductance and transpiration from porometer data. *HortScience, 25*, 1538–1548.

McDonald, M. G., & Harbaugh, A. W. (1988). A modular three-dimensional finite-difference groundwater flow model. In *Techniques of water resources investigations* (Book 6, Chap. A1). Denver: U.S. Geological Survey.

McDonald, C. C., & Hughes, G. H. (1968). *Studies of the consumptive use of water by phreatophytes and hydrophytes near Yuma, Arizona* (U.S. Geological Survey Professional Paper 486-F, 24 p). Washington, DC: U.S. Government Printing Office.

McElrone, A. J., Pockman, W. T., Martínez-Vilalta, J., & Jackson, R. B. (2004). Variation in xylem structure and function in stems and roots of trees to 20 m depth. *The New Phytologist, 163*, 507–517.

McFarlane, C., Pfleeger, T., & Fletcher, J. (1990). Effect, uptake and disposition of nitrobenzene in several terrestrial plants. *Environmental Toxicology and Chemistry, 9*, 513–520.

McKay, S. E., Kluitenberg, G. J., Butler, J. J., Jr., Zhan, X., Aufman, M. S., & Brauchler, R. (2004). *In-situ determination of specific yield using soil moisture and water level changes in the riparian zone of the Arkansas River, Kansas* (EOS Transactions, AGU Vol. 85). Fall Meeting Supplement, Abstract H31D-0425.

McLinn, E. L., Vondracek, J. E., & Aitchison, E. A. (2001). Monitoring remediation with trembling leaves: Assessing the effectiveness of a full-scale phytoremediation system. In A. Leeson, E. A. Foote, M. K. Banks, & V. S. Magar (Eds.), *Phytoremediation, wetlands, and sediments. Proceedings of the 6th International In Situ and On-Site Bioremediation Symposium,* San Diego, CA, June 4–7, 2001 (Vol. 6, pp. 121–127). Columbus: Battelle Press.

McQueen, I. S., & Miller, R. F. (1972). *Soil-moisture and energy relationships associated with riparian vegetation near San Carlos, Arizona* (U.S. Geological Survey Professional Paper 655-E, 51 p). Washington, DC: U.S. Government Printing Office.

Mead, D. W. (1904). *Notes on hydrology and the application of its laws to the problems of hydraulic engineering*. Chicago: S. Smith & Company.

Mead, D. W. (1919). *Hydrology: The fundamental basis of hydraulic engineering:*. New York: McGraw-Hill Book Company, Inc.

Meinzer, O. E. (1923). *Outline of groundwater hydrology, with definitions* (U.S. Geological Survey Water-Supply Paper 494). Washington, DC: U.S. Government Printing Office.

Meinzer, O. E. (1926). Plants as indicators of groundwater. *Journal of the Washington Academy of Sciences, 16*, 553–564.

Meinzer, O. E. (1927). *Plants as indicators of groundwater* (U.S. Geological Survey Water-Supply Paper 577, 95 p). Washington, DC: U.S. Government Printing Office.

Meinzer, O. E., & Hare, R. F. (1915). *Geology and water resources of the Tularosa Basin, New Mexico* (U.S. Geological Survey Water-Supply Paper 343). Washington, DC: U.S. Government Printing Office.

Meinzer, O. E., & Kelton, F. C. (1913). *Geology and water resources of Sulphur Spring Valley, Arizona* (U.S. Geological Survey Water-Supply Paper 320, pp. 171–187). Washington, DC: U.S. Government Printing Office.

Meinzer, F. C., Fownes, J. H., & Harrington, R. A. (1996). Growth indices and stomatal control of transpiration in Acacia koa stands planted at different densities. *Tree Physiology, 16*, 607–615.

Meyboom, P. (1964). Three observations on streamflow depletion by phreatophytes. *Journal of Hydrology, 2*, 248–261.

Meyboom, P. (1966). Unsteady groundwater flow near a willow ring in hummocky moraine. *Journal of Hydrology, 4*, 38–62.

Michaud, S. C., & Richardson, C. J. (1989). Relative radial oxygen loss in five wetland plants. In D. A. Hammer, (Ed.), *Constructed wetlands for wastewater treatment: Municipal, industrial and agricultural* (Chap. 38, pp. 501–507). Chelsea: Lewis Publishers, Inc.

Miller, E. C. (1938). Plant Physiology, McGraw-Hill Book Company, Inc., New York, 1,201 p.

Miller, R. R. (1996). *Phytoremediation.* (Groundwater Remediation Technologies Analysis Center Technology Overview Report TO-96-03).

Miller, M. H., Mamaril, C. P., & Blair, G. J. (1970). Ammonium effects on phosphorus absorption through pH changes and phosphorus precipitation, at the soil-root interface. *Journal of Agronomy, 62*, 524–527.

Mirck, J., & Volk, T. A. (2010). Seasonal sap flow of four Salix varieties growing on the Solvay wastebeds in Syracuse, NY, USA. *International Journal of Phytoremediation, 12*, 1–23.

Monteith, J. L. (1965). Evaporation and environment. In G. E. Fogg (Ed.), *The state and movement of water in living organisms* (Symposium of the society of experimental biology, pp. 205–234). San Diego: Academic.

Moran, M. J., Zogorski, J. S., & Squillace, P. J. (2007). Chlorinated solvents in groundwater of the United States. *Environmental Science & Technology, 41*, 74–81.

Moreau, C., Aksenov, N., Lorenzo, M. G., Segerman, B., Funk, C., Nilsson, P., Jansson, S., & Tuominen, H. (2005). A genomic approach to investigate developmental cell death in woody tissues of *Populus* trees. *Genome Biology, 6*, R34.

Moreo, M. T., Laczniak, R. J., & Stannard, D. I. (2007). *Evapotranspiration rate measurements of vegetation typical of groundwater discharge areas in the Basin and Range Carbonate-Rock Aquifer System,* Nevada and Utah, September 2005–August 2006 (U.S. Geological Survey Scientific Investigations Report 2007-5078, 36 p).

Mower, R. W., & Feltis, R. D. (1968). *Groundwater hydrology of the Sevier Desert, Utah* (U.S. Geological Survey Water-Supply Paper 1854, 75 p). Washington, DC: U.S. Department of the Interior.

Mower, R. W., & Nace, R. L. (1957). *Water consumption by water-loving plants in the Malad Valley, Oneida County, Idaho* (U.S. Geological Survey Water-Supply Paper 1412, 33 p). Washington, DC: U.S. Government Printing Office.

Mower, R. W., Hood, J. W., Cushman, R. L., Borton, R. L., & Galloway, S. E. (1964). *An appraisal of potential groundwater salvage along the Pecos River between Acme and Artesia, New Mexico* (U.S.Geological Survey Water-Supply Paper 1659, 98 p). Washington, DC: U.S. Government Printing Office.

Muratova, A., Hübner, Th, Tischer, S., Turkovskaya, O., Möder, M., & Kuschk, P. (2003). Plant-rhizosphere-microflora association during phytoremediation of PAH-contaminated soil. *International Journal of Phytoremediation, 5*, 137–151.

Nadezhdina, N. (1999). Sap flow as an indicator of plant water status. *Tree Physiology, 19*, 885–891.

Naff, R. L., Baker, A. A., & Gross, G. W. (1975). *Environmental controls in groundwater chemistry in New Mexico, Part I—The effects of phreatophytes* (New Mexico Water Resources Research Institute Report 052, 102 p).

Nagler, P. L., Glenn, E. P., & Thompson, T. L. (2003). Comparison of transpiration rates among saltcedar, cottonwood and willow trees by sap flow and canopy temperature methods. *Agricultural and Forest Meteorology, 116*, 73–89.

Narayanan, M., Russell, N. K., Davis, L. C., & Erickson, L. E. (1999). Fate and transport of trichloroethylene in a chamber with alfalfa plants. *International Journal of Phytoremediation, 1*, 387–411.

National Research Council. (1994). *Alternatives for groundwater cleanup*. Washington, DC: National Academy of Sciences.

National Research Council. (2000). *Natural attenuation for groundwater remediation* (274 p). Washington, DC: National Academy of Sciences.

Nativ, R. (2004). Can the desert bloom? Lessons learned from the Israeli case. *Ground Water, 42*, 651–657.

Nearing, H., & Nearing, S. (1971). *The maple sugar book* (273 p). New York: Schocken Books.

Neitch, C. T., Morris, J. T., & Vroblesky, D. A. (1999). Biophysical mechanisms of trichloroethene uptake and loss in bald cypress growing in shallow contaminated groundwater. *Environmental Science & Technology, 33*, 2899–2904.

Newman, L. A., & Arnold, C. W. (2003). Phytoremediation of MTBE—A review of the state of the technology. In E. E. Moyer & P. T. Kostecki (Eds.), *MTBE remediation handbook* (pp. 279–287). Amherst: Amherst Scientific Publishers.

Newman, L. A., Strand, S. E., Duffy, J., Ekuan, G., Raszaj, M., Shurtleff, B., Wilmoth, J., Heilman, P., & Gordon, M. (1997). Uptake and biotransformation of trichloroethylene by hybrid poplars. *Environmental Science & Technology, 31*, 1062–1067.

Newman, L. A., Wang, X., Doty, S. L., Gery, K. L., Heilman, P. E., Muiznieks, I., Shang, T. Q., Siemieniec, S. T., Strand, S. E., Wilson, A. M., & Gordon, M. P. (1998). Phytoremediation of organic contaminants: A review of phytoremediation research at the University of Washington. *Journal of Soil Contamination, 7*, 532–542.

Newman, L. A., Gordon, M. P., Heilman, P., Cannon, D. L., Lory, E., Miller, K., Osgood, J., & Strand, S. E. (1999a). Phytoremediation of MTBE at a California naval site. *Soil & Groundwater Cleanup, Feb/Mar.*, 42–45.

Newman, L. A., Doty, S. L., Muiznieks, I., Ekuan, G., Ruszaj, M., Cortellucci, R., Domroes, D., Karscig, G., Newman, T., Crampton, R. S., Hashmonay, R. A., Yost, M. G., Heilman, P. E., Duffy, J., Gordon, M. P., & Strand, S. E. (1999b). Remediation of trichloroethylene in an artificial aquifer with trees: A controlled field study. *Environmental Science & Technology, 33*, 2257–2265.

Nichols, W. D. (1993). Estimating discharge of shallow groundwater by transpiration from greasewood in the Northern Great Basin. *Water Resources Research, 29*, 2771–2778.

Nichols, T. D., Wolf, D. C., Rogers, H. B., Beyrouty, C. A., & Reynolds, C. M. (1997). Rhizosphere microbial populations in contaminated soils. *Water, Air, and Soil Pollution, 95*, 165–178.

Nilsen, E. T., Sharifi, M. R., Rundel, P. W., & others. (1983). Diurnal and seasonal water relations of the desert phreatophyte Prosopis Glandulosa (Honey Mesquite) in the Sonoran Desert, California. *Ecology*, 64, 1381–1393.

Nilson, S. E., & Assmann, S. M. (2007). The control of transpiration. Insights from Arabidopsis. *Plant Physiology, 143*, 19–27.

Nimick, D. A., Gammons, C. H., Cleasby, T. E., Madison, J. P., Skaar, D., & Brick, C. M. (2003). Diel cycles in dissolved metal concentrations in streams—Occurrence and possible causes. *Water Resources Research, 39*, 1247.

Nimmo, J. R. (2007). Simple predictions of maximum transport rate in unsaturated soil and rock. *Water Resources Research*, 43, 11.

Noctor, G., Arisi, A. M., Jouanin, L., Kunert, K. J., Rennenberg, H., & Foyer, C. H. (1998). Glutathione: Biosynthesis, metabolism and relationship to stress tolerance explored in transformed plants. *Journal of Experimental Botany, 49*, 249–270.

Nolan, B. T., Bayless, E. R., Green, C. T., Garg, S., Voss, F. D., Lampe, D. C., Barbash, J. E., Capel, P. D., & Bekins, B. A. (2005). *Evaluation of unsaturated-zone solute-transport models for studies of agricultural chemicals* (U.S. Geological Survey Open-File Report 2005-1196, 16 p).

Norton, W. H. (1897). *Artesian wells of Iowa* (Vol. 6). Des Moines: Iowa Geological Survey.

Nowack, B., Schulin, R., & Robinson, B. H. (2006). Critical assessment of chelant-enhanced metal phytoextraction. *Environmental Science & Technology, 40*, 5225–5232.

Nnyamah, J. U., & Black, T. A. (1977). Rates and patterns of water uptake in a Douglas-Fir forest. *Journal of the Soil Science Society of America, 41*, 972–979.

Nzengung, V. A., Wang, C., & Harvey, G. (1999). Plant-mediated transformation of perchlorate into chloride. *Environmental Science & Technology, 33*, 1470–1478.

Oborn, E. T. (1962). Iron content of selected water and land plants. In *Chemistry of iron in natural water* (U.S. Geological Survey Water-Supply Paper 1459, Chap. G, pp. 191–213). Washington, DC: U.S. Government Printing Office.

O'Conner, T. G. (1985). *A synthesis of field experiments concerning the grass layer in the Savanna Regions of South Africa* (Report No. 114). Pretoria: South African National Scientific Programmes, Foundation for Research Development.

Olson, P. E., & Fletcher, J. S. (1999). Field evaluation of mulberry root structure with regard to phytoremediation. *Bioremediation Journal, 3*, 27–34.

Orchard, B. J., Doucette, W. J., Chard, J. K., & Bugbee, B. (2000). Uptake of trichloroethylene by hybrid poplar trees grown hydroponically in flow-through plant growth chambers. *Environmental Toxicology and Chemistry, 19*, 895–903.

Oren, R., Phillips, N., Ewers, B. E., Pataki, D. E., & Megonigal, J. P. (1999). Sap-flux-scaled transpiration responses to light, vapor pressure deficit, and leaf area reduction in a flooded Taxodium distichum forest. *Tree Physiology, 19*, 337–347.

Ouvrard, S., LaPole, D., & Morel, J. L. (2006). Root exudates impact on phenanthrene availability. *Water, Air, and Soil Pollution: Focus, 6*, 343–352.

Ouyang, Y. (2002). Phytoremediation: Modeling plant uptake and contaminant transport in the soil-plant-atmosphere continuum. *Journal of Hydrology, 266*, 66–82.

Pallardy, S. G., & Kozlowski, T. T. (1979). Stomatal response of *Populus* clones to light intensity and vapor pressure deficit. *Plant Physiology, 64*, 112–114.

Pallardy, S. G., & Kozlowski, T. T. (1981). Water relations of *Populus* clones. *Ecology, 62*, 159–169.

Palissy, B. (1580). *Discourse admirables* (Aure lie La Rocque Trans. 1957, 264 p). Urbana: University of Illinois Press.

Palmroth, S., Katul, G. G., Hui, D., McCarthy, H. R., Jackson, R. B., & Oren, R. (2010). Estimation of long-term basin scale evapotranspiration from streamflow time series. *Water Resources Research, 46*, 13 pp. Published online 9 October 2010.

Paramelle. (1856). The art of finding springs. Dalmont et Dunod, Paris.

Park, R., & Epstein, S. (1960). Carbon isotope fractionation during photosynthesis. *Geochimica et Cosmochimica Acta, 21*, 110–126.

Parshall, R. L. (1937). Laboratory measurements of evapotranspiration losses. *Journal of Forestry, 35*, 1033–1040.

Passioura, J. B. (1988). Water transport in and to roots. *Annual Review of Plant Physiology and Plant Molecular Biology, 39*, 245–265.

Pate, J. S., Jeschke, W. D., & Aylward, M. J. (1995). Hydraulic architecture and xylem structure of the dimorphic root systems of south-west Australian species of *Proteaceae*. *Journal of Experimental Botany, 46*, 907–915.

Paterson, S., & Mackay, D. (1995). Interpreting chemical partitioning in soil-plant-air systems with a fugacity model. In S. Trapp & C. McFarlane (Eds.), *Plant contamination—Modeling and simulation of organic chemical processes*. Boca Raton: CRC Press, Inc.

Patric, J. H. (1961). The Sans Dimas larger lysimeters. *Journal of Soil Water Conservation, 16*, 13–17.

Paul, E. A., & Clark, F. E. (1989). Soil Microbiology and Biochemistry, Academic Press, San Diego, 273 p.

Peel, M. C., McMahon, T. A., & Finlayson, B. L. (2010). Vegetation impact on mean annual evapotranspiration at a global catchment scale. *Water Resources Research*, 46. Published online 4 September 2010.

Penman, H. L. (1948). Natural evaporation from open water, bare soil, and grass. *Proceedings Royal Society London Series A, 193*, 120–146.

Perrault, P. (1674). *De l'origine des fountains* [Treatise on the origin of springs] (353 p). Paris: Pierre le Petit.

Persson, G. (1995). Willow stand evapotranspiration simulated for Swedish soils. *Agricultural Water Management, 28*, 271–293.

Pessin, L. J. (1939). Root habits of longleaf pine and associated species. *Ecology, 20*, 47–57.

Phipps, R. L. (1967). *Annual growth of a suppressed chestnut oak and red maple, a basis for hydrologic inference* (U.S. Geological Survey Professional Paper 485-C, 27 p). Washington, DC: U.S. Government Printing Office.

Phipps, R. L. (1985). *Collecting, preparing, crossdating, and measuring tree increment cores* (U.S. Geological Survey Water-Resources Investigations Report 85-4148, 48 p).

Phipps, R. L., Ireley, D. L., & Baker, C. P. (1978). *Tree rings as indicators of hydrologic change in the Great Dismal Swamp, Virginia and North Carolina* (U.S. Geological Survey Water-Resources Investigations Report 78-136, 26 p).

Pinay, G., Fabre, A., Vervier, P., & Gazelle, F. (1992). Landscape Ecology. *Control of C,N, P distribution in soils of riparian forests, 6*, 121–132.

Pitterle, M. T., Andersen, R. G., Novak, J. T., & Widdowson, M. A. (2005). Push-pull tests to quantify in situ degradation rates at a phytoremediation site. *Environmental Science & Technology, 39*, 9317–9323.

Plewa, M. J., & Wagner, E. D. (1993). Activation of promutagens by green plants. *Annual Review of Genetics, 27*, 93–113.

Plummer, C. C., & McGeary, D. (1985). *Physical geology*. Dubuque: Wm. C. Brown Publishers.

Plutarch. (1914). *The parallel lives* (Vol. 1). New York: Loeb Classical Library Edition.

Pollan, M. (2001). *The botany of desire; a plant's-eye view of the world*. New York: Random House, Inc. 271 p.

Pollock, D. W. (1994). *User's guide for MODPATH/MODPATH-PLOT, version 3: A particle tracking post-processing package for MODFLOW, the U.S. Geological Survey finite-difference groundwater flow model* (U.S. Geological Survey Open-File Report 94-464). Reston: The Survey.

Poulsen, D. L., Simmons, C. T., Le Galle La Salle, C., & Cox, J. W. (2006). Assessing catchment-scale spatial and temporal patterns of groundwater and stream salinity. *Hydrogeology Journal, 14*, 1339–1359.

Powell, J. W. (1885). *Report of the Director (1883–84)* (U.S. Geological Survey 5th Annual Report, pp. xvii–xxxvi). Washington, DC.

Preston, R. (2008). *The wild trees: Random house trade paperback edition* (284 p). New York: Random House, Inc.

Preston, G. M., & McBride, R. A. (2004). Assessing the use of poplar tree systems as a landfill evapotranspiration barrier with the SHAW model. *Water Management Research, 22*, 291–305.

Preston, G. M., McBride, R. A., Bryan, J., & Candido, M. (2004). Estimating root mass in young hybrid poplar trees using the electrical capacitance method. *Agroforestry Systems, 60*, 305–309.

Proctor, R. J. (1968). Geology of the Desert Hot Springs-Upper Coachella Valley Area, California. California Division of Mines. *Geol. Spec. Report, 94*, 50 pp.

Purvis, A. C. (1997). Role of the alternative oxidase in limiting superoxide production by plant mitochondria. *Plant Physiology, 100*, 165–170.

Quinn, J. J., & Johnson, R. L. (2005). Continuous water-level monitoring in the assessment of groundwater remediation and refinement of a conceptual site model. *Remediation Journal, 15*, 49–61.

Quinn, J. J., Negri, M. C., Hinchman, R. R., Moos, L. P., Wozniak, J. B., & Gatliff, E. G. (2001). Predicting the effect of deep-rooted hybrid poplars on the groundwater flow system at a large-scale phytoremediation site. *International Journal of Phytoremediation, 3*, 41–60.

Radwan, S., Sorkhoh, N., & El-Nemr, I. (1995). Oil biodegradation around roots. *Nature, 376*, 302.

Raes, D. (2004). *UPFLOW: Water movement in a soil profile from a shallow water table to the topsoil (capillary rise)* (18 p). Reference manual, version 2.2. Leuven: K.U. Leuven University.

Raes, D., & Deproost, P. (2003). Model to assess water movement from a shallow water table to the root zone. *Agricultural Water Management, 62*, 79–91.

Ramaswami, A., Rubin, E., & Bonola, S. (2003). Non-significance of rhizosphere degradation during phytoremediation of MTBE. *International Journal of Phytoremediation, 5*, 315–331.

Randall, A. (1987). *Resource economics: An economic approach to natural resource and environmental policy* (434 p). New York: Wiley.

Rantz, S. E. (1968). *A suggested method for estimating evapotranspiration by native phreatophytes*. Geological Survey Research. Chapter D, U.S. Geological Survey Professional Paper 600-D, pp 10–13.

Raskin, I. (1996). Phytoremediation. In *Phytoremediation, Proceedings of International Business Communications Conference*, Arlington, VA. 5/8–10/96.

Raven, J. A., Handley, L. L., Macfarlane, J. J., McInroy, S., McKenzie, L., Richlands, J. H., & Samuelsson, G. (1987). Tansley Review No. 13: The role of CO_2 uptake by roots and CAM in acquisition of inorganic C by plants of the isoetid life-form: A review, with new data on Ericaulon decangulare l. *The New Phytologist, 108*, 125–148.

Ray Benayas, J. M., Bernaldez, F. G., Levassor, C., & Peco, B. (1990). Vegetation of groundwater discharge sites in the Douro basin, central Spain. *Journal of Vegetation Science, 1*, 461–466.

Read, D. B., & Gregory, P. J. (1997). Surface tension and viscosity of axenic maize and lupin root mucilages. *The New Phytologist, 137*, 623–628.

Read, D. B., Bengough, A. G., Gregory, P. J., Crawford, J. W., Robinson, D., Scrimgeour, C. M., Young, I. M., Zhang, K., & Zhang, X. (2003). Plant roots release phospholipids surfactants that modify the physical and chemical properties of soil. *The New Phytologist, 157*, 315–326.

Reicosky, D. C., Smith, R. C. G., & Meyer, W. S. (1985). Foliage temperature as a means of detecting stress of cotton subjected to a short-term water-table gradient. *Agricultural and Forest Meterology, 35*, 193–203.

Reilley, K. A., Banks, M. K., & Schwab, A. P. (1996). Dissipation of polycyclic aromatic hydrocarbons in the rhizosphere. *Journal of Environmental Quality, 25*, 212–219.

Reilly, T. E., & Harbaugh, A. W. (2004). *Guidelines for evaluating groundwater flow models* (U.S. Geological Survey Scientific Investigations Report 2004-5038, 30 p). Reston: U.S. Geological Survey.

Rentz, J. A., Chapman, B., Alvarez, J. J., & Schnoor, J. L. (2003). Stimulation of hybrid poplar growth in petroleum-contaminated soils through oxygen addition and soil nutrient amendments. *International Journal of Phytoremediation, 5*, 57–72.

Reporter, M., Robideaux, M., Wickster, P, Wagner, J., & Kapustka., L. (1991). Ecotoxicological assessment of toluene and cadmium using plant cell cultures. In J. W. Gorsuch, W. R. Lower, W. Wang, & M. A. Lewis (Eds.), *Plants for toxicity assessment: Second volume* (ASTM STP 1115, pp. 240–249). Philadelphia: American Society for Testing and Materials.

Rheinhardt, R. D., & Hershner, C. (1992). The relationship of below-ground hydrology to canopy composition in five tidal freshwater swamps. *Wetlands, 12*, 208–216.

Richards, J. H., & Caldwell, M. M. (1987). Hydraulic lift: Substantial nocturnal water transport between soil layers by *Artemisia tridenta* roots. *Oecologia, 73*, 486–489.

Ricklefs, R. E. (1979). *Ecology*. New York: Chiron Press Incorporated.

Richter, J. P. (1888). *The Notebooks of Leonardo da Vinci* (English translation). CreateSpace, 2010, 294 p.

Riederer, M. (1995). Partitioning and transport of organic chemicals between the atmospheric environment and leaves. In S. Trapp & C. McFarlane (Eds.), *Plant contamination: Modeling and simulation of organic chemical processes*. Boca Raton: CRC Press, Inc.

Robinson, T. W. (1958). *Phreatophytes* (U.S. Geological Survey Water–Supply Paper 1423, 84 p). Reston: U.S. Geological Survey.

Robinson, T. W. (1965). *Introduction, spread and areal extent of saltcedar (Tamarix) in the Western States* (U.S. Geological Survey Professional Paper 491-A, 12 p). Washington, DC: U.S. Government Printing Office.

Robinson, T. W. (1970). *Evapotranspiration by woody phreatophytes in the Humboldt River Valley near Winnemucca, Nevada* (U.S. Geological Survey Professional Paper 491-D, 41 p). Washington, DC: U.S. Government Printing Office.

Robinson, T. W., & Donaldson, D. (1967). Pontacyl brilliant pink as tracer dye in the movement of water in phreatophytes. *Water Resources Research, 3*, 203–211.

Robinson, N. J., Procter, C. M., Connolly, E. L., & Guerinot, M. L. (1999). A ferric-chelate reductase for iron uptake from soils. *Nature, 397*, 694–697.

Robinson, B., Green, S., Mills, T., Clothier, B., van der Velde, M., Laplane, R., Fung, L., Deuer, M., Hurst, S., Thayalakumaran, T., & van der Dijssel, C. (2003). Phytoremediation: Using plants as biopumps to improve degraded environments. *Australian Journal of Soil Research, 41*, 599–611.

Roberts, K. (1951). *Henry gross and is dowsing rod* (310 p). Garden City: Doubleday & Company, Inc.

Rock, S. A. (2003). Vegetative covers for waste containment. *Advances in Biochemical Engineering/Biotechnology, 78*, 157–170.

Rodell, M., & Famiglietti, J. S. (2002). The potential for satellite-based monitoring of groundwater storage changes using GRACE: The High Plains aquifer, Central US. *Journal of Hydrology, 263*, 245–256.

Roelle, J. E., & Hagenbuck, W. W. (1995). Surface cover changes in the Rio Grande floodplain, 1935–89. In E. T. LaRoe, G. S. Farris, C. E. Puckett, P. D. Doran, & M. J. Mac, (Eds.), *Our living resources: A report to the nation on the distribution, abundance, and health of U.S. plants, animals, and ecosystems* (pp. 290–292). Washington DC: U.S. Department of the Interior, National Biological Service, U.S. Geological Survey.

Rohrer, W. L., Newman, L., Sharp, M., Heilman, P., & Wallis, B. R. 2000. Monitoring site constraints at NUWC Keyport's hybrid poplar phytoremediation plantation. In G. B. Wickramanayake, A. R. Gavaskar, B .C. Alleman, & V. S. Magar (Eds.), *Bioremediation and phytoremediation of chlorinated and recalcitrant compounds* (Vol. C2–4, pp. 467–477). Columbus: Battelle Press.

Römheld, V., & Marschner, H. (1986). Evidence for a specific uptake system for iron phytosiderophores in roots of grasses. *Plant Physiology, 80*, 175–180.

Rosenberry, D. O., & Menheer, M. A. (2006). *A system for calibrating seepage meters used to measure flow between ground water and surface water* (U.S. Geological Survey Scientific Investigations Report 2005-5053, 21 p).

Rosenberry, D. O., Striegl, R. G., & Hudson, D. C. (2000). Plants as indicators of focused groundwater discharge to a northern Minnesota lake. *Ground Water, 38*, 296–303.

Rosenshein, J. S., Moore, J. E., Lohman, S. W., & Chase E. B. (Eds.). (1986). *Two-hundred years of hydrogeology in the United States* (U.S. Geological Survey Open–File Report 86–480, 110 p).

Ross, C. P. (1922). *Routes to desert watering places in the Lower Gila Region, Arizona* (U.S. Geological Survey Water-Supply Paper 490-C, 315 p). Washington, DC: U.S. Government Printing Office.

Ross, W. D. (1927). *Aristotle, selections* (359 p). New York: Charles Scribner's Sons.

Ross, B. (2007). Phreatophytes in the Bible. *Ground Water, 45*, 652–654.

Ross, M. S., Mitchell-Bruker, S., Sah, J. P., Stothoff, S., Ruiz, P. L., Reed, D. L., Jayachandran, K., & Coultas, C. L. (2006). Interaction of hydrology and nutrient limitation in the Ridge and Slough landscape of the southern Everglades. *Hydrobiologia, 569*, 37–59.

Rotteveel, T., Al-ahmad, H., & Gressel, J. (2006). Assessing risks and containing or mitigating gene-flow of transgenic and non-transgenic phytoremediating plants. In M. Mackova, D. N. Dowling, & T. Macek (Eds.), *Phytoremediation rhizoremediation* (pp. 259–284). Dordrecht: Springer.

Royce, C. L., Fletcher, J. S., Risser, P. G., McFarlane, J. C., & Benenati, F. E. (1984). PHYTOTOX: A database dealing with the effect of organic chemicals and terrestrial vascular plants. *Journal of Chemical Information and Modeling, 24*, 7–10.

Rubin, E., & Ramaswami, A. (2001). The potential for phytoremediation of MTBE. *Water Research, 35*, 1348–1353.

Rundel, P. W. (1972). Habitat restriction in Giant Sequoia: The environmental control of grove boundaries. *The American Midland Naturalist, 87*, 81–89.

Running, S. W. (1979). *Environmental and physiological control of water flux through Pinus contorta*. Ph.D. dissertation, Colorado State University, Fort Collins.

Rural Industries Research and Development Corporation. (2000). *Sustainable hardwood production in shallow watertable areas* (Water and salinity issues in agroforestry no. 6, RIRDC publication no. 00/163, 105 p). RIRDC.

Safir, G. R. (Ed.). (1987). *Ecophysiology of VA mycorrhizal plants*. Boca Raton: CRC Press.

Sakuratani, T. (1981). A heat balance method for measuring water flux in the stem of intact plants. *Journal of Agricultural Meteorology, 37*, 9–17.

Sala, A., & Smith, S. D. (1996). Water use by *Tamarix ramosissima* and associated phreatophytes in a Mojave desert floodplain. *Ecological Applications, 6*, 888–898.

Sánchez-Pérez, J. M., Lucot, E., Bariac, T., & Trémolières, M. (2008). Water uptake by trees in a riparian hardwood forest (Rhine floodplain, France). *Hydrological Processes, 22*, 366–375.

Sandermann, H., Jr. (1994). Higher plant metabolism of xenobiotics: The "green liver" concept. *Pharmocogenetics, 4*, 225–241.

Sandermann, H., Diesperger, H., & Scheel, D. (1977). Metabolism of xenobiotics by plant cell cultures. In W. Barz, E. Reinhard, & M. H. Zenk (Eds.), *Plant tissue culture and its biotechnological application* (178 p). Berlin: Springer.

Schaeffer, S., Williams, D. G., & Goodrich, D. C. (2000). Transpiration of cottonwood/willow forest estimated from sap flux. *Agricultural and Forest Meteorology, 105*, 257–270.

Schnabel, W., & White, D. (2001). The effect of mycorrhizal fungi on the fate of PCBs in two vegetated systems. *International Journal of Phytoremediation, 3*, 203–220.

Schnabel, W. E., Dietz, A. C., Burken, J. G., Schnoor, J. L., & Alvarez, P. J. (1997). Uptake and transformation of trichloroethylene by edible garden plants. *Water Research, 31*, 816–824.

Schneider, W. H., Hirsh, S. R., Compton, H. R., Burgess, A. E., & Wrobel, J. G. (2000). *Analysis of hydrologic data to evaluate phytoremediation system performance*. Columbus: Battelle Press.

Schnoor, J. L. (1997). *Phytoremediation* (Ground-Water Remediation Technology Analysis Center Technology Evaluation Report TE-98-01).

Schnoor, J. L., Licht, L. A., McCutcheon, S. C., Wolfe, N. L., & Carreira, L. H. (1995). Phytoremediation of organic and nutrient contaminants. *Environmental Science & Technology, 29*, 318–323.

Schroeder, P. R., Dozier, T. S., Zappi, P. A., McEnroe, B. M., Sjostrom, J. W., & Peyton, R. L. (1994). *The hydrologic evaluation of landfill performance (HELP) model engineering documentation for version 3* (EPA/600/R-94/168B) Cincinnati: U.S. EPA Risk Reduction Engineering Laboratory.

Schumacher, J. G., Sutley, S. J., & Cathcart, J. D. (1993). *Geochemical data for the Weldon Spring training area and vicinity property, St. Charles County, Missouri—1990–92* (U.S. Geological Survey Open-File Report 93-153, 86 p).

Schumacher, J. G., Stuckhoff, G. C., & Burken, J. G. (2004). *Assessment of subsurface chlorinated solvent contamination using tree cores at the Front Street Site and a former dry cleaning facility at the Riverfront Superfund Site, New Haven, Missouri, 1999–2003* (U.S. Geological Survey Scientific Investigations Report 2004-5049, 35 p).

Schwarz, O. J., & Eisele, G. R. (1984). Food chain transport of synfuels: experimental approaches for acquisition of baseline data. In K. E. Cowser (Ed.), *Synthetic fossil fuel technologies: Result of health and environmental studies. Proceedings of the Fifth Life Sciences Symposium, Gatlinburg, TN* (pp. 441–462). Boston: Butterworth.

Schwarz, O. J., & Jones, L. W. (1997). Bioaccumulation of xenobiotic organic chemicals by terrestrial plants. In W. Wang, J. W. Gorsuch, & J. S. Hughes (Eds.), *Plants for environmental studies* (pp. 417–449). Boca Raton: CRC Press.

Scholander, P. F., van Dam, L., & Scholander, S. I. (1955). Gas exchange in the roots of mangroves. *American Journal of Botany, 42*, 92–98.

Scholander, P. F., Hammel, H. T., Bradstreet, E. D., & Hemmingsen, E. A. (1965). Sap pressure in vascular plants. *Science, 148*, 339–346.

Schwennesen, A. T. (1918). *Groundwater in the Animas, Playas, Hachita, and San Luis basins, New Mexico* (U.S. Geological Survey Water-Supply Paper 422). Washington, DC: U.S. Government Printing Office.

Scott, M. L., Shafroth, P. B., & Auble, G. T. (1999). Responses of riparian cottonwoods to alluvial water table declines. *Environmental Management, 23*, 347–358.

Schröder, P. (2006). Enzymes transferring biomolecules to organic foreign compounds: A role for glucosyltransferase and glutathione s-transferase in phytoremediation. In M. Mackova, D. N. Dowling, & T. Macek (Eds.), *Phytoremediation rhizoremediation* (pp. 133–142). Dordrecht: Springer.

Schröder, P., & Collins, C. (2002). Conjugating enzymes involved in xenobiotic metabolism of organic xenobiotics in plants. *International Journal of Phytoremediation, 4*, 247–265.

Schwab, A. P., & Banks, M. K. (1994). Biologically mediated dissipation of polyaromatic hydrocarbons in the root zone. In T. A. Anderson & J. R. Coats (Eds.), *Bioremediation through rhizosphere technology* (ACS Symposium Series No. 563, pp. 132–141). Washington, DC: American Chemical Society.

Scott, R. L., Watts, C., Payan, J. G., Edwards, E., Goodrich, D. C., Williams, D., & Shuttleworth, W. J. (2003). The understory and overstory partitioning of energy and water fluxes in an open canopy, semiarid woodland. *Agricultural and Forest Meteorology, 114*, 127–139.

Scott, R. L., Huxman, T. E., Williams, D. G., & Goodrich, D. G. (2006). Ecohydrological impacts of woody-plant encroachment: Seasonal patterns of water and carbon dioxide exchange within a semiarid riparian environment. *Global Change Biology, 12*, 311–324.

Scrimgeour, C. M. (1995). Measurement of plant and soil water isotope composition by direct equilibrium methods. *Journal of Hydrology, 172*, 261–274.

Seuss, Dr. (1971). The Lorax, Random House, Inc., New York.

Seyfferth, A. L., & Parker, D. R. (2007). Effects of genotype and transpiration rate on the uptake and accumulation of perchlorate (ClO_4^-) in lettuce. *Environmental Science & Technology, 41*, 3361–3367.

Shafroth, P. B., Stromberg, J. C., & Patten, D. T. (2000). Woody riparian vegetation response to different alluvial water table regimes. *Western North American Naturalist, 60*, 66–76.

Shah, N., Nachabe, M., & Ross, M. (2007). Extinction depth and evapotranspiration from groundwater under selected land covers. *Groundwater, 45*, 329–338.

Shang, T. Q., Doty, S. L., Wilson, A. M., Howald, W. N., & Gordon, M. P. (2001). Trichloroethylene oxidative metabolism in plants: The trichoroethanol pathway. *Phytochemistry, 58*, 1055–1065.

Shang, T. Q., Newman, L. A., & Gordon, M. P. (2003). Fate of trichloroethylene in terrestrial plants. In S. C. McCutcheon & J. L. Schnoor (Eds.), *Phytoremediation: transformation and control of contaminants* (pp. 529–560). Hoboken: Wiley-Interscience.

Shannon, M. C., Bañuelos, G. S., Draper, J. H., Ajwa, H., Jordahl, J., & Licht, L. (1999). Tolerance of hybrid poplar (*Populus*) trees irrigated with varied levels of salt, selenium, and boron. *International Journal of Phytoremediation, 1*, 273–288.

Sharp, J. M., Jr., & Simmons, C. T. (2005). The compleat Darcy: New lessons learned from the first English translation of *Les Fountaines Publiques de la Ville Dijon. Ground Water, 43*, 457–460.

Shaw, G., Nichols, E. G., Cook, R., Fetzer, B., Messier, J. P., & Atkinson, B. (2010). The impact of trees on groundwater hydrology and jet-fuel contamination. In *Proceedings of the 7th International Conference on Remediation of Chlorinated and Recalcitrant Compounds*, Monterey, CA, May 2010.

Sheppard, J. C., & Funk, W. H. (1975). Trees as environmental sensors monitoring long-term heavy metal contamination of Spokane River, Idaho. *Environmental Science & Technology, 9*, 638–642.

Shim, H., Chauhan, S., Ryoo, D., Bowers, K., Thomas, S. M., Canada, K. A., Burken, J. G., & Wood, T. K. (2000). Rhizosphere competitiveness of trichloroethylene-degrading, poplar-colonizing recombinant bacteria. *Applied and Environmental Microbiology, 66*, 4673–4678.

Shinozaki, K., Yoda, K., Hozumi, K., & Kira, T. (1964). A quantitative analyses of plant form—The pipe model theory. I. Basic analyses. II. Further evidence of the theory and its implications in forest ecology. *Japanese Journal of Ecology,* 14, 97–105, 133–139.

Shoenmuth, B. W., & Pestemer, W. (2004a). Dendroremediation of trinitrotoluene (TNT) Part I: Literature overview and research concept. *Environmental Science & Pollution Research, 11*, 273–278.

Shoenmuth, B. W., & Pestemer, W. (2004b). Dendroremediation of trinitrotoluene (TNT) Part 2: Fate of radio-labelled TNT in trees. *Environmental Science & Pollution Research, 11*, 331–339.

Shone, M. G. T., & Wood, A. V. (1974). A comparison of the uptake and translocation of some organic herbicides and a systemic fungicide by barley: I. Absorption in relation to physico-chemical properties. *Journal of Experimental Botany, 25*, 390–400.

Shrout, J. S., Struckhoff, G. C., Parkin, G. F., & Schnoor, J. L. (2006). Stimulation and molecular characterization of bacterial perchlorate degradation by plant-produced electron donors. *Environmental Science & Technology, 40*, 310–317.

Siciliano, S. D., Fortin, N., Mihoc, A., Wisse, G., Labelle, S., Beaumier, D., Ouellette, D., Roy, R., Whyte, L. G., Banks, M. K., Schwab, P., Lee, K., & Greer, C. W. (2001). Selection of specific endophytic bacterial genotypes by plants in response to soil contamination. *Applied and Environmental Microbiology, 67*, 2469–2475.

Simunek, J., van Genuchten, M. Th., & Segina, M. (2005). *The HYDRUS-1D software package for simulating the movement of water, heat, and multiple solutes in variably saturated media* (270 p). Riverside, California: Department of Environmental Science, University of California Riverside.

Singer, A. C. (2006). The chemical ecology of pollutant biodegradation. In M. Mackova, D. N. Dowling, & T. Macek (Eds.), *Phytoremediation rhizoremediation* (pp. 5–21). Dordrecht: Springer.

Skaggs, R. W., Wells, L. G., & Ghate, S. R. (1978). Predicted and measured drainage porosities for field soils. *Transactions of ASAE, 22*, 522–528.

Skelton, N. J., & Allaway, W. G. (1996). Oxygen and pressure changes measured in situ during flooding in roots of the Grey Mangrove *Avicennia marina* (Forssk.) Vierh. *Aquatic Botany, 54*, 165–175.

Smith, G. E. P. (1915). *Transactions of the American Society* of *Civil Engineers*, 78. University of California Press, various pagination.

Smith, J., & Szathamáry, E. (1999). *The origins of life: From the birth of life to the origin of language* (180 p). New York: Oxford University Press, Inc.

Snyder, K. A., & Williams, D. G. (2000). Water sources used by riparian trees varies among stream types on the San Pedro River, Arizona. *Agricultural and Forest Meteorology, 105*, 227–240.

Snyder, K. A., & Williams, D. G. (2003). Defoliation alters water uptake by deep and shallow roots of *Prosopis velutina* (Velvet Mesquite). *Functional Ecology, 17*, 363–374.

Somma, F., Hopmans, J. W., & Clausnitzer, V. (1998). Transient three-dimensional modeling of soil water and solute transport with simultaneous root growth, root water and nutrient uptake. *Plant and Soil, 202*, 281–293.

Sorek, A., Atzmon, N., Dahan, O., Gerstl, Z., Kushisin, L., Laor, Y., Mingelgrin, U., Nasser, A., Ronen, D., Tsechansky, L., Weisbrod, N., & Graber, E. R. (2008). "Phytoscreening": The use of trees for discovering subsurface contamination by VOCs. *Environmental Science & Technology, 42*, 536–542.

Sorey, M. L., Farrar, C. D., Gerlack, T. M., McGee, K. A., Evans, W. C., Colvard, E. M., Hill, D. P., Bacley, R. A., Rogie, J. D., Hendley, J. W., II. & Stauffer, P. H. (2000). *Invisible CO_2 gas killing trees at Mammoth Mountain, California* (U.S. Geological Survey Fact Sheet 172-96, version 2.0, 2 p).

Spalding, V. M. (1909). *Distribution and movements of desert plants* (Carnegie Institution Washington Publication No. 113, pp. 9–12). Tortugas: Carnegie Institution Washington.

Sparks, J. P., & Black, R. A. (1999). Regulation of water loss in populations of *Populus trichocarpa*: The role of stomatal conductance in preventing xylem cavitation. *Tree Physiology, 19*, 453–459.

Speiran, G. K. (2010). Effects of groundwater-flow paths on nitrate concentrations across two riparian forest corridors. *Journal of the American Water Resources Association, 46*, 246–260.

Speiran, G. K., Hamilton, P. A., & Woodside, M. D. (1998). *Natural processes for managing nitrate in groundwater discharged to Chesapeake Bay and other surface waters: more than forest buffers* (U.S. Geological Survey Fact Sheet FS-178-97, 6 p).

Spittlehouse, D. L., & Black, T. A. (1981). A growing season water balance model applied to two Douglas fir stands. *Water Resources Research, 17*, 1651–1656.

Stearns, N. D. (1927). *Laboratory tests on physical properties of water-bearing materials* (U.S. Geological Survey Water-Supply Paper 596-F, pp. 121–176). Washington, DC: U.S. Government Printing Office.

Steinbeck, J. (1962). *Travels with Charley. In search of America*. New York: Viking Penguin Inc.

Steinberg, S., van Bavel, C. H. M., & McFarlane, M. J. (1989). A gauge to measure mass flow rate of sap in stems and trunks of woody plants. *Journal of the American Society for Horticultural Science, 114*, 466–472.

Steinberg, S. L., van Bavel, C. H. M., & McFarland, M. J. (1990a). Improved sap flow gauge for woody and herbaceous plants. *Agronomy Journal, 82*, 851–854.

Steinberg, S. L., McFarland, M. J., & Worthington, J. W. (1990b). Comparison of trunk and branch sap flow with canopy transpiration in pecan. *Journal of Experimental Botany, 41*, 653–659.

Stephens, D. B., & Ankeny, M. D. (2004). A missing link in the historical development of hydrogeology. *Groundwater, 42*, 304–309.

Sterky, F., et al. (2004). A *Populus* EST resource for plant functional genomics. *Proceedings of the National Academy of Sciences of the United States of America, 101*, 13951–13956.

Sternberg, L., DeNiro, M. J., & Johnson, H. B. (1986). Oxygen and hydrogen isotope ratios of water from photosynthetic tissues of CAM and C_3 plants. *Plant Physiology, 82*, 428–431.

Stewart, J. B. (1984). Measurement and prediction of evaporation from forested and agricultural catchments. In M. L. Sharma (Ed.), *Evapotranspiration from plant communities* (pp. 1–28). New York: Elsevier Science.

Strand, S. E., Walter, G. A., & Stensel, H. D. (1995). Effect of trichloroethylene loading on mixed methanotrophic community stability. In R. E. Hinchee, A. Leeson, & L. Semprini (Eds.), *Bioremediation of chlorinated solvents* (pp. 161–168). Columbus: Battelle Press.

Strand, S. E., Wang, X., Newman, L. A., Martin, D., Choe, N., Shurtleff, B., Wilmoth, J., Muiznieks, I., Gordon, M. P., Ekuan, G., Heilman, P., Massman, J., & Duffy, J. (1998). *Phytoremediation of chlorinated solvents: Laboratory and pilot-scale results*. Groundwater quality: Remediation and protection, Proceedings of the GQ98 Conference held at Tübingen, Germany, September 1998, IAHS Publication no. 250, pp. 245–248.

Struckhoff, G. C., Burken, J. G., & Schumacher, J. G. (2005). Vapor-phase exchange of perchloroethene between soil and plants. *Environmental Science & Technology, 39*, 1563–1568.

Susarla, S., Bacchus, S. T., Harvey, G., & McCutcheon, S. C. (2000). Phytotransformation of perchlorate contaminated waters. *Environmental Technology, 21*, 1055–1065.

Szilágyi, J., Gribovszki, Z., Kalicz, P., & Kucsara, M. (2008). On diurnal riparian zone groundwater-level and streamflow fluctuations. *Journal of Hydrology, 349*, 1–5.

Tabacchi, E., Lambs, L., Guilloy, H., Planty-Tabacchi, A., Muller, E., & Décamps, H. (2000). Impacts of riparian vegetation on hydrological processes. *Hydrological Processes, 14*, 2959–2976.

Taghavi, S., Barac, T., Greenberg, B., Borremans, B., Vangronsveld, J., & van der Lelie, D. (2005). Horizontal gene transfer to endogenous endophytic bacteria from poplar improves phytoremediation of toluene. *Applied and Environmental Microbiology, 71*, 8500–8505.

Tanaka, S., Brentner, L. B., Merchie, K. M., & Schnoor, J. L. (2007). Analysis of gene expression in poplar trees (*Populus Deltoides* x Nigra, DN34) exposed to the toxic explosive hexahydro-1,3,5-trinitro-1,3,5-triazine (RDX). *International Journal of Phytoremediation, 9*, 15–30.

Tao, S., Xu, F., Liu, W., Cui, Y., & Coveney, R. (2006). A chemical extraction method for mimicking bioavailability of polycyclic aromatic hydrocarbons to wheat grown in soils containing various amounts of organic matter. *Environmental Science & Technology, 40*, 2219–2224.

Taylor, H. M., Upchurch, D. R., & McMichael, B. L. (1990). Applications and limitations of rhizotrons and minirhizotrons for root studies. *Plant and Soil, 129*, 29–35.

Teuling, A. J., Uijenhoet, R., Hupet, F., & Troch, P. A. (2006). Impact of plant water uptake strategy on soil moisture and evapotranspiration dynamics during drydown. *Geophysical Research Letters, 33*, 1–5.

The Bible, English Revised Standard Version. (1971). Dallas: The Melton Book Company.

The Economist. (2005). *Ears of plenty* (Vol. 377, No. 8458). December 24, 2005. Westminster: The Economist Newspapers Limited.

Theis, C. V. (1935). The relation between the lowering of the piezometric surface and the rate and duration of discharge of a well using ground-water storage. *American Geophysical Union Transactions, 16*, 519–524. 16th Annual Meeting.

Thomas, H. E. (1952). *Hydrologic reconnaissance of the Green River in Utah and Colorado* (U.S. Geological Survey Circular 129), U.S. Geological Survey, 32 p.

Thomas, W. A. (1967). Dye and calcium ascent in dogwood trees. *Plant Physiology, 42*, 1800–1802.

Thompson, P. L., Ramer, L. A., Guffey, A. P., & Schnoor, J. L. (1998a). Decreased transpiration in poplar trees exposed to 2,4,6-trinitrotoluene. *Environmental Toxicology & Chemistry, 17*, 902–906.

Thompson, P. L., Ramer, L. A., & Schnoor, J. L. (1998b). Uptake and transformation of TNT by hybrid poplar trees. *Environmental Science & Technology, 32*, 975–980.

Thoreau, H. D. (1854). *Walden and civil disobedience*. New York: Signet Classic.

Thornthwaite, C. W., & Holzman, B. (1939). The determination of evaporation from land and water surfaces. *Monthly Weather Review, 67*, 4–11.

Thygesen, R. S., & Trapp, S. (2002). Phytotoxicity of polycyclic aromatic hydrocarbons to willow trees. *Journal of Soils and Sediments, 2*, 77–82.

Tillman, F. D. (2009). *Results of the analyses for 1,4-dioxane of groundwater samples collected in the Tucson Airport Remediation Project area, south-central Arizona, 2006–2009* (U.S. Geological Survey Open-File Report 2009–1196, 14 p).

Tindall, J. A., & Kunkel, J. R. (1999). *Unsaturated zone hydrology for scientists and engineers* (624 p). Englewood Cliffs: Prentice Hall.

Tognetti, V. B., Monti, M. R., Valle, E. M., Carrillo, N., & Smania, A. M. (2007). Detoxification of 2,4-dinitrotoluene by transgenic tobacco plants expressing a bacterial flavodoxin. *Environmental Science & Technology, 41*, 4071–4076.

Transeau, E. N. (1926). The accumulation of energy by plants. *The Ohio Journal of Science, 26*, 1–10.

Trapp, S. (1995). Model for uptake of xenobiotics into plants. In S. Trapp & J. C. McFarlane (Eds.), *Plant contamination modeling and simulation of organic chemical processes* (pp. 107–152). Boca Raton: CRC Press, Inc.

Trapp, S. (2002). Dynamic root uptake model for neutral lipophilic organics. *Environmental Toxicology and Chemistry, 21*, 203–206.

Trapp, S. (2004). Plant uptake and transport models for neutral and ionic chemicals. *Environmental Science & Pollution Research, 11* (1), 33–39.

Trapp, S., & Christiansen, H. (2003). Phytoremediation of cyanide-polluted soils. In S. C. McCutcheon & J. L. Schnoor (Eds.), *Phytoremediation: Transformation and control of contaminants* (pp. 829–862). Hoboken: Wiley.

Trapp, S., Matthies, M., Scheunert, I., & Topp, E. M. (1990). Modeling the bioconcentration of organic chemicals in plants. *Environmental Science & Technology, 24*, 1246–1252.

Trapp, S., McFarlane, J. C., & Matthies, M. (1994). Model for uptake of xenobiotics into plants—Validation with bromocil experiments. *Environmental Toxicology and Chemistry, 13*, 413–422.

Trapp, S., Zambrano, K. C., Kusk, K. O., & Karlson, U. (2000). A phytotoxicity test using transpiration of willows. *Archives of Environmental Contaminant Toxicity, 39*, 154–160.

Trapp, S., Miglioranza, K. S. B., & Mosbæk, H. (2001a). Sorption of lipophilic organic compounds to wood and implications for their environmental fate. *Environmental Science & Technology, 35*, 1561–1566.

Trapp, S., Köhler, A., Larsen, L. C., Zambrano, K. C., & Karlson, U. (2001b). Phytotoxicity of fresh and weathered diesel and gasoline to willow and poplar trees. *Journal of Soils and Sediments, 2*, 71–76.

Trapp, S., Rasmussen, D., & Samsøe-Petersen, L. (2003). Fruit tree model for uptake of organic contaminants from soil. *SAR and QSAR in Environmental Research, 14*, 17–26.

Trapp, S., Cammarano, A., Capri, E., Reichenberg, F., & Mayer, P. (2007). Diffusion of PAH in potato and carrot slices and application for a potato model. *Environmental Science & Technology, 41*, 3103–3108.

Travis, C., & Arms, A. (1988). Bioconcentration in beef, milk and vegetation. *Environmental Science & Technology, 22*, 271–274.

Travis, E. R., Hannink, N. K., Van Der Gast, C. J., Thompson, I. P., Rosser, S. J., & Bruce, N. C. (2007). Impact of transgenic tobacco on trinitrotoluene (TNT) contaminated soil community. *Environmental Science & Technology, 41*, 5854–5861.

Trousdell, K. B., & Hoover, M. D. (1955). A change in groundwater level after clearcutting of loblolly pine in the coastal plain. *Journal of Forestry, 53*, 493–498.

Tsao, D. T. (2003). Overview of phytotechnologies. *Advances in Biochemical Engineering/Biotechnology, 78*, 1–50.

Tsao, D. T. (2005). *Cost comparisons of phytotechnologies to other remedial approaches*. Presentation at the Third International Phytotechnologies Conference, Atlanta, GA, April 20–22, 2005. Accessed at January 31, 2006, http://www.clu-in.org on.

Tuominen, H., Puech, L., Fink, S., & Sundberg, B. (1997). A radial concentration gradient of Indole-3-acetic acid is related to secondary xylem development in hybrid aspen. *Plant Physiology, 115*, 577–585.

Tuskan, G. A. (2006). The genome of the Black Cottonwood, *Populus trichocarpa* (Torr. & Gray). *Science, 313*, 1596–1604.

Tyree, M. T., Kolb, K. J., Rood, S. B., & Patino, S. (1994). Vulnerability to drought-induced cavitation of riparian cottonwoods in Alberta: A possible factor in the decline of the ecosystem? *Tree Physiology, 14*, 455–466.

U.S. Environmental Protection Agency. (1993). *Guidance for evaluating the technical impracticability of groundwater restoration* (Interim final, EPA Directive 9234.2-25).

U.S. Environmental Protection Agency. (1994). *Land application of sewage sludge—A guide for land appliers on the requirements of the Federal Standards for the use or disposal of Sewage sludge, 40 CFR Part 503* (EPA/831-B-93-002b).

U.S. Environmental Protection Agency. (1995). *Standards for the use and disposal of sewage sludge, 40 CFR Part 503*. February 19, 1993, and amendments dated February 25, 1994, October 25, 1995.

U.S. Environmental Protection Agency. (1999). *Phytoremediation resource Guide* (EPA 542-B-99-003).

U.S. Environmental Protection Agency. (2000a). *Introduction to phytoremediation* (EPA/600/R-99/107).

U.S. Environmental Protection Agency. (2000b). *J-Field phytoremediation project field events and activities through July 31, 2000*. Aberdeen Proving Ground, Edgewood, Maryland. August 31, 2000.

U.S. Environmental Protection Agency. (2001a). *Brownfields technology primer: selecting and using phytoremediation for site cleanup* (EPA 542-R-01-006, Various pagination).

U.S. Environmental Protection Agency (2001b). *Cost analysis for selected groundwater cleanup projects: Pump and treat systems and permeable reactive barriers* (EPA 542-R-00-013).

U.S. Environmental Protection Agency. (2003). *Phytoremediation of groundwater at Air Force Plant 4, Carswell, Texas, Innovative Technology Evaluation Report* (EPA/540/R-03/506).

U.S. Environmental Protection Agency. (2004). In Green, C., & Hoffnagle, A. (Eds.), *Phytoremediation field studies database for chlorinated solvents, pesticides, explosives, and metals.* Washington, DC: U.S. Environmental Protection Agency Office of Superfund Remediation and Technology Innovation.

U.S. Environmental Protection Agency. (2005a). *Evaluation of phytoremediation for management of chlorinated solvents in soil and groundwater* (EPA 542-R-05-001, 34 p).

U.S. Environmental Protection Agency. (2005b). *Use of field-scale phytotechnology for chlorinated solvents, metals, explosives and propellants, and pesticides* (EPA 542-R-05-002, 14 p).

U.S. Environmental Protection Agency. (2008). *A guide for assessing biodegradation and source identification of organic ground water contaminants using compound specific isotope analysis (CSIA)* (EPA 600/R-08/148).

U.S. Geological Survey. (1998). *Innovative technologies join Superfund cleanup of groundwater at Fort Lewis.* Washington, DC: U.S. Geological Survey Fact Sheet 082–98.

U.S. Geological Survey. (Variously dated). *National field manual for the collection of water-quality data* (U.S. Geological Survey Techniques of Water-Resources Investigations, Book, 9, Chap. A1). http://pubs.water.usgs.gov/twri9A2/

Van Aken, B., & Schnoor, J. L. (2002). Evidence of perchlorate (ClO_4^-) reduction in plant tissues (poplar tree) using radio-labeled $^{36}ClO_4^-$. *Environmental Science & Technology, 36*, 2783–2788.

Van Aken, B., Moon Yoon, J., Just, C. L., & Schnoor, J. L. (2004a). Metabolism and mineralization of hexahydro-1,3,5-trinitro-1,3,-5-triazine inside poplar tissues (*Populus deltoides* x nigra DN-34). *Environmental Science & Technology, 38*, 4572–4579.

Van Aken, B., Moon Yoon, J., & Schnoor, J. L. (2004b). Biodegradation of nitro-substituted explosives 2,4,6-trinitrotoluene, hexahydro-1,3,5-trinitro-1,3,5-triazine, and octahydro-1,3,5,7-tetranitro-1,3,5-tetrazocine by a phytosymbiotic *Methylobacterium* sp. Associated with poplar tissues (*Populus deltoides x nigra* DN-34). *Applied and Environmental Microbiology, 70*, 508–517.

Van Bavel, C. H. M. (1966). Potential evapotranspiration, the combination concept and its experimental verification. *Water Resources Research, 2*, 455–467.

Van Bavel, C. H. M., & van Bavel, M. (1988). *ETP software.* Houston: Dynamax.

Van Bavel, C. H. M., & van Bavel, M. (1990). *ETP Penman-Van Bavel Software, version 2.1.* Houston: Dynamax, Inc.

Van Bavel, C. H. M., Nakayama, F. S., & Ehrler, W. L. (1965). Measuring transpiration resistance of leaves. *Plant Physiology, 40*, 535–540.

Van Den Bos, A. (2002). Phytoremediation of volatile organic compounds in groundwater: Case studies in plume control. Draft report prepared for the U.S. EPA Technology Innovation Office under a National Network for Environmental Management Studies Fellowship.

Van der Leeden, F., Troise, F. L., & Todd, D. K. (1990a). *The water encyclopedia* (808 p). Chelsea: Lewis Publishers, Inc.

Van der Leeden, F., Troise, F. L., & Todd, D. K. (1990b). *The water encyclopedia.* Chelsea: Lewis Publishers.

Van der Lelie, D., Schwitzguebel, J.-P., Glass, D. J., Vangronsveld, J., & Baker, A. (2001). Assessing phytoremediation's progress in the United States and Europe. *Environmental Science & Technology, 35*, 447A–452A.

Van Doren, M. (1955). *Travels of William Bartram.* New York: Dover publications, Inc.

Van Epps, A. (2006). *Phytoremediation of petroleum hydrocarbons.* Prepared for EPA Office of Solid Waste and Emergency Response, Office of Superfund Remediation and Technology Innovation, Washington, DC, August, http:.//www.clu-in.org

Van Hylckama, T. E. A. (1968). Water-level fluctuations in evapotranspirometers. *Water Resources Research, 4*, 761–768.

Van Hylckama, T. E. A. (1970). Water use by saltcedar. *Water Resources Research, 6*, 728–735.

Van Hylckama, T. E. A. (1974). *Water use by saltcedar as measured by the water budget method* (U.S. Geological Survey Professional Paper 491-E, 30 p). Washington, DC: U.S. Government Printing Office.

Verry, E. S. (2003). Estimating groundwater yield in small research basins. *Groundwater, 41*, 1001–1004.

Viessman, W., Jr., Knapp, J. W., Lewis, G. L., & Harbaugh, T. E. (1977). *Introduction to hydrology* (704 p). New York: Harper & Row.

Vieweg, G. H., & Ziegler, H. (1960). Thermoelektrische Registrierung der Geschwindigkeit des Transpirationsstromes. *Berichte. Deutsche Botanische Gesellschaft., 73*, 221–226.

Virginia Natural Heritage Program. Accessed at November 11, 2009, http://www.dcr.virginia.gov/natural_heritage/mission.shtml

Von Caemmerer, S., & Baker, N. (2007). The biology of transpiration: From guard cells to globe. *Plant Physiology, 143*, 3.

von Sachs, J. (1875). *Textbook of botany* (English translation ed.). Oxford: Oxford University Press.

Vonnegut, K. (1973). *Breakfast of champions.* New York: Delacorte Press.

Vosátka, M., Rydlová, J., Sudová, R., & Vohník, M. (2006). Mycorrhizal fungi as helping agents in phytoremediation of degraded and contaminated soils. In M. Mackova, D. N. Dowling, & T. Macek (Eds.), *Phytoremediation rhizoremediation* (pp. 237–257). Dordrecht: Springer.

Vose, J. M., Swank, W. T., Harvey, G. J., Clinton, B. D., & Sobek, C. (2000). Leaf water relations and sapflow in eastern cottonwood (*Populus deltoids* Bartr.) trees planted for phytoremediation of a groundwater pollutant. *International Journal of Phytoremediation, 2*, 53–73.

Voss, C. I. (1984). *A finite-element simulation model for saturated-unsaturated, fluid-density-dependent groundwater flow with energy transport or chemically-reactive single-species solute transport* (U.S. Geological Survey Water-Resources Investigations Report 84-4369, 409 p).

Vroblesky, D. A. (2008). *User's guide to the collection and analysis of tree cores to assess the distribution of subsurface volatile organic compounds* (U.S. Geological Survey Scientific Investigations Report 2008-5088, 59 p).

Vroblesky, D. A., & Yanosky, T. M. (1990). Use of tree-ring chemistry to document historical groundwater contamination events. *Groundwater, 28*, 677–684.

Vroblesky, D. A., Yanosky, T. M., & Siegel, F. R. (1992). Increased concentrations of potassium in heartwood of trees in response to groundwater contamination. *Environmental Geology and Water Science, 19*, 71–74.

Vroblesky, D. A., Nietch, C. T., & Morris, J. T. (1999a). Chlorinated ethenes from groundwater in tree trunks. *Environmental Science & Technology, 33*, 510–515.

Vroblesky, D. A., Nietch, C. T., Robertson, J. F., Bradley, P. M., Coates, J., & Morris, J. T. (1999). *Natural attenuation potential of chlorinated volatile organic compounds in groundwater, TNX flood plain, Savannah River Site, South Carolina* (U.S. Geological Survey Water-Resources Investigations Report 99-4071, 43 p).

Vroblesky, D. A., Clinton, B. D., Vose, J. M., Casey, C. C., Harvey, G. J., & Bradley, P. M. (2004). Groundwater chlorinated ethenes in tree trunks: Case studies, influence of recharge, and potential degradation mechanisms. *Groundwater Monitoring & Remediation, 24*, 124–138.

Vroblesky, D. A., Willey, R. E., Clifford, S., & Murphy, J. J. (2007). *Real-time and delayed analysis of tree and shrub cores as indicators of subsurface volatile organic compound contamination, Durham Meadows, Superfund Site, Durham, Connecticut, August 19, 2006* (U.S. Geological Survey Scientific Investigations Report 2007-5212, 12 p).

Vroblesky, D. A., Canova, J. L., Bradley, P. M., & Landmeyer, J. E. (2009). *Tritium concentrations in environmental samples and transpiration rates from the vicinity of Mary's Branch Creek and background areas, Barnwell, South Carolina, 2007–2008* (U.S. Geological Survey Scientific Investigations Report 2009-5245, 12 p).

Vuorinen, A. H., Vapaavuori, E. M., & Lapinjoki, S. (1989). Time-course of uptake of dissolved inorganic carbon through willow roots in light and in darkness. *Physiologia Plantarum, 77*, 33–38.

Wagner, P. A. (1916). The geology and mineral industry of Southwest Africa. *South Africa Geological Survey Memorandum* 7.

Waisel, Y., Liphschitz, N., & Kuller, T. (1972). Patterns of water movement in trees and shrubs. *Ecology, 53*, 520–523.

Walton, B. T., & Anderson, T. A. (1992). Plant-microbe treatment systems for toxic waste. *Biotechnology, 3*, 267–270.

Walton, B. T., Guthrie, E. A., & Hoylman, A. M. (1994). Toxicant degradation in the rhizosphere. In T. A. Anderson & J. R. Coats (Eds.), *Bioremediation through rhizosphere technology* (ACS Symposium Series No. 563) Washington, DC: ACS.

Walton, B. T., Hoylman, A. M., Perez, M. M., Anderson, T. A., Johnson, T. R., Guthrie, E. A., & Christman, R. F. (1994). Rhizosphere microbial communities as a plant defense against toxic substances in soils: In T. A. Anderson & J. R. Coats (Eds.), *Bioremediation through rhizosphere technology* (ACS Symposium Series No. 563, pp. 82–92). Washington, DC: ACS.

Wan, C., Sosebee, R. E., & McMichael, B. L. (1993). Growth, photosynthesis, and stomatal conductance in *Gutierrezia sarothrae* associated with hydraulic conductance and soil water extraction by deep roots. *International Journal of Plant Science, 154*, 144–151.

Wardsman, P., & Candeias, L. P. (1996). Fenton Centennial Symposium, Fenton Chemistry: An Introduction. *Radiation Research, 145*, 523–531.

Wayment, D. G., Bhadra, R., Lauritzen, J., Hughes, J. B., & Shanks, J. V. (1999). A transient study of formation of conjugates during TNT metabolism by plant tissues. *International Journal of Phytoremediation, 1*, 227–239.

Webb, R. H., Boyer, D. E., & Berry, K. H. (2001). *Changes in riparian vegetation in the Southwestern United States: Historical changes along the Mojave River, California* (U.S. Geological Survey Open-File Report 01-245).

Weeks, E. P., Weaver, H. L., Campell, G. S., & Tanner, B. D. (1987). *Water use by saltcedar and by replacement vegetation in the Pecos River floodplain between Acme and Artesia, New Mexico* (U.S. Geological Survey Professional Paper 491-G, 33 p).

Weigel, D., & Jürgens, G. (2002). Stem cells that make stems. *Nature, 415*, 751–754.

Weidemeier, T. H., Swanson, M. A., Wilson, J. T., Kampbell, D. H., Miller, R. N., & Hansen, J. E. (1996). Approximation of biodegradation rate constants for monoaromatic hydrocarbons (BTEX) in ground water. *Ground Water Monitoring & Remediation, 16*, 186–194.

Westover, K. M., Kennedy, A. C., & Kelley, S. E. (1997). Patterns of rhizosphere microbial community structure associated with co-occurring plant species. *Journal of Ecology, 85*, 863–873.

Westbrooks, R. G., & Preacher, J. W. (1986). *Poisonous plants of eastern North America* (226 p). Columbia: University of South Carolina Press.

Weyers, J., & Meidner, H. (1990). *Methods in stomata research: Longman scientific and technical*. London: Essex.

White, W. N. (1932). A method of estimating groundwater supplies based on discharge by plants and evaporation from soil-results of investigations in Escalante Valley, Utah (Part A). In *Contributions to the hydrology of the United States, U.S. Geological Survey Water-Supply Paper, 659*, 1–106.

White, J. W., Cook, E. R., Lawrence, J. R., & Broecker, W. S. (1985). The D/H ratios of sap in trees: Implications for water sources and tree ring D/H ratios. *Geochimica et Cosmochimica Acta, 49*, 237–246.

Whiting, G. J., & Chanton, J. P. (1996). Control of the diurnal pattern of methane emission from emergent aquatic macrophytes by gas transport mechanisms. *Aquatic Botany, 54*, 237–253.

Whittaker, R. H., & Feeny, P. P. (1971). Allelochemics: Chemical interactions between species. *Science, 171*, 757–770.

Wickliff, C., & Fletcher, J. S. (1991). Tissue culture as a method for evaluating the biotransformation of xenobiotics by plants. In J. W. Gorsuch, W. R. Lower, W. Wang, & M. A. Lewis (Eds.), *Plants for toxicity assessment: Second volume* (ASTM STP 1115, pp. 250–257). Philadelphia: American Society for Testing and Materials.

Widdowson, M. A., Shearer, S., Andersen, R. G., & Novak, J. T. (2005a). Remediation of polycyclic aromatic hydrocarbon compounds in groundwater using poplar trees. *Environmental Science & Technology, 39*, 1598–1605.

Widdowson, M. A., Al-Sayed, A., Hester, E., & Landmeyer, J. E. (2005b). *SEAM3D—Plant Uptake Package (PUP). A numerical model for 3-D transport coupled to sequential electron-acceptor-based biodegradation reactions in groundwater, Documentation and Users Guide*. Virginia Tech.

Wießner, A., Kuschk, P., Kästner, M., & Stottmeister, U. (2002). Abilities of helophyte species to release oxygen into rhizospheres with varying redox conditions in laboratory-scale hydroponic systems. *International Journal of Phytoremediation, 4*, 1–15.

Wilcke, W. (2000). Polycyclic aromatic hydrocarbons (PAH) in soil: A review. *Journal of Plant Nutrition and Soil Science, 163*, 229–248.

Wilcox, B. P., Owens, M. K., Dugas, W. A., Ueckert, D. N., & Hart, C. R. (2006). Shrubs, streamflow, and the paradox of scale. *Hydrological Processes, 20*, 3245–3259.

Wild, E., Dent, J., Barber, J. L., Thomas, G. O., & Jones, K. C. (2004). A novel analytical approach for visualizing and tracking organic chemicals in plants. *Environmental Science & Technology, 38*, 4195–4199.

Wild, E., Dent, J., Thomas, G. O., & Jones, K. C. (2005a). Real-time visualization and quantification of PAH photodegradation on and within plant leaves. *Environmental Science & Technology, 39*, 268–273.

Wild, E., Dent, J., Thomas, G. O., & Jones, K. C. (2005b). Direct observation of organic contaminant uptake, storage, and metabolism. *Environmental Science & Technology, 39*, 3695–3702.

Williams, S. (1789). *The natural and civil history of Vermont*. Burlington: Samuel Mills Printing.

Williams, J. B. (2002). Phytoremediation in wetland ecosystems: Progress, problems, and potential. *Critical Reviews in Plant Science, 21*, 607–635.

Willstätter, R., & Stoll, A. (1913). *Untersuchung über Chlorophyllen, methoden, und ergebnisse*: Berlin: Julius Springer.

Wilson, K. J. (1995). Molecular techniques for the study of rhizobial ecology in the field. *Soil biology & Biochemistry, 27*, 501–514.

Wilson, J. T., & Wilson, B. H. (1985). Biotransformation of trichloroethylene in soil. *Applied and Environmental Microbiology, 56*, 1012–1016.

Wilson, J. T., Ross, R. P., & Acree, S. (2005). Using direct-push tools to map hydrostratigraphy and MTBE plume diving. *Groundwater Monitoring & Remediation, 25*, 93–102.

Winter, T. C. (1999). Relation of streams, lakes, and wetlands to groundwater flow systems. *Hydrogeology Journal, 7*, 28–45.

Winter, T. C., & Rosenberry, D. O. (1995). The interaction of ground-water with prairie pothole wetlands in the Cottonwood Lake area, east-central North Dakota, 1979–1990. *Wetlands, 15*, 193–211.

Winter, T. C., Harvey, J. W., Franke, O. L., & Alley, W. M. (1998). *Groundwater and surface water—A single resource* (U.S. Geological Survey Circular 1139, 79 p).

Wisler, C. O., & Brater, E. F. (1956). *Hydrology* (419 p). Wiley.

Witherspoon, J.P., Jr. (1964). Cycling of cesium-134 in white oak trees. *Ecological Monograph, 34*, 403–420.

Witte, H., Langenohl, T., & Offenbacher, G. (1988). Investigation of the entry of organic pollutants into soils and plants through the use of sewage sludge in agriculture. *Korrespndenz Abwasser, 13*, 118–136.

Wium-Anderson, S. (1971). Photosynthetic uptake of free CO_2 by the roots of *Lobelia dortmanna*. *Physiologia Plantarum, 25*, 245–248.

Woodall, G. S., & Ward, B. H. (2002). Soil water relations, crop production and root pruning of a belt of trees. *Agricultural Water Management, 53*, 153–169.

Woodruff, D. R., Bond, B. J., & Meinzer, F. C. (2004). Does turgor limit growth in tall trees? *Plant, Cell & Environment, 27*, 229–236.

Worth, G. D., Holawe, F., & McIntyre, G. N. (1994). An inexpensive and reliable atmometer for estimating evaporation from sandy surfaces. *Theoretical and Applied Climatology, 49*, 263–265.

Wullschleger, S. D., Meinzer, F. C., & Vertessy, R. A. (1998). A review of whole-plant water use studies in trees. *Tree Physiology, 18*, 499–512.

Xiao-Zhang, Y., & Gu, Ji-Dong. (2006). Uptake, metabolism, and toxicity of methyl tert-butyl ether (MTBE) in weeping willows. *Journal of Hazardous Materials, 137*, 1417–1423.

Yanosky, T. M., & Vroblesky, D. A. (1992). Relation of nickel concentrations in tree rings to groundwater contamination. *Water Resources Research, 28*, 2077–2083.

Yanosky, T. M., & Vroblesky, D. A. (1995). Element analysis of tree rings in groundwater contamination studies. In T. E. Lewis (Ed.), *Tree rings as indicators of ecosystem health* (pp. 177–205). Boca Raton: CRC Press.

Yee, D. C., Maynard, J. A., & Wood, T. K. (1998). Rhizoremediation of trichloroethylene by a recombinant, root-colonizing *Pseudomonas fluorescens* strain expressing toluene ortho-monooxygenase constitutively. *Applied and Environmental Microbiology, 64*, 112–118.

Yeh, T.-C. J., & Guzman-Guzman, A. (1995). Tensiometry. In L. G. Wilson, L. G. Everett, & S. J. Cullen (Eds.), *Handbook of vadose zone characterization and monitoring* (pp. 319–328). Ann Arbor: Lewis Publishers.

Yeh, T.-C. J., & Liu, S. (2000). Hydraulic tomography: Development of a new aquifer test method. *Water Resources Research, 36*, 2095–2105.

Yifru, D. D., & Nzengung, V. A. (2006). Uptake of N-nitrosodimethylamine (NDMA) from water by phreatophytes in the absence and presence of perchlorate as a co-contaminant. *Environmental Science & Technology, 40*, 7374–7380.

Yoon, J. M., van Aken, B., & Schnoor, J. L. (2006). Leaching of contaminated leaves following uptake and phytoremediation of RDX, HMX, and TNT by poplar. *International Journal of Phytoremediation, 8*, 81–89.

Young, A. A., & Blaney, H. F. (1942). *Use of water by native vegetation* (Division of Water Resources, Bulletin 50, 160 p). California: Department of Public Works.

York, J. P., Person, M., Gutowski, W. J., & Winter, T. C. (2002). Putting aquifers into atmospheric simulation models: An example from the Mill Creek Watershed, northeastern Kansas. *Advances in Water Resources, 25*, 221–238.

Yu, X., & Gu, J. (2006). Uptake, metabolism, and toxicity of methyl tert-butyl ether (MTBE) in weeping willows. *Journal of Hazardous Materials, 137*, 1417–1423.

Zalesney, R. S., Jr., Bauer, E. O., Hall, R. B., Zalesney, J. A., Kunzman, J., Rog, C. J., & Riemenschneider, D. E. (2005). Clonal variation in survival and growth of hybrid poplar and willow in an in situ trial on soils heavily contaminated with petroleum hydrocarbons. *International Journal of Phytoremediation, 7*, 177–197.

Zhang, H., Morison, J. I. L., & Simmonds, L. P. (1999). Transpiration and water relations of poplar trees growing close to the water table. *Tree Physiology, 19*, 563–573.

Ziegler, H. (1995). Stable isotopes in plant physiology. *Progress in Botany, 56*, 1–24.

Zimmermann, R. C. (1969). *Plant ecology of an arid basin Tres Alamos-Redington area Southeastern Arizona* (U.S. Geological Survey Professional Paper 485-D, 51 p).

Zimmermann, U. (1989). Water relations of plant cells: Pressure probe technique. *Methods in Enzymology, 174*, 338–366.

Zimmermann, U., Schneider, H., Wegner, L. H., & Haase, A. (2004). Water ascent in tall trees: Does evolution of land plants rely on a highly metastable state? *The New Phytologist, 162*, 575–615.

Zogorski, J. S., Carter, J. M., Ivahnenko, T., Lapham, W. W., Moran, M. J., Rowe, B. L., Squillace, P. J., & Toccalino, P. L. (2006). *The quality of our Nation's waters- volatile organic compounds in the Nation's ground water and drinking-water supply wells* (U.S. Geological Survey Circular 1292, 101 p). Reston: US Geological Survey.

Index

A

Aberdeen Proving Ground, 206–207, 332, 361
Aboveground storage tanks (AST), 137, 172, 195, 311, 316, 340
Abscisic acid, 55
Abscission, 55, 185
Absorption, 5, 11, 48, 49, 59, 60, 67–68, 79, 99, 168, 170, 175, 247, 248, 251, 252, 259, 277, 280, 283, 289, 295, 309, 313, 318, 319, 326, 345, 348
Acetone, 280
Acetyl-CoA, 52, 252, 257, 301
Acid, 45, 52, 67, 77, 178, 258, 260, 272, 291, 292, 295, 302, 318, 320, 344, 366, 369, 374
Acid rain, 24, 374
Activation, 248, 253, 298, 370–372
Actual evapotranspiration, 39, 116, 200
Adak National Forest, 141
Additives, 265, 320–321, 325
Adenosine diphosphate (ADP), 49, 88, 177, 252, 292
Adenosine triphosphate (ATP), 46–50, 52, 64, 70, 88, 177, 249–257, 274, 292, 302, 303
ADP. *See* Adenosine diphosphate
Adsorption, 277
Aerenchyma, 47, 51, 52, 62, 64, 251, 259, 294, 297, 322, 327
Africa, 14, 16, 155, 171, 267
Agent Orange, 56, 308
Air temperature, 19, 35, 36, 40, 41, 48, 56, 81, 82, 96, 124, 140–142, 167, 170, 192, 197, 198, 228–230, 239, 269, 294
Alchemy, 5, 6
Alder, 6, 66, 98, 120, 155, 160, 162, 255, 293
Alfalfa, 14–16, 20–22, 39, 65, 81, 156, 157, 173, 180, 181, 197, 201, 256, 290, 293, 311, 318–320, 328, 335, 338, 339. *See also* *Medicago sativa*
Algae, 45, 46, 60, 61, 67, 73, 93, 264, 265, 274
Alkaline, 161, 257
Alkalinity, 14, 126, 138
Alkaloids, 5, 56, 253, 261, 263, 264, 291, 292, 365
Alkane monooxygenase *(alkB)*, 290, 305
Allelopathy, 260–264
Alpha-ketoglutarate, 252
Amino acids, 45, 47, 55, 74, 253, 255, 258, 262, 302, 352, 368
Ammonia, 66, 176, 177, 254–256
Ammonia carbonate, 11, 25
Ammonium, 254–256, 291, 333
Amphistomatous, 78
Amygdalin, 303
Amyloplasts, 58
Anatome Plantarum, 7, 23
The Anatomy of Plants, 7, 24, 43–93
Anaxagoras of Clazomenae, 28
Angiosperms, 9, 23, 62, 74, 158, 159, 166
Anhydride linkages, 50
Anisocytic, 78
Anisotropic, 99
Anisotropy, 99, 114, 193
Annual ring, 62, 75, 115, 354, 358, 363
Anomocytic, 78
Anoxic, 49, 50, 64, 66, 67, 75, 85, 139, 164, 168, 170, 175, 182, 183, 196, 232, 249–251, 256, 258, 259, 267, 269, 270, 273, 292–296, 303, 313, 321, 322, 360, 361
Anthocyanins, 48
Anthracene, 286, 287, 313, 320
Anthraquinone, 287
Antimicrobial, 187, 262, 263
Antioxidants, 249, 299, 301
Apical dominance, 55
Apical meristems, 24, 46, 170
Apoplastic pathway, 63, 71, 211, 258, 281, 282, 285
Appalachian Mountains, 120, 159, 366
Applicable or Relevant and Appropriate Requirements (ARAR), 239
Aquatic plants, 11, 43, 50–51, 60, 62, 73, 253, 256, 257, 265, 288, 294, 335, 337
Aquifer, 12, 31, 78, 97, 116, 120, 135, 159, 190, 218, 233, 250, 275, 311, 341, 355, 373, 377
ARAR. *See* Applicable or Relevant and Appropriate Requirements
Arbuscular mycorrhizae, 66, 289
Archimedes, 23
Argonne National Laboratory, 204–206
Aristotle, 4–6, 10, 23, 28, 33, 76
Aromatic, 45, 248, 279, 300, 302, 310–320, 333, 339, 363
Arrowweed, 124
Artesian pressure, 19
Artesian well, 25, 106
Artificial irrigation, 14, 165, 171, 215, 225, 331
Artois, 106
Asexual reproduction, 166, 167
Ashland Research Site, 124
Aspartic acid, 51
Assessment, 14, 83, 88, 104, 131–153, 156, 160, 166, 175, 176, 179, 180, 183, 187–195, 197, 233–236, 240, 263, 265–266, 269, 355, 357, 370
AST. *See* Aboveground storage tanks
Atmometers, 39
Atoms, 44, 47, 49, 51, 52, 64, 65, 76, 90, 252, 254, 256, 299–301, 325, 352, 371
ATP. *See* Adenosine triphosphate
Atrazine, 309, 310
Australia, 54, 55, 63, 121, 125, 126, 166, 177, 185, 186, 217, 227, 267, 293

J.E. Landmeyer, *Introduction to Phytoremediation of Contaminated Groundwater*,
DOI 10.1007/978-94-007-1957-6, © Springer Science+Business Media B.V. 2012

Automatic water-level recorder, 18, 19, 118, 123, 161
Autotrophs, 43, 63, 255, 273
Auxin, 55–57, 74, 167, 168, 170, 177, 300, 308
Av gas, 311
Axenic cell culture, 332, 364
Azospirillum, 255, 304
Azotobacter, 255

B

Bacillus, 304
Bacillus thuringiensis, 370
Backfill, 169–171, 176, 179, 180, 183, 201, 205, 206, 208, 324
Bacon, F., 6, 23
Bacteria, 45, 47, 50, 64–67, 75, 88, 139, 175, 215, 255–259, 262–263, 272–274, 288–292, 294, 296–298, 300, 303–305, 310, 313, 318, 319, 322, 325, 328, 330–333, 336–338, 351, 352, 355, 363, 364, 371, 377
Baldcypress, 65, 140, 163, 182, 187, 224, 293, 327, 354, 355
Bare-root cutting, 168–170, 196, 197
Bark, 7–9, 45, 46, 53, 56, 59, 64, 66, 67, 74, 76, 77, 84–85, 88, 160, 167, 176, 186–188, 214, 228, 261, 262, 293, 294, 303–305, 329, 330, 352, 355–359, 372
Barley, 180, 260, 278, 281
Barometric efficiencies, 207, 218
Barometric pressure, 113, 123, 148, 197, 218, 219
Bartram, W., 262
Baseflow, 110
Basin, 12, 17, 18, 23, 27, 29–32, 35, 38, 42, 54, 99, 114, 115, 119, 120, 121, 123, 126, 126, 181, 183, 189, 191, 209, 214, 220, 318, 346, 348, 373
Bauhin, G., 4
Beaupré, *P. trichocarpa* Torr & A. Gray x *P. deltoides* Bartr. *ex* Marsh, 117
Beavers, 136, 194
Bedrock, 111, 149, 150, 175, 202, 208, 230
Beer, 262
Beggiatoa, 67, 259
Benzene, 34, 173, 236, 240, 265, 275, 281, 284, 289, 290, 292, 297, 300, 310, 312–317, 324, 354, 366, 371
Benzene, toluene, ethylbenzene and xylene (BTEX), 236, 271, 273, 290, 297, 310–313, 321, 323, 324, 354, 360, 364, 366
Benzo[*a*]pyrene, 312, 313
Benzo[*e*]pyrene, 312
Bermuda grass, 156, 197
Bernoulli, D., 102
Bernoulli equation, 102, 106
Best management practices (BMP), 270
Betacarotene, 60
Bible, 28, 54, 155, 157
Bicarbonate, 51, 60, 62, 251, 253, 257, 291, 359
Bigwoods Experimental Forest, 123
Binomial classification, 10
Bioaccumulate, 276, 308, 309, 325
Bioaccumulation, 276, 277, 279, 280, 302, 339, 368, 372
Bioassay, 265
Bioavailable, 37, 65, 67, 69, 71, 92, 93, 109, 165, 174, 176, 181, 192, 209, 246, 253, 257, 258, 279, 300, 313, 318, 319, 335, 372
Bioconcentration factor, 281, 286, 372
Bioflavonoids, 301
Biofouling, 136
Biogeochemical, 245–274
Biogeochemical prospecting, 266
Bioindicator, 264
Biological oxygen demand, 170
Biomass, 5, 40, 53, 73, 81, 82, 117, 138, 156, 158, 166, 170, 176, 186, 239, 259, 271, 288, 290, 295, 297, 319, 323, 340, 352, 368, 372, 374, 376
Biosolids, 272
Biotransformation, 3, 279, 305, 338
Birch tree, 84
Black, J., 250
Blackwater, 262
Blaney-Criddle method, 190
Blood, 7, 8, 29, 102, 248, 251, 298, 299, 303, 375
Blood pressure, 102
Blue-green algae, 51, 60, 93
Blue Ridge Mountains, 366
BMP. *See* Best management practices
Bog, 11, 269, 295, 313
Bonnet, C., 24, 43, 76
Borehole, 104, 105, 133, 134, 146, 147, 149, 165, 169, 171, 183, 201, 202, 204–206, 208
Borer, 171, 174, 314, 352, 354
Boron, 177
Botanicom Gallicum, 9
Botany, 4, 9
Bottled water, 233
Boundary layer, 49, 51, 79, 86, 215
Bowen ratio method, 39, 190
Boyle, R., 250
Branches, 8, 29, 52, 53, 56, 59, 77, 78, 80, 81, 84, 86, 87, 168, 174, 185–188, 211–213, 216, 284, 330, 334, 355–357, 359, 366, 369
Bristlecone pine, 53
Bromide, 345, 362
Brownfields, 150, 373
Brownian motion, 46
Brown, R., 24, 46
BTEX. *See* Benzene, toluene, ethylbenzene and xylene
Buffer, 122, 165, 179, 256, 272, 274
Bulk density, 144, 175, 176, 274, 286, 343
Burkholderia, 304, 305

C

C_3, 50–51, 80, 215, 223, 360, 376
C_4, 50–51, 215, 223, 224, 360
CAA. *See* Clean Air Act
Cabbages, 37
Cairo, 112
Caisson, 204
Calcium oxalate, 89, 165, 263
California, 13–18, 40, 50, 54, 72, 74, 78, 119, 124, 141, 142, 160, 161, 177, 190, 194, 229, 268, 271, 275, 290, 323
Calvin-Benson cycle, 50
Calvin cycle, 25, 50, 252
CAM. *See* Crassulacean acid metabolism
Cambium, 46, 53, 56, 57, 64, 73–75, 84, 85, 183, 188, 287, 294, 298, 304, 308, 357, 372
Camerarius, R.J., 9, 23
Canadian Plains, 123
Candle, 43, 250, 251
Cannon, W.A., 11, 115
CAP. *See* Corrective Action Plan
Capillaries, 7, 68
Capillary fringe, 3, 11, 13, 16, 19, 21, 22, 23, 52, 58, 60, 68, 71, 76, 81, 83, 103–105, 109, 113, 114, 124–126, 143–146, 156, 159, 172–174, 178, 182, 183, 190, 192, 194, 195, 209, 211, 219, 226, 232, 250, 259, 291, 327, 338
Capture zone, 154, 195, 220, 221

Carbohydrate, 49–51, 56, 64, 89, 143, 168, 249, 250, 253, 278, 291, 295, 337, 367
Carbon, 25, 43, 50–52, 59, 65, 66, 80, 85, 88, 157, 182, 223, 224, 239, 249–255, 256, 270, 274, 279, 292, 295, 297, 300, 312, 325, 333, 337, 340, 344, 352, 354, 359, 360, 363, 365, 366, 374
Carbonate, 51
Carbon dioxide (CO_2), 36, 43, 115, 139, 176, 215, 245, 283, 309, 344, 351, 372
Carbonic acid, 51, 100
Carboniferous period, 61
Carbon sequestration, 239, 340, 374
Carbon tetrachloride, 304, 332, 356–359
Carolina Bays, 269
Carson, R., 264. *See also* Silent Spring
Casparian strip, 59, 71, 72, 258, 259, 278, 279, 281, 282, 285, 304, 305, 321, 331
Catabolic genotypes, 290
Catalyst, 47, 177, 253
Catalytic converter, 254, 320
Cation-exchange capacity, 138, 178, 196
Catnip, 9, 261
Cattails, 156, 157, 270, 271, 288, 294, 330
Caventou, J.B., 24, 47
Cavitation, 74, 76, 81, 82, 357
Cell, 6, 43, 98, 158, 205, 210, 240, 248, 278, 308, 344, 352, 365, 377
More than 300 instances. We have picked first instance page number from each chapter
Cella, 6
Cell division, 52–54, 56, 57, 85, 157, 168, 308
Cell membrane, 45–47, 62, 63, 68–70, 98, 177, 210, 211, 256, 258, 278–282, 284–286, 299
Cell theory of Life, 7, 24
Cellulose, 45, 51, 55, 274, 283, 289, 300, 304
CERCLA. *See* Comprehensive Environmental Response, Compensation and Liability Act
Cesalpino, A., 5, 23
Characterization, 131–153, 156, 159, 165, 166, 175, 176, 179, 180, 183, 189, 234, 235, 238–240, 341
Charleston, S.C., 84, 180, 195–200, 212, 213, 219, 240, 313–315, 317, 356
Chase, J.S., 14
Chelate, 67, 258, 292
Chemical oxygen demand, 170
Chemical potential, 90, 91, 282, 287
Chemolithotrophic, 259
Chinese, 28, 112
Chloride, 22, 25, 37, 70, 111, 126, 162, 177, 183, 194, 215, 267, 332, 334, 337, 345, 352–354, 363
Chlorinated solvents, 3, 153, 171, 202, 203, 206–208, 235, 237, 253, 265, 275, 285, 289, 296, 299, 301, 302, 304, 307, 314, 325–334, 339, 340, 346, 348, 354, 356, 361, 362, 366, 369, 376
Chlorobenzene, 313
Chloroform, 245, 263, 307
Chlorophyll *a*, 47, 48, 60, 177, 248, 337, 359, 360
Chlorophyll *b*, 47, 248
Chloroplast, 24, 44, 45, 47–50, 57, 61, 62, 79, 85, 88, 188, 247, 259, 301, 312, 359
Christ, 4
Chrysene, 281, 313
cis–1,2-DCE. *See* *cis*–1,2-dichloroethylene
cis–1,2-dichloroethylene (*cis*–1,2-DCE), 362, 367
Civil Action, 135
Civil war, 11, 12
Class A evaporation pan, 36. *See also* Evaporation pan
Clay, 11, 14, 15, 18, 59, 71, 97, 98, 100, 104, 106, 108, 109, 138, 143, 146, 149, 156, 165, 169, 173, 175, 176, 180, 195, 203, 204, 206, 207, 219, 221, 259, 270, 273, 316, 324, 330, 356
Clean Air Act (CAA), 239
Cleanup, 150, 220, 239–241, 270, 305, 364
Clean Water Act (CWA), 132, 239, 270, 351
Clear cutting, 122, 123, 163
Climate change, 57, 374
Climax community, 61, 164, 188
Clone, 9, 25, 119, 158, 159, 196, 204, 235, 260, 263, 304, 330, 332, 338, 374
Closed loop, 7, 8, 208
CO_2. *See* Carbon dioxide (CO_2)
Coachella Valley, 14, 15
Coal, 61, 176, 254, 264, 310, 312, 320
Coal tar, 195, 196, 312–314, 316, 317, 373, 374
Coastal Plain, 63, 104, 116, 120, 141, 175, 183, 234, 262
Cobalt chloride, 37, 215
Coenzyme A, 52
Cohesion-tension theory, 76–78, 80, 84, 210
Collenchyma, 47
Colorado River, 12, 124, 160, 174
Cometabolism, 292. *See also* Co-metabolism
Co-metabolism, 311, 330, 371
Common goods, 233
Compartmentalization, 259, 298, 303, 304, 346, 355
Compound-specific isotope analysis (CSIA), 360. *See also* Isotopes
Comprehensive Environmental Response, Compensation and Liability Act (CERCLA), 132, 137, 234, 239, 351
Comte (Count) of Buffon, 10. *See also* Georges-Louis Leclerc
Conduction, 211, 247, 278
Cone of depression, 112, 220
Confined aquifer, 78, 103, 105–109, 112, 114, 135, 145, 149, 183, 204, 205
Confining unit, 106, 195, 204, 207
Congaree National Park, 65, 119
Conifers, 61, 72–75, 84, 86–88, 138, 163, 187, 294, 323
Conjugation, 258, 298, 301–303, 335, 368, 370, 372
Consumptive use, 12, 31, 39, 42, 99, 119, 120, 155, 191, 373
Convection, 211, 247, 294, 345
Copper, 82, 177
Cork, 6, 23, 44, 46, 53, 84, 187, 262
Corn, 39, 73, 80, 86, 223, 240, 241, 249, 310, 325, 370, 372
Corrective Action Plan (CAP), 132, 366
Cortex, 52, 56, 57, 59, 62, 64, 65, 67, 69–72, 74, 84, 251, 253, 278, 280, 282, 287, 293, 304, 327, 344, 357, 362, 372, 376
Cottonwood, 15, 19, 29, 54, 55, 81, 82, 119–121, 124, 155, 158–161, 173, 181–184, 204, 214, 218, 219, 223, 294, 296, 297, 330, 356, 366, 371. *See also* Populus fremontii
Cotyledon, 9
Coweeta Experimental Forest, 120
Crassulacean acid metabolism (CAM), 50–51, 223–224, 295
Creosote bush, 17–19, 359
Cretaceous Period, 62, 158, 159, 260
Crop coefficients, 39, 200
Crude oil, 251, 310, 312, 320, 360
Crystal idioblasts, 89
Crystals, 47, 70, 89, 263, 267
Cultivar, 9, 158, 264
Cuticle, 49, 51, 79, 85–87, 248, 283, 286
Cutting, 9, 55, 58, 62, 92, 115, 121, 122, 146, 147, 160, 163–172, 177–180, 183, 186, 187, 196, 197, 202, 204, 207, 208, 235, 268, 282, 297, 300, 304, 309, 310, 312, 321–323, 329, 330, 334, 336–338, 358, 362, 364, 367, 368
CWA. *See* Clean Water Act (CWA)
Cyanide, 263, 264, 303

Cyanobacteria, 47, 88, 255
Cypress, 59, 64, 184, 293, 312, 354
Cypress knees, 64
Cysteine, 262, 301, 303
Cytochrome, 47
Cytochrome P–450, 299, 302, 332, 333, 335, 336, 368, 370, 371
Cytoplasm, 45–47, 49, 51, 53, 55, 58, 61, 63, 69–71, 73, 74, 79, 89, 177, 211, 252, 263, 283, 286, 299, 304
Cytoplasmic membrane, 45, 47
Cytosol, 252, 300

D

2,4-D, 56, 256, 308
Dalton, J., 30, 35
Dalton's law of partial pressures, 35
Darcy, H., 95–101
Darcy's law, 95–101, 108, 113, 114, 143, 148, 193, 202, 222, 342, 343, 345, 349
Dark reaction, 50, 252
Darwin, C., 10, 11, 25
da Vinci, Leonardo, 29, 74, 77
DBCP. *See* Dibromochloropropane
DDT. *See* Dichlorodiphenyltrichloroethane
Dead Sea, 14
Dealkylation, 298, 299, 301, 309
De Architectura Libri Decem, 28
de Candolle, A., 9
Decarboxylation, 298, 299, 301
Decaying, 5–7, 53, 59, 61, 67, 75, 231, 246, 253, 294, 304, 310, 374
Dechlorination, 295, 296, 328, 330
Deciduous, 40, 48, 55, 57, 61, 74, 75, 84, 85, 87, 88, 138, 163, 186–188, 215, 217, 228, 245, 268, 359, 364, 366
Decision tree, 151, 152, 371
Decomposers, 249
Defensive chemical, 245, 253, 260, 365
Dehalogenation, 298, 299, 302
De l'origine des fountains (Treatise on the Origin of Springs), 29
Democritus of Abdera, 246
Denaturing gradient gel electrophoresis (DGGE), 351–352
Denitrification, 66, 122, 255, 270, 271
Dense nonaqueous phase liquid (DNAPL), 196, 202, 314–317, 343, 362
Deoxyribonucleic acid (DNA), 45–47, 58, 60, 240, 257, 300, 301, 303, 352, 363, 370
Department of Defense (DOD), 22, 154, 203, 237, 335
Department of Energy (DOE), 153, 156, 158, 204, 208, 237
de Saussure, N.T., 11, 24, 44
Desert Botanical Laboratory of the Carnegie Institution of Washington, 11
Desert, deserts, 11, 13–15, 17, 40, 51, 82, 89, 117, 118, 137, 141, 142, 155–157, 162, 181, 229, 247, 268, 295, 334, 359
Des Plantis libri, 5, 23
Determinate growth, 53
Detoxification, 245, 251, 263, 276, 282, 290, 292, 297–305, 307–309, 325, 335, 336, 338, 357, 365, 368–369, 371, 372, 375, 376
Deuterium, 224
Dew, 30, 61, 89, 99, 122, 224, 229
DFA. *See* Direct fluorescent antibody
DGGE. *See* Denaturing gradient gel electrophoresis
Diacytic, 78
Dialysis, 361–362
Dibromochloropropane (DBCP), 278, 307
Dibromoethylene, 278, 281, 307, 320, 321, 325. *See also* Ethylene dibromide (EDB)
Dichlorodiphenyltrichloroethane (DDT), 245
1,2-Dichloroethane, 320
Dicotyledons, 9, 23, 52, 223, 258
Differentiation, cells, 8, 45
Diffuse porous
Diffusion, 8, 35, 36, 38, 49, 51, 56, 58–65, 68–72, 75, 79, 81, 83, 85, 104, 142, 165, 168, 170, 175, 176, 211, 215, 223, 224, 230, 250–253, 256, 258, 267, 275, 277, 278, 280–286, 291–296, 299, 304, 305, 327–330, 343–345, 354, 356–358, 361–362, 369, 370
Diffusion traps, 322, 323, 329, 358–359
1,3-Dinitrobenzene (DNB), 264
Dioscorides, 5
Dioxane, 333–337
Dioxygenase, 290, 299, 300
Direct fluorescent antibody (DFA), 352
Direct push, 146–147
Discharge, 14, 27, 54, 96, 117–123, 135, 155, 189, 218, 235, 270, 296, 330, 341, 351, 373
Discourse Admirables, 23, 29
Dispersion, 343–345, 349
Dissolved organic matter (DOM), 279
Diurnal, 113, 122, 124–125, 161, 193, 218, 219, 221, 223, 361
Divining rod, 27
DN34, 297, 312, 321, 334
DN–34, 9, 268, 310, 322, 324, 329, 334, 336, 337, 367. *See also* DN34
DNA. *See* Deoxyribonucleic acid
DNAPL. *See* Dense nonaqueous phase liquid
DNT, 333, 336
DOD. *See* Department of Defense
DOE. *See* Department of Energy
Dormancy, 43–56, 123, 136, 141, 151, 163, 167–170, 186, 188, 222, 245, 268, 269, 376
Dorschkamp, *P. deltoides* x *P nigra* L., 117
Drought, 13, 28, 33, 54, 65, 68, 74, 76, 81, 82, 110, 117, 118, 120, 134, 136, 143, 157, 171, 180, 182, 199, 201, 204, 206, 211, 216, 228, 234, 340, 363, 374
Druids, 56, 346, 347
Duckweed, 62, 83, 265
Dupuit, 24, 104
Dupuit-Forchheimer, 104
Dust bowl, 54, 57
Dutrochet, H.H., 11, 24
Dutrochet, J.H., 68
Dyes, 76, 77, 231, 233

E

Earth tides, 78, 113, 218
Eastern cottonwood, 158, 204, 356. *See also Populus deltoides*
Eastern red cedar, 162, 163
Ecohydrology, 115
Ecology, 121, 127, 133, 158, 164, 185, 207, 216, 246–259, 365, 376
Ectomycorrhizae, 67, 289, 371
Ectotrophic mycorrhizae, 67, 69
Eddy covariance, 230
EDTA, 258
Effective porosity, 98, 100, 105, 106, 198, 219–221, 342, 343, 345, 363
Egypt, 14
Egyptian, 112
Eh, 296
Elat Gulf, 14
Electric Power Research Institute (EPRI), 313
Electrochemical gradients, 70
Electromagnetic energy, 141, 231, 247, 266
Electron acceptor, 252, 256, 271, 296, 302, 322, 337, 338, 344, 371
Electron donor, 253, 337

Electrons, 33, 49, 51, 60, 64, 78, 82, 88, 98, 149, 246–249, 252–255, 257, 258, 292, 299, 301, 325, 328, 331, 333, 336, 359
Electron transport, 49, 253, 257, 259, 303
Elements, 3, 4, 10, 45, 49, 60, 66, 82, 122, 174, 176–177, 223, 246, 249, 250, 257, 266, 352–354, 363, 367
Elevation head, 102, 103, 106
Elimination reaction, 298
ELISA. *See* Enzyme-linked immunosorbent assay
Elkington, J., 24, 95
Embden-Meyerhof-Parnas pathway (EMP), 252
Embolism, 74, 76, 78, 84, 168
EMP. *See* Embden-Meyerhof-Parnas pathway
Empedocles of Agrigentum, 4
Endodermal, 11, 279, 282, 285
Endomycorrhizae, 289
Endophytes, 263, 289, 292, 304–305
Endoplasmic reticulum, 46, 299
Endotrophic mycorrhizae, 66, 67
Energy, 8, 27, 43, 101, 117, 141, 156, 189, 222, 237, 245, 286, 312, 359, 374 More than 300 instances. We have picked first instance page numberfrom each chapter
Energy balance, 38, 222, 227–228
Enquiry into Plants, 4, 5, 23, 47
Environments, 4, 43, 50, 51, 61, 63, 64, 67, 111, 115, 119, 150, 173, 245–274, 277, 294–296, 326, 361, 364
Enzyme-linked immunosorbent assay (ELISA), 352
Ephesus, 121
Epidermal cells, 46, 58, 59, 61, 71, 79, 187, 258, 280, 286, 287
EPRI. *See* Electric Power Research Institute
Equilibrium, 30, 34, 51, 68, 83, 90, 92, 101, 142, 149, 174, 184, 185, 190, 276–278, 280–282, 284–287, 344, 356
Equipotential lines, 108, 109, 111, 195, 221
Eratosthenes, 112
Ericoid mycorrhizae (ERM), 289
Erosion, 12, 54, 57, 121, 123, 136, 138, 149, 197, 274
Esherichia coli, 45
Essential oils, 261, 263, 375
ET_a. *See* Actual evapotranspiration
Ethanol, 34, 250, 299, 321, 325, 330
Ether, 321, 354
Ethylbenzene, 236, 281, 310, 314
Ethylene dibromide (EDB), 278, 281, 307, 320, 321, 325. *See also* Dibromoethylene
ET Package, 231, 347
Eucalyptus, 54, 121, 140, 177, 184–186, 214, 217, 261, 264, 268, 321, 332
Eukaryote, 47, 299
Eutrophication, 274
Evaporation, 3, 8, 11, 16, 18, 19, 21, 24, 28, 30–42, 59, 68, 69, 71–74, 76, 81, 83, 88–90, 92, 109, 110, 115, 119, 122, 126, 144, 161, 171, 177–179, 182, 183, 189–190, 206, 209, 215, 216, 218, 222, 224, 226, 229, 247, 268, 307, 330, 334, 359, 363
Evaporation pan, 36–38, 41
Evaporite deposits, 11
Everglades, 55, 118, 126, 256, 294
Evolution, 10, 28, 43, 60–62, 66, 166, 245, 371, 375
Experimentation, 4, 28–30
Experiments and observations, 44
Experiments on vegetables, 24, 44, 246
Explosives, 3, 305, 333, 335, 336, 339, 367, 370, 371
Exposure, 4, 47, 56, 67, 86, 87, 109, 140, 187, 212, 240, 255, 259, 261, 264–267, 271, 273, 290, 298, 299, 303, 304, 309, 310, 312, 313, 322, 330, 331, 334, 336, 337, 354, 365, 367–370, 372
Extinction depth, 205, 231, 348
Extract, 4, 57, 144, 153, 261, 308, 353, 362
Exudates, 71, 288, 290–292, 295, 300, 318, 319, 328, 337, 344, 359, 376

F

FACT. *See* Flow and contaminant transport
Facultative phreatophytes, 11, 12, 27, 62, 91, 155, 159, 171, 172, 180, 192, 202
FAD. *See* Flavin adenine diphosphate
Fairchild, D., 55
Fan palm, 14. *See also Washington filifera*
Faults, 14, 15
FDA. *See* Food and Drug Administration
Federal Insecticide, Fungicide, and Rodenticide Act (FIFRA), 265
Federal Water Pollution Control Act (FWPCA), 270, 351
Federal Wetland Delineation Manual, 269, 351. *See also* FWPCA
Fen, 118, 269, 295
Fermentation, 250, 325
Ferns, 61, 73, 172
Ferredoxin, 255, 257
Ferric, 175, 257–259
Ferrireductase, 258
Ferrous, 175, 256–259, 292, 296
Fertilizer, 65, 132, 157, 165, 176, 176, 179, 238, 257, 272, 324, 373
Fibonacci series, 87
Fibrous roots, 291
Fick's diffusion equation, 358
Field capacity, 109, 144, 210
FIFRA. *See* Federal Insecticide, Fungicide and Rodenticide Act
First-order degradation, 240, 288, 342
Flavin adenine diphosphate (FAD), 52, 257
Flavonoids, 301
Flood, 274
Flood plain, 12, 117, 119–122, 124, 160, 174, 180, 269, 271, 294, 348, 354
Flow and contaminant transport (FACT), 348
Flowing fluid, 102
Flow net, 108
Fluctuations, groundwater, diurnal, 18–23, 112–113, 198–200
Fluid mechanics, 98
Fluoranthene, 312, 313
Fluorescence, 359–360
Fluorine, 313
Fluorometric analysis, 231
Flux, 203–204
Foliage volume, 191
Fontana, F., 46
Food and Drug Administration (FDA), 265
Forest, 18, 57, 61, 64, 67, 81, 89, 122, 123, 141, 149, 158, 164, 165, 183, 185, 203, 204, 216, 217, 227, 230, 249, 296, 335, 365
Formaldehyde, 261, 325, 366
Fossilization, 10
Fossils, 29, 60–62, 66, 72, 157, 159, 251, 310. *See also* Fossilization
Fossil water, 61
Four elements, 4
Fourier's law, 211
Fractionation, 223, 224, 363

Framework, 30, 44, 135, 145, 150, 189, 193, 200, 201, 203–208, 246, 249, 276, 307, 308, 341–343, 345–346
France, D., 95
Frankia, 66, 255
Free energy, 49, 90, 91
Freeman method, 37
Free product, 133, 275, 325, 341
Free radicals, 257, 299, 301
Free-surface water evaporation, 189–190
French drain, 204, 314
Frost, 19, 20, 89, 136, 169, 170, 248
Fruits, 9, 245, 266, 301, 369, 372
Fuel additives, 325. *See also* Additives
Fuel oxygenates, 3, 320–321. *See also* Oxygenates
Fumigant, 261, 278, 307, 325
Functionalization, 298, 299
Fungi, 5, 7, 45, 47, 61, 65–67, 88, 258, 262, 263, 273, 288, 289, 291, 300, 304, 313, 331, 374
FWPCA. *See* Federal Water Pollution Control Act

G

Gaining stream, 124
Gas, 5, 24, 34, 35, 43, 44, 51, 52, 55, 56, 60, 61, 64, 74, 76, 78–81, 84–86, 88, 92, 107, 133, 134, 139, 142, 164, 176, 188, 195, 196, 198–200, 210, 215, 218, 223, 224, 231, 240, 245, 246, 249–254, 264, 266, 271, 275–278, 280, 284, 285, 293–297, 311, 320–323, 326, 327, 329, 331, 354, 356, 357, 359, 360, 363–364, 366, 376
Gas bags, 231, 359
Gas-flow porometer, 215, 360. *See also* Porometer
Gasoline, 133, 137, 139, 183, 195, 234, 236, 259, 270–271, 275, 299, 300, 307, 310–314, 320, 321, 323, 324, 328
Gatewood, J.S., 17, 22, 25, 54, 118, 159, 160
Gene, 60, 66, 86, 99, 240, 241, 260, 290, 305, 336, 352, 371
Gene clock, 60
Gene flow, 371
Genera Plantarum, 9
Genome, 80, 159, 240, 371, 376
Genus, 9, 140, 158–160, 166, 264, 303, 334
Geochemistry, 89, 99, 122, 126, 183, 223, 277, 353, 361
Geology, 12, 14, 24, 52, 95, 104, 207, 266
Geotextiles, 136
Geothermal water, 14
Germination, 4, 47, 54, 61, 67, 72, 158, 163, 183, 227, 265, 303, 364
Gibberellin, 57
Gibbs, J.W., 90
Gila River, 22, 118, 121, 353
Girdling, 7
Glacier, 245
Global warming, 61, 374
Glucose, 45, 50–52, 64, 66, 249, 250, 252, 261, 274
Glutathione, 300, 301, 303, 363, 368, 370
Glutathione-S-transferase (GST), 300, 302, 336
Glycolysis, 50–52, 250, 252
Glycoside, 261, 263, 264, 302, 303, 368
Göethe, 24, 52, 84. *See also* Johann Wolfgang von Göethe
Golgi bodies, 46
Goodding willow, 119. *See also Salix gooddingii*
GORE-TEX, 68, 364
Goths, 121
GRACE. *See* Gravity Recovery and Climate Experiment
Gravity, 8, 20, 21, 28, 29, 58, 59, 73, 75, 76, 83, 91, 95, 101–105, 107, 109, 112, 143, 144, 175, 179, 190, 195, 221, 230, 271, 275, 297, 325, 326, 345
Gravity Recovery and Climate Experiment (GRACE), 230
Greeks, 28, 29, 112
Greenhouse, 164, 239, 332, 335, 369, 374
Green liver, 25, 299–300, 336
Green wood, 161, 188
Grew, N., 7, 9, 24
Ground-penetrating radar (GPR), 149–150. *See also* Radar
Groundwater, 3, 27, 43, 95, 117, 131, 155, 189, 209, 233, 245, 275, 307, 341, 351, 365, 375
Groundwater Decision Tree, 151, 152. *See also* Decision Tree
Groundwater flux, 203–204. *See also* Flux
Groundwater model, 113–114, 220, 231–232. *See also* Model
Groundwater system, 18, 111, 113, 145, 192, 193, 199, 205, 272, 333
Grow bags, 167
Growing degree days (GDD), 40, 190
Growth, 4, 29, 43, 99, 117, 137, 156, 205, 226, 233, 246, 285, 308, 343, 351, 371, 376
Growth Rings, 256, 303
GST. *See* Glutathione-S-transferase
Guano, 254
Guard cells, 55, 78–80, 85
Gulf of Mexico, 116, 118
Guttation, 69, 353
Gymnosperms, 61, 74, 159, 166

H

Haber-Bosch process, 254
Hack-Slack Cycle, 51
Haeckel, E., 246
Hagen, 96
Hail, 55, 136
Hales, S., 8, 10, 24, 31, 37, 99, 102, 250
Half-life, 302, 303, 309, 321, 334, 337, 366
Halley, E., 24, 30–32
Halophytes, 11, 12, 265, 353
Hand auger, 104, 105, 146, 147, 165, 175, 331, 351
Hardwood cutting, 167, 207. *See also* Cutting
Hardwoods, 74, 184
Harvey, W., 7, 8, 102, 118
Hawaii, 58
Headspace, 92, 314, 319, 324, 326, 327, 354, 356, 369, 370
Heartwood, 186, 262, 266, 294, 304, 354, 355
Heat flux, 38, 39, 247
Heat-pulse method, 37
Heavy water, 188
Helium, 247
HELP. *See* Hydrologic Evaluation of Landfill Performance model
Hemlock, 4, 263
Hemoglobin, 47, 248, 303
Henry's law, 277, 282–284, 326, 327, 358
Herbaceous, 17, 18, 40, 46, 57, 67, 84, 87, 89, 156–157, 163, 164, 167, 173, 269, 289, 310, 338
Herbal, 259
Herbicides, 165, 261, 265, 281, 282, 300, 308–309, 362, 364, 370
Herbivores, 89, 260, 261, 303, 365
Herbivory, 46, 52, 73, 245, 260, 262, 297
Heterogeneous, 98, 196, 207
Heterosis, 158
Heterotrophs, 66, 273
Hidden hunger, 257
High-melting-point explosive (HMX), 333, 335, 336, 339, 367, 368
Hippocrates, 5
Histoire Naturelle, 10, 24
Historia Plantarum, 9, 24. *See also* John Ray and Theophrastus
Hokkaido, 122
Homer, 28
Homogeneous, 44, 98, 99

Hooke, R., 6–7, 23, 44
Hormones, 52–57, 74, 80, 158, 167, 253, 263, 308
Horton overland flow, 136. *See also* Overland flow
Hubbard Brook Experimental Station, 121
Humid, 11–14, 33, 38, 39, 48, 50, 81, 82, 89, 103, 109, 111, 118–120, 123, 126, 145, 148, 157, 159–160, 162–163, 170, 172, 181, 184, 192, 194, 200, 215, 228, 229, 261, 334, 335, 371
Humidity, 14, 19, 34–36, 40, 62, 68, 69, 79–82, 84, 91, 99, 124, 140, 142, 168, 170, 183, 184, 192, 194, 196–198, 213–215, 228–229, 283, 373. *See also* Humid
Humus theory, 4, 5, 10, 23
Hvorslev method, 222
Hybrid, 24, 58, 62, 76, 84, 156, 158, 164–166, 168, 169, 171–173, 183, 184, 196–198, 201, 202, 204–207, 213, 223, 230, 234, 239, 284, 288, 303–305, 309, 310, 312, 314–316, 321, 322, 324, 325, 330, 332, 334, 336–339, 342, 367, 370, 371, 374, 376
Hybridization, 158
Hybrid poplars, 37, 76, 156, 158, 165, 187, 197, 202, 207, 239, 288, 309, 325, 332, 338. *See also* Poplars
Hydrathodes, 86
Hydraulic conductivity, 11, 35, 36, 62–63, 71, 74, 76, 81, 83, 91, 96–100, 106–108, 114, 116, 125, 143, 145, 147–149, 176, 193, 196, 198, 201–204, 206, 210, 211, 220, 222, 223, 232, 236, 270, 342, 343, 345–347, 361–363
Hydraulic gradient, 21, 74, 96–98, 101, 102, 104, 108, 112, 120, 125, 137, 180, 198, 222, 235, 342, 345, 361, 363
Hydraulic head, 96, 101, 123, 148
Hydraulic lift, 71–72, 221, 226, 227
Hydraulic push, 72
Hydraulic tomography, 108
Hydroecology, 13, 115, 137–144, 156
Hydrogen, 33, 34, 43, 49, 50, 64, 67, 70, 76, 89–91, 139, 176, 178, 223–225, 227, 245, 247, 248, 251, 252, 254, 255, 259, 263, 267, 291, 292, 295, 301, 320, 325, 334, 337, 352, 359, 363
Hydrogen sulfide, 34, 50, 67, 259, 267, 295. *See also* Sulfide
Hydrogeologists, 3, 11–14, 16, 27, 102, 113, 123, 132, 151, 361
Hydrogeology, 4, 11, 12, 16, 23, 25, 83, 95–114, 145, 153, 173, 194, 196, 207, 210, 220
Hydrologic barrier, 136, 151, 171–173, 236, 237
Hydrologic control, 18, 114, 132, 135–137, 142, 144, 145, 148, 150, 151, 153, 156, 164–166, 171, 173, 180, 184, 185, 189, 190, 192–194, 197, 205, 209, 225, 234–240, 338, 377
Hydrologic cycle, 27–33, 42, 43, 82, 95, 115, 179, 189, 206, 245–247, 256, 259, 272, 284, 288, 334
Hydrologic Evaluation of Landfill Performance (HELP), 232
Hydrologists, 10, 97, 113, 352
Hydrology textbook, 25, 110
Hydrolysis, 50, 51, 255, 298, 299, 302, 309
Hydropher, 102, 103
Hydrophilic, 45, 46, 276, 279, 281, 299, 303, 321
Hydrophobic, 45, 46, 175, 258, 276, 278, 279, 283, 299, 303, 369
Hydrophytes, 11, 12
Hydroponic, 62, 83, 281, 289, 295, 296, 304, 308, 310, 321, 323, 325, 329, 330, 334, 336–338, 359
Hydroquinones, 313
Hydrostatic pressure, 22, 91, 92, 190, 210, 219
Hydroxylation, 298–301, 312
Hydroxyl radical, 259, 301, 366
HYDRUS, 349
HYDRUS–1D, 171
HYDRUS–2D, 349
Hygroscopic, 144
Hyperspectral sensing, 231
Hysteresis, 109, 112

I

Idioblasts, 89
Igneous rocks, 173
Iliad, 28
Illustrated Exposition Of Plants, 4
Imperial Carolina, 171, 196, 201, 310, 312
Increment borer, 314, 352, 354. *See also* Borer
Indeterminate growth, 53
Index Herbariorum, 5
Indolbutyric acid (IBA), 167
Indoleacetic acid (IAA), 55
Infiltration, 13, 31, 72, 82, 83, 109, 110, 117, 124, 126, 136, 142, 143, 145–147, 165, 168, 170, 171, 173, 175, 176, 178, 182, 190, 193, 205, 206, 219, 222, 312, 334, 349
Infiltrometers, 146
Infrared gas analyzers (IRGA), 359
Ingemarsson, C., 9. *See also* Linnaeus, C.
Ingemarsson, N., 9
Ingenhousz, J., 24, 44, 247
Inorganics, 43, 44, 64, 65, 70, 82, 139, 165, 175–176, 182, 249, 251, 256, 356
Insecticide, 184, 245, 261, 265, 308
Insolation maps, 88, 141
Intermolecular forces, 33, 34
Internode cutting, 167. *See also* Cutting
Interstate Technology & Regulatory Council (ITRC), 150–153, 156, 157, 160, 161, 163, 187, 234
Intrinsic permeability, 97
Ion pump, 70, 354
Ions, 8, 49, 60, 70, 122, 179, 253–255, 258, 267, 291, 292, 353
Iron, 47, 49, 64, 67, 110, 175, 177, 180, 183, 245, 246, 250, 251, 255–259, 264, 266, 272, 292, 294–297, 300, 303, 353, 354, 366
Iron oxidation, 295
Iron-reducing bacteria, 175
Iron reduction, 49, 183, 258, 297
Irrigation, 14, 16, 18, 143, 145, 151, 156, 162, 165, 171, 178–180, 201, 204, 207, 215, 225, 235, 238, 324, 329, 331, 346, 369, 376. *See also* Natural and artificial
Isocitrate, 252
Isotopes, 44, 72, 88, 110, 115, 155, 159, 163, 174, 223–227, 231, 232, 256, 258, 360, 363
Isotopic composition, 44, 223, 224, 226, 363
Israel, 14, 112, 160
ITRC. *See* Interstate Technology & Regulatory Council

J

Jesuits, 9
Jet fuel, 172, 259, 311
Johann Wolfgang von Göethe, 52
Joseph's well, 112
JP–8, 311
Juglone, 261, 263, 303
Jurassic Period, 72

K

Kamen, M., 44
Kaustsky effect, 359
K_d, 276–278, 343, 345
Kinetic energy, 43, 50, 89, 102, 247
King Solomon, 28
Kioritz, 308
K_{oc}, 277, 310, 321
Köelreuter, J.G., 24, 158

K_{ow}, 85, 276–287, 304, 305, 307, 310, 313, 321, 323, 326, 334, 341, 344, 348, 349, 372
Kreb's cycle, 52, 252, 301
K-T boundary, 159
K-T extinction, 159
Kudzu, 54, 55

L
LAI. *See* Leaf area index
Lakes, 27, 28, 31, 33, 38, 73, 78, 111, 115, 117, 121, 135, 141, 145, 147, 159, 163, 218, 227, 269, 271, 274
Lamarck, J.B., 12, 102
Landfill covers, 136
Langmuir isotherm, 332
Lapland, 74
Lateral bud, 53, 55, 56, 168, 169, 185, 187
Laurent de Jussieu, A., 9
Lava tubes, 58
Lavoisier, A., 33, 250, 251
Law of Laplace, 75
Leachate, 131, 134, 136, 145, 151, 171, 334–336, 367, 373
Lead, 34, 36, 56, 57, 75, 76, 91, 98, 102, 104, 116, 118, 121, 122, 131, 137–139, 141–143, 150, 165, 174, 175, 177–178, 182, 185, 186, 189, 194, 210, 221, 239, 252, 271, 272, 279, 280, 290, 291, 293, 296, 298, 301, 302, 305, 311, 320, 324, 338, 344, 371, 372
Leaf, 5, 34, 43, 116, 136, 163, 191, 211, 245, 278, 307, 352, 365
Leaf area index (LAI), 182, 184, 185, 187, 200, 201, 213, 216, 217, 232, 239, 274, 360
Leaf drop, 20, 55, 84, 85, 87, 88, 168, 170, 186, 188, 216, 228, 245, 330, 336, 365–367
Leaf litter, 59, 67, 116, 126, 184, 253, 366–367
Leaves, 3, 30, 43, 99, 115, 144, 155, 191, 209, 245, 276, 308, 352, 365, 375
Leclerc, G.L., 8, 10, 24. *See also* Comte (Count) of Buffon
Leghemoglobin, 66, 255
Legumes, 65, 66, 255, 335
Lens, grinding, 6
Lenticels, 52, 56, 62, 74, 76, 81, 84–85, 188, 252–253, 262, 293, 294
Leucoplasts, 47
Lichen, 61, 67, 73, 89, 174
Lidar, 230
Life-cycle approach, 234, 236, 237
Light reaction, 251–252, 308
Lignin, 45, 47, 73, 82, 274, 282, 283, 289, 300, 302, 304
Lime, 179, 207
Limestone, 28, 99, 100, 163, 179, 203, 251
Limonene, 262, 301, 366
Linden tree, 9
Linnaeus, C., 8, 10, 24, 74. *See also* Ingemarsson, C.
Liquid culture, 36, 83
Little San Bernardino Mountain, 14
Liver, 25, 276, 288, 298–300, 336
Loblolly pine, 123, 163, 184, 326, 354, 356. *See also* Pine
Locust, 260, 338
Log K_{ow}, 277–283, 285–287, 304, 305, 307, 309, 310, 313, 321, 323, 326, 334, 344, 372
Long Island, New York, 116, 191
Longleaf pine, 119, 141, 163. *See also* Pine
Losing stream, 121
Love Canal, 273
Low-flow groundwater sampling, 361. *See also* Sampling
Lucas, J., 12, 25, 102
Lysergic acid, 263
Lysimeter, 37, 39, 361, 363
Lysosomes, 46

M
Magnesium, 47, 48, 177, 196, 256, 353
Malate, 252
Malic acid, 51, 224, 318
Malpighi, M., 7, 8, 23
Manganese, 125, 177, 294, 296
Mangrove, 89, 266–268, 293
Manometers, 96, 102, 147
Manufactured gas plant (MGP), 195–200, 219, 240, 303, 313–315, 317
Mariotte, E., 10, 29, 30, 32, 102
Marsh, 20, 117, 126, 207, 267, 271, 292, 293
Marsh marigold (*Caltha palustris* L.), 126
Mass flux, 38, 203, 341, 342
Matric potential, 91
Mayer, R., 6–7, 44, 46, 274, 352, 355, 374
Mayow, J., 250
Medicago sativa, 14–16, 156, 181, 197, 201, 256, 290, 311, 318, 320, 335, 338, 339
Mediterranean Sea, 30, 54
Meinzer, O.E., 13–19, 25, 155–157, 159, 161, 162, 166, 172, 181, 183, 218, 266, 361
Melaleuca, 54, 55
Meltwater, 121
Meniscus, 35
Mercury, 34, 35, 96, 102, 258, 374
Mesophyll, 49, 51, 79, 80, 85, 188, 223, 247, 283
Mesophytes, 11, 12
Mesquite, 12, 17–20, 117, 121, 161–163, 173, 181, 218, 227, 268. *See also Prosopis* spp.
Metabolism, 45, 50–53, 83, 84, 140, 143, 224, 247, 249–251, 255, 262, 274, 275, 282, 285, 287, 289, 292, 295, 298, 299, 302, 303, 309, 311, 312, 322, 330, 331, 345, 353, 369, 371, 372
Metabolites, 245, 280, 292, 309, 318, 332, 336, 365, 367, 370, 372
Metalloenzymes, 47
Metals, 3, 25, 73, 122, 173, 245, 258, 264, 266, 272, 276, 307, 314, 339, 353
Metamorphic rocks, 100
Meteorologica, 28
Methane, 75, 82, 139, 245, 253, 264, 292–295, 320, 328, 362
Methane monooxygenases, 328
Methanogenesis, 51, 294, 296, 297
Methanol, 34, 283, 324, 331, 354
Methyl *tert*-butyl ether (MTBE), 354
MGP. *See* Manufactured gas plant
Michaelis-Menten, 345
Microbes, 25, 180, 246, 255–257, 259, 260, 265, 269–274, 288–290, 292, 296, 299, 300, 305, 310, 318, 319, 330, 333, 338, 351, 352, 364, 371, 374, 377. *See also* Microorganisms
Microfibrils, 45
Micrographia, 6, 23
Micronutrients, 64, 82, 138, 177, 246, 256
Microorganisms, 4, 49, 65, 144, 176, 180, 273, 289, 294, 338
Microscope, 5–7, 9, 10, 23, 44, 46
Middle ages, 28–29
Mint, 43, 160
Mississippi River, 13, 159
Mistletoe, 56, 188
Mitochondria, 46, 49, 51, 52, 252
MNA. *See* Monitored natural attenuation
Model, 8, 29, 71, 97, 137, 159, 204, 221, 239, 285, 313, 342, 357, 373
MODFLOW, 114, 205, 230, 231, 346, 347
MODPATH, 114, 205, 346
Molds, 136, 194
Molecular clock, 60
Molecular mass, 278
Molybdenum, 177, 255

Mona Lisa, 158
Monitored natural attenuation (MNA), 136, 365
Monocotyledons, 9, 23, 52
Monoculture, 66, 166, 261, 267,
Monooxygenases, 299, 300, 328, 368
Monteith method, 38, 39
Morel, G., 25, 158
Moss, 60, 61, 165, 224
Most probable number (MPN), 289, 296, 310, 351, 352
Mouse, 43, 250
MPN. *See* Most probable number
MS4, 270
MTBE, 271, 275, 279, 280,. 281, 284, 292, 297, 302, 307, 320–325, 328, 346, 354, 366. *See also* Methyl *tert*-butyl ether
Mucigel, 57, 66, 71, 288, 291
Municipal storm-sewer system, 270. *See also* MS4
Mushrooms, 67, 289
Mutagen, 370–372
Mycellium, 67
Mycorrhizae, 64–67, 69, 197, 266, 289, 294

N
NADP. *See* Nicotinamide adenine dinucleotide phosphate
NADPH, reduced NADP, 49, 50, 250, 252, 299
nahG, 290
Naphthalene, 281, 290, 297, 301, 305, 312–320, 362, 364, 370
Naphthalene dioxygenase (ndoB), 290, 305
NAPL. *See* Non-aqueous phase liquid
National Pollutant Discharge Elimination System, 270
National Research Council, 145, 220, 272, 273
National Secondary Drinking Water Standard (NSDWS), 259
National Solar Radiation Data Base, 141
National Water-Quality Assessment Program (NAWQA), 263
Native Americans, 14, 29
Natural attenuation, 135, 136, 238, 269, 272–274, 307, 320, 326, 340, 341, 343, 344, 348, 355, 377
Natural History, 4, 23
Natural irrigation,
Naval Air Station, 203
Naval Undersea Warfare Center (NUWC), 207
Navier-Stokes equation, 98
NAWQA Program, *See* National Water-Quality Assessment Program
NDMA. *See* N-nitrosodi methylamine
Negev desert, 14
Neolithic, 112
Nero, 5
Nested wells, 147, 221
Net Present Value (NPV), 236, 237
Neutron probes, 124, 220
Neutrons, 44
Nevada, 13, 40, 54, 72, 121, 142, 191, 334, 373
Newton, I., 96, 101, 102
Nicotinamide adenine dinucleotide phosphate (NADP), 49, 52, 250, 252
Nicotine, 52, 261, 308
Nicot, J., 261
Nighttime transpiration, 214, 215
Nitrate gardens, 256
Nitrogen, 54, 122, 138, 157, 253, 291, 319
Nitrogenase, 254, 255
Nitrogen fixation, 65–66, 254, 255
Nitrogen-fixing bacteria, 65, 66, 255, 256
Nitrogen stable isotopes, 256
Nitrotoluene monooxygenase (*ntnM*), 290, 305
2-Nitrotoluene reductase *(ntdAa)*, 290, 305
N-nitrosodi methylamine, (NDMA), 333–335
Nocturnal sap flow, 214–215
Nodule cell cultures, 332, 364
Non-aqueous phase liquid (NAPL), 314
Nonexclusive resources, 233
Nonpoint source, 307
NPDES. *See* National Pollutant Discharge Elimination System
NSRDB. See National Solar Radiation Data Base
Nucleus, 24, 46, 49
Nursery, 166, 168, 170, 204, 338
Nutrients, 3, 59, 61, 65, 67, 82, 83, 118, 122, 126, 126, 138, 165, 166, 173–177, 183, 201, 216, 246, 251, 253, 257, 264, 271, 272, 274, 297, 366, 374
Nuts, 9, 245, 303, 369, 370
NUWC. *See* Naval Undersea Warfare Center

O
Oak, 52, 140, 162, 163, 199, 202, 232, 262, 296, 324, 330, 331, 354–356
Oasis, 15, 38
Obligate phreatophytes, 11, 12, 54, 57, 155, 222, 225, 232
Observation well, 15
Ocean, 28, 29, 31, 35, 60, 69, 109, 156, 171, 218, 225, 259, 263, 323
Octanol, 85, 276, 278, 284, 287
Octanol-water partition coefficient, 277
Offensive chemicals, 245, 260
Office of Research and Development and Technology Innovation Office, 153
Ogden, U.T., 201
Ohm's law, 98
Okeefanokee swamp, 111
O&M. *See* Operation and maintenance
On the Causes of Plants, 4, 5, 23
OP–367, 183, 197, 223, 267–268
Operation and maintenance (O&M), 236, 238–239
Orchids, 45, 47
Organic matter (OM), 4, 43, 138, 165, 194, 223, 251, 275, 278, 313, 344, 367.*See also* Decaying
Orindole–3 butyric acid, 55
Osmosis, 11, 24, 68–71, 73, 76, 83, 89, 92, 104, 143, 144, 176, 177, 210, 286, 308, 327, 344
Overland flow, 206, 270
Oxaloacetic acid, 51
Oxic, 50, 66, 85, 164, 194, 255, 270, 274, 296, 311, 321, 325, 328, 360, 366
Oxidation, 51, 64, 139, 175, 176, 202, 245, 252, 255, 257, 274, 292–296, 298–302, 311, 312, 320–322, 325, 328, 330, 333, 368, 370–372
Oxygen, 3, 33, 43, 115, 133, 165, 194, 223, 245, 275, 307, 344, 354, 366
Oxygenated gasoline, 314
Oxygenates, 3, 320–325
Oxygen, dissolved, 257, 266, 314, 362
Oxygen releasing compound, 297
Ozone, 50, 261, 301
Palisade parenchyma, 79–81, 84
Palissy, B., 23, 29, 95, 146
Palm, 14, 15, 53, 118, 162, 181, 264, 267
Palm Springs, 14
Paracelsus, 259
Paracytic, 78
Paramelle, 98, 160
Parenchyma, 46, 79–81
Parenchymal cells, 6, 59, 75, 84, 187
Particle-size analysis, 138

Partition coefficient, 85, 276–278, 282–285, 322, 326, 327, 345, 358
Pascal, 23, 79, 90, 101, 143, 185, 210, 217, 229
PCB. *See* Polychlorinated biphenyls
PCE. *See* Perchloroethylene
Peat, 15, 61, 82, 109, 144, 165, 183, 202, 205, 295, 313
Peat bogs, 313
Pectin, 45, 47, 304
Pelletier, P.J., 24, 47
Pencil tree, 163
Penman method, 38
Penman-Monteith method, 38, 39
Peptides, 262–263
Peptidoglycan, 45
Perchlorate, 271, 333–337
Perchloroethylene (PCE), 207, 282, 284, 285, 292, 295, 307, 325–327, 332, 334, 337, 338, 355–357, 366
Percolation, 136, 146
Pericycle, 57, 71
Peridermal layer, 46
Permeability, 11, 13, 14, 24, 59, 61, 97, 136, 137, 145, 149, 165, 170, 171, 175, 176, 195, 200, 204–206, 226, 266, 270, 282, 285, 311, 316, 345
Permeases, 70
Peroxidase, 300, 337
Perrault, P., 29
Perylene, 312
Pesticide root zone model (PRZM), 362
Pesticides, 173, 260, 261, 265, 277, 278, 281, 286, 302, 308–310, 339, 362
Petiole, 73, 79, 85, 86
Petroleum hydrocarbons, 3, 171, 183, 201, 235, 251265, 286, 295, 297, 307, 309, 310, 312–314, 332, 338–340, 354, 362
PGA. *See* Phosphoglyceric acid
Phase I reaction, 298–302, 307, 309, 312, 325, 335–338, 357, 368, 376
Phase II reaction, 298, 301–304, 309, 335, 336, 338
Phase III reaction, 298, 304–305, 336
Phellogen, 85
Phenanthrene, 286, 287, 312, 313, 318, 319
Philosophers, 4–5
Phinizy swamp, 271
Phloem, 7, 10, 52, 53, 55, 59, 62, 64, 73–75, 80, 81, 84, 85, 92, 171, 187, 304, 309, 345, 357, 366, 369, 372
Phlogiston, 33, 250, 251
Phosphatidylcholines, 71
Phosphoglyceric acid (PGA), 50
Phospholipid bilayer, 46
Phospholipid fatty acid analysis (PLFA), 351, 352
Phosphorus, 64, 118, 126, 138, 177, 196, 256, 259, 271, 274, 366
Phosphorylation, noncyclic, 49
Photo-ionization detector, 354
Photon, 79, 247–249, 286
Photon-induced X-ray emission (PIXE), 354
Photoperiodism, 55
Photosynthesis, 25, 36, 43–57, 60, 63, 64, 74, 79–82, 84, 88–91, 115, 141, 176, 182, 187, 215, 223, 228, 246–255, 260, 265, 292, 294, 295, 301, 302, 308, 320, 336, 337, 359, 360
Photosynthetically active radiation (PAR), 49, 124, 199, 228
Phototropins, 252
Phototropism, 55, 248
Phreatophyte facies, 17–19
Phreatophytes, 11, 27, 51, 99, 117, 140, 155, 190, 219, 234, 255, 293, 310, 349, 352, 373
Phytochromes, 252
Phytoextraction, 3
Phytomining, 266
Phytoplankton, 60, 263
Phytoremediation, 3, 37, 43, 95, 115, 131, 155, 189, 245, 275, 307, 341, 351
Phytoremediation, defined, 3, 311
Phytosiderophores, 258
PhytoSociety, 131
Phytostabilization, 3
PHYTOTOX, 266
Phytotoxicity, 265
Phytovolatilization, 3
Pickleweed, 162, 183, 223
Piezometers, 102
Piezopit, 110
Pine, 37, 53, 59, 74, 86, 119, 123, 141, 163, 164, 184, 196, 225, 260, 261, 326, 328, 331, 339, 354, 356, 366
Pine Barrens, 141
Pistil, 9, 166
Pitcher plants, 254, 256
PIXE. *See* Photon-induced X-ray emission
Plant-nutrient availability, 138–139
Plant physiologists, 3, 10, 11, 13, 14, 27, 76, 83, 105, 143, 151, 215, 216, 360
PlantX, 285, 344–345
Plasmadesmata, 45, 62, 210
Plasmalemma, 45, 47, 69
Plastids, 47
Plato, 28
PLFA. *See* Phospholipid fatty acid analysis
Pliny the Elder, 4, 24, 140, 260
Point source, 133, 307
Poiseuille, 96
Poiseuille's equation, 93
Poison ivy, 260, 264
Poles, 169, 183, 235
Pollen, 9, 61, 158, 166
Pollination, 9, 158, 240
Pollinators, 261
Polychlorinated biphenyls (PCB), 265, 289, 325, 331
Polysaccharides, 45
Polyvinyl chloride (PVC), 110, 146, 332
Pond, 62, 73, 135, 163, 184, 185, 271, 316
Pope Clement VIII, 5
Poplars, 54, 55, 76, 80–82, 99, 155, 156, 158–160, 163, 165, 168, 170, 182, 184–187, 197, 202, 204, 207, 215, 227, 232, 235, 239, 255, 268, 269, 288, 297, 309, 325, 330–332, 338, 339, 356, 362, 370, 371, 376
Populus
  P. deltoids, 9, 119, 158, 163, 182, 201, 204, 206, 297, 310, 312, 321, 334, 336, 337, 342, 356, 367
  P. fremontii, 119, 174, 366
  P. trichocarpa, 158, 159, 206, 264–265, 330, 332
Populusdb, 159
Populus deltoides var. DN–34, 9. *See also* DN–34
Pore water, 175, 178, 183, 223, 258, 259, 267, 268, 273, 279, 282, 286, 292–295, 330, 361, 362
Porometer, 215, 360
Porosity, 11, 36, 52, 57, 59, 75, 77, 82, 98–101, 104–106, 112, 116, 145, 148, 165, 168, 174–176, 180, 193, 194, 198–200, 203, 219–221, 285, 297, 329, 342, 343, 345, 363
Porphyrin, 47, 48, 248
Potassium, 70, 138, 176, 177, 196, 255, 259, 354
Potato, 8, 263, 374
Potential energy, 49, 50, 89–91, 102, 252
Potential evapotranspiration, 37–39, 116, 118, 141–142, 171, 190–195, 197, 198, 200
Potentiometric surface, 106, 107
Potometer, 37

Powell, J.W., 12
Precipitation, 11, 27, 52, 95, 115, 137, 157, 189, 209, 235, 246, 284, 316, 346, 356, 367
Pre-dawn water potential, 92
Pressure
bomb, 76, 210
head, 96, 102, 103, 106, 125, 147
probe, 78, 92, 210
transducer, 219, 324
Priestley, J., 33, 43, 44, 250
Primary growth, 46, 52, 53, 57, 73, 75, 157, 158
Primary porosity, 100
Primary production, 216, 247, 249, 274, 391
Prokaryote, 46
Property rights, 233
Prosopis spp., 17
Protein, 55, 66, 254, 255, 259, 263, 302, 308, 352
Protons, 258
Protozoa, 272
Pruning, 53, 70, 185–187, 304, 356
PRZM. *See* Pesticide root zone model
PRZM-2, 362
Pseudomonas, 304
Psilophyta, 61
Psychrometer, 92, 209, 210
Pump-and-treat, 131, 136, 145, 150, 151, 164, 172, 173, 185, 202, 203, 220, 234–238, 273
Pump test, 204, 220, 345, 363
Push-pull tests, 362–363
PVC. *See* Polyvinyl chloride
Pyrenees, 54
Pyrrole rings, 248
Pyruvate, 51, 52, 252, 301
Pyruvic acid, 51, 52, 252

R

Rabbitbrush, 41, 161–163
Radar, 149–150
Radiation, 35, 36, 38, 48–50, 55, 80, 83, 86, 88, 116, 124, 137, 138, 141, 166, 192, 194, 196–199, 211, 212, 216, 219, 228–230, 245, 247–249, 255, 266, 286, 301, 313, 314, 374, 375
Radiation balance, 83
Radioactive isotopes, 231, 304, 334
Radioisotopes, 37
Radiotracer, 295
Raoult's law,
Raphides, 89, 365
Raskin, I., 25
Raunkiaer, C., 25, 89
Ray, J., 9
Rays, 75, 124, 187, 216, 246, 301, 366
RCRA. *See* Resource Conservation and Recovery Act
RDX. *See* Royal demolition explosive
Recharge, 3, 31, 68, 95, 115, 132, 157, 189, 218, 234, 250, 292, 311, 347, 360, 373
Record of Decision (ROD), 27, 132, 146, 147, 179, 207, 321, 340, 362
Redox, 63–64, 164, 203, 252, 257, 258, 259, 270, 271, 273, 291, 292, 294–296, 299, 302, 314, 355, 392
Red Sea, 157
Reduction, 8, 49–51, 64, 65, 67, 117–120, 122, 131, 133–136, 143–145, 150, 151, 157, 171, 172, 175, 179, 183, 187, 188, 190, 191, 195–200, 202, 221, 223, 234, 238, 248, 252, 255, 256, 258, 266, 269, 271, 272, 292–302, 314, 316, 326, 333, 335–337, 341–343, 370, 373
Reductive dechlorination, 295, 296, 328, 330
Redwoods, 72, 73, 76, 78, 80, 179
Reeds, 155, 157, 160, 319
Reformulated gasoline (RFG), 320
Relative humidity, 14, 35, 36, 69, 79, 81, 82, 91, 124, 142, 184, 192, 194, 196–198, 213, 214, 228–229
Remediation Technologies Development Forum (RTDF), 153, 237
Remote sensing, 230–231, 248
Renaissance, 29, 158
Research Site, 124, 207, 220, 359
Resource Conservation and Recovery Act (RCRA), 132, 136, 137, 234, 237, 239, 351
Respiration, 43–56, 62–68, 74, 81, 85, 138, 139, 142–144, 157, 165, 168–173, 175, 182, 206, 215, 245, 249–253, 255, 259, 261, 266, 267, 269, 273, 274, 291–295, 301, 308, 312, 320, 327, 337, 338, 362
RFG. *See* Reformulated gasoline
Rhizobia, 65, 66
Rhizobium, 65, 255
Rhizoferrin, 258
Rhizofiltration, 3
Rhizoids, 60
Rhizome, 46, 62, 73
Rhizoremediation, 3
Rhizosphere, 25, 52, 64–67, 71, 85, 139, 149, 157, 168, 176, 180, 251, 253, 255, 263, 267, 270, 273–275, 282, 288–292, 294, 295, 298, 300, 303, 310, 318–320, 322, 326–328, 331, 337, 338, 341, 344, 351, 352, 361–363, 367, 371, 376
Rhizotron cameras, 149
Rhodamine, 231
Ribonucleic acid (RNA), 46, 47, 56, 303
Ribosomes, 46, 47
Ribulose 1,5-biphosphate carboxylase/oxygenase, 252
Ribulose bisphosphate (RuBP), 50
Richards equation, 210, 232, 345, 346, 349
Ricin, 263
Ring cleavage, 300, 312
Ring porous, 75, 329, 353–355, 357, 358
Riparian, 12, 41, 54, 62, 80, 110, 118–127, 155, 159, 160, 165, 177, 180–182, 191, 198, 214, 219, 226, 227, 230–231, 267, 274, 293, 335, 353, 371, 373, 376
Riparian buffers, 122–123, 274
Riparian Evapotranspiration Package (RIP-ET), 231
RNA. *See* Ribonucleic acid
Rock, S., 136
Rocky Mountains, 25
ROD. *See* Record of Decision
Ronald R., 366
Root concentration factor, 280
Root dip, 197
Root hairs, 36, 51, 57–59, 62–76, 78, 79, 82, 83, 104, 143, 144, 168, 170, 176, 182, 224, 226, 227, 246, 255, 261, 262, 280, 287, 307, 340, 356
Root hydraulic conductivity, 62–63, 98, 210, 211
Roots, fibrous, tap, adventitious, suberized, hydraulic conductivity, 56–72, 84, 93, 98, 119, 124, 141, 157, 160, 168–170, 186, 210, 211, 224, 288, 291, 294, 326, 335, 348
Root tips, 46, 53, 64, 149, 258
Root water potential, 143
Root zone water quality model (RZWQM), 349
Rose, 29, 64, 82, 96, 102, 123, 162, 176
Rosette, 295
Rotenone, 261
Royal demolition explosive (RDX), 333, 335, 336, 367, 368
RTDF. *See* Remediation Technologies Development Forum
Rub' Al Khali, 14
Ruben, S., 44

Rubisco, 50, 223, 224, 252. *See also* RuBP
RuBP. *See* Ribulose bisphosphate
Rushes, 155, 156, 160, 271
Rye grass, 157
RZWQM. *See* Root zone water quality model

S
Safe Drinking Water Act (SDWA), 239, 351
Sahara, 14, 155
Sahel, 155
Salicaceae, 157–159
Saline seep, 178
Salix, 12, 55, 88, 119, 123, 159, 161, 163, 181, 227, 313, 323, 325, 331, 334, 337, 339
Salix gooddingii, 119, 174
Salt, 11, 16–18, 20, 25, 27, 35, 69, 70, 98, 113, 116, 125–126, 138, 157, 160–162, 173, 177, 183, 197, 223, 224, 232, 257, 267–268, 352, 353
Saltcedar, 22, 39, 54, 113, 117–118, 121, 123, 124, 161–162, 180, 182, 183, 223, 353. *See also* Tamarisk, Tamarix
Salt glands, 353
Saltgrass, 41, 156, 157, 181
Salton Sea, 17, 18
Saltwater, 177, 196, 267, 269, 373
Sampling, 111, 202, 217, 238, 314–316, 324, 334, 354, 360–362, 364
Sand, 15, 17, 18, 35, 82, 95–100, 102, 104, 106, 108, 109, 117, 135, 138, 143, 144, 146, 148, 149, 173, 175, 176, 180, 193, 194, 196, 201–208, 221, 234, 253, 271, 272, 288, 309, 316, 323, 324, 330, 332, 346
Sand hills, 104, 116, 119
San Jacinto Mountain, 14
Sap, 7, 8, 34, 48, 69, 74–76, 82, 84, 92, 99, 102, 120, 123–124, 159, 171, 186, 201, 202, 204, 206, 209–217, 224, 225, 228, 230, 238, 258, 314, 315, 324, 335, 355, 358, 366, 369, 372
Sapogenins, 263
Sapwood, 186, 202, 354, 355
Saudi Arabia, 14
Sawgrass, 271
Saw palmetto, 330, 331
Scheele, K.W., 33, 250
Schleiden, M.J., 7, 24, 44
Schlerenchyma, 47
Scholander, P.F., 76, 84, 92, 210, 293
Scholander Pressure Bomb, 210
Schwann, T., 7, 24
Screened interval, 108, 147, 220, 363
SDWA. *See* Safe Drinking Water Act
Sealants, 312
Sea level, 73, 101, 105, 184, 196, 199, 207, 247, 267
SEAM3D, 348
SEAM3D-PUP, 348
Secondary growth, 46, 52, 53, 57, 58, 62, 156, 157
Secondary porosity, 100, 101, 175
Sedimentation, 12, 249
Seedlings, 7, 55, 62, 65, 158, 166–167, 183, 204, 265, 327, 332, 364
Seeds, 4–7, 9, 16, 29, 47, 52, 54–57, 61, 62, 67, 68, 89, 117, 158, 160, 163, 166, 170, 183, 240, 260, 261, 262, 265, 266, 267, 303, 364, 365, 369–371
Seep, 109, 111, 174, 316
Seine River, 29–31
Selenium, 177, 266, 272, 308, 352
Semipermeable membrane devices (SPMD), 358
Senebier, J., 44
Seneca, 95, 109
Septic system, 175, 192, 272
Sequoias, 72
Sessile, 79, 260
Sewage sludge, 272, 369, 370
Sexual reproduction, 24, 57, 166, 167, 235, 260, 263, 268
SHAW. *See* Simultaneous heat and water model
Short rotation coppice (SRC), 117
Short-rotation wood culture (SRWC), 158, 165, 166, 239
Shrews, 136
Siderophores, 258, 292
Silent Spring, 263–264
Silurian period, 61
Simultaneous Heat and Water model (SHAW), 231–232
Sinai Peninsula, 14
Sinkholes, 28
Site history, 132–134, 222
SITE Program. *See* Superfund Innovative Technology Evaluation Program
Site visit, 135, 137–139, 151
Slash, 126, 141, 184, 186
Slough, 118, 123, 126, 176, 269, 288, 310
Slug test, 147–149, 196, 204, 345
Smith, G.E.P., 18–19, 25, 33, 123, 161, 218, 361
Smith, W., 24, 95
Snakeweed, 57
Snow, 28, 88, 136, 141, 356
Socrates, 4, 263
Sodium, 70, 126, 162, 177, 178, 183, 196, 256–257, 267, 352, 353
Sodium chloride (NaCl), 70, 266, 267, 308
Software, 108, 214, 228
Softwood cutting, 167
Softwoods, 75, 167, 354
Soil formation, 174
Soil-gas samples, 133, 331, 364
Soil moisture, 20, 32, 35–37, 39, 40, 56, 57, 59, 61, 64, 71, 72, 76, 83, 91, 92, 104, 109, 124, 137, 138, 140, 143, 144, 156, 157, 163, 165, 170, 171, 174, 177–180, 183, 185, 186, 190, 192, 198, 200, 209, 213, 217, 220–226, 228–232, 268, 274, 349
Soil-plant-atmosphere continuum (SPAC), 82–83, 174
Soil water potential, 83, 91, 92, 143, 176, 180, 209, 268
Solar panels, 88, 197
Solid-phase microextraction (SPME), 358
Solubility, 64, 67, 70, 142, 175, 194, 195, 229, 240, 245, 250, 254, 255, 257, 258, 261, 265, 275–277, 279, 282, 284, 287, 288, 291–293, 298, 302, 303, 305, 307, 310, 313, 321–323, 325, 326, 331, 343, 372
Sonic drilling, 146
Source area, 114, 131, 132, 134, 136, 137, 145, 148, 171, 192, 194, 195, 205, 208, 235, 237, 273, 287, 310–312, 323, 324, 342, 343, 346–348
Southington, CT, 202
SPAC. *See* Soil-plant-atmosphere continuum
Spanish conquistadors, 54,
Spartina, 70, 89, 177, 259, 266–268, 292, 294
Species, 9–13, 17, 24, 39, 43, 54, 56, 57, 59, 65, 67, 73, 74, 77, 83, 85–87, 110, 117, 119, 123, 137–139, 147, 155, 156, 158–160, 163, 164, 166, 172, 183–185, 188, 202, 206, 214, 217, 234, 235, 249, 262, 264–268, 273, 276, 289, 294, 301–303, 308, 316, 335–338, 349, 352, 354, 366, 370, 371, 374
Species Plantarum, 10, 24
Specific capacity, 56
Specific conductance, 124, 126, 139, 361
Specific discharge, 96, 97, 193, 198
Specific retention, 105, 142, 143
Specific yield, 20, 105, 108, 109, 112, 136, 145, 164, 190, 202, 206, 220, 221, 240

Spiral ascent, 77
SPMD. *See* Semipermeable membrane devices
SPME. *See* Solid-phase microextraction
Spongy parenchyma, 79–81
Spontaneous generation, 7
Spores, 61, 187, 262, 289, 303, 355
Springs, 8, 14, 19, 23, 28–31, 53, 55, 59, 75, 95, 98, 99, 106, 109, 111, 115, 116, 135, 159, 162, 163, 167, 170, 172, 186, 192, 201, 202, 216, 219, 222, 235, 296, 315, 316, 357
SRC. *See* Short rotation coppice
SRWC. *See* Short-rotation wood culture
Stable hydrogen isotopes, 224, 225, 227
Stable isotopes, 44, 72, 110, 115, 155, 159, 163, 174, 223–227, 232, 256, 258, 360, 363
Stable oxygen isotopes, 44, 225
Stakeholders, 151, 233, 263, 365, 377
Stamen, 9, 10, 166
Static fluid, 101, 102
Steady state, 30, 38, 68, 96, 107, 110, 113, 189, 193, 205, 211, 215, 221, 232, 239, 240, 246, 250, 251, 286–288, 308, 342, 344, 348
Stem cells, 53, 54, 57, 73, 74
Stolon, 46
Stomata, 39, 48, 49, 51, 52, 55, 57, 60, 66, 74, 76, 78–82, 84–86, 91, 92, 127, 184, 214, 215, 219, 224, 252–253, 262, 267, 278, 283, 286, 294, 295, 305, 330, 360, 369
Stomatal conductance, 74, 79, 80, 98, 166, 215–216, 232, 283, 294, 360
Stomatal resistance, 39, 78–81, 228
Storage, 28, 31, 38, 45–47, 49–51, 54, 62, 69, 82, 83, 85, 88, 89, 106–108, 110, 112, 113, 124, 167–170, 172, 175, 190, 193, 198, 199, 204–206, 211, 212, 214, 219, 222, 230, 248, 272, 275, 286, 303, 332, 342, 368
Storage coefficient, 106–108, 193
Storm drain, 192
Straight ascent, 77
Strasburger, E., 73
Strategy I, 257–259
Strategy II, 257, 258
Stream, 17, 18, 25, 28–31, 33, 74, 76, 82, 84, 96, 109–111, 116–123, 124–126, 135, 145, 147, 155, 159, 160–161, 163, 181, 194, 211, 215, 219, 226, 227, 235, 239, 258, 265, 267, 271, 274, 278, 279, 281–283, 285–288, 292, 298, 299, 305, 313, 314, 316, 326, 330, 342, 347, 352, 356–358, 366, 369, 373
Strobaeus, K., 9
Stroma, 47, 50
Suberin, 46, 59, 63, 84, 187, 258
Suberized, 59, 66, 168, 211, 224, 348
Succinate, 252, 257
Suez Gulf, 14
Sugar, 7, 48–52, 57, 66, 69, 71, 74, 75, 80, 84, 186, 187, 223, 245, 247, 252, 253, 266, 273, 291, 302, 303, 335, 355, 368, 369, 372
Sulfide, 259, 314
Sulfur, 50, 67, 177, 259, 320
Sumerians, 4
Sun, 24, 27–29, 33, 34, 38, 40, 43, 48–50, 54, 56, 72, 76, 79, 83, 88, 90, 124, 141, 144, 174, 184, 216, 217, 222, 246–249, 252, 273, 286, 301, 359
Sun dances, 29
Sunflower, 99, 158
Superfund, 25, 132, 173, 201–203, 206–207, 234, 237, 239, 340, 351
Superfund Innovative Technology Evaluation (SITE) Program, 201, 203, 204
Superoxide anions, 301
Suquamish Tribe, 207
Surface tension, 33–35, 65, 70, 71, 91, 101, 104, 105, 109, 143, 144, 307
Surfactants, 65, 71
SUTRA, 348
Swamp, 85, 99, 111, 117, 123, 138, 159, 163, 180, 182, 183, 224, 235, 256, 259, 269, 271, 293, 294
SWATRE model, 232
Symplastic pathway, 62, 63, 71, 211, 258, 259, 281, 285
Syrian-African rift, 14
System Naturae, 9, 24

T

2,4,5-T, 308
Tamarisk, 54,–119, 155, 160, 177, 182
Tamarix, 22, 54, 81, 113, 117–119, 124, 160, 161, 174, 182, 227, 353, 370
TAME. *See* Tertiary-amyl methyl ether
Tanks, metal, 16
Tannins, 47, 187, 262, 291, 303, 321
TCA. *See* Tricarboxylic acid cycle
TCAA. *See* Trichloroacetic acid
TCE. *See* Trichloroethylene
TCLP. *See* Toxicity Characteristics Leaching Procedure
Teardrop, 34
Technical impracticability, 372–373
Tensiometer, 92, 93, 104, 179, 209, 222, 361, 363
Terminal buds, 53
Terpenes, 261, 291, 301, 366
Terpenoids, 261, 331
Tertiary-amyl methyl ether (TAME), 321, 346
Tertiary period,
Theis, C.V., 102, 149, 373
Theophrastus, 4, 9, 23, 47, 260
Theophrastus of Eresus, 4
Thermocouple, 209, 211, 212, 214
Thermocouple dissipation probes, 211, 212
Thermodynamics, 89, 90
Thiobacillus, 259
Thionins, 262–263
Thoreau, H. D., 27
Thornthwaite and Holzman method, 34, 37, 38, 100
Thornthwaite, C.W., 37, 190
Thylakoids, 47
Tidal efficiencies, 218
Time-domain reflectometry, 222
Tissues, 3, 6, 8, 10, 12, 37, 45–47, 51–53, 55, 59–62, 72–89, 92, 159, 162, 174, 183, 209, 210, 224, 225, 228, 234, 247, 250, 251, 259, 262, 264, 267, 275, 276, 279, 280, 283, 284, 286, 287, 292, 294, 298, 301–304, 309, 313, 326, 330–332, 334–337, 341, 352–358, 369, 370, 372
TNT, 292, 302, 333–336, 338, 339, 367, 368, 371
Toadstools, 67
Tobacco, 25, 86, 261, 264, 308, 371
Toluene, 236, 240, 281, 283, 289, 290, 300, 302, 305, 312–317, 333, 335, 336, 354, 372
Tonoplast, 47, 69
Tortuosity, 91, 100, 170, 343
Total dissolved solids, 122, 138, 170, 194
Total petroleum hydrocarbon, 297, 364
Toxic, 3, 4, 47, 49, 56, 67, 73, 132, 133, 136, 177, 183, 222, 234, 239, 245, 249, 258–261, 263–267, 295, 297–300, 302, 303, 307–309, 325, 332, 333, 336, 337, 340, 342, 364, 365, 368, 372
Toxicity Characteristics Leaching Procedure (TCLP), 368, 369
Toxics Substances Hydrology Program, 307
Toxic Substances Control Act (TSCA), 325
Trace elements, 3, 122, 176, 257, 352–354
Trace metals, 122
Tracer, 77, 114, 211, 281, 304, 343, 345, 359, 362
Trachieds,
Transeau, E., 249

Transformation, 52, 91, 224, 248, 261, 264, 270, 274, 275, 286, 287, 298–300, 302, 305, 309, 312, 325, 330, 332–333, 335–338, 355, 369, 370
Transgenics, 166, 240–241, 333, 335, 336, 370–372, 374
Transient, 31, 110, 113, 193, 205, 240
Translocation, 36, 80, 245, 259, 263, 265, 278, 281, 285, 287, 304, 309, 310, 313, 331, 334, 336, 348, 354, 366, 369
Transmissivity, 106, 107, 112, 145
Transpiration, 11, 30, 48, 99, 115, 136, 156, 189, 209, 234, 249, 278, 307, 341, 352, 366, 376
Transpiration efficiency (TE), 48, 49
Transpiration ratio, 73
Transpiration Stream Concentration Factor (TSCF), 239, 240, 280–282, 285, 287, 288, 313, 321, 327, 330, 332, 334, 336, 339, 342, 348, 370
Tree cores, 174, 284, 304, 324, 326, 329, 331, 335, 352–358
Tree Islands, 118, 126
Tree rings, 115, 125, 256, 314, 353, 354, 363
TreeWell, 204, 205
Tree-wells, 145, 150
Trench, 17, 99, 131, 149, 169, 170, 180, 201, 202, 204, 206, 208, 222, 234–238, 316, 317
Tricarboxylic acid cycle (TCA), 52, 252–253, 257, 281, 312, 332
Trichloroacetic acid (TCAA), 300, 304, 305, 332
Trichloroethanol, 332
Trichloroethylene, 325
Trichomes, 86
Trihalomethane, 307
Tritium, 231, 307, 333–337, 359
TSCA. *See* Toxic Substances Control Act
TSCF. *See* Transpiration stream concentration factor
Tull, J., 6, 165
Turgor, 53, 54, 69, 78, 80, 92, 168, 210

U

Ultraviolet, 50, 202, 248, 266, 286, 301, 313
Underground storage tank (UST), 133, 137, 195, 275, 323, 340
United states (US), 11–18, 21, 24, 25, 27, 31, 33, 37, 40, 41, 53–57, 59, 60, 63, 88, 99, 104, 112, 116, 118, 119, 120, 131–133, 135, 140–142, 153, 156–163, 169–175, 177, 178, 182, 183, 191, 192, 194, 196, 199, 200, 202, 206, 213, 217, 233, 239, 241, 261–263, 266, 269–272, 275, 309, 311, 324, 325, 339, 361, 366, 371–374
University of Arizona, 18, 218
Unsaturated zone, 32, 40, 52, 57, 63, 71, 91, 98, 103–105, 109, 110, 114, 115, 116, 124–128, 133, 137–139, 143, 145, 171, 174, 175, 198, 218, 221, 222, 226, 230–232, 250, 251, 273, 275, 277, 284, 287, 290, 291, 295, 297, 303, 312, 317, 318, 324, 326, 328, 331, 334, 346, 348–349, 356, 361, 363
UPFLOW model, 232
Uptake rate, 148, 190, 193, 239, 240, 282, 288, 303, 324, 326, 342
US. *See* United states
U.S. Congress, 320
USDA. *See* U.S. Department of Agriculture
U.S. Department of Agriculture (USDA), 140
U.S. Environmental Protection Agency (USEPA), 25, 131, 132, 150, 153, 200–202, 206, 207, 220, 237, 238, 265, 271, 279, 303, 310, 312, 333, 351, 362
USEPA. *See* U.S. Environmental Protection Agency
U.S. Geological Survey (USGS), 12–17, 19, 22, 24, 27, 40, 99, 102, 118, 121, 123, 135, 149, 157, 160, 166, 181, 191, 197, 203, 205, 207, 230, 231, 263, 307, 316, 317, 334, 353, 355, 359, 361
USGS. *See* U.S. Geological Survey
UST. *See* Underground storage tank
Utah, 15, 40, 54, 119, 120, 160, 222, 369
Utilities, 134, 146, 166, 238
UV radiation, 301, 313

V

Vacuole, 45, 47, 53, 57, 69, 71, 88–89, 177, 224, 261–263, 283, 284, 287, 295, 300, 302–304, 312, 368
VAM. *See* Vesicular arbuscular mycorrhizae
Vanadium bush, 156, 161, 266
Van Bavel method, 38, 39, 190
van Helmont, J.B., 5–6, 10, 23, 176
van Leeuwenhoek, A., 7, 23, 89
Van Niel, C.B., 25, 44
Vapor pressure, 34–36, 38, 79, 92, 142, 144, 184, 209, 224, 228, 229, 275, 277, 284, 304, 305, 310, 326, 328, 331, 344, 358
Vapor pressure deficit (VPD), 38, 79, 81, 82, 84, 86, 124, 142, 159, 166, 184–185, 196, 197, 214–217, 223, 228, 229, 304, 318, 360
Variety, 3, 9, 22, 32, 38, 65, 85, 124, 138, 147, 148, 165, 167, 171, 172, 176, 196, 206, 222, 232, 239, 245, 260, 262, 265, 266, 269, 298, 299, 307, 310, 312, 314, 333, 337, 354, 359, 361, 362, 369, 374, 376
Vascular plants, 57, 61, 66, 73, 75, 82, 174, 264, 265, 289, 304
Vascular system, 53, 57, 59, 60, 62, 72, 75, 85, 168, 231, 265, 282, 287, 298, 304, 356
VC. *See* Vinyl chloride
Vegetable Staticks, 8, 24
Vegetated covers, 136, 150
Vegetated scarps, 14
Venus fly traps, 254, 256
Vesicular arbuscular mycorrhizae (VAM), 66
Vessel cells, 73
Vikings, 28
Vineyard, 95, 185, 187
Vinyl chloride (VC), 307, 371
Virgin Islands, 116
Viruses, 66, 262, 272, 355
Vitruvius, 28, 155
VOC. *See* Volatile organic compound
Volatile organic compound (VOC), 136, 147, 202, 204, 238, 245, 260–263, 277, 278, 283, 285, 307, 326, 332, 347, 354–357, 359, 361, 363–364, 366, 367
Voles, 136, 194
von Mohl, H., 24, 44
von Sachs, J., 11, 25
VPD. *See* Vapor pressure deficit
VS2D, 349

W

Walden, 27
Walnut, 162, 261, 263, 303
Washington filifera, 14, 161
Wastewater, 117, 263, 265, 269–272, 274, 348
Water budget, 16, 17, 22, 23, 30–33, 38, 40, 42, 112–114, 117, 121, 126, 142, 145, 153, 166, 181, 189, 191, 204, 206–209, 215, 216, 221, 226, 231, 247, 286, 324, 330, 346, 347
Water holes, 14
Water hyacinth, 38
Water lily, 11, 62, 73
Water potential, 3, 8, 11, 14, 53, 58, 59, 63–65, 69, 71, 74, 76, 78–83, 86, 89–93, 104, 125, 143, 144, 176, 177, 180, 181, 190, 209–211, 214, 215, 219, 221, 226, 227, 232, 268, 288, 293, 304, 337, 361
Water status, 11, 86, 89–93, 209, 211, 214, 229
Water-Supply Paper 577, 13, 25
Water table, 3, 35, 49, 103, 116, 131, 156, 190, 214, 234, 250, 275, 311, 345, 353, 374, 376

Water-table aquifer, 104, 105, 107, 110, 112, 114, 135, 193, 196, 198–201, 221, 275, 316, 324, 346
Water-use efficiency (WUE), 71, 82
Water wheel, 50
Water witch, 27
Wavelengths, 48, 49, 247–249, 286, 359
Wax myrtle, 87, 255
Weather station, 192, 197, 206, 219, 228, 238
Weighing lysimeter, 37–39
Well, 7, 27, 46, 95, 115, 131, 155, 190–191, 212, 234, 249, 275, 311, 341, 352, 365
Western black cottonwood,158. *See also Populus trichocarpa*
Wetlands, 111, 118, 122, 126, 135, 138, 145, 159, 175, 231, 235, 250, 256, 269–271, 273, 293, 294, 327, 330, 348
  constructed, 269–271
  natural, 111, 269–271
Wetting front, 109, 226
Wetwood, 294
Whips, 164, 165, 168–170, 186, 197, 199, 201, 204, 211, 213, 214, 216, 228, 324, 364
White's equation, 20
Williams, S., 10, 24
Willow, 5, 6, 12, 20, 27, 29, 41, 54, 55, 64, 81, 87, 119, 123, 140, 155, 157–158, 160–163, 168, 171, 180–185, 205, 214, 215, 223, 227, 231, 269, 270, 283, 291, 295, 303, 312, 313, 323, 325, 331, 334, 337–339, 356
Willow tree, 5, 10, 20, 87, 123, 185, 202, 214, 218, 313, 336
Wilted, 7, 19, 69
Wilting, 34, 36, 69, 82, 83, 93, 105, 109, 143, 144, 170, 180, 210, 219, 222, 337
Winogradsky, S.N., 255
Witch hazel, 27
Wood, 7, 27, 45, 47, 49, 54, 59, 75, 83, 86, 112, 121, 158, 161, 176, 186, 188, 196, 204, 214, 223, 239, 240, 250, 276, 282–284, 294, 303, 304, 320, 332, 358, 366
Woodward, J., 6, 24, 36
Woody, 6, 17, 18, 21, 40, 46, 52, 53, 55–60, 67, 73, 81, 82, 84, 86, 89, 128, 155, 156, 157–164, 167, 168, 181, 224, 231, 261, 267, 269, 274, 288, 293, 295, 304, 310, 327, 334, 338, 374
World Tree, 28, 29
WUE. *See* Water-use efficiency

X

Xenobiotic, 64, 176, 245, 263, 264, 271, 274–276, 281, 289, 290, 298–304, 307–309, 338, 339, 341, 344, 348, 349, 353, 363, 368, 369, 375
Xerohalophytes, 11
Xerophytes, 11–13, 126, 225
Xylem, 7, 8, 10, 23, 44, 51–53, 56, 59–64, 69–78, 80–82, 84–86, 92, 93, 115, 174, 179, 186, 188, 210, 211, 214, 215, 224–227, 258, 259, 261, 267, 278, 279, 281–287, 294, 298, 304, 305, 309, 329–331, 337, 344, 345, 354, 355, 357, 358, 363, 369, 372
Xylenes, 236, 281, 289, 310, 314

Y

Yeast, 47, 253
Yggdrasil, 28

Z

Zeaxanthin, 48, 266
Zinc, 125, 177, 266, 300
Zygote, 52, 166

Printed by Publishers' Graphics LLC USA

2012